Watershed Assessment of River Stability and Sediment Supply (WARSSS)

Dave Rosgen
Wildland Hydrology
Fort Collins, Colorado

Illustrations
Hilton Lee Silvey
Western Hydrology
Lakewood, Colorado

Technical Editing
Darcie Frantila
Wildland Hydrology
Fort Collins, Colorado

Cover photograph of the Grand Tetons with Snake River in foreground, Copyright © 2006 Alpen Glow Images.

Cover design by Sandy Cochran, Fort Collins, Colorado.

Rosgen, Dave, 1942 –
Watershed Assessment of River Stability and Sediment Supply (WARSSS)

ISBN-13: 978-0-9791308-0-9
ISBN-10: 0-9791308-0-8

Includes References and Index
1. Geomorphology;
2. Hydrology;
3. Environmental Engineering;
4. Fisheries Management;
5. Watershed Science; and
6. Sedimentology.

Prepress (including page layout, design and cover design):
Sandy Cochran Graphic Design, Fort Collins, Colorado.
Printing: Friesens, Altona, Manitoba, Canada

Copyright © 2006 by Wildland Hydrology
11210 N County Rd 19
Fort Collins, Colorado

All Illustrations copyright © by Hilton Lee Silvey unless indicated otherwise

No part of this book may be reproduced by any mechanical, photographic or electronic process, nor can it be stored in a retrieval system, transmitted or otherwise copied for public or private use without written permission from the publisher.

Printed in Canada

Dedication

This book is dedicated to two outstanding individuals
who have both made a difference to field practitioners:

To the memory of

Luna B. Leopold,

who for nearly six decades contributed to the river science and land ethic. Luna instilled a legacy of deductive reasoning based on field observation in the study of river process and form. His contributions to rivers and to those who study them will never be diminished by time. Luna will be forever heard in the quantitative expression of the flow, only to be read from the river. We still see Luna as the campfire dims, with his hearty laugh and constant challenge…an impression never forgotten.

And to

Hilton Lee Silvey,

hydrologist extraordinaire, my river partner and close friend for 35 years…whose good judgment, keen insight, inspiration, thoughtfulness and great artistic talent has earned him the highest respect and admiration of his peers. I am indebted to Lee for his substantial contributions to *Applied River Morphology* and *WARSSS*.

Preface

This book is designed for the field practitioner. The assessment methodology presented is a compendium of 40+ years in the field. The questions posed to me early in my career included: How much timber can we cut without adversely affecting watersheds and streams? How will this type of road at this location affect the streams? How will continued, heavy, season-long grazing affect this stream? In other words…How much is too much? These were complex questions, and I was not alone in the search for answers. A small group of hydrologists from the USDA Forest Service, along with Hilton Lee Silvey and myself, were instrumental in the early work of watershed assessment. Opinions were as common as the problems but even more variable. To be effective, any management advice offered needed to be backed up according to the principles of physical science. Observations were common, but data was not. In seeking out answers to these questions, I sought out the advice of Dr. Luna B. Leopold, the world-renowned geomorphologist. His advice can be succinctly stated: "There are no models . . . you must find the answer on the landscape and in the river. Your daily observations and measurements must be chronicled into a permanent record. You will refer back to this record and draw conclusions from it over time." Luna's advice was followed over many years and the conclusions from such observations formed the basis for the majority of the procedures presented in *WARSSS*.

A scientist must first observe then document such observations. A rigorous analysis must follow to test and to substantiate initial interpretations. Such interpretations can be of the most help in conserving and properly managing our precious water resources if we put our collective understanding into the hands of those who can make a difference. Technology transfer of complex subjects through the applied science is necessary if we are to put sound principles into practice.

This book represents efforts in the applied science of watershed management. The results of the physical and biological impairment has led us to better understand and predict the cumulative watershed impacts designed to prevent reoccurring stream channel instability and related sediment problems. The methods presented do not skip the fact of uncertainty of prediction, the variability of nature and its ever-present complexity. One has to best use and continue to refine the tools necessary in watershed assessment, but monitoring is the best long-term solution. It gives us the confidence of prediction and builds an understanding and appreciation for river response. With changed management and/or mitigation, the efforts in assessment and monitoring will reward the considerable efforts expended toward improving our watersheds and river systems.

Preface

This system is designed to be a framework for various predictive tools. It does not require one to use only one model or method, but rather requires one to access those tools of greatest familiarity and confidence to best represent the processes at work. The method allows a rapid, initial assessment to set priorities for more detailed study on smaller parcels and river reaches.

The overriding objective of *WARSSS* is to understand the cause and consequence of change. Process- and site-specific mitigation can then be prescribed to secure the physical stability and the natural potential functions of river systems.

Acknowledgements

To mention all those who influenced this book would require another chapter; however, special recognition is deserving of the following individuals who were key contributors to *WARSSS*.

Hilton Lee Silvey, illustrator, provided excellent graphical artwork, technical guidance and helped with the field form creation. His skilled portrayals of the science help many see who might not otherwise.

Darcie Frantila, technical editor, provided exceptional technical editing, great insight, organization skills and dedication to make the words and document more understandable. Darcie helped make the book "user friendly," especially the worksheets and flowcharts, and provided technical consistency.

The outstanding contributions by both Lee Silvey and Darcie Frantila are deeply appreciated.

The EPA supported the earlier work providing initial funding and peer review. Don Brady was instrumental in initiating this project and Doug Norton followed it through on an interactive web page (http://www.epa.gov/WARSSS) making the *WARSSS* procedure available worldwide. The technical support of the EPA is much appreciated in helping make *WARSSS* available to everyone.

In appreciation of Dr. Richard Hey, whose insight and exchange of ideas made a positive influence on *WARSSS*. Also, Jim Nankervis provided the water yield analysis, statistical inference and contributions to the dimensionless sediment relation analysis. Josh Kurz assisted in the sediment rating curve testing, data collection, analysis and graphical presentation. Both Jim's and Josh's efforts significantly contributed to *WARSSS*.

Bridget Julian also contributed to the formatting, editing and overall improvement of the book. Sandy Cochran provided the cover design and prepress layout. Editorial draft reviews were improved by the following contributors: Lee Chavez, Doug Norton, Lee Silvey, Angela Greene, Elain Opdahl, Becky Talley and George Athanasakes. Field measurements, data collection and laboratory analysis used in the various relations were contributed by Lee Chavez, Lee Silvey, Brandon Rosgen, Jim Nankervis, Christine Jones-Lucero, Tracie Sylte, Bob Kassun, Keith Candaleria and Tiffany Thompson. Technical typing was provided by Sandy Mion.

And finally, to my wife and business partner, Kae Rosgen, whose coordination, perseverance and kind persuasion encouraged and supported the completion of *WARSSS*.

Table of Contents

Chapter 1: The WARSSS Assessment Methodology 1-1
    Introduction 1-1
    Basic assumptions for the use of WARSSS 1-6
        The importance of field measurement 1-6
        Identification, prediction and measurement of sediment sources 1-6
    River stability, sediment and land uses 1-7
        Land use influences on river stability and sediment 1-7
        The influence of major flooding on river stability and sediment 1-9
        Using existing spatial data: GIS and remote sensing 1-11
        Potential uses of WARSSS 1-11
    Conclusion 1-12

**Chapter 2: A Review of Hillslope, Hydrologic and
Channel Process Relations** 2-1
    Hillslope erosional processes 2-1
        Surface erosion 2-1
        Mass erosion 2-3
    Channel processes 2-6
        Bedload transport 2-7
            Sediment entrainment 2-8
            Sediment competence 2-14
            Sediment transport 2-14
        Suspended sediment transport 2-19
        Total bed-material load 2-19
        River classification 2-20
        Valley classification 2-26
            Hydraulic geometry 2-27
        Management applications 2-30
            Classification versus river stability 2-30
            Sediment relations 2-33

Table of Contents

    Streambank erosion 2-35
        Streambank erosion prediction 2-37
    River stability 2-37
        River stability concepts 2-38
        Aggradation 2-38
        Degradation 2-40
        Channel enlargement 2-42
    Gully erosion 2-43
    Stream channel succession 2-44
Hydrologic processes 2-53
    Streamflow 2-53
    Bankfull discharge 2-55
Integration of hydrologic, sediment and stability relations 2-57
    Stream stability, reference condition and sediment rating curves 2-57
Application of basic principles 2-60
    Dimensionless bedload and suspended sediment rating curves 2-60
    Stability test/sediment rating curves 2-67
    Entrainment/competence calculation approach 2-70
        West Fork San Juan River 2-72
    Bedload transport 2-73
**Weminuche Creek: A case study of the channel succession
   sequences of an unstable river** 2-76
    ***Hydraulic, sedimentological and morphological consequences of
    stream type change*** 2-84
    Sediment effects 2-87
    Streambank erosion prediction results: The BANCS model 2-87
    ***Comparison of predicted-to-observed values for the BANCS
    streambank erosion model*** 2-92

Chapter 3: The Reconnaissance Level Assessment (*RLA*) 3-1

    Step 1. **Compile existing data** 3-2
    Step 2. **Review the landscape history** 3-6
    Step 3. **Summarize activities that potentially affect sediment supply
                 and channel stability** 3-6
    Step 4. **Identify specific process relations** 3-8
    Step 5. **Review the landscape overview and map the watershed** 3-10

Table of Contents

Step 6.	Identify hillslope processes	3-13
Step 7.	Document surface erosion	3-14
Step 8.	Document mass erosion	3-16
Step 9.	Assess hydrologic processes	3-16
Step 10.	Identify streamflow changes	3-17
Step 11.	Analyze channel processes	3-18
Step 12.	Detect direct impacts to streambanks and channels	3-19
Step 13.	Summarize problem verification: recognition of places, processes and sources	3-20
Step 14.	Eliminate areas, sub-watersheds and/or river reaches that are not contributors to either sediment or stability problems	3-20
Step 15.	Select areas, sub-watersheds and river reaches for further assessment	3-20

Chapter 4: The Rapid Resource Inventory for Sediment and Stability Consequence (*RRISSC*) 4-1

RRISSC assessment concept 4-2
Sequential assessment steps 4-2

Step 1.	Identify land use activities	4-4
Step 2.	Perform landscape and river inventory	4-7
Step 3.	Determine variables influenced	4-10
Step 4.	Compile data for risk rating system	4-12

Hillslope processes: Steps 5–8 4-14

Step 5.	Select appropriate models and compile data for hillslope processes	4-14
Step 6.	Assess mass erosion risk	4-14
Step 7.	Evaluate road impact risk	4-21
Step 8.	Determine surface erosion risk	4-26

Hydrologic processes: Steps 9–10 4-34

Step 9.	Compile maps and data for hydrologic processes	4-34
Step 10.	Assess potential for streamflow changes	4-35

Channel processes: Steps 11–15 4-43

Step 11.	Compile data for channel processes	4-43
Step 12.	Assess broad-level channel stability	4-43
Step 13.	Determine streambank erosion risk	4-44
Step 14.	Assess in-channel mining	4-50
Step 15.	Assess direct channel impacts	4-53

Table of Contents

    Riparian vegetation change risks 4-55

    Debris blockage risk 4-57

Risk potential assessment: Steps 16–20 4-59

    Step 16.   Calculate channel enlargement risk potential 4-59

    Step 17.   Calculate aggradation/excess sediment risk 4-62

        Depositional patterns adjustment 4-66

        Decrease in streamflow due to diversions and/or flow regulation adjustment 4-66

    Step 18.   Determine channel evolution potential 4-67

    Step 19.   Calculate degradation risk 4-69

    Step 20.   Summarize total potential sediment and stream channel stability risk .4-73

    Step 21.   Create overall risk rating summary 4-73

    Step 22.   Discard processes with low overall risk ratings 4-73

    Step 23.   Create management change recommendations for processes
              with moderate overall risk ratings 4-74

        Mitigation 4-74

        Monitoring 4-74

    Step 24.   Advance high-risk and/or high-consequence processes to the
              PLA assessment phase 4.74

Chapter 5: The Prediction Level Assessment (*PLA*) 5-1

PLA analytical concepts 5-4

  Initial preparation 5-4

Steps 1–4: Bankfull discharge and hydraulic relations 5-8

    Step 1.   Develop and/or obtain regional curves of bankfull discharge versus
              drainage area and bankfull dimensions channel versus drainage area ..5-9

    Step 2.   Delineate the watershed boundary on a USGS 7.5′ Quad map.
              Calculate drainage area in square miles 5-12

    Step 3.   Field calibrate bankfull discharge estimates by either direct discharge
              measurements or field calibration procedures at local gage stations
              to verify regional hydrology curves applied to ungaged sites 5-14

        Calibrating bankfull stage to known streamflows 5-17

    Step 4.   Calculate bankfull discharge and dimensions 5-20

Steps 5–6: Level II stream classification and dimensionless ratios

              of channel features 5-27

    Step 5.   Classify streams 5-28

    Step 6.   Calculate detailed morphological descriptions
              (including dimensionless ratios) 5-30

Table of Contents

Step 7. Identify stream stability indices 5-34
    a. Riparian vegetation 5-36
    b. Flow regime 5-38
    c. Stream order and stream size 5-39
    d. Meander patterns 5-40
    e. Depositional patterns 5-41
    f. Channel blockages 5-42
    g. Width/depth ratio state 5-43
    h. Modified Pfankuch stability rating 5-45
    i. Degree of channel incision 5-47
    j. Degree of channel confinement (lateral containment) 5-50

Steps 8–9: Streambank erosion (BANCS model) 5-53
    Step 8. Calculate Bank Erosion Hazard Index (BEHI) and Near-Bank Stress (NBS) ratings 5-54
    The Bank Erosion Hazard Index (BEHI) 5-55
    Near-Bank Stress (NBS) 5-65
    Step 9. Predict annual streambank erosion rate using BEHI and NBS ratings .5-78

Steps 10–15: Total annual sediment yield prediction using the FLOWSED model ... 5-82
    Step 10. Develop dimensionless flow-duration curve 5-89
    Snowmelt-dominated hydrograph 5-89
    Stormflow-dominated hydrograph 5-89
    Step 11. Collect bankfull discharge, bedload sediment and suspended sediment 5-92
    Step 12. Obtain or establish dimensionless bedload and suspended sediment rating curves 5-93
    Step 13. Convert dimensionless bedload and suspended sediment rating curves to dimensioned sediment rating curves 5-94
    Step 14. Convert dimensionless flow-duration curve to dimensioned flow-duration curve 5-96
    Step 15- Calculate total annual sediment yield for both bedload and suspended sediment 5-97

Steps 16–17: Calculate flow-related changes in annual sediment yield using water yield and FLOWSED models 5-98
    Step 16. Select and run a water yield model 5-100
    Snowmelt models 5-100
    Stormflow models 5-100
    Pre-treatment condition 5-101

xiii

Table of Contents

  Post-treatment condition: Changes in runoff 5-101

  Proposed condition 5-101

 Step 17. Calculate flow-related changes in annual sediment yield 5-102

Steps 18–19: Sediment transport capacity model (POWERSED) 5-103

 Step 18. Evaluate channel characteristics that change hydraulic and morphological variables and develop hydraulic geometry relations for a wide range of flows 5-110

 Step 19. Calculate bedload and suspended sand bed-material load transport capacity (stream power) 5-111

Steps 20–21: Sediment delivery from hillslope processes 5-112

 Step 20. Determine potential sediment delivery from roads, surface erosion and mass erosion 5-114

  Sediment delivery from roads 5-114

  Sediment delivery from surface erosion 5-118

  Sediment delivery from mass erosion 5-125

 Step 21. Summarize total annual sediment yield (tons/year) from hillslope processes 5-126

 Step 22. Calculate sediment entrainment/competence .5-127

 Step 23. Predict channel response based on sediment competence and transport capacity 5-137

  Sediment transport capacity (Steps 18 and 19) 5-138

  Sediment competence (Step 22) 5-138

Steps 24–27: Channel stability ratings by various processes and source locations 5-139

 Step 24. Calculate potential stream channel successional stage shift 5-142

 Step 25. Calculate lateral stability rating 5-144

 Step 26. Calculate vertical stability ratings 5-146

  Aggradation or excess deposition 5-146

  Degradation/channel incision 5-146

 Step 27. Calculate potential channel enlargement 5-149

 Step 28. Determine overall sediment supply rating from individual and combined stability ratings 5-150

Steps 29–31: Summary evaluations 5-153

 Step 29. Calculate total sediment yield (all sources) 5-156

 Step 30. Compare potential accelerated or increased sediment supply above a reference or baseline condition 5-157

Step 31. Evaluate potential consequences of increased sediment supply and/or channel stability changes 5-159

    A. Channel stability considerations and summary interpretations 5-159

    B. Sediment supply considerations and summary interpretations 5-160

Discussion and application of *PLA* 5-163

Conclusion 5-164

Chapter 6: Monitoring 6-1

Effectiveness monitoring 6-2

    Temporal measurements 6-2

    Spatial measurements 6-3

    Scale measurements 6-3

    Streamflow variation measurements 6-3

    Site or reach variation measurements 6-3

    Implementation variation measurements 6-3

    Design concepts for effectiveness monitoring 6-4

    Relating physical to biological monitoring 6-5

Calibration and validation monitoring 6-6

    Calibration monitoring 6-6

    Validation monitoring 6-6

Field methods and procedures 6-6

    (1) Frame and pin 6-8

    (2) Bar sample 6-9

    (3) Scour chains 6-10

    (4) Bank pins 6-13

    (5) Bank profile 6-13

    (6) Replicate annual survey 6-16

Conclusion 6-18

Chapter 7: *The WARSSS* Methodology:
A Case Study of Wolf Creek, Colorado 7-1

The *Reconnaissance Level Assessment* (*RLA*):
Wolf Creek Case Study 7-1

Problem identification 7-3

Table of Contents

Step 1. Compile existing data 7-3
Step 2. Review the landscape history 7-6
Step 3. Summarize activities that potentially affect sediment supply and channel stability 7-7
Step 4. Identify specific process relations 7-10
Step 5. Review the landscape overview and map the watershed 7-12
Step 6. Identify hillslope processes 7-14
Step 7. Document surface erosion 7-14
Step 8. Document mass erosion 7-15
Step 9. Assess hydrologic processes 7-15
Step 10. Identify streamflow changes 7-16
Step 11. Analyze channel processes 7-17
Step 12. Detect direct impacts to streambanks and channels 7-18
Step 13. Summarize problem verification: recognition of places, processes and sources 7-19
Step 14. Eliminate areas, sub-watersheds and/or river reaches that are not contributors to either sediment or stability problems 7-19
Step 15. Select areas, sub-watersheds and river reaches for further assessment 7-19

The *Rapid Resource Inventory for Sediment and Stability Consequence* (*RRISSC*) Assessment Level: Wolf Creek Case Study 7-21

Sequential assessment steps 7-21
Step 1. Identify land use activities 7-23
Step 2. Perform landscape and river inventory 7-26
Step 3. Determine variables influenced 7-31
Step 4. Compile data for risk rating system 7-33

Hillslope processes: Steps 5–8 7-35
Step 5. Select appropriate models and compile data for hillslope processes 7-35
Step 6. Assess mass erosion risk 7-35
Step 7. Evaluate road impact risk 7-40
Step 8. Determine surface erosion risk 7-46

Hydrologic processes: Steps 9–10 7-47
Step 9. Compile maps and data for hydrologic processes 7-47

Table of Contents

    Step 10.   Assess potential for streamflow changes 7-48

Channel processes: Steps 11–15 7-51

    Step 11.   Compile data for channel processes 7-51

    Step 12.   Assess broad-level channel stability 7-51

    Step 13.   Determine streambank erosion risk 7-51

    Step 14.   Assess in-channel mining 7-57

    Step 15.   Assess direct channel impacts 7-57

Risk potential assessment: Steps 16–20 7-62

    Step 16.   Calculate channel enlargement risk potential 7-62

    Step 17.   Calculate aggradation/excess sediment risk 7-65

    Step 18.   Determine channel evolution potential 7-69

    Step 19.   Calculate degradation risk 7-71

    Step 20.   Summarize total potential sediment and stream
                    channel stability risk 7-75

    Step 21.   Create overall risk rating summary 7-75

              Narrative Summary 7-77

    Step 22.   Discard processes with low overall risk ratings 7-80

    Step 23.   Create management change recommendations for
                    processes with moderate overall risk ratings 7-80

    Step 24.   Advance high-risk and/or high-consequence processes
                    to the *PLA* assessment phase 7-81

The *Prediction Level Assessment (PLA)*: Wolf Creek Case Study 7-83

Steps 1–4: Bankfull discharge and hydraulic relations 7-87

    Step 1.   Develop and/or obtain regional curves of bankfull discharge versus
                  drainage area and bankfull dimensions versus drainage area 7-89

    Step 2.   Delineate the watershed boundary on a USGS 7.5' Quad map.
                  Calculate drainage area in square miles 7-90

    Step 3.   Field calibrate bankfull discharge estimates by either direct discharge
                  measurements or field calibration procedures at local gage stations
                  to verify regional hydrology curves applied to ungaged sites 7-90

    Step 4.   Calculate bankfull discharge and dimensions 7-90

Steps 5–6: Level II stream classification and dimensionless ratios
                of channel features 7-93

    Step 5.   Classify streams 7-94

    Step 6.   Calculate detailed morphological descriptions (including
                  dimensionless ratios) 7-97

xvii

Table of Contents

Step 7. Identify stream stability indices 7-106

    a. Riparian vegetation 7-108

    b. Flow regime 7-111

    c. Stream order and stream size 7-112

    d. Meander patterns 7-113

    e. Depositional patterns 7-114

    f. Channel blockages 7-115

    g. Width/depth ratio state 7-116

    h. Modified Pfankuch stability rating 7-117

    i. Degree of channel incision 7-120

    j. Degree of channel confinement (lateral containment) 7-121

Steps 8–9: Streambank erosion (BANCS model) 7-123

    Step 8. Calculate Bank Erosion Hazard Index (BEHI) and Near-Bank Stress (NBS) ratings 7-125

    Step 9. Predict annual streambank erosion rate using BEHI and NBS ratings 7-131

Steps 10–17: Total annual sediment yield prediction and flow-related increases in sediment using water yield and FLOWSED models 7-137

    Step 10. Develop dimensionless flow-duration curve 7-145

    Step 11. Collect bankfull discharge, bedload sediment and suspended sediment 7-145

    Step 12. Obtain or establish dimensionless bedload and suspended sediment rating curves 7-146

    Step 13. Convert dimensionless bedload and suspended sediment rating curves to dimensioned sediment rating curves 7-146

    Step 14. Convert dimensionless flow-duration curve to dimensioned flow-duration curve 7-146

    Step 15. Calculate total annual sediment yield for both bedload and suspended sediment 7-147

    Step 16. Select and run a water yield model 7-147

    Step 17. Calculate flow-related changes in annual sediment yield 7-148

Steps 18–19: Sediment transport capacity model (POWERSED) 7-149

    Step 18. Evaluate channel characteristics that change hydraulic and morphological variables and develop hydraulic geometry relations for a wide range of flows 7-151

    Step 19. Calculate bedload and suspended sand bed-material load transport capacity (stream power) 7-152

Steps 20–21: Sediment delivery from hillslope processes 7-159

Table of Contents

Step 20. Determine potential sediment delivery from roads, surface erosion and mass erosion ... 7-161

Sediment delivery from roads ... 7-161

Sediment delivery from surface erosion ... 7-165

Sediment delivery from mass erosion ... 7-165

Step 21. Summarize total annual sediment yield (tons/year) from hillslope processes ... 7-167

Step 22: Sediment competence prediction ... 7-168

Step 23. Predict channel response based on sediment competence and transport capacity ... 7-174

Steps 24–27: Channel stability ratings by various processes and source locations ... 7-175

Step 24. Calculate potential stream channel successional stage shift ... 7-177

Step 25. Calculate lateral stability rating ... 7-179

Step 26. Calculate vertical stability ratings ... 7-182

Aggradation or excess deposition ... 7-182

Degradation/channel incision ... 7-182

Step 27. Calculate potential channel enlargement ... 7-187

Step 28: Overall sediment supply rating by individual and combined stability ratings ... 7-190

Steps 29–31: Summary evaluations ... 7-193

Step 29. Calculate total sediment yield (all sources) ... 7-194

Step 30. Compare potential accelerated or increased sediment supply above a reference or baseline condition ... 7-196

Total annual sediment yield (Step 15) ... 7-196

Sediment transport capacity (Step 19) ... 7-200

Streambank erosion (Steps 8–9) ... 7-200

Hillslope processes (Steps 20–21) ... 7-200

Sediment competence (Step 22) ... 7-200

Channel stability and sediment supply summaries (Steps 24–28) ... 7-201

Step 31. Evaluate potential consequences of increased sediment supply and/or channel stability changes ... 7-202

A. Channel stability considerations and summary interpretations ... 7-202

B. Sediment supply considerations and summary interpretations ... 7-209

Appendix A ... I

Table of Contents

References VII

Subject Index XXI

"The answer is not in the model or in the book...it is in the river."
— Luna B. Leopold

CHAPTER **1**

The WARSSS Assessment Methodology

Introduction

The cumulative effects of watershed development can have serious adverse consequences on water resources. Because of the diversity of river types and land uses, however, not all systems respond similarly to treatment. The challenge for land managers and researchers is to identify impacts from current land use practices, as compared to past land uses, and to separate natural geologic erosion rates from anthropogenic influences.

The 1996 National Water Quality Inventory (USEPA, 1997a) ranks sediment as a leading cause of water quality impairment in assessed rivers and lakes. Accelerated erosion and deposition from hillslope and channel processes can impair designated uses in rivers in many ways. Stream channel instability caused by excess sediment deposition can have severe biotic impacts on food chains, spawning and rearing habitat, instream cover and water temperatures. Adverse sediment impacts also include damage to recreation, navigation, conveyance, water treatment systems and water storage. Excessive sediment deposition and transport problems in surface waters have resulted in annual damage costs of approximately $16 billion (Osterkamp et al., 1992).

Watershed Assessment of River Stability and Sediment Supply (*WARSSS*) is a geomorphology-based procedure for quantifying the effects of land uses on sediment relations and channel stability. Candidates for the *WARSSS* assessment include, but are not limited to, Clean Water Act Section 303(d)-listed streams with a variety of impairments that may be caused by sediment imbalances or channel instability. Streams that have ceased to be functional for either physical and/or biological purposes may also require watershed assessments. The *WARSSS* methodology provides the following:

- A mechanism to put fundamental geomorphic principles into practice;

- A consistent, quantitative and comparative analysis that minimizes subjective bias;

- Watershed-based and specific river reach assessments;

- Linkages between various land uses and their associated sources of accelerated sediment supply;

- A procedure that rapidly prioritizes high-risk sub-watersheds and river reaches at broad "screening" levels yet provides for more detailed assessments;

- Methods to assess the probability, risk and potential consequences of sediment problems despite the inevitable gaps in knowledge, large uncertainty and spatial and temporal variability;

- A basis for mitigation and/or restoration plans that isolates processes responsible for high-risk/adverse sediment-related consequences of watersheds and river systems;

- A companion assessment that can be related to aquatic habitat;

- A summary of output parameters useful for assessment of sediment consequences; and

- A time-integrated procedure to assess sediment and stability problems of the past and present and to help set environmentally sound management direction for the future.

WARSSS identifies the hillslope, hydrologic and channel processes responsible for significant changes in erosion, sedimentation and related stream channel instability. It uses a three-phase assessment process to quickly separate areas into low-, moderate- and high-risk landscapes and/or river reaches. This process allows large watersheds to be assessed efficiently without sacrificing the ability to identify specific problem sites in need of mitigation. *Mitigation* is associated with a combination of potential activities including, but not limited to, rehabilitation, stabilization, restoration, enhancement and changes in land use management to reduce the adverse consequences of sediment-related impairment.

The results of the *WARSSS* assessment reveal significant, adverse influences of land uses on stream channel stability, sediment sources and sediment yield that may affect the material beneficial uses of rivers and streams. *WARSSS* data can be used for watershed planning, "clean sediment" Total Maximum Daily Load (TMDL) assessments for non-point source pollution and stability analysis for river restoration. The *WARSSS* methodology is a flexible framework that permits the substitution of other appropriate prediction methods for specific processes based on the user's familiarity and experience with the models.

This book provides a detailed explanation of the *WARSSS* procedure, along with guides for field research, study design and worksheets to complete the assessment levels. It presents examples and a final case study of a full *WARSSS* implementation to assist others who are undertaking watershed-based assessments.

Chapter 1 presents a summary of objectives, a discussion of the potential impacts of various practices on hillslope and channel processes and an overview of the *WARSSS* methodology. The *WARSSS* process is shown in **Flowchart 1-1**. The colors used to indicate the three assessment phases are also used to quickly distinguish related material in the following chapters. For example, the *Reconnaissance Level Assessment (RLA)* material is shown in "green." A green line will appear at the top of every page of Chapter 3 (*RLA*) and again in the *RLA* section of the case study in Chapter 7.

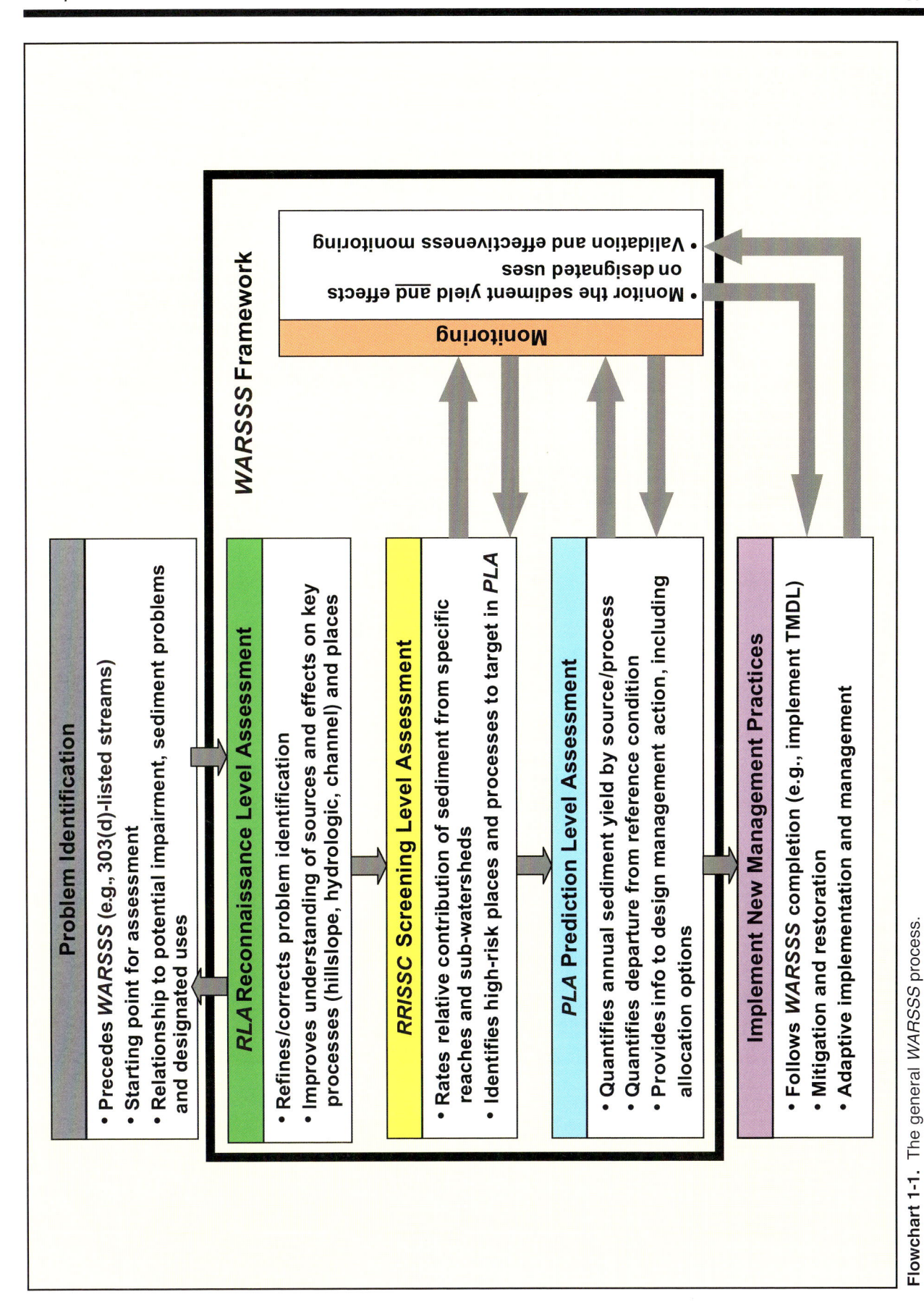

Chapter 1 — The WARSSS Assessment Methodology

WARSSS Framework

Monitoring
- Monitor the sediment yield and effects on designated uses
- Validation and effectiveness monitoring

Problem Identification
- Precedes WARSSS (e.g., 303(d)-listed streams)
- Starting point for assessment
- Relationship to potential impairment, sediment problems and designated uses

RLA Reconnaissance Level Assessment
- Refines/corrects problem identification
- Improves understanding of sources and effects on key processes (hillslope, hydrologic, channel) and places

RRISSC Screening Level Assessment
- Rates relative contribution of sediment from specific reaches and sub-watersheds
- Identifies high-risk places and processes to target in PLA

PLA Prediction Level Assessment
- Quantifies annual sediment yield by source/process
- Quantifies departure from reference condition
- Provides info to design management action, including allocation options

Implement New Management Practices
- Follows WARSSS completion (e.g., implement TMDL)
- Mitigation and restoration
- Adaptive implementation and management

Flowchart 1-1. The general WARSSS process.

1-3

Chapter 2 presents general principles related to hillslope and channel processes and provides examples of specific assessment applications. Chapter 3 presents the *Reconnaissance Level Assessment* (*RLA*), the first of three major assessment phases. This is the broadest screening level, designed to separate low- and high-risk landscapes and river reaches. *RLA* identifies those areas that do not appear to contribute to a sedimentological, hydrological or morphological problem. These low-risk areas can be eliminated from more detailed, time-consuming field data collection and analysis.

Chapter 4 describes the second assessment phase, the *Rapid Resource Inventory for Sediment and Stability Consequence* (*RRISSC*). This phase subjects the moderate- and high-risk landscapes and river reaches identified in *RLA* to an intermediate level of assessment. Rapid assessments of generalized, process-specific nomographic relations are developed and explained in this chapter. *RRISSC* identifies and prioritizes low-, moderate- and/or high-risk slopes, sub-watersheds and river reaches using a comparative analysis based on process/land use and erosional/depositional relations. Low-risk areas do not require mitigation. Moderate-risk areas identified at this level may require recommended changes in management and/or site- and process-specific mitigation, along with subsequent effectiveness monitoring. The identified high-risk areas advance to the third and final assessment phase.

Chapter 5 describes the most detailed assessment phase, the *Prediction Level Assessment* (*PLA*). This assessment level utilizes a consistent, quantitative, comparative analysis of potential impairment compared to a reference condition. The analysis employs models of hillslope, hydrologic and channel processes for specific locations and individual land uses. This allows land managers to address the causes and consequences of impairment for explicit land uses at a designated location. The outputs are designed to indicate spatially sensitive, diverse and disproportionate sediment sources. Mitigation recommendations at this level can effectively target identified problems, processes and places. Monitoring is recommended not only to validate a given model but also to determine mitigation effectiveness.

Chapter 6 describes validation and effectiveness monitoring methods. Validation monitoring reduces the uncertainty of prediction, while effectiveness monitoring determines the success of management changes and/or specific mitigation practices, including project implementation. This chapter also provides study design guidance and methods for field monitoring.

Chapter 7 is a case study of the Wolf Creek Watershed in Southwestern Colorado. The case study details the implementation of all three *WARSSS* assessment levels, including discussion, interpretation and outcomes for sediment supply and stability.

The conceptual watershed assessment model for *WARSSS*, including implementation and monitoring, is shown in **Flowchart 1-2**.

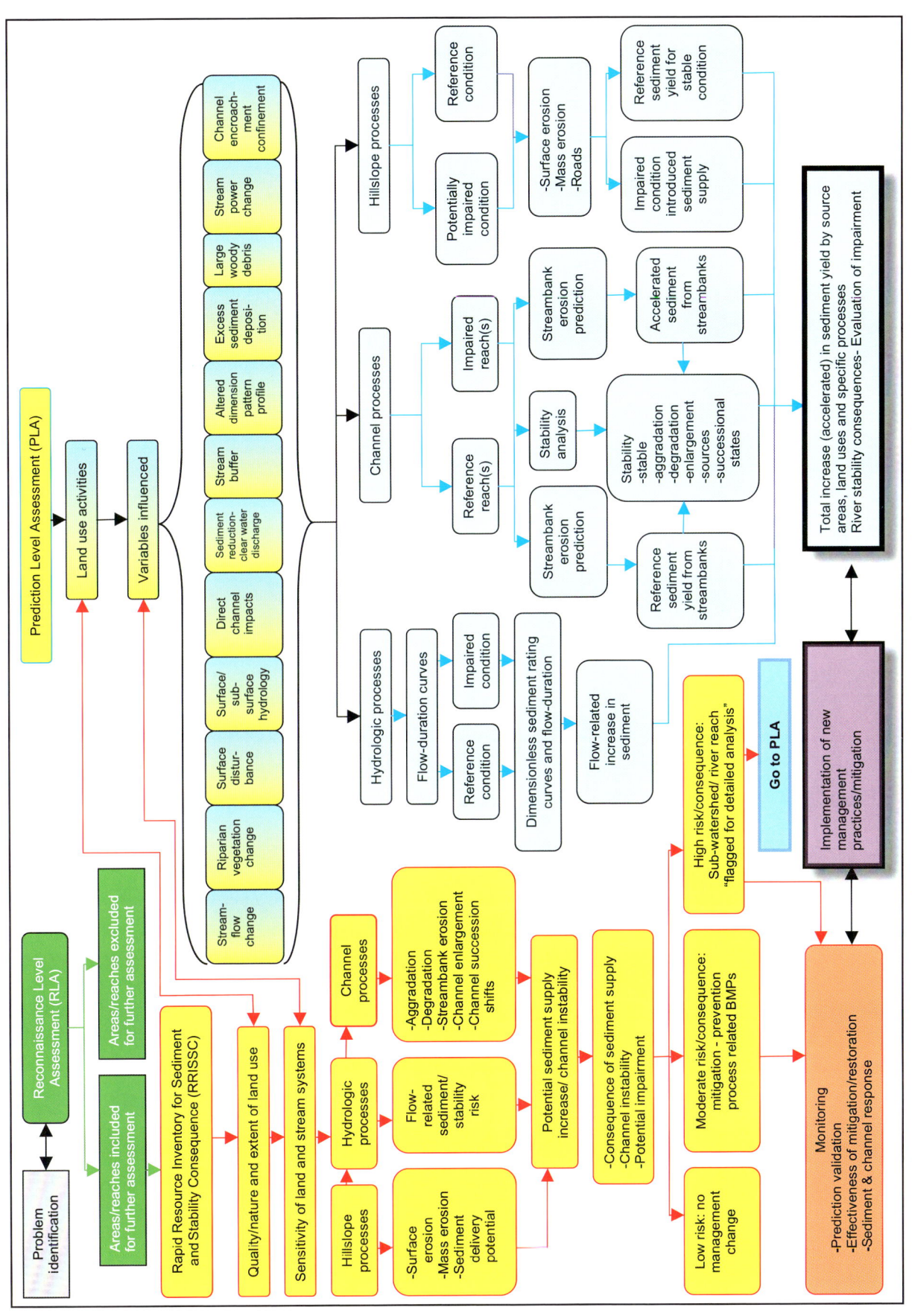

Flowchart 1-2. Conceptual model for *Watershed Assessment of River Stability and Sediment Supply* (WARSSS).

Basic assumptions for the use of WARSSS

The importance of field measurement

The *WARSSS* methodology is not exclusively a desktop modeling procedure. It relies heavily on field data collection and corresponding analysis.

> **Caution: Due to the nature of this methodology, it is essential that assessments be conducted by individuals with training and experience in geomorphology, hydrology, engineering, geology, soil science, plant science and other related scientific disciplines. Individuals should be specifically trained and experienced in hillslope, hydrologic and channel processes.**

Identification, prediction and measurement of sediment sources

The most definitive approach to quantify sediment contributions is to make direct measurements at known source locations. Because sediment is associated with a great variety of temporal and spatial variability, it requires considerable effort to:

- Identify natural variability with flow/season/geology;
- Determine contributing erosional processes;
- Document land use influence on magnitude/duration;
- Understand consequences of sediment changes;
- Assess departure from a stable reference condition;
- Compare sediment from past versus existing land use;
- Quantify the influence of sediment source/stability on upstream and downstream reaches; and
- Conduct measurements of stream discharge, velocity, slope and channel characteristics to accompany sediment data analysis.

Thus, measurements must be accompanied by an assessment of hillslope, hydrologic and channel processes that identifies the nature, extent and location of specific erosional processes linked to past and present land uses.

The assessment of potential departure from a reference condition requires measurements of sediment and channel stability from several sources, including:

- Concurrent channel measurements above and below specific stability/sediment source locations;
- Measurements before and after the onset of contributing source activities;
- Controlled or paired watersheds with concurrent measurements; and
- Reference streams that represent the stable form of the same valley type, stream type and vegetative type.

Measurements must be obtained in multiple locations to address the beneficial uses of the whole water body. Without data obtained over a wide range of flow conditions (including bankfull discharge or normal high flow, along with the rising and recession limb of both stormflow and snowmelt hydrographs), it is difficult to correctly interpret sediment data. High costs, specialized equipment, the difficulty of high water measurements and a lack of training and experience in standardized measurement techniques often preclude proper sediment data collection and analysis. Sediment measurements, however, can and should be obtained by trained individuals to determine the reliability of prediction, to calibrate specific models over large areas and to determine the effectiveness of management, mitigation and/or restoration efforts at specific locations. Sediment measurements are vital in areas where controversy, risk and consequences are high.

Stability measurements should include an annual resurvey of permanent cross-sections/profiles, streambanks and the channel material size study sites. This information can validate stability predictions of dimension, pattern, profile and channel materials. Chapter 6 describes the design and methods for effectiveness and validation monitoring.

River stability, sediment and land uses

Land use influences on river stability and sediment

Surface disturbance activities, roads, streamflow changes, channel alteration and major vegetation impacts can affect multiple processes leading to stream channel instability and corresponding changes in sediment supply and transport. Stream channel adjustments due to channel instability (disequilibrium) often result in accelerated sediment yields and associated adverse consequences.

The terms *equilibrium* and *stream channel stability* are often used interchangeably. *Stability* is defined as a river or stream's ability in the present climate to transport the streamflows and sediment of its watershed over time in such a manner that the channel maintains its dimension, pattern and profile without either aggrading or degrading (Rosgen, 1996, 2001b). *Aggradation* is a raising of local base level due to sediment depositional processes, whereas *degradation* is a lowering of local base level due to channel incision processes.

WARSSS includes a river stability component that addresses processes of flow and sediment changes below reservoirs and streamflow modifications due to urban and non-urban changes in the hydrologic response to various land uses. The products of accelerated erosion and deposition, as well as their associated impacts on river stability and on corresponding physical and biological function, are key issues. The natural, geologic, erosional and depositional processes need to be understood in order to determine acceleration of these processes or departure from natural, stable conditions due to poor land use practices.

WARSSS is designed to isolate the location of specific problem practices that contribute disproportionate amounts of sediment. The identification of such problems will help the user design process-specific treatment. *WARSSS* is also designed to help land use planners avoid potential future problems and prioritize mitigation through changes in land use practices, restoration and/or rehabilitation.

The extent to which land use practices might generate river-impairing sediment levels needs to be assessed in a rigorous, consistent and quantitative manner. The stability of the affected lands and streams and the type of land use practices affect the risks associated with potential sediment problems. The major categories of erosional/depositional processes potentially influenced by a wide range of land uses include:

- Surface erosion;
- Mass erosion (slump/earthflow, debris torrents);
- Stream channel instability/channel adjustments;
- Accelerated streambank erosion;
- Channel degradation/incision;
- Channel aggradation/excess deposition;
- Lateral channel containment/confinement;
- Accelerated channel meander migration; and
- Channel enlargement.

Current sediment problems are not only a product of the various geologic erosional rates but are also a legacy of historical land use activities. Many past land uses, such as placer mining at the turn of the last century, are responsible for channel adjustments that continue to add sediment to stream systems (Pickup et al., 1983; Pickup, 1988; James, 1991; Riefenberger and Baird, 1991). Accelerated sediment deposition following poor farming practices in the Piedmont province of the Eastern United States aggraded many rivers. Today, with improved farming practices resulting in reduced hillslope erosion, these rivers are routing excess sediments out of the system.

To correct past sediment and stability problems, many projects to restore aquatic ecosystems have been implemented in recent years (National Research Council, 1992). For restoration projects to succeed, the design must separate natural from anthropogenic influences on the fluvial system. An assessor must understand natural recovery following disturbance, evolution of morphological states, response of systems to imposed change and the natural stable state.

Channel changes can be adversely affected by reduced sediment loads, as well as by excess sediment. Either condition may lead to degradation, loss of aquatic habitat and long-term stream channel instability. This has been demonstrated in the literature regarding "hungry water" degradation below reservoirs and clear water discharge from storm water drains in urban rivers (López, 2004; Kondolf, 1997). An equilibrium state, or a balance of energy and sediment supply, is central to the evolution of stable morphological forms and associated beneficial water uses.

The influence of major flooding on river stability and sediment

The assessment of land use erosional impacts on sediment yields and stream ecosystems is often complicated by the episodic nature of major flood events. Such events generate large amounts of sediment in river systems. How well the landscape and streams accommodate natural events is influenced by modifications to surface and internal drainage via vegetative changes, direct channel impacts and road systems. Vegetative conversion from woody to grass/forb riparian communities, or direct alterations to stream channels, such as channelization, abandonment of floodplains and confinement of river systems, can impair a stream's ability to deal with flooding. For example, the 1982 Lawn Lake flood in Larimer County, Colorado, affected both the Roaring River and the Fall River, two different stream types, in very different ways. The breach of a man-made reservoir resulted in a flood magnitude 30 times that of the 500-year flood on both of the rivers. The Lawn Lake flood, however, resulted in far more damage on the Roaring River than on the Fall River.

The Roaring River, an A3a+ stream type (Rosgen, 1994), is a steep-gradient, entrenched, low width/depth ratio, large cobble-bed channel associated with heterogeneous, unconsolidated materials of glacial till. The geomorphic consequences of the 500-year flood on the Roaring River were catastrophic. Channels increased in width from 10–16 ft to 70–500 ft and increased in depth from 1–2 ft to 5–50 ft (Jarrett and Costa, 1986). The geomorphological nature of these A3a+ stream types is associated with infrequent debris torrents/avalanches that respond similarly to major events. The A3a+ stream types are associated with high energy, high sediment supply channel systems that transport enormous sediment loads. An extensive alluvial fan of 394,600 yds^3 was created at the mouth of the Roaring River, where seven-foot-diameter boulders were deposited. The flood waters were then routed into the low-gradient Fall River valley downstream of the alluvial fan.

The Fall River, whose bankfull discharge is approximately 380 cfs, was estimated at 12,000 cfs after the Lawn Lake flood. The U.S. Geological Survey estimated that this flood was the largest since the end of the last glacial retreat at the start of the Holocene period, 10,000 years ago (Jarrett and Costa, 1986). The Fall River is a meandering, gravel-bed alluvial channel in a broad valley with a well-developed floodplain. It classifies as a C4 stream type with a sinuosity of 2, slope of 0.005, D_{50} of 35 mm, a width/depth ratio of 16 and an entrenchment ratio of 24 (Rosgen, 1994). The riparian willow vegetation is very dense throughout several miles of its length.

Following the Lawn Lake flood, the dimension, pattern and profile of the Fall River remained unchanged. The tight meanders of the river remained intact, with accumulations of sand in the bed that took several years to redistribute (Pitlick, 1985). The Fall River represents a stable reference reach channel with a well-developed floodplain that had not been extensively grazed. A *reference reach* is a stable stream type, not necessarily a pristine stream, that represents the same stream and valley type as an impaired stream reach. Large elk populations had been utilizing the willows along the Fall River without changing the density and/or composition of the riparian vegetation. As a result, the stream remained intact despite the high sediment supply and rare, extreme-magnitude flood. **Figure 1-1**, a photograph of the Fall River taken in 2001, shows this stable alluvial channel reach downstream of the alluvial fan. Many incorrectly believed that an alluvial channel would not maintain the same dimension, pattern and profile following such a catastrophic flood.

Figure 1-1. Fall River, C4 stream type (2001), looking downstream within 1/4 mile of the alluvial fan from the Lawn Lake flood. Note the dense willow riparian vegetation maintaining natural stability. This channel is in the same location as before the flood.

As streams are self-formed, they should also be self-maintaining. A stable system, such as the Fall River, presents a good example of the process of self-maintenance.

The downstream reach of the Fall River extends into the town of Estes Park. This reach had been altered, straightened and was confined by residential development. The direct alterations of the channel created river instability, making this reach more susceptible to flood effects. The instability of the lower reach of the Fall River set it up for failure. The flood damage through this reach was extensive, causing millions of dollars worth of damage. The stream channel was severely eroded and required extensive flood rehabilitation using federal and state agency flood-relief funds. None of these funds were expended on the upper Fall River, as this reach was a naturally stable, self-formed and self-maintained river system. Had the upper Fall River been altered and destabilized as the lower reach, however, the flood would have likely resulted in similar severe and expensive adverse consequences.

Landscapes and stream systems like the lower reaches of the Fall River are often prone to failure, depending on the extent of land uses that affect their natural function. *WARSSS* assesses the relative risks and consequences of sediment generation and transport influenced by past, current and potential future land use activities that contribute to erosional and depositional processes, including their effects on floods.

Chapter 1 *The WARSSS Assessment Methodology*

Using existing spatial data: GIS and remote sensing

Researchers should obtain and interpret inventory information that identifies potential land use impacts on a diverse landscape at all assessment levels. Much of this initial inventory can be accomplished by using aerial photography, which is widespread, or by using multi-spectral and hyper-spectral imagery where available. GIS LiDAR (Light Detection and Ranging) images should also be used in areas with heavy vegetation. LiDAR imaging removes vegetation effects for a more accurate depiction of stream and ground features. In addition, recent satellite data has improved resolution, frequency of coverage and product availability and is often suitable for landscape analyses in large watersheds. Appropriate systems for specific applications are summarized by Lillesand and Kiefer (1999).

It is increasingly commonplace to find GIS data, such as land use/land cover or detailed drainage mapping from completed remote sensing analyses. Such data can help provide a unique analysis more suited to sediment computations. For example, suspended sediment concentrations sufficient to affect spectral reflectance can be identified on aerial photography or imagery and validated. A sediment source study conducted on the West Fork Madison River in Southwestern Montana used color-infrared photography taken from helicopter, supplemented with ground truth, to quantify sediment source areas during snowmelt runoff (Rosgen, 1973, 1976). Photo density readings from the transparencies of the color-infrared photographs were correlated with actual suspended sediment measurements taken over a wide range of concentrations, channel sizes and elevations. The relations produced a correlation coefficient R^2 of 0.95, which was highly significant at the 99% level (Rosgen, 1973, 1976). Once this correlation was established, photo density readings were obtained for 100% of the streams during snowmelt runoff to determine sediment contributions by specific reach and sub-watershed. This approach proved to be a cost-effective and reasonable means of determining sediment source locations over a large area in a short period of time.

Potential uses of WARSSS

Future watershed development plans must contain assessments of past and current land uses and their cumulative watershed impacts. Numerical water quality standards for non-toxic or clean sediment have not been set in many states given the great complexity and uncertainty of prediction and the extreme natural variability over temporal and spatial scales. This complexity has also precluded the establishment of universal sediment standards. Nonetheless, widespread adverse sediment impacts on beneficial water uses require consistent, quantitative assessment methods.

Rather than attempting to predict or develop a numeric sediment standard, *WARSSS* uses hillslope, hydrologic and channel erosional/depositional process relations to identify specific disproportionate sediment sources that impact beneficial water uses. When used to implement clean sediment TMDLs, *WARSSS* emphasizes the potential proportional contribution of sediment sources influenced by land uses rather than focusing on model outputs that predict "absolute" total annual sediment yield. Although annual sediment yields are calculated at the *PLA* analysis level, the values are used to apportion relative source contributions and assess the likely consequences for beneficial uses. This emphasis leads directly to appropriate management, improved design specifications and alleviation prescriptions.

WARSSS is appropriate for evaluating physical water resources and their condition for watershed planning. It is also compatible with the integration of biological assessments, such as fish habitat indices, and with restoration designs that require a stability assessment to address the causes, consequences and corrections of river impairments. The majority of the *WARSSS* procedures contained herein have been successfully applied in river assessment and restoration for over 15 years.

Conclusion

Watershed management has historically placed more emphasis on project planning and design than on project response or evaluation. Sediment and stream reach stability problems have traditionally been treated by "patching symptoms" rather than by treating the *cause* of the problem.

The *WARSSS* methodology provides a procedure to assess large watersheds with a practical, rapid screening component that integrates hillslope, hydrologic and channel processes. It is designed to identify the location, nature, extent and consequence of various past, as well as proposed, land use impacts. Before changes in land use management are implemented, it is of utmost importance to first understand the cause of the impairment. A good assessment can lead to more effective solutions, including ***preventive strategies*** that may be cheaper and more effective than restoration.

The *WARSSS* procedure also helps scientists to integrate existing resource inventory information into a structured, reproducible analysis. The underlying purpose of the method is to better understand the four "C's" of river/watershed assessment:

1. The *cause* of the impairment (problem);
2. The *consequence* of change;
3. The *correction* (prevention, mitigation, stabilization, restoration) of the problem; and
4. The *communication* of assessment results and recommendations.

"We must consider and decide for ourselves, what hydrology is, where it lies in the spectrum of human knowledge and endeavor, between the extremes of mere technical application and the deeper understanding of science."
— Nash (1992)

CHAPTER 2

A Review of Hillslope, Hydrologic and Channel Process Relations

Those conducting watershed assessments must determine the location, nature and magnitude of geologic versus anthropogenic sediment supply and/or channel instability, and must understand their associated causes and consequences. The hillslope process component of watersheds often results in accelerated erosion and delivered sediment, causing river instability and impairment. Identification of the sources, magnitude and sediment delivery potential of mass erosion, surface erosion and/or road source erosion leads to specific treatment at identified locations. Understanding hillslope processes may also help prevent accelerated sediment supply.

Changes in sediment supply can, in turn, greatly affect a stream's hydrologic and channel processes, including its ability to transport sediment and maintain channel stability. Before proceeding with the *WARSSS* methodology, it is important to review some fundamental relations among hillslope, hydrologic and channel processes. This chapter discusses currently available models for calculating the effects of these processes and offers recommendations for their use. Specific formulas used in subsequent risk assessments are also examined.

Hillslope erosional processes

Surface erosion

Surface erosion is the wearing away of the surface soil by water, wind, ice or other geological processes. The potential for surface erosion is associated in part with the amount of bare, compacted soil exposed to rainfall and runoff. An example of overland flow, potentially detaching soil for transport on a compacted slope created by log skidding activity in Idaho, is shown in **Figure 2-1**.

WATERSHED ASSESSMENT OF RIVER STABILITY AND SEDIMENT SUPPLY

Figure 2-1. Overland flow associated with surface erosion on compacted areas due to logging, Idaho.

The two main types of processes associated with surface erosion are as follows:

1. Those that generate erosion, such as rill, sheet wash, dry ravel or freeze-thaw; and
2. Those that route eroded soil to drainageways (sediment delivery).

A general approach to predicting sediment yields from surface erosion is the Pacific Southwest Inter-Agency Committee (PSIAC) method (Pacific Southwest Inter-Agency Committee, 1968). It involves rating a watershed on nine factors: surface geology, soil, climate, runoff, topography, land use, upland erosion, channel erosion and sediment transport. There are additional soil loss models, including the Universal Soil Loss Equation (USLE) (Wischmeier and Smith, 1978), the Modified Soil Loss Equation as applied in WRENSS (USEPA, 1980) and the Water Erosion Prediction Procedure (WEPP), as well as the methods of Lane and Nearing (1989), Nearing et al. (1989) and Flanagan and Nearing (1995). Elliot and Hall (1997) adapted the WEPP model for forest soils (FSWEPP), and a recent version of the Revised Universal Soil Loss Equation (RUSLE) (Renard et al., 1997) also provides estimates of surface soil loss.

Due to the inherent complexity of sediment routing on hillslopes, sediment delivery is difficult to predict and verify. Direct observation over short lengths of slope is possible if soil has been transported beyond its location of initial detachment. The quantification of a sediment delivery ratio has been extrapolated per unit area of watershed from downstream reservoir surveys (Wischmeier and Smith, 1978); however, this generally overestimates sediment due to surface erosion as sediment from streambank erosion is not separately expressed. The USLE rainfall erosivity factor used in place of the storm runoff parameter by Williams (1975) is designed to substitute for a sediment delivery ratio and is favored over the above drainage area relation. The sediment delivery relation in WRENSS (USEPA, 1980) evaluates site conditions related to sediment delivery potential. According to Reid and Dunne (1996), Williams and Berndt (1972) provide the best predictions of measured sediment delivery, using:

$$S_{dr} = 63\, S_m^{0.40} \qquad \text{(Equation 2-1)}$$

where:

S_{dr} = sediment delivery ratio

S_m = average slope of the mainstream channel

The WEPP model adaptation by Toy and Foster (1998) is applied in mined-lands erosion studies. Prediction methods associated with road-related sediment using a USLE-based approach are shown by Reid and Dunne (1984), and the FSWEPP-based approach is shown by Elliot and Foltz (2001). Caution should be exercised in the use of this relation if there is slope discontinuity and/or changes in slope shape, surface debris or lack of rills, as **Equation 2-1** would potentially over-predict delivered sediment from onsite surface erosion in these cases. This relation may be appropriate for hillslopes in very close proximity to the channel slopes used in the prediction.

These surface erosion models rely on the assumptions of *Hortonian overland flow*, where precipitation rate exceeds infiltration rate, and many such models were developed for low-gradient, agricultural applications. Use of these models often requires local calibration. These models were designed for site-specific prediction rather than watershed-wide levels of application. The *RRISSC* phase of *WARSSS* uses more general risk relations to screen for potential surface erosion/sediment yield contributions. The *PLA* phase of assessment for specific high-risk areas, as identified at the *RRISSC* screening level, uses appropriate models as recommended in *WARSSS* or selected by the assessor.

Vegetative ground cover density is generally the most sensitive variable in these models. Ground cover includes live plants as well as surface organic litter. Increased risk of erosion and sediment delivery is associated with high soil erodibility; little ground cover; steep, long, continuous slopes; high-intensity storms; high drainage density of the slope and close proximity to streams (e.g., road fills closer than 61 meters or 200 ft). Mitigation for surface erosion (surface drainage diversion, terracing, alternate row crops, lop and scatter logging debris, surface mulch and seeding, etc.) generally provides an increase in ground cover or surface protection and breaks up continuous slope length. Best management practices for roads involve road and ditch surfacing, vegetating ditches, out-sloping the road surface, cross-drains, vegetating cut banks and road fills and proper culvert and bridge designs.

Mass erosion

There are two primary types of mass erosional processes:

1. Shallow, fast movements of debris avalanches/debris torrents and mudflows that generally move in response to extensive (high intensity and/or long duration) precipitation events; and

2. Slow, deep-seated slump/earthflow erosional processes that move intermittently over varying time scales in response to infrequent events and/or disturbance factors.

Evidence of mass erosion can be detected on aerial photographs; thus, slide frequency and magnitude can be observed using time-sequential aerial photos. Examples of debris torrents are shown in **Figure 2-2** in the Idaho batholith and in **Figure 2-3** on an A3a+ stream type in Colorado. Mass erosion slides are shown in **Figure 2-4** (Willow Creek, Colorado) and in the marine shale geology of **Figure 2-5** (Blue River, Colorado).

Sediment delivery is often estimated by measuring the concave slope remnant less that portion of the slide mass removed by fluvial entrainment. Annual sediment yield associated with mass erosion processes is extremely difficult to predict due to the episodic nature of climatic events that initiate movement. Often, landslides occur many years after vegetation change and road construction because of the complex interactions of root mass decay and soil saturation from major storms. The short-term impact to stream channels from landslides triggered by roads or other imposed land use activities may be of more importance than their long-term contribution to annual sediment yield.

Models to predict annual sediment yield from mass erosion do not presently exist. Any predictive assessment of these erosional processes must rely

Figure 2-2. Debris torrent form of mass erosion, North Fork Clearwater River, Idaho.

Figure 2-3. An A3a+ stream type depicting debris torrent mass erosion, Colorado.

Chapter 2   *A Review of Hillslope, Hydrologic and Channel Process Relations*

primarily on the experienced individual who can recognize the relative stability/instability of an area from soil and geology maps, aerial photographs, vegetation indicators and field observations. GPS, digital terrain models and other appropriate tools can speed up the mapping process of these features. An overlay of existing and proposed road systems placed over landslide risk maps provides valuable warning indicators of past and/or impending potential high-risk areas.

Nilsen et al. (1976), Swanson and Swanson (1976), Dietrich and Dunne (1978), Caine (1980), WRENSS (USEPA, 1980), Reid (1981), Kelsey (1982), Rice and Pillsbury (1982), Burroughs (1984), Wilson (1985) and Benda and Dunne (1987) have methodologies that assess landslide potential related to controlling variables, triggering events and potential sediment contributions. A prediction methodology based on terrain mapping and landslide occurrence for quantitatively predicting landslide susceptibility following timber harvest was developed by Howes (1987). *WARSSS* users can select from this range of applied research methods for the identification and prediction of specific landslide processes appropriate to their assessment site.

Figure 2-4. Slump/earthflow erosion process adjacent to stream, Colorado.

Figure 2-5. Slump/earthflow erosional processes adjacent to Blue River, Colorado. Exposed erosional faces are now influenced by fluvial erosion.

Sediment delivery potential estimated from the methods presented in WRENSS (USEPA, 1980) may be compared to direct field measurements of slide activity. Sediment delivery may be estimated by calculating a delivery ratio by dividing the slide mass removed from the slide path or toe of slide into the channel by the initial mass involved in the landslide. Relations may be back-calculated from actual slides based on failure type, slope gradient, channel type, proximity to channel and vegetative cover to empirically generate values for sediment delivery potential.

Mitigation associated with various land uses in landslide-prone areas involves the following strategies:

- Avoid mapped "high hazard" slump/earthflow areas with large-scale vegetation conversions, such as clearcut silvicultural treatments;
- Construct roads in areas with a low hazard for debris torrent and avoid slump/earthflow terrain;
- Concentrate surface or intercepted sub-surface runoff from existing roads onto low-risk slopes and disperse runoff;
- Keep cut bank heights to a minimum and avoid through-cuts; and
- Stabilize abandoned roads by redistributing road fill on road prism and cut bank and out-sloping the road.

These mitigation procedures reduce sub-surface flow interception, disperse surface runoff, discourage traffic and allow for the revegetation of disturbed areas. Drainage crossings should be cleared to prevent fill erosion and debris torrent "dams."

Channel processes

Transport physics, sediment size, sediment load, increases in the magnitude and duration of streamflow and the stability of streambanks and beds all influence the contribution of sediment from channel processes. Hillslope processes reflect the potential sediment delivered to the channel. How the stream accommodates delivered sediment is dependent on the particle size and magnitude of the delivered sediment load, as well as the stream's competence and capacity.

Measurements of suspended and bedload sediment are not routinely obtained; thus, available U.S. data is limited. Relations based on measured sediment and flow data over a longer period are routinely extrapolated to rivers without sediment measurements. Bedload data, which is key for assessment of stream stability and habitat change, is rare due to the difficulty of obtaining measurements during high flows. As a consequence, the analysis of measured bedload transport magnitude and frequency has traditionally relied on modeled transport rates. The difficulty in prediction is often selecting the appropriate model for the stream type. Models are frequently used without calibration, which requires *measured* bedload data. Without calibration, predicted values can vary from actual rates by many orders of magnitude. Bedload prediction is essential to maintain the natural stability and fish habitat features of rivers (e.g., glides, pools and spawning redds) and the conveyance of sediment, especially the coarser component of total sediment transport. To protect a river's designated water uses, such as recreation or aquatic life support, it is often as important to ascertain the channel response to a given sediment load and/or size as it is to predict its response to total annual yield or export. Total annual sediment yield, however, is valuable to compare proportionate contributions from various watershed sources and processes.

Bedload transport

Bedload is that portion of the total sediment in transport that is carried by intermittent contact with the streambed by rolling, saltating, sliding and bouncing. Bedload transport prediction has two broad approaches: those calculated by force, and those calculated by power. The summary by Yang (1996) describes nine specific bedload formula approaches:

1. Shear stress
2. Energy slope
3. Discharge
4. Velocity
5. Bed form
6. Probabilistic
7. Stochastic
8. Regression
9. Equal mobility

There appear to be as many approaches to determining bedload transport in the literature as there are stream conditions.

Vanoni (1975, p. 190) offered an earlier, but still relevant, assessment of the status of sediment transport formulas:

> "Unfortunately, available methods or relations for computing sediment discharge are far from satisfactory, with the result that plans for works involving sediment movement by water cannot be based strongly on such relations. At best these relations serve as guides to planning, and usually the engineer is forced to rely on experience and judgment in such work."

An excellent summary of sediment transport equations, including bedload relations, is presented in Reid and Dunne (1996). The application of any sediment transport relation relies on an understanding of its assumptions and limitations and the availability of measured data to properly calibrate the model. Some basic concepts used in most sediment transport models are discussed in the following sections.

Sediment entrainment

When the flow conditions exceed the criteria for incipient motion, sediment particles on the streambed start to move. The transport of bed particles in a stream is a function of the fluid forces per unit area—the tractive force or shear stress, τ, acting on the streambed. Under steady, uniform flow conditions, the shear stress is:

$$\tau = \gamma RS \qquad \text{(Equation 2-2)}$$

where:
    γ = specific weight of the fluid

    R = hydraulic radius (ratio of cross-sectional area to wetted perimeter, often the same as mean depth for wide channels)

    S = water surface slope

The gravitational force resisting particle entrainment, f_g, is proportional to:

$$f_g \propto (\gamma_s - \gamma/\gamma)D \qquad \text{(Equation 2-3)}$$

where:
    γ_s = specific weight of sediment

    D = particle diameter

The Shields relation (Shields, 1936) with modifications by Graf (1971) developed relations associated with initiation of particle movement using the ratio of the fluid forces to the gravitational force proportional to the dimensionless quantity, called the critical dimensionless shear stress, τ^*_c, where:

$$\tau^*_c \propto \frac{dS}{(\gamma_s - \gamma/\gamma)D} \qquad \text{(Equation 2-4)}$$

where:
    d = mean depth

The dimensionless bed-material transport rate per unit width of streambed, Q^*_B is:

$$Q^*_B = \frac{Q_s}{[(\gamma_s - \gamma/\gamma)gD]^{1/2}D} \qquad \text{(Equation 2-5)}$$

where:
    g = gravitational acceleration

    Q_s = volumetric transport rate per unit width of streambed determined from bedload samples

The empirical function developed by Parker (1979) is:

$$Q^*_B = \frac{11.2(\tau^* - \tau^*_c)^{4.5}}{(\tau^*_c)^3}$$ (Equation 2-6)

where:
τ^* = dimensionless shear stress

τ^*_c = threshold value of τ^* required to initiate particle motion

The widely used application of critical shear stress by particle size class as initially developed by Shields (1936) is shown in **Figure 2-6**.

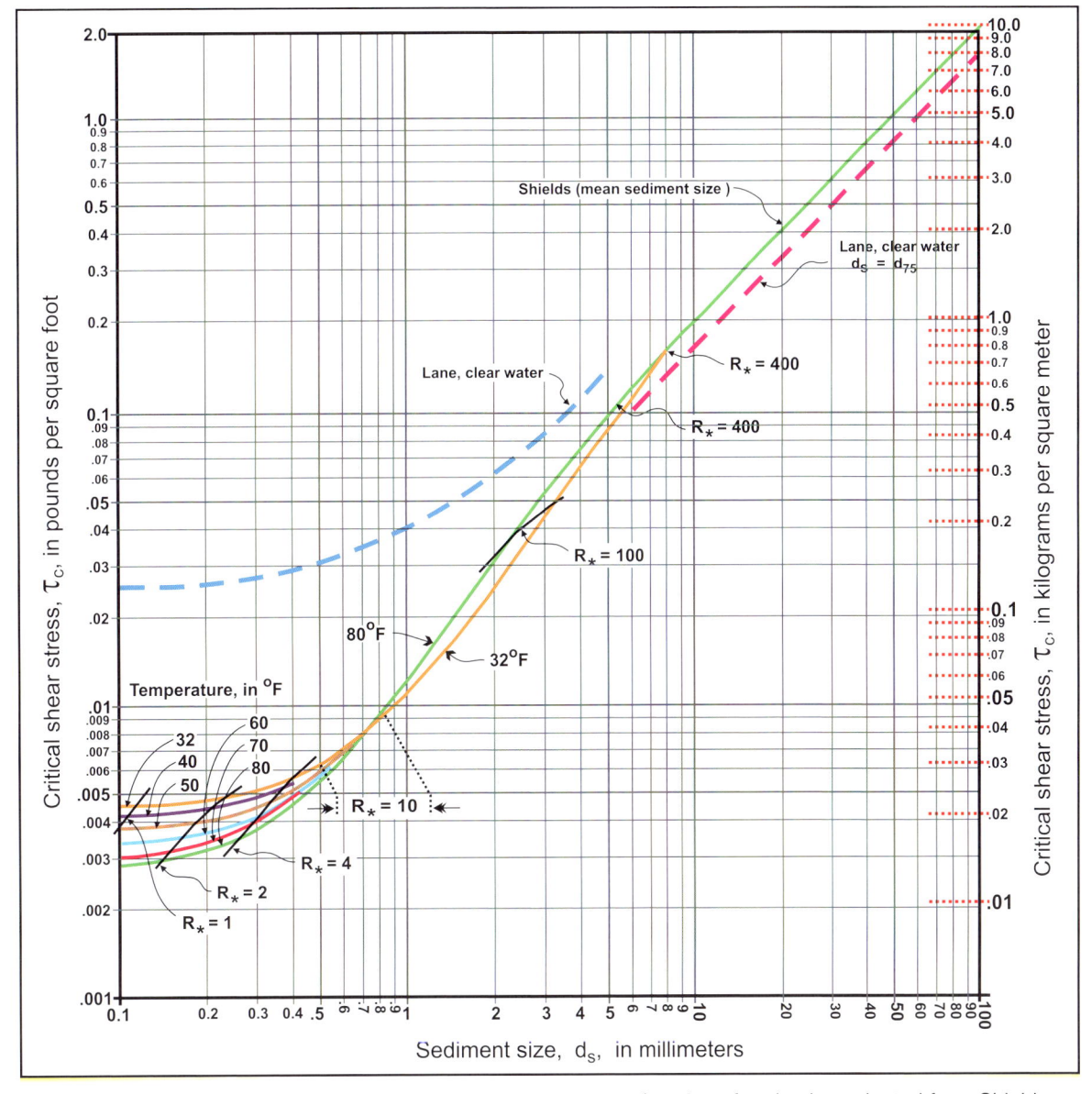

Figure 2-6. Critical shear stress for quartz sediment in water as function of grain size; adapted from Shields (1936), Lane (1955) and Leopold et al. (1964).

Field measurements by the author have shown that for heterogeneous bed materials, larger particles are entrained at shear stress values much lower than indicated in the relation in **Figure 2-6**. A revised relation based on measured stream data from Colorado is presented in Chapter 5 (**Figure 5-54**).

As reported by Erman et al. (1988), **Equation 2-5** is identical to the Einstein bedload function (Einstein, 1950) but has a much simpler mathematical form. For the bed-material transport curve plotted in **Figure 2-7**, threshold dimensionless shear stress was computed by a function derived by Andrews (1983) and was found to be in excellent agreement with observed threshold conditions in Sagehen Creek (Andrews and Erman, 1986).

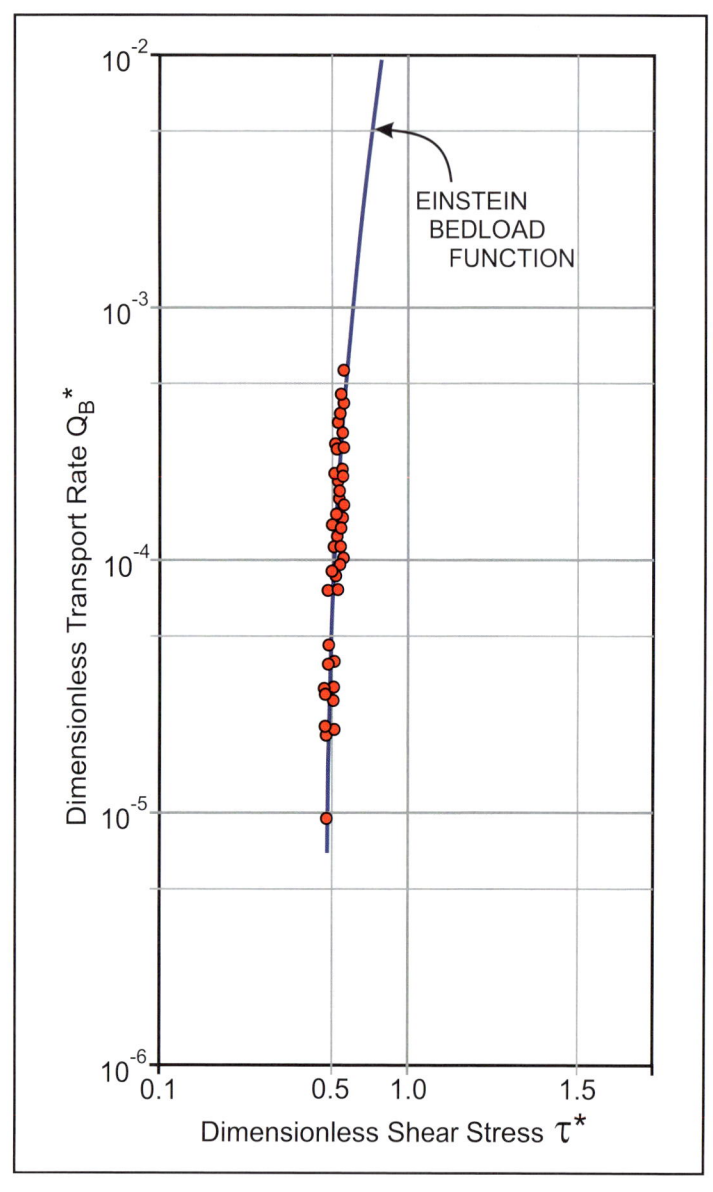

Figure 2-7. Dimensionless transport rate of bed material in Sagehen Creek in relation to dimensionless shear stress. The 58 daily totals from 1982, 1983 and 1984 are in excellent agreement with the curve drawn from a function developed by Einstein (1950) and later modified by Parker (1979) (Andrews and Erman, 1986, reproduced/modified by permission of the American Geophysical Union).

The computed value for τ^*_c for Sagehen Creek is 0.043. The limits of the curve shown in **Figure 2-8** are within the range of Einstein's original data used by Parker (1979).

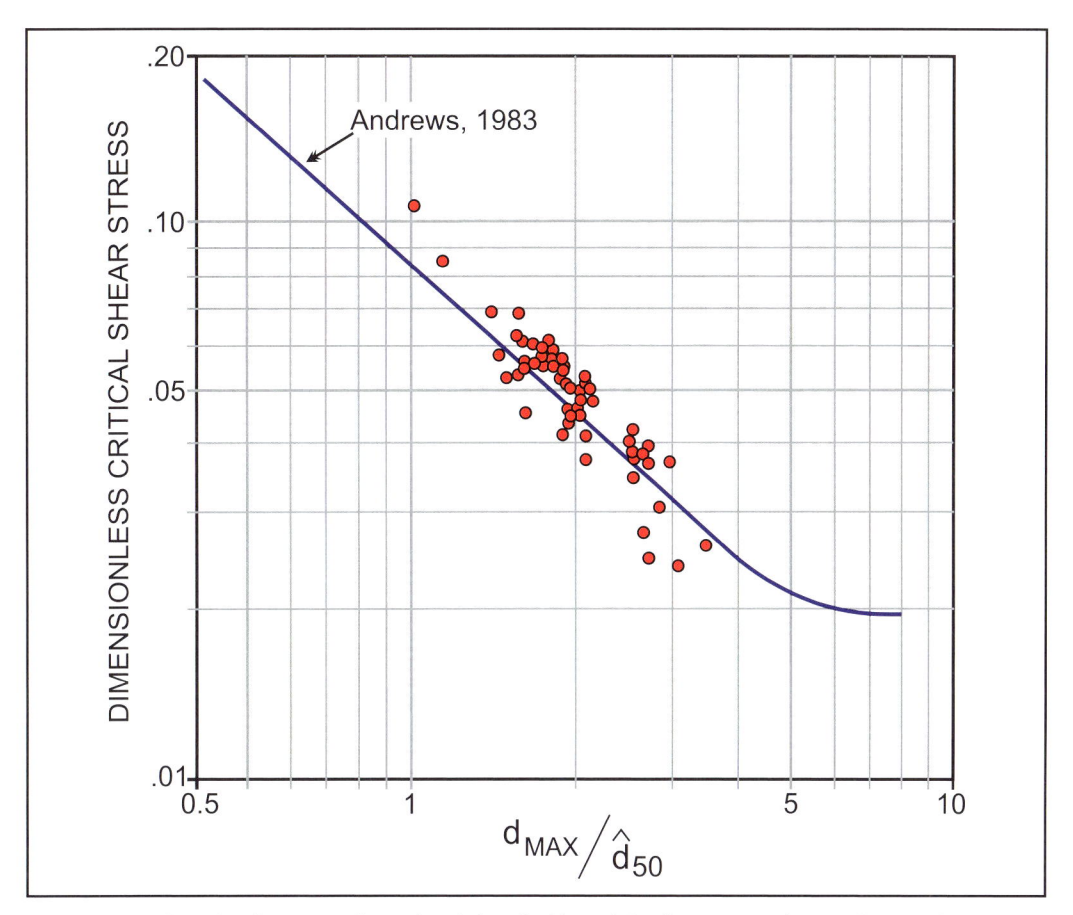

Figure 2-8. Relation between the ratio of threshold particle diameter to the median particle diameter of sub-surface bed material and the critical dimensionless shear stress (Andrews and Erman, 1986, reproduced/modified by permission of the American Geophysical Union).

Andrews (1984) investigated critical dimensionless shear stress by developing a relationship using the ratio of a given size particle, d_i, to sub-surface, $\hat{d}_{50}$, where τ^*_c is equal to:

$$\tau^*_c = 0.0834 \left(d_i / \hat{d}_{50}\right)^{-0.872} \qquad \textbf{(Equation 2-7)}$$

where:
  τ^*_c = critical dimensionless shear stress
  d_i = size of a given particle on riverbed surface
  $\hat{d}_{50}$ = median size particle of sub-surface sample

The values of τ^*_c derived from measured data are shown in **Figure 2-8** (Andrews and Erman, 1986). The threshold shear stress for median size particles in the riverbed surface can be calculated using **Equation 2-7**. Surface and sub-surface material sampling procedures are summarized by Bunte and Abt (2001).

Work by Andrews and Erman (1986) on Sagehen Creek, California, and additionally reported by Andrews and Nankervis (1995), produced the equation:

$$\tau^*_c = 0.0384 \left(d_i / d_{50}\right)^{-0.887} \quad \textbf{(Equation 2-8)}$$

where:
d_{50} = median size particle of the bed surface

The entrainment calculation is significant not only for bedload transport but also for calculating the competence of the river to transport the largest sediment clasts made available by its catchment.

The equations described by Bagnold (1980), Erman et al. (1988), Einstein (1950), Parker (1979) and Andrews and Erman (1986) require some form of entrainment relation for the initiation of particle motion. Various computational methods for τ^*_c are summarized in **Table 2-1** (Reid and Dunne, 1996). Their summary includes the statement: "The uncertainties in estimating this critical stress…therefore seriously degrade efforts to predict transport rates and make worthwhile the expenditure of some effort to constrain these rates using local field evidence."

Table 2-1. Equations for initiation of motion. τ_c (dynes-cm^{-2}) is calculated for particles of size D_i (cm) lying in a pavement or on a substrate with median grain size D_{50} (cm) or geometric mean diameter D_g (cm) (Reid and Dunne, 1996, reprinted with permission from Catena Verlag).

Source	Equation	D_{50} type	Method***
Shields, 1936	$\tau_c = 0.056(\rho_s - \rho)gD_i$		
Miller et al., 1977	$\tau_c = 0.045(\rho_s - \rho)gD_i$		
Parker et al., 1982	$\tau_c = 0.09(\rho_s - \rho)gD_{50}$	Sub-surface	Bedload
Diplas, 1987	$\tau_c = 0.087(\rho_s - \rho)gD_i^{0.06}D_{50}^{0.94}$	Sub-surface	Bedload
Parker, 1990*	$\tau_c = \tau_{rg}(\rho_s - \rho)gD_i^{0.10}D_g^{0.90}$	Surface	Bedload
Komar, 1987a**	$\tau_c = 0.045(\rho_s - \rho)gD_i^{0.35}D_{50}^{0.65}$	Sub-surface	Clast
Andrews, 1983	$\tau_c = 0.083(\rho_s - \rho)gD_i^{0.13}D_{50}^{0.87}$	Sub-surface	Clast
Ashworth & Ferguson, 1989	$\tau_c = 0.072(\rho_s - \rho)gD_i^{0.35}D_{50}^{0.65}$	Surface	Bedload
" " "	$\tau_c = 0.054(\rho_s - \rho)gD_i^{0.33}D_{50}^{0.67}$	Surface	Bedload
" " "	$\tau_c = 0.087(\rho_s - \rho)gD_i^{0.08}D_{50}^{0.92}$	Surface	Bedload
Komar & Carling, 1991	$\tau_c = 0.059(\rho_s - \rho)gD_i^{0.36}D_{50}^{0.64}$	Sub-surface	Clast
" " "	$\tau_c = 0.039(\rho_s - \rho)gD_i^{0.18}D_{50}^{0.82}$	Sub-surface	Clast
Costa, 1983	$\tau_c = 26.6 D_i^{1.21}$		Deposit
Williams, 1983	$\tau_c = 12.9 D_i^{1.34}$		Deposit

* D_g is the surface geometric mean particle diameter and is calculated as: $\ln D_g = \Sigma F_i \ln D_i$, where F_i is the fraction of particles in size class i and having a geometric mean of D_i

$\tau_{rg} = 0.836 (D_g/D_{50sub})^{-0.905}$ and is the reference dimensionless shear stress for D_g

**   Equation was derived using surface grain parameter but was applied using sub-surface.

*** Method...
     Bedload: Initiation defined by the occurrence of a small but finite amount of transport.
     Clast: Initiation threshold defined from the largest particle moving at a given τ.
     Deposit: Initiation threshold defined from the largest particles deposited during an event with given τ at crest stage.

As obtaining a sub-pavement sample can be difficult, Rosgen (1996, pp. 7-7 to 7-9; 2001b) identifies a modified procedure that requires a core sample at specific elevations and locations of depositional features, such as point bars, to determine τ^*_c at the bankfull stage. This approach substitutes for pavement/sub-pavement samples by obtaining D_{max} (largest particle size of the bar sample) and $\hat{D}_{50}$ (median size particle from the bar sample). In field practice, the values of $d_i/\hat{d}_{50}$ by Andrews (1984) in **Equation 2-7** are often substituted by using $D_{50}/\hat{D}_{50}$, where D_{50} is the median diameter of bed material on the bed of the riffle and $\hat{D}_{50}$ is the median diameter of a point bar core sample. The range of these ratio values applicable for the calculation of bankfull dimensionless shear stress (τ^*_c) are between 3.0–7.0.

If the resultant ratio of the $D_{50}/\hat{D}_{50}$ is outside the acceptable range, then substituting the largest particle on the bar (D_{max}) for d_i in **Equation 2-8** is recommended to calculate the bankfull dimensionless shear stress value (τ^*_c). The acceptable range of the ratio of the largest particle on the bar (D_{max}) to the median diameter particle on the bed of the riffle (D_{50}) is between *1.3–3.0* using this adaptation of **Equation 2-8** (Rosgen, 2001b). The bankfull dimensionless shear stress value (τ^*_c) determined in this manner can be used for bedload transport relations and for calculation of competence.

Sediment competence

A rearrangement of **Equation 2-4**, shown as **Equation 2-9**, provides an estimate of competence, or a calculation of the depth of flow and/or water surface slope required to move the largest particle of sediment at the bankfull stage made available for the dimensionless shear stress calculated using **Equation 2-7** or **2-8**. The largest particle required to be moved using this method is the largest clast (D_{max}) measured on the surface of the lower third position of the point bar.

$$d = (\gamma_s - 1) D_{max} \tau^*_c / S \qquad \text{(Equation 2-9)}$$

where:

γ_s = immersed specific gravity of sediment

d = depth at the bankfull stage

D_{max} = largest clast to be moved as located on point bar site

Sediment transport

Bagnold's approach (1980) involves the use of *stream power*, the mean rate of kinetic energy supply and dissipation along a stream channel.

Stream power (ω) is defined as:

$$\omega = \gamma QS \qquad \text{(Equation 2-10)}$$

where:
Q = stream discharge

Unit stream power, or power per unit of streambed area (ω_a), is defined as:

$$\omega_a = \tau u \qquad \text{(Equation 2-11)}$$

where:

τ is defined in **Equation 2-2**

u = mean velocity

Yang (1996) uses unit stream power per unit weight (ω_w), where:

$$\omega_w = uS \qquad \text{(Equation 2-12)}$$

The relation between unit stream power (ω_a) (**Equation 2-11**) and measured sediment transport rate by size classes is shown in **Figure 2-9** (Dunne and Leopold, 1978).

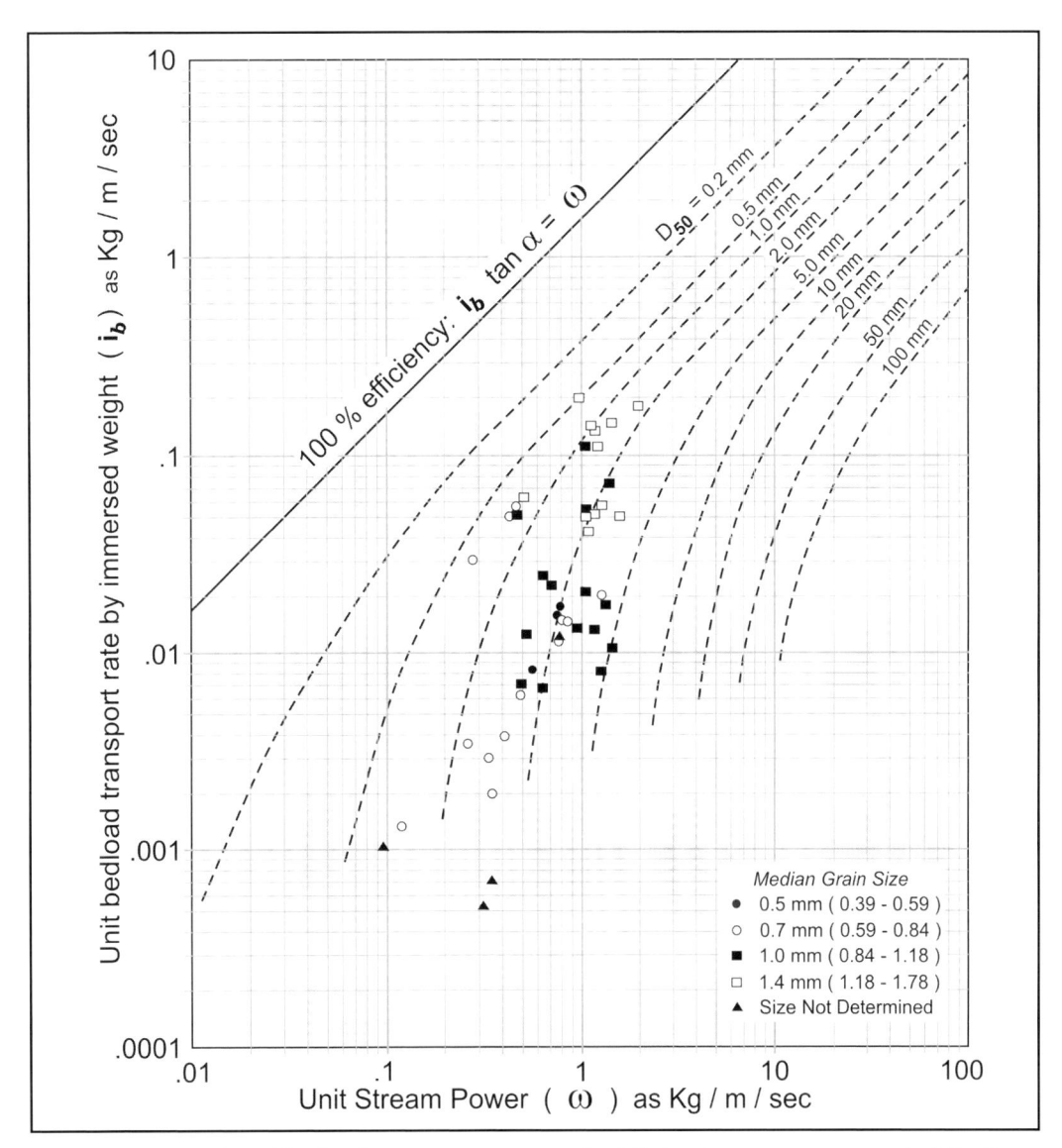

Figure 2-9. Relation of bedload transport rate per unit width to stream power per unit width, the family of curves being the D_{50} for the moving load. Plotted points are bedload measurements at the East Fork bedload trap, Wyoming (Dunne and Leopold, 1978, reprinted with permission from W.H. Freeman and Company).

The application of this relation is related to corresponding changes in sediment transport by particle size as related to changes in stream channel depth, velocity and/or slope—the variables influencing unit stream power. Thus, increases in width/depth ratio due to streambank instability or direct disturbance would show potential sediment deposition, filling of pools and other related channel consequences.

Bagnold's (1980) transport relation (i_b) using stream power is shown as:

(Equation 2-13)
$$i_b/(i_b)_* = \left(\frac{D}{D_*}\right)^{-1/2} \left[\frac{\omega - \omega_o}{(\omega - \omega_o)_*}\right]^{3/2} \left(\frac{Y}{Y_*}\right)^{-2/3}$$

where:
$$\omega_o = 5.75 \left[(\gamma D \Theta_o)^{3/2} \left(\frac{g}{\rho}\right)^{0.5} \log\left(12\frac{Y}{D}\right) \right]$$

and where:

i_b = transport rate of bedload by immersed weight per unit width

$(i_b)_*$ = a reference value of i_b (0.1 kg m⁻¹s⁻¹)

ω = stream power per unit bed area (**Equation 2-11**)

ω_o = threshold unit stream power

$(\omega - \omega_o)_* = 0.5$ kg m⁻¹s⁻¹

Y = depth of flow

Y_* = a reference value of Y (0.1 m)

D = mode size of bed material

D_* = reference value of D (0.0011 m)

γ = excess density of solids

Θ_o = Shields threshold criterion

g = gravitational acceleration

ρ = density of liquid

The dimensionless shear stress value τ^*_c, as shown in **Equation 2-8**, is essentially the same as Θ_o in **Equation 2-13** above.

After testing seven bedload equations on 22 streams with measured hydraulic and sediment data, Lopes et al. (2001a) determined that the best overall equations were those of Schoklitsch (1962) and Bagnold (1980). Their comparison of various bedload models for the Chippewa River in Wisconsin against measured bedload data is shown in **Figure 2-10** (Lopes et al., 2001a). Gomez and Church (1989) concluded that prediction of bedload under limited hydraulic information is best accomplished by using equations based on the stream power concept. However, after testing 12 various bedload equations in gravel-bed rivers for 410 bedload events, they concluded that none of the equations performed consistently, due to the limitations of the data and the complexity of the sediment transport phenomena. The following are the conclusions resulting from this work:

- The need to collect localized bedload and suspended sediment rating curve data to establish sediment supply values; and
- The need to calibrate sediment transport models based on observed values.

Without these observed sediment values, predicted transport rates will continue to differ significantly from actual transport rates.

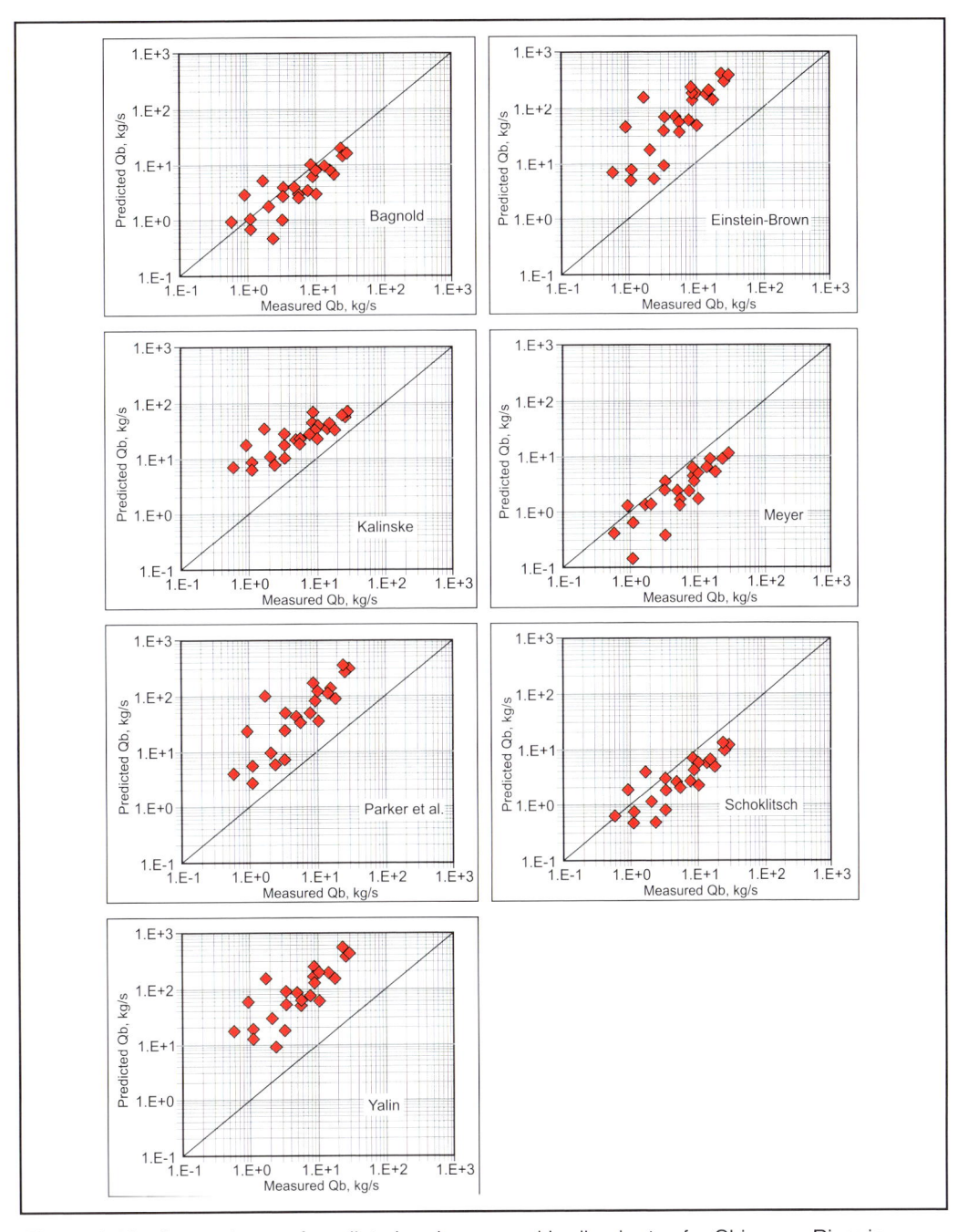

Figure 2-10. Comparisons of predicted and measured bedload rates for Chippewa River in Durand, Wisconsin (Lopes et al., 2001a).

A reference dimensionless bedload sediment rating curve developed for a specific stream type/stability (**Figure 2-11**) (Troendle et al., 2001) can be used to develop a localized bedload sediment rating curve. To convert from dimensionless values, measured bedload and concurrent streamflow discharge must be obtained at the bankfull stage for a stream reach of the same type/stability relation. The reference curve should represent a range of stable stream types (Rosgen, 1994) and stability ratings (Pfankuch, 1975; Rosgen, 2001b). The values of both bankfull stage bedload and stream discharge are multiplied by the appropriate bedload sediment ratios (S_B/S_{Bbkf}) and corresponding discharge ratios (Q/Q_{bkf}) for the appropriate reference curve to generate a dimensioned bedload rating curve. The bankfull discharge values in the equation are converted from an instantaneous discharge to mean daily discharge (**Figure 2-11**). Reference dimensionless bedload rating curves are constructed from actual bedload and streamflow measurements over a wide range of flows for specific stream types and stability ratings in the hydro-physiographic province being assessed.

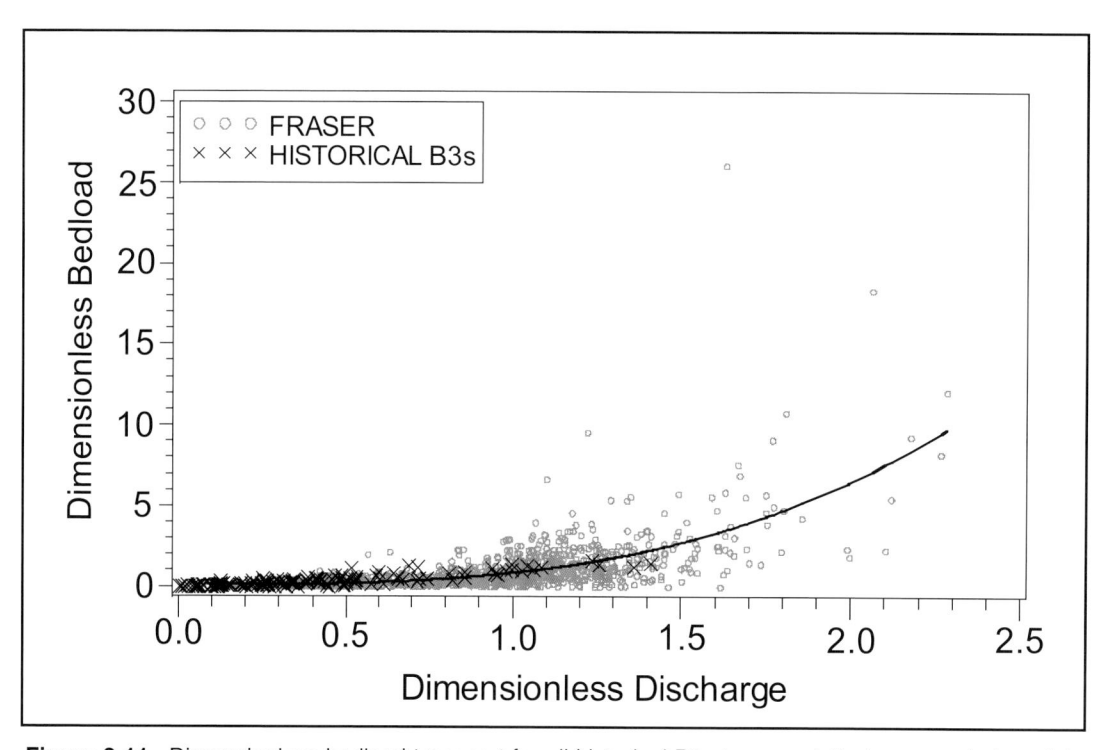

Figure 2-11. Dimensionless bedload transport for all historical B3 streams plotted over pooled model for reference streams (Troendle et al., 2001).

Suspended sediment transport

Suspended sediment is that portion of the total sediment load of rivers that is carried in the water column. It is measured with a depth-integrated sampler that proportionately represents streamflow and reports suspended sediment as a concentration in mg/l. Field and lab procedures are documented in USGS (1999). It contains *washload*, or that portion of the suspended load not represented in the bed material. In practice, washload usually comprises the silt/clay or colloidal fraction that is more controlled by supply than energy. Most of the suspended sediment transport and washload transport relations are derived from measured sediment rating curves and flow-duration curves.

Total bed-material load

Bedload and bed-material suspended load transport are often termed *total bed-material load equations*, where a portion of the bed material is composed of sand that is transported as both bedload and suspended load. Total bed-material load equations do not include washload. A detailed listing and evaluation of total bed-material load equations appears in Reid and Dunne (1996). Ackers and White (1973), Bagnold (1966) and Yang (1979) commonly used total bed-material load or suspended bed-material load equations. GSTARS 2.0 is a computer program that offers the user a selection of various sediment transport equations (Yang et al., 1998).

Dimensionless suspended sediment rating curves were developed and tested against an independent data set of the same stream type/stability class, as in the bedload example in Troendle et al. (2001) (**Figure 2-12**). Once reference dimensionless suspended sediment rating curves are established for a region, dimensioned sediment rating curves can be established by the methods described previously.

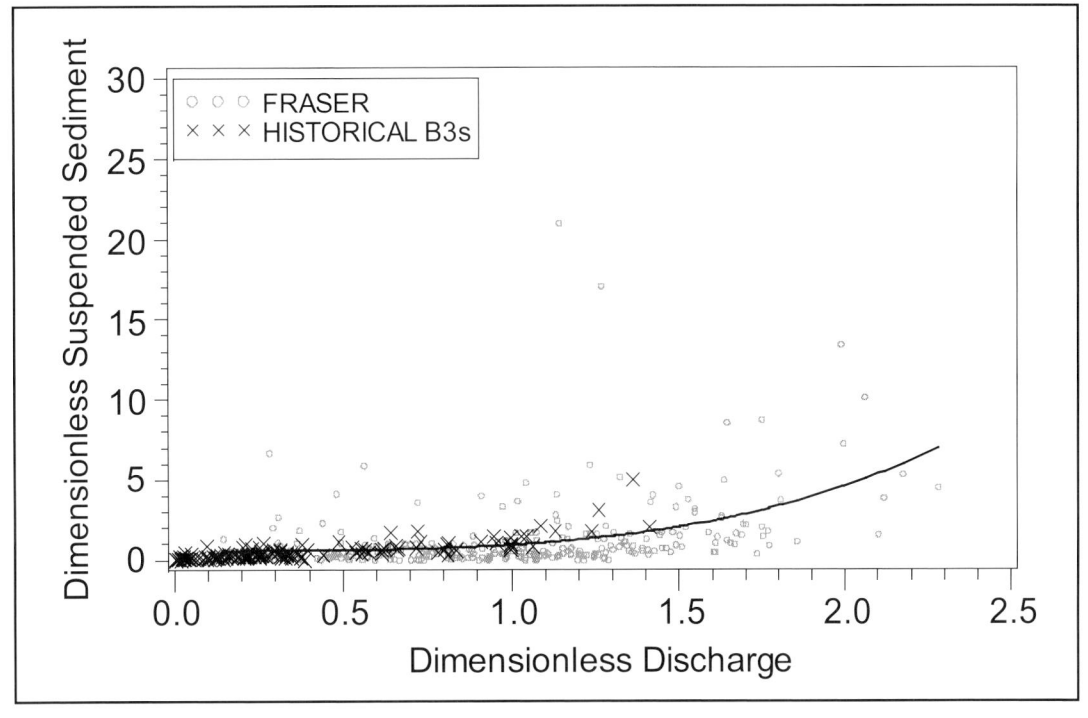

Figure 2-12. Dimensionless suspended sediment transport for all historical B3 streams plotted over pooled model for reference streams (Troendle et al., 2001).

River classification

Stable streams take on many combinations of dimension, pattern, profile and materials amidst a wide range of valley slopes, sediment sizes, sediment, sediment loads and streamflows. Because of the great diversity of morphological features among rivers, a stream classification system was developed to stratify and describe various river types (Rosgen, 1994, 1996). This river classification system integrates individual variables into morphological descriptions that combine various forms of the existing as well as "probable state" variables. The following are the five objectives of this stream classification scheme:

1. To predict a river's behavior from its appearance, based on documentation of similar response from similar types for imposed conditions;

2. To stratify empirical hydraulic and sediment relations by stream type by state (condition) to minimize variance;

3. To provide a mechanism to extrapolate site-specific morphological data;

4. To describe physical stream relations to complement biological inventory and assist in establishing potential and departure states; and

5. To provide a consistent frame of reference for communicating stream morphology and condition among a variety of disciplines.

Examples of broad-level stream morphology (broad-level geomorphic characterization) are shown in **Figure 2-13** and summarized in **Table 2-2** (Rosgen, 1994, 1996). This basic level of classification is utilized in the primary phase of the *WARSSS* assessment, the *RLA* evaluation. It is often rapidly obtained from aerial photographs and topographic maps by valley types (Rosgen, 1994, 1996). Field checks are generally recommended to ensure proper remote sensing interpretations. The broad-level stream classification is based on morphological features associated with stream patterns, shape (width/depth ratio) and vertical containment (entrenchment ratio). Width/depth ratios are determined from cross-sections based on bankfull surface width divided by bankfull mean depth. High width/depth ratios are associated with wide, shallow streams. Conversely, low width/depth ratios are associated with narrow, deep channels.

Entrenchment ratio is a measure of vertical containment by a ratio of the flood-prone area width divided by the bankfull width. The flood-prone area width is obtained at an elevation two times the maximum bankfull depth. If the entrenchment ratio is less than 1.4 (± 0.2 to allow for the continuum of channel form), the stream is classified as entrenched or ***vertically*** contained (A, G and F stream types). If the entrenchment ratios are 1.4–2.2, the stream is moderately entrenched (B stream types). If the ratio is greater than 2.2, the stream is not entrenched (C, E and DA stream types). The general description of each broad-level stream type is provided in **Table 2-2**.

A more detailed stream classification, including dimension, pattern and profile measurements as well as channel materials, is required in subsequent assessments and is shown in **Figure 2-14** (Rosgen, 1994, 1996).

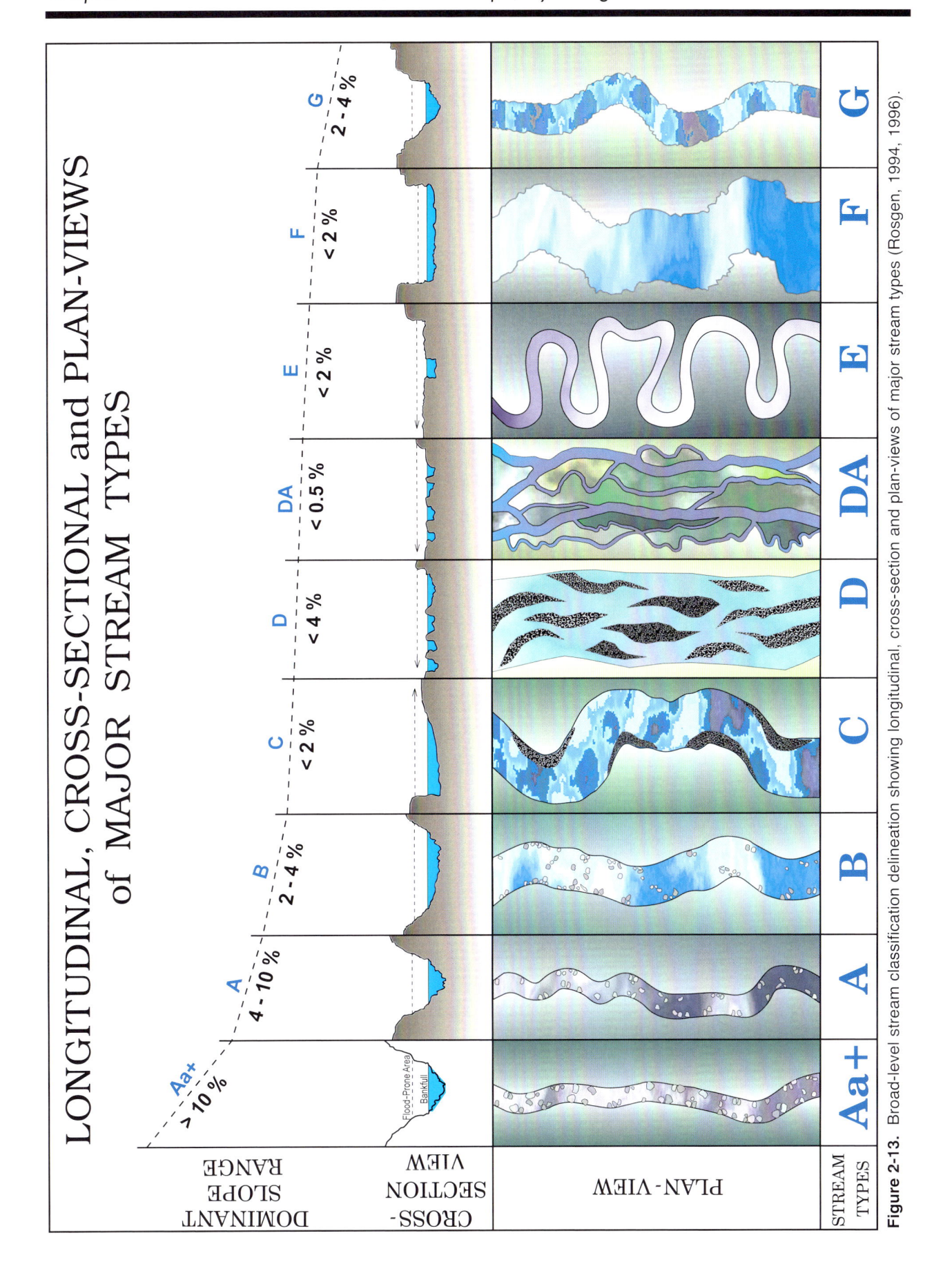

Chapter 2 — A Review of Hillslope, Hydrologic and Channel Process Relations

Figure 2-13. Broad-level stream classification delineation showing longitudinal, cross-section and plan-views of major stream types (Rosgen, 1994, 1996).

Table 2-2. General stream type descriptions and delineative criteria for broad-level classification (Rosgen, 1994, 1996).

Stream Type	General Description	Entrenchment Ratio	W/D Ratio	Sinuosity	Slope	Landform/Soils/Features
Aa+	Very steep, deeply entrenched, debris transport, torrent streams.	<1.4	<12	1.0 to 1.1	>.10	Very high relief. Erosional, bedrock or depositional features; debris flow potential. Deeply entrenched streams. Vertical steps with deep scour pools; waterfalls.
A	Steep, entrenched, cascading, step/pool streams. High energy/debris transport associated with depositional soils. Very stable if bedrock- or boulder-dominated channel.	<1.4	<12	1.0 to 1.2	.04 to .10	High relief. Erosional or depositional and bedrock forms. Entrenched and confined streams with cascading reaches. Frequently spaced, deep pools in associated step/pool bed morphology.
B	Moderately entrenched, moderate gradient, riffle dominated channel, with infrequently spaced pools. Very stable plan and profile. Stable banks.	1.4 to 2.2	>12	>1.2	.02 to .039	Moderate relief, colluvial deposition and/or structural. Moderate entrenchment and width/depth ratio. Narrow, gently sloping valleys. Rapids predominate with scour pools.
C	Low gradient, meandering, point-bar, riffle/pool, alluvial channels with broad, well-defined floodplains.	>2.2	>12	>1.2	<.02	Broad valleys with terraces, in association with floodplains, alluvial soils. Slightly entrenched with well-defined meandering channels. Riffle/pool bed morphology.
D	Braided channel with longitudinal and transverse bars. Very wide channel with eroding banks.	n/a	>40	n/a	<.04	Broad valleys with alluvium, steeper fans. Glacial debris and depositional features. Active lateral adjustment, with abundance of sediment supply. Convergence. Divergence of bed features, aggradational processes, high bedload and bank erosion.
DA	Anastomosing (multiple channels) narrow and deep with extensive, well-vegetated floodplains and associated wetlands. Very gentle relief with highly variable sinuosities and width/depth ratios. Very stable streambanks.	>2.2	highly variable	highly variable	<.005	Broad, low-gradient valleys w/ fine alluvium and/or lacustrine soils. Anastomosed (multiple channel) geologic control creating fine deposition w/ well-vegetated bars that are laterally stable w/ broad wetland floodplains. Very low bedload, high wash load sediment.
E	Low gradient, meandering riffle/pool stream with low width/depth ratio and little deposition. Very efficient and stable. High meander width ratio.	>2.2	<12	>1.5	<.02	Broad valley/meadows. Alluvial materials with floodplains. Highly sinuous with stable, well-vegetated banks. Riffle/pool morphology with very low width/depth ratios.
F	Entrenched meandering riffle/pool channel on low gradients with high width/depth ratio.	<1.4	>12	>1.2	<.02	Entrenched in highly weathered material. Gentle gradients with a high width/depth ratio. Meandering, laterally unstable with high bank erosion rates. Riffle/pool morphology.
G	Entrenched "gully" step/pool and low width/depth ratio on moderate gradients.	<1.4	<12	>1.2	<.039	Gullies, step/pool morphology with moderate slopes and low width/depth ratio. Narrow valleys, or deeply incised in alluvial or colluvial materials; i.e., fans or deltas. Unstable, with grade control problems and high bank erosion rates.

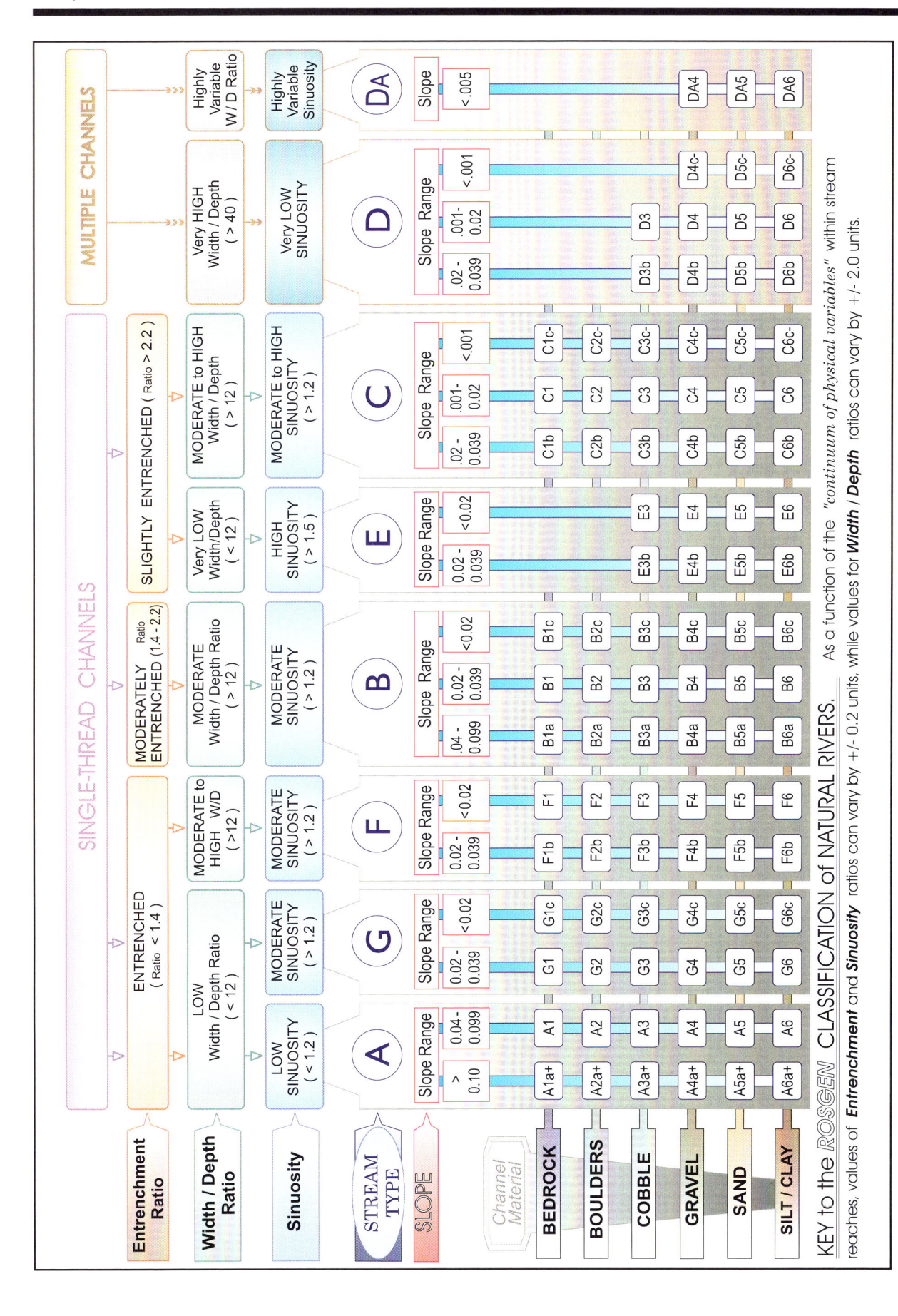

Chapter 2 A Review of Hillslope, Hydrologic and Channel Process Relations

Figure 2-14. Stream classification key for natural rivers (Rosgen, 1994, 1996).

2-23

The classification key in **Figure 2-14** is used much like a dichotomous key in plant identification. The A→ G letter designation follows the same broad categories as shown in **Figure 2-13**. Slope categories are broken into smaller ranges, as designated by subscripted small letters, while numbers 1–6 reflect channel material. Stream classification must also be stratified by valley type (Chapter 4, Rosgen, 1996). This level of information requires field measurements to describe and verify morphological relations.

The stream classification system was developed with an understanding that stability evaluation must be conducted at a higher degree of resolution than morphological groupings. Channel stability assessments, however, must be stratified by stream type for extrapolation purposes. Barbour et al. (1991) commented on the role of stream classification in bio-assessment as follows:

> "One function of classification is to increase the resolving power or sensitivity of biological surveys to detect impairment by partitioning variation within selected environmental parameters or among sites. The importance of minimizing variation… clearly it is easier to distinguish impairment if the parameters have low variability. Formal statistical tests (parametric and nonparametric) indicate greater resolution and power exist if there is low variance within elements being compared. Effective classification leads to improving resolving power by partitioning and accounting for variability. A coarse classification yields higher variance and therefore lower resolving power; vice versa for finer classifications."

Reference reach data describe the *stable* morphological form of dimension, pattern and profile for a range of valley slopes, channel materials, riparian vegetation and other measurable variables (Rosgen, 1998). Dimensionless relations are developed from reference reach data as stratified by stream type in order to quantitatively describe specific channel attributes. Examples are the dimensionless parameters of depth and slope of bed features such as pools, runs, riffles and glides. Dimensionless relations of meander and hydraulic geometry are also developed by stream type to determine departure and provide information for natural channel design (Rosgen, 1998). The advantage of developing dimensionless relations is that they may be extrapolated to rivers of similar channel and valley types but of different sizes.

Applications of regime equations are most appropriate if the river where the data was empirically derived is similar to the river to which it is applied. Often, stratification of such relations by stream type and condition provides insight as to how the relations were developed. For example, the relation of stream width/discharge as influenced by bank stability, shown in **Figures 2-15** and **2-16** (Copeland et al., 2001), assumes there are no differences in the width/discharge relation due to morphological stream types.

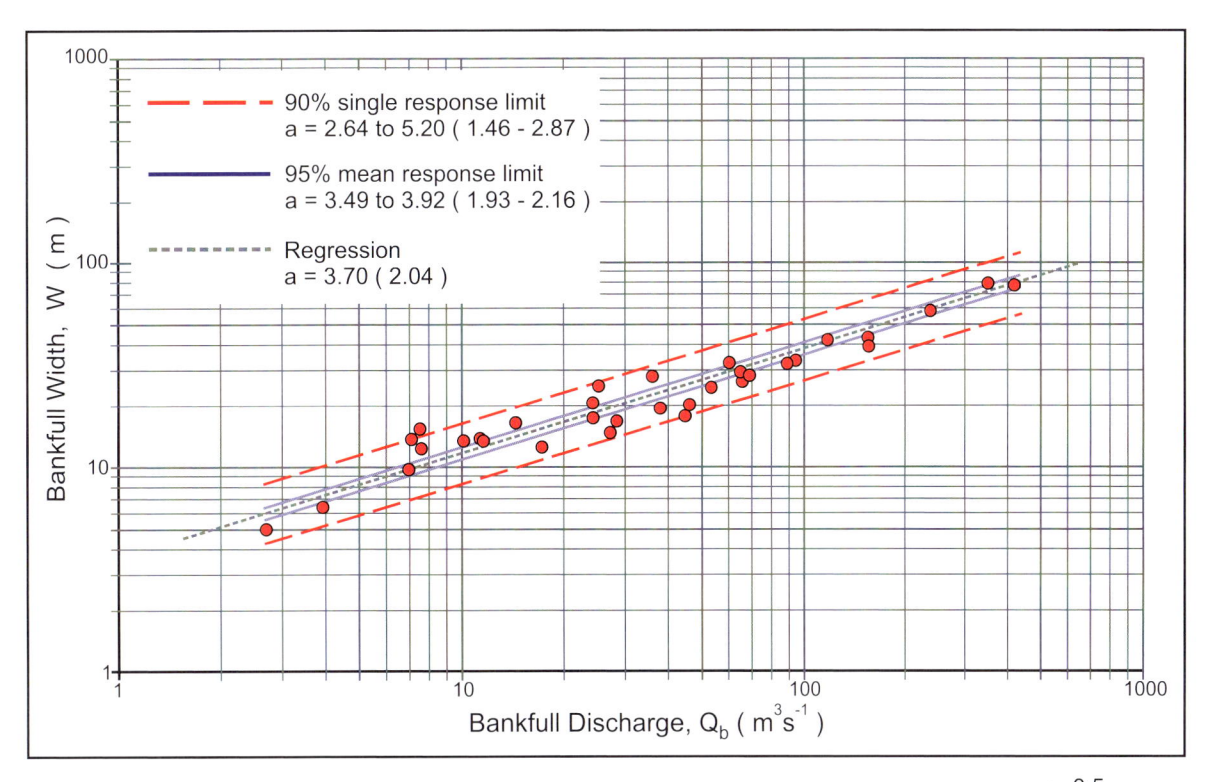

Chapter 2 — A Review of Hillslope, Hydrologic and Channel Process Relations

Figure 2-15. Downstream width hydraulic geometry for United Kingdom gravel-bed rivers, $W = a\, Q_b^{0.5}$ with confidence bands. Based on 36 sites in the United Kingdom with erodible banks. S.I. units, m and m^3/sec (English units, ft and ft^3/sec) (Copeland et al., 2001).

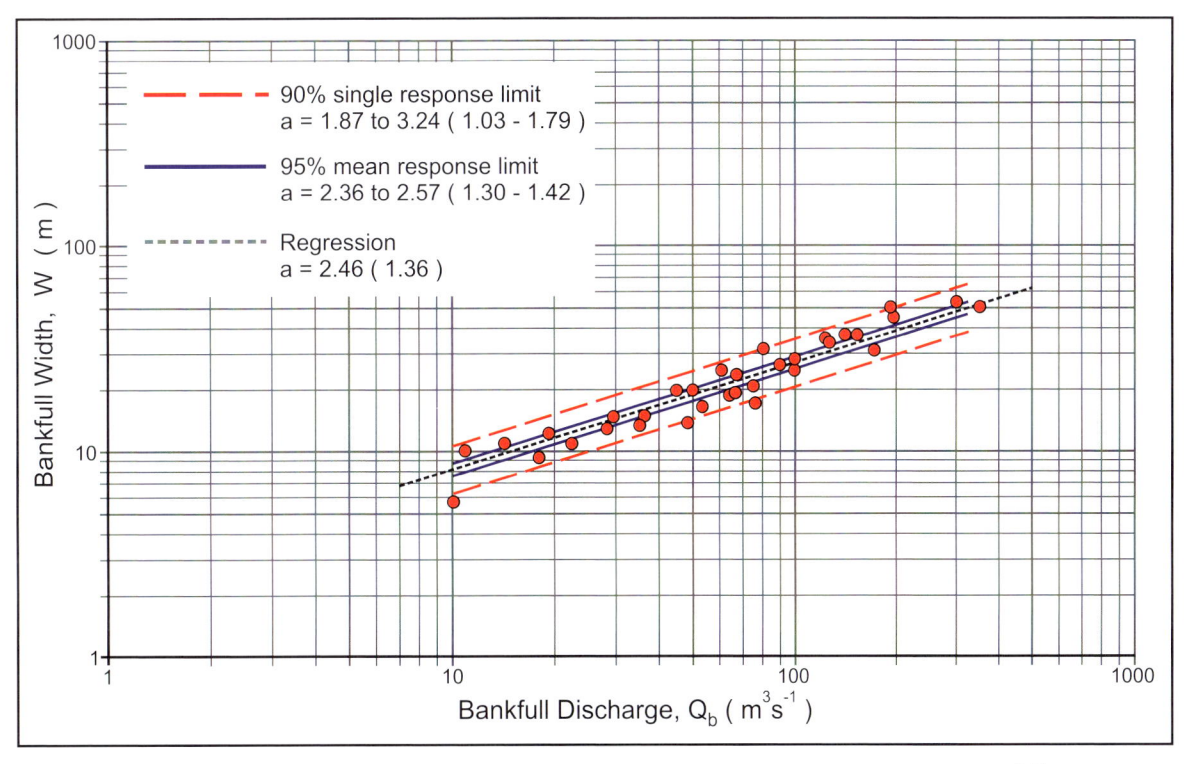

Figure 2-16. Downstream width hydraulic geometry for United Kingdom gravel rivers, $W = a\, Q_b^{0.5}$ with confidence bands. Based on 43 sites in the United Kingdom with resistant banks. S.I. units, m and m^3/sec (English units, ft and ft^3/sec) (Copeland et al., 2001).

Valley classification

The stability and activity of a stream is also dependent on the stream's location or valley type. **Table 2-3** summarizes valley types and associated characteristics. Identification of valley type is required in subsequent assessments.

Table 2-3. Valley types used in geomorphic characterization (Rosgen, 1996).

Valley Types	Summary Description of Valley Types
I	Steep, confined, "V" notched canyons, rejuvenated sideslopes
II	Moderately steep, gentle-sloping side slopes often in colluvial valleys
III	Alluvial fans and debris cones
IV	Gentle gradient canyons, gorges and confined alluvial and bedrock-controlled valleys
V	Moderately steep, "U" shaped glacial-trough valleys
VI	Moderately steep, fault, joint, or bedrock (structural) controlled valleys
VII	Steep, fluvial dissected, high-drainage density alluvial slopes
VIII	Wide, gentle valley slope with well-developed floodplain adjacent to river and/or glacial terraces — alluvial valley fills
IX	Broad, moderate to gentle slopes, associated with glacial outwash and/or eolian sand dunes
X	Very broad and gentle valley slope, associated with glacio- and non-glacio-lacustrine deposits
XI	Deltas

Reference reach data from the more detailed stream classification that represents the stable form of various morphological types is used throughout the *WARSSS* methodology to compare the assessed channel's degree of departure from the reference reach's dimension, pattern and profile (Rosgen, 1998, 1999, 2001b, 2001c).

Chapter 2   *A Review of Hillslope, Hydrologic and Channel Process Relations*

Hydraulic geometry

Hydraulic geometry relations, as predicted by regime equations, can be improved with stream type data. The variability of hydraulic geometry relations, such as width versus discharge, can be minimized by stream type stratification. For example, a stable E4 stream type that has a width/depth ratio of 3 (**Figure 2-17**) has a much lower width for the same discharge than a C4 stream type that has a stable width/depth ratio of 20 (**Figure 2-18**). Both streams in **Figures 2-17** and **2-18** are associated with valley type X.

Figure 2-17. Example of a typical E4 stream type, South Fork, South Platte River, Colorado (Kurz, 1999).

Figure 2-18. Example of a C4 stream type similar in size to the river in **Figure 2-17**, tributary to the Payette River, central Idaho.

2-27

Both streams are gravel-bed, low-gradient, meandering streams with floodplains, and both are stable, but their widths for the same discharge are much different. Smaller rivers that exhibit the same small width/depth ratios for similar flows are shown in the E4 stream type (**Figure 2-19**) and, for a higher width/depth ratio, the C4 stream type (**Figure 2-20**).

Figure 2-19. An E4 stream type, valley type X, with a bankfull width of 4 ft for a bankfull discharge of 75 cfs, small tributary in North-central Idaho.

Figure 2-20. A stable C4 stream type, valley type VIII, with a bankfull width of 15 ft (w/d ratio 14) for approximately 80 cfs, upper Willow Creek, Colorado.

The data points shown in **Figure 2-21** for North American gravel-bed rivers (Copeland et al., 2001) express great variability and a range of widths from 5–30 meters for a bankfull discharge of 10 cfs. If the data points were stratified by stream type, or by width/depth ratio, this variance could be explained or at least minimized.

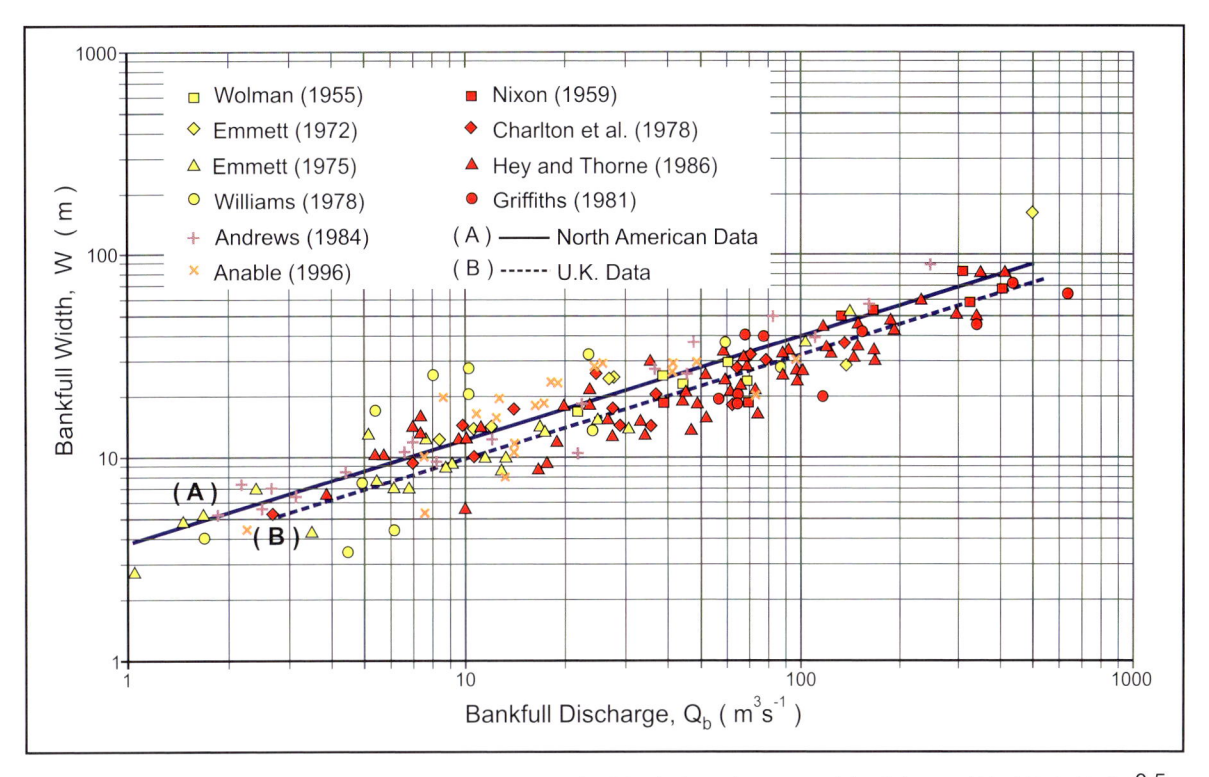

Figure 2-21. Downstream width hydraulic geometry for North American gravel-bed rivers, (A): $W = 3.68\, Q_b^{0.5}$ and (B): U.K. gravel-bed rivers, $W = 2.99\, Q_b^{0.5}$ (Copeland et al., 2001).

Much of the observed differences in width for the same discharge have been explained by morphological stream type. The relations of width/discharge in **Figures 2-15, 2-16** and **2-21** may be improved if they are stratified by stream type, or at least by width/depth ratio, one of the key delineative criteria for stream classification.

Hydraulic geometry relations, including width, are presented for various stream types of the same size and depict large differences in width for the same discharge (**Figure 2-22**) (Rosgen, 1994, 1996). This relation underscores the differences in width for the same discharge between E3 and C3 stream types.

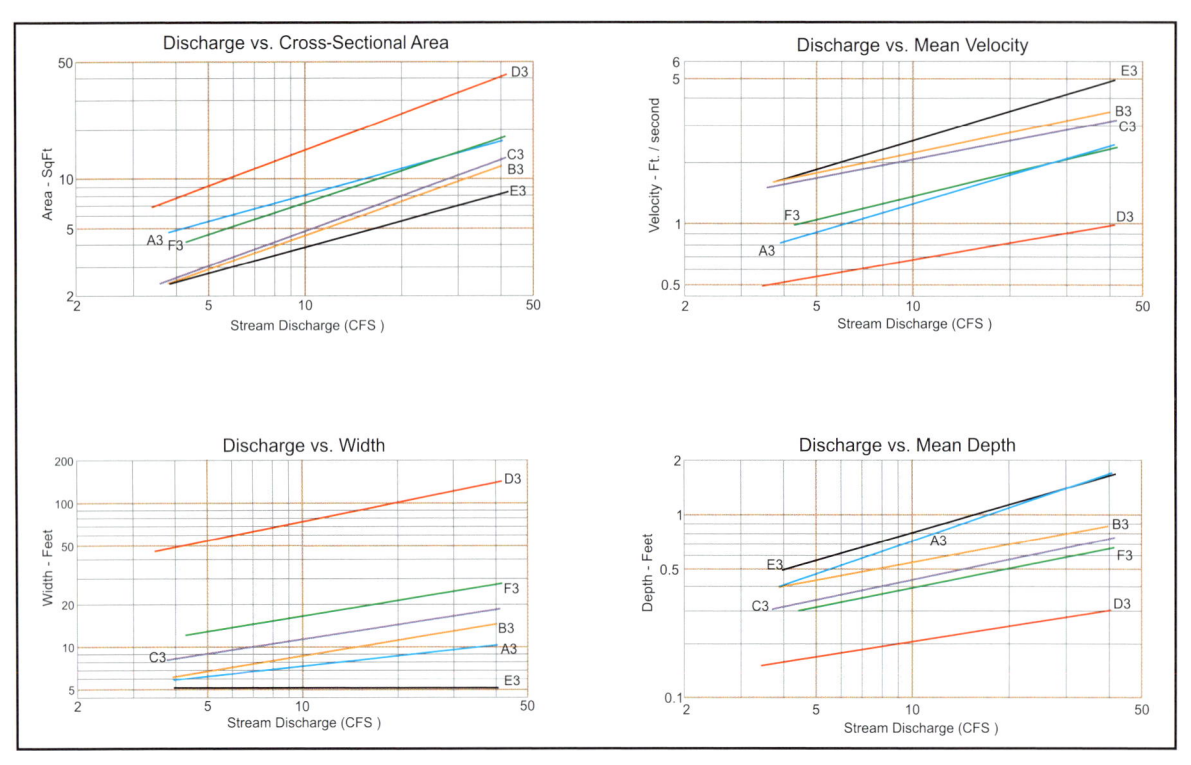

Figure 2-22. Hydraulic geometry relations for selected stream types of uniform size (Rosgen, 1994, 1996).

Management applications

Classification versus river stability

The Rosgen classification system does not assess stability but rather describes various river types and quantifies their morphological parameters. When the values of these quantitative morphological variables depart from value ranges typical of a stable state based on dimension, pattern, profile and materials, the channel may exceed a "stability threshold," resulting in degradation, aggradation, accelerated lateral extension, avulsion and other instability consequences. The corresponding instability or disequilibrium often results in a morphological shift (evolution) to a new stream type.

The sensitivity of streams to imposed changes, such as increase in flow, direct disturbance and riparian vegetation influence, varies by stream type. These relations are shown in **Table 2-4** and reflect differences in potential sensitivity, recovery, sediment supply, streambank erosion and vegetative controlling influence (Rosgen, 1994, 1996). Contrary to some interpretations, this data does not imply that D, G and F stream types cannot be stable. Rather, they represent the stable but geologically active forms found in many valley types. A departure analysis must be conducted to understand the stable reference reach form within a valley type.

Table 2-4. Broad-level/generalized management interpretations by stream type (Rosgen, 1994, 1996).

Stream type	Sensitivity to disturbance	Recovery potential	Sediment supply	Streambank erosion potential	Vegetation controlling influence
A1	very low	excellent	very low	very low	negligible
A2	very low	excellent	very low	very low	negligible
A3	very high	very poor	very high	very high	negligible
A4	extreme	very poor	very high	very high	negligible
A5	extreme	very poor	very high	very high	negligible
A6	high	poor	high	high	negligible
B1	very low	excellent	very low	very low	negligible
B2	very low	excellent	very low	very low	negligible
B3	low	excellent	low	low	moderate
B4	moderate	excellent	moderate	low	moderate
B5	moderate	excellent	moderate	moderate	moderate
B6	moderate	excellent	moderate	low	moderate
C1	low	very good	very low	low	moderate
C2	low	very good	low	low	moderate
C3	moderate	good	moderate	moderate	very high
C4	very high	good	high	very high	very high
C5	very high	fair	very high	very high	very high
C6	very high	good	high	high	very high
D3	very high	poor	very high	very high	moderate
D4	very high	poor	very high	very high	moderate
D5	very high	poor	very high	very high	moderate
D6	high	poor	high	high	moderate
DA4	moderate	good	very low	low	very high
DA5	moderate	good	low	low	very high
DA6	moderate	good	very low	very low	very high
E3	high	good	low	moderate	very high
E4	very high	good	moderate	high	very high
E5	very high	good	moderate	high	very high
E6	very high	good	low	moderate	very high
F1	low	fair	low	moderate	low
F2	low	fair	moderate	moderate	low
F3	moderate	poor	very high	very high	moderate
F4	extreme	poor	very high	very high	moderate
F5	very high	poor	very high	very high	moderate
F6	very high	fair	high	very high	moderate
G1	low	good	low	low	low
G2	moderate	fair	moderate	moderate	low
G3	very high	poor	very high	very high	high
G4	extreme	very poor	very high	very high	high
G5	extreme	very poor	very high	very high	high
G6	very high	poor	high	high	high

Applications of **Table 2-4** might have improved interpretations that appeared in a recent Minnesota Department of Natural Resources publication (Devore, 1998), wherein a landowner claimed that streambanks and riparian vegetation were improved by grazing, while the exclusion of livestock led to accelerated erosion. While the article did not distinguish differences in stream type, the grazed reach that exhibited great improvement due to a properly applied grazing system (**Figure 2-23**) was an E4 stream type, whereas the reach excluded from grazing (**Figure 2-24**) was an F4 stream type.

Figure 2-23. E4 stream type, Minnesota (from Kluckhohn, 1998).

Figure 2-24. F4 stream type, Minnesota (from Kluckhohn, 1998).

The summary in **Table 2-5** depicts the major differences between these two gravel-bed streams (Rosgen, 1994, 1996).

Table 2-5. Comparison of management interpretations between E4 and F4 stream types (Rosgen, 1994, 1996).

Stream type	Sensitivity to disturbance	Recovery potential	Sediment supply	Streambank erosion potential	Vegetation controlling influence
E4	very high	good	moderate	high	very high
F4	extreme	poor	very high	very high	moderate

As this table demonstrates, the exclusion of livestock from the F4 stream type would not produce a rapid vegetative response due to the poor recovery potential and reduced vegetative controlling influence. If the F4 stream type evolved (through active lateral erosion of the high banks) to an E4 (the stable reference reach type) inside of the previous F4 type, then the recovery potential and vegetative influence would change—but then, so would the stream type. Without the benefit of the interpretations of stream classification, the conclusions that were drawn—that the grazed reach was more stable than the livestock-excluded area—may be correct, but for the wrong reasons.

Although good grazing practices improved the E4 stream type, good grazing practices have little influence on the F4 stream type due to the poor recovery potential of the high banks. The author has observed F4 stream types in North-central Nevada where fences excluded cattle for a 10-year period without any visible difference between the grazed and un-grazed stream sections. If a stream eventually evolves from an F4 to a C4 or E4 stream type, then vegetation begins to play a key role as sediment supply decreases due to reduced bed and bank erosion.

Sediment relations

Bedload sediment rating curve data representing 55 individual Colorado rivers (**Figure 2-25**) indicate that bedload transport rates may vary over six orders of magnitude for the same discharge (Williams and Rosgen, 1989; Rosgen, 1996). When individual streams are stratified by stream type from the same population (**Figure 2-26**), much of the variability in the bedload transport relation is explained (Rosgen, 1996, 2001b). Stream classification and channel stability ratings (Pfankuch, 1975) are coupled to further stratify differences in measured sediment rating curves and to establish reference dimensionless bedload and suspended sediment rating curve data (Troendle et al., 2001). These methods are recommended for application in *WARSSS* and will be described in more detail later in this chapter.

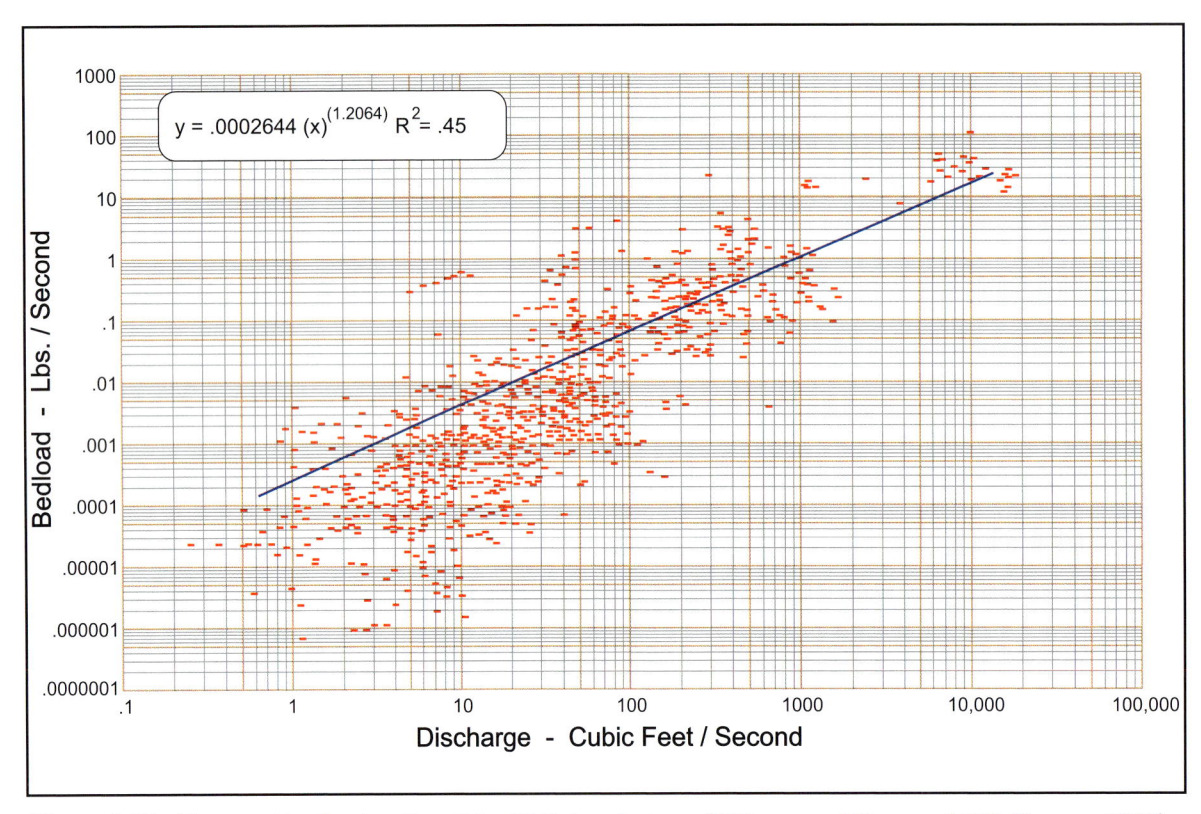

WATERSHED ASSESSMENT OF RIVER STABILITY AND SEDIMENT SUPPLY

Figure 2-25. Measured bedload sediment for 55 Colorado rivers (Williams and Rosgen, 1989; Rosgen, 1996).

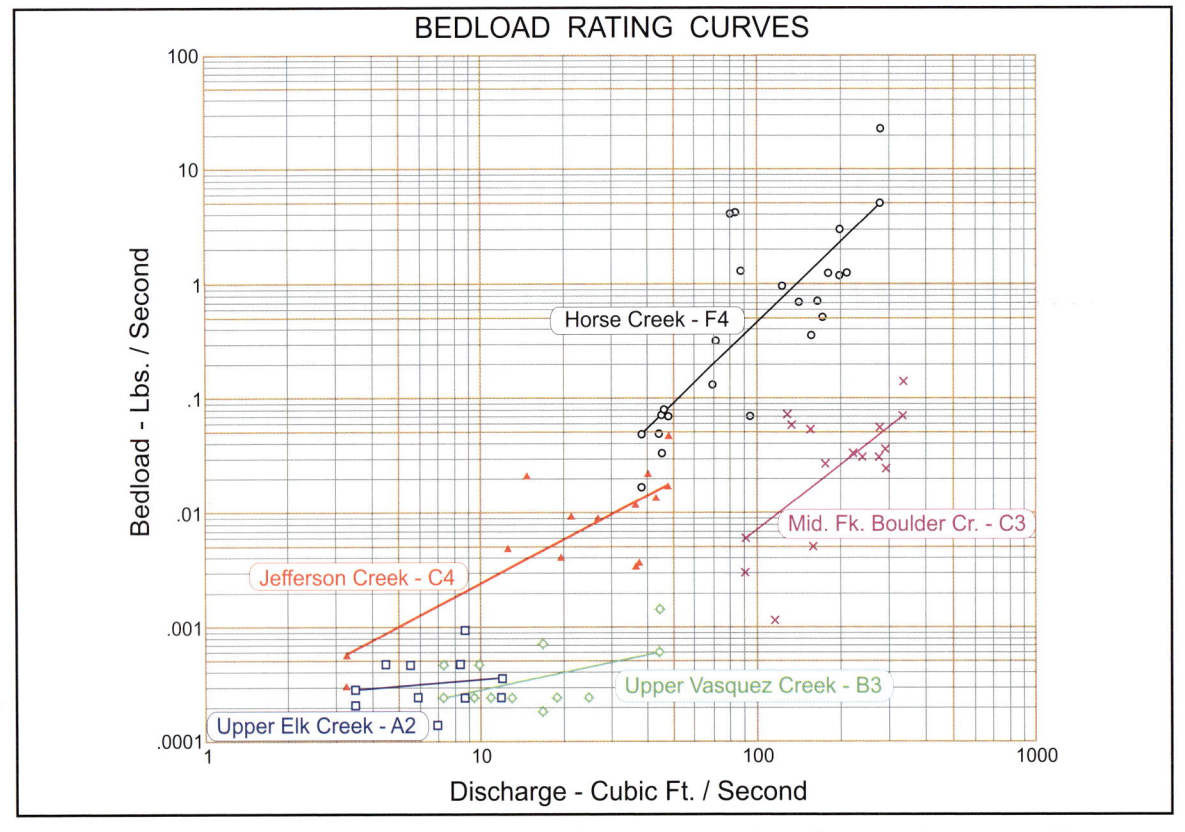

Figure 2-26. Bedload sediment rating curves stratified by stream type; from the same data set used for **Figure 2-25** (Rosgen, 1996).

Streambank erosion

Streambank erosion is a natural process. Acceleration of this natural process, however, leads to a disproportionate sediment supply, stream channel instability, land loss, aquatic habitat loss and other adverse effects. Streambank erosion processes, although complex, are driven by two major components: streambank characteristics (erodibility) and hydraulic/gravitational forces. Many land use activities affect both of these components and lead to accelerated bank erosion.

Vegetation rooting characteristics can protect banks from fluvial entrainment and collapse and also can provide internal bank strength. When riparian vegetation is changed from woody species to annual grasses and/or forbs, the internal strength is weakened, accelerating mass erosion processes. Streambed aggradation or degradation is often a response to stream channel instability. Because bank erosion is often a symptom of larger, more complex problems, the long-term solutions often involve much more than bank stabilization.

Erosion rates can sometimes be estimated from sequential, time-trend aerial photos on larger rivers. Lateral erosion rates from aerial photos and field-measured bank heights have been used to estimate sediment production rates (Lehre et al., 1983). Aerial photographs were also used to compare various riparian vegetative types to bank retreat rates (Pizzuto and Mecklenburg, 1989). Annual bank erosion rates were increased by three orders of magnitude due to willow eradication and conversion from woody species to a grass/forb riparian community on Wolf Creek in Colorado (Rosgen, 2001a).

Numerous studies have demonstrated that streambank erosion contributes a large portion of the annual sediment yield. Over 90% of the total suspended sediment yield during bankfull discharge of a snowmelt runoff event on the West Fork of the Madison River in Montana was associated with unstable streambanks and channel instability (Rosgen, 1973, 1976). Simon (1989a) reported lateral bank erosion rates from the Forked Deer River system in West Tennessee of 1.5 million tons per year (m/yr), of which 82% was contributed by bank erosion and 18% by bed degradation. Streambank erosion rates of 14 m/yr were measured in the Cimmaron River in Kansas (Schumm and Lichty, 1963), 50 m/yr in the Gila River in Arizona and 100 m/yr on some reaches of the Toutle River in Washington (Simon, 1992).

Streambank erosion contributed 46% of the total annual sediment yield from three miles of unstable channel on the East Fork San Juan River in Colorado (Rosgen, 2001a). The accelerated sediment supply contributing to the annual sediment budget was caused by the conversion of a stable C4 stream type (meandering channel, width/depth ratio of 25, gravel-bed river with a well-developed floodplain) to an unstable D4 stream type (gravel-bed, braided channel with a width/depth ratio of 250). The conversion and associated instability was caused by willow eradication in the 1930s (Rosgen, 2001a). As a result, the stream widened from 60 ft to locations resulting in channel widths from 400 to 840 ft in a 70-year period. The "river pedestals" of the East Fork of the San Juan River shown in **Figure 2-27** and **Figure 2-28** indicate the high rate of lateral erosion from channel widening. The pedestals are remnants of the streambank of the river terrace left standing approximately four feet above the gravel bar.

Figure 2-27. "River pedestals" of the East Fork San Juan River, remnant of the previous river terrace bank, indicating a high rate of lateral erosion.

Figure 2-28. "River pedestals" of the East Fork San Juan River, showing extensive, rapid bank retreat.

Streambank erosion prediction

The complexity of predicting annual streambank erosion rates has limited the application of available models. The *factor of safety approach* is a detailed, process-specific quantitative prediction of the likelihood or relative risk of streambank erosion as described by Thorne (1999) and Simon et al. (1999). The factor of safety approach has many valuable applications, but it may not be applicable to *WARSSS* due to the detailed data collection requirements and the inability of the model to predict annual streambank erosion rates. A process-integration approach is recommended as an alternative that provides a means for the field practitioner to estimate annual bank erosion (Rosgen, 2001a). This method involves collecting field data on streambank and channel characteristics to calculate a Bank Erosion Hazard Index (BEHI) and Near-Bank Stress (NBS) rating using the Bank Assessment for Non-point source Consequences of Sediment (BANCS) model. The user then relates these ratings to measured erosion. The advantage of predicting streambank erosion rates is to document and quantify the sediment load from this source. This is accomplished by multiplying the lateral erosion rate by the bank height by the length of bank associated with a given BEHI and NBS. These values are converted from cubic ft/yr to tons/yr then compared to total annual loads from measured values. Both BEHI and NBS play significant roles in *WARSSS* field measurement and sediment budget analysis. These relations can be further verified by a comparison with lateral erosion rates from time-trend aerial photography.

River stability

Within the scientific community, the terms *channel stability, equilibrium, quasi-equilibrium, dynamic equilibrium* and *regime channels* evoke various interpretations. It is necessary for those working with river systems to have a consistent, working definition of what constitutes a stable river. Davis (1902) defined a *graded stream* as the condition of "balance between erosion and deposition attained by mature rivers." Mackin (1948), as reported by Leopold et al. (1964), defined a *graded stream* as "…one in which, over a period of years, slope is delicately adjusted to provide, with available discharge and with prevailing channel characteristics, just the velocity required for the transport of the load supplied from the drainage basin." The *graded stream* is a system in equilibrium; its diagnostic characteristic is "…that any change in any of the controlling factors will cause a displacement of the equilibrium in a direction that will tend to absorb the effect of the change." The controlling factors described by Leopold et al. (1964) were width, depth, velocity, slope, discharge, size of sediment, concentration of sediment and roughness of the channel. If any one of these variables changes, it sets up a series of concurrent adjustments of the other variables to seek a new equilibrium.

Strahler (1957) and Hack (1960) used the term *dynamic equilibrium* to refer to an open system in a steady state in which there is a continuous inflow and output of materials, in which the form or character of the system remains unchanged. Equations showing river variables as a function of discharge were derived by Leopold and Maddock (1953) and by Langbein (1964). These hydraulic geometry relations described adjustable characteristics of open channel systems in terms of independent and dependent variables in quasi-equilibrium (neither aggrading nor degrading). Streams described as *in regime* are synonymous with *stable channels*. Equations describing the three-dimensional geometry of stable, mobile, gravel-bed rivers were presented by Hey and Thorne (1986). Additional equations and discussion on stable river morphology were presented in Hey (1997). Regime channels, as discussed by Hey (1997), allow for some erosion and deposition but no net change in dimension, pattern or profile for a period of years. Processes of stream channel scour and/or deposition do occur in a natural stable channel, but over time, if this leads to degradation or aggradation, respectively, then the stream is not stable.

These concepts of stream channel stability, the graded river, regime, quasi-equilibrium and/or dynamic equilibrium form a central theme in the WARSSS methodology. The central tendency, amidst the complex variables that shape channels and provide for erosion and deposition, is that *rivers progress toward their most probable form* (Leopold, 1994). As described in Chapter 1, a *stable channel* is one whose most probable form among rivers is the channel morphology that, over time, in the present climate, transports the water and sediment produced by its watershed in such a manner that the stream maintains its dimension, pattern and profile without aggrading or degrading (Rosgen, 1996, 2001b).

River stability concepts

Balanced sediment input and output is central to river channel equilibrium. Stream channel instability, or disequilibrium, is often associated with an excess load of sediment and/or size of delivered sediment beyond the carrying capacity of the river, resulting in storage and potential aggradation. Disequilibrium can also occur due to streambank instability, which increases the width/depth ratio and changes channel slope and pattern. An increase in the bankfull channel width changes the distribution of energy such that, for the same flow, there is a marked decrease in both shear stress (τ) and in unit stream power (ω_a). The increased width results in shear stress and stream power reduction due to the decrease in depth, velocity and slope. These channel and hydraulic changes decrease sediment transport competence and capacity, leading to excess sediment deposition. The adjustment of the channel in the presence of the high width/depth ratio and excess sediment deposition results in an increase in sediment supply from channel enlargement (widening) and accelerated streambank erosion from lateral adjustments. During major floods, channels with a high width/depth ratio can aggrade, causing loss of capacity, filling of pools and significant loss of aquatic habitat. Calculating the competence and capacity of sediment transport is a critical requirement for maintaining natural stable channels.

Aggradation

Aggradation involves the raising of the streambed elevation, an increase in width/depth ratio and a corresponding decrease in channel capacity. Over-bank flows occur more frequently with less than high-water events. Excess sediment deposition in the channel and coarse particle deposits on floodplains are characteristic of the aggrading river. Often, aggradation is the result of an increase in upstream sediment load and/or sediment size that exceeds the transport competence and/or capacity of the channel. Aggradation may also result from instability caused by over-widening of the channel, with a resultant decrease in stream power and shear stress. In other instances, aggradation can be caused by tectonic processes downstream of an affected reach, inducing deposition from a channel slope decrease. Aggradation can also be associated with tributary deltas due to disproportionate sediment supply, change in timing of flows or change in sediment transport.

Based on the relations in **Figure 2-9**, a wider, shallow depth channel with a corresponding decrease in unit stream power indicates a corresponding dramatic reduction in both size and transport rate of bedload. Adverse consequences associated with aggradation include channel *avulsion* (complete abandonment and initiation of a new channel) and major changes in stream type evolution. The adverse effects on beneficial uses can be very high due to the corresponding adjustments of the channel. Examples of aggradation processes are shown in **Figures 2-29, 2-30** and **2-31**. Common adverse consequences include increased flood risk, land loss through accelerated streambank erosion, the decline of fish habitats, elevated stream temperatures and loss of biological functions.

Bedload models and entrainment calculations can be useful in predicting aggradation potential as long as the models are locally calibrated. Bedload models using stream power equations are preferred, as they accommodate the velocity and shear stress changes often associated with the altered dimension, pattern and profile caused by channel disturbance.

Figure 2-29. Aggradation of coarse gravel and cobble on an over-wide C3 stream type on lower West Fork San Juan River, Colorado.

Figure 2-30. Aggradation of sand and fine gravel in a C4 stream type on Blue Joe Creek, Idaho.

Figure 2-31. Aggradation on Willow Creek, Colorado, due to excess sediment supply from upstream sources.

Degradation

Degradation is the lowering of the local base level of rivers through the process of excess bed scour and channel incision. The lowering of the streambed abandons floodplains, lowers the water table and increases bank height, which adds to bank erosion and often leads to long-term instability. The causes of degradation are complex and may be related to many sources. Clear water discharge below reservoirs, urban storm drains, excess shear stress due to changes in flow regime, straightening of the channel alignment that alters slope, headward advancement (headcut) of base-level shifts due to downstream alteration, excess shear through contraction and bed scour from bridges and culverts can all contribute to channel degradation. Examples of degradation in stream channels are shown in **Figures 2-32** and **2-33**.

Field evidence of degradation is a combination of a lowered width/depth ratio and an increased Bank-Height Ratio (BHR). *Bank-height ratio* is a measure of the degree of channel incision (Rosgen, 2001b) leading toward channel entrenchment (vertical containment).

$$BHR = LBH/d_{bmx} \qquad \textbf{(Equation 2-14)}$$

where:
BHR = Bank-Height Ratio

LBH = Lowest Bank Height

d_{bmx} = bankfull maximum depth

If the bank-height ratio is greater than 1, then flows greater than bankfull discharge are required to over-top the bank. The longitudinal profile survey, where elevations of the bed, water surface, bankfull stage and bank-height are plotted in the downstream direction, provides valuable information as to the direction and extent of the incision. If the "wedge," or the bank-height ratio, is *increasing* in the downstream direction, then an incision is advancing headward. If the wedge is *decreasing* in the downstream direction, then the source of the incision is from upstream. This field technique helps to determine the source, extent and direction of channel incision. The bank-height ratio is also an "early warning" indicator of incision, prior to eventual entrenchment or vertical containment.

An increase in bank-height ratio generally results in increased shear stress and stream power, and has the potential to lower the streambed and enlarge the channel. Degradation is also associated with the undermining of the root mass of riparian species and the collapse of the upper bank with trees and other vegetation. Generally, streams do not incise unless the width/depth ratio is less than 10 (Rosgen, 2001b). A lowering of the width/depth ratio generally corresponds with higher shear stress and unit stream power; thus, the incision is generally associated with a shift in stream type from E or C to G (gully). E stream type channels are often straightened in meadow environments to drain the meadow for agricultural purposes. The results of the increased slope and lowered width/depth ratio are to increase shear stress and stream power, causing incision. Any disruption of the natural energy balance and sediment transport that is reflected in the morphological variables of dimension, pattern and profile may lead to serious long-term adjustments, such as incision and/or degradation.

Chapter 2 A Review of Hillslope, Hydrologic and Channel Process Relations

Figure 2-32. Example of a gully created due to degradation caused by high shear stress and stream power below the "double-barrel shotgun" culverts, Maryland.

Figure 2-33. Headward advancement of a degraded gully in a meadow, Colorado.

Prediction of potential degradation requires competence and critical depth computations (**Equations 2-7**, **2-8** and **2-9**). If the shear stress and depth/slope relations are sufficient to move particle sizes larger than what is contained in the bed, there is a high probability of excess shear and bed scour, leading to channel incision and eventual degradation. Degradation potential becomes an essential stability evaluation due to the long-term and severe adverse adjustments of stream channels once they become incised or entrenched (vertically contained). The sediment supply associated with degradation can be extremely high, as presented by Simon (1989a).

Channel enlargement

Channel enlargement can be caused by combined processes of incision, bank erosion, direct modification by construction activities and large floods in incised channels. Lateral erosion may occur in stable streams, but the point bar follows at the same rate; thus, some streams do not get wider over time. This contrasts with channel enlargement, where the width of the stream increases due to lateral erosion, often concurrently on both banks. The results of enlargement are increased sediment supply from the bed and banks, increased deposition due to decreased shear stress and stream power, loss of habitat, increased water temperatures and a shift in stream types. Often, a C4 stream type (single-thread, meandering channel) is shifted to a D4 stream type (multiple-thread, braided channel). Increased flows due to watershed changes, trans-basin diversions (imported water), storm drains from urban runoff, power generation due to "ramping flows" from reservoir releases and contraction scour below culverts and bridges can all contribute to channel enlargement. Combined processes of incision, degradation, aggradation and lateral accretion may also be associated with enlargement.

Channels are frequently mechanically enlarged to increase flood flow capacity for bridge construction, for flood-control projects and often for emergency flood-relief work. These over-widening projects generally result in eventual excess deposition in the channel, actually decreasing the channel's capacity to transport sediment. These projects result in increased bank erosion; thus, many of these projects require extensive maintenance, including dredging sediment and stabilizing streambanks. An example of channel enlargement is shown in **Figure 2-34**.

Figure 2-34. An over-wide gravel-bed stream evolving from a C4 to D4 stream type. Enlargement is due to the effects of combined bank erosion on both banks and excess coarse sediment deposition from an upstream source, Wolf Creek, Northeast California.

Gully erosion

Local base-level shifts in ephemeral drainageways can accelerate incision processes and excess erosion of both the channel bed and banks of the ephemeral gully. The Rosgen stream classification system (1994, 1996) is used to classify the ephemeral gully and to obtain the dimension, pattern, profile and channel materials as a function of drainage area and width/depth ratio associated with the entrenched A, F or G stream types. Rates of incision and bank erosion (enlargement) use the same computational methods, as do streambank erosion and channel stability (i.e., aggradation, degradation, confinement). An example of gully erosion is shown in **Figure 2-35**.

Figure 2-35. Example of a G5 gully, Florida.

Mitigation in gully systems generally involves creating grade control structures to contain the headcut advancement, back-sloping and vegetating exposed streambanks to decrease streambank erodibility. In descending priority order, other restoration/stabilization methods for entrenched rivers involve re-establishing the original surface of the incised channel, constructing a stable stream type in place and creating a confined but stable stream type, such as converting a G stream type to a B stream type or an F stream type to a Bc- stream type (Rosgen, 1997).

Stream channel succession

The adverse consequences of accelerated sediment supply, accelerated bank erosion rates, degradation, aggradation from channel disturbance, streamflow changes and sediment budget changes can lead to channel change. These changes result in stability shifts and adjustments leading to stream channel morphological changes and eventual stream type changes, as shown in **Figure 2-36** (Rosgen, 1996).

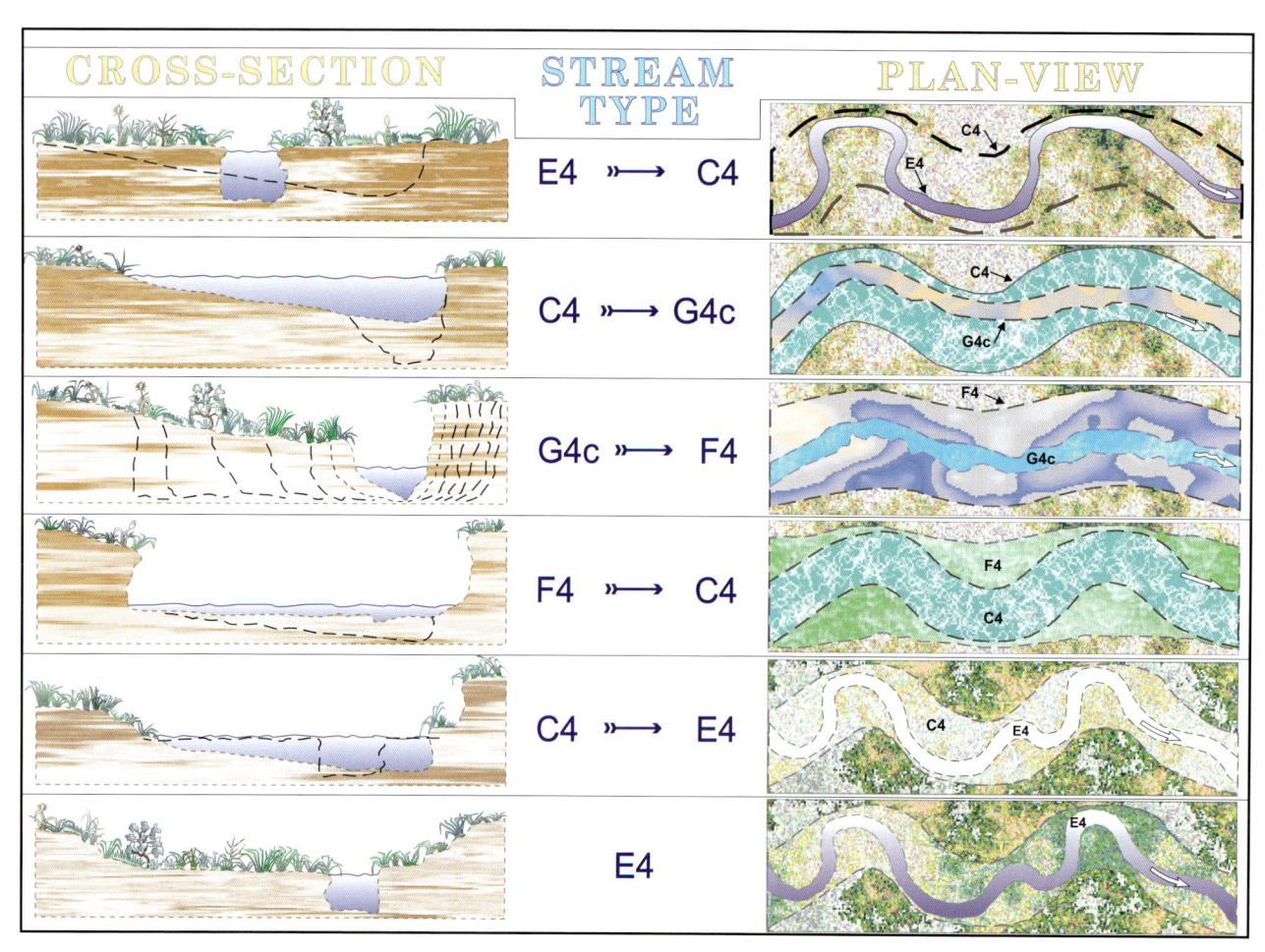

Figure 2-36. Adjustments of channel cross-section and plan-view patterns, as stream types shift through successional cycles (Rosgen, 1996).

The stream type succession sequences in **Figures 2-36**, **2-37** and **2-38** are similar to the stages of channel evolution as reported by Schumm et al. (1984) and Simon and Hupp (1986), but these figures present more change scenarios than the original channel evolution concept. The relation between successional stages and stream classification is shown in **Figure 2-37** (Rosgen, 1999). The nine scenarios in **Figure 2-38**, depicting the successional sequences of stream type change, indicate a larger range of possible morphological shifts and their tendencies toward stable endpoints than previously published (Rosgen, 1999, 2001b).

Chapter 2 — A Review of Hillslope, Hydrologic and Channel Process Relations

Sequence of Stream Type Occurrence Due to Morphological Change
ALLUVIAL VALLEY — Gravel Bed — Valley Slope .009

Channel Evolution Model (Simon and Hupp, 1986)

Stage 1: PreModified
Stage 2: Constructed

Stream Type C4
Entrenchment Ratio... 20
Width / Depth Ratio... 18
Sinuosity 1.8
Channel Slope ...... .005
Meander Width Ratio ... 8
Simon & Hupp Evolutionary Stage ....1 & 2

Stage 3: Degradation

Stream Type G4
Entrenchment Ratio... 1.1
Width / Depth Ratio.......5
Sinuosity 1.3
Channel Slope ...... .007
Meander Width Ratio ... 4
Simon & Hupp Evolutionary Stage 3

Stage 4: Degradation and widening
Stage 5: Aggradation and widening

Stream Type F4
Entrenchment Ratio...1.0
Width / Depth Ratio...150
Sinuosity 1.8
Channel Slope ...... .008
Meander Width Ratio..1.5
Simon & Hupp Evolutionary Stage ....4 & 5

Stage 6: Quasi equilibrium

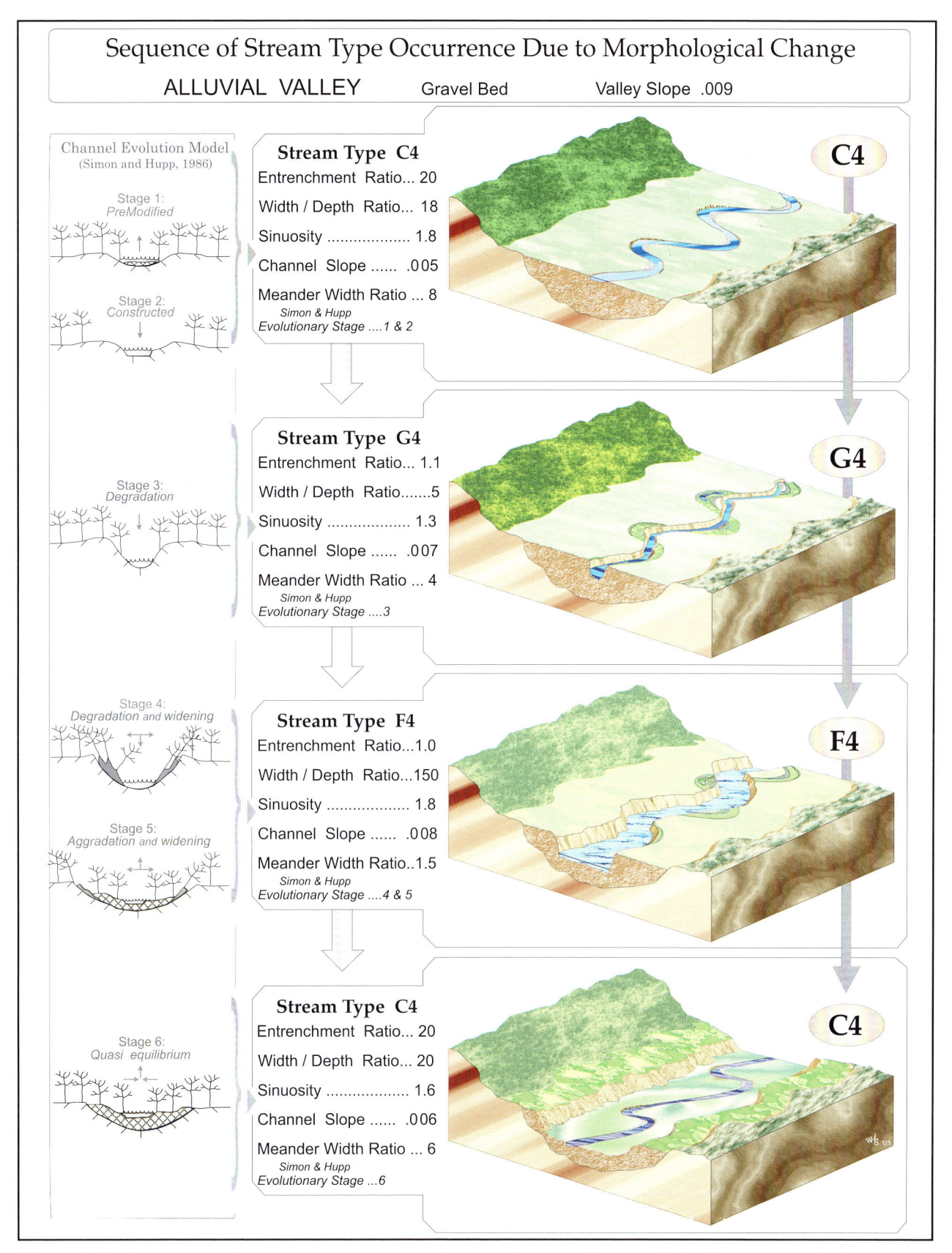

Stream Type C4
Entrenchment Ratio... 20
Width / Depth Ratio... 20
Sinuosity 1.6
Channel Slope ...... .006
Meander Width Ratio ... 6
Simon & Hupp Evolutionary Stage ...6

Figure 2-37. Comparison of channel evolution model stages of Simon and Hupp (1986) with one morphological sequence of Rosgen stream types (adapted from Rosgen, 1999).

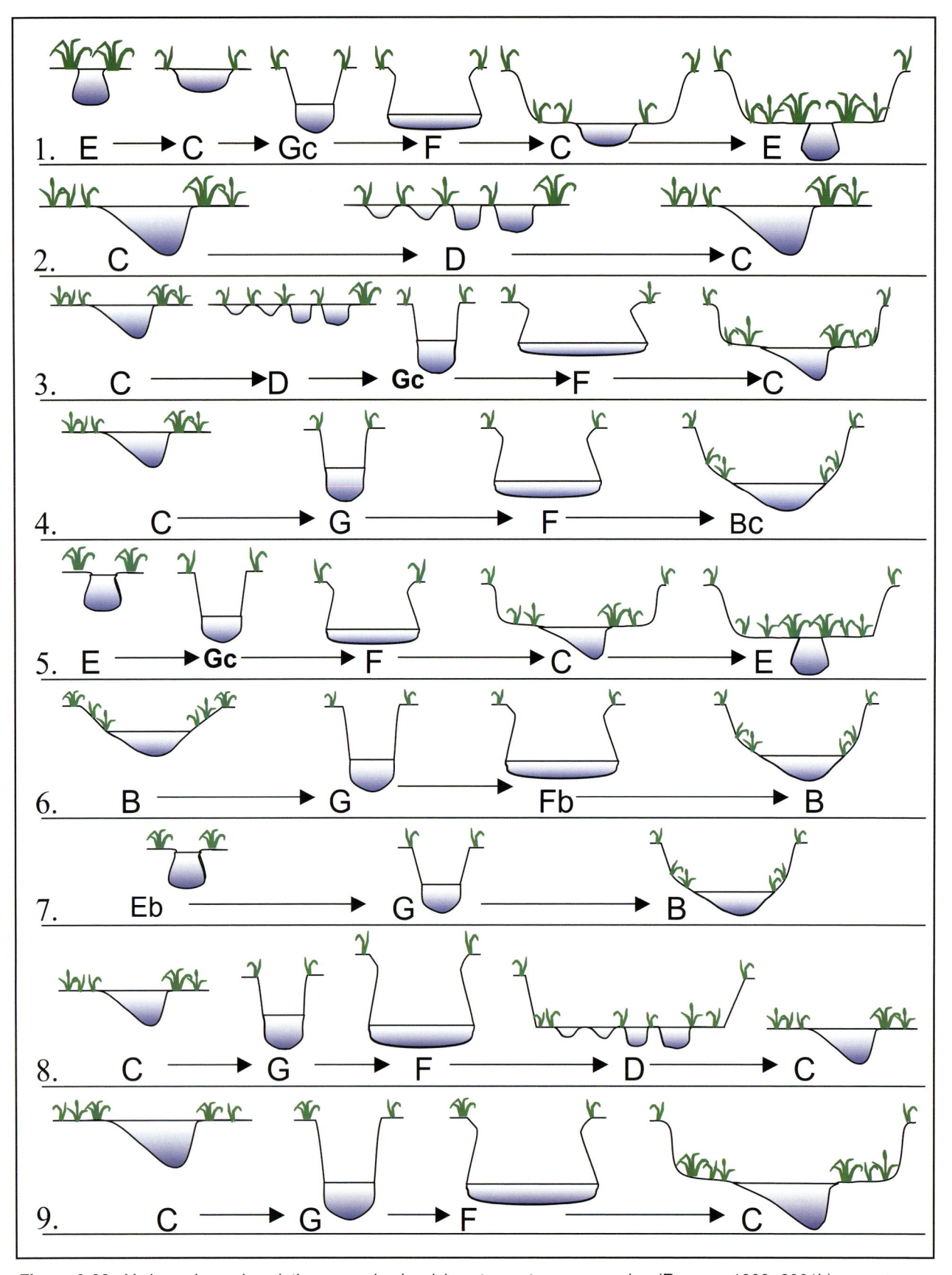

Figure 2-38. Various channel evolution scenarios involving stream type succession (Rosgen, 1999, 2001b).

Each stage of the individual sequences shown in **Figures 2-37** and **2-38** is associated with unique quantitative relations of morphological, hydrological, sedimentological and biological functions. Adverse adjustments due to shifts in equilibrium can create accelerated sediment yields, loss of land, lowering of the water table, decreased land productivity, loss of aquatic habitat and diminished recreational and visual values. The *WARSSS* procedure identifies the existing stream type, associated channel adjustment scenario and the current state of a stream reach in the successional sequence. In assessing stream reaches where excess sediment is associated with a disequilibrium condition of a particular stability state or stage of adjustment, the succession relations assist in determining the following information:

- The appropriate morphological scenario;
- Within a scenario, the current state or stage of the successional stage of the existing stream type;
- The various stages generally associated with a succession endpoint;
- The series of natural changes that occur prior to reaching stability (quasi-equilibrium); and
- The potential stable form of the channel type.

Once the appropriate sequence is identified within a range of scenarios (**Figure 2-38**), the current successional stage of the existing stream type within that sequence is identified.

Figure 2-38 indicates the stream type changes that can occur as the stream seeks a morphological form with greater stability or quasi-equilibrium. For each stage, there are large arrays of morphological variables that describe the current dimension, pattern and profile of the stream type. An example of a stable C4 reference reach, with excellent woody riparian vegetation and negligible channel erosion, is shown in **Figure 2-39**. An unstable C4 reach, with poor riparian vegetation and accelerated streambank erosion, appears in **Figure 2-40**. These examples represent the same stream type, but the latter indicates potential channel change from a C4 stream type to a D4 stream type, as depicted in Scenario #2 in **Figure 2-38**. Bank erosion can often continue with channel enlargement and aggradation, resulting in a D4 or braided stream type (**Figure 2-41**).

The sediment consequences associated with disequilibrium are evident in the previous photographs, as the channel types change from a stable C4 to an unstable C4, and then to a D4 stream type. The beneficial uses of water for fish habitat, recreation and aesthetics are all adversely impacted with these changes. With proper riparian vegetation, the stream will eventually restore itself to a C4 stream type. The East Fork of the San Juan River, described previously, changed from a C4 to a D4 stream type following willow eradication. The channel succession pattern was typical of Scenario #2 in **Figure 2-38**.

Figure 2-39. A stable C4 stream type, valley type VIII, with excellent riparian vegetation.

Figure 2-40. An unstable C4 stream type, valley type VIII, showing higher w/d ratio due to accelerated streambank erosion. Note grass/forb vegetation.

Figure 2-41. An unstable D4 stream type, valley type VIII, exhibiting multiple thread channels, an extremely high w/d ratio (254) and accelerated bank erosion.

Another series of channel changes due to instability is related to Scenario #1 (**Figure 2-38**), involving degradation and lateral extension processes. This sequence is shown in **Figures 2-42** through **2-45**.

The consequences of instability due to sediment sources from both the bed and banks of the G and F stream types are quite evident in these photographs. It is important to recognize the central tendency of rivers to seek stability, as shown in the diminished sediment supply associated with the more stable C and E stream types. Sediment consequence data will be presented in Chapter 7 for Wolf Creek as an example of the application of *WARSSS*.

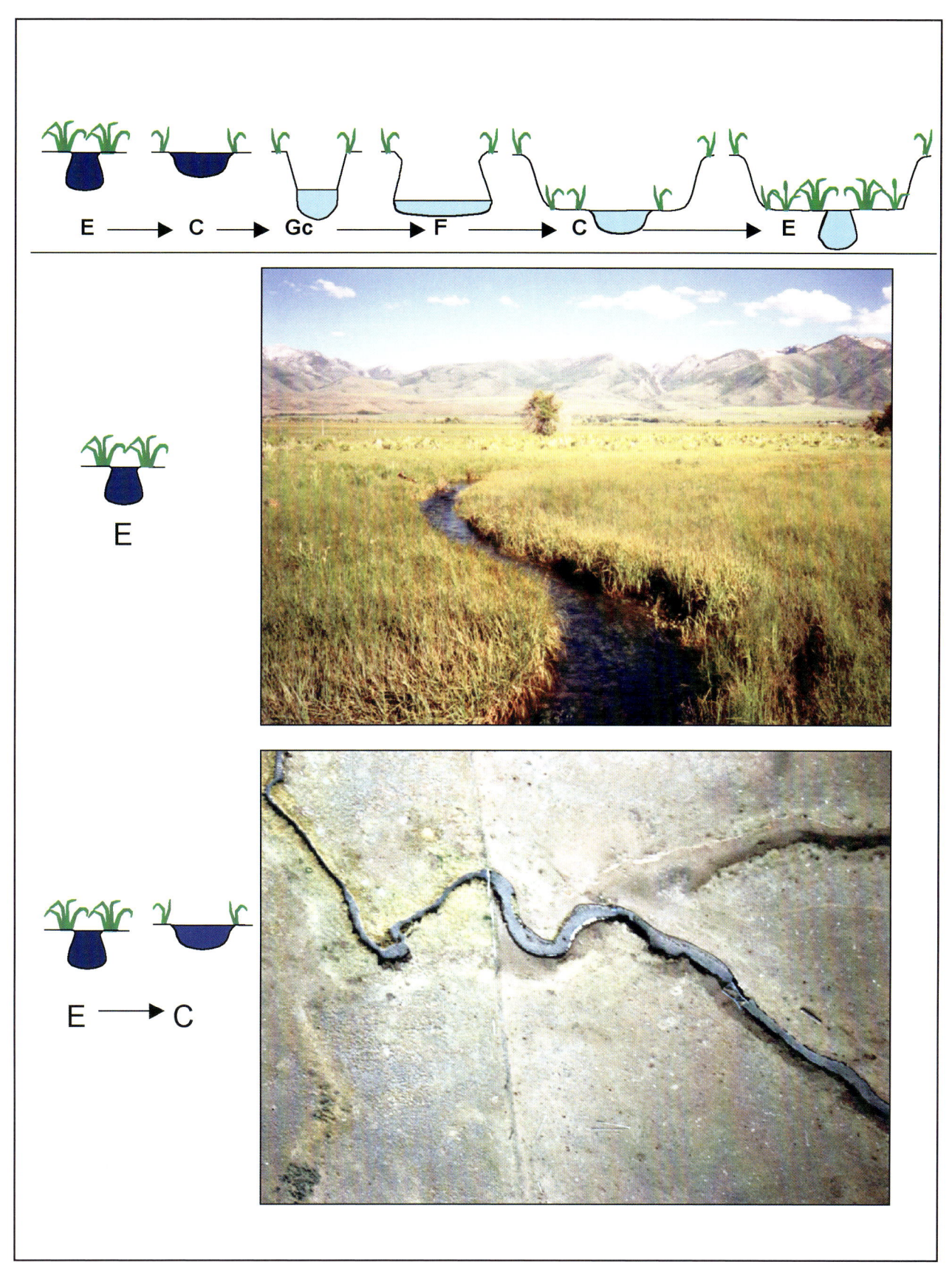

Chapter 2 — A Review of Hillslope, Hydrologic and Channel Process Relations

Figure 2-42. Channel succession stage from an E to an unstable C stream type; note increase in w/d ratio.

2-49

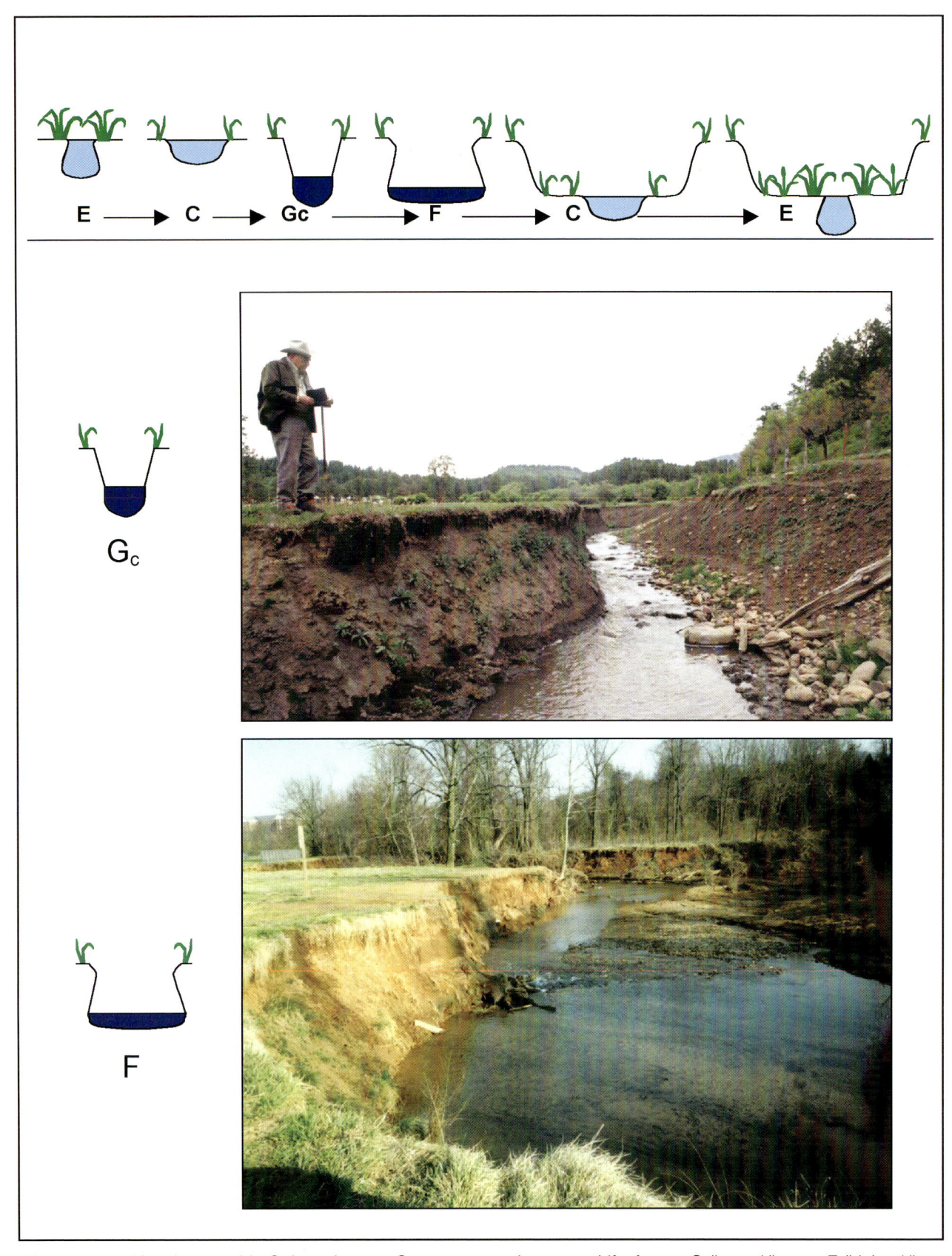

WATERSHED ASSESSMENT OF RIVER STABILITY AND SEDIMENT SUPPLY

Figure 2-43. After the unstable C degrades to a G stream type, the stage shifts from a G (low w/d) to an F (high w/d) stream type. Note Luna B. Leopold making field note entries on the G_c channel.

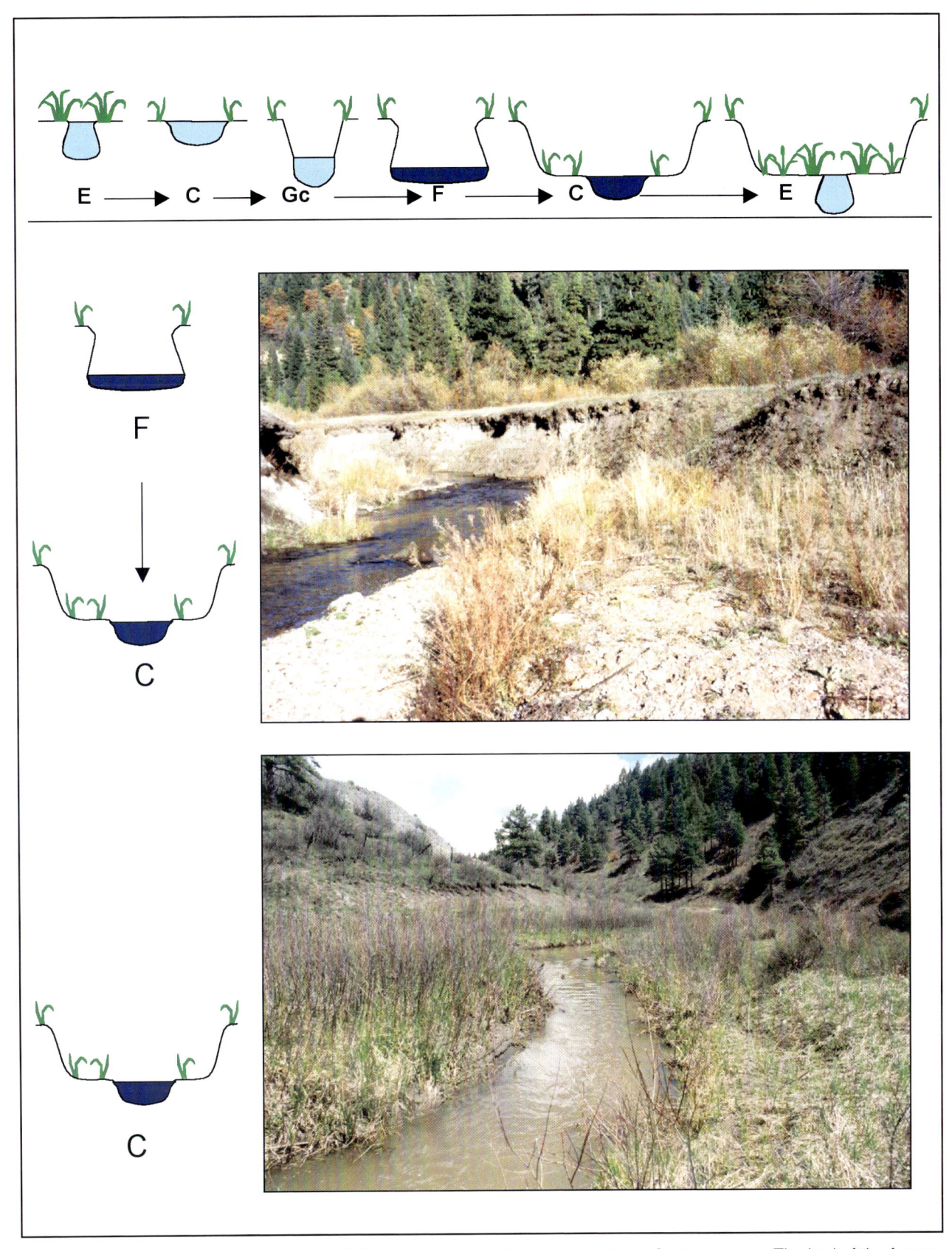

Chapter 2 *A Review of Hillslope, Hydrologic and Channel Process Relations*

E → C → Gc → F → C → E

Figure 2-44. Channel succession stage shift from an unstable F to a more stable C stream type. The bed of the former F stream type is the new floodplain for the C stream type.

2-51

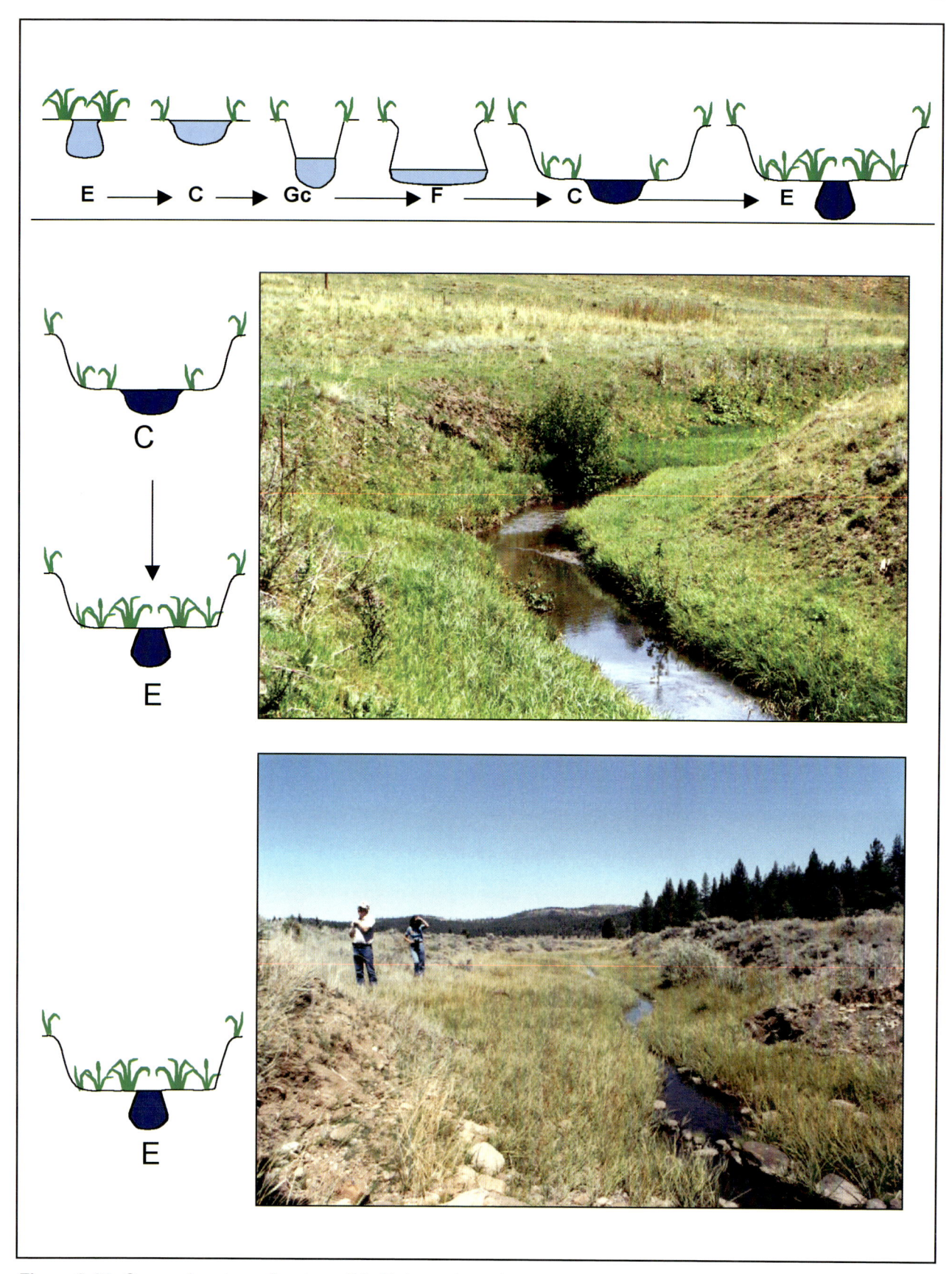

Figure 2-45. Succession stage showing a C to E stream type change as vegetation reduces width/depth ratio.

Hydrologic processes

Streamflow

Many changes in stability are associated with changes in the magnitude, duration and timing of streamflows. Research conducted by the USDA Forest Service has demonstrated significant long-term increases in streamflow following timber harvest and road construction. These changes in flow are associated with reductions in forest evapo-transpiration, rainfall interception loss and an increase in snowpack water equivalent, making excess water available for runoff. A snowmelt hydrograph for Fool Creek in the Central Rockies Province of Colorado shows changes in peak and duration of flows following a 50% stand removal by patch clearcutting on a second-order stream (**Figure 2-46**; Troendle and Olsen, 1994). Hydrologic recovery in the Rocky Mountain snowmelt-dominated runoff region is estimated at approximately 60 years (Porth, 2006). Based on those estimates, the full hydrologic recovery of Fool Creek would be extended to the year 2016 rather than 1971, as shown in **Figure 2-46**.

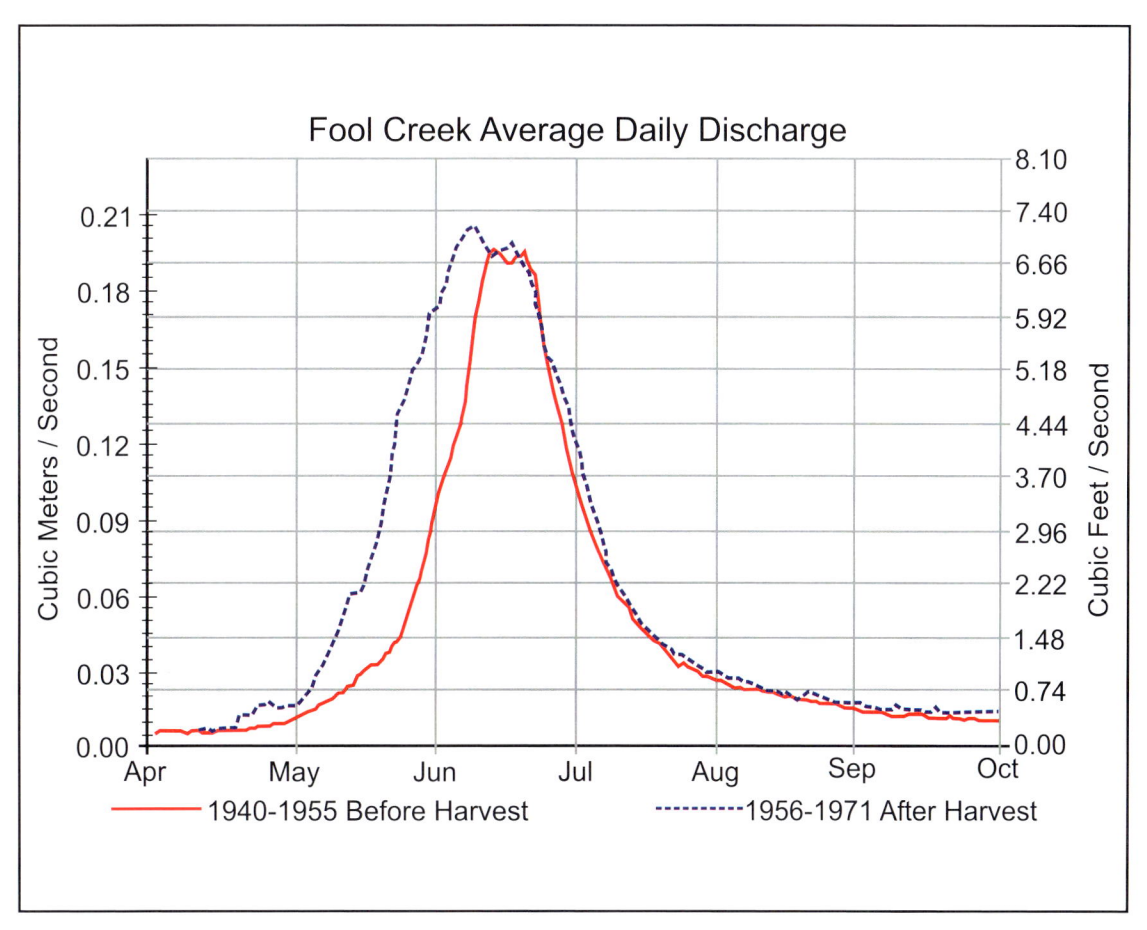

Figure 2-46. Water yield increase following patch clearcutting, Fool Creek, Colorado (Troendle and Olsen, 1994).

Streamflow increases can potentially alter the sediment transport relations and morphological character of rivers. The stream's response to increased streamflow is a function of both the stream type and the stability of the channel, as well as the magnitude and duration of flow changes. Numerous studies have demonstrated that forest disturbances can increase the amount of introduced channel sediment. Bosch and Hewlett (1982) summarized the effect of timber harvest on increased streamflow for over 100 experiments worldwide. Following partial clearcutting on Deadhorse Creek, Colorado, a significant increase in both sediment export and flow was observed (Troendle and Olsen, 1994). The analysis of Deadhorse Creek indicated that the increase in sediment was from within the channel and not the result of increased sediment introduction following road building and timber harvest. The timber harvest and roads shown in the Willow Creek Watershed, Colorado (**Figure 2-47**) depict high potential for increased streamflow and corresponding sediment yields due to the extent and nature of the forest disturbance. Flow-related increases in sediment, however, have not been widely reported. Cumulative effects of such practices need to be assessed for their respective influence on increased sediment yields and channel instability.

Figure 2-47. Cumulative effects of clearcutting and road construction, Willow Creek Watershed, Colorado.

Water yield models for both snowmelt and stormflow-generated runoff have been used by hydrologists for many years. The snowmelt model in WRENSS (USEPA, 1980) was validated nationwide and is recommended in *WARSSS* for snowmelt-dominated regions. An improvement based on the WRENSS model was documented by Swanson (2004) and Troendle and Swanson (in press), which is being supported by the USDA Forest Service Stream Team in Fort Collins, Colorado. The Unit Hydrograph and flow-duration curve methods, such as TR-20, TR-55, WRENSS (USEPA, 1980) and those recently described in Haan et al. (1994), are models that predict a runoff response based on a given storm. These approaches are typically applied in urban watersheds and for stormflow-dominated regions. The snowmelt version by Sheppard et al. (1991) is presently being updated for a Windows™ software-based program by Troendle and Swanson (in press) and is included in Swanson (2004).

Diversions also affect flow and sediment dynamics. Flow increases due to imported water can create excess shear stress and stream power in channels leading to enlargement, degradation and accelerated bank erosion. Decreases in flow or dewatering of mainstem channels can reduce the channel's ability to maintain transport capacity; thus, the sediment delivered to the regulated trunk stream by unregulated tributaries leads to aggradation. Flow releases by reservoirs and diversions can be determined by the operational hydrology of the system or back-calculated from high-stage scour lines in the channel using indirect methods to estimate discharge.

Bankfull discharge

The term *bankfull* was originally used to describe the incipient elevation on the bank where flooding begins. In many stream systems, the bankfull stage is associated with the flow that fills the channel to the top of its banks and at a point where the water begins to overflow onto a floodplain (Leopold et al., 1964). The bankfull stage and its attendant discharge serve as consistent morphological indices, which can be related to the formation, maintenance and dimensions of the channel as it exists under the modern climatic regime. The terms *effective* and *dominant discharge* are synonymous with bankfull discharge as used in this procedure; see the federal manual *Stream Corridor Restoration: Principles, Processes, and Practices* (Federal Interagency Stream Restoration Working Group, 1998) for a detailed discussion of these concepts. Stream dimensions, patterns and bed features associated with the longitudinal river profile are generally described as a function of channel width as measured at the bankfull stage. Streams are self-formed and self-maintained; thus, it is important to relate measurable features one can identify in the field to a corresponding bankfull discharge. This definition of bankfull discharge applies primarily to stream types that have an observable floodplain feature, no matter how wide. Floodplains can be quite small in certain stream types where they may be naturally indistinct or currently being developed.

Streams that are incising or deeply entrenched in the landform do not exhibit significant changes in channel width as flood flows increase. With increasing flood stage, stream depth generally increases at a more rapid rate than channel width. Bankfull stage can be observed and determined within the F (entrenched) stream types by using a series of common stage indicators generally located along the boundary of the active channel. It is very difficult, however, to field determine the bankfull stage of actively incising G channels. Bankfull discharge must be determined indirectly using regional hydrology curves developed from streamgage data, as discussed in Chapter 5. Similarly, for recently aggrading channels, flows less than bankfull discharge can occupy floodplains. Again, regional hydrology curves for a specific hydro-physiographic province must be used (see **Figure 5-1** in Chapter 5 as an example).

A commonly accepted and universally applicable definition of *bankfull* was provided by Dunne and Leopold (1978): "The bankfull stage corresponds to the discharge at which channel maintenance is the most effective, that is, the discharge at which moving sediment, forming or removing bars, forming or changing bends and meanders, and generally doing work results in the average morphologic characteristics of channels." It is this discharge, along with the range of flows that make up an annual hydrograph, that governs the shape and size of the active channel. Bankfull discharge is associated with a momentary maximum flow that has an average recurrence interval of 1.5 years, as determined using a flood frequency analysis (Dunne and Leopold, 1978).

Although great erosion and enlargement of steep, incised channels may occur during extreme fluvial events, it is the modest flow regimes that transport the greatest quantity of sediment material over time due to the higher frequency of occurrence for such events (Wolman and Miller, 1960). An example of the relationship between flow magnitude and frequency of flow occurrence is shown in **Figure 2-48**.

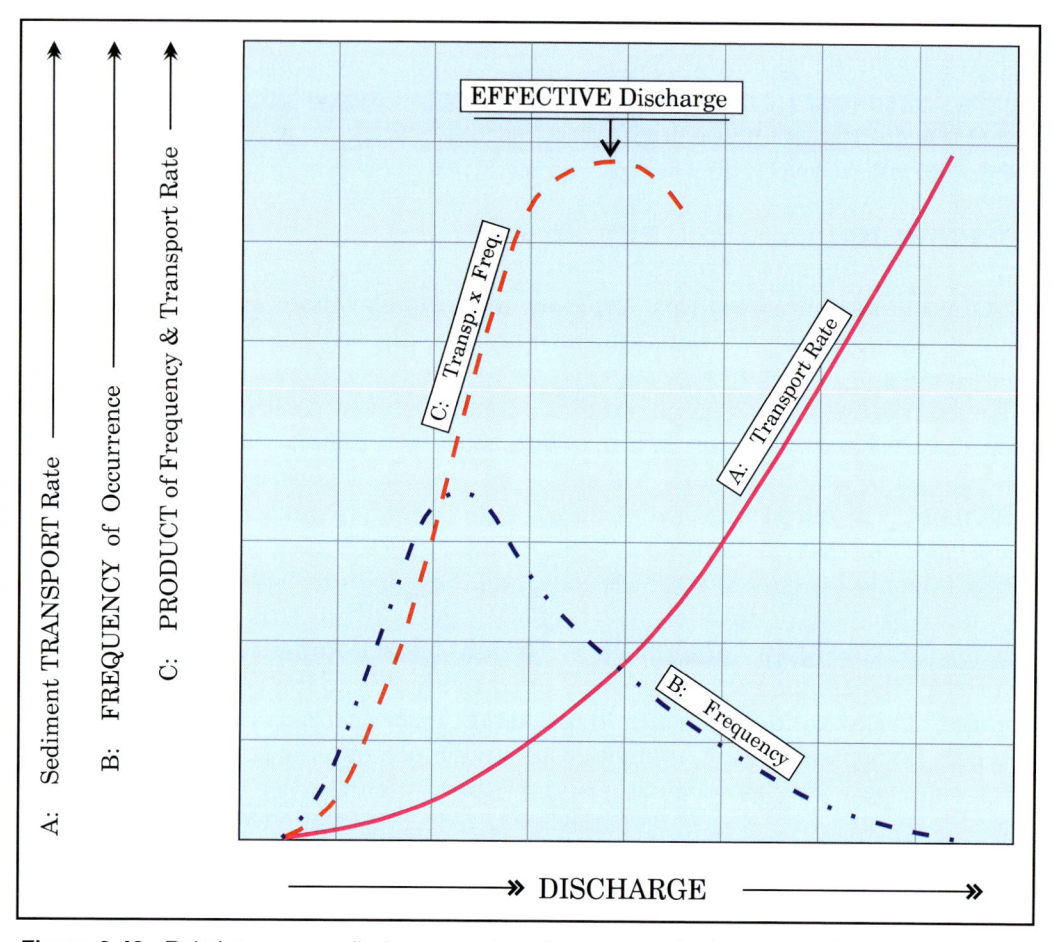

Figure 2-48. Relations among discharge, sediment transport rate, frequency of occurrence, and the product of frequency and transport rate (after Wolman and Miller, 1960).

The dominant, effective or bankfull discharge is associated with the peak of cumulative sediment transport for a given streamflow magnitude and frequency of occurrence. The majority of work over time is accomplished at moderate flow rates, as shown in **Figure 2-48**. The effectiveness of bankfull discharge and a discussion of dominant or effective discharge theory is summarized by Andrews (1980).

Recent analyses of peak flow data for gage stations on 47 rivers located in Ontario, Canada, indicated that their bankfull discharges have an average return interval of 1.6 years, with a range from 1.5–1.7 years (Annable, 1995). A review of the literature by Williams (1978) indicates return periods for "bankfull discharge" ranging from 1–25 years. In the Williams study, however, no clear distinction was made between the elevations of the low terrace and the active floodplain, which serves as the indicator of bankfull stage. A low terrace is, by definition, an abandoned floodplain.

The flows necessary to over-top the low terrace bank must be associated with a flood of large magnitude, much larger than the actual bankfull discharge. A low terrace feature is often mistaken for an active floodplain by an untrained field observer.

An analysis of return periods related to field-determined bankfull discharge, conducted by the author over the past 40 years and using data from gage stations located on rivers throughout North America, indicates a range in return interval from 1.05–1.8 years. Exceptions to this finding appear to be associated with highly developed urban watersheds, where the return period of the bankfull discharge is closer to 1.1 years. Often, the U.S. Army Corps of Engineers field interpretation of "ordinary high water" and the bankfull stage are synonymous. Thus, one may conclude that the flow regime associated with bankfull discharge is a relatively frequent event. Using a partial duration series (multiple peaks per year) for flood probability analysis, it is common for bankfull discharge to occur several times per year. The annual series method (the largest magnitude single peak flow per year), however, is used to calculate the recurrence interval for flood probability analysis and to establish regional curves (Dunne and Leopold, 1978).

Bankfull discharge is a key concept for properly classifying streams and determining departure from reference conditions, such as in the calculation of width/depth ratio. The width and mean depth measured for the width/depth ratio calculation are those dimensions at the stage related to bankfull discharge.

Integration of hydrologic, sediment and stability relations

Stream stability, reference condition and sediment rating curves

Stream stability shifts are reflected in sediment rating curves where measured sediment values are regressed against measured discharge. The upward shift in the slope and/or intercept values of the sediment rating curve are due to increased sediment supply as a result of a variety of sources. The upward shift in the sediment rating curve exponentially increases the sediment yield for selected increments of streamflow. Land uses that increase streamflow magnitude and duration can be instrumental in accelerating flow-related increases in sediment due to characteristically unstable channels with elevated sediment rating curves. Examples of sediment rating curves influenced by land use are shown in Missouri gully erosion (Piest et al., 1975), silvicultural impacts (USEPA, 1980; Rosgen, 2001c), Colorado silvicultural and reservoir impacts (Rosgen, 1996), West Tennessee channelization for land drainage and flood control (Simon and Hupp, 1986) and Arizona silvicultural impacts (Lopes et al., 2001b).

Many ongoing stream system adjustments involve events and/or past perturbations associated with the various stages of morphological types. It is essential to identify the potentially stable stream type of the existing river. A reference stream type provides not only the morphologically stable form of dimension, pattern and profile, but also the corresponding rates of sediment supply, bank erosion rates and other characteristics representing a stable (quasi-equilibrium) geomorphic condition. Stream types and channel stability relations are used concurrently in the use of sediment rating curve relations.

An example of the use of a reference stream and stability interpretation to assess changes in sediment yields is that of the Hatchie and South Fork Forked Deer Rivers in West Tennessee. Simon (1989b) presented the sediment consequence associated with a shift in stability and evolution stages. The Hatchie River, a stable, meandering, sand-bed, E5 stream type with a well-developed floodplain and a low width/depth ratio, had an annual sediment yield of 57 metric tons/yr/km^2 (163 tons/yr/mi^2). Channelization for land drainage was applied to the South Fork Forked Deer River in the same basin as the Hatchie River. The channelization entrenched the river and initiated a channel evolution sequence from stage I to III/IV (Simon and Hupp, 1986). The changes in channel evolution through successional stages corresponded to a shift from stream types E5→ G5→ F5, or Scenario #5 in **Figure 2-38** (Rosgen, 1994, 1999, 2001b). The F5 stream type is a relatively straight, incised, high width/depth ratio, sand-bed stream. The resultant unstable, entrenched (F5 stream type) was producing 872 metric tons/yr/km^2 (2490 tons/yr/mi^2) (Simon, 1989b). The majority of the sediment supply was from streambank erosion. The U.S. Geological Survey's suspended sediment rating curves for the Hatchie and South Fork Forked Deer Rivers appear in **Figure 2-49** (Simon, 1989b).

The sediment rating curves shown in **Figure 2-49** were converted to dimensionless relations (**Figure 2-50**). In this relation, the mg/l data from Simon (1989b) were converted to tons/day to match the published rating curve. The Hatchie E5 stream type is a stable reach, whereas the South Fork Forked Deer F5 stream type is an unstable reach. A statistical analysis performed on the two dimensionless sediment rating curves resulted in a p value of less than *0.0001*, indicating that the South Fork Forked Deer River curve is statistically significantly different from the reference Hatchie River curve. Because the sediment rating curves did not collapse into the same curve when made dimensionless, this indicates the change in stability and corresponding stream type change from an E5 to an F5 stream type, resulting in a new sediment rating curve reflecting the higher sediment supply of approximately two orders of magnitude.

The direct disturbance that created the accelerated sediment supply has existed since the 1950s, but the river has yet to reach an equilibrium state. These channel adjustments have resulted in long-term adverse sediment effects. The channel stability shifts and corresponding changes in stream types are associated with measured quantitative morphological relations involving channel dimension, pattern, profile and channel materials. The corresponding sediment consequences due to streambank erosion are obvious.

The prediction of a shift from a reference condition using measured sediment rating curves and ranking good to poor stability ratings (Pfankuch, 1975) by stream type (Rosgen, 1994, 1996) was conducted by Troendle et al. (2001). This work established dimensionless sediment rating curves for suspended sediment and bedload sediment from measured data, using measured bankfull discharge and sediment values for normalization. Study results demonstrated that there was a statistically significant difference (departure) of the curve compiled from streams with poor stability ratings as compared to the dimensionless reference curve for stream types with good/fair stability. Stream channel stability analysis by stream type can thus be used to show departure from the reference sediment rating curve.

Other sediment studies have obtained similar results. Measured suspended sediment rating curves in the Redwood Creek drainage were plotted by channel stability rating categories, indicating that sediment concentration increases exponentially with discharge as a function of channel stability ratings (**Figure 2-51**) (Leven, 1977; USEPA, 1980; Rosgen, 2001b). Measured bedload sediment rating curves by stream type from Colorado rivers show the variation in sediment rating curves associated with various stream types (Rosgen, 1996, 2001b). Stream type change often is associated with instability and corresponding changes in sediment supply; therefore, ***sediment relations must encompass both stream type and stability assessments.***

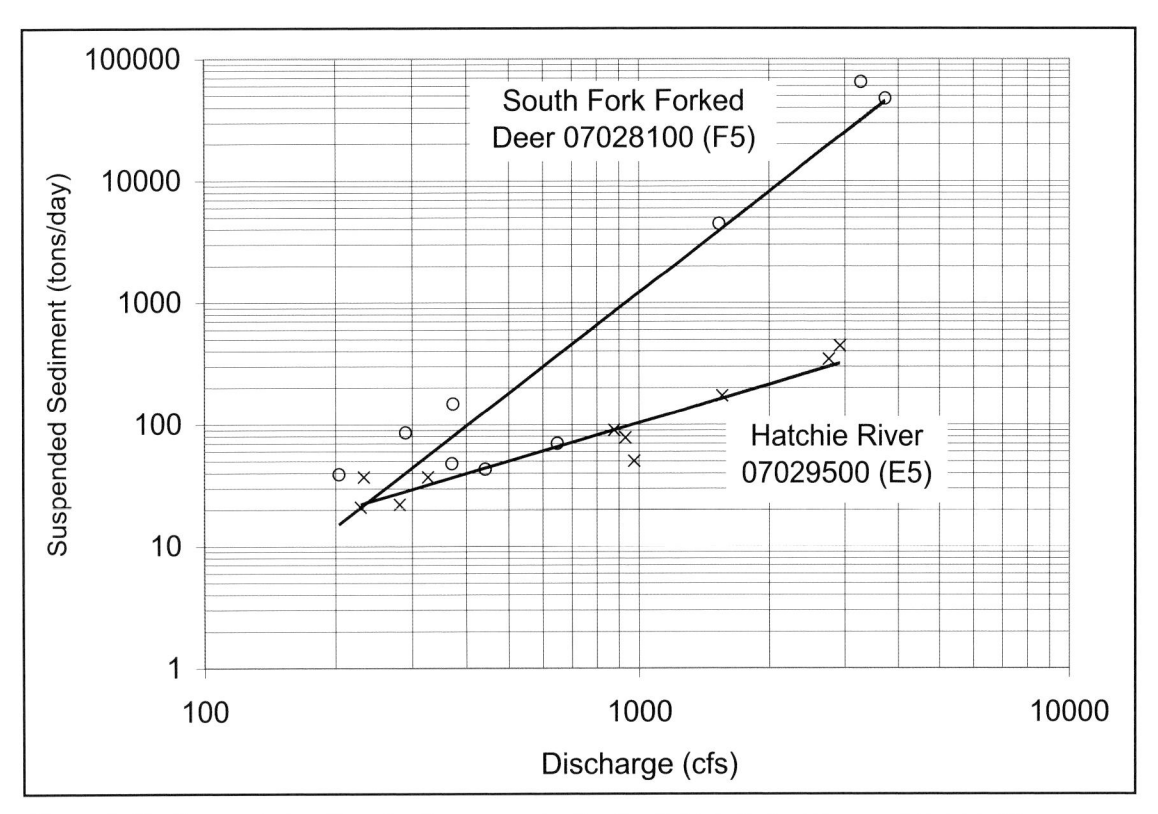

Chapter 2 *A Review of Hillslope, Hydrologic and Channel Process Relations*

Figure 2-49. Suspended sediment rating curves for the South Fork Forked Deer and Hatchie Rivers, Tennessee (data from USGS stations, Simon, 1989b).

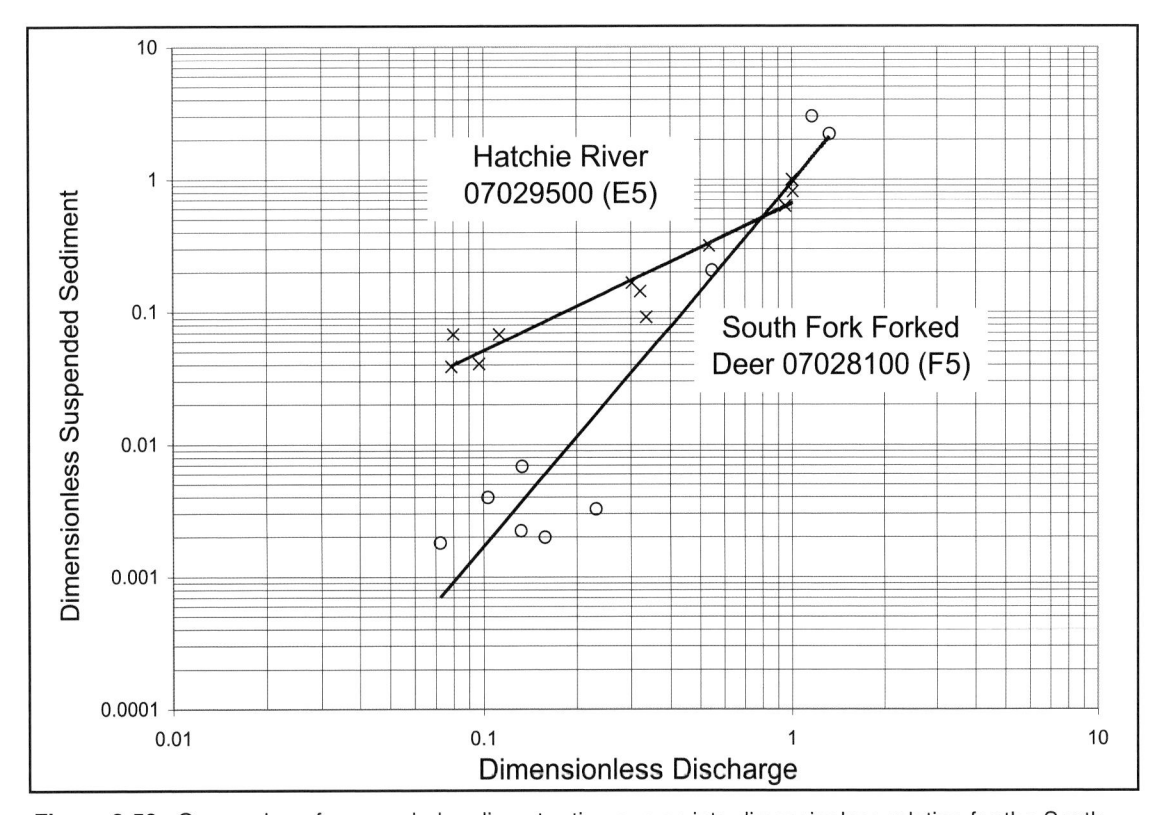

Figure 2-50. Conversion of suspended sediment rating curves into dimensionless relation for the South Fork Forked Deer and Hatchie Rivers, Tennessee (data from USGS stations, Simon, 1989b).

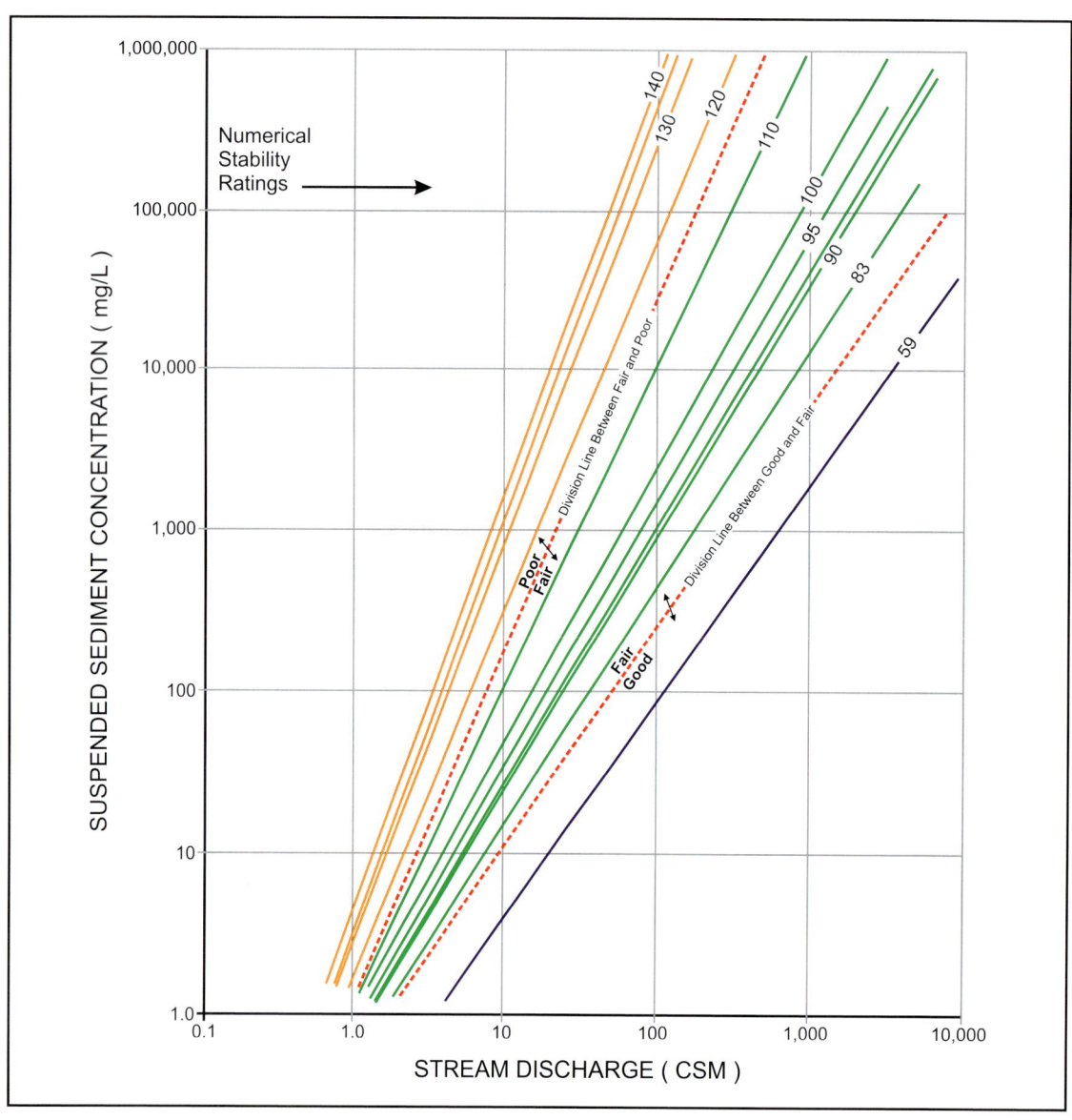

Figure 2-51. Suspended sediment rating curves by channel stability ratings of various reaches of Redwood, California (Leven, 1977; USEPA, 1980; Rosgen, 2001b).

Application of basic principles

Dimensionless bedload and suspended sediment rating curves

Establishing bedload and suspended sediment rating curves requires measuring flow and sediment over a wide range of temporal and spatial variability. It is expensive and time-consuming. It is also difficult, based on past experience, to expect a bedload or total-load equation to predict actual sediment rating curves. However, by normalizing actual measured bedload and suspended sediment relations with measured bankfull values of bedload, suspended sediment and streamflow, reference sediment rating curves can be established. The "reference" is associated with streams of similar type and stability ratings that reflect a certain sediment supply for a range of flows. When departure from reference occurs in stream channel stability, increased sediment supply often results in a shift in the sediment rating curves (Rosgen, 2001c). This shift may be detected by establishing a river-specific sediment rating curve, making it dimensionless and then testing against the reference curve, as in Troendle et al. (2001).

Flow-related increases in sediment are generated from a flow model and a sediment rating curve as described in WRENSS (USEPA, 1980). Because sediment rating curves are difficult to establish, sediment transport equations have been developed to provide rating curves. Unfortunately, without actual measured sediment data to calibrate these models, the predicted values often vary many orders of magnitude from actual data. The application of the reference dimensionless sediment rating curve provides a new method for establishing sediment rating curves. This is made possible by collecting bedload, suspended sediment and discharge data at the bankfull stage and converting the dimensionless relations to actual values as described previously.

Examples of the application of dimensionless sediment rating curves are shown below, where locally derived bedload and suspended sediment dimensionless relations were tested elsewhere on a variety of rivers using bankfull discharge and sediment values. The majority of the suspended sediment and bedload sediment data tested is from U.S. Geological Survey measurements in Idaho (Emmett, 1975, 2001). Model validation tested 20 rivers whose data were independent from those rivers used in the development of the relation. Rivers ranging from Alaska (Tanana) to Tennessee (South Fork Forked Deer) were evaluated using the dimensionless ratio relations from Colorado. The reference dimensionless bedload and suspended rating curve data was developed from stable snowmelt runoff streams in Southwestern Colorado (**Figures 2-52** and **2-53**).

Computations for the test involved obtaining the bankfull discharge and bankfull suspended and bedload sediment values for each river, then applying the appropriate reference equation from dimensionless bedload sediment (**Equation 2-15**) and dimensionless suspended sediment (**Equation 2-16**).

(Equation 2-15)

$$\text{Bedload (good/fair): } y = -0.0113 + 1.0139 x^{2.1929}$$

(Equation 2-16)

$$\text{Suspended sediment (good/fair): } y = 0.0636 + 0.9326 x^{2.4085}$$

where:
y = dimensionless sediment

x = dimensionless discharge

Multiplying the bankfull values for the test rivers by the calculated dimensionless ratios generated a sediment rating curve. Comparisons of predicted to measured values for suspended sediment and bedload rating curves for 20 rivers are shown in **Figures 2-54** through **2-57**.

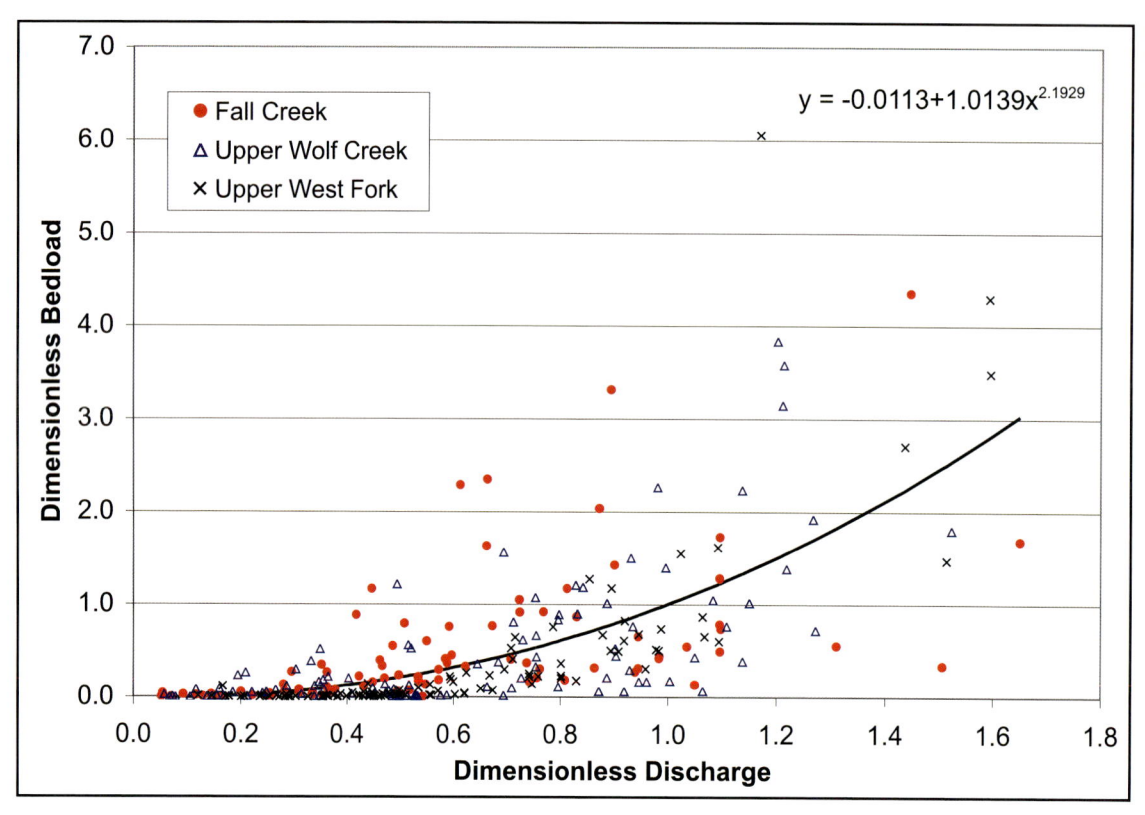

Figure 2-52. Dimensionless bedload sediment rating curve for "good/fair" streams/stability, Pagosa Springs, Colorado.

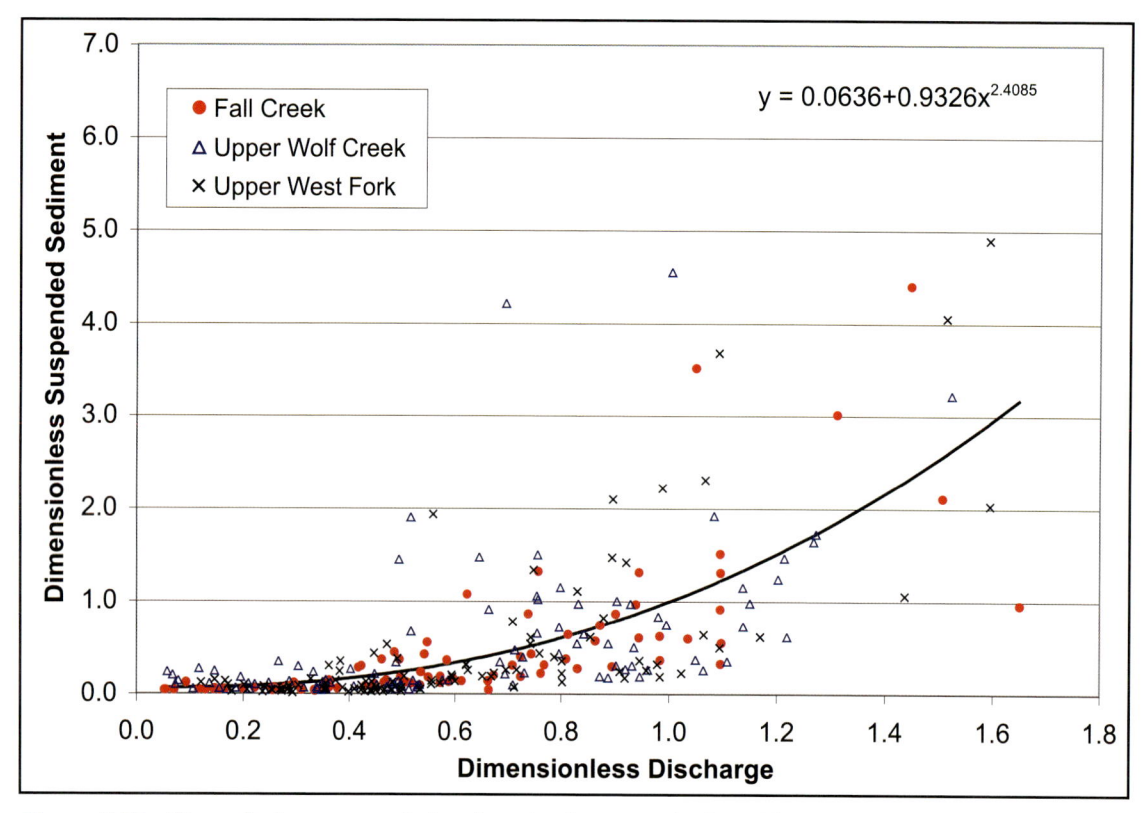

Figure 2-53. Dimensionless suspended sediment rating curve for "good/fair" streams/stability, Pagosa Springs, Colorado.

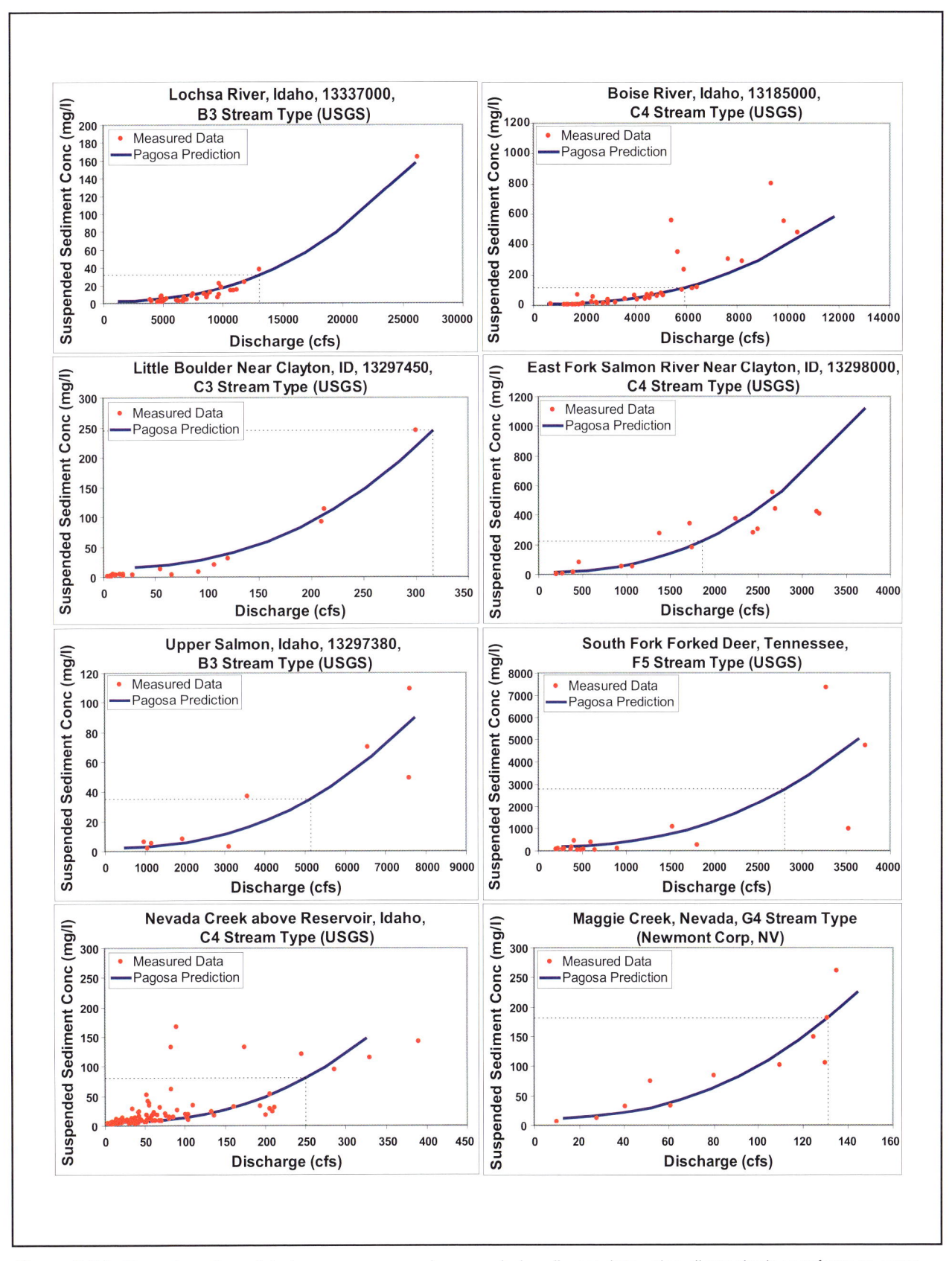

Chapter 2 — A Review of Hillslope, Hydrologic and Channel Process Relations

Figure 2-54. Examples of predicted versus measured suspended sediment data using dimensionless reference curve (dashed line represents the bankfull condition).

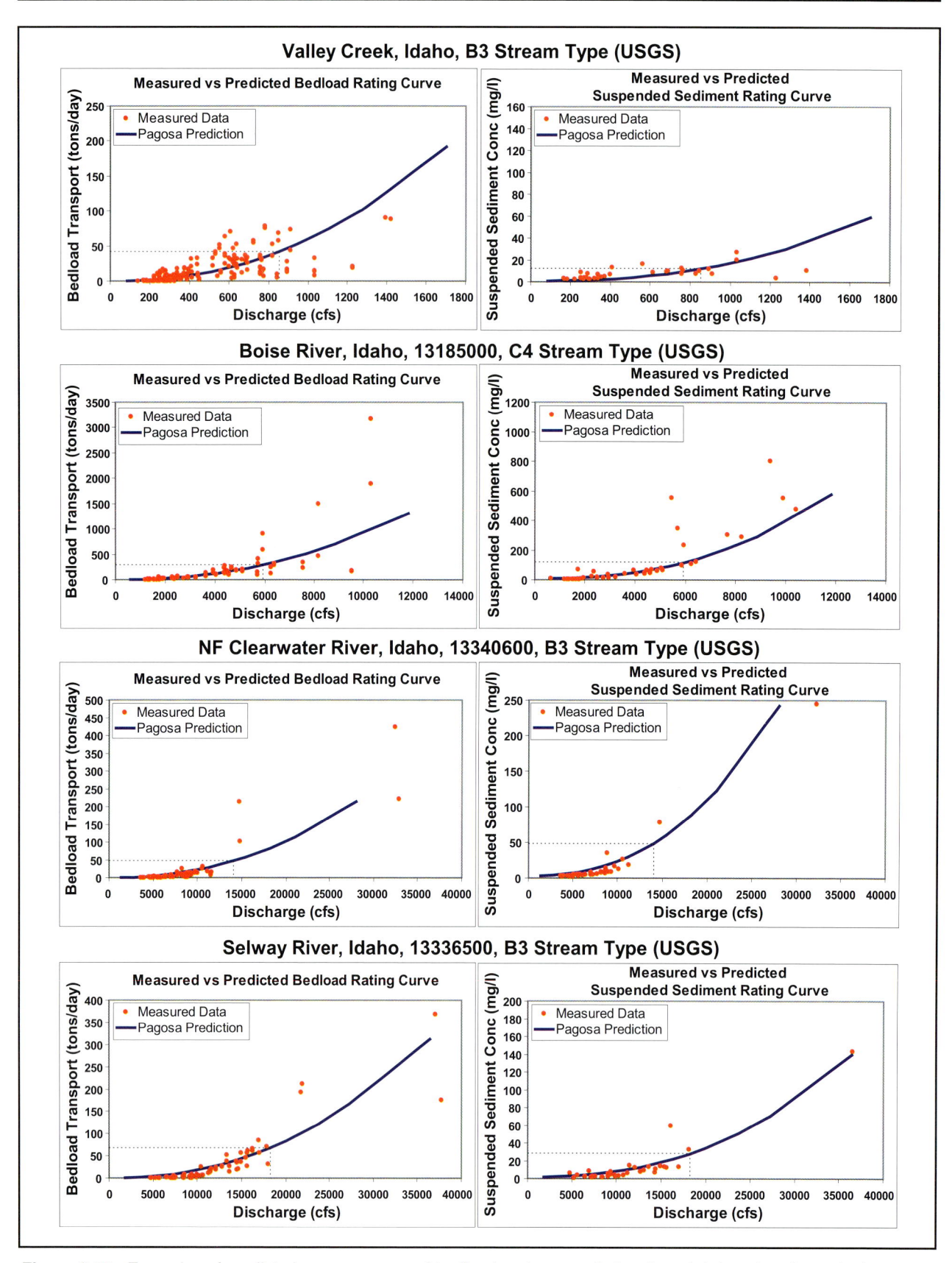

Figure 2-55. Examples of predicted versus measured bedload and suspended sediment data using dimensionless reference curve (dashed line represents the bankfull condition).

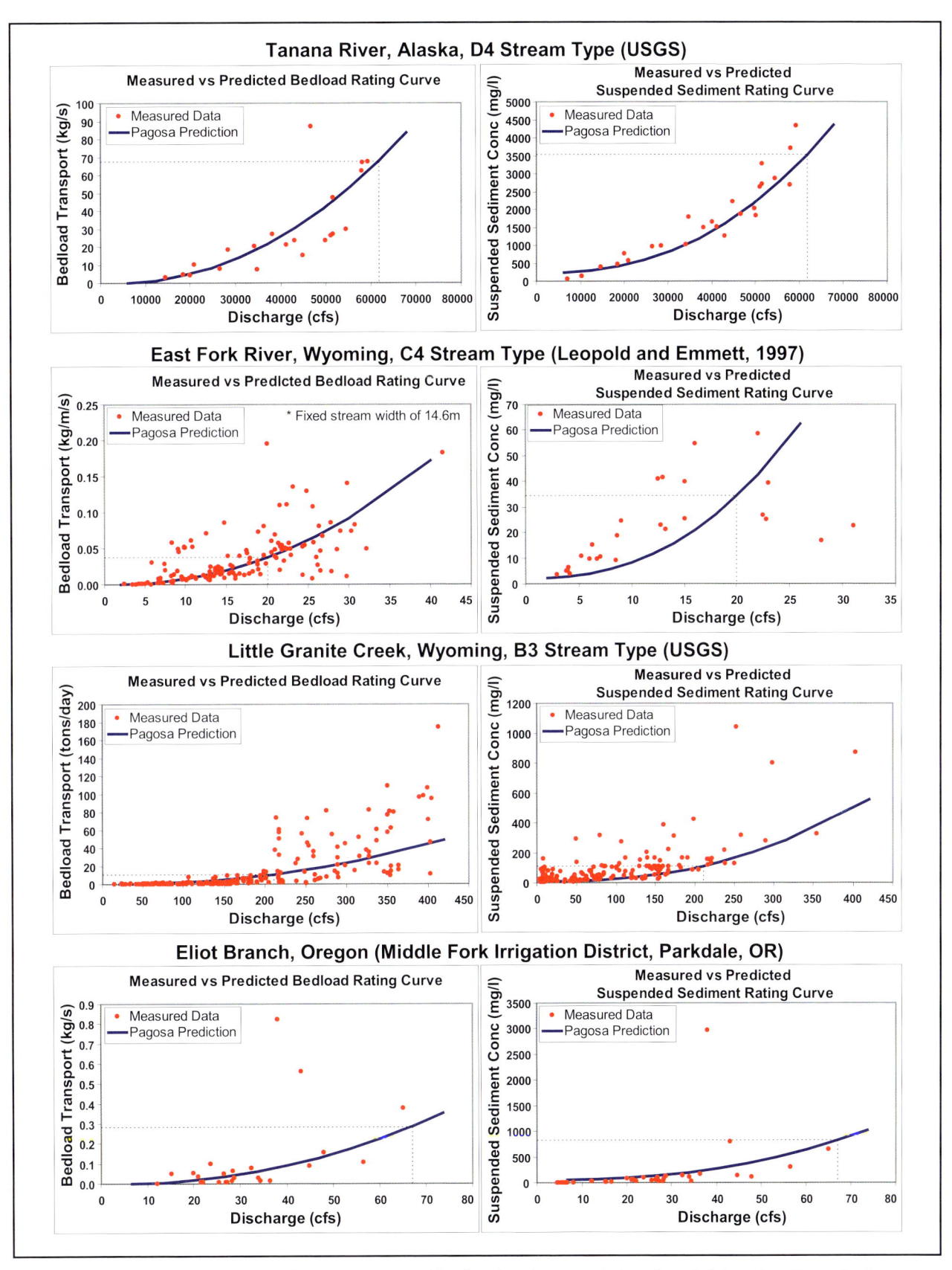

Figure 2-56. Examples of predicted versus measured bedload and suspended sediment data using dimensionless reference curve (dashed line represents the bankfull condition).

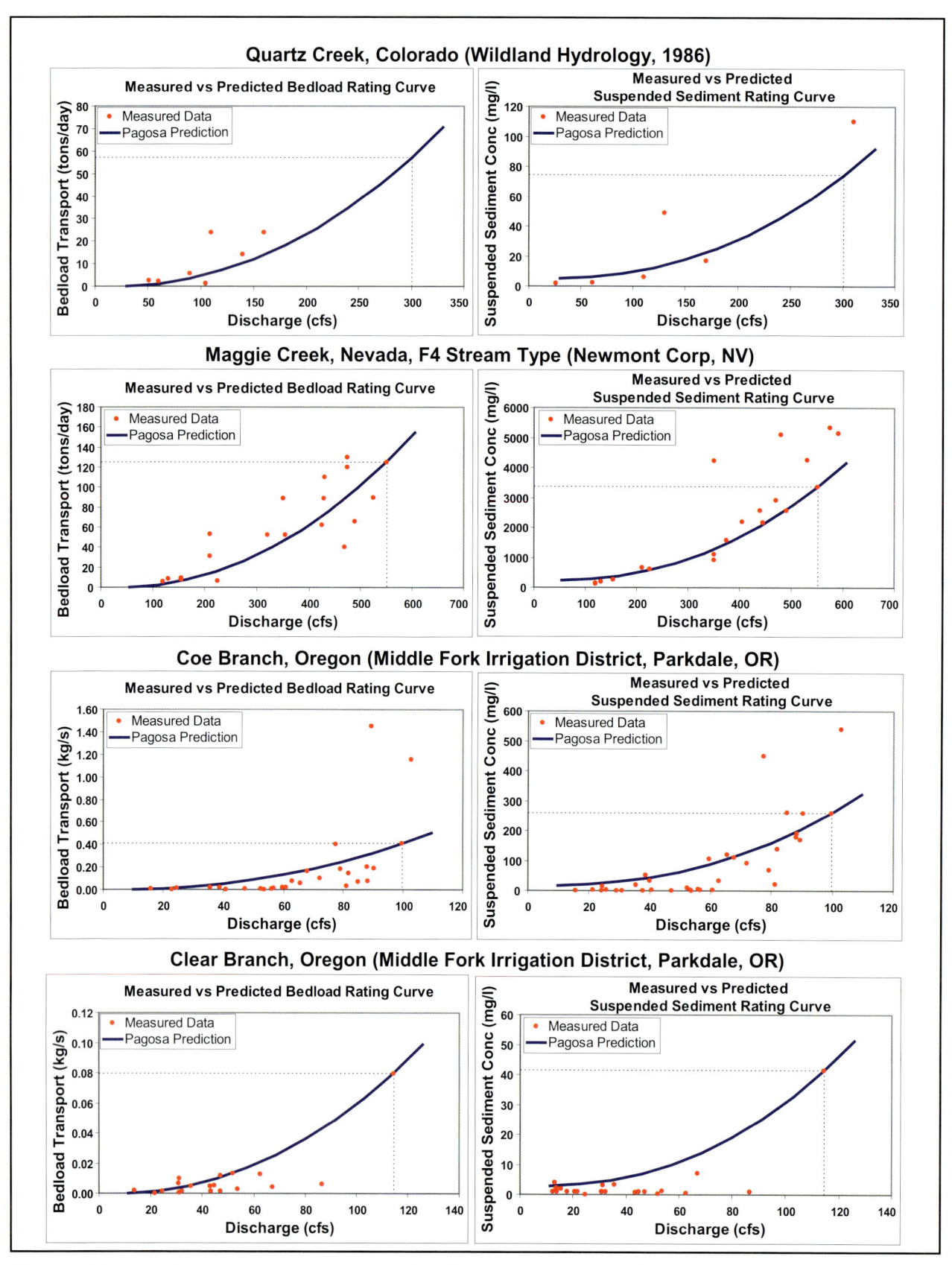

Figure 2-57. Example of predicted versus measured bedload and suspended sediment data using dimensionless reference curve (dashed line represents the bankfull condition).

The relations developed and tested appear to be exceptionally good, especially considering the difficulty of predicting sediment rating curves from bedload and suspended sediment transport formulas. The applications of this approach make it feasible to develop both suspended sediment and bedload sediment rating curves from bankfull measurements. This assumes that the bankfull measurements are properly obtained and accurately represent the bankfull condition. If great variability in sediment occurs during the same discharge at the bankfull stage, then several measurements should be taken to establish average values. Nonetheless, these results are very encouraging. Coupled with a water yield analysis converted to flow-duration curves, flow-related sediment increases can be reasonably obtained (FLOWSED model, Rosgen, 2006). The contributions of sediment integrated in the sediment rating curves can be proportionally determined by more process-specific predictions. These sediment prediction procedures are presented in the *Prediction Level Assessment* (*PLA*) in Chapter 5 and are demonstrated in Chapter 7.

Stability test/sediment rating curves

Applying the same test as in Troendle et al. (2001), dimensionless suspended and bedload rating curves were tested against the measured sediment data associated with three poor stability stream reaches. These reaches were lower Wolf Creek, lower West Fork San Juan River and Weminuche Creek. All were C stream types with poor stability ratings. The upper Wolf Creek, upper West Fork San Juan River and Fall Creek (C stream types with good to fair stability ratings) were also compared with the reference dimensionless suspended and bedload sediment rating curves. The results showed significant differences between the reference model curves and both the bedload and suspended sediment curves for the poor stability streams. The three good to fair streams were not statistically different from the good to fair reference relations, as was shown previously (Troendle et al., 2001).

The statistical tests were made for all the streams at the 95% confidence level. The results demonstrate that poor stability rivers have a statistically significant increase in measured sediment supply over good and fair stability reaches, as indicated by the sediment rating curves for each river type and stability. This information is important for prediction, as it associates changes in sediment rating curves with instability. The linkage between the two is intuitively obvious; however, there has been limited data from past studies to indicate statistical significance. For this reason, the *WARSSS* methodology integrates sediment supply based on channel stability and stream type into each assessment level.

Stream channel stability indicators can be consistently applied by various observers. It is advantageous to identify a potential problem early in the channel adjustment phase in order to apply mitigation and/or restoration rather than waiting until the instability shifts the sediment rating curve of the river. This particular test demonstrated that instability can significantly increase sediment supply and shift the suspended and/or bedload rating curves of the river. The bedload and suspended sediment (poor) relations were based on 544 data points for each of the bedload and suspended sediment rating curves. These data points represent observations of poor stability for three rivers in Southwestern Colorado near Pagosa Springs. The poor stability bedload and suspended sediment rating curves are shown in **Figure 2-58** and **Figure 2-59**, with corresponding equations of:

(Equation 2-17)

$$\text{Bedload (poor): } y = 0.07176 + 1.02176x^{2.3772}$$

(Equation 2-18)

$$\text{Suspended sediment (poor): } y = 0.0989 + 0.9213x^{3.659}$$

where:
y = dimensionless sediment

x = dimensionless discharge

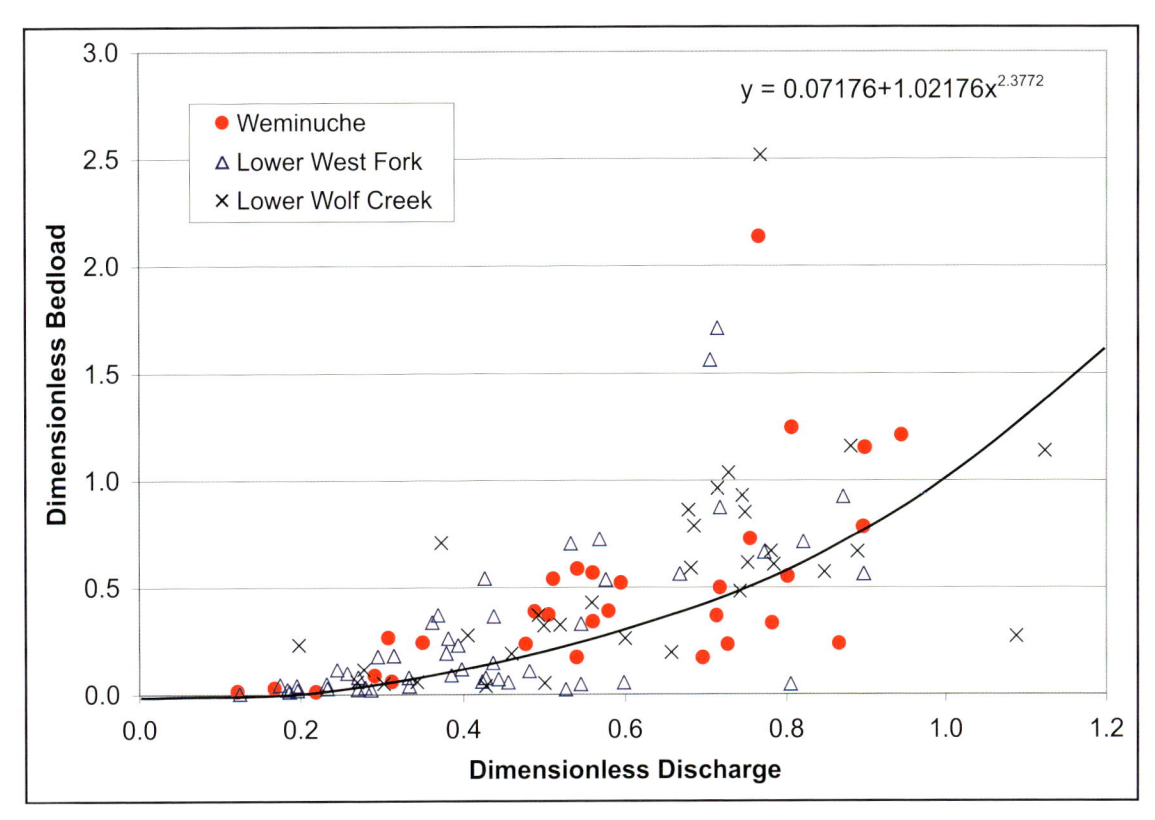

Chapter 2 — A Review of Hillslope, Hydrologic and Channel Process Relations

Figure 2-58. Dimensionless bedload rating curve for three unstable "poor" streams, Pagosa Springs, Colorado.

$$y = 0.07176 + 1.02176 x^{2.3772}$$

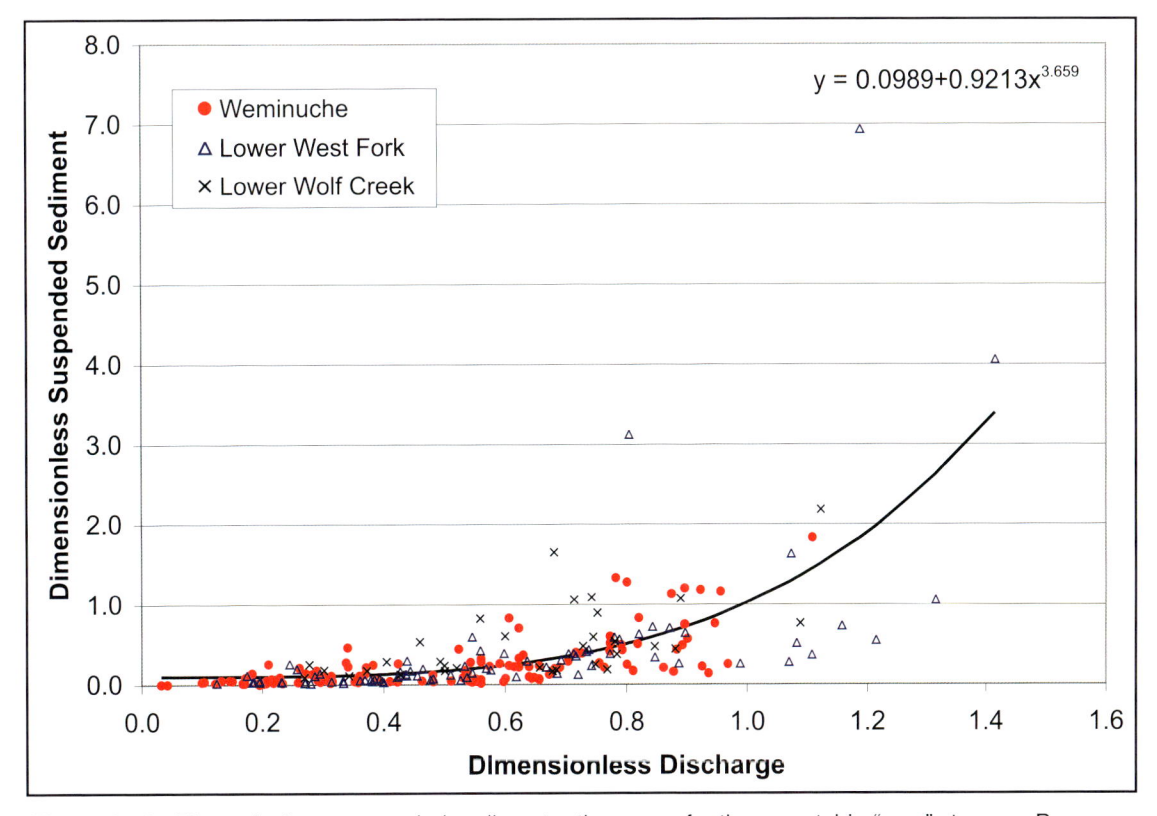

Figure 2-59. Dimensionless suspended sediment rating curve for three unstable "poor" streams, Pagosa Springs, Colorado.

$$y = 0.0989 + 0.9213 x^{3.659}$$

Entrainment/competence calculation approach

The importance of critical shear stress calculations for sediment transport models was discussed earlier. The conditions of incipient motion for various flow depths are not only important for sediment transport formulas, but also for sediment competence calculations. Modifications to previous approaches were tested to provide observers with improved field methods for computation. Bar core samples were obtained on point bars on the lower third plan-view position through a bend and 0.5 elevation of the bankfull stage (Rosgen, 1996, Chapter 7, p. 8; 2001b). This location substituted for the sub-surface sample described by Andrews and Erman (1986). The core sample is sieved in the field, with the largest-size particle (D_{max}) recorded along with the D_{50} of the sample. The size distribution of the bar sample is similar to the gradation of particles associated with bedload transport at the bankfull stage (Rosgen, 1996, pp. 7–9).

The relations using bar samples agree quite well with the original relation presented by Andrews and Erman (1986) (**Figure 2-60**).

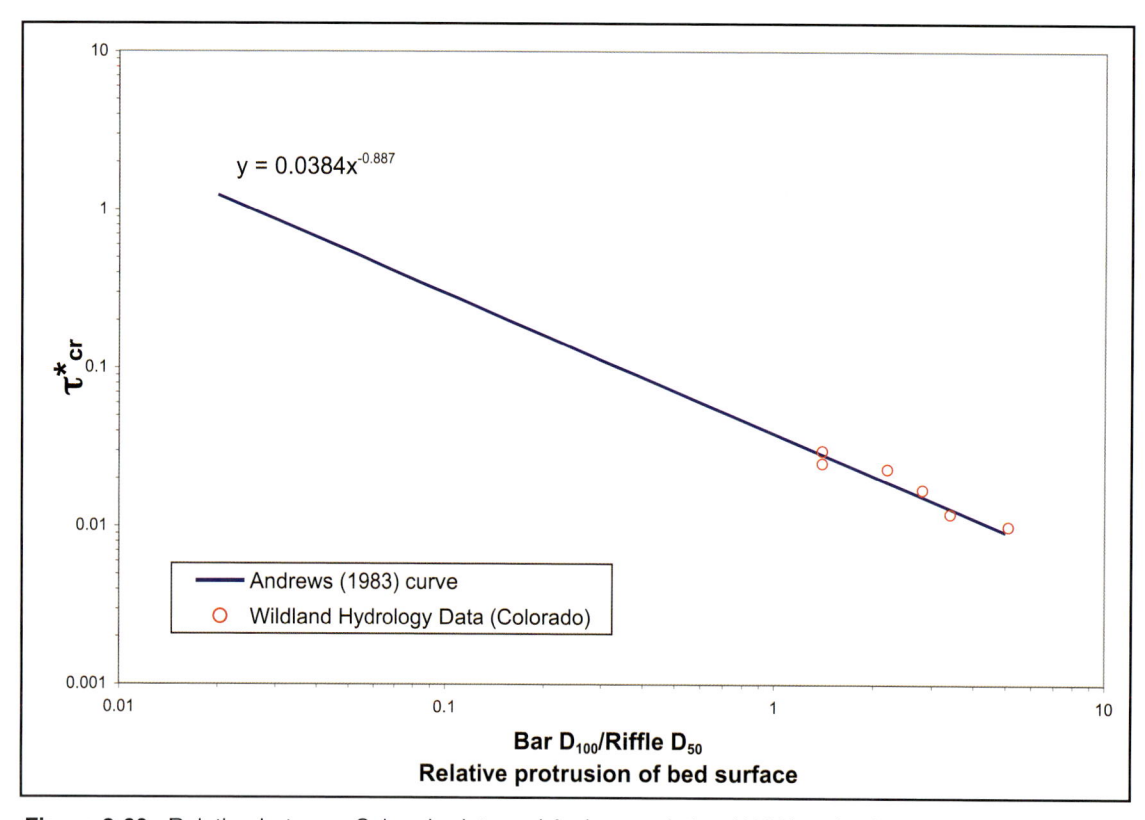

Figure 2-60. Relation between Colorado data and Andrews relation (1983), using bar sample and bed-material data.

Once the bankfull critical dimensionless shear stress is determined, the competence calculation assesses the depth and/or slope necessary to move the largest particle made available on the point bar, D_{max}, using **Equation 2-9**.

West Fork San Juan River

Entrainment and sediment competence calculations using **Equation 2-7** were made on a stable reach at a USGS streamgage location at the West Fork San Juan River, Colorado, in order to compare the predictions with observed data. An example of the entrainment and competence calculations is shown in **Table 2-6**.

Table 2-6. Entrainment and sediment competence computation for lower West Fork of the San Juan River, Colorado.

Stream: West Fork San Juan
Drainage Area: 85.4 mi^2
Bankfull Discharge: 1150 cfs

Step 1: Calculate bankfull critical dimensionless shear stress

$$\tau_c^* = 0.0834 \left(D_{50} / \hat{D}_{50} \right)^{-0.872} \quad \text{(eq. 2-7)}$$

where:

τ_c^*: critical dimensionless shear stress

D_{50}: bed material D_{50} (from riffle pebble count) = **70mm**

$\hat{D}_{50}$: sub-surface $\hat{D}_{50}$ or bar sample $\hat{D}_{50}$ = **20mm**

$$\tau_c^* = 0.0834 \left(\frac{70}{20} \right)^{-0.872}$$

$$\tau_c^* = 0.028$$

Step 2: Calculate the mean bankfull depth required to move the largest particle from bar sample

$$\tau_c^* = \frac{dS}{(\gamma_s - \gamma/\gamma) D_{max}} \quad \text{(eq. 2-4)}$$

transformed to:

$$d = (\gamma_s - 1) D_{max} \tau_c^* / S \quad \text{(eq. 2-9)}$$

where:
d: mean bankfull depth at riffle (ft)
τ_c^*: critical dimensionless shear stress = **0.028**
S: bankfull water surface slope = **0.007 ft/ft**
γ_s: specific weight of sediment = **2.65**
$\gamma_s - 1$: submerged specific weight of sediment = **1.65**
D_{max}: Largest particle in Bar Sample 130mm = **0.43 ft**

$$d = \frac{(0.028)(1.65)(0.43)}{0.007}$$

$$d = 2.8 \text{ ft}$$

Comparison of the particle protrusion ratio of the $D_{50}/\hat{D}_{50}$, using **Equation 2-7** with measured bedload and scour chain data, indicated that the equation predicts well within the ratio values of 3.0–7.0. The prediction of the largest size on the bar (D_{max}) of 130 mm was very close to the largest particle of 140 mm measured by the 6-inch Helley-Smith bedload sampler at the bankfull stage. The predicted depth of 2.8 ft necessary to move a 130 mm particle, using the relations shown in **Table 2-6** and **Equation 2-9**, closely approximates the measured bankfull depth of 2.6 ft moving a 140 mm particle. The predicted depth overestimated the measured depth by 0.2 ft; however, this is well within an acceptable range of prediction.

Using **Equation 2-8**, the largest particle on the bar (D_{max}) was 130 mm (0.43 ft) and the D_{50} (median diameter) of the active bed was 70 mm. The ratio of 130 mm/70 mm equals 1.8. This is also within the acceptable range of the relation in **Equation 2-8** of 1.3–3.0. Thus, the results of the depth prediction to move a 130 mm particle on a slope of 0.007 are:

$$\tau^*_c = 0.0384 \left(D_{max} / D_{50} \right)^{-0.887} \quad \textbf{(Equation 2-8)}$$

$$\tau^*_c = 0.023$$

$$d = \left(\gamma_s - 1 \right) D_{max} \tau^*_c / S \quad \textbf{(Equation 2-9)}$$

$$d = (1.65)(0.43)(0.023) / (0.007)$$

d = 2.3 ft

The measured bankfull depth at the gage site was 2.6 ft; thus, the equation underpredicted by 0.3 ft. This is also well within an acceptable range of estimation for stability examinations. Both **Equations 2-7** and **2-8** bracketed the measured value of bankfull depth. In field practice, **Equation 2-8** would be used first. If the ratio is within the 1.3–3.0 range, it would not be necessary to field-sieve the bar sample as required for the use of **Equation 2-7**. If neither equations are within the range, then **Equation 2-2** estimating bankfull dimensional shear stress is used with an empirical particle entrainment relation, as shown in **Figure 5-54** in Chapter 5 (Rosgen and Silvey, 2005).

Bedload transport

In addition to sediment competence, or size of sediment moved, sediment load and transport capacity must be considered. Bedload formulas (e.g., Bagnold, 1980) predict sediment load, generally by size fraction for the energy available for each flow stage on an annual basis. The Bagnold equation was tested against measured data, as it uses unit stream power (ω_a) (**Equation 2-11**), or shear stress times velocity. Because the discharge between upper and lower Wolf Creek is similar, the unit stream power would differ at the two reaches due to reductions of depth, slope and velocity in the braided reach. A plot showing the relation for discharge and unit stream power for the two river types is presented in **Figure 2-61**.

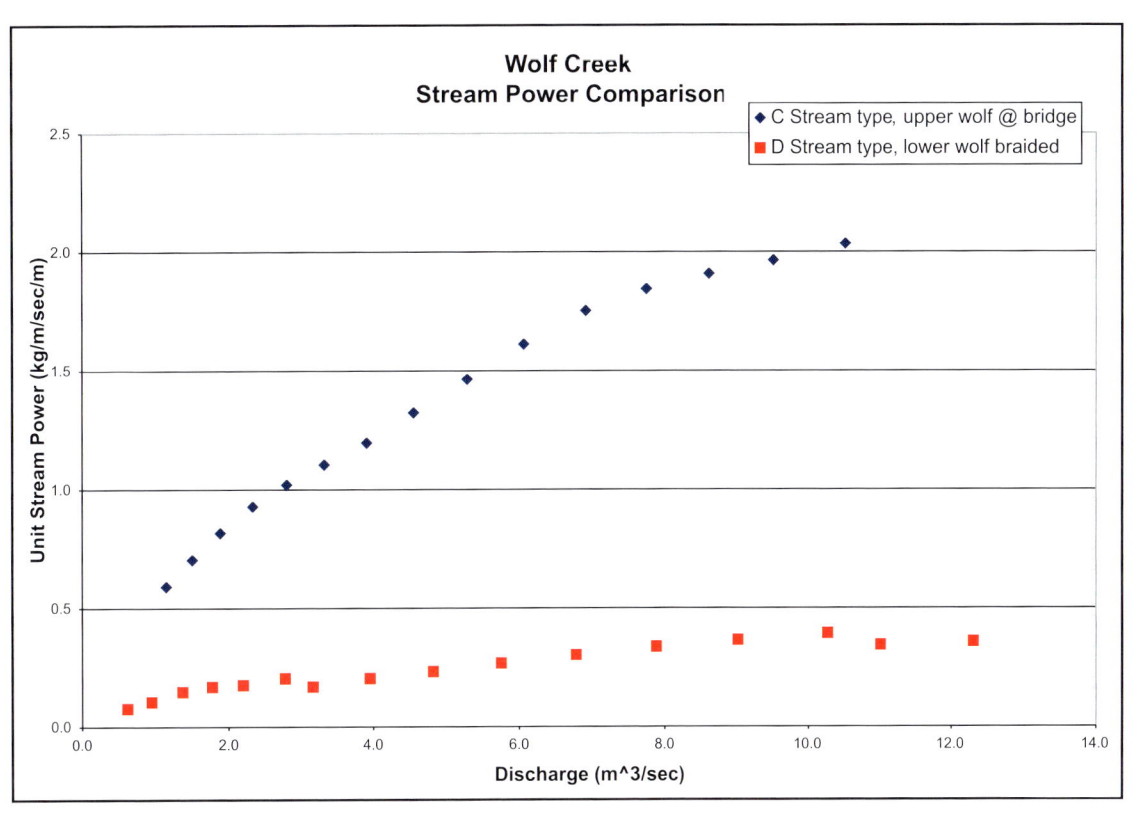

Figure 2-61. Relationship of unit stream power versus discharge for upper Wolf Creek (C4 stream type) and lower Wolf Creek (D4 stream type).

For the same discharge between the two rivers, the unit stream power is an order of magnitude lower for the braided (D4) reach at the bankfull stage. The two variables in the Bagnold relation that are generally not changed are the Θ_o value of 0.04 and the use of D_{50} of bed material for transport size. In this example, however, the D_{max} of the bar sample was used in place of D_{50} for the calculation of total bedload transport, and the dimensionless shear stress value (τ^*_c), computed as in **Table 2-6**, was substituted for the Shields threshold criterion (Θ_o) value in **Equation 2-13**. The application of the Bagnold equation to upper Wolf Creek is shown in **Figure 2-62**, where predicted values and actual values are in close agreement.

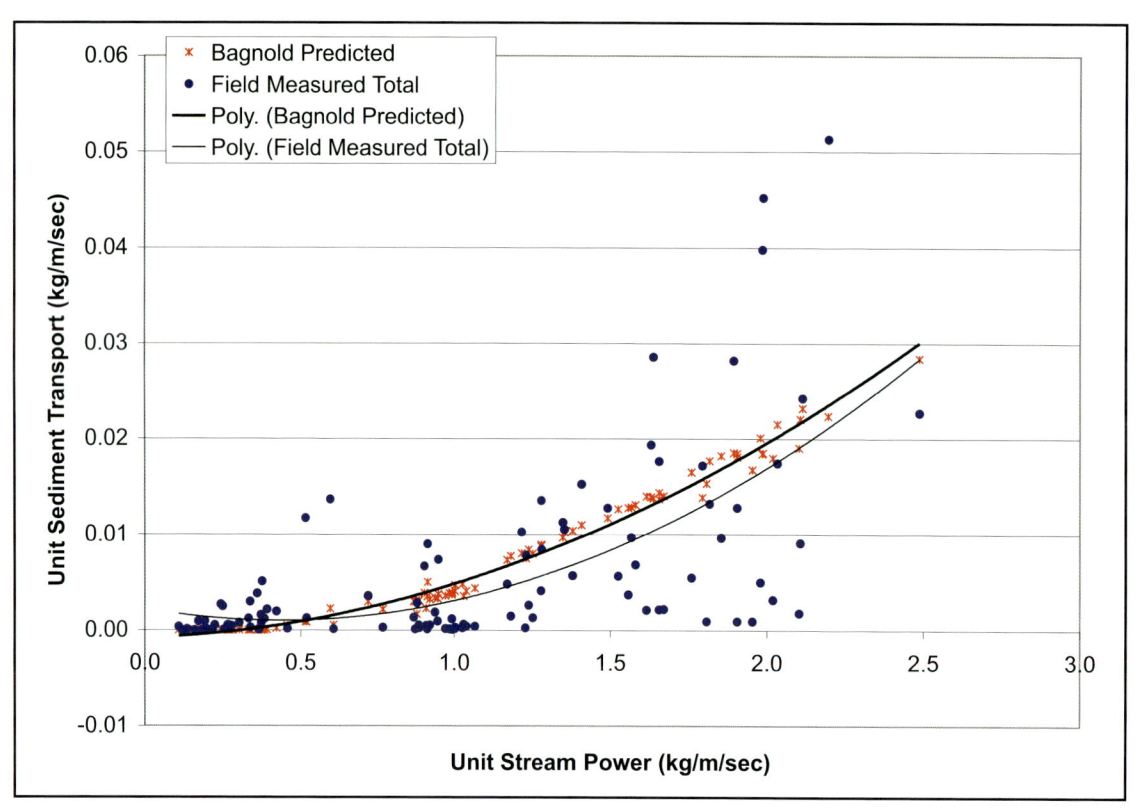

Figure 2-62. Comparison of predicted bedload using the Bagnold formula versus measured values for upper Wolf Creek, Colorado.

A bedload and suspended sand material load (total bed-material load) prediction model using dimensionless sediment rating curves and flow-duration curves (FLOWSED) is described in Rosgen (2006). The sediment rating curves from FLOWSED are converted from discharge to stream power using the POWERSED model (Rosgen, 2006). This model predicts the total sediment transport and also the bed stability based on transport capacity. Both the FLOWSED and POWERSED models are used in RIVERMorph™ software (version 4.0) and are further described in Chapters 5 and 7.

The total bedload transport for lower Wolf Creek (**Figure 2-63**) shows an approximately 50% reduction in unit transport rate compared to the upstream reach. The Bagnold relation, using a revised dimensionless shear stress and D_{max} of the bar sample as previously discussed, over-predicted the total bedload transport of the lower reach compared to measured values, but it accurately predicted a reduction in the coarse sediment transport rate.

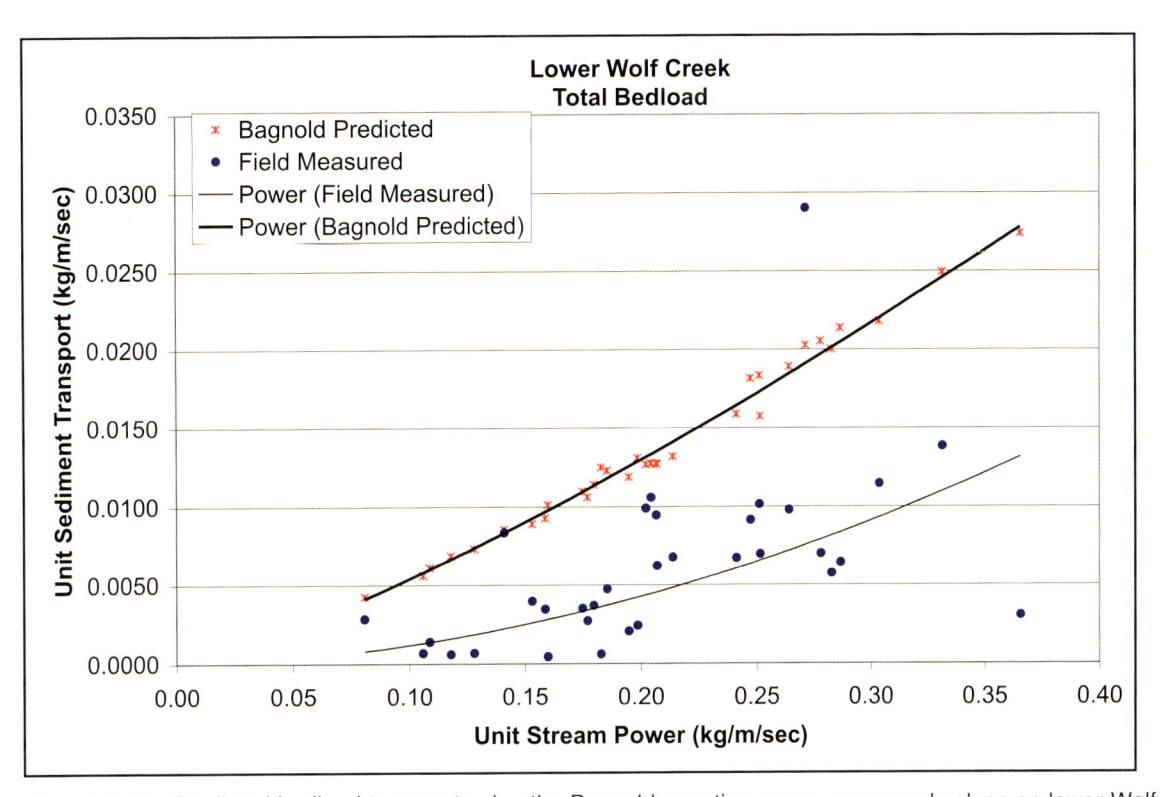

Figure 2-63. Predicted bedload transport using the Bagnold equation versus measured values on lower Wolf Creek, Colorado (D4 stream type).

The measured values indicate the amount of deposition material compared to upper Wolf Creek, a C4 stream type (**Figure 2-62**). Measured bed-material size at this cross-section indicates a shift to finer bed material and excess sediment deposition. The stream type is associated with a braided form and is contributing to excess sediment storage in the reach. An increase in bank erosion, however, has increased sediment supply due to a channel width enlargement from 37 ft to approximately 203 ft. Additional and more comprehensive assessments are presented for Wolf Creek in Chapter 7.

Weminuche Creek: A case study of the channel succession sequences of an unstable river

Willow removal from herbicide spraying and heavy grazing pressure caused Weminuche Creek in Southwestern Colorado to change over a 12-year period from a stable, meandering, single-thread C4 stream type to a braided (D4) stream type in a terraced, alluvial valley (valley type VIII) (**Figure 2-64**).

The spraying changed the vegetative composition from a willow/cottonwood-dominated riparian community to a grass/forb community. Both streambank erosion and width/depth ratio increased, and this change resulted in aggradation, as indicated in the upstream end of the photograph in **Figure 2-64**. The aggradation process led to a channel avulsion where 2,100 ft of channel was abandoned and created a major headcut. The headcut resulted in an increase in slope and a decrease in width/depth ratio as the stream type shifted from a D4 to a G4 (Scenario #3, **Figure 2-38**). Thus, the aggradation process was reversed and the stream degraded due to the avulsion. The downstream end of the photograph indicates the avulsion potential as the streamflow abandoned the main channel at streamflows less than bankfull and the flow is parallel with the fall line of the valley.

Figure 2-64. Weminuche Creek, showing conversion of a C4 stream type to a D4 stream type. Note the aggradation process (Rosgen, 2003).

The initial adjustments in Weminuche Creek, as a result of the streambank destabilization by riparian vegetation change and subsequent accelerated erosion, converted a C4 (meandering) channel to a D4 (braided) channel (**Figures 2-64** and **2-65**). The typical consequences of this stream type change include:

- Increases in sediment supply from streambank erosion;
- Decreases in shear stress;
- Decreases in unit stream power;
- Reduced sediment transport capacity;
- Decreases in competence (ability to move largest-sized sediment made available);
- Filling in of pools and bi-modal distribution of bed material with a resultant increase composition of sand;
- Decreases in fish habitat;
- Major land loss; and
- Loss of visual value.

Another Weminuche Creek example is shown in **Figure 2-65**. Change in the rooting depth and density made the banks more susceptible to fluvial entrainment, triggering upper bank collapse. The lack of sediment transport capacity resulted in excess deposition, redirecting high shear stress to the near-bank region and causing accelerated streambank erosion and down-valley meander migration. The deposition from natural, geologic upstream sources, as well as increased supply from streambank erosion, add to the long-term instability of Weminuche Creek.

Figure 2-65. Weminuche Creek, Colorado, depicting widening, conversion from a C4 stream type to a D4 stream type, down-valley meander migration and fishery habitat loss.

WATERSHED ASSESSMENT OF RIVER STABILITY AND SEDIMENT SUPPLY

The photographs in **Figure 2-66** document the conversion of a meandering, single-thread, C4 stream type to a braided, D4 stream type over a 2-year period. The streambank erosion rates are evident in **Figure 2-66**, as the channel widened from 45 ft to over 120 ft.

C4 stream type

D4 stream type

Figure 2-66. Replicate photos between 1999 and 2001, showing Weminuche channel change from a C4 to D4 stream type due to willow spraying.

Chapter 2 *A Review of Hillslope, Hydrologic and Channel Process Relations*

Heavy grazing pressure and willow spraying on Weminuche Creek created instability, with corresponding channel evolutionary Scenario #3 (**Figure 2-38**). The stream shifted in a C4→ D4→ G4→ F4→ C4 series, as depicted in the aerial oblique photos (**Figures 2-64** and **2-65**). The graphical succession sequence for Weminuche Creek is shown in **Figures 2-67, 2-68** and **2-69**. This sequence depicts the morphological change from C4→ D4→ G4→ F4→ C4 that occurred over a 12-year period.

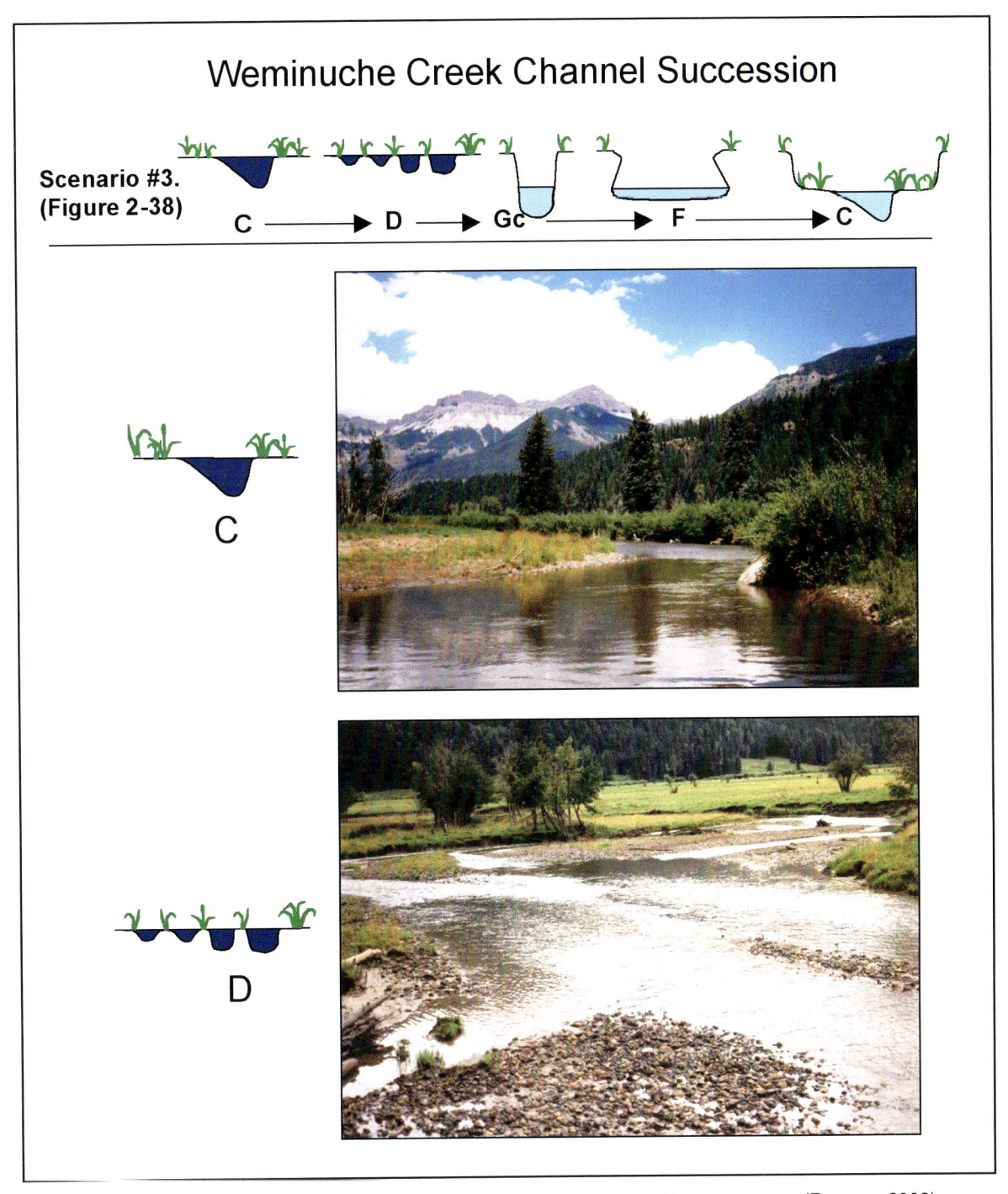

Figure 2-67. Conversion of a C4 stream type on Weminuche Creek to a D4 stream type (Rosgen, 2003).

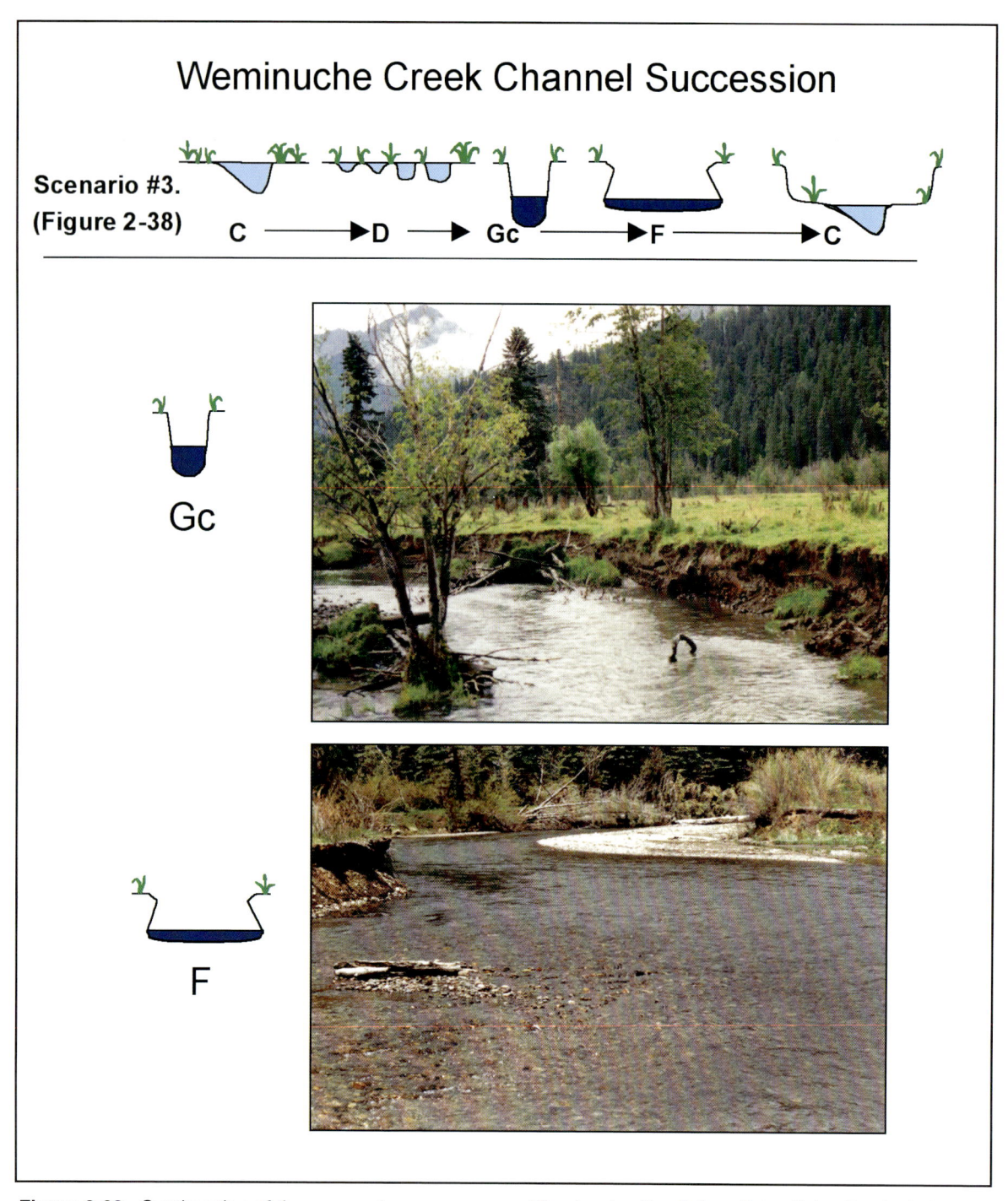

Figure 2-68. Continuation of the succession sequence on Weminuche Creek from D4→ G4→ F4, depicting the G4 and F4 stream types (Rosgen, 2003).

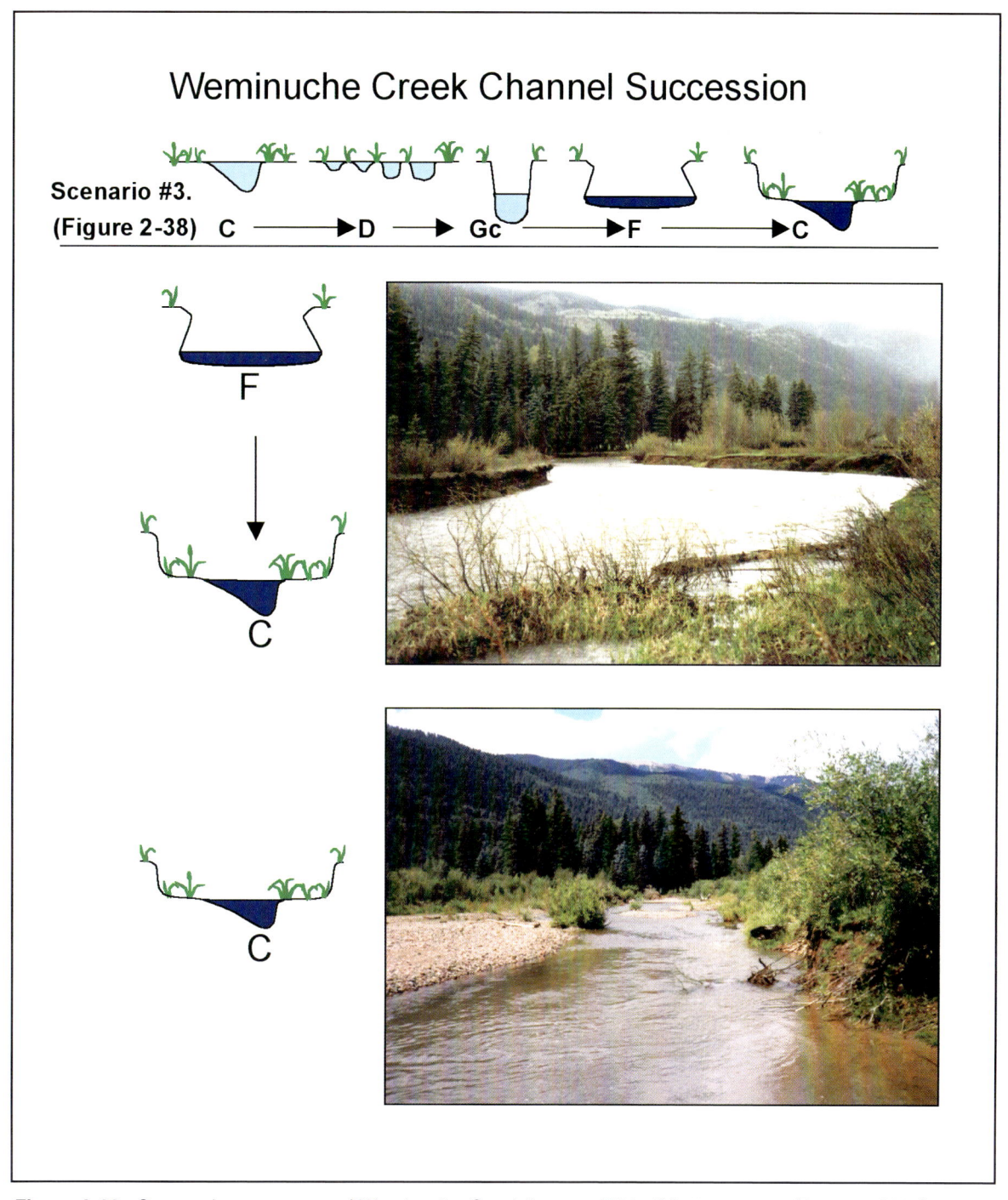

Chapter 2   A Review of Hillslope, Hydrologic and Channel Process Relations

Figure 2-69. Succession sequence of Weminuche Creek from an F4 to C4 stream type (Rosgen, 2003).

Stream research studies conducted in Southwestern Colorado since 1985 were summarized with results that documented the processes responsible for sediment and stability change. Some results documented stable reference reaches, while other reaches illustrated the consequences of instability and associated sediment problems. The example of a G4 headcutting into a D4 stream type on Weminuche Creek is depicted in **Figure 2-70**. As the headcut advanced, it deepened and widened, resulting in a conversion from a G4 to an F4 stream type (**Figure 2-71**). The stream continued to widen and then constructed a new floodplain within the F4 stream type at an elevation two feet lower than the previous channel (**Figure 2-72**). The difference in elevation from the streambed to the new floodplain was the same as the previous channel, representing the tendency for the stream to maintain a "bankfull" channel, even though the stream reach was still in disequilibrium.

Figure 2-70. Weminuche Creek channel avulsion, looking upstream at an advancing headcut as the stream type is at the incipient point of changing headward from a D4 to G4 stream type.

Chapter 2   A Review of Hillslope, Hydrologic and Channel Process Relations

G4 stream type

F4 stream type

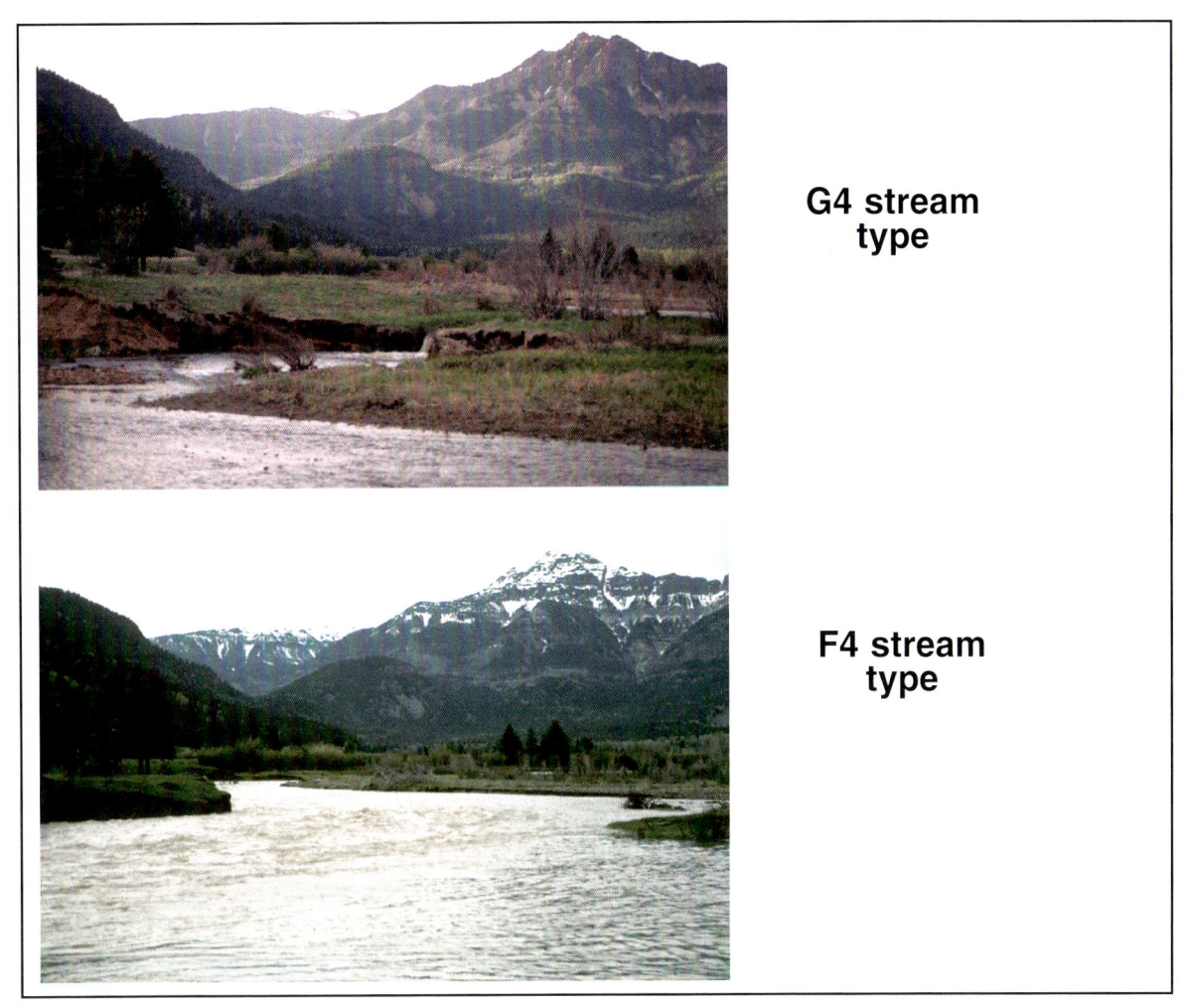

Figure 2-71. Weminuche Creek channel avulsion, looking upstream. The F4 stream type is occupied with a bankfull discharge, 1999—2001.

Previous floodplain

New floodplain

Figure 2-72. Cross-section 32+00 on Weminuche Creek, showing a transition from C4→ D4→ G4→ F4→ C4 at a new base level, 1999—2001.

The change in size distribution resulting from a conversion of a C4 to D4 stream type is reflected in the bed-material size distribution data (**Figure 2-73**).

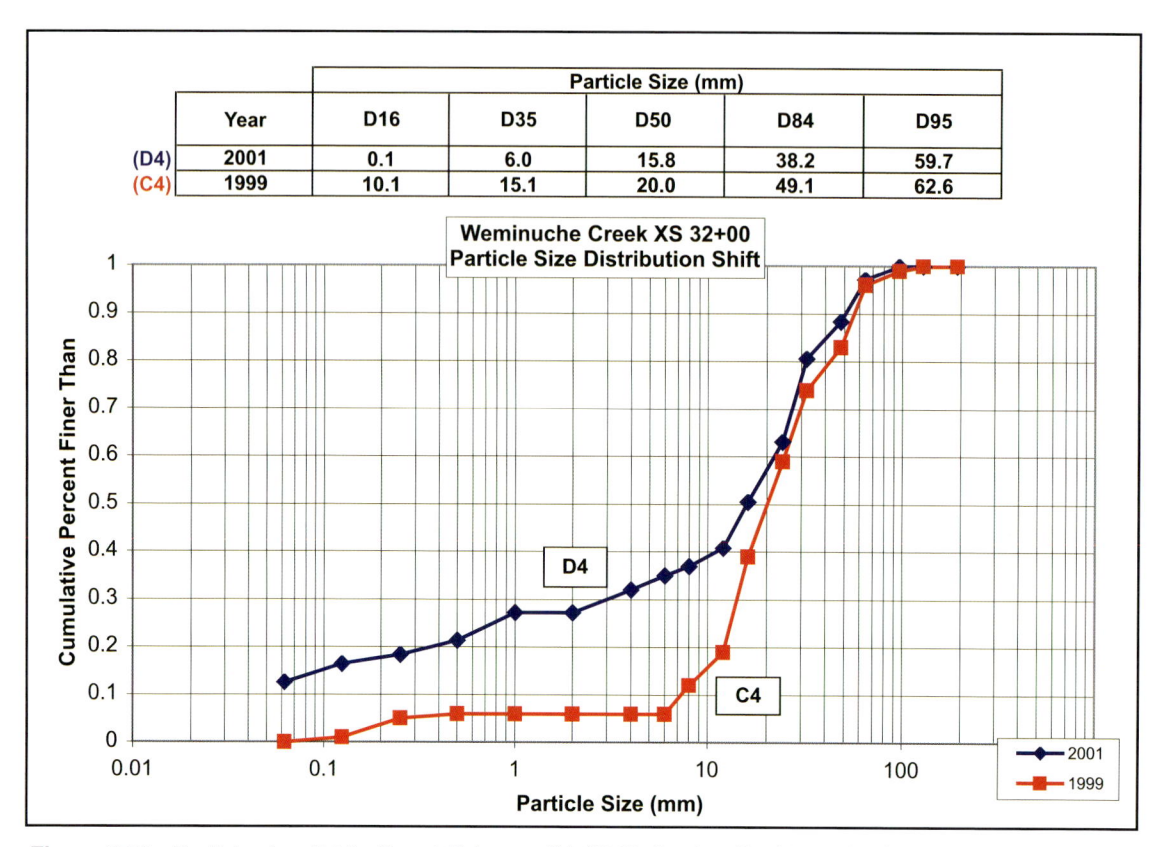

		Particle Size (mm)				
	Year	D16	D35	D50	D84	D95
(D4)	2001	0.1	6.0	15.8	38.2	59.7
(C4)	1999	10.1	15.1	20.0	49.1	62.6

Figure 2-73. Particle size distribution shift from a C4 (1999 data) to D4 (2001 data) stream type at cross-section 32+00 (Rosgen, 2003).

Hydraulic, sedimentological and morphological consequences of stream type change

As a result of willow spraying and poor grazing practices, Weminuche Creek went through a series of successional stages, changing stream types from C4→ D4→ G4→ F4 and eventually back to C4. The consequence summary based on measured data is depicted in **Tables 2-7** and **2-8**, which reflect the magnitude and direction of corresponding change. Besides the obvious sedimentological, morphological and hydraulic changes, serious losses of land, fish habitat and aesthetic value occurred.

Table 2-7. Stream succession and corresponding stream type sequence showing comparisons of channel morphology, hydraulics and bank erosion from reference condition above avulsion at cross-section 32+00: 1999–2001 (Rosgen, 2003).

Stream Type	C4	C4	G4	F4
Stream Condition	Good	Poor	Poor	Poor
Bankfull Discharge (cfs)	400	400	400	400
Bankfull Width (ft)	39.7	54.9	18.0	92.9
Mean Bankfull Depth (ft)	1.78	1.8	3.6	1.15
Width/Depth Ratio	22.3	30.0	5.0	80.8
Bankfull Cross-Sectional Area (ft^2)	70.6	100.2	66.0	106.5
Max Bankfull Depth (ft)	2.9	2.7	6.0	2.4
Width Floodprone Area (ft)	500	500	27	103
Entrenchment Ratio	12.6	9.1	1.5	1.1
D_{15} (mm)	10.6	9.4	20.0	0.1
D_{35} (mm)	19.7	15.0	40.0	6.0
D_{50} (mm)	26.9	19.8	52.0	15.8
D_{84} (mm)	52.5	48.2	70.0	38.2
D_{95} (mm)	68.8	61.3	95.0	59.7
Slope (ft/ft)	0.0045	0.0050	0.0150	0.0051
Bank Erosion (ft/year)	0.015	1.9	10	24.8
Mean Bankfull Velocity (ft/s)	5.7	4.0	6.1	3.8
Shear Stress (lb/ft^2)	0.50	0.57	3.37	0.37
Unit Stream Power (lb/ft/s)	2.8	2.3	20.8	1.4
D_{50} bar (mm)	20	12	52	7
Bedload (kg/s) at Bankfull Stage	0.65	1.3	4.2	3
(tons/day)	61.9	124	400	286
Suspended Sediment (mg/l)	325	500	750	720
(tons/day)	351	540	821	778

Table 2-8. Stream succession sequence above avulsion at cross-section 45+51: 1999–2001 (Rosgen, 2003).

Stream Type	C4	C4	D4
Stream Condition	Good	Poor	Poor
Bankfull Discharge (cfs)	400	400	400
Bankfull Width (ft)	39.7	94.9	207.2
Mean Bankfull Depth (ft)	1.8	1.3	0.9
Width/Depth Ratio	22.3	75.3	222.8
Bankfull Cross-Sectional Area (ft^2)	70.6	119.7	192.6
Max Bankfull Depth (ft)	2.9	3.7	2.3
Width Floodprone Area (ft)	500	500	500
Entrenchment Ratio	12.6	5.3	2.4
D_{15} (mm)	10.6	1.4	0.1
D_{35} (mm)	19.7	10.2	0.1
D_{50} (mm)	26.9	17.6	6.9
D_{84} (mm)	52.5	43.4	28.8
D_{95} (mm)	68.8	64.0	50.0
Slope (ft/ft)	0.0045	0.0038	0.0032
Bank Erosion (ft/year)	0.015	7.1	2.4
Mean Bankfull Velocity (ft/s)	5.7	3.3	2.1
Shear Stress (lb/ft^2)	0.50	0.30	0.18
Unit Stream Power (lb/ft/s)	2.8	1.0	0.4
D_{50} bar (mm)	20	8	7
Bedload at Bankfull Stage (kg/s)	1.5	1.3	0.25
(tons/day)	143	124	24
Suspended Sediment (mg/l)	325	500	520
(tons/day)	351	540	562

Sediment effects

The sediment effects of a change in stream channel stability leading to stream type changes are reflected in the bedload and suspended sediment rating curves in **Figure 2-74**. The upstream change from C4→ G4→ F4 occurred with the avulsion between 1999 and 2001. The sediment rating curves reflect the highest sediment supply relations due to the shift in stream type. For example, bedload transport increased from 0.65 kg/s near bankfull discharge to 1.25 kg/s for the same flow (**Figure 2-74**). This would result in an increase of 57 tons/day in sediment yield. The suspended sediment increased from 325 mg/l to 500 mg/l near the bankfull discharge, or an increase from 351 tons/day to 540 tons/day—an increase of 189 tons/day (**Figure 2-74**).

The change due to the avulsion downstream resulted in a conversion from a C4→ D4 stream type (**Table 2-8** and **Figure 2-66**). The sediment consequence resulted in a reduction in bedload transport from a bankfull transport of 1.5 kg/s (143 tons/day) for the C4 stream type to 0.25 kg/s (24 tons/day) for the D4 stream type (**Table 2-8**). This was due to the reduction in shear stress of 0.50 lbs/ft^2 for the C4 stream type to 0.18 lbs/ft^2 for the D4 stream type, and the reduction in velocity of 5.7 ft/sec for the C4 stream type to 2.1 ft/sec for the D4 stream type. The resultant change in stream type associated with a high width/depth ratio reduced unit stream power from 2.8 lbs/ft/sec to 0.4 lbs/ft/sec (**Table 2-8**, Rosgen, 2003). The change in the morphologic and hydraulic character of this reach of Weminuche Creek decreased sediment transport capacity and led to excess deposition. The aggradation thus caused a change in stream type from a C4→ D4 (**Figure 2-66**).

The suspended sediment concentration, however, increased from 325 mg/l (351 tons/day) to 520 mg/l (562 tons/day). This was due to the increase in streambank erosion of 0.015 ft/yr for the C4 stream type compared to 2.4 ft/yr for the D4 stream type (**Table 2-8**). This increase is due to the finer particles associated with the streambank materials resulting in high suspended sediment concentrations. The enlargement of the channel (39.7 ft to 207.2 ft) also verifies the high rate of streambank erosion. *This demonstrates the dynamic linkage between process and form variables that help shape the present river.*

Streambank erosion prediction results: The BANCS model

A process integration model, BANCS (Bank Assessment for Non-point source Consequences of Sediment), was developed (Rosgen 1996, 2001a) to predict annual streambank erosion rates. The model was developed for field practitioners and uses measurements that can be rapidly obtained to represent streambank erodibility variables and the energy distribution in the "near-bank" region. This model was statistically validated via comparison to measured annual streambank erosion rates (Rosgen 1996, 2001a). The BANCS model converts bank erodibility field measurements to a Bank Erosion Hazard Index, or BEHI rating, and energy distribution measurements to a Near-Bank Stress, or NBS rating. The procedure is described in more detail in Chapter 5 and is demonstrated in Chapter 7.

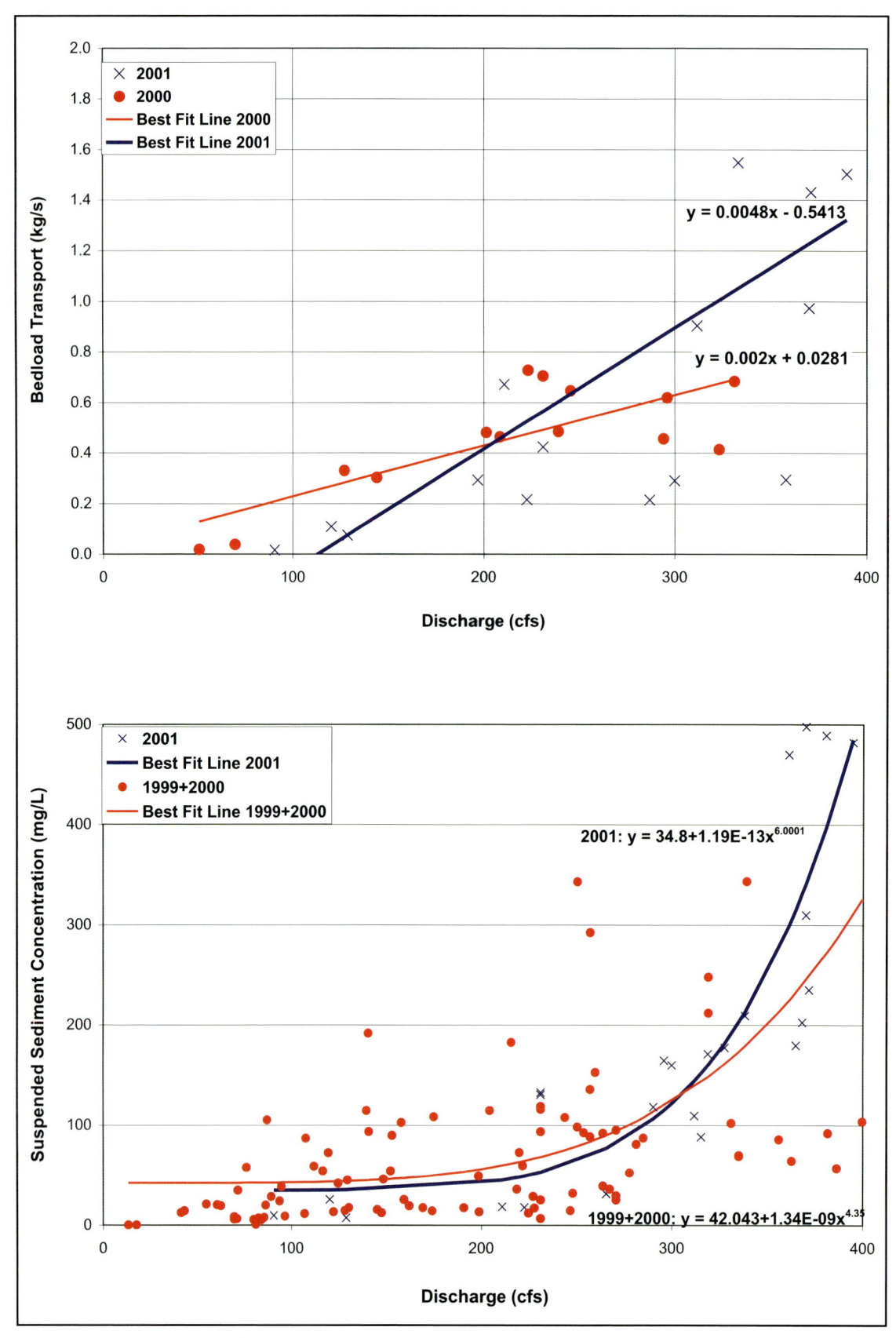

Figure 2-74. Weminuche Creek bedload and suspended sediment rating curves; changes between 1999 and 2000 vs. 2001, reflecting changes from a C4 to F4 stream type.

Chapter 2 A Review of Hillslope, Hydrologic and Channel Process Relations

The instability and consequential shift in succession sequence of Weminuche Creek is associated with accelerated streambank erosion. It is important to predict annual streambank erosion rates. Examples of measured erosion on Weminuche Creek are shown in **Figures 2-75, 2-76** and **2-77**. Prediction of streambank erosion using the BANCS model (BEHI, NBS) must be compared with observed data to validate or develop new (empirical) relations for a local or regional area.

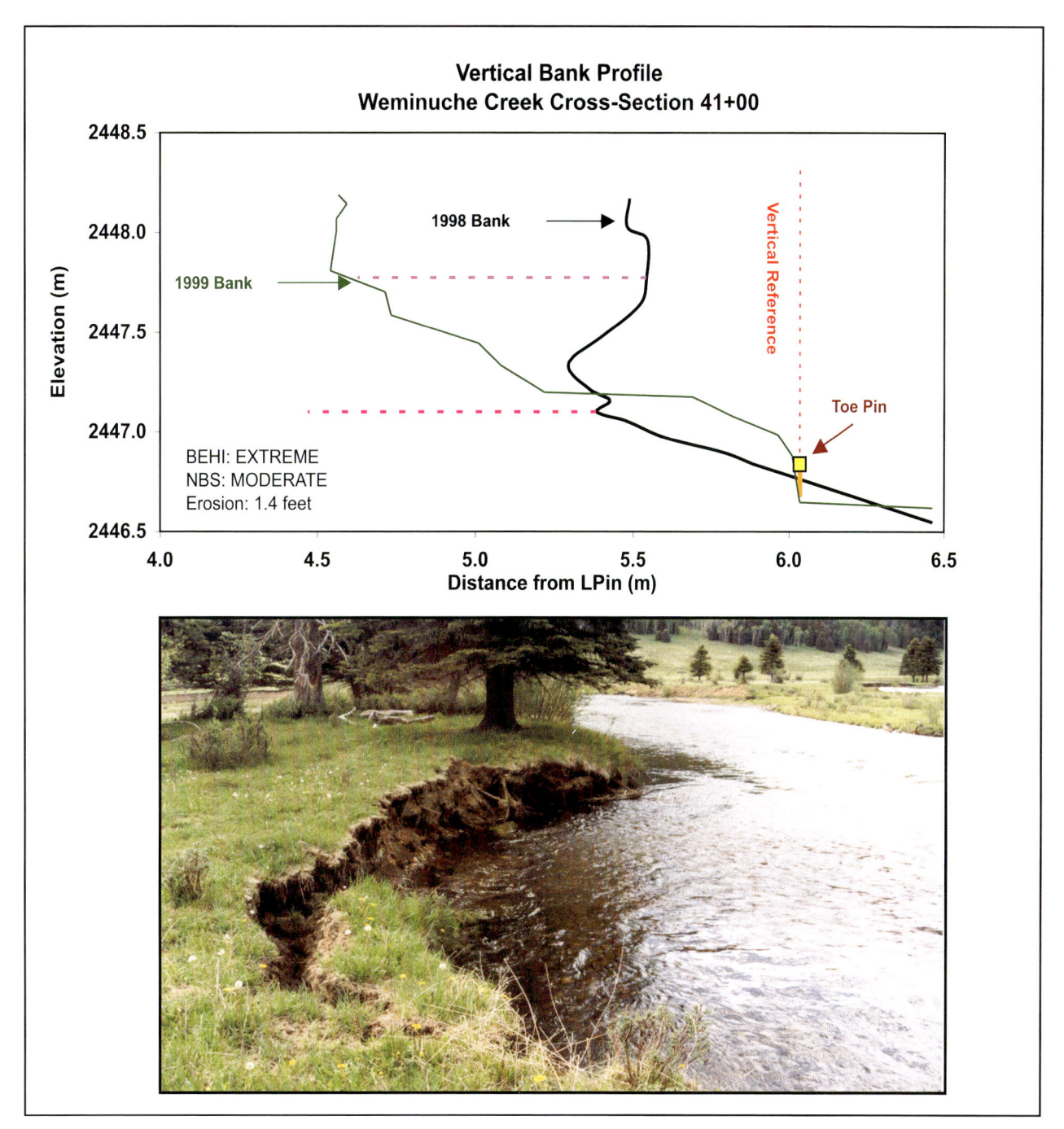

Figure 2-75. Example of accelerated streambank erosion on Weminuche Creek cross-section 41+00 with corresponding Extreme BEHI and Moderate NBS ratings.

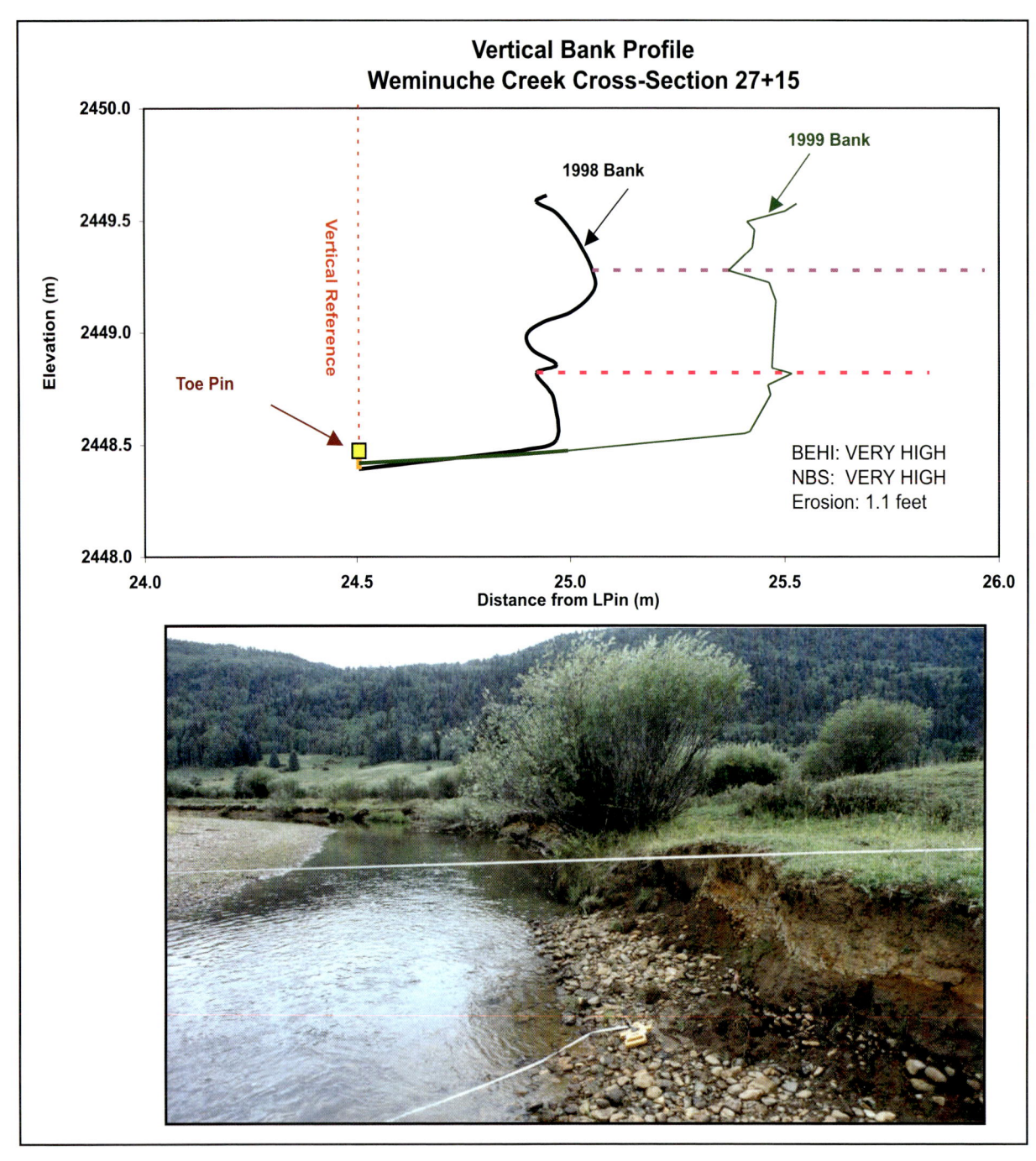

WATERSHED ASSESSMENT OF RIVER STABILITY AND SEDIMENT SUPPLY

Figure 2-76. Example of accelerated streambank erosion on Weminuche Creek cross-section 27+15 with corresponding Very High BEHI and Very High NBS ratings.

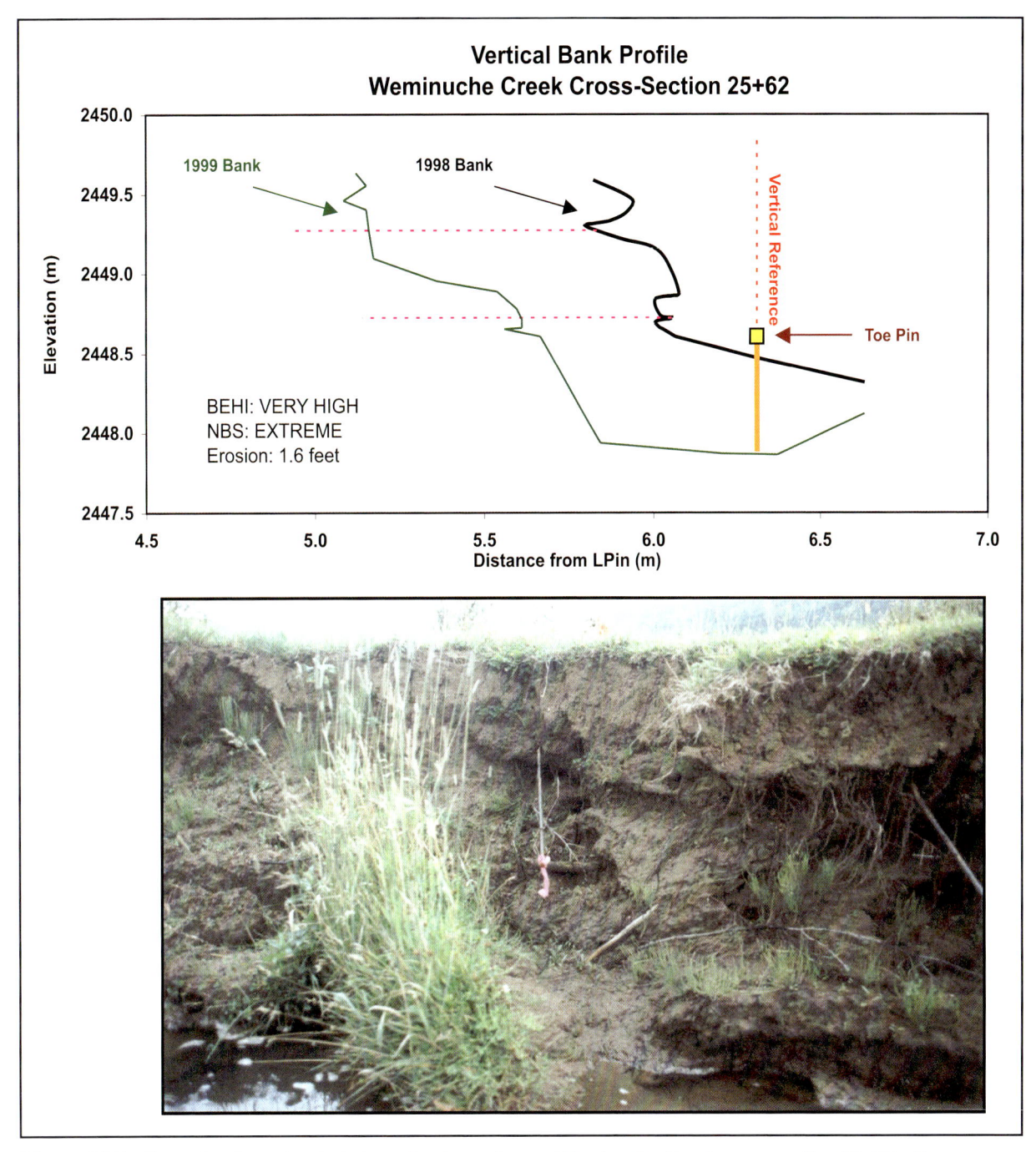

Chapter 2 — A Review of Hillslope, Hydrologic and Channel Process Relations

Figure 2-77. Example of accelerated streambank erosion on Weminuche Creek cross-section 25+62 with corresponding Very High BEHI and Extreme NBS ratings.

The annual lateral streambank erosion rates on the upper reaches of Weminuche Creek associated with this succession averaged 1.8 ft/yr, contributing over 1,370 tons/mi/yr. Obviously, not all of this sediment is routed out of the basin. Much of the coarser material goes into temporary storage in the channel. The bank erosion, however, contributes a large portion of the available sediment and increases the loss of habitat and beneficial uses. The reference streambank erosion rate measured for the stable C4 stream type condition is approximately 3.8 tons/mi/yr.

It is important to remember that, rather than using the derived empirical values of erosion rate in ft/yr for a given BEHI and NBS, users must calibrate the model for the specific soil types and conditions within a local region. The relations used for Weminuche Creek were developed by the author for volcanic and alpine glaciation geology from Yellowstone National Park (Rosgen, 2001a).

Comparison of predicted-to-observed values for the BANCS streambank erosion model

Because the recommended BANCS streambank erosion prediction model (Rosgen, 2001a) was developed in Yellowstone and on the Front Range of Colorado, it was important to test this procedure for local adoption. Independent data sets were used to test the relation on Weminuche Creek in Colorado, the East Fork San Juan River in Colorado and the South Fork Mitchell River in North Carolina (Rosgen, 2001a). The curve that was used to predict annual bank erosion rates for the Mitchell River was adapted from the Front Range of Colorado that represented metamorphic and sedimentary geologic processes. The results showed that the model performed well for predicting streambank erosion; a Chi-squared test of the sites indicated that the populations tested did not differ significantly from the model. Comparisons of predicted rates using the BANCS model versus observed annual bank erosion rates for Weminuche Creek are shown in **Table 2-9**. The East Fork San Juan River, Colorado and the South Fork Mitchell River, North Carolina are shown in **Table 2-10**. The research presented in **Tables 2-9** and **2-10** represents an independent validation of the method's potential applicability beyond its initial development area (Rosgen, 2001a).

Table 2-9. Comparison of predicted versus measured annual streambank erosion rates for selected reaches of Weminuche Creek, Colorado (Rosgen, 2001a).

Cross-Section Location	Bank Erosion Hazard Index (BEHI)	Near-Bank Stress (NBS)	Predicted Erosion m/yr	Predicted Erosion ft/yr	Measured Erosion m/yr	Measured Erosion ft/yr
25+62 (**Figure 2-77**)	Very High	Extreme	0.46	1.50	0.48	1.60
27+15 (**Figure 2-76**)	Very High	Very High	0.27	0.80	0.34	1.10
40+26.5	Very High	Moderate	0.06	0.20	0.06	0.20
41+00 (**Figure 2-75**)	Extreme	Moderate	0.34	1.10	0.43	1.40
44+25	Low	Very High	0.08	0.26	0.09	0.30

Table 2-10. Comparison of predicted versus measured annual streambank erosion rates for the Mitchell River, North Carolina (Harmon and Jessup, 1999), and for the braided reach of East Fork San Juan River, Colorado (Rosgen, 2001a).

River Location	Bank Erosion Hazard Index (BEHI)	Near-Bank Stress (NBS)	Predicted Erosion m/yr	Predicted Erosion ft/yr	Measured Erosion m/yr	Measured Erosion ft/yr
Mitchell River	Moderate	High	0.12	0.38	0.09	0.30
	Moderate	Extreme	0.45	1.50	0.21	0.70
	High	Extreme	0.76	2.5	0.85	2.80
	Extreme	Extreme	4.27	14.0	3.35	11.0
East Fork San Juan River	Extreme	Extreme	0.85	2.80	0.73	2.40
	Extreme	High	0.55	1.80	0.59	1.95
	Moderate	High	0.19	0.62	0.22	0.73
	Low	High	0.06	0.20	0.06	0.20
	High	Low	0.14	0.45	0.12	0.40
	High	Low	0.14	0.45	0.14	0.47

Once the adverse consequences of change due to instability are understood, the impetus to prevent the cause(s) and restore the stream becomes a reality. Watershed and river assessment methodologies help to provide an understanding of the location, nature, extent and consequence of change. Chapter 3 explains the initial assessment process necessary to predict instability and adverse consequences.

"Reading the river makes one look beyond the water."

CHAPTER **3**

The Reconnaissance Level Assessment (RLA)

The *Reconnaissance Level Assessment (RLA)* of the *WARSSS* assessment methodology has three primary objectives:

1. To identify sediment sources and channel stability problems linked to specific processes influenced by land and river management activities;

2. To refine, clarify or, if necessary, redirect problem identification; and

3. To locate potential problem areas and reaches within a large watershed that require a more detailed level of assessment.

The *Reconnaissance Level Assessment* (*RLA*) is the first and most general phase of the three *WARSSS* assessment phases. It provides a broad overview of the landscape while focusing on processes that may affect sediment supply and channel stability. The *RLA* identifies erosional or depositional processes and locations that are influenced by a variety of existing and past land use practices. This initial screening eliminates stable, low-risk slopes, sub-watersheds and river reaches from further analysis. By briefly evaluating a large assortment of processes, practices and places, the *RLA* reveals specific locations that require more detailed analyses at the *RRISSC* or *PLA* levels. This reduces the time and cost of the *WARSSS* assessment. Conducting a more detailed assessment of targeted sites is justified if the user consistently applies the *RLA* methodology and documents the initial results and recommendations. Even though field measurements are generally not required for this level, a site visit is necessary to verify aerial photograph interpretations and the valley and stream type mapping, as well as to confirm, reject or redirect the initial problem identification. The time required for one person to assess a third-order watershed using *RLA* is measured in days (rather than weeks or months), depending on availability of maps, photographs, soils mapping and access to land use history.

In summary, this broad-level assessment method provides the following:

- A basis for selecting obvious sediment supply sources;

- The location of stable slopes, sub-watersheds and stream channels not requiring additional assessments;

- Verification of perceived problems;

- Familiarity with the watershed being assessed, including preparation of maps and photographs to be used for later analysis; and

- The opportunity to identify sources and causes of problems not intuitively obvious, and a preliminary database for use in other applications.

The *RLA* flowchart (**Flowchart 3-1**) illustrates the general assessment process using a sequence of numerical steps. The remainder of this chapter discusses each step in detail. A summary worksheet provides the guidance criteria and analysis summary for hillslope, hydrologic and channel processes to determine which areas and/or stream reaches require a more detailed assessment. This worksheet also documents the location and justification for areas and/or river reaches not requiring further assessment. For a completed version of worksheets and tables associated with *RLA*, see the *RLA* section of Chapter 7.

The importance of problem identification

Before beginning *RLA*, it is essential to revisit the initial information on the nature of sediment impairment and existing data sources. Because some form of Problem Identification (Problem ID) usually already exists, it is a starting point for the *RLA* phase of the *WARSSS* assessment rather than a formal part of *WARSSS*. The details of an existing Problem ID can vary considerably and affect the amount of effort put into *RLA*.

In most cases where a *WARSSS* assessment is planned, some form of Problem ID has taken place and is responsible for the current recognition of the waterbody's sediment problems. State or other water quality monitoring data may have revealed non-attainment of water quality criteria, and probable causes for the impairment may have been identified. The waterbody may already appear as sediment-impaired for non-point source pollution on a recent State 303(d) list or Integrated Report. A TMDL or a watershed plan may be scheduled for development. Results of a sediment model may even be available.

In the better-documented settings, several of the data sources and reconnaissance-level analytical steps of the *RLA* phase may have been completed. However, some assessments begin with limited information and substantial uncertainty as to the suspected sediment problem, its sources and effects. In these situations, the *RLA* phase needs to become the Problem ID and play a critical role in developing a more specific identification rather than simply verifying the existing problem. At a minimum, existing problem identification should be reviewed thoroughly and used as a data source for the *RLA* phase.

If a TMDL is planned as part of a *WARSSS* assessment, more detailed guidance on developing a Problem ID appears in pages 3-1 through 3-9 of the *Protocol for Developing Sediment TMDLs* (USEPA, 1999).

Step 1. Compile existing data

The first *RLA* step assembles data sources that will be used recurrently in *WARSSS*. The user compiles all available information from existing monitoring data then obtains the information necessary to initially identify likely sediment sources.

An existing water quality monitoring and reporting program may have already established the type of sediment impairments found, identified their relationship to water quality criteria and designated uses, compiled the monitoring data and sample locations, and explained how this information indicates impairment in all or part of the waterbody. Sources of sediment may be known or suspected, but this can vary; water quality monitoring programs generally document the effects of a stressor with less attention to its origins. *WARSSS* was designed to improve the

Chapter 3 　　　　　　　　　　　　　　　　　　　　　　　　　　　The Reconnaissance Level Assessment (RLA)

Reconnaissance Level Assessment (RLA)

Start with problem identification

1) Compile existing data

What sources and effects?
2) Review the landscape history
3) Summarize activities that potentially affect sediment supply and channel stability
4) Identify specific process relations

Where do they occur?
5) Review the landscape overview and map the watershed
6) Identify hillslope processes
7) Document surface erosion
8) Document mass erosion
9) Assess hydrologic processes
10) Identify streamflow changes
11) Analyze channel processes
12) Detect direct impacts to streambanks and channels

What places should be targeted?
13) Summarize problem verification: recognition of places, processes and sources
14) Eliminate areas, sub-watersheds and reaches that are not contributors to impairment
15) Select areas, sub-watersheds and reaches for further assessment

Proceed to RRISSC

Refine problem identification if necessary

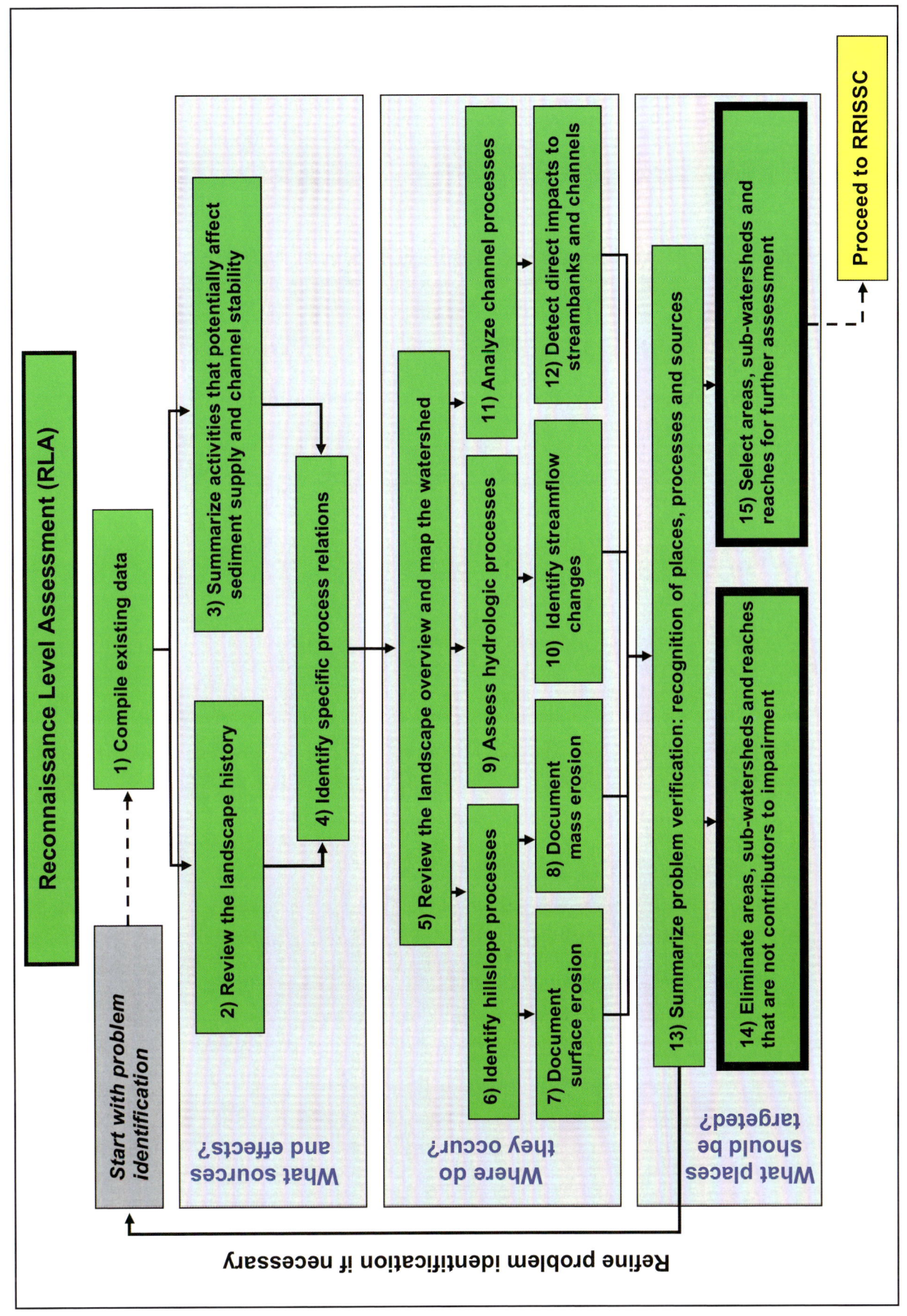

Flowchart 3-1. The *Reconnaissance Level Assessment (RLA)* step-wise sequence.

3-3

understanding of sediment sources and their relative contributions, thereby informing the choice of sediment control options.

Geographic information about the watershed plays a major role in the *RLA* phase's initial focus on sediment sources. A GIS, if available, can accelerate several *RLA* steps, add new findings to existing data and indicate reaches and sub-watersheds for further assessment. Nevertheless, the *RLA* can be completed in a few days without a GIS.

The following GIS data sources, if available, are recommended (note the several online sources):

- *Topographic maps:* Essential for delineating sub-watersheds, identifying steep slopes, planning field access routes and many other logistical tasks. If you have no local source, go online to http://topomaps.usgs.gov/ordering_maps.html for 1:24,000 scale topo maps.

- *Aerial photographs:* They are inexpensive, available nationwide since the 1940s and still heavily used in environmental analyses. Current and historical (30+ years) coverage should be acquired to note channel migration, bank erosion and land use changes. They are best analyzed using quality equipment, such as mirror stereoscopes or binocular optics on a light table mount. Black-and-white, color or color-infrared photos are acceptable. Color-infrared aerial photographs, in particular, can help to identify sediment sources (**Figure 3-1**). Obtain the largest scale possible (i.e., the *smallest* number in the scale fraction's denominator— 1:5,000 scale is better than 1:40,000), preferably 1:24,000 or better. Allow several weeks for delivery time. Assistance on searching airphoto holdings from most public and private sources (the APSRS database) is available through USGS by providing a map of the area where coverage is desired. Check online at http://erg.usgs.gov/isb/pubs/factsheets/fs22096.html

- *Soil and geology maps:* Essential for noting the natural factors affecting erosion potential, etc. Soil surveys (check online at http://soils.usda.gov/survey/printed_surveys/) are available in print if not digitally available in GIS.

- *GIS LiDAR (Light Detection and Ranging):* LiDAR images are helpful in areas with heavy vegetation. The images depict terrain and stream features by removing vegetation effects.

- *USGS flood history data:* Real-time flow information (check online at http://waterdata.usgs.gov/nwis/rt) is essential if the waterbody is gaged; local sources of information can also be sought for ungaged waters.

- *Current land cover (land use and vegetation) maps:* Land cover information is commonly available on a local-scale GIS or from state or national land cover information sources at http://www.mrlc.gov/index.asp Note the date of the images used to develop any land cover product and check current aerial photography or field notes for significant changes.

- *Historical land use documentation and special activities on site:* Examples include SCS/NRCS land drainage and emergency flood-relief projects, U.S. Army Corps of Engineers channel projects (flood-control projects associated with channel dredging, straightening, flood levees, etc.), navigation projects, grazing history/reports, timber harvest and transport history and urbanization patterns.

Chapter 3   *The Reconnaissance Level Assessment (RLA)*

This information is used throughout the *WARSSS* assessment and helps users to make an initial assessment of possible hillslope, hydrologic and channel processes that may affect sediment supply and river stability. Color-infrared aerial photographs can quantitatively determine the concentration of sediment and sources with aerial coverage of a watershed during a runoff event (Rosgen, 1973, 1976). This method involves ground truth of measured sediment at the time of the flight and densinometric analysis of the photo transparencies. The color-infrared photo shown in **Figure 3-1** indicates the higher sediment concentration of the smaller tributary (West Fork Madison River) compared to the larger receiving reach of the Madison River (Rosgen, 1973, 1976).

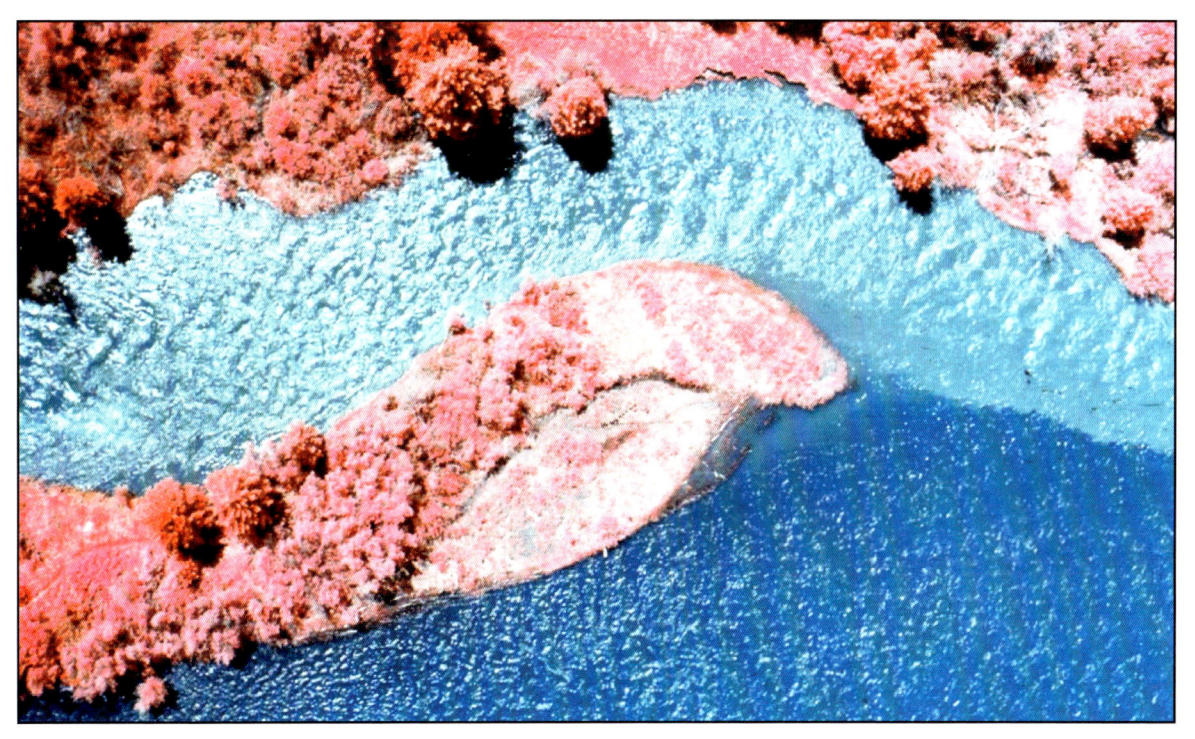

Figure 3-1. Color-infrared photograph at the confluence of the West Fork with the Madison River, Montana (Rosgen, 1973).

Step 2. Review the landscape history

A review of land use history helps to:

- Develop an understanding of the activities that have occurred, their extent and nature; and
- Characterize trends and associated impacts for various land uses over time.

For example, in the Piedmont region of the Eastern United States, tobacco farming on steep, erodible slopes caused excessive sediment delivery to stream channels during the 19th and 20th centuries. Since then, the landscape has been converted to deciduous forests, reducing sediment delivery. Subsequently, the streams have been down-cutting in the erosional debris associated with land use practices that have long been discontinued. Channel evolution associated with successional stages of channel adjustment and flow-related sediment increases needs to be assessed for such streams. In the West, many riparian areas are recovering along major rivers due to significant reductions in livestock numbers since the late 1800s.

Step 3. Summarize activities that potentially affect sediment supply and channel stability

In this step, the watershed is broken into discrete sub-watersheds and river reaches of similar types and uses. The summary of watershed land use history in **Step 2** is vital to linking land and river management activities to the sources that influence sediment supply and river stability. The user may note potential land use impacts for the overall watershed using **Table 3-1**. This table summarizes the direct, indirect and negligible effects of fundamental land and river management activities. If these activities are associated with the land use history of the watershed, this step identifies them for later evaluation during the more detailed portion of this assessment (**Steps 5–12**). If there is no evidence of these activities, the user can eliminate additional evaluation. In this way, the information helps to determine whether more data is needed to identify potential problem areas and/or reaches.

Chapter 3 — The Reconnaissance Level Assessment (RLA)

Table 3-1. Direct and indirect potential influences of land use variables on stream channels and sediment supply.

Land Uses	(1) Streamflow changes (magnitude/ timing/duration)	(2) Riparian vegetation change (composition/ density)	(3) Surface disturbance (% bare ground/ compaction)	(4) Surface/ sub-surface slope hydrology	(5) Direct channel impacts that destabilize channel	(6) Clear water discharge	(7) Loss of stream buffers, surface filters, ground cover	(8) Altered dimension, pattern and profile	(9) Excess sediment deposition/ supply (all sources)	(10) Large woody debris in channel	(11) Stream power change (energy distribution)	(12) Floodplain encroachment channel confinement (lateral containment)
Urban development	D	D	D	D	D	D	D	D	I	D	D	D
Silvicultural	D	D	D	D	D		D	I	D	D	I	D
Agricultural	D	D	D	D	D		D	D	D	D	D	D
Channelization	D	D		D	D		D	D	D	D	D	D
Fires	D	D	D	D	I		D	D	D	D	D	
Flood control, clearing, veg. removal, dredging, levees	D			I	D		D	I	I	D	D	D
Reservoir storage, hydropower	D	I		I	D	D			I/D	I	D	
Diversions, depletions (-) Imported (+)	D	I		I	D	D			I/D			
Grazing	I	D	D	D	D		D	D	D	D	D	D
Roads	D		D	D	D		I	D	D	D	D	D
Mining	D	D	D	D	D		D	D	D	D	D	D
In-channel mining	D	D		D	D		D	D	D	D	D	D

D = Direct potential impact
I = Indirect potential impact
Blank = Little to no impact

Step 4. Identify specific process relations

This step involves a review of the linkages among variables influenced by land use activities and potential erosional process impacts. This assessment is conducted by sub-watershed and by major river reach segments. **Table 3-2** is a summary of stream and channel variables influenced by land uses and the erosional processes potentially impacted. Process relations information needs to be:

1. Checked against the land use history that specifies past and present land use and channel impacts, as compiled in **Table 3-1**; and

2. Related to potential impacts on water quality and designated uses using **Table 3-2**. **Table 3-2** relates the direct impacts of land uses, such as streamflow change, to potential indirect and direct erosional impacts, including mass erosion, gully erosion, streambank erosion, channel enlargement, channel aggradation or degradation, change in stream type and change in sediment delivery capabilities.

Many places and processes within a watershed have a low risk of impact on sediment supply and channel stability. It is not necessary to proceed with a more detailed assessment of these areas. Areas identified as potential contributors to sediment and stability problems, however, require further assessment.

Chapter 3 — The Reconnaissance Level Assessment (RLA)

Table 3-2. Relation of stream and channel variables to erosional processes.

Variables Influenced	Surface erosion	Mass erosion	Gully erosion	Streambank erosion	Channel enlargement	Aggradation	Degradation	Channel succession state	Sediment delivery efficiency
(1) Streamflow changes (magnitude/ timing/ duration)		I	D	D	D	D	D	D	I
(2) Riparian vegetation change (composition/ density)			D	D	D	D	D	D	I
(3) Surface disturbance (% bare ground/ compaction)	D	I (debris torrents)	D (rills-gully)						D
(4) Surface/ sub-surface slope hydrology	D	D	D						D
(5) Direct channel impacts that destabilize channel			D	D	D	D	D	D	I
(6) Clear water discharge			D	D	D	I	D	D	
(7) Loss of stream buffers, surface filters, ground cover	D		I						D
(8) Altered dimension, pattern and profile				D	D	D	D	D	
(9) Excess sediment deposition/ supply				D	D	D	D	D	
(10) Large woody debris in-channel		D	D	D	D	D	D	D	
(11) Stream power change (energy redistribution)			D	D	D	D	D	D	
(12) Floodplain encroachment channel confinement (lateral containment)		I	I	D	D	D	D	I	D

D = Direct potential contribution
I = Indirect potential contribution
Blank = Little to no influence

Step 5. Review the landscape overview and map the watershed

Now that basic locations with sediment problems have been identified, the next step is to review soil types, topographic maps, aerial photographs and land use maps to locate and document obvious sediment source areas and/or processes that may adversely affect sediment supply and river stability. This assessment is performed without field measurements but requires a site visit to verify uncertain interpretations from aerial photographs, topographic maps and other data used in the initial delineation.

The site visit can be accomplished via aerial reconnaissance (fixed wing or helicopter) or, if permitted access is not a problem, by vehicle and/or on foot. Ideally, the procedure involves looking at the entire watershed from the mouth of the drainage to the watershed divide, but it may be limited to checking key sites indicated by the landscape overview map/photo analysis. If sources are not clear from the maps and photos, initial field observations of instream problems are often explained by looking further upstream for source areas and causes.

Where a GIS is available, the landscape overview can involve single-factor and composited overlay analysis of land uses on soil maps, topographic maps and aerial photography. At this point, a major watershed is broken into sub-watersheds based on second- to third-order drainage basins. Stream order is delineated using the Strahler method of ordering where two first orders at their confluence create a second order, two second orders at their confluence create a third order, etc. Stream and valley types must also be identified and mapped, based on the Level I stream classification system (Rosgen, 1994, 1996). Valley type designations (**Table 2-3**) are important as they are an initial stratification at a broad level, which helps identify potential stream types associated with a unique valley type. Key data for valley types are gradient, width and dominant geologic/geomorphic processes. A detailed summary of valley types is available in Rosgen (1996).

Various stream type delineations are shown in **Figures 3-2**, **3-3** and **3-4**. **Figure 3-4** also shows an example of valley type delineation; for review, see **Figures 2-13** and **2-14** and **Table 2-2**.

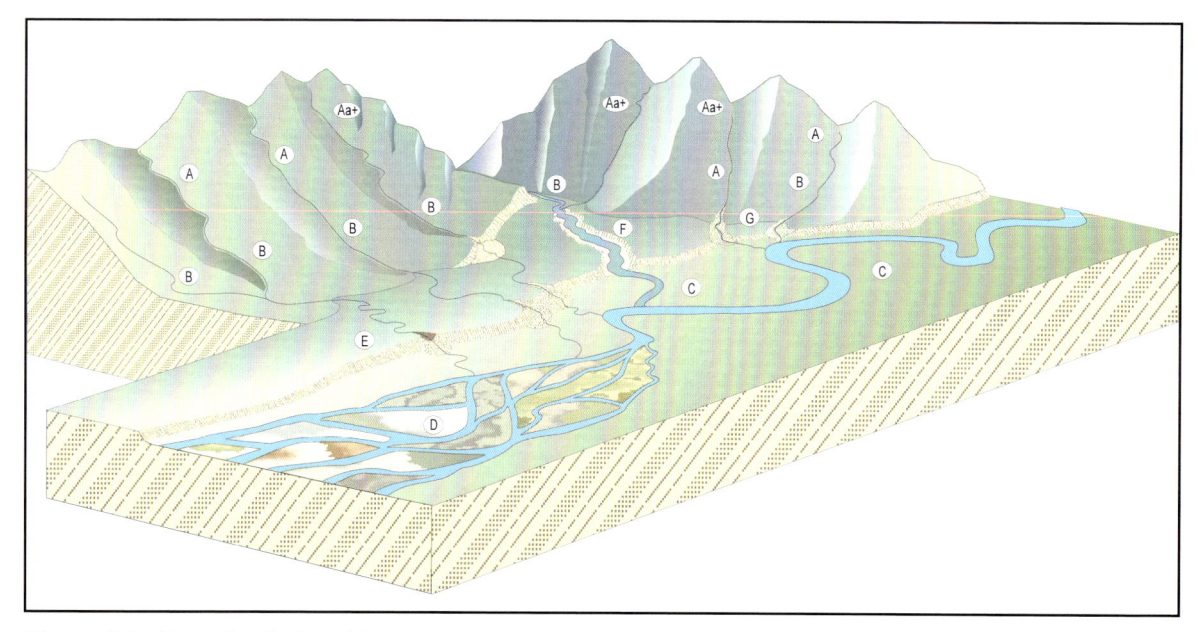

Figure 3-2. Example of a Level I stream type delineation.

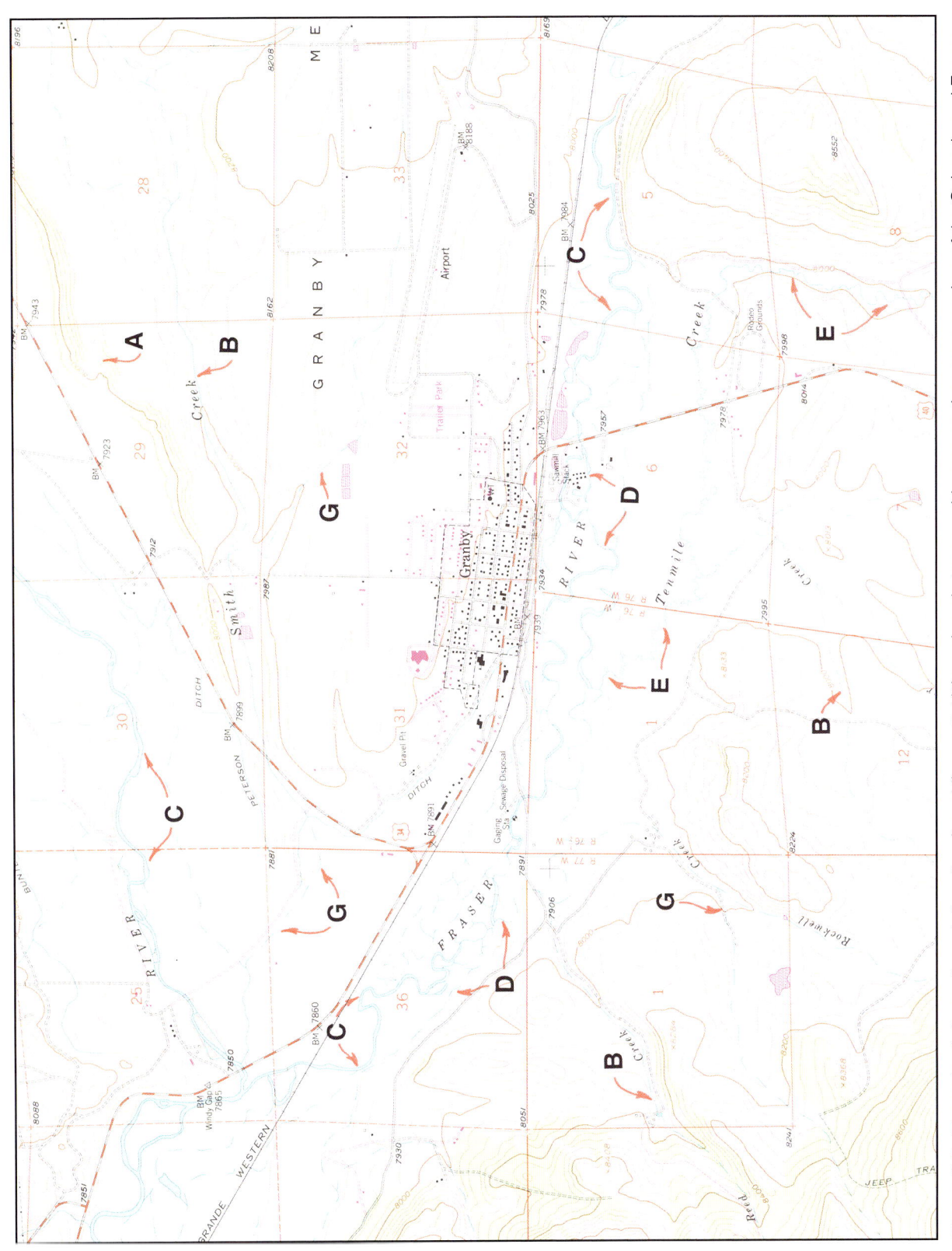

Chapter 3
The Reconnaissance Level Assessment (RLA)

Figure 3-3. Example of stream type delineation (Level I) on a 7-1/2' quadrangle topographic map on the upper reaches of the Colorado and Fraser Rivers, Colorado (Rosgen, 1996).

3-11

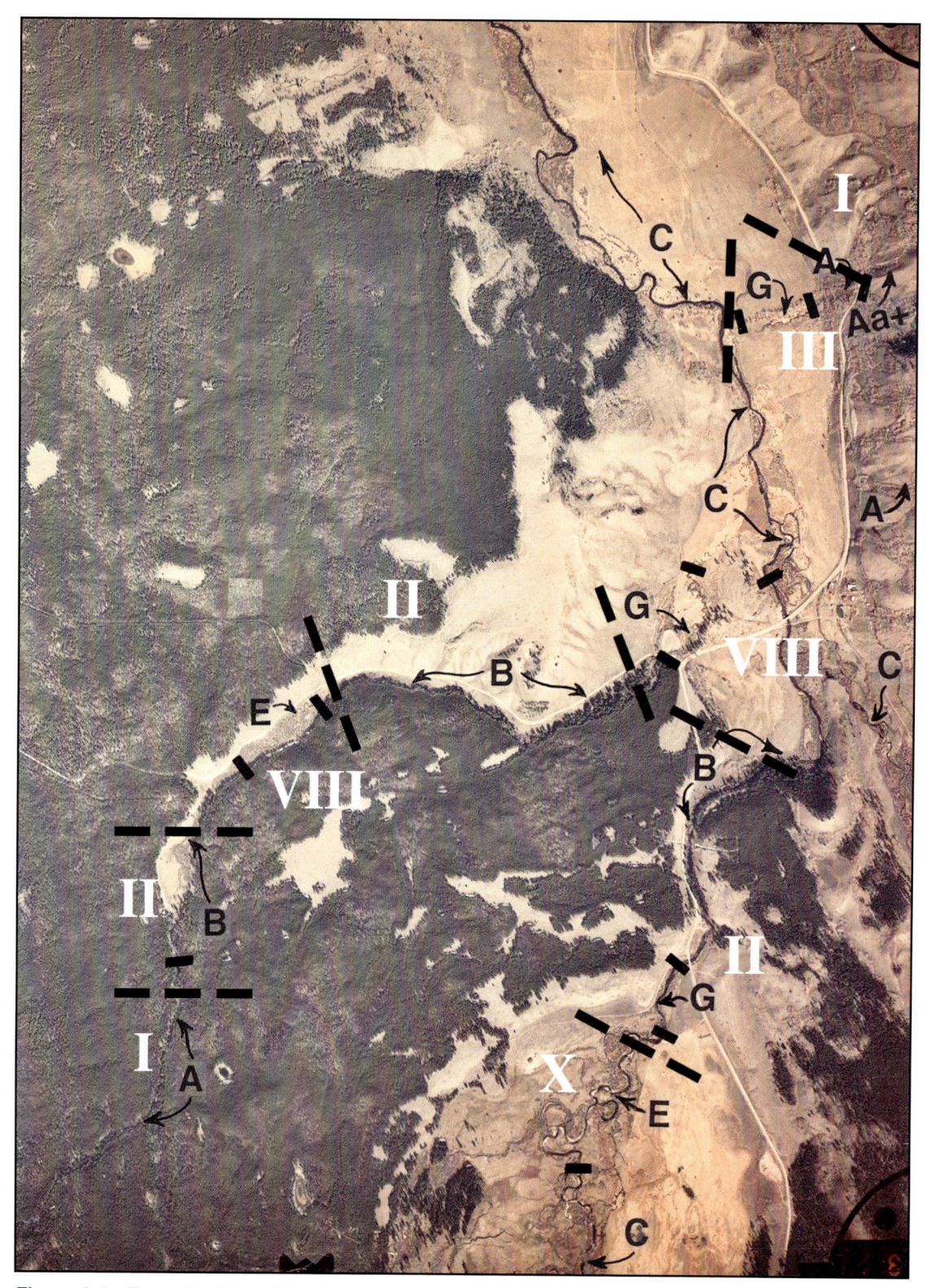

Figure 3-4. Example of broad-level stream and valley type delineation using aerial photography, upper Laramie River, Colorado (adapted from Rosgen, 1996).

Step 6. Identify hillslope processes

The interpretations of the response of various stream types A–G to increases in streamflow will vary by channel materials. Although the broad Level I stream type mapping utilized at the *RLA* phase does not require site-specific, detailed measurements, channel materials can often be successfully inferred from interpretations from soil and geology maps, landforms/land types and valley types. Historical mapping and other studies can also provide interpretations of channel materials; for example, a tributary stream cut through an alluvial terrace as well as those incised in an alluvial fan would be characteristic of a 3 (cobble) or 4 (gravel) material. It is appropriate to insert a number that infers a dominant particle size without a detailed pebble count at this level of assessment (refer to **River classification** section in Chapter 2 and Rosgen 1994, 1996). An A1 or A2 stream type will ***not*** be sensitive to streamflow increases; however, this is just the opposite of A3–A6 stream types. The A stream types are high energy streams, but coupled with a heterogeneous, unconsolidated mixture of cobble, gravel, sand or silt/clay, these streams become high energy and high sediment supply. Later guidance criteria used in *RLA* requires this distinction of a general, dominant channel material. A more detailed Level II classification and field verification of the preliminary mapping conducted at this level is required at the *RRISSC* level (Chapter 4).

This step identifies evidence of erosional processes on landscapes prone to erosion and/or where land use activities have accelerated these processes. The procedure involves overlaying soil types susceptible to erosion or failure with land uses at specific locations within the watershed. Some land use data, such as "percent of land area in clearcut condition," can also be used in hydrologic and channel process evaluations (**Steps 9–11**). Direct observations and recent aerial photographs are often very helpful in identifying erosional problems in conjunction with roads, land clearing and other surface disturbance activities. If the activities occur on steep terrain, highly dissected landscapes or lower slopes in proximity to streams, then there is a higher likelihood of sediment delivery in the form of erosional debris. The approximate dates of historical disturbances should be noted to build understanding of landscape recovery from similar impacts. Review **Tables 3-1** and **3-2** and, where necessary, indicate the sub-watersheds where potential mass erosion is likely. The remaining steps in this assessment use the information gathered in the preceding steps, along with area maps, to determine which areas are at highest risk and, therefore, are in need of additional evaluation at the *RRISSC* assessment level.

Step 7. Document surface erosion

Soil maps, slope dissection and gradient from topographic maps, as well as vegetation cover and presence of rills or gullies, etc. from aerial photographs, are used to identify areas associated with particular land uses that may contribute sediment due to surface erosion processes. Analyzing potential conversion of surface erosion to sediment supply helps to determine which areas may require more detailed assessments.

For example, if there was evidence of erosion due to land clearing, but the slopes were gentle with good internal drainage, had low drainage dissection and were located on an upper slope position far removed from drainageways, then the likelihood of sediment delivery would be low; thus, this site could be excluded from further analysis of surface erosion prediction. However, the effects of land clearing on water yield, potentially affecting hydrologic and channel processes, would have to be considered in a separate analysis. Use the following guidance criteria (**Table 3-3**) to review maps and information in **Tables 3-1** and **3-2**, and complete the appropriate columns in summary **Worksheet 3-1**. Information should be recorded by both geographic location and stream type. If any one of several conditions is present, it is recommended that those areas and/or river reaches advance to the *RRISSC* assessment.

Table 3-3. Guidance criteria for advancement to the *RRISSC* assessment based on surface erosion.

Surface erosion guidance criteria for advancement to *RRISSC*
1. If surface erosion is evident on steep, dissected slopes.
2. If surface erosion is evident on unstable soils at lower slope positions in close proximity to drainageways.
3. If activities such as skid trails are continuous down-slope indicating a high potential of surface erosion converted to sediment delivery to a drainageway.
4. If surface disturbance activities occur on rill-dominated slopes.

The above information is entered in **Worksheet 3-1** by sub-watershed and reach location. Areas recommended for advancement to *RRISSC* are documented, along with those areas requiring no further assessment.

Chapter 3 — The Reconnaissance Level Assessment (RLA)

Worksheet 3-1. Evaluation and summary of guidance criteria for selection of sub-watershed to proceed to *RR/SSC* or to exclude from further assessment.

Sub-watershed/ reach location ID	Step 7: Surface erosion — Circle selected guidance criteria number (Table 3-3)*	Step 7: Reason for exclusion	Step 8: Mass erosion — Circle selected guidance criteria number (Table 3-4)*	Step 8: Reason for exclusion	Step 10: Streamflow change — Circle selected guidance criteria number (Table 3-5)* Roads	Step 10: Reason for exclusion	Step 11: Channel processes — Circle selected guidance criteria number (Table 3-6)*	Step 11: Reason for exclusion	Step 12: Direct channel impacts — Circle selected guidance criteria number (Table 3-7)*	Step 12: Reason for exclusion	Step 15: ✓ location selected for advancement to *RR/SSC***
1.	(1) (2) (3) (4)		(1) (2) (3) (4) (5)		(1) (2) (3) (4) (5) (6)		(1) (2) (3) (4) (5) (6)		(1) (2)		
2.	(1) (2) (3) (4)		(1) (2) (3) (4) (5)		(1) (2) (3) (4) (5) (6)		(1) (2) (3) (4) (5) (6)		(1) (2)		
3.	(1) (2) (3) (4)		(1) (2) (3) (4) (5)		(1) (2) (3) (4) (5) (6)		(1) (2) (3) (4) (5) (6)		(1) (2)		
4.	(1) (2) (3) (4)		(1) (2) (3) (4) (5)		(1) (2) (3) (4) (5) (6)		(1) (2) (3) (4) (5) (6)		(1) (2)		
5.	(1) (2) (3) (4)		(1) (2) (3) (4) (5)		(1) (2) (3) (4) (5) (6)		(1) (2) (3) (4) (5) (6)		(1) (2)		
6.	(1) (2) (3) (4)		(1) (2) (3) (4) (5)		(1) (2) (3) (4) (5) (6)		(1) (2) (3) (4) (5) (6)		(1) (2)		
7.	(1) (2) (3) (4)		(1) (2) (3) (4) (5)		(1) (2) (3) (4) (5) (6)		(1) (2) (3) (4) (5) (6)		(1) (2)		
8.	(1) (2) (3) (4)		(1) (2) (3) (4) (5)		(1) (2) (3) (4) (5) (6)		(1) (2) (3) (4) (5) (6)		(1) (2)		
9.	(1) (2) (3) (4)		(1) (2) (3) (4) (5)		(1) (2) (3) (4) (5) (6)		(1) (2) (3) (4) (5) (6)		(1) (2)		
10.	(1) (2) (3) (4)		(1) (2) (3) (4) (5)		(1) (2) (3) (4) (5) (6)		(1) (2) (3) (4) (5) (6)		(1) (2)		
11.	(1) (2) (3) (4)		(1) (2) (3) (4) (5)		(1) (2) (3) (4) (5) (6)		(1) (2) (3) (4) (5) (6)		(1) (2)		
12.	(1) (2) (3) (4)		(1) (2) (3) (4) (5)		(1) (2) (3) (4) (5) (6)		(1) (2) (3) (4) (5) (6)		(1) (2)		
13.	(1) (2) (3) (4)		(1) (2) (3) (4) (5)		(1) (2) (3) (4) (5) (6)		(1) (2) (3) (4) (5) (6)		(1) (2)		
14.	(1) (2) (3) (4)		(1) (2) (3) (4) (5)		(1) (2) (3) (4) (5) (6)		(1) (2) (3) (4) (5) (6)		(1) (2)		
15.	(1) (2) (3) (4)		(1) (2) (3) (4) (5)		(1) (2) (3) (4) (5) (6)		(1) (2) (3) (4) (5) (6)		(1) (2)		
16.	(1) (2) (3) (4)		(1) (2) (3) (4) (5)		(1) (2) (3) (4) (5) (6)		(1) (2) (3) (4) (5) (6)		(1) (2)		

*Criteria based on overall review of the list in Tables 3-1 and 3-2.

**Locations that meet one or more selection criteria should proceed to the *RR/SSC* assessment level.

Step 8. Document mass erosion

Mass erosion, or mass wasting (debris torrents, debris avalanches and/or slump/earthflow), is caused by various sources. Extensive vegetation removal on steep slopes can add to soil saturation and reduce the internal strength afforded by roots, accelerating mass erosion. Roads can affect slope equilibrium and result in accelerated mass erosion. Log skidding that concentrates rather than disperses flow can accelerate debris torrent/debris flow processes. If landslides occur on discontinuous slopes, there is less likelihood of erosional debris becoming sediment; thus, these areas may be excluded from additional assessment. Evident land movement on steep or concave slopes within the lower one-third of the slope or in close proximity to drainageways requires advancement to the *RRISSC* level of assessment. The user should review maps and **Table 3-2** to complete the evaluation according to the guidance criteria in **Table 3-4**.

Table 3-4. Guidance criteria for advancement to the *RRISSC* assessment for mass erosion.

Mass erosion guidance criteria for advancement to *RRISSC*
1. If evidence exists of recent (within last 10 years) slump/earthflow and/or debris flow/debris avalanche activity.
2. If slide activity is located on steep, concave, continuous slopes.
3. If there is a high percentage of vegetation clearing in proximity to landslide prone terrain.
4. If the location of slide activity is in or adjacent to drainageways.
5. If evidence exists of slump/earthflow and/or debris flow/debris avalanche caused by road location.

The summary of this evaluation is entered in **Worksheet 3-1** by sub-watershed and reach location. Areas recommended for advancement to *RRISSC* are documented, along with those areas requiring no further assessment.

Step 9. Assess hydrologic processes

Changes in the vegetative cover that influence evapo-transpiration, interception, snow depositional patterns, etc. can change the magnitude, duration and timing of runoff. Urban watershed development can result in a high percentage of impervious surface that changes the probability, frequency, magnitude and duration of runoff events. These flow increases have varying outcomes, depending on the morphology of the stream channels and their stability. For example, increased flood flows to incised stream types (G and F) below urban development are contained within the channel, causing excess bed and bank erosion, land loss and increased sedimentation.

Vegetation and urban changes should be evaluated within second- to third-order basins to determine potential flow changes. Road density associated with slope steepness needs to be observed by sub-watershed and reach at this level. This assessment determines potential changes in drainage density due to sub-surface flow interception and change in evapo-transpiration without hydrologic recovery.

Step 10. Identify streamflow changes

This step identifies sub-watersheds where flow-related sediment and channel instability is potentially influenced by increased streamflow. The observer must determine whether there are sufficient potential changes in streamflow to warrant further investigation. The user should review **Table 3-1** and watershed maps to complete the guidance criteria in **Table 3-5**.

Table 3-5. Guidance criteria for advancement to the *RRISSC* assessment for potential streamflow changes.

Streamflow change guidance criteria for advancement to *RRISSC*
1. If rural (non-urban) watersheds have a percentage of bare ground, hydrologic modification due to change in vegetative type and clearcutting timber stands that exceed 20% of first- to third-order watershed areas in the presence of A3–A6, C, D, E, F and G stream types.
2. If urban watersheds have impervious conditions that exceed 10% of second- to third-order watershed areas in the presence of A3–A6, C, D, E, F and G stream types. No hydrologic recovery is recognized.
3. *Time-trend of vegetation* (rural or non-urban). If the vegetative conversions occurred within the last 15–20 years for rain-dominated or temperate climates, or 80 years or less for snowmelt-dominated montane and/or sub-alpine climatic regions, there likely has not been sufficient time for full hydrologic recovery. These recovery times are based on revegetating sites and the time necessary to regain pre-treatment evapo-transpiration, snow deposition patterns and other similar processes reflecting consumptive water loss.
4. *Diversions, imported water, water depletion and/or return flows.* If the recipient or depleted stream types are alluvial and susceptible to degradation, aggradation, streambank erosion or enlargement (stream types A3–A6, C, D, E, F and G).
5. *Reservoirs.* All reservoirs located on alluvial channel types or those incised in landslide debris, glacial tills, etc. need to be assessed at the *RRISSC* or *PLA* level. This is due to the complexity of potential impacts, the nature of the stream type, the variation in the operational hydrology of the reservoir, potential ramping flows due to power generation (rapid raising and lowering of flow stage), timing of releases with downstream unregulated tributaries and clear water discharge effects. Temperature and other water quality parameters may also need to be assessed.
6. *Roads.* If roads are located in the lower one-third of slope position on moderate to steep slopes (sub-surface flow interception). Road densities over 10% of watershed area of first- and second-order watersheds. Roads traversing highly dissected slopes or with multiple stream crossings. Drainageway crossings associated with floodplain fill blockages, and base-level changes above and/or below culverts and/or bridges.

The summary of the guidance criteria evaluation is entered in **Worksheet 3-1** by sub-watershed and reach location. Areas recommended for advancement to *RRISSC* are documented, along with those areas requiring no further assessment.

Step 11. Analyze channel processes

Stream channels convey both water and sediment from their watersheds. The channel morphology (form), stability and past land use history influence sediment supply and potential impairment to designated uses.

The user should review watershed maps and photographs to complete the guidance criteria in **Table 3-6**.

Table 3-6. Guidance criteria for advancement to the *RRISSC* assessment for channel processes.

Channel process guidance criteria for advancement to *RRISSC*
1. If there are potential increases in streamflow within the sub-watershed associated with A3–A6, C, D, E, F and/or G stream types.
2. If there appear to be stream types that are of the unstable form for a given valley type, i.e., G and F types in valley types II, IX, and X, then proceeding to the *RRISSC* assessment level is recommended. The observer is reminded to compare reference to existing conditions to determine if the existing stream type is appropriate for the valley type being studied. For example, if a D stream type was mapped in a valley type IX (glacial outwash valley), it would be indicative of the stable form for that valley type. However, if a D stream type was mapped in valley types II, IV, VI, VIII or X, it would not represent the typical stable form and should be flagged to require the *RRISSC* assessment.
3. If the current stream type departs from the stable form as indicated in the potential channel evolution or successional stage of channel adjustment relations (**Figure 2-38**), then proceed to the *RRISSC* assessment level (see also *Succession* background information in Chapter 2).

4. If aerial photographs or site visits reveal the following channel-destabilizing processes:
  a. aggradation (excess deposition, wide/shallow)
  b. degradation (incision, floodplain abandonment)
  c. lateral accretion (excess bank erosion)
  d. avulsion (abandonment of previous channels)
  e. enlargement
  f. meandering to braided channels

5. If time-trend aerial photography analysis indicates little recovery of apparent channel condition associated with the magnitude, extent and/or obvious consequence of channel change.
6. If road drainage, stream crossings and/or lack of floodplain drains (through-fill crossings) cause adverse channel adjustment.

The summary of the guidance criteria evaluation is entered in **Worksheet 3-1** by sub-watershed and reach location. Areas recommended for advancement to *RRISSC* are documented, along with those areas requiring no further assessment.

Step 12. Detect direct impacts to streambanks and channels

This step detects potentially adverse effects on stream channels due to obvious direct changes. These potential impacts can be observed from aerial photography, existing data (such as flood-control projects, land drainage/channelization records) and/or direct field observations. Straightening rivers, lining river channels, levees, lowering, dredging, emergency flood-relief excavation and other direct impacts are readily observable. Detailed documentation of such "river improvement works" often exists. Land uses that convert riparian vegetation from woody to grass/forb communities accelerate streambank erosion on C and E stream types. Heavy concentrations of livestock can also adversely affect streambank erosion and channel stability. For example, feedlots using live streams for watering would be flagged for potential adverse effects. Roads can have adverse effects on river stability and sediment supply by fill encroachment, realignment, reducing stream length, cutting off floodplains by through-fills and direct modifications of base level and grade by stream crossings. The user should review **Table 3-1** to complete the guidance criteria in **Table 3-7**.

Table 3-7. Guidance criteria for advancement to the *RRISSC* assessment due to direct channel impacts.

Direct channel impact guidance criteria for advancement to *RRISSC*
1. If the stream's dimension, pattern and profile have been altered due to direct impacts from various sources, then the influence of time of disturbance on channel recovery must be determined at a more advanced level of assessment.
2. If evidence exists of riparian vegetation alteration from woody plants to a grass/forb community or annuals.

The summary of the guidance criteria evaluation is entered in **Worksheet 3-1** by sub-watershed and reach location. Areas recommended for advancement to *RRISSC* are documented, along with those areas requiring no further assessment.

Step 13. Summarize problem verification: recognition of places, processes and sources

Steps 1–12 culminate in this step. Use **Worksheet 3-1** to locate areas at potential risk for instability and disproportionate sediment supply on a watershed map. The map organizes the *RLA* results by specific sub-watershed and river reach, documenting which sites are eliminated from further assessment and which sites advance to the *RRISSC* assessment level. Verification of initial problem identification can be accomplished using supplemental information, including past reports, timber harvest dates, interviews with landowners, flood records, road plans and older aerial photographs to observe evidence of changes over time. This step compiles the results needed for **Steps 14** and **15**.

Step 14. Eliminate areas, sub-watersheds and/or river reaches that are not contributors to either sediment or stability problems

There are many places and processes within a watershed that have a low risk of impact on sediment supply and channel stability, making it unnecessary to proceed. Users can safely disregard these sites in their subsequent assessments. **Tables 3-1** and **3-2** and summary **Worksheet 3-1** document the basis for excluding these areas from further assessment. It is also recommended to prepare a narrative summary explaining the reasons for exclusion.

Step 15. Select areas, sub-watersheds and river reaches for further assessment

It is important to recognize those areas that, based on the initial screening and the supplied guidance criteria, may potentially contribute to sediment and stability problems. These selected sites should be evaluated at the *RRISSC* level of assessment discussed in Chapter 4. **Tables 3-1** and **3-2** and summary **Worksheet 3-1** document the basis for selecting these areas for a more detailed assessment. It is also recommended to prepare a narrative summary explaining the reasons for pursuing additional evaluation at the *RRISSC* assessment level.

"The answers are often found by taking that extra step beyond one's immediate reach."

CHAPTER **4**

The Rapid Resource Inventory for Sediment and Stability Consequence (RRISSC)

Budget, time and scale considerations are realistic constraints in stability assessments. The greatest efforts should be directed toward the most critical areas. Sensitive landscapes, potentially unstable stream systems and sediment-generating land use activities need to be identified, prioritized and evaluated for potential impacts at a level of detail beyond their initial *RLA* analysis. The *Rapid Resource Inventory for Sediment and Stability Consequence* (*RRISSC*) provides this finer level of analysis. The time required for one person to assess a third-order watershed using *RRISSC* is one week (rather than days or a month), depending on availability of land use cover maps, aerial photographs, soil information and access to land use history.

Many land use activities occur on stable landscapes associated with sound land management practices. Using *RRISSC* to screen out such areas based on their limited risk of significant sediment yield reduces the level of effort needed during the final assessment phase, the *Prediction Level Assessment* (*PLA*). The risk rating system within the *RRISSC* phase of *WARSSS* requires analysis of the type and extent of land uses, the erosion potential of the landscape and channels and the relationship of potential sediment sources to hillslope, hydrologic and channel processes.

These rapid assessment methods are not designed to overlook problems but rather to isolate those land and stream systems with poor conditions and other variables that may be observed in a consistent, objective and, hopefully, reproducible manner. When in doubt, or in the event of high risks and high values, one should advance to the Prediction Level Assessment (PLA) to reduce uncertainty and to obtain sufficient detail to prescribe process-based, effective mitigation.

The *RRISSC* phase builds on the methods and results of the *RLA* phase in a number of ways. *RRISSC* begins with the subset of key areas (second- to third-order streams, sub-watersheds and specific landscapes) identified during *RLA* as potentially significant sediment sources. Many of the process relationships used in *WARSSS* are empirically derived from smaller watersheds and stream reaches. Using generalized process relations, *RRISSC* again reduces the number of these key areas that will advance to the next, most intensive *WARSSS* phase. Like *RLA*, *RRISSC* can be applied in a consistent manner over large areas in a relatively short time. The *RRISSC* assessment's more detailed screening requires an inventory of the type, location, nature and extent of land uses, as well as the sensitivity or erosion potential of the landscape and streams and the potential sediment sources. Again like *RLA*, *RRISSC* highlights hillslope, hydrologic and channel processes.

RRISSC assessment concept

Each of the hillslope, hydrologic and channel process assessments is used to create a summary risk rating by specific location. These ratings determine if a given sub-watershed and/or river reach is:

- "Tagged" for a further, more detailed assessment (*high* or *very high* risk);
- Scheduled to have a management action change (*moderate* risk);
- Planned for monitoring (*moderate* risk); or
- Excluded from further assessment (*very low*, *low* or *moderate* risk).

Site-specific mitigation can be prescribed at this level for *moderate* risk ratings, along with a monitoring plan.

It is not within the scope of this assessment methodology to provide detailed land use management prescriptions or site- and process-specific mitigation recommendations to reduce sediment and channel stability risk. The user must have the necessary experience and familiarity with a region to assign proper land use management practices and/or effective mitigation. *RRISSC*, however, is designed to assist the user in identifying multiple processes and associated potential impacts/risk ratings contributing to impaired waters. With such an assessment, the user not only can identify and specify land use management changes and mitigation, but can also document the reasons for such recommendations. The *high* and *very high* risk ratings also justify advancing areas to *PLA* to obtain more detailed information. The *PLA* analysis provides the foundation for site- and process-specific mitigation.

Sequential assessment steps

Flowchart 4-1 shows the procedural sequence of analysis used for the *RRISSC* assessment. The numerical sequence in the chart corresponds to the procedural steps described throughout this chapter.

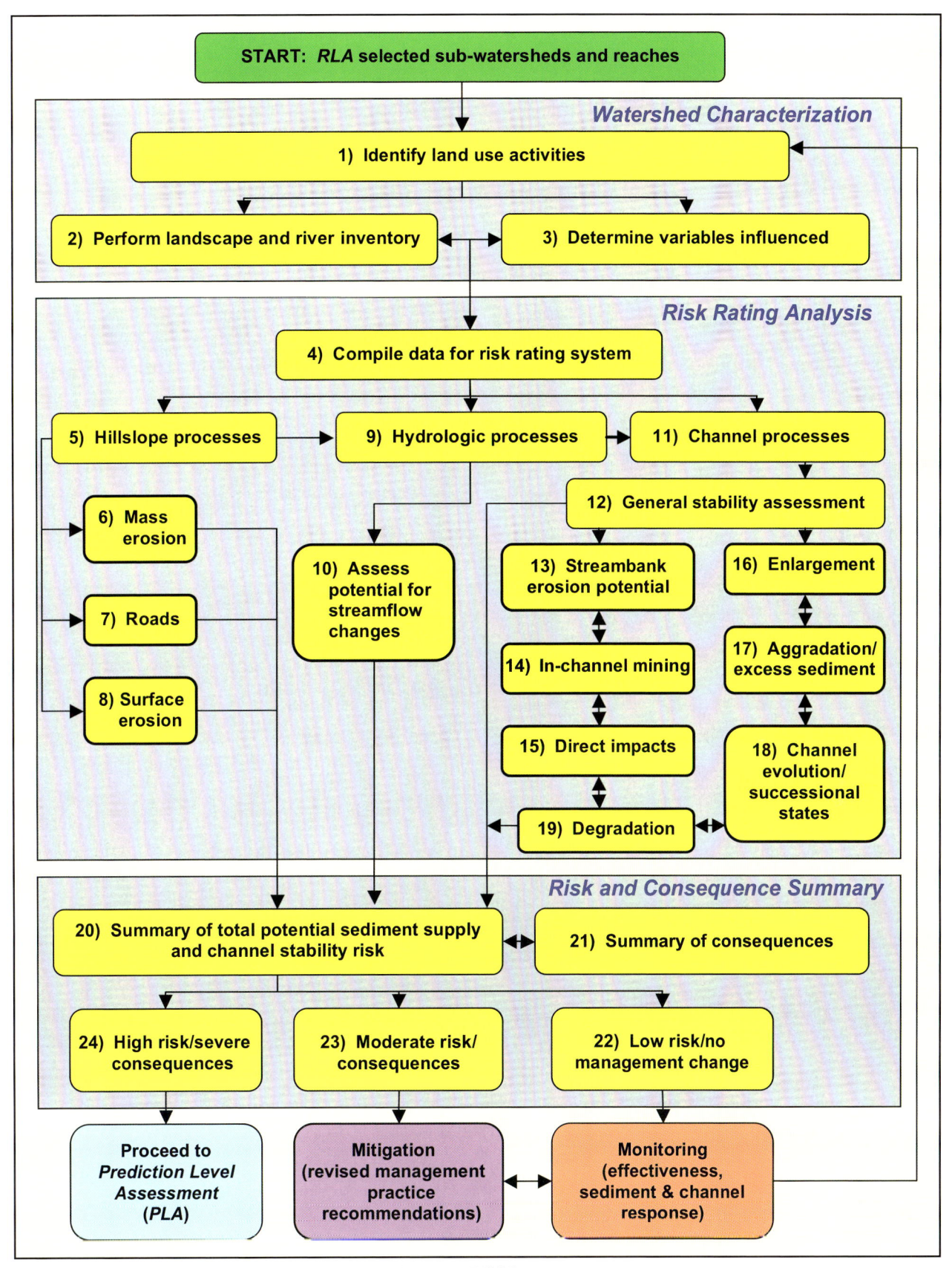

Chapter 4   The Rapid Resource Inventory for Sediment and Stability Consequence (RRISSC)

START: *RLA* selected sub-watersheds and reaches

Watershed Characterization
1) Identify land use activities
2) Perform landscape and river inventory
3) Determine variables influenced

Risk Rating Analysis
4) Compile data for risk rating system
5) Hillslope processes
6) Mass erosion
7) Roads
8) Surface erosion
9) Hydrologic processes
10) Assess potential for streamflow changes
11) Channel processes
12) General stability assessment
13) Streambank erosion potential
14) In-channel mining
15) Direct impacts
16) Enlargement
17) Aggradation/excess sediment
18) Channel evolution/successional states
19) Degradation

Risk and Consequence Summary
20) Summary of total potential sediment supply and channel stability risk
21) Summary of consequences
22) Low risk/no management change
23) Moderate risk/consequences
24) High risk/severe consequences

Proceed to *Prediction Level Assessment* (*PLA*)

Mitigation (revised management practice recommendations)

Monitoring (effectiveness, sediment & channel response)

Flowchart 4-1. Procedural sequence of analysis for the *RRISSC* assessment.

Step 1. Identify land use activities

Land use practices within the watershed area were identified in a general fashion in **Step 2** of the *RLA* phase. Land uses are also key ingredients for the *RRISSC* phase. *RRISSC* documents in greater detail the nature, extent, location and dates of land use practices within the sub-watersheds selected in *RLA*. This step begins with a detailed review of the guidance criteria summarized in **Worksheet 3-1**. The user should review the relevant *RLA* inventory information for links between the worksheet and the land use list in **Table 4-1**. This table identifies various land use activities and their potential influence on variables related to adverse water quality and channel change. The land use list is stratified by hillslope, hydrologic and channel processes, which enables an assessor with GIS access to make three thematic sub-analysis maps from what is usually one overall land use/land cover map. The land use information is then organized by major process type to conform to the process-based analysis in *RRISSC*. Unmapped information, such as historical land use patterns, may need to be considered separately from the mapped data or, where possible, added as attributes to mapped features.

The list included in **Table 4-1** specifies the inventory information necessary to assess risk. For example, a road inventory requires not only a map of road locations but also the number of stream crossings, acres of road (including cut bank, fill slope and road surface to calculate road density), road location in relation to slope position and the drainage area of second- to third-order sub-watersheds containing roads. This information is then used to determine risk of sediment delivery and road sediment source.

Table 4-1. A list of various land and river management activities and inventory information by hillslope, hydrologic and channel processes.

Land and river management activities and inventory information

Hillslope processes
 Roads
   1. Road density
   2. Location within the drainage network
   3. Number and type of stream crossings (culverts, bridges, fords)
   4. Type and quality of construction
   5. Road surfacing
   6. Type of drainage (culvert spacing, in-sloped, out-sloped, lined ditches)
   7. Drainage area of second- or third-order watershed containing roads
 	
 Surface disturbance
   1. Compaction potential
   2. Percent bare ground
   3. Vegetation composition
   4. Vegetation density change
   5. Location in regards to unstable soils
 	
 Mining
   1. Altered vegetation (composition, density)
   2. Altered surface and sub-surface soils
   3. Altered surface and sub-surface hydrology
   4. Spoil piles, tailings
   5. Altered channel dimension, pattern and profile

Table 4-1 - *Continued*. A list of various land and river management activities and inventory information by hillslope, hydrologic and channel processes.

Land and river management activities and inventory information - *Continued*

Silvicultural activities
 1. Silvicultural treatments (clearcutting, partial cutting)
 2. Log landings
 3. Log skidding
 4. Site preparation (scarification, burning)
 5. Fuel hazard reduction (burning, tractor bunching)
 6. Haul roads (density, location, stream crossings)

Agricultural applications
 1. Type of site preparation (plowing, no-till, summer fallow)
 2. Concentration of animals
 3. Vegetation type conversions

Urban development
 1. Percent impervious of land surface changed due to development
 2. Encroachment on drainageways/floodplains
 3. Alteration of the drainage network (reducing infiltration, percolation and use of vegetation for reducing and dispersing urban runoff)
 4. Stormwater management

Hydrologic processes
Roads
 1. Affecting drainage density
 2. Change in evapo-transpiration (excess runoff)
 3. Change in time of concentration of flow
 4. Increasing snowpack storage and water availability for runoff
 5. Interception of sub-surface water, increased time of concentration
 6. Drainageway crossings: provide direct sediment source from fill erosion and ditch-line source, as well as local channel instability

Surface disturbance
 1. Compaction, change in time of concentration of flow, reduced infiltration, potential flow increase
 2. Percent vegetative cover, type of vegetative cover and age class distribution affecting evapo-transpiration and water yield
 3. Wildfires, controlled burns

Reservoirs, hydropower dams, diversions, imported water
 1. Change in sediment regime (clear water discharge)
 2. Change in flow magnitude, duration and timing
 3. Change in water quality attributes affecting designated uses (water temperature)
 4. Ramping flows affecting streambank stability

Many hillslope processes, including urban development, mining, agricultural and silvicultural applications, influence hydrologic processes.

Table 4-1 - *Continued*. A list of various land and river management activities and inventory information by hillslope, hydrologic and channel processes.

Land and river management activities and inventory information - *Continued*

Channel processes

Roads
1. Encroachment on active channels and floodplains, and/or straightening of channels
2. Elimination of riparian buffers, and encroachment of road fills
3. Drainage density, increased sub-surface interception and surface runoff leading sediment directly into stream systems
4. Time of concentration of flows potentially increased (nature of drainage)
5. Source of direct sediment introduction
6. Number and type of drainageway crossings
7. Age of road

Surface and/or direct disturbance activities
1. Direct disturbance to stream channels by straightening, raising, lowering, lining and levee construction
2. Vegetation clearing and type conversion
3. Large woody debris clearing and tree removal
4. Dams for recreation, flood control and/or hydropower generation, in-channel structures (grade control, check dams)

Mining
1. Reworked channel materials
2. Altered channel dimension, pattern and profile
3. Vertical and lateral containment
4. Riparian vegetation change
5. Tailings; location and extent

Silvicultural activities
1. Timber harvest within riparian (stand removal/management)
2. Surface disturbance within riparian from skidding, site preparation or fuel hazard reduction
3. Roads or skid trails encroached on channel and multiple stream crossings

Agricultural applications
1. Dams and irrigation diversions
2. Poor grazing practices that change vegetation composition and density and add to hoof shear on streambanks
3. Direct disturbance to stream channels due to land drainage
4. Riparian vegetation change to cultivated crops
5. Feedlots and other animal concentration activities

Chapter 4   *The Rapid Resource Inventory for Sediment and Stability Consequence (RRISSC)*

Table 4-1 - *Continued*. A list of various land and river management activities and inventory information by hillslope, hydrologic and channel processes.

Land and river management activities and inventory information - *Continued*

Urban development
     1. Encroachment on the drainage network
     2. Elimination of floodplains
     3. Concentration of flows (storm drain outfalls leading to stream degradation)
     4. Direct alteration of channels (dimension, pattern, profile and materials)

Step 2. Perform landscape and river inventory

This step reviews and upgrades the existing map and database information needed for the upcoming *RRISSC* steps. The soil maps, aerial photographs, topographic maps, vegetation mapping and much of the inventory information gathered in **Step 1** of the *RLA* level can be used in the *RRISSC* phase. Locating existing inventory data reduces costs. A list of the information needed for *RRISSC* is included in **Table 4-2**. Much of the data needed for *RRISSC* is obtained from existing sources including:

- Previous inventories;
- Regional curves for hydrology;
- Reference reach data; and
- Records from federal, state, county and private land sources.

Obtaining road data from those who located, designed and constructed roads will speed up the data collection process. Recent aerial photographs are also extremely valuable. Vegetative changes for both hillslope and stream channels, direct disturbance indicators, road data and surficial geology and associated slope stability can be rapidly documented by a trained aerial photograph interpreter. Much of the data used in **Worksheet 3-1** can also be applied at this inventory step.

Bankfull discharge with field checks is best obtained from regional curves by drainage area for a representative hydro-physiographic province. If regional curves are not available, the procedures for development of such curves are discussed in Chapter 5.

Reference reach data at the *RRISSC* level is not as extensive as that required at the *PLA* level. The bankfull width/depth ratio data for representative stream types by valley type are used at this level. Level II stream classification often obtains reference reach data and is required for reaches of concern identified in *RLA*. Channel materials are measured using a pebble count frequency distribution method that validates interpretations from soils, geology and landforms made in *RLA*. Measurements to define width/depth ratio, entrenchment ratio, slope, sinuosity and channel materials are obtained at this level. Users may review classification basics in **Figures 2-13** and **2-14**, **Tables 2-2** and **2-3**, and Chapter 5 in Rosgen (1996). Use **Worksheet 4-1** for Level II stream classification.

Table 4-2. Basic minimum information needed for *RRISSC*.

Information needed for *RRISSC*

• Topographic maps (7.5' USGS quadrangle)
• Aerial photography
    - current (within 5 years)
    - past (50 years +)
• Soil survey map (hazards interpretations)
• Surficial geology map
• Summary of vegetative change
    - nature of change (estimated % change by major growth form)
    - extent (acres) and location
    - any known dates of change
    - riparian vegetation composition/density
• Level I valley types
• Level II stream types
• Roads (need road system map)
    - acres of road
    - number of stream crossings
    - location indicating slope position
    - acres of sub-watershed (second- and/or third-order) containing such roads
    - specific mitigation previously implemented (such as surfacing)
    - age of roads
    - class of road (main, secondary, unimproved skid roads, etc.)
    - vegetative cover (% bare soil) of cut banks, fill slopes
    - road and/or ditch-line surfacing
• Surface waters base map (National Hydrography Data: see http://www.epa.gov/waters/)
• Drainage area
• Regional hydrology curves
• Streambank heights
• Radius of curvature of river channel bends
• Bankfull discharge
• Bankfull width
• Bankfull depth
• Width/depth ratio for reference reach and existing condition
• Length of river with direct impacts
• Length of river with mining impacts
• Operational hydrology of reservoirs, diversions
• Percent impervious surfaces from urban development

Chapter 4 — The Rapid Resource Inventory for Sediment and Stability Consequence (RRISSC)

Worksheet 4-1. Field form for Level II stream classification (Rosgen, 1996; Rosgen and Silvey, 2005).

Stream Name:			
Basin Name:	Drainage Area:	acres	mi²
Location:			
Twp.&Rge:	Sec.&Qtr.:	Valley Type:	
Cross-Section Monuments (Lat./Long.):			
Observers:		Date:	

Bankfull WIDTH (W_{bkf})
WIDTH of the stream channel at bankfull stage elevation, in a riffle section. — ft

Bankfull DEPTH (d_{bkf})
Mean DEPTH of the stream channel cross-section, at bankfull stage elevation, in a riffle section ($d_{bkf} = A / W_{bkf}$). — ft

Bankfull X-Section AREA (A_{bkf})
AREA of the stream channel cross-section, at bankfull stage elevation, in a riffle section. — ft²

Width/Depth Ratio (W_{bkf} / d_{bkf})
Bankfull WIDTH divided by bankfull mean DEPTH, in a riffle section. — ft/ft

Maximum DEPTH (d_{mbkf})
Maximum depth of the bankfull channel cross-section, or distance between the bankfull stage and Thalweg elevations, in a riffle section. — ft

WIDTH of Flood-Prone Area (W_{fpa})
Twice maximum DEPTH, or ($2 \times d_{mbkf}$) = the stage/elevation at which flood-prone area WIDTH is determined in a riffle section. — ft

Entrenchment Ratio (ER)
The ratio of flood-prone area WIDTH divided by bankfull channel WIDTH (W_{fpa} / W_{bkf}) (riffle section). — ft/ft

Channel Materials (Particle Size Index) D_{50}
The D_{50} particle size index represents the mean diameter of channel materials, as sampled from the channel surface, between the bankfull stage and Thalweg elevations. — mm

Water Surface SLOPE (S)
Channel slope = "rise over run" for a reach approximately 20–30 bankfull channel widths in length, with the "riffle-to-riffle" water surface slope representing the gradient at bankfull stage. — ft/ft

Channel SINUOSITY (k)
Sinuosity is an index of channel pattern, determined from a ratio of stream length divided by valley length (SL / VL); or estimated from a ratio of valley slope divided by channel slope (VS / S).

Stream Type ⟵ (See **Figure 2-14**)

4-9

Step 3. Determine variables influenced

A list of land use activities and their influences on process variables and potential consequences, along with the corresponding *RRISSC* prediction methods, appears in **Table 4-3**. The user should review the relevant *RLA* inventory information for links among **Tables 3-1**, **3-2**, **4-1** and **4-3** for variables and specific land uses. For example, a reservoir located on an unstable G4 (entrenched) stream type that released increased streamflows would help to explain channel enlargement and degradation below the dam. Some of the variables influenced by this land use are magnitude, duration and timing of streamflow, clear water discharge, increased stream power, degradation and energy slope. This combination of reservoir use and G4 stream type represents a high risk of both excess sediment supply and channel instability. Specific mitigation effects on the operational hydrology of the dam would require a more detailed *PLA* assessment. If the reservoir mentioned above was on a B2 stream type, however, the response to the change in flow and sediment would be entirely different and would require very little mitigation for channel stability.

Chapter 4 *The Rapid Resource Inventory for Sediment and Stability Consequence (RRISSC)*

Table 4-3. Relationship among land uses/activities, process influences, consequences and assessment methods.

Potential change from land uses/activities	Processes influenced	Potential consequences	RRISSC prediction method
Streamflow decrease in magnitude, duration and altered timing due to reservoirs or diversions	Shear stress ↓ Stream power ↓ Sediment transport competency and capacity ↓	Excess sediment deposition Aggradation Accelerated bank erosion Widening channel Successional state	Worksheet 4-11 Worksheet 4-11 Worksheet 4-7 Worksheet 4-10 Table 4-5
Streamflow discharge increase due to high % impervious and storm water drains from urban development. Clear water discharge "ramping flows" from reservoir releases	Shear stress ↑ Stream power ↑ Sediment transport capacity ↑	Degradation Channel enlargement Bank erosion Channel succession shift Increased sediment load (supply)	Worksheet 4-12 Worksheet 4-10 Worksheet 4-7 Table 4-5 Worksheet 4-11
Streamflow increase from vegetative alteration, clearcutting, land clearing and roads	Shear stress ↑ Stream power ↑ Magnitude of flow ↑ Duration of flows ↑	Channel enlargement Bank erosion Degradation Channel succession shift Increased sediment load (supply) Surface erosion	Worksheet 4-10 Worksheet 4-7 Worksheet 4-12 Table 4-5 Worksheet 4-11 Worksheet 4-5
Riparian vegetation alteration (% of channel length by stream type)	Bank erodibility ↑ Sediment transport capacity ↓ Stream power ↓ Shear stress ↓	Bank erosion Aggradation Enlargement Channel succession shift	Worksheet 4-7 Worksheet 4-11 Worksheet 4-10 Table 4-5
Surface disturbances (% of ground cover) and roads	Surface runoff ↑ Sub-surface flow interception (roads) ↑ Deposition ↑ Sediment transport capacity (aggradation) ↓ Excess scour (degradation) ↑	Surface erosion delivered to stream Road source sediment Gully erosion Aggradation Degradation Streambank erosion	Worksheet 4-5 Worksheet 4-4 Worksheets 4-7, 9, 10, 12 Worksheet 4-11 Worksheet 4-12 Worksheet 4-7
Water yield – harvest and roads – add to soil water influencing slope stability	Surface/sub-surface hydrology ↑ Soil saturation ↑ Internal strength by roots ↓ Slope equilibrium ↓	Mass erosion: - slump earthflow ↑ - debris torrent ↑ - sediment supply delivered to channel ↑ Aggradation ↑ Channel succession shift Enlargement ↑ Surface erosion ↑	Table 4-4 Worksheet 4-3 Worksheet 4-11 Table 4-5 Worksheet 4-10 Worksheet 4-5

Direct channel impacts
  Channelization
  Levees
  Straightening
  Dredging
	Shear stress ↑↓
	Stream power ↑↓
	Width ↑
	Confinement ↑
	Incision ↑
		Gully erosion ↑
		Bank erosion ↑
		Channel enlargement ↑
		Degradation ↑
		Aggradation ↑
		Channel succession shift
			Worksheets 4-7, 9, 10, 12
			Worksheet 4-7
			Worksheet 4-10
			Worksheet 4-12
			Worksheet 4-11
			Table 4-5

Channel clearing, cleaning, grubbing, large woody debris removal	Stream power ↑ Shear stress ↑ Sediment transport capacity ↓ Competence ↑ Degradation ↑ Energy dissipation ↓	Sediment deposition ↑ Degradation ↑ Bank erosion ↑ Channel enlargement ↑ Sediment supply ↑ Aggradation ↑	Worksheet 4-11 Worksheet 4-12 Worksheet 4-7 Worksheet 4-10 Worksheet 4-11 Worksheet 4-11

Note: Potential consequences column is directly related to *RRISSC* prediction method column; for example, potential excess sediment deposition is assessed in **Worksheet 4-11**.

4-11

Step 4. Compile data for risk rating system

The risk rating system evaluates the key variables for hillslope, hydrologic and channel processes representing erosional/depositional characteristics that are influenced by land uses. The potential to contribute to disproportionate sediment supply and/or stream channel instability is evaluated using a series of relations developed for *RRISSC*. The following steps of the risk rating system identify low-, moderate- and high-risk conditions that are additionally assessed to determine areas that warrant further analysis. At this step, the soil and geology maps, vegetative cover changes, aerial photographs, topographic maps and inventory list in **Table 4-2** are compiled.

During subsequent risk rating steps (**Steps 6–19**), the user obtains overall risk ratings sorted by hillslope, hydrologic and channel processes for various land uses. These overall risk ratings for multiple sub-watersheds and river reaches (i.e., entire watershed results) are recorded in the *RRISSC* summary **Worksheet 4-2** after each risk rating step. In the summary worksheet, overall risk ratings for mass erosion, roads, surface erosion and streamflow changes are recorded by geographic land location within the watershed; e.g., location 1, location 2, etc. The remaining variables are recorded by stream type reach; e.g., E3, C4, etc. See Chapter 7 *RRISSC* for an example of a completed worksheet.

Chapter 4   The Rapid Resource Inventory for Sediment and Stability Consequence (RRISSC)

Worksheet 4-2. *RRISSC* summary worksheet for multiple sites/river reaches within a study watershed. Insert both the adjective and numeric overall risk rating.

Watershed Name:					Date:		Observer(s):						
Location code/ river reach I.D.	**Geographic Location**					**Stream Type Location**						Processes identified by step for advancement to *PLA*	✓ Location selected for advancement to *PLA*
	Step 6: Mass erosion (Worksheet 4-3)	Step 7: Roads (Worksheet 4-4)	Step 8: Surface erosion (Worksheet 4-5)	Step 10: Streamflow change (Worksheet 4-6)	Step 13: Streambank erosion (Worksheet 4-7)	Step 14: In-channel mining (Worksheet 4-8)	Step 15: Direct channel impacts (Worksheet 4-9)	Step 16: Channel enlargement (Worksheet 4-10)	Step 17: Aggradation/ excess sediment (Worksheet 4-11)	Step 18: Channel evolution/ successional states (Table 4-5)	Step 19: Degradation (Worksheet 4-12)		
1.													
2.													
3.													
4.													
5.													
6.													
7.													
8.													
9.													
10.													
11.													
12.													
13.													
14.													
15.													

4-13

Hillslope processes: Steps 5–8

The *RRISSC* analysis of hillslope processes is oriented toward land uses that influence sediment supply. The extent of certain land use activities is analyzed and linked to erosional risk categories. These land uses are overlain on soil/geology hazard maps to identify potential erodibility risks, revealing specific locations with potentially accelerated erosion and increased sediment levels. Sediment delivery potential will only be quantified in the upcoming *PLA* phase for those sites that have a *high* to *very high* risk of erosion. *Moderate*-risk areas may need mitigation or stabilization to reduce erosion/sediment levels when associated with a relatively low consequence of impact. These *moderate*-risk and low-consequence areas probably do not warrant the level of detail required in *PLA*. If serious questions arise, or if extremely high resource value is involved and/or serious potential consequences may result, then the analysis should advance to the highly detailed *PLA* phase. The *PLA* analysis allows for the design of process-specific mitigation measures backed by quantitative field data rather than the application of generic or "laundry list" measures that may or may not be effective.

Step 5. Select appropriate models and compile data for hillslope processes

In this step, land uses, including road systems, are overlain on soil/geology hazard maps to identify potential erodibility risks. Overall sediment delivery potential risk ratings for areas identified in *RLA* are obtained for mass erosion, roads and surface erosion in **Steps 6**, **7** and **8**. Sediment delivery potential is further evaluated for only those sites that have a *high* or *very high* risk rating of mass and/or surface erosion, as determined in **Steps 6** and **8**. *Very low*- and *low*-risk areas will be excluded from further analysis. *Moderate*-risk areas may need specific mitigation or stabilization recommendations and may be adequately addressed at the completion of the *RRISSC* phase. Areas in which serious risks arise, high value water resources are involved and/or borderline results are accompanied by substantial uncertainty should advance to the more detailed (*PLA*) assessment level discussed in Chapter 5.

Note: Sub-watersheds that did not meet any guidance criteria in RLA for mass erosion or surface erosion do not require a RRISSC assessment in Steps 6 or 8, and are automatically given a risk rating of "Very Low (1)" to be inserted in the RRISSC summary worksheet; these ratings will be used in subsequent risk rating steps. Also, all roads identified in RLA (Table 3-5, criterion 6) are evaluated in the RRISSC assessment level. Roads are also evaluated if shown to have direct impact as indicated in Table 3-1 in RLA, requiring a road impact risk calculation in Step 7.

Step 6. Assess mass erosion risk

The processes associated with mass erosion include two primary types: 1) shallow, fast movements of debris avalanche/debris torrents and mudflows that generally move in response to a major precipitation event, and 2) slow, deep-seated slump/earthflow erosional processes that move intermittently over a wide range of time scales in response to infrequent events and/or disturbance factors. The evidence of recent mass erosion can be observed on aerial photographs; frequency and magnitude of the slides can be observed over time-sequential aerial photos.

Examples of debris avalanche/debris torrent slides are shown in Chapter 2 in **Figure 2-2** (in the Idaho batholith) and **Figure 2-3**, associated with an A3a+ stream type (in Colorado). Slump/earthflow (slow deep-seated mass erosion) examples are shown in **Figure 2-4** (Willow

Creek, Colorado) and in the Marine shale geology in **Figure 2-5** (Blue River, Colorado). Sediment delivery is often computed by measuring the concave slope remnant, less that portion of the slide mass removed by fluvial entrainment. Prediction of annual rates of sediment associated with mass erosion processes is extremely difficult due to the episodic nature of events that initiate movement and time-lapse releases. The impact on stream channels and associated beneficial uses by landslides triggered by roads or other imposed land use activities may be of more importance to channel stability assessment than the long-term contribution to annual sediment yield.

Current models do not accurately predict annual sediment yield from mass erosion. Any assessment of these erosional processes must rely primarily on an experienced individual who can recognize the relative stability or instability of an area using soil and geology maps, aerial photographs, vegetation indicators and field observations. GPS, digital terrain models and other appropriate tools can speed up the mapping process of these features. An overlay of existing and proposed road systems over the landslide risk maps provides valuable warning indicators of past and/or impending potential high risk for failure. The user should select tested methods for the identification and prediction of specific landslide processes for the appropriate geologic type of the assessment site and also should review the soil mass movement chapter (Chapter V) in WRENSS (USEPA, 1980).

The general guidelines to identify high risk of debris avalanche/debris torrent or slump/earthflow processes are shown in **Table 4-4**. The characteristics of landform features, soils, rock type, vegetative cover and precipitation associated with high risk of debris avalanche/debris torrent and slump/earthflow processes are used to identify mass erosion potential.

Note: The user must be familiar with mass erosion processes and must be able to use photographic interpretations to identify high-risk features to properly conduct this step.

Table 4-4. General guidelines for broad-level high risk of mass erosion potential (debris avalanche/debris torrent and slump/earthflow) (USEPA, 1980).

Guideline for various slopes	Debris avalanche/ debris torrent	Slump/earthflow
Landform features • slope gradient • slope shape	**Steep >34°** • concave • continuous/uniform	**2° → steep** • hummocky • discontinuous/irregular
Soil characteristics • depth • type	**Shallow** • residual or colluvium • glacial till deposition	**Deep** • deeply weathered • fine-grained • presence of mica • high clay content (pyroclastic soils)
Hydrologic characteristics of site • concentration of ground water	• linear depressions parallel to the slope	• saturated depressions • bowls • springs • "elk wallows" (undermined by road at toe of slide activity)
Rock type/geology	• hard, resistant volcanic rock • granites • diorites • alpine glaciation • highly weathered volcanic rock	• volcanic ash • breccias • silt sandstones • mudstones • highly weathered rock • attitude (dip of beds, parallel to slope)
Vegetative cover	• rooting depth impacts	• indication of clearcuts/roads increase risk • old age vs. new stands (indication of activity) • leaning trees

Precipitation	• varies
	  - snowmelt
	  - convectional storms
		• high intensity
		• short-duration storms

Risk rating: If none of the characteristics from **Table 4-4** are met, the risk for mass erosion is low and the user can proceed to **Step 7**. "Low (2)" can be recorded in the summary **Worksheet 4-2**. Any surface disturbance activities associated with high risk for mass erosion (slump/earthflow) or debris avalanche/debris torrent that influence vegetative cover, flow interception and routing, such as stand conversion, roads, skidding and yarding, land clearing and related activities, are to be considered a high risk for erosion potential. **Worksheet 4-3** is used to record data and to determine the overall risk rating for mass erosion using **Figures 4-1** and **4-2**. The data needed to complete this step includes:

- Slope gradient;
- Slope shape; and
- Slope position.

Sediment delivery potential is determined using general relations of slope gradient, shape irregularity and slope position of landslide activity. These relations determine the likelihood and subsequent risk of converting erosion to sediment. Sediment delivery potential (converting erosional debris to sediment in a conveyance channel) is very important for this analysis as the combination of a high risk for erosion, coupled with a high risk for sediment delivery, requires the most detailed investigation and quantitative prediction of sediment yield at the *PLA* level of *WARSSS*.

Watershed Assessment of River Stability and Sediment Supply

Worksheet 4-3. Risk rating worksheet for mass erosion sediment delivery.

(1) Sub-watershed location (I.D.)	(2) Slope gradient (degrees)	(3) Slope shape (discontinuous or continuous)	(4) Risk rating: slope gradient by slope shape (**Figure 4-1**)	(5) Slope position (lower 1/4, mid to lower 1/4, mid to upper 1/4, upper 1/4 or stream adjacent)	(6) Risk rating: slope position (**Figure 4-2**)	(7) Total risk rating points by sub-watershed Σ[(4)+(6)]	(8) Overall mass erosion risk rating (use column (7) points; insert adjective and numerical risk rating) VL(1) = 2–3 L(2) = 3–4 M(3) = 5–6 H(4) = 7–8 VH(5) = 9–10
1.							
2.							
3.							
4.							
5.							
6.							
7.							
8.							
9.							
10.							
11.							
12.							
13.							
14.							
15.							

| Chapter 4 | *The Rapid Resource Inventory for Sediment and Stability Consequence (RRISSC)* |

Sediment delivery potential is first assessed using the relationships among sediment delivery risk, slope gradient and slope shape, as shown in **Figure 4-1**. Use this figure to determine the risk rating and insert the adjective and numeric risk rating in Column (4) in **Worksheet 4-3**.

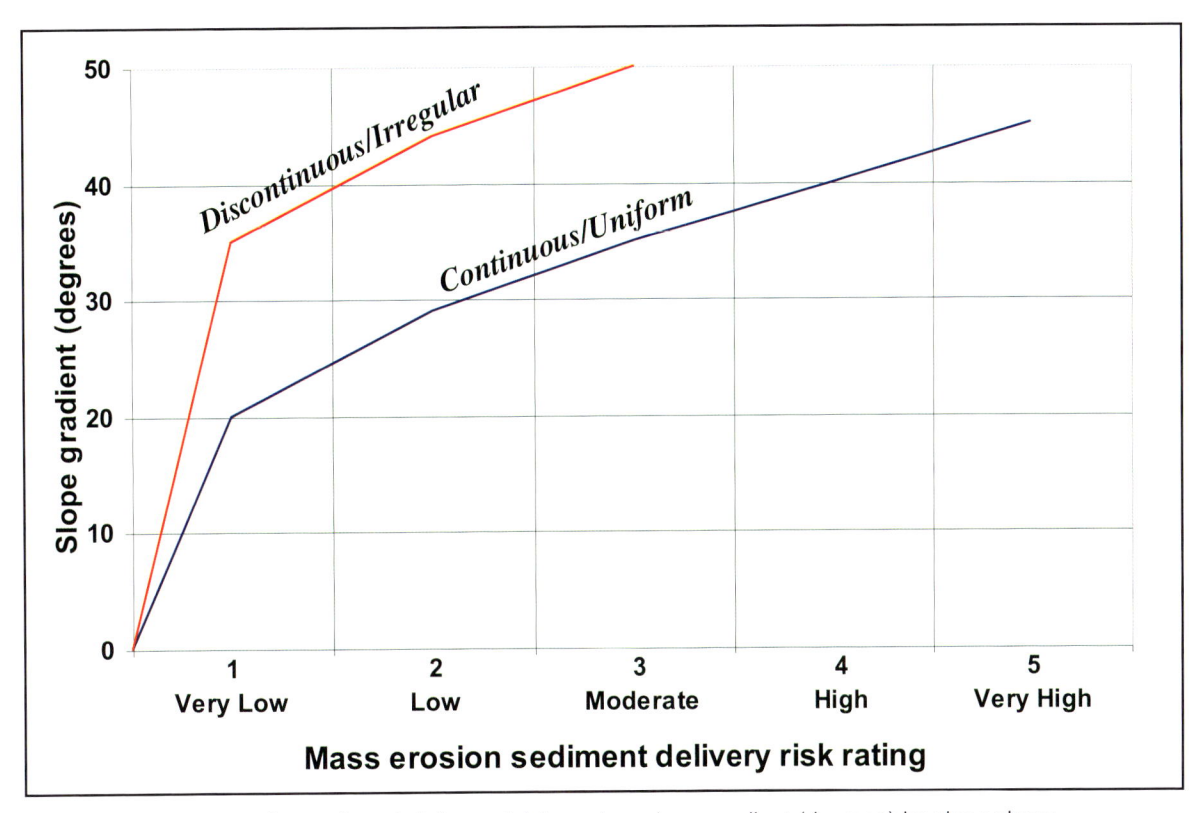

Figure 4-1. Mass erosion sediment delivery risk based on slope gradient (degrees) by slope shape.

Figure 4-2 shows the relationship of delivery potential and slope position and is also used to determine a numerical risk rating. Insert the numeric risk rating in Column (6) in **Worksheet 4-3**.

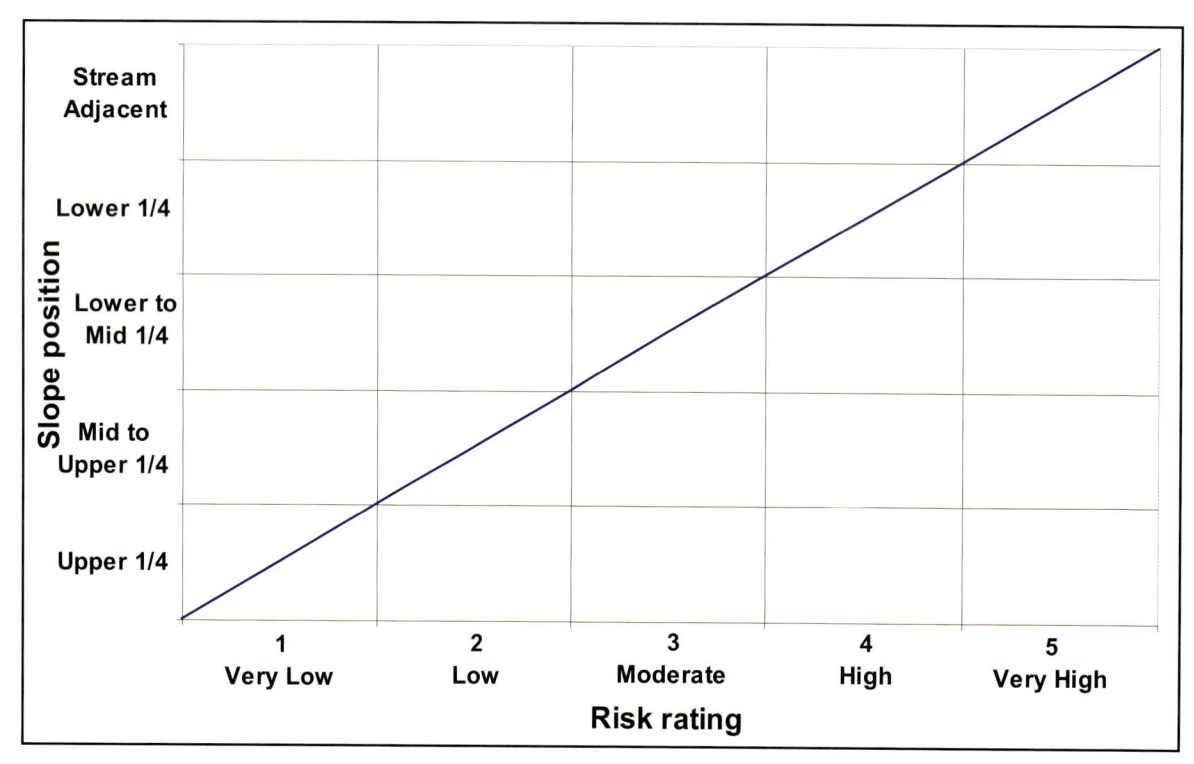

Figure 4-2. Mass erosion sediment delivery risk based on slope position.

The overall risk rating for mass erosion can now be calculated using **Worksheet 4-3**. This risk rating is recorded in the *RRISSC* summary **Worksheet 4-2** for each location assessed.

Land use activities, such as roads, clearcuts, land clearing, control burns, etc., that occur on *high-* or *very high*-risk landscapes associated with these erosional processes and have a high sediment delivery potential will likely require a *PLA* analysis. Similar activities on *moderate*-risk landscapes normally require mitigation and/or management recommendations to reduce potential adverse impacts.

Step 7. Evaluate road impact risk

Roads can directly introduce sediment from exposed cut banks, road fills, road surfaces and ditch-line erosion. Roads can also increase available water for runoff by sub-surface interception, decreased evapo-transpiration and increased snowpack storage of water equivalent in the road prism. Changes in time of flow concentration can occur, which increase drainage density. These flow changes can also increase flow-related channel erosion. Direct disturbance to streams at road crossings and contraction scour below culverts and bridges often affect local base level, upstream aggradation and downstream degradation. Floodplain blockages are common on stream crossings, reducing floodplain function and increasing the risk of channel instability and adverse changes in sediment transport. This assessment step involves an inventory of roads that requires the following information:

- Acres of sub-watershed;
- Acres of surface disturbance of roads including road surface, cut, fill and ditch line;
- Number of stream crossings;
- Slope position;
- Slope of road;
- Age of road;
- Mitigation such as road surfacing, ditch-line surfacing, etc.;
- Vegetative cover of cut banks and road fills; and
- Presence of unstable terrain associated with mass erosion processes.

Information by first- and second-order sub-watersheds is entered in **Worksheet 4-4** to determine the overall risk rating of potential sediment delivery from roads. This overall risk rating is calculated by determining individual risk ratings for:

- Road impact on sediment delivery by slope position (**Figure 4-3**);
- Potential sediment delivery from distance of road fill to channel (**Figure 4-4**); and
- Slope of road surface (**Figure 4-5**).

The risk for sediment supply and sediment delivery potential using the road impact index and slope position is determined from **Figure 4-3**. Insert the numeric risk rating in Column (7) in **Worksheet 4-4**.

WATERSHED ASSESSMENT OF RIVER STABILITY AND SEDIMENT SUPPLY

Worksheet 4-4. Risk rating worksheet for potential sediment delivery from roads.

(1)	(2)	(3)	(4)	(5)	(6)	(7)	(8)	(9)	(10)	(11)	(12)	(13)	(14)			(15)	(16)	(17)
Location of sub-watershed (I.D.)	Acres of sub-watershed (200–5000 acres)	Acres disturbance of road (include cut bank, fill slope, road surface)	Number of stream crossings	Calculate road impact index [(3)/(2)X(4)] *If crossings = 0, multiply by 1.	Slope position (lower or mid-upper)	Risk rating: road impact index (5) by slope position (Fig. 4-3)	Distance of road fill to stream (ft)	Risk rating: distance of road fill to stream (ft) (Fig. 4-4)	Slope of road (%)	Risk rating: slope of road (%) (Fig. 4-5)	Total individual risk rating points Σ[(7)+(9)+(11)]	Overall risk rating for potential sediment from roads (Fig. 4-6)	Adjustments for construction, design and age of road			Risk rating adjustments for mass erosion potential slump/ earthflow*** (Table 4-4, Figures 4-1, 4-2)	Debris torrent/ avalanche: If erosion risk and sediment delivery potential is High, raise final road risk rating to Very High (Table 4-4, Figures 4-1, 4-2)	Final risk rating of potential sediment from roads
													Age of road: If > 7 yrs and sediment delivery potential = Low, reduce one risk category*	Road surfacing: If gravel/ asphalt, then reduce one risk category**	Ditch line: If surfacing out-sloped, reduce one risk category	Vegetative condition of cut banks, road fills: If > 50% ground cover, reduce one risk category		
1.																		
2.																		
3.																		
4.																		
5.																		
6.																		
7.																		
8.																		
9.																		
10.																		
11.																		
12.																		
13.																		
14.																		
15.																		

*Unless: road has not recovered; poor maintenance; poor vegetative cover on cut bank and fill slopes - ditch line is still leading water into stream.

**Unless: road cut bank, fills and ditch line continue to provide sediment source to stream.

***If risk is *high* for potential sediment delivery of mass erosion (Worksheet 4-3), then adjust overall risk up one category.

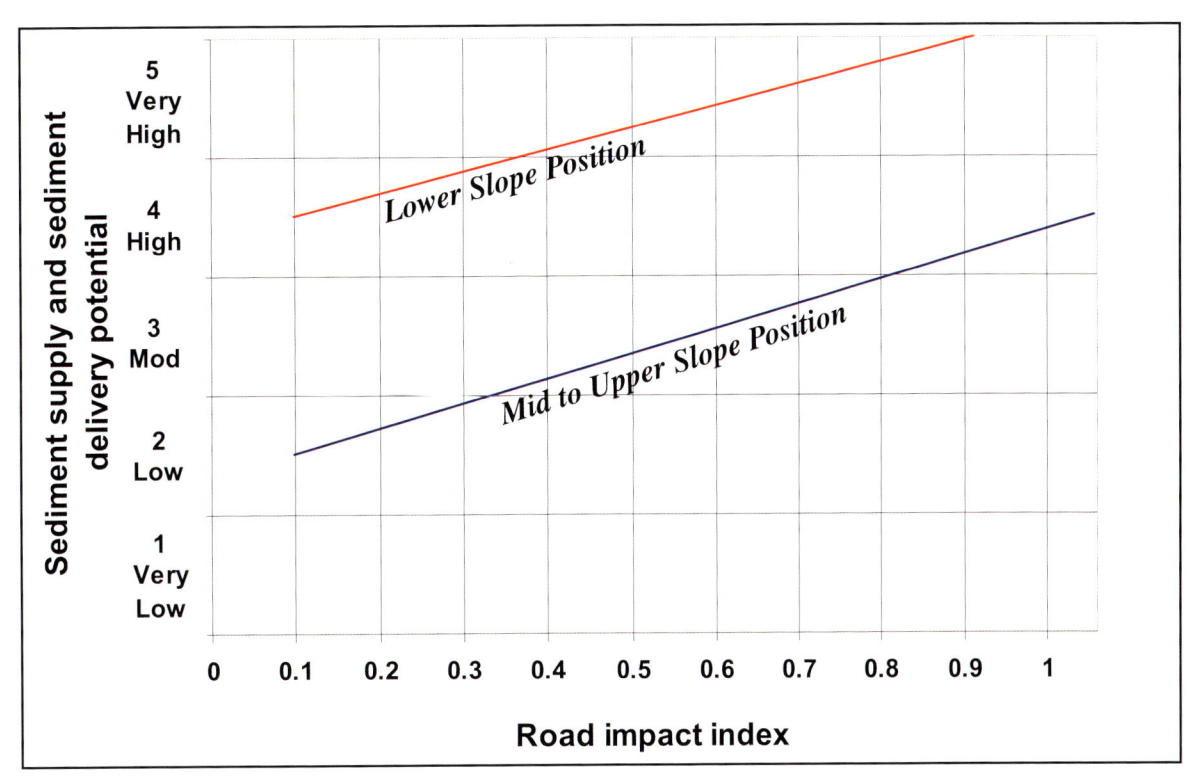

Chapter 4   *The Rapid Resource Inventory for Sediment and Stability Consequence (RRISSC)*

Figure 4-3. Road sediment delivery risk based on road impact index by slope position. Figure modified from Rosgen (2001b) based on measured delivered road sediment to debris basins in Horse Creek Watershed, Idaho and Fool Creek, Colorado using experimental watershed data from USDA Forest Service.

Additional risk is estimated based on the distance of the road fill from the stream using **Figure 4-4**. Insert the numeric risk rating in Column (9) in **Worksheet 4-4**.

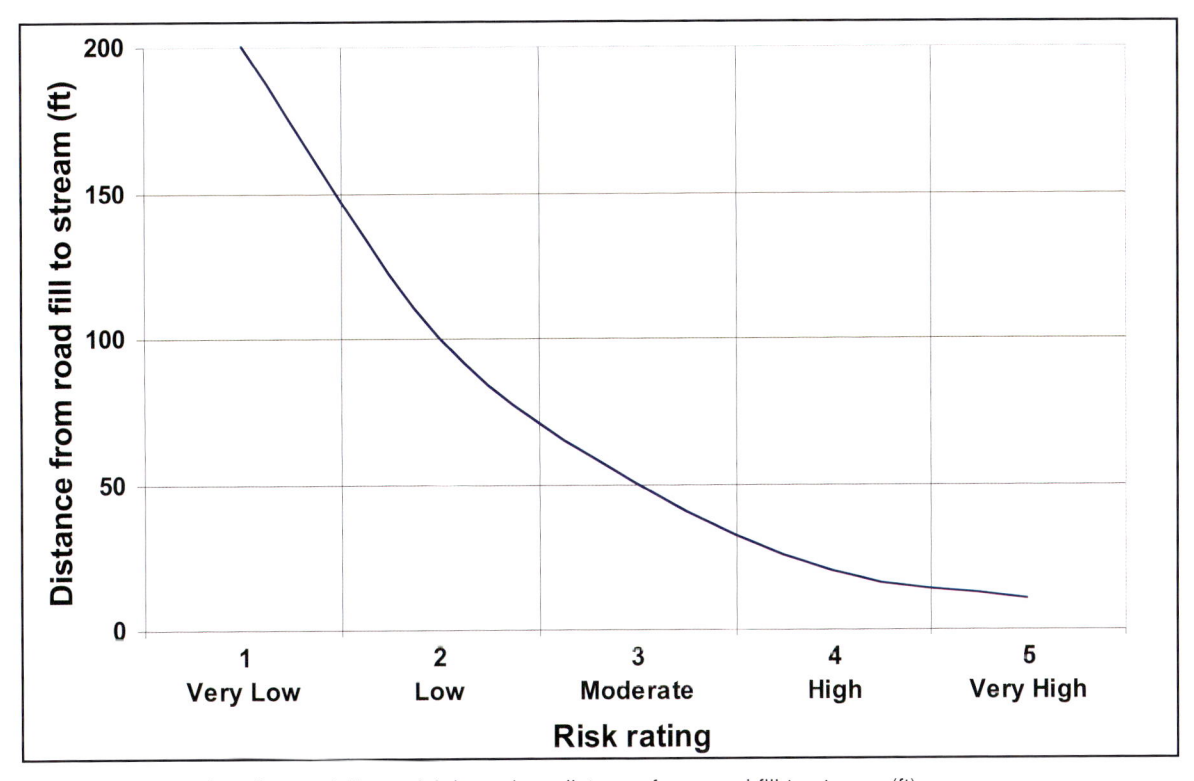

Figure 4-4. Road sediment delivery risk based on distance from road fill to stream (ft).

4-23

The risk of delivered sediment from roads is also determined by the gradient of the road surface using **Figure 4-5**. Insert the numeric risk rating in Column (11) in **Worksheet 4-4**.

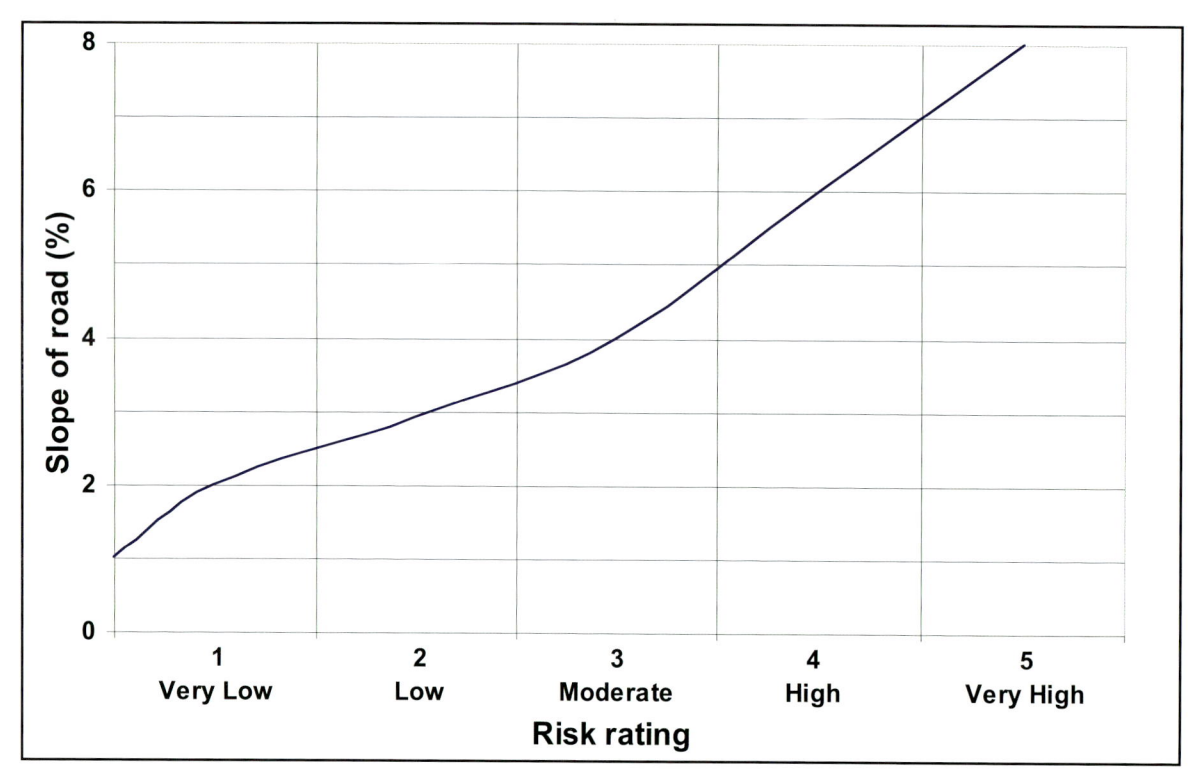

Figure 4-5. Road sediment delivery risk based on slope of road (%) (curve derived from data from Reid and Dunne, 1984).

Chapter 4 *The Rapid Resource Inventory for Sediment and Stability Consequence (RRISSC)*

The overall risk rating from roads is calculated by taking the sum of individual road risk ratings (Column (12) in **Worksheet 4-4**) and using **Figure 4-6**. Record the adjective and numeric risk ratings in Column (13) in **Worksheet 4-4**.

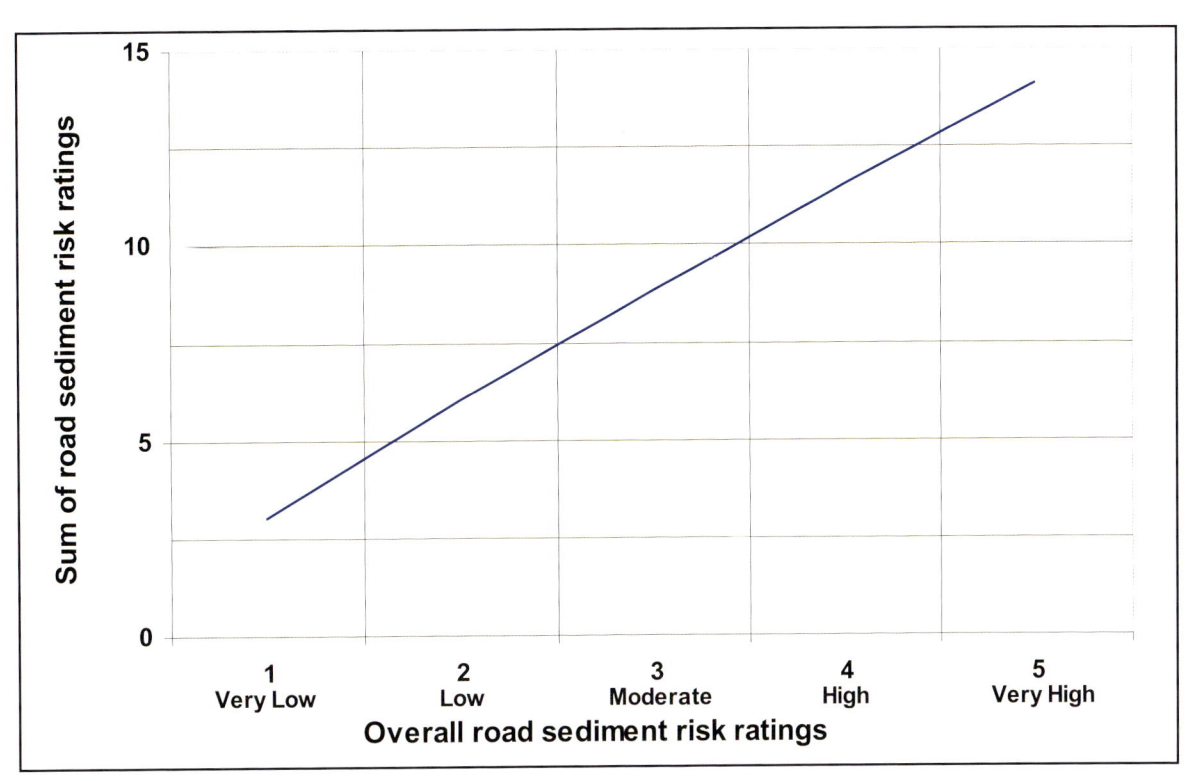

Figure 4-6. Overall road sediment delivery risk based on the sum of individual sediment risk ratings.

After determining the overall risk rating for potential sediment from roads, use **Worksheet 4-4** to adjust this risk rating if necessary. For example, if the road is older than seven years and if the sediment delivery potential is low, decrease the overall risk by one full category as shown in Column (14) in **Worksheet 4-4**. The final risk rating of potential sediment from roads is entered in the *RRISSC* summary **Worksheet 4-2**.

A *moderate* risk rating for a sub-watershed indicates a disproportionate source of sediment or channel stability problems associated with a particular road system and justifies site-specific mitigation. Mitigation would typically be recommended to repair, stabilize, revegetate, change drainageway crossings, out-slope or even "hydrologically" close certain roads. Sub-watersheds that have *high* or *very high* risk ratings require analysis at the *PLA* level and eventually will need site-specific and process-specific mitigation.

Step 8. Determine surface erosion risk

Accelerated erosion may occur from surface erosion processes due to exposed bare soil, compaction and poor conservation practices (activities associated with agriculture, surface mining, land clearing and silviculture). Surface erosion that occurs on steep, highly dissected slopes has a higher potential for conversion to sediment delivery due to rill and gully erosion. **Flowchart 4-2** depicts the detailed assessment methodology for this step.

Areas subject to rill erosion and Hortonian overland flow are mapped for further analysis. The summary of the surface erosion assessment is shown in **Worksheet 4-5**. Finally, candidate areas for mitigation, Best Management Practices (BMPs) and/or areas needing a more detailed assessment at the *PLA* level are mapped and targeted.

The information needed for this risk assessment includes the following:

- Total acres being evaluated for a sub-watershed and soil type erodibility potential;
- Acres impacted;
- Percent bare ground of impacted acres;
- Drainage density of impacted slope or width of interfluve spacing;
- Slope position and gradient (identify if highly dissected slope);
- Distance of disturbance to nearest stream; and
- Buffer width of riparian corridor.

The data entry and individual risk ratings are recorded and used to determine the overall surface erosion/sediment delivery risk rating in **Worksheet 4-5**. First, a risk rating for surface erosion potential is calculated using **Worksheet 4-5** and **Figure 4-7** to determine the percent of acres impacted with more than 50% bare ground. If the risk rating for surface erosion potential is *very low, low* or *moderate*, then that rating is the overall risk rating for surface erosion and the rating can be entered in the summary **Worksheet 4-2**; the user can then proceed to **Step 9**.

If a sub-watershed has a *high* or *very high* risk rating for surface erosion potential, then a sediment delivery risk rating needs to be calculated. The overall sediment delivery risk rating is determined from individual risk ratings for:

- Drainage density (**Figure 4-8**);
- Slope position (**Figure 4-9**);
- Percent ground cover (**Figure 4-10**);
- Distance from disturbance to stream (**Figure 4-11**); and
- Stream buffer (**Figure 4-12**).

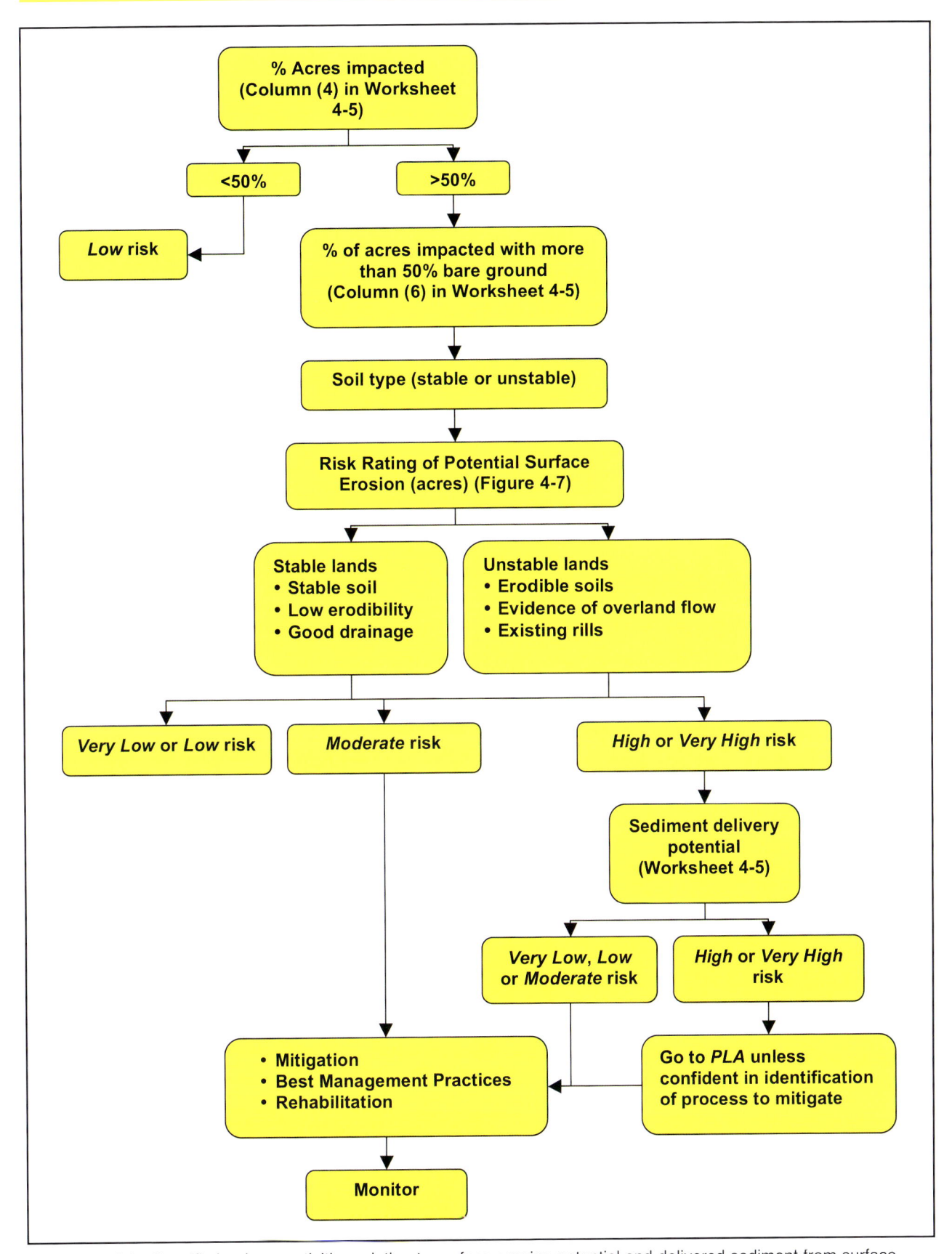

Chapter 4 — *The Rapid Resource Inventory for Sediment and Stability Consequence (RRISSC)*

Flowchart 4-2. Specific land use activities relating to surface erosion potential and delivered sediment from surface disturbance.

```
% Acres impacted (Column (4) in Worksheet 4-5)
├── <50% → Low risk
└── >50% → % of acres impacted with more than 50% bare ground (Column (6) in Worksheet 4-5)
            ↓
            Soil type (stable or unstable)
            ↓
            Risk Rating of Potential Surface Erosion (acres) (Figure 4-7)
            ├── Stable lands
            │    • Stable soil
            │    • Low erodibility
            │    • Good drainage
            │    ├── Very Low or Low risk
            │    └── Moderate risk → Mitigation / Best Management Practices / Rehabilitation → Monitor
            └── Unstable lands
                 • Erodible soils
                 • Evidence of overland flow
                 • Existing rills
                 └── High or Very High risk
                      ↓
                      Sediment delivery potential (Worksheet 4-5)
                      ├── Very Low, Low or Moderate risk → Mitigation / Best Management Practices / Rehabilitation → Monitor
                      └── High or Very High risk → Go to PLA unless confident in identification of process to mitigate
```

4-27

Watershed Assessment of River Stability and Sediment Supply

Worksheet 4-5. Risk rating worksheet for surface erosion and sediment delivery potential.

(1)	(2)	(3)	(4)	(5)	(6)	(7)	(8)	(9)	(10)	(11)	(12)	(13)	(14)	(15)	(16)
	Surface Erosion Potential							**Sediment Delivery Potential**						
								Continue only if rating in Column (8) is *High* or *Very High*						
								Converted ratios or conditions for numerical risk ratings of sediment delivery potential						
Location of sub-watershed (I.D.)	Total acres of sub-watershed	Acres impacted*	% acres impacted [(3)/(2)X100]	Acres impacted (3) with more than 50% bare ground	% of acres impacted with more than 50% bare ground [(5)/(3)X100]	Landscape type (stable or unstable)	Overall risk rating: surface erosion (**Fig. 4-7**)	Risk rating: drainage density by slope gradient (%) (**Fig. 4-8**)	Risk rating: slope position (**Fig. 4-9**)	Risk rating: % ground cover (**Fig. 4-10**)	Risk rating: distance of disturbance to stream (ft) (**Fig. 4-11**)	Risk rating: stream buffer (ft) (**Fig. 4-12**)	Total individual risk rating points ∑[(9) through (13)]	Overall risk rating: sediment delivery potential: use (14) points (**Fig. 4-13**)	% of sub-watershed with *H* or *VH* erosion potential, and with *H* or *VH* sediment delivery potential (see map)
1.															
2.															
3.															
4.															
5.															
6.															
7.															
8.															
9.															
10.															
11.															
12.															
13.															
14.															
15.															

*Do not include road acres.

Chapter 4 *The Rapid Resource Inventory for Sediment and Stability Consequence (RRISSC)*

The basic relationship among bare ground, soil type and risk rating is shown in **Figure 4-7**. Insert both the adjective and numeric risk rating in Column (8) in **Worksheet 4-5**.

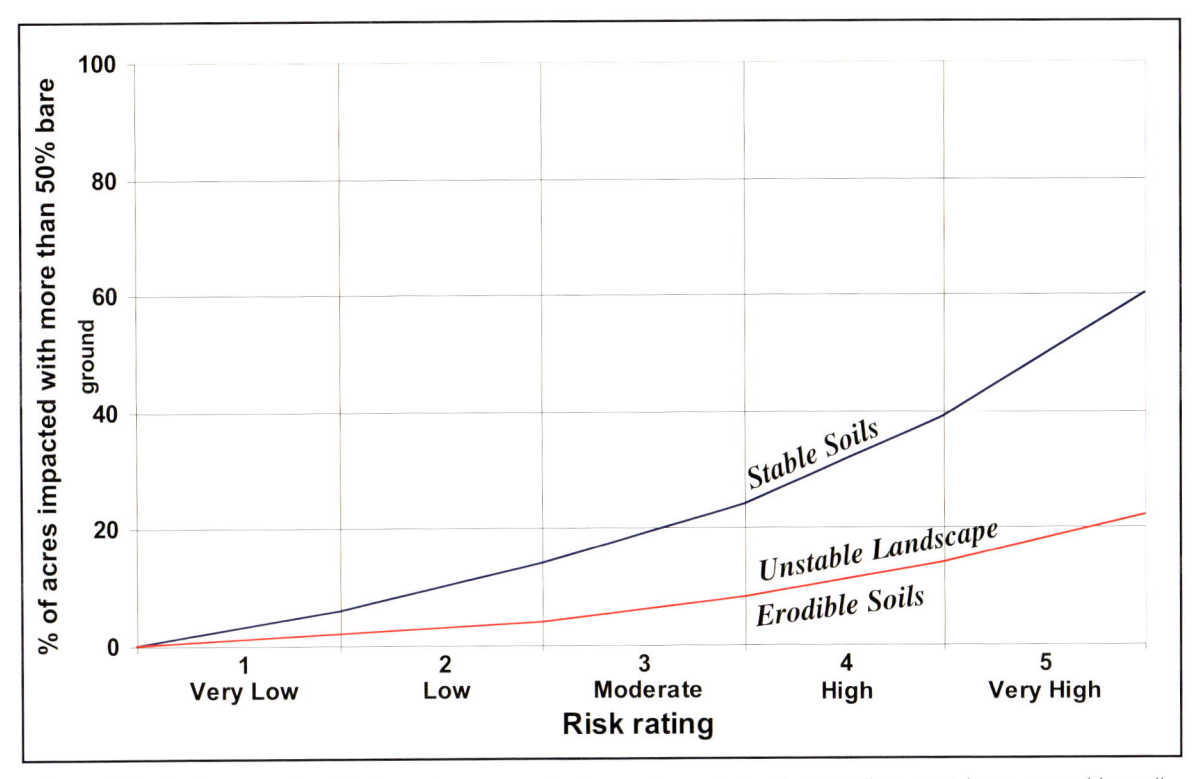

Figure 4-7. Surface erosion risk based on percent of acres impacted with more than 50% bare ground by soil type.

If the risk rating for percent area disturbed with greater than 50% bare soil exposed is *very low, low* or *moderate*, then sediment delivery potential does not need to be calculated. The risk rating can be entered in the summary **Worksheet 4-2** and the user can proceed to **Step 9**.

If the risk rating is *high* or *very high*, then proceed with **Step 8** using **Worksheet 4-5** and **Figures 4-8** through **4-13** to calculate sediment delivery potential.

Sediment delivery potential risk ratings are obtained through the relationship between drainage density and slope gradient, using **Figure 4-8**. Insert the numeric risk rating in Column (9) in **Worksheet 4-5**.

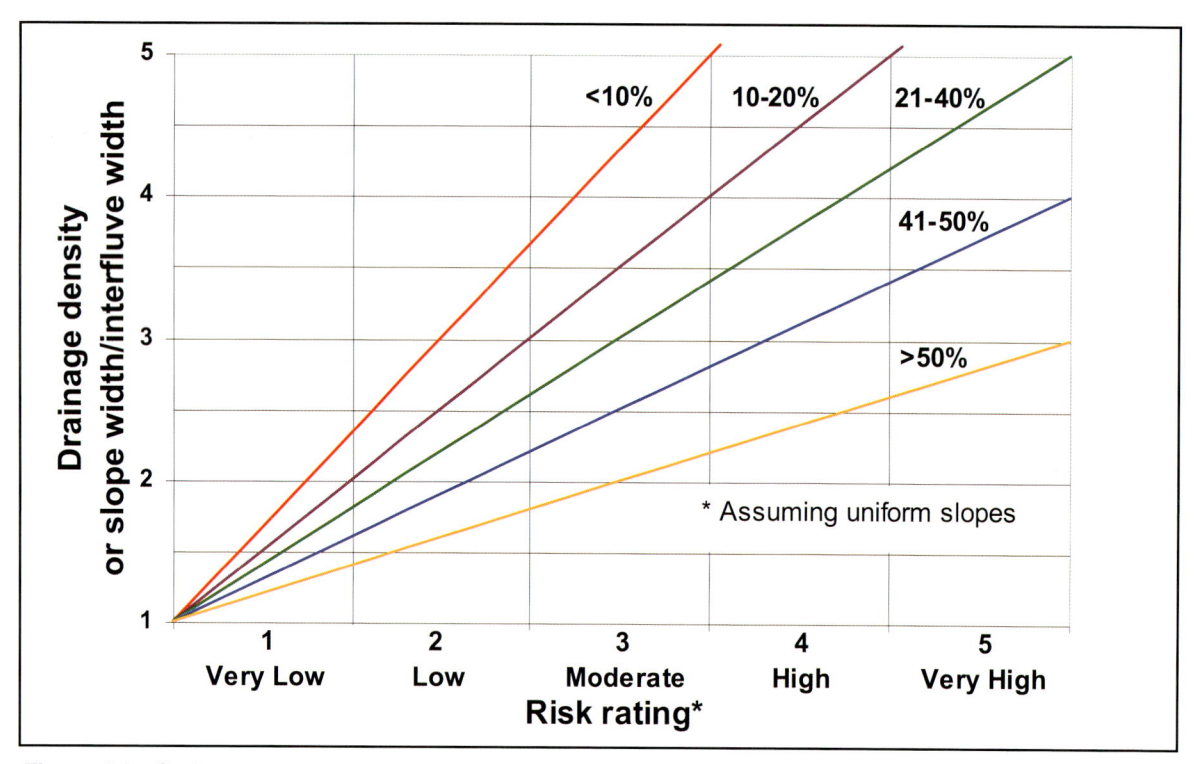

Figure 4-8. Surface erosion sediment delivery risk based on drainage density by slope gradient (%).

The relationship between slope position and sediment delivery potential is shown in **Figure 4-9**. Insert the numeric risk rating in Column (10) in **Worksheet 4-5**.

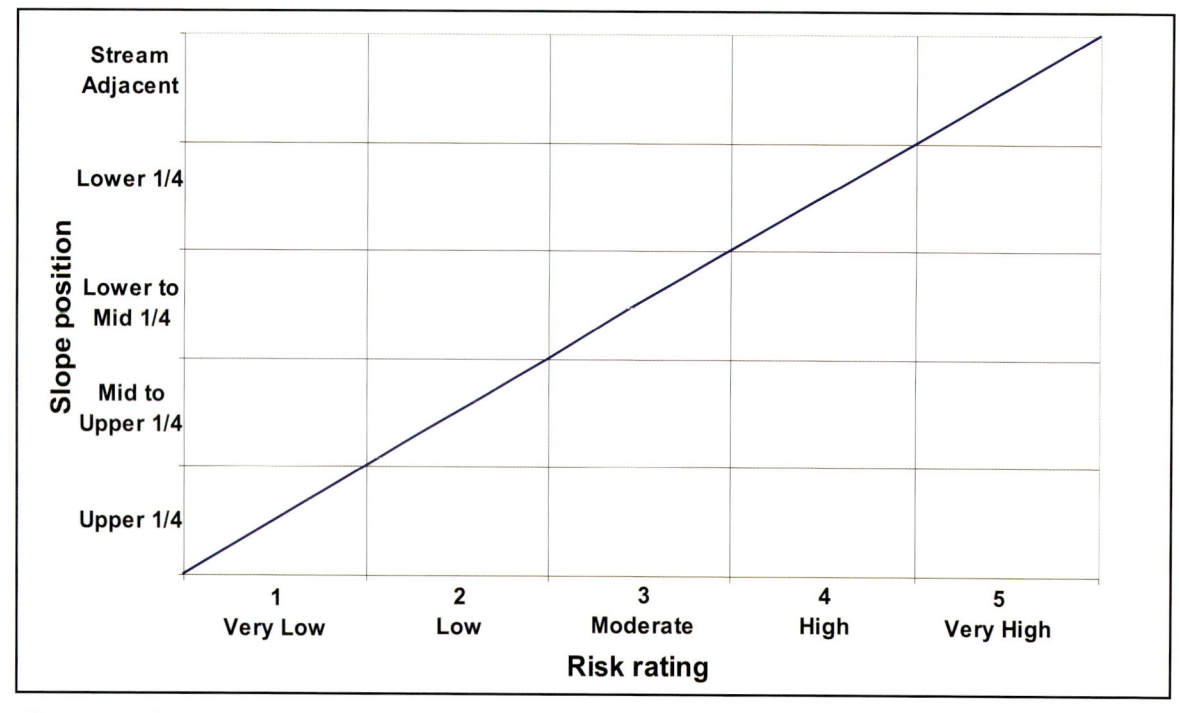

Figure 4-9. Surface erosion sediment delivery risk based on slope position.

4-30

Chapter 4 — The Rapid Resource Inventory for Sediment and Stability Consequence (RRISSC)

The effect of ground cover on sediment delivery risk is determined according to the relation shown in **Figure 4-10**. Insert the numeric risk rating in Column (11) in **Worksheet 4-5**.

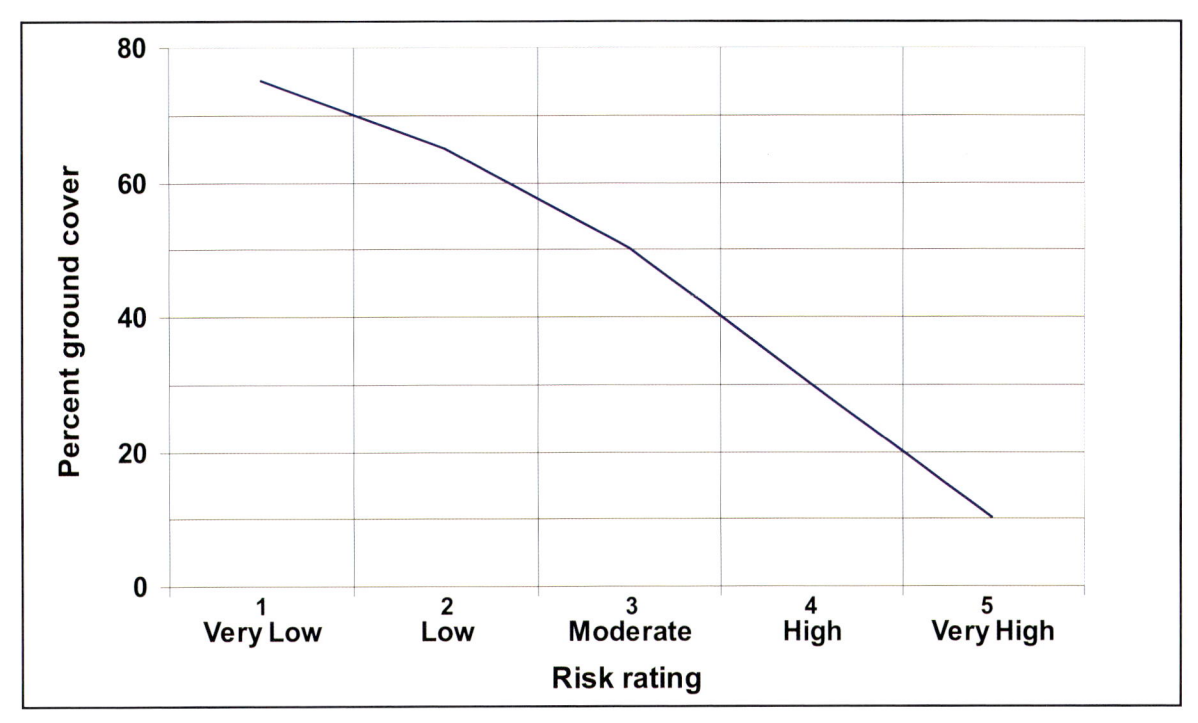

Figure 4-10. Surface erosion sediment delivery risk based on percent ground cover.

Sediment delivery risk is also affected by the distance of a disturbance from the stream, as shown in **Figure 4-11**. Insert the numeric risk rating in Column (12) in **Worksheet 4-5**.

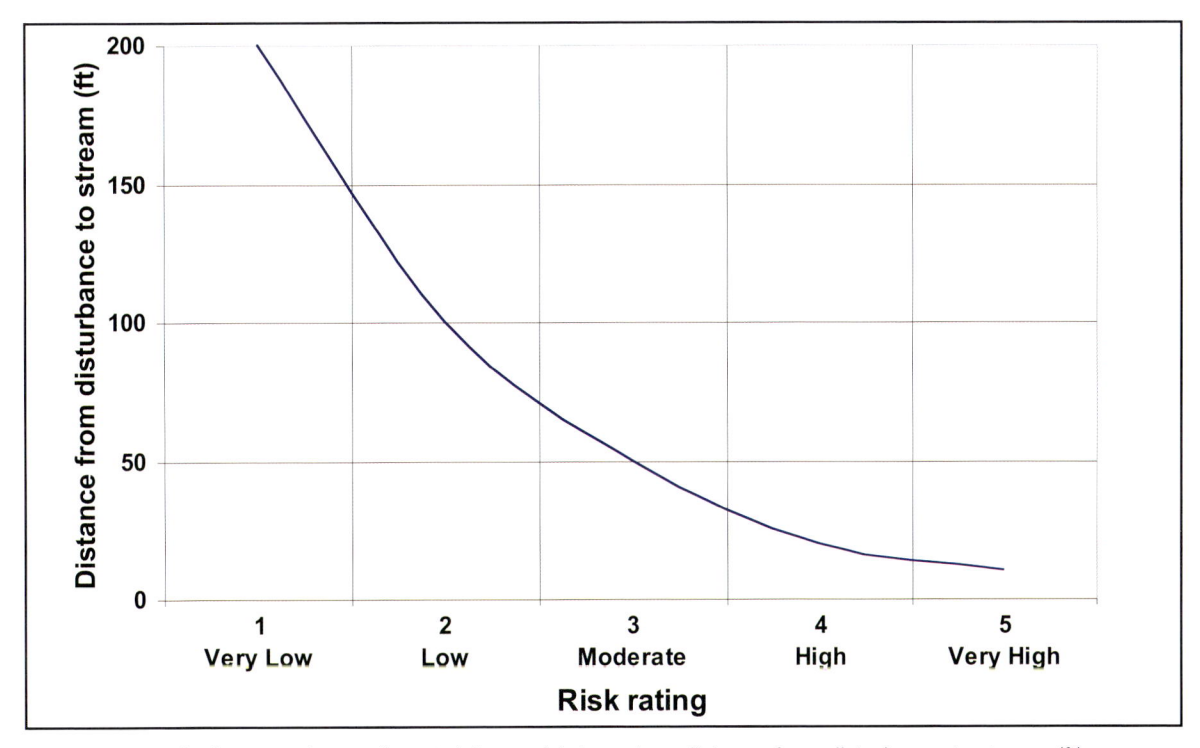

Figure 4-11. Surface erosion sediment delivery risk based on distance from disturbance to stream (ft).

Stream buffer widths from disturbance sites are also used to determine risk of sediment delivery with risk ratings derived from the relation shown in **Figure 4-12**. Insert the numeric risk rating in Column (13) in **Worksheet 4-5**.

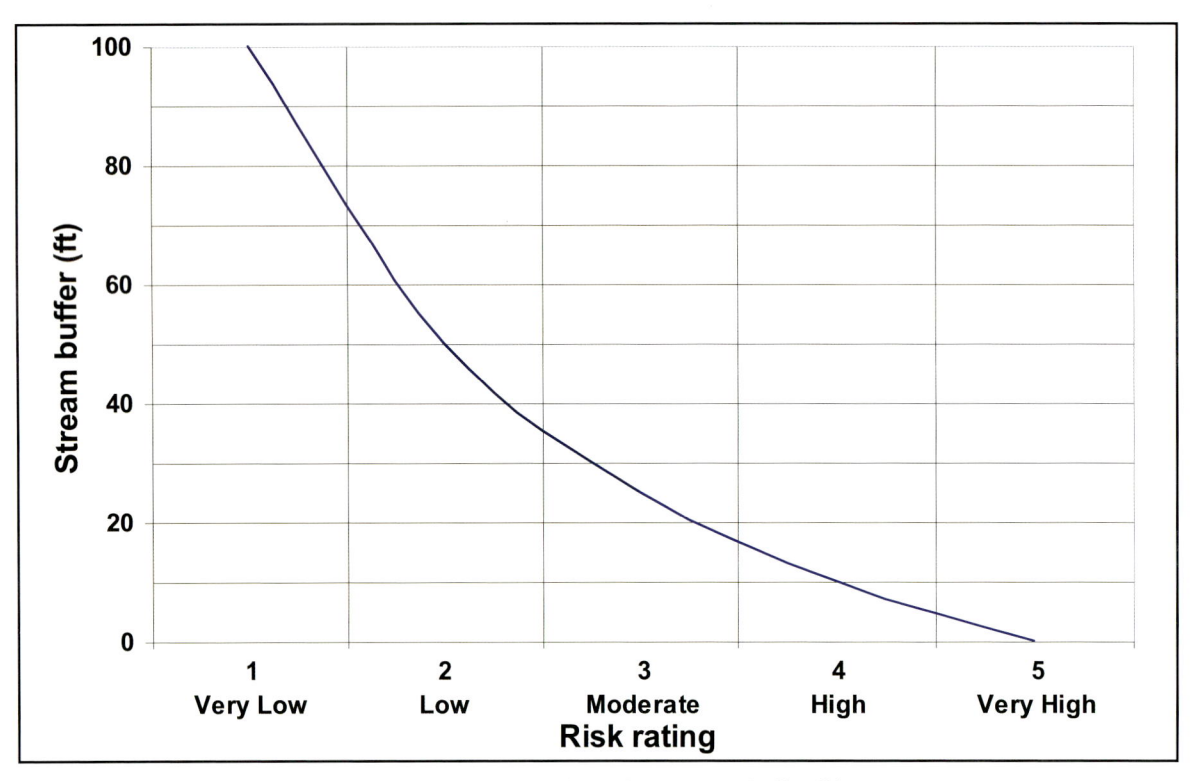

Figure 4-12. Surface erosion sediment delivery risk based on stream buffer (ft).

Figure 4-13 is used in conjunction with **Worksheet 4-5** to determine the overall surface erosion risk rating. Insert the adjective and numeric risk ratings in Column (15) in **Worksheet 4-5**.

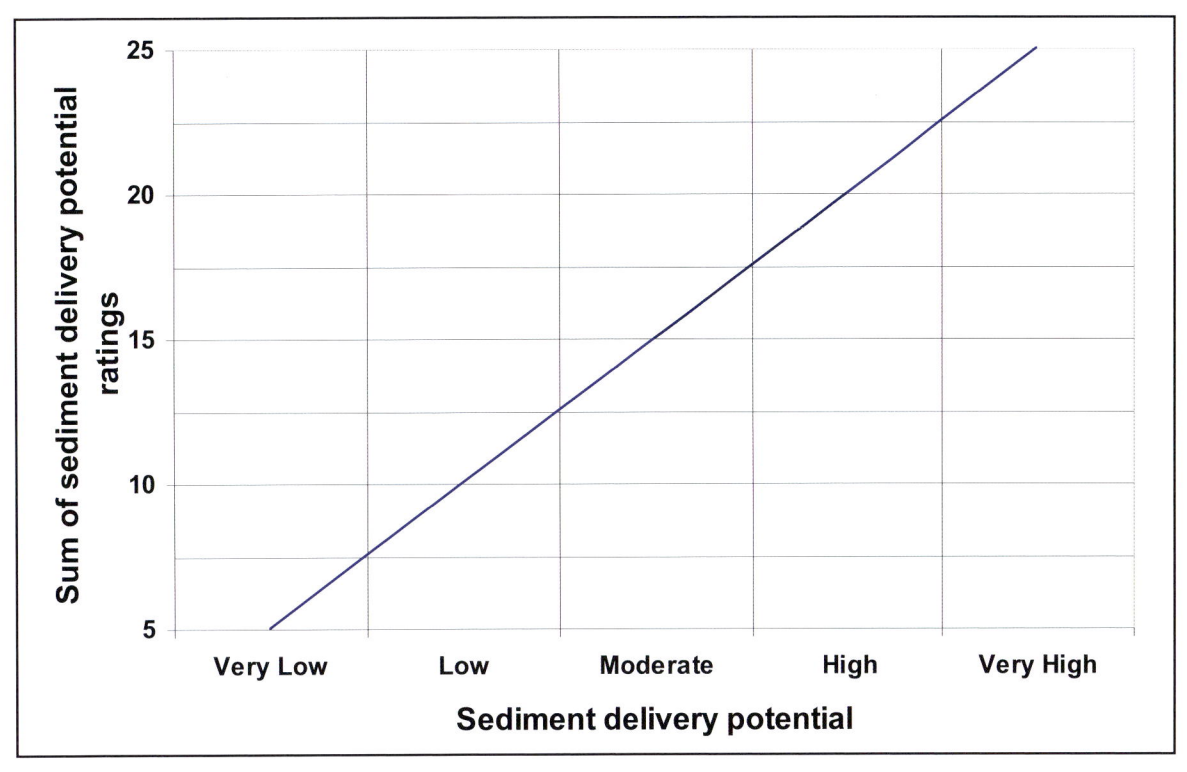

Figure 4-13. Overall sediment delivery risk based on the sum of individual sediment delivery risk ratings.

The overall surface erosion risk rating is entered in the *RRISSC* summary **Worksheet 4-2**.

Moderate risk ratings for surface erosion processes can be lowered for specific land uses with site-specific mitigation. Locations with *high* to *very high* surface erosion risk ratings should advance to the *PLA* phase.

Hydrologic processes: Steps 9–10

RRISSC evaluates the potential for increased water yield and associated flow-related sediment increases. Rural watersheds are assessed in terms of percentage of the watershed in a modified vegetative condition (i.e., clearcuts or non-forested land cover types), while urban watersheds are evaluated mainly for percentage of impervious areas. The higher the altered percentage, the higher the potential for increased flow-related sediment supply due to channel enlargement and incision. Magnitude, duration and timing of flow are also important, including both natural and human-altered patterns such as operational hydrology of reservoirs and diversions. In addition, risk ratings of increased flow are affected by stream types, which are grouped into risk categories according to the degree of susceptibility for increased sediment due to flow increases.

Step 9. Compile maps and data for hydrologic processes

Extensive changes in vegetative cover, reservoirs, diversions, return flows from irrigation and inter- and intra-basin water transfers can change river stability and sediment supply. Channel and sediment processes are sensitive to potential variations in magnitude (too much or too little), timing and duration of flows. The likelihood and risk of hydrologic change is evaluated in **Step 10**.

Note: Sub-watersheds that did not meet any guidance criteria in RLA for streamflow changes do not require a RRISSC assessment in Step 10 and are automatically given a risk rating value of "Very Low (1)" to be inserted in the RRISSC summary worksheet for further assessment summaries.

Often, mitigation assigned to streambank stabilization results in "patching the symptoms" rather than dealing with the cause of the destabilized streambanks. Not all streams respond the same way to changes in flow; thus, it is critical to map the stream types below areas of flow change to evaluate the effects of potential changes. For example, bedrock and boulder-bed channel types are not susceptible to base-level changes due to flow. These would be considered low risk, as there is a low probability that a flow change would lead to adverse channel adjustment and/or sediment problems. If the streams were entrenched (vertically contained) types, or incised in erodible bed and bank material, then the stream response and corresponding risk rating would be different.

It is imperative to obtain the operational hydrology of reservoirs, diversions and/or hydropower facilities. This data should include the duration of seasonal and diurnal flows. In the following steps, it is important to determine changes in timing, magnitude and duration of regulated flows. This data is necessary to assess potential channel impacts and unregulated tributary impacts.

Chapter 4   The Rapid Resource Inventory for Sediment and Stability Consequence (RRISSC)

Step 10. Assess potential for streamflow changes

The following information is needed for this step:

- Stream types (Level II);
- Vegetative modification history of the sub-watershed, percent of watershed altered;
- Operational hydrology of reservoirs, diversions and inter- and intra-basin water transfers;
- "Ramping flows" fom hydro-electric projects;
- Presence of significant agricultural return flows;
- Road densities on steep/dissected slopes (potential for converting sub-surface to surface runoff, and change in time of concentration of flow);
- Wildfire locations and history; and
- Percent of the urban watershed that is impervious.

Data to determine the overall risk rating for streamflow changes based on sub-watershed location is entered in **Worksheet 4-6** for both rural and urban conditions. The *Rural sub-watershed risk* portion of **Worksheet 4-6** is used to determine risk for rural sub-watersheds based on the percent of acres in a vegetative-altered state (**Figure 4-14**). The *Urban sub-watershed risk* portion of **Worksheet 4-6** is used to determine risk for urban sub-watersheds based on percent impervious (**Figure 4-15**).

If flows are regulated due to diversion, reservoir releases and "ramping flows" for hydropower generation, the operational hydrology is obtained to ascertain the percent change in increase (**Figure 4-16**) or decrease (**Figure 4-17**) that may correspond to increased risk ratings for rural and urban watersheds.

The flow increase risk summary identifies which processes may be responsible for channel instability and/or sediment supply problems. This identification helps to isolate processes with which to prescribe appropriate mitigation of *moderate*-risk watersheds. For *high*- or *very high*-risk watersheds, the assessment identifies specific sub-watersheds in need of a more detailed *PLA* assessment.

In addition to individual sub-watershed streamflow change, the total watershed flow increase potential is also assessed. The overall risk rating for rural or urban sub-watersheds is influenced by the dominant "weak link" stream type most likely to show a response to flow-related sediment increases (**Worksheet 4-6**). Furthermore, overall risk ratings for streamflow change by stream type are also necessary for subsequent risk assessment steps.

For rural sub-watersheds, use **Worksheet 4-6** to calculate percent of sub-watershed cleared/harvested, and use **Figure 4-14** to determine risk. For total watershed summary, use **Worksheet 4-6** to calculate percent of watershed cleared/harvested. Total watershed acres in Column (2) are a sum of all sub-watersheds assessed; total watershed acres cleared/harvested in Column (3) are a sum of sub-watershed and road acres cleared/harvested. The "weak link" stream type most susceptible to change for the watershed in Column (5) needs to be determined.

For urban sub-watersheds, use **Worksheet 4-6** to calculate percent of the sub-watershed that is in impervious condition, and use **Figure 4-15** by stream type in Column (9) to determine risk. For total watershed summary, use **Worksheet 4-6** to calculate percent of watershed that is in impervious condition. The "weak link" stream type most susceptible to change for the watershed in Column (9) needs to be determined.

Chapter 4   The Rapid Resource Inventory for Sediment and Stability Consequence (RRISSC)

Worksheet 4-6. Risk rating worksheet for streamflow changes.

(1) Sub-watershed location/river reach I.D. (include cumulative total watershed following sub-watershed I.D.s)	(2) Total acres	Rural Sub-watershed Risk			Urban Sub-watershed Risk				Adjustments		(13) Overall risk rating: streamflow changes (insert adjective and numeric rating)
		(3) Acres cleared/harvested (include roads) [roads + clearcut = total]	(4) % cleared/harvested of total [(3)/(2)X100]	(5) Stream type most susceptible to change or "weak link"	(6) Risk rating: rural sub-watershed risk (Fig. 4-14) (4) by stream type (5)	(7) Total impervious acres	(8) % impervious [(7)/(2)X100]	(9) Stream type most susceptible to change or "weak link"	(10) Risk rating: urban sub-watershed risk (Fig. 4-15) (8) by stream type (9)	(11) Risk rating: % increase over bankfull discharge (Fig. 4-16)*	(12) Risk rating: % reduction in bankfull discharge (Fig. 4-17)*	
1.												
2.												
3.												
4.												
5.												
6.												
7.												
8.												
9.												
10.												
11.												
12.												
13.												
14.												
15.												

* Describe source of increased or decreased bankfull discharge adjustment, i.e., operational hydrology of reservoir.

Figure 4-14 is used to determine the risk of increased flow for rural watersheds based on the area in a vegetative-altered state by the dominant "weak link" stream type. Insert the risk rating in Column (6) in **Worksheet 4-6**.

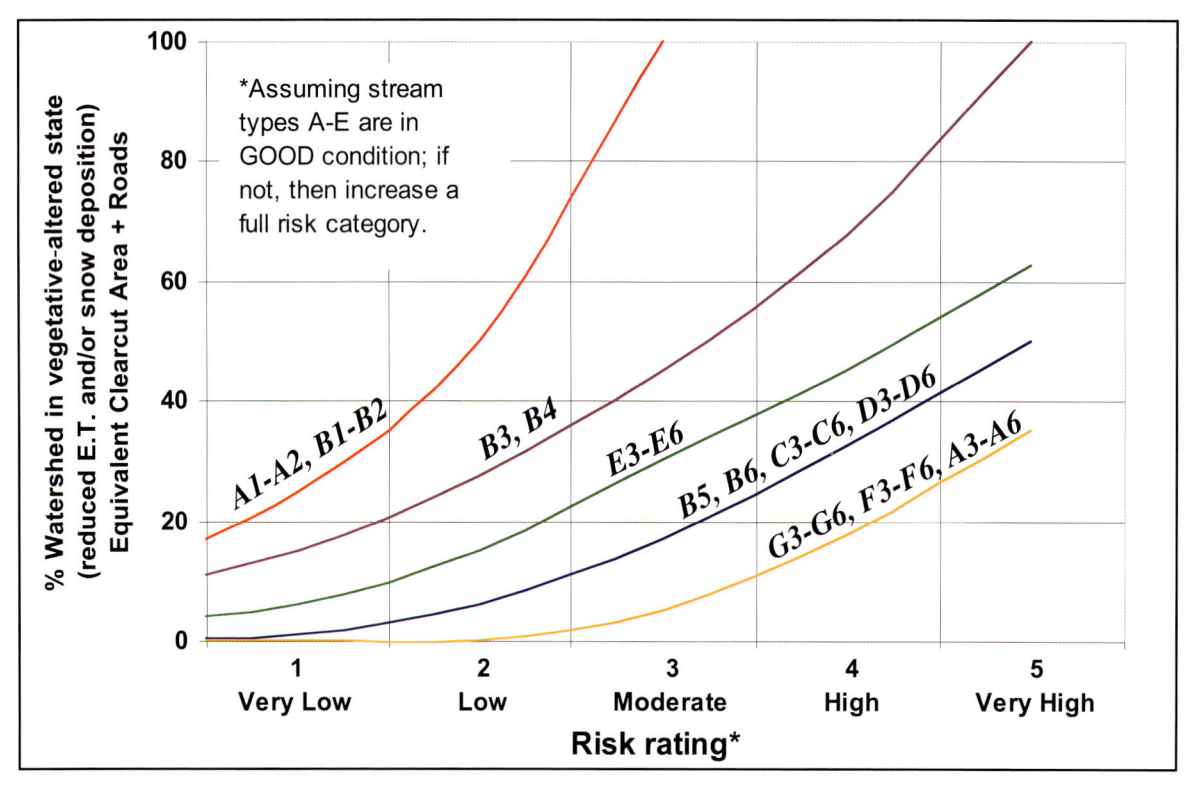

Figure 4-14. Rural watershed flow-related sediment increase risk based on percent of watershed in vegetation-altered state by stream type.

This rating assumes that stream types A through E are in good condition; if not, then the risk rating should be increased one category. If the vegetation recovery is sufficient to provide for hydrologic recovery, adjust the risk rating accordingly.

Chapter 4   The Rapid Resource Inventory for Sediment and Stability Consequence (RRISSC)

The risk of increased streamflow for urban watersheds based on percent imperviousness by the dominant "weak link" stream type is determined using **Figure 4-15**. Hydrologic recovery is not factored into urban development unless extensive, large-scale stormwater management has been sufficient to reduce flood peaks. Insert the risk rating in Column (10) in **Worksheet 4-6**.

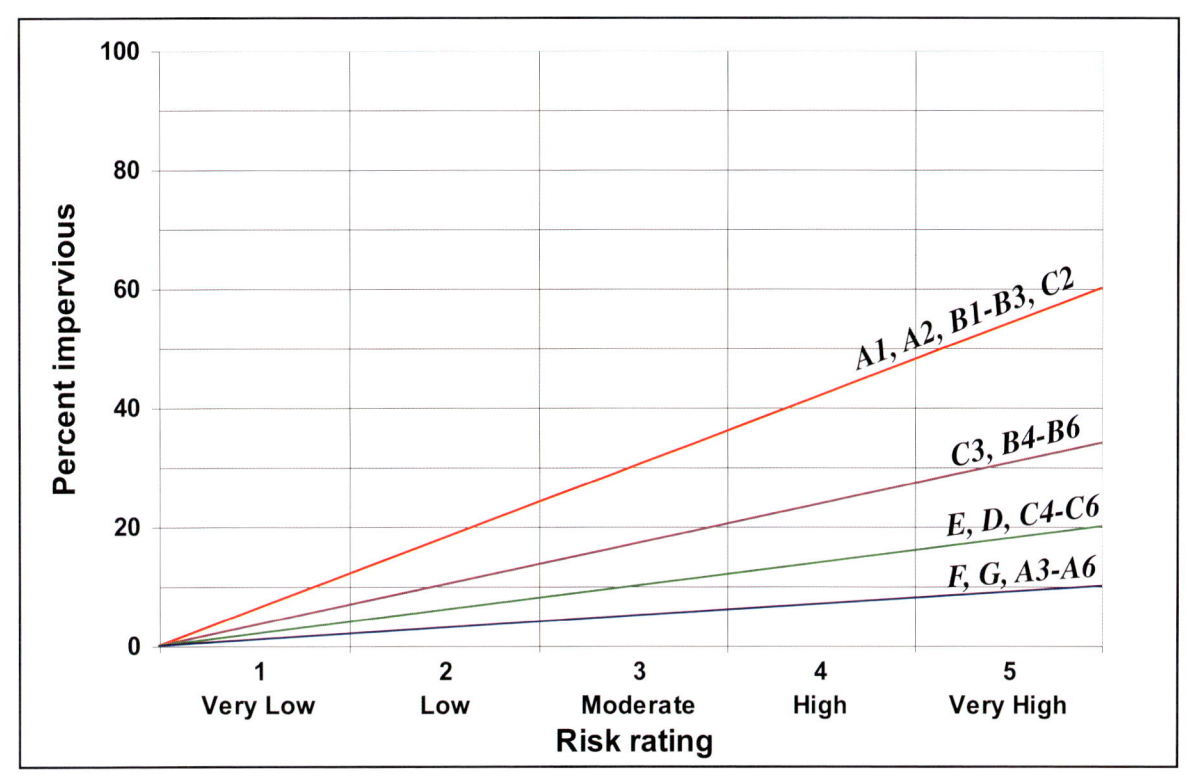

Figure 4-15. Urban development flow-related sediment increase risk based on percent impervious by stream type.

A risk rating of *high* or *very high* for either a rural or urban sub-watershed will advance to *PLA*, and the rating is recorded in the *RRISSC* summary **Worksheet 4-2**.

Urban or rural watershed locations with risk ratings of *very low*, *low* or *moderate* should apply **Worksheet 4-6** and **Figures 4-16** and **4-17** to determine if upward adjustments to these risk ratings are necessary due to bankfull discharge increases or decreases. In both cases, the highest rating should be transferred to the overall risk rating. For example, if the initial rating of *moderate* corresponds to a bankfull discharge increase with a rating of *very high*, then the overall risk rating for that location is *very high*.

To assess the risk of change in bankfull discharge increases, use **Figure 4-16**. The relation in **Figure 4-16** is based on the risk of flow increase by stream type categories and the potential for increased magnitude, duration, timing and frequency of flooding. This increase in flow is due to direct releases from reservoirs, diversion, trans-basin imported water and/or "ramping flows" for hydropower generation. Operational hydrology from the diversions/reservoirs, etc. is used to determine percent increase from bankfull discharge. Sediment increase and channel instability risks are increased due to changes in time of concentration of flow, changes in runoff curve

Watershed Assessment of River Stability and Sediment Supply

numbers (stormflow runoff), high road density, wildfires, equivalent clearcut areas exceeding 40% of basins, excessive storm drains and changes in streamflow interception and streamflow routing. Operational hydrology of reservoirs, diversions and/or hydropower facilities needs to be understood to assess flow changes and channel response. Hydrology changes can adversely affect stream channels and sediment relations. Some streams are very resilient to changes in flow, while others are very sensitive to similar flow changes.

The risk rating using **Figure 4-16** is inserted in Column (11) in **Worksheet 4-6**. The final, adjusted streamflow change rating from Column (13) is entered in the *RRISSC* summary **Worksheet 4-2**.

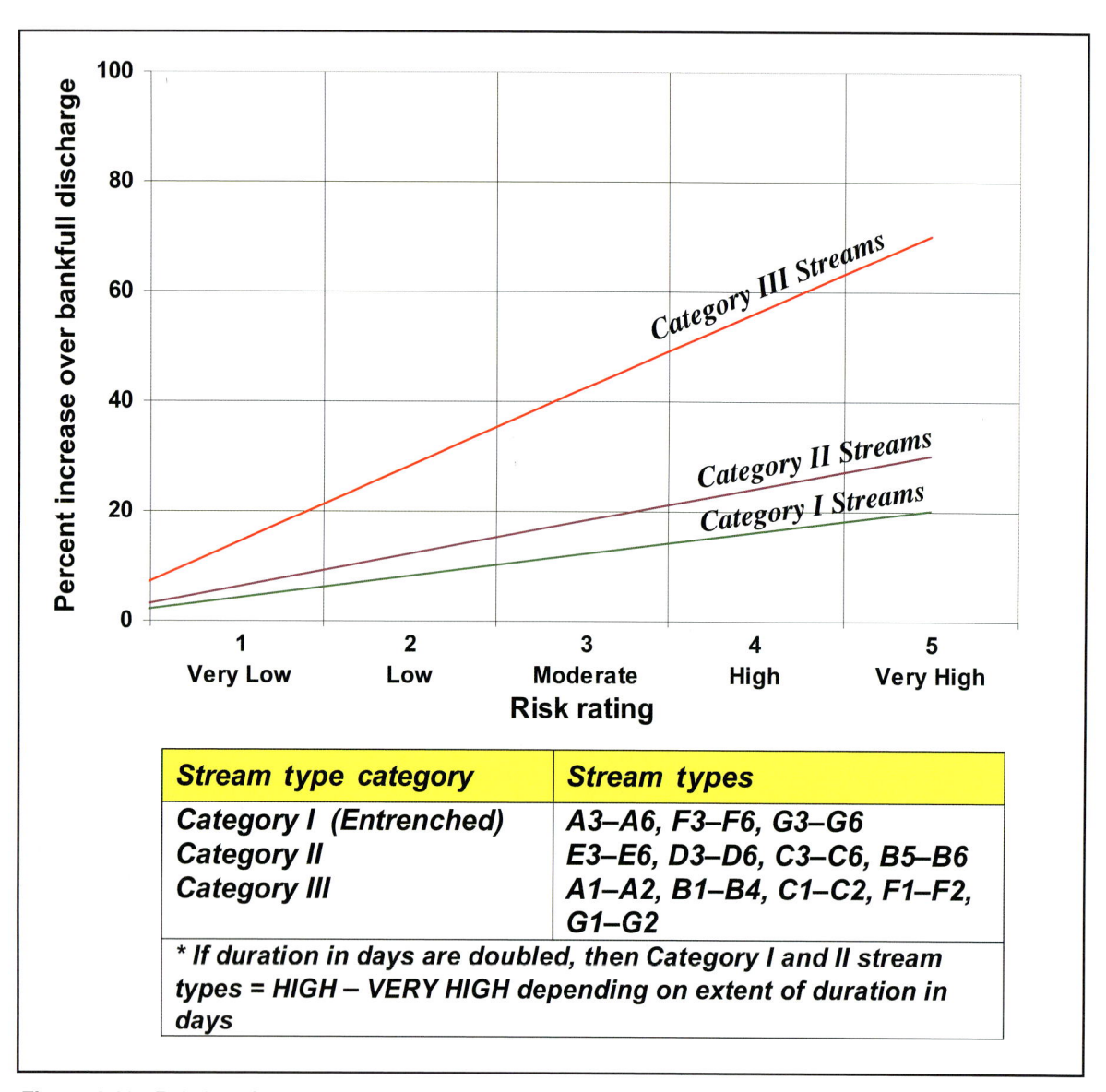

Stream type category	Stream types
Category I (Entrenched)	A3–A6, F3–F6, G3–G6
Category II	E3–E6, D3–D6, C3–C6, B5–B6
Category III	A1–A2, B1–B4, C1–C2, F1–F2, G1–G2
* If duration in days are doubled, then Category I and II stream types = HIGH – VERY HIGH depending on extent of duration in days	

Figure 4-16. Relation of potential risk for channel adjustment/sediment supply due to increase in bankfull discharge from increased streamflow from imported water or reservoir releases by stream type category. Category I stream types are the most sensitive or subjective to rapid adverse change due to flow increases.

Chapter 4 The Rapid Resource Inventory for Sediment and Stability Consequence (RRISSC)

Decreases in bankfull and/or changes in timing, duration and magnitude can have an adverse effect on river stability. If unregulated tributaries deliver sediment to a de-watered or regulated main stream reach with diminished flows, excess sediment deposition will occur. The operational hydrology of such flow regulation must be assessed by "weak link" stream type using **Figure 4-17**. This relation is used only under direct streamflow reduction due to regulation conditions. If there are no regulated structures, then the risk would be *very low* for this condition.

The risk rating using **Figure 4-17** is inserted in Column (12) in **Worksheet 4-6**. The final, adjusted streamflow change rating from Column (13) is entered in the *RRISSC* summary **Worksheet 4-2**.

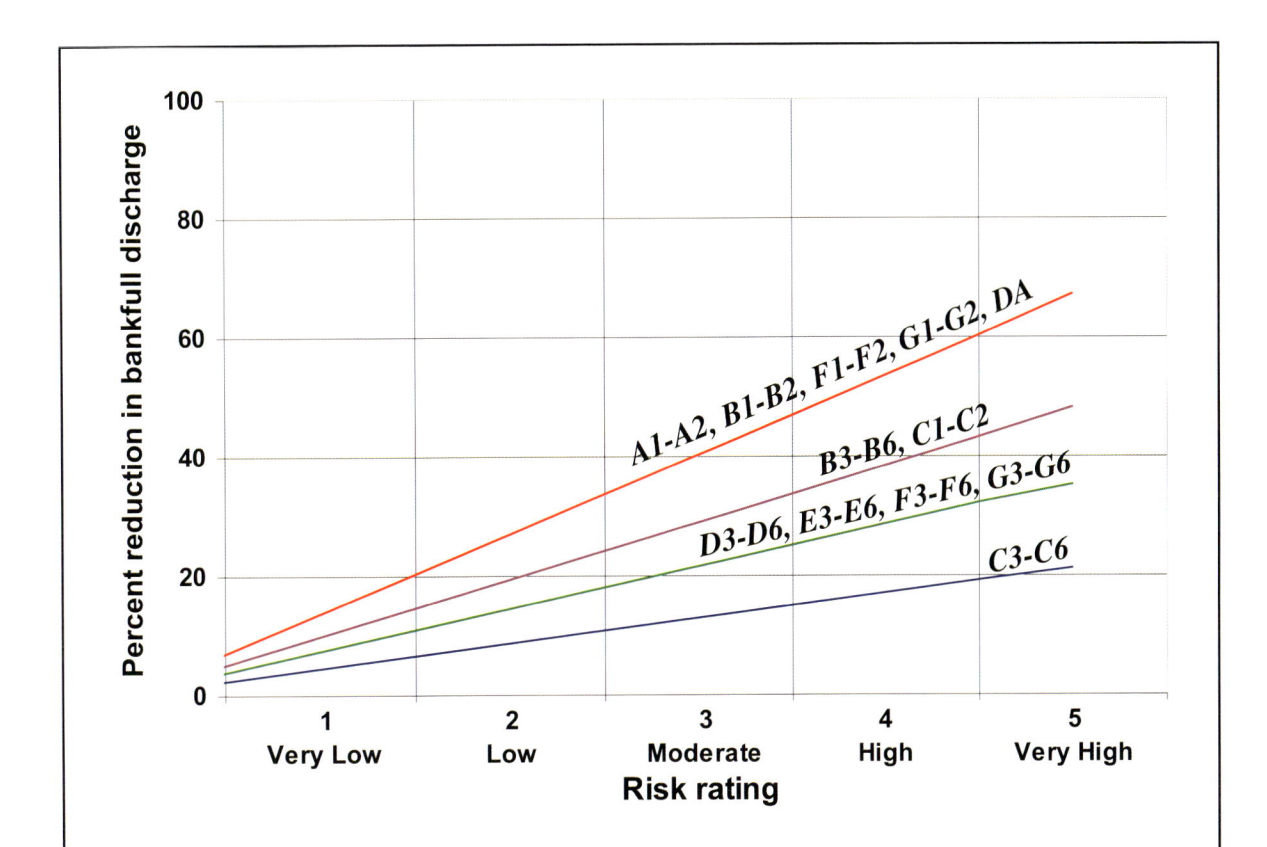

Note: Risk rating assumes timing of depletions induce sediment deposition from the unregulated tributaries. If reduction in natural sediment supply due to depletions causes hungry water and/or reduction of replacement gravel for spawning and maintenance of bed-material size distribution, then increase the risk rating to Very High for all C3–C6, E3–E6, G3–G6 and A3–A6 stream types.

Figure 4-17. Relation of potential risk of adverse channel adjustment due to flow depletion/timing change by stream type.

In addition to the streamflow change risk ratings for sub-watersheds, roads and the total watershed, risk also needs to be assessed by stream type to be used in the channel enlargement and degradation assessment steps, **Steps 16** and **19**, respectively. To properly assess overall stream type risk to enter in **Worksheet 4-6**, the following guidance is offered:

1. For hillslope processes, select the highest risk rating of any location and/or process for the potential impact of a "receiving" stream reach. A *receiving stream reach* is defined as the location of a stream reach in proximity or below an impacting activity.

2. The cumulative effects of increased sediment supply from total watershed sources are used to rate the most representative "weak link" stream type that is most susceptible or sensitive to the combined watershed processes. The proximity of the "weak link" reach or receiving streams to the impact is important in properly assessing risk.

3. The user can spatially locate specific reaches influenced by specific processes relating to potential impact. This will help specify only those reaches/processes/locations where mitigation will be proposed or recommendations can be made to advance to *PLA*.

4. If the overall risk for watershed streamflow risk is *high* or *very high*, then the downstream, receiving "weak link" reach would have the same ratings. This condition would advance to *PLA* due to the high potential risk of adverse cumulative effects.

The streamflow change risk ratings for stream types are entered in both **Worksheet 4-6** and the *RRISSC* summary **Worksheet 4-2**.

The sub-watersheds that have an overall rating of *very low*, *low* or *moderate* for increases or decreases in streamflow do not require a water yield analysis at the *PLA* level. However, the sub-watersheds rating *high* or *very high* require some potential mitigation and monitoring. Mitigation may involve revegetation, dispersing runoff from roads, stormflow management for urban stream systems, flow release modification from hydropower facilities or reservoir release schedules. If river restoration is needed to offset the problem of having the channel match new flow regimes, then these conditions generally warrant conducting a water yield model as part of a *PLA* analysis to ensure proper design.

Other considerations with flow-related increases in sediment include restoring streams, changing a stream type to its most stable potential morphology and stabilizing streambeds and banks to reduce accelerated erosion and sediment yields. A *PLA* analysis is required to successfully plan these efforts.

Chapter 4   The Rapid Resource Inventory for Sediment and Stability Consequence (RRISSC)

Channel processes: Steps 11–15

Stream type delineation plays an important role in *RRISSC*. For example, if a stream was described as a gravel-bed stream and a question arose as to the response of this stream to increased flow, there would be a big difference in the flow response of a B4 versus an F4 or G4 stream type. The interpretations and influence of flow-related impacts, vegetation-controlling influence, recovery potential and sediment supply all vary with channel morphology and stability. The risk rating system used for this part of the assessment relies on an understanding of differential response of stream types to imposed change (**Table 2-2**).

An individual could spend a great amount of time and effort running very complex bedload transport equations and doing factor of safety analysis on streambanks where the potential for instability and/or disproportionate sediment supply problems may be minimal. Rapid assessment methods will systematically eliminate these low-risk areas from further analysis.

Step 11. Compile data for channel processes

Streams are often seen as barometers of the health and function of their watersheds. Impairment of the beneficial uses of water often involves changes in fisheries habitat, flood stage changes, land loss, accelerated sediment deposition and lowering of local baseflow levels that decrease agricultural field productivity. Stream channels influence all of these consequences; their condition is key to any watershed assessment. **Steps 12–15** assess the impairment risk associated with sediment supply and/or stream channel stability.

One key to a broad-level assessment of channel process is to obtain aerial photos or documented evidence of channel change over time. Time-trend analysis can be rapidly performed with documentation. Often, aerial photos from the late 1930s are available to compare with the present river location. Photographs, bridge design compared to present-day characteristics, old maps of channel location, riparian vegetative changes and similar documents are compiled for channel process analysis. If riparian vegetation has changed, then the assessment should look for bank erosion processes. If water tables have dropped, then be alert for channel incision and stream type changes. Channel pattern changes can be readily observed leading to assessment of potential channel instability and successional stages that indicate a change in stream type. The aerial photos and documentation help to direct the assessment toward processes that have been altered and/or are presently undergoing change.

Step 12. Assess broad-level channel stability

The following analysis employs general stability associations rather than a site-specific field measurement approach. This general approach identifies river reaches that are typically not indicative of stability/sediment problems. Delineation of stream types helps to define areas that may require or justify a more detailed assessment.

Step 13. Determine streambank erosion risk

Streambank erosion can be a major contributor to total sediment supply and is associated with river instability, land loss, fish habitat loss, flood issues and other related problems. Appropriate mitigation strategies require the sources of sediment and stability problems to be identified. Because streambanks can be directly observed and measured, inferences from observed relations are used to provide a risk assessment. Data required for this assessment is as follows:

- Stream types (Level II);
- Aerial photographs and drainage area maps;
- Regional curves (bankfull dimensions vs. drainage area);
- Bankfull width;
- Radius of curvature;
- Riparian species composition;
- Bank height; and
- Bankfull depth.

Worksheet 4-7 is used for data entry and to determine streambank erosion risk by computing individual risk ratings for:

- Vegetation composition (**Figure 4-18**);
- Bank-height ratio (**Figure 4-19**); and
- Radius of curvature divided by bankfull width (**Figure 4-20**).

The overall risk rating for streambank erosion is then determined from the integration of these various individual risk ratings using **Figure 4-21**.

Chapter 4 The Rapid Resource Inventory for Sediment and Stability Consequence (RRISSC)

Worksheet 4-7. Risk rating worksheet for streambank erosion.

(1) Location code/ river reach I.D.	(2) Vegetation composition	(3) Risk rating: vegetation composition (Fig. 4-18)	(4) Bank-height ratio	(5) Risk rating: bank-height ratio (Fig. 4-19)	(6) Radius of curvature divided by bankfull width	(7) Risk rating: radius of curvature divided by bankfull width (Fig. 4-20)	(8) Total individual risk rating points by reach $\sum[(3)+(5)+(7)]$	(9) Overall risk rating by stream type (Fig. 4-21)
1.								
2.								
3.								
4.								
5.								
6.								
7.								
8.								
9.								
10.								
11.								
12.								
13.								
14.								
15.								

The relation developed for riparian vegetation composition represents the susceptibility of streambanks to erosion. The rooting depth and density of the riparian plant community both directly affect bank failure. The internal strength afforded by plants with dense and deep root structure reduces mass erosion due to shear and cantilever bank failures. Plants can also reduce fluvial entrainment by protecting exposed soil from detachment. Use **Figure 4-18** to obtain the risk rating for vegetative influence on bank erosion. Insert the numeric risk rating in Column (3) in **Worksheet 4-7**.

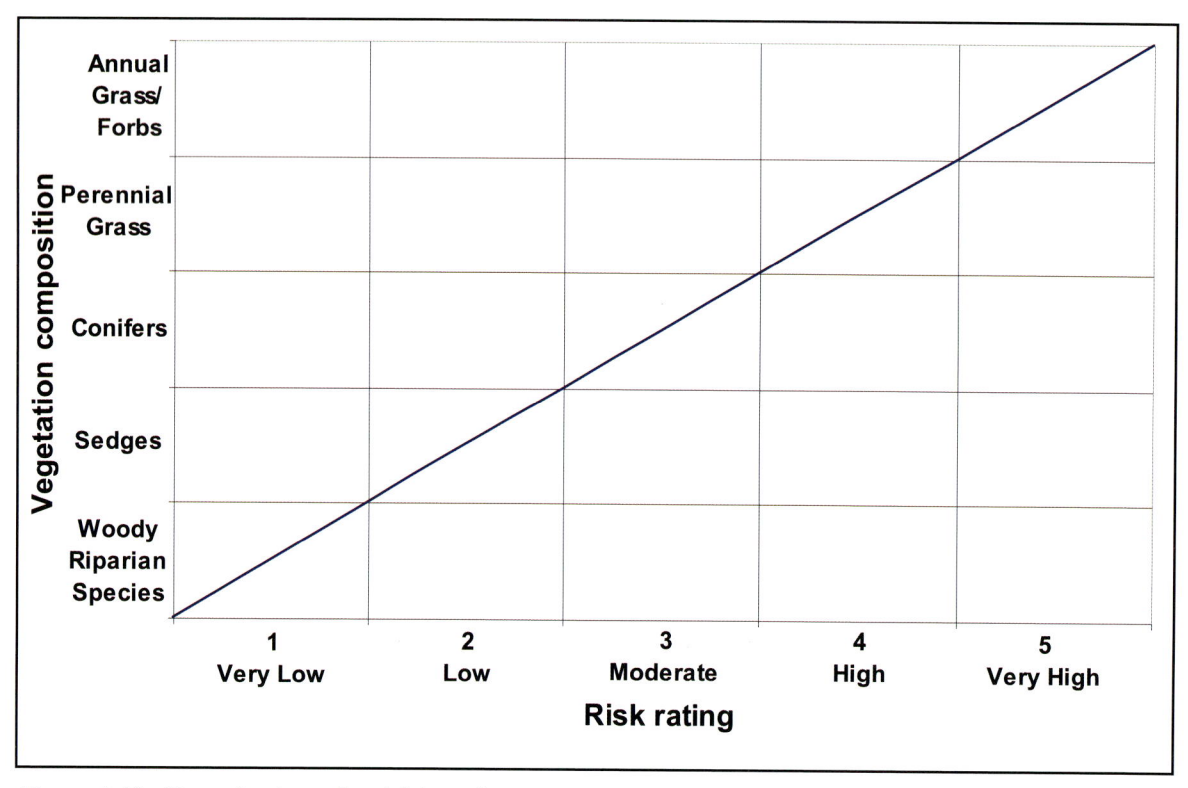

Figure 4-18. Streambank erosion risk based on vegetation composition.

Chapter 4   The Rapid Resource Inventory for Sediment and Stability Consequence (RRISSC)

The bank-height ratio is the height of the study bank divided by the bankfull depth at the toe of the bank. The greater the value is above 1.0, the greater the likelihood of increased erosion due to various erosional processes, including mass erosion, dry ravel, freeze-thaw and rill erosion. The relation of bank-height ratio to risk rating is shown in **Figure 4-19**. Insert the bank-height ratio in Column (4) in **Worksheet 4-7** and the corresponding risk rating in Column (5).

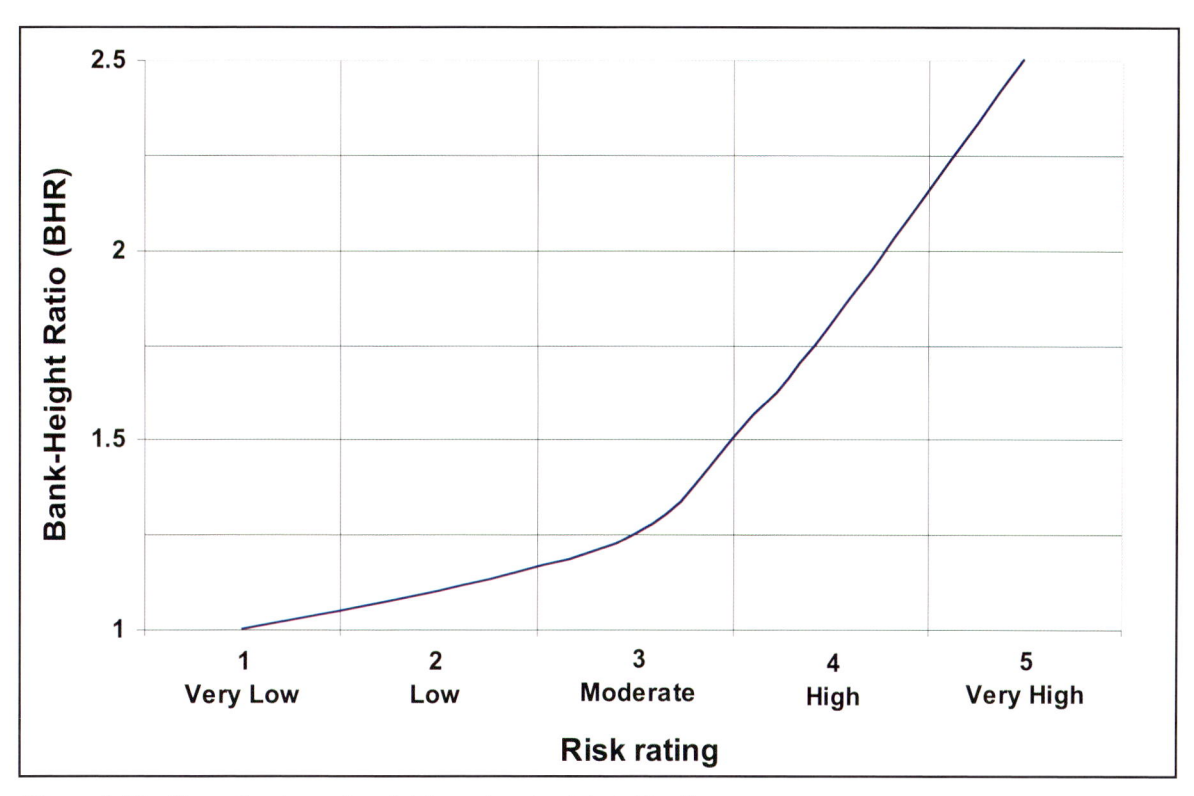

Figure 4-19. Streambank erosion risk based on bank-height ratio.

The tighter the radius of curvature, the greater the boundary shear stress and velocity gradient directed to the bank, causing accelerated erosion. Radius of curvature can be obtained from aerial photography or rapidly measured in the field. The radius of curvature is divided by the bankfull width to obtain the risk rating in **Figure 4-20**. Insert the radius of curvature divided by bankfull width in Column (6) in **Worksheet 4-7** and the corresponding risk rating in Column (7).

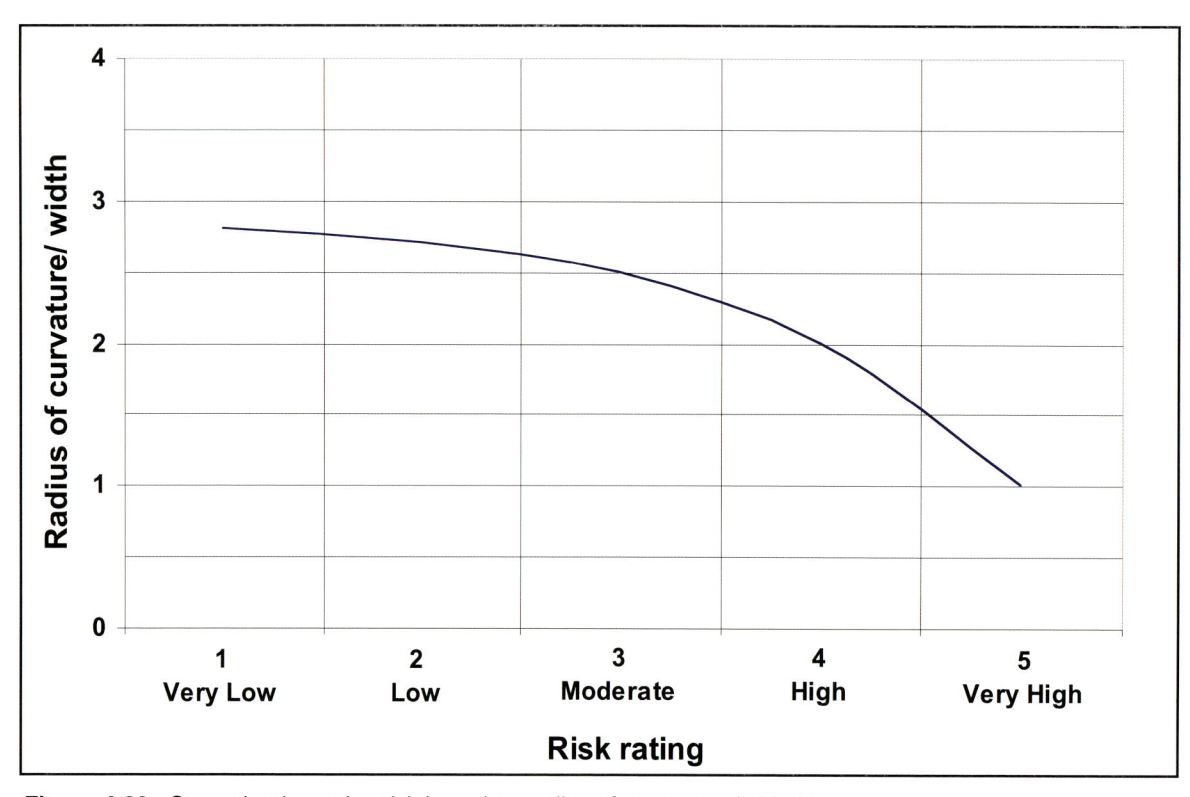

Figure 4-20. Streambank erosion risk based on radius of curvature divided by width.

Chapter 4   The Rapid Resource Inventory for Sediment and Stability Consequence (RRISSC)

The overall risk rating is calculated in **Worksheet 4-7**, using **Figure 4-21**.

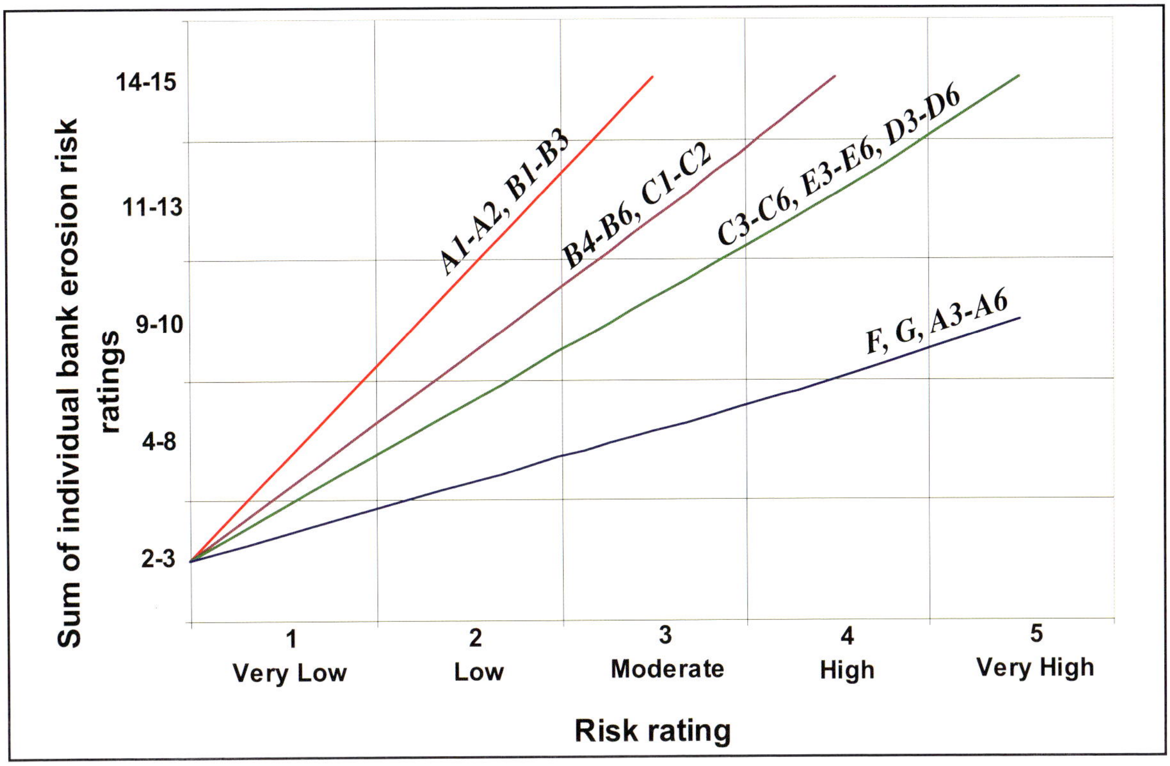

Figure 4-21. Overall streambank erosion risk based on the sum of individual risk ratings by stream type.

The overall risk rating for streambank erosion is recorded in the *RRISSC* summary **Worksheet 4-2**. If the streambank erosion risk rating is *very low* or *low*, then no mitigation or further assessment is needed. If the risk rating is *moderate*, then it is necessary to do the following:

- Determine major processes/causes of bank erosion risk for the following variables: radius of curvature divided by width, riparian vegetation change and bank-height ratio;

- Identify specific mitigation that could potentially reduce the streambank erosion risk from *moderate* to *low*. For example, a riparian grazing strategy could be designed to help recover the key species (such as woody vegetation for a C4 stream type). Riparian management zones could be established where surface disturbance activities could be minimized; riparian vegetative communities improved; debris management could be specified; and buffer widths and vegetative management could be specified (e.g., even-aged versus uneven-aged management for overstory species and composition/density of understory community); and

- Provide general restoration guidance to reduce "tight radius" curves and to add grade control to help incision processes.

For those reaches rating *high* or *very high*, a *PLA* assessment is recommended. This more detailed assessment level will:

- Identify more detailed, specific processes responsible for accelerated bank erosion;

- Quantify bank erosion rates and compare to total annual sediment yield. Mitigation can then be directed at disproportionately high sediment sources, specific locations and processes; and

- Obtain data for the design of stream restoration/streambank stabilization and enhancement projects.

Step 14. Assess in-channel mining

In-channel mining affects channel stability by 1) lowering local baseflow level, 2) causing headcuts, 3) incising channels, 4) changing dimension, pattern and profile, and 5) accelerating streambank erosion. The nature, extent and consequence of channel adjustments are influenced by stream type, and risk ratings are adjusted accordingly. The data needed for this assessment includes:

- Stream types (Level II);
- Length of affected stream; and
- Nature and extent of mining (including base-level lowering).

Note: If no in-channel mining, a RRISSC assessment is not necessary, and "Very Low (1)" is automatically inserted in the RRISSC summary worksheet to be used for further assessment summaries.

Worksheet 4-8 is used for data entry and to determine the percent of channel length impacted by in-channel mining. The overall in-channel mining risk is then calculated using **Worksheet 4-8**. This risk is based on the direct disturbance in relation to the percent of channel length impacted, as shown in **Figure 4-22**.

Worksheet 4-8. Risk rating worksheet for in-channel mining.

(1) Location code/ reach I.D.	(2) Total acres of reach	(3) Total acres impacted by in-channel mining	(4) % of channel length impacted by in-channel mining [(3)/(2)X100]	(5) Overall adjective and numeric risk rating (Fig. 4-22) (4) by stream type
1.				
2.				
3.				
4.				
5.				
6.				
7.				
8.				
9.				
10.				
11.				
12.				
13.				
14.				
15.				

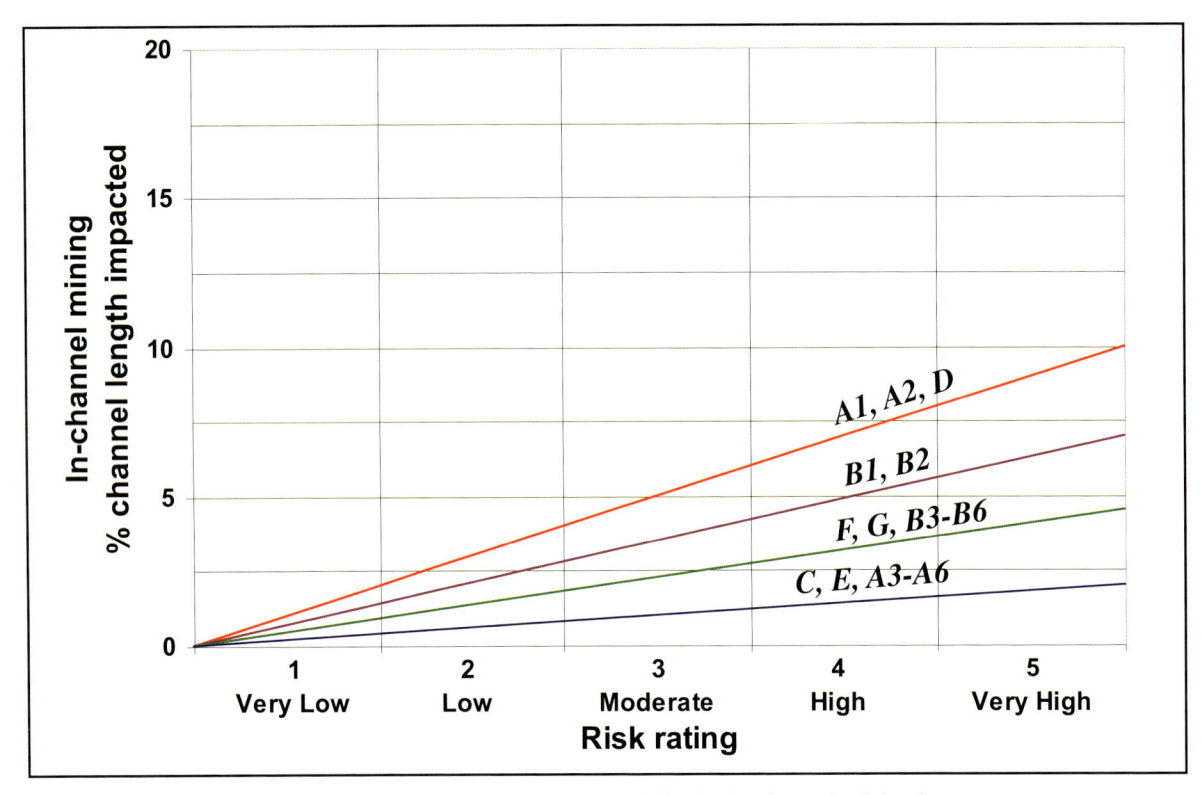

Figure 4-22. Risk rating for potential sediment/channel stability for in-channel mining impacts.

The overall adjective risk rating is recorded in the *RRISSC* summary **Worksheet 4-2**. This risk rating is also used to evaluate channel enlargement and degradation potential in **Steps 16** and **19**, respectively.

If the risk rating for in-channel mining is *moderate*, alternatives to reduce such risk should be evaluated. For example, if depositional features (bars) are excavated, then only the center should be removed, leaving the boundary of the bar to properly distribute flows on the channel margin. Excess sediment may also be transported into an off-channel storage area using a vortex-tube design. This would prevent equipment from re-entering the channel on a frequent basis but still provide a product. These are only a few examples of potentially effective mitigation strategies. Monitoring would be helpful to determine the effectiveness of such mitigation.

If the risk rating for in-channel mining is *high* or *very high*, then the level of detail required to implement restoration/stabilization designs necessitates a *PLA* analysis.

Step 15. Assess direct channel impacts

Flood control, land drainage, emergency flood relief, vegetative conversions, heavy grazing pressure, livestock concentrations, straightening, levees, dredging, clearing vegetation, grubbing streamside zones and other related "river engineering" works have caused major instability and sediment problems.

The information needed to assess direct channel impacts includes the following:

- Stream types (Level II);
- Time-trend aerial photos;
- Percent of riparian vegetation changed;
- Length of channel with changed riparian vegetation, including:
    - nature of direct disturbance (hoof shear, straightening, etc.)
    - percent of channel directly impacted (straightened, dredged, levees, etc.)
- Percent of channel blockage, including woody debris, structures, etc.

Worksheet 4-9 is used for data entry and to determine the overall direct channel impact risk by computing individual risk ratings for:

- Percent riparian vegetation changed by direct disturbance (**Figure 4-23**);
- Percent channel length impacted by vegetation utilization as described above (**Figure 4-24**); and
- Channel blockage from large woody debris (**Figure 4-25**).

The overall risk rating for direct channel impacts is determined by using the highest individual risk rating rather than by summarizing total points.

Watershed Assessment of River Stability and Sediment Supply

Worksheet 4-9. Risk rating worksheet for direct channel impacts.

(1) Location code/ river reach I.D.	(2) Total channel length in feet	(3) Riparian vegetation change in feet	(4) % of total length impacted [(3)/(2)X100]	(5) Risk rating: % riparian vegetation change (**Fig.** 4-23) (4) by stream type	(6) Length impacted by direct channel disturbance in feet	(7) % of total length impacted [(6)/(2)X100]	(8) Risk rating: % channel length impacted (**Fig.** 4-24) (7) by stream type	(9) Length impacted by large woody debris in feet	(10) % of length of debris blockage [(9)/(2)X100]	(11) Risk rating: debris blockage (**Fig.** 4-25)	(12) Overall risk rating for direct channel impacts (Insert highest risk rating from Columns 5, 8 and 11)
1.											
2.											
3.											
4.											
5.											
6.											
7.											
8.											
9.											
10.											
11.											
12.											
13.											

Direct disturbances to banks are associated with straightening, scarification, levees, lining of banks with "sacrificial" rip rap (river gravel or cobble), heavy grazing pressure causing hoof shear or other similar impacts.

Riparian vegetation change risks

Use **Worksheet 4-9** to calculate channel length acres impacted due to significant riparian vegetation change. A significant percentage of riparian vegetation composition change represents a 75% reduction in key riparian species. Use the relations in **Figure 4-23** to determine the percentage of channel length at risk due to these significant changes in riparian vegetation by stream type. Insert the numeric risk rating in Column (5) in **Worksheet 4-9**.

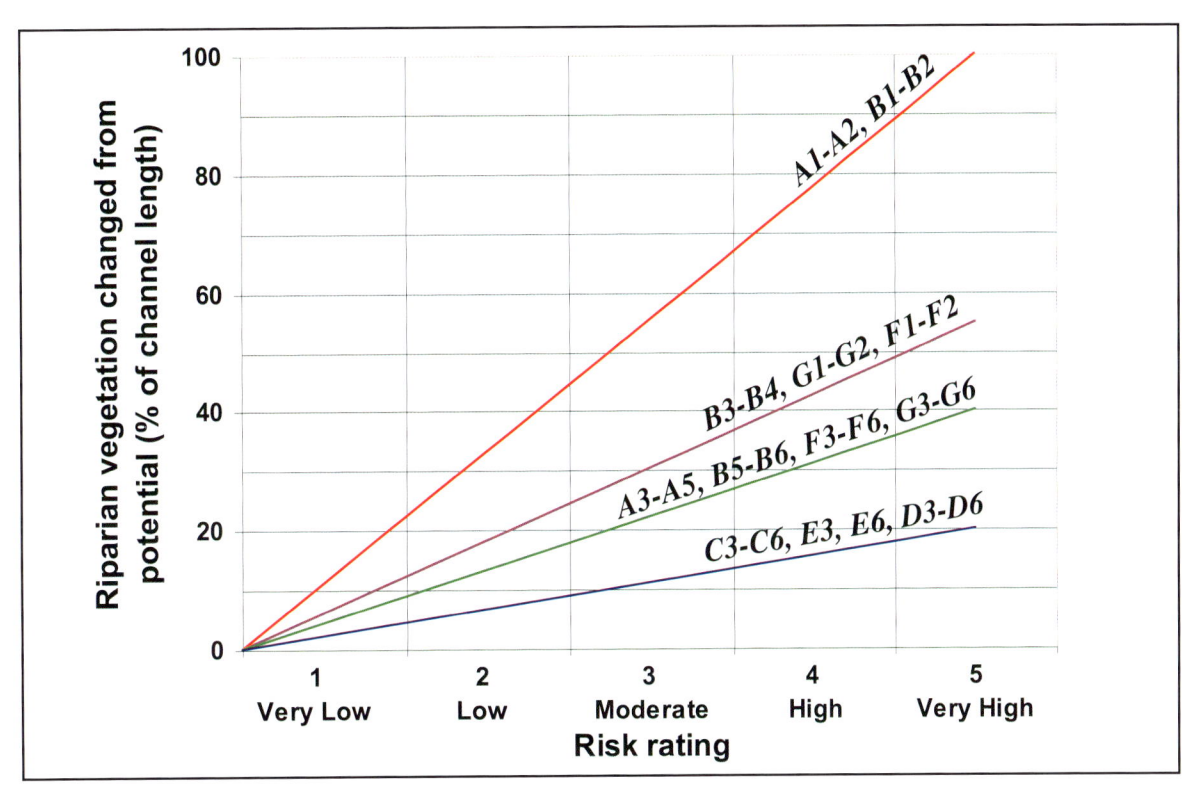

Figure 4-23. Risk rating for potential introduced sediment and channel instability by stream type based on percentage of channel length affected by vegetation change.

The second rating associated with riparian vegetation involves a relation of the percent of channel impacted with direct disturbance, adjusted by stream type category. Use **Worksheet 4-9** to determine percent of total acres impacted. The risk rating derived from this process is shown in **Figure 4-24**. Heavy riparian vegetation use represents 15–20% forage use for woody "key species" plants and/or 10% hoof shear drainage to streambanks. This risk rating also represents additional direct impacts such as channelization, straightening, lining and levees. Insert the numeric risk rating in Column (8) in **Worksheet 4-9**. It is important to retain the mapping of impacts by channel length/location in order to specify future management changes/mitigation or to select reaches for the *PLA* phase.

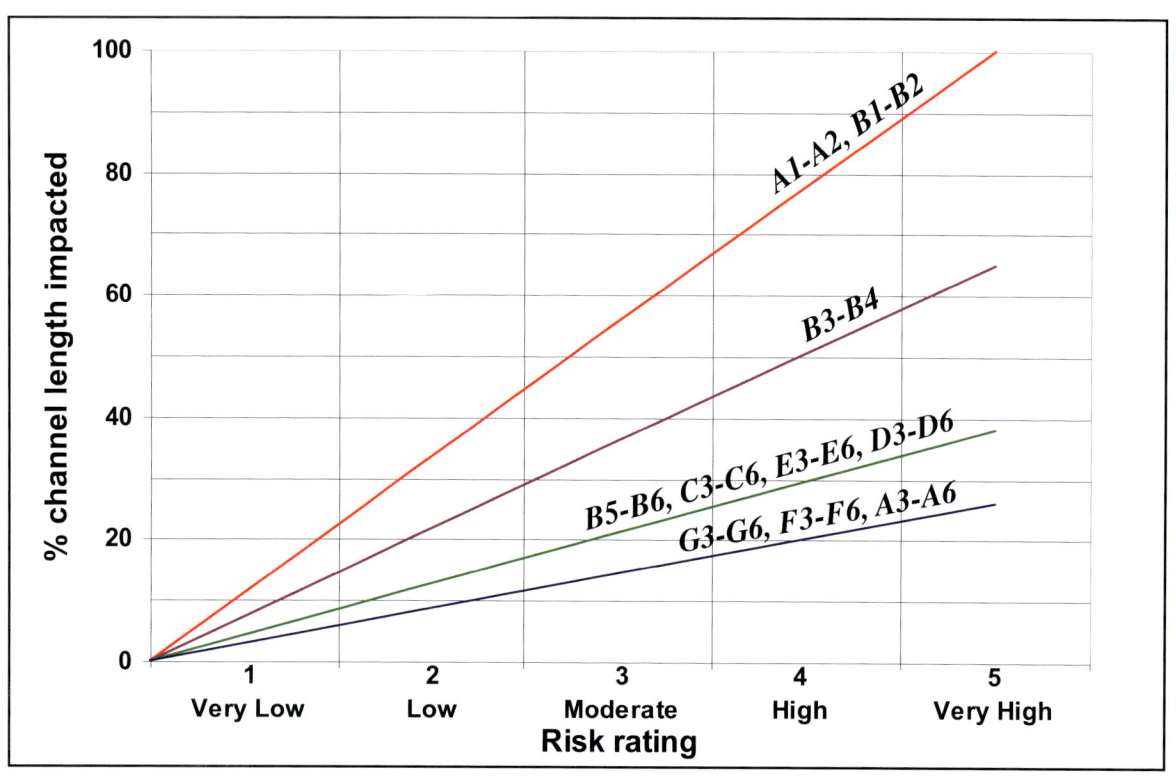

Figure 4-24. Risk rating relation of percent of channel length impacted by vegetation utilization and bank impacts according to stream type.

Debris blockage risk

Large wood or other debris may cause channel blockages and create a lack of sediment transport competence and/or capacity. Use **Worksheet 4-9** to determine percent of debris blockage, and then use **Figure 4-25** to calculate risk ratings by percent of channel impaired due to large wood or other debris. Insert the numeric risk rating in Column (11) in **Worksheet 4-9**.

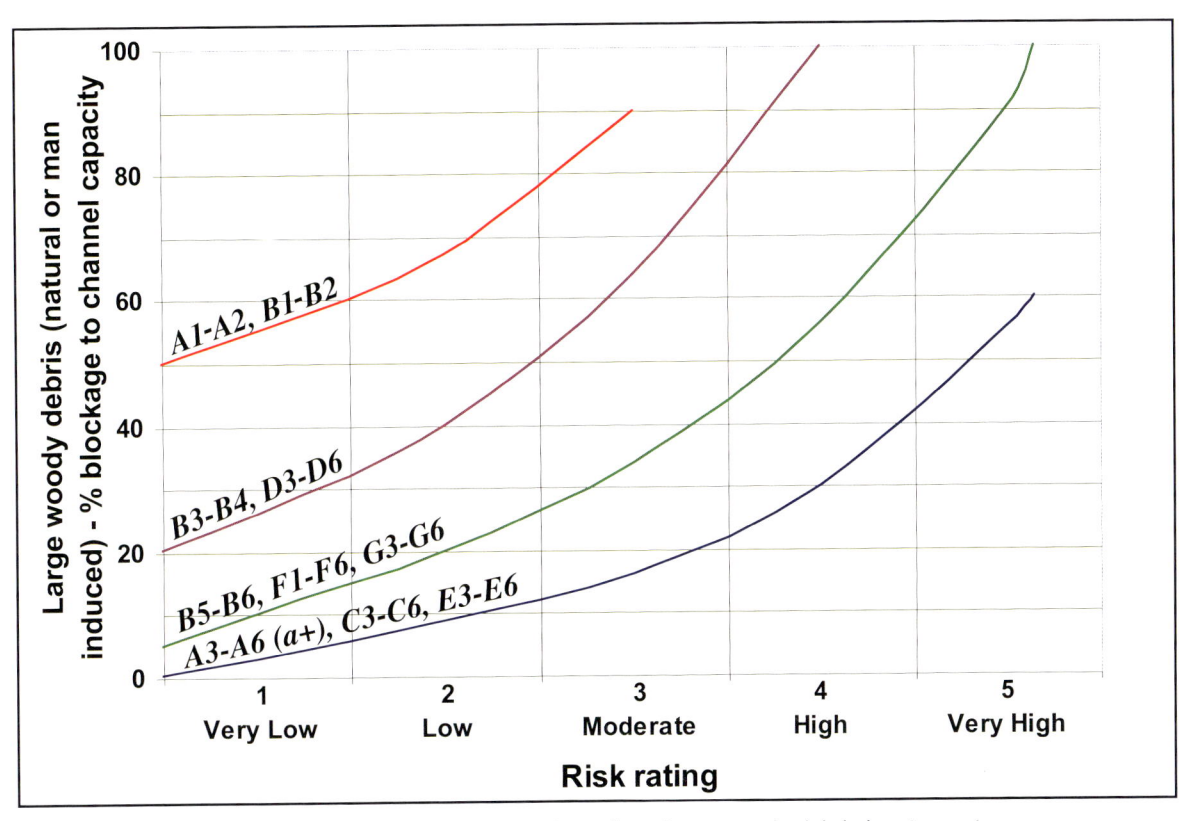

Figure 4-25. Risk rating in relation to channel blockage from large woody debris by stream type.

Worksheet 4-9 can now be used to determine the overall risk rating for direct channel impacts. The overall rating reflects the highest individual risk rating from Columns (5), (8) and (11). The overall risk rating is recorded in the *RRISSC* summary **Worksheet 4-2**.

Due to the multiple criteria used to determine risk for direct channel impacts, accurate documentation on maps and aerial photographs of the "cause" and location of *moderate* and greater risk ratings helps the user to:

- Identify the nature and extent of potential impacts such as:
    - riparian vegetation change
    - livestock concentrations
    - direct change in channel dimension, pattern and profile
    - channel blockages/debris/structures
    - in-channel mining
- Locate and specify land use management changes related to the "cause" of moderate risks;
- Identify mitigation recommendations specific to potential impacts;
- Design a monitoring system to evaluate effectiveness of management/mitigation implementation; and
- Advance specific locations of river reaches to the *PLA* phase.

For a *moderate* risk rating of large woody debris or channel blockages from various causes, the following example of mitigation considerations may apply:

- Inventory large woody debris by location to manage by 1) clearing to an acceptable cross-sectional area, 2) stabilizing/locating large woody debris at the channel margins (banks), and 3) specifying riparian vegetation to provide natural recruitment over time; and
- Identify locations of blockages such as structures to 1) modify structures to decrease width/depth ratio, and 2) alter structures to provide for fish passage, sediment transport and channel stability.

If the risk rating is *high* or *very high* for large woody debris or channel blockages from various causes, then proceed to *PLA*. A more detailed riparian/woody debris management plan can be implemented following the *PLA* analysis.

Chapter 4   *The Rapid Resource Inventory for Sediment and Stability Consequence (RRISSC)*

Risk potential assessment: Steps 16–20

Step 16. Calculate channel enlargement risk potential

Enlargement relations are based on the susceptibility of streams to incise and/or widen at an accelerated rate due to changes in flow, clear water discharge, direct disturbance and streambank erosion. Information needed for this summary assessment includes the following:

- Stream types (Level II);
- Streamflow changes risk (**Step 10** in **Worksheet 4-2**; **Worksheet 4-6**);
- Streambank erosion risk (**Step 13** in **Worksheet 4-2**; **Worksheet 4-7**);
- In-channel mining impact risk (**Step 14** in **Worksheet 4-2**; **Worksheet 4-8**); and
- Direct channel disturbance risk (**Step 15** in **Worksheet 4-2**; **Worksheet 4-9**).

The relations used to obtain risk for channel enlargement involve a summary of the risk ratings from the preceding steps. The summary risk ratings used to determine enlargement risk are found in **Worksheet 4-2**. The combination of these risk scores are then stratified by stream type category for an overall risk rating in **Worksheet 4-10**, using the relations in **Figure 4-26**.

Note: If any in-channel mining exists, the reach will rate as "high risk" for enlargement and will advance to PLA.

Worksheet 4-10. Risk rating worksheet for channel enlargement.

(1) Location code/ river reach I.D.	(2) Overall risk rating: streamflow changes **(Step 10 in Worksheet 4-2; Worksheet 4-6)**	(3) Overall risk rating: streambank erosion **(Step 13 in Worksheet 4-2; Worksheet 4-7)**	(4) Overall risk rating: direct channel impacts **(Step 15 in Worksheet 4-2; Worksheet 4-9)**	(5) Total numeric score $\sum[(2)+(3)+(4)]$	(6) Overall risk rating for channel enlargement **(Fig. 4-26)** (5) by stream type	(7) Adjustment due to in-channel mining*
1.						
2.						
3.						
4.						
5.						
6.						
7.						
8.						
9.						
10.						
11.						
12.						
13.						
14.						
15.						

*Any in-channel mining automatically raises reach to *high* risk for enlargement and advances reach to *PLA*.

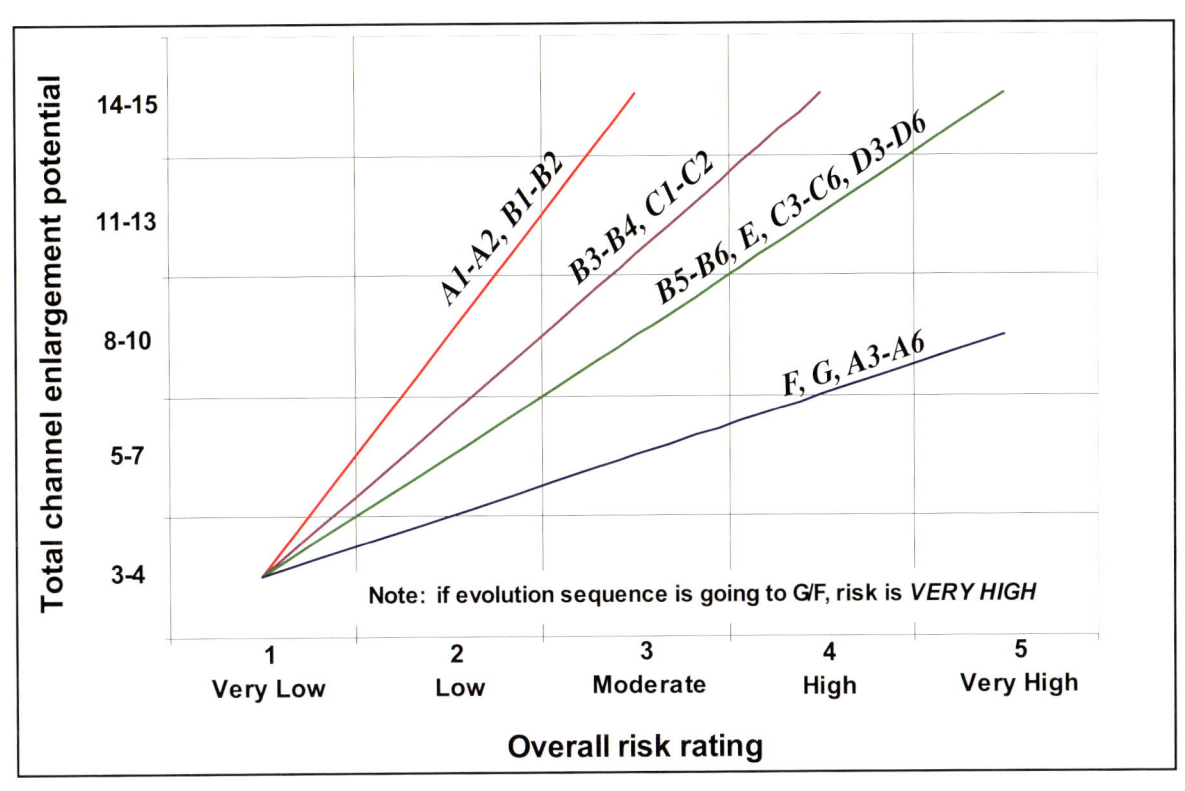

Chapter 4 *The Rapid Resource Inventory for Sediment and Stability Consequence (RRISSC)*

Figure 4-26. Increased sediment and channel instability risk based on channel enlargement potential by stream type.

Insert the overall risk rating for channel enlargement in the *RRISSC* summary **Worksheet 4-2**.

Moderate risk ratings for enlargement require a summary of the relations responsible for the rating in order to direct mitigation (e.g., channel evolution, flow regime changes, streambank erosion risk, in-channel mining and direct channel impacts).

Step 17. Calculate aggradation/excess sediment risk

Excess sediment deposition and reductions in sediment competence and capacity due to increases in width/depth ratio and/or slope changes are often responsible for aggradation. Direct disturbances to stream channels, such as over-widening due to bridges, channelization projects, riparian vegetation changes and excessive hoof shear from poor grazing practices, "back water" from road crossings, accelerated streambank erosion and excess sediment supply from hillslope processes, can all potentially lead to aggradation. Information needed to complete this risk assessment using **Worksheet 4-11** includes the following:

- Aerial photographs;
- Stream types;
- Overall hillslope processes sediment supply risk:
    - mass erosion risk (**Step 6** in **Worksheet 4-2**; **Worksheet 4-3**)
    - roads risk (**Step 7** in **Worksheet 4-2**; **Worksheet 4-4**)
    - surface erosion risk (**Step 8** in **Worksheet 4-2**; **Worksheet 4-5**)
- Bankfull width/depth ratio of existing and reference (stable) reaches (**Figure 4-27**);
- Channel enlargement risk (**Step 16** in **Worksheet 4-2**; **Worksheet 4-10**);
- Streambank erosion risk (**Step 13** in **Worksheet 4-2**; **Worksheet 4-7**); and
- Depositional pattern evaluation (**Figure 4-28**).

To determine aggradation/excess sediment supply risk using **Worksheet 4-11**, the *Hillslope Risk Ratings (Sediment Supply)* section is completed for each sub-watershed and road analyzed. The *Channel Process Response to Excess Sediment* section is completed for each reach. Overall risk ratings for potential aggradation/excess sediment deposition using **Worksheet 4-11** for stream types are evaluated in the following manner:

1. For each location sub-watershed and road analyzed, use **Worksheet 4-11** to determine an overall hillslope processes risk rating by using the individual risk ratings for:
    - mass erosion (**Step 6** in **Worksheet 4-2**; **Worksheet 4-3**)
    - roads (**Step 7** in **Worksheet 4-2**; **Worksheet 4-4**)
    - surface erosion (**Step 8** in **Worksheet 4-2**; **Worksheet 4-5**)

2. The dominant and/or overall highest hillslope sediment supply source by process (mass erosion, roads, etc.) is used in conjunction with the receiving stream type;

3. The overall hillslope risk ratings are then used in Column (7); for each reach, select the most representative stream type and use the associated hillslope risk rating where the stream type occurs. If a given hillslope process that rated *high* or *very high* is below a given sensitive stream type, the rating would not be adjusted due to such risk. However, if the same stream type was adjacent to or below the activity generating the sediment supply, then the stream type would indicate the corresponding risk from one or more of columns (2), (3) or (4) and summarized in column (7);

4. The location of the sediment supply is identified in Column (7) in **Worksheet 4-11**;

5. The remaining channel process risk ratings by stream type reach are completed following the references in **Worksheet 4-11**, Columns (8)–(10) for:

    - width/depth ratio departure (**Figure 4-27**)

    - channel enlargement (**Step 16** in **Worksheet 4-2**; **Worksheet 4-10**)

    - streambank erosion (**Step 13** in **Worksheet 4-2**; **Worksheet 4-7**)

6. Adjustments may be necessary to reach a final aggradation/excess sediment supply risk rating by following the references in Columns (13) and (14) that assess aggradation/excess sediment supply indicators, including depositional patterns (**Figure 4-28**) and reduction in discharge due to flow regulation.

The stream type risk ratings identified by location and erosional depositional processes allow for the user to adequately specify either mitigation or advancement to *PLA*.

For an example of a completed aggradation/excess sediment deposition worksheet, see **Worksheet 7-10** in Chapter 7.

Watershed Assessment of River Stability and Sediment Supply

Worksheet 4-11. Summary of risk ratings for potential aggradation and/or excess sediment deposition.

(1)	Hillslope Risk Ratings (Sediment Supply)					Channel Process Response to Excess Sediment						(14)	(15)
	(2)	(3)	(4)	(5)	(6)	(7)	(8)	(9)	(10)	(11)	(12)	(13)		
Location code/ river reach I.D.	Risk rating: mass erosion **(Step 6 in Worksheet 4-2; Worksheet 4-3)**	Risk rating: roads **(Step 7 in Worksheet 4-2; Worksheet 4-4)**	Risk rating: surface erosion risk/ delivered sediment risk **(Step 8 in Worksheet 4-2; Worksheet 4-5)**	Point subtotal $\Sigma[(2)+(3)+(4)]$	Hillslope summary overall rating; use points from column (5) (Insert both numeric and adjective ratings) VL(1) = 3 L(2) = 4–7 M(3) = 8–10 H(4) = 11–14 VH(5) = >14	Representative location and associated rating points from column (6)*	Risk rating: width/depth ratio departure **(Fig. 4-27)** VL(1) = HS L(2) = S M(3) = MU H(4) = U VH(5) = HU	Risk rating: channel enlargement **(Step 16 in Worksheet 4-2; Worksheet 4-10)**	Risk rating: streambank erosion **(Step 13 in Worksheet 4-2; Worksheet 4-7)**	Point subtotal $\Sigma[(7)+(8)+(9)+(10)]$	Risk rating: use points from column (11) (Insert adjective risk rating) VL(1) < 5 L(2) = 5–8 M(3) = 9–12 H(4) = 13–16 VH(5) >16	Adjustments: aggradation/excess sediment Indicators** **a.** obvious excess deposition **b.** filling of pools **c.** deposition of sand or larger material on floodplain **d.** bi-modal depositional patterns B3, B5-B7 **e.** (Fig. 4-28) (note categories that apply)	Adjustment: reduction in flow due to regulation**	Final aggradation/ excess sediment deposition risk rating (insert adjective risk rating)
1.														
2.														
3.														
4.														
5.														
6.														
7.														
8.														
9.														
10.														
11.														
12.														
13.														
14.														
15.														

* To apply risk rating from Hillslope Processes for aggradation risk, it is important to identify the location of the sediment supply in relation to the most representative or "weak link" stream type.
** Adjust a full risk category upward if streamflow decrease and/or indicators provide evidence appropriate to the observed condition such as aggradation indicators on categories a,b,c,d and e.

Chapter 4   The Rapid Resource Inventory for Sediment and Stability Consequence (RRISSC)

A reference reach width/depth ratio (surface width at bankfull stage divided by mean bankfull depth) relation is compared to the existing condition stream type. If the existing stream has a higher value than the reference, the value will be greater than 1. The higher the departure ratio (existing w/d divided by reference w/d) is above 1, the more likely the stream is to induce excess sediment deposition and potentially lead to aggradation and channel enlargement. Increases in width/depth ratio above reference create reduction in shear stress, mean velocity and unit stream power. The corresponding reduction in the hydraulic variables leads to excess sediment deposition. The degree of w/d departure at this level of assessment that creates unstable conditions or high risk is associated with values greater than 1.4 (**Figure 4-27**). The stability rating using **Figure 4-27** is inserted in Column (8) in **Worksheet 4-11**.

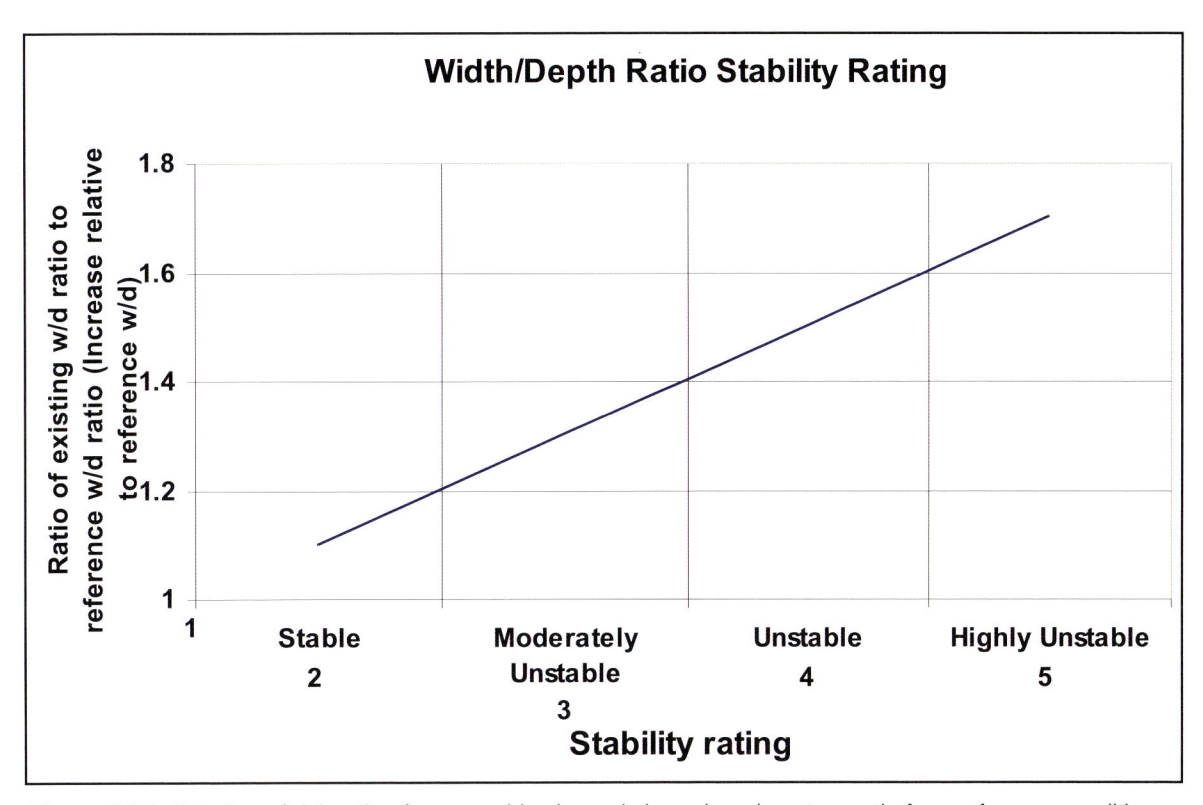

Figure 4-27. Relation of risk rating for over-wide channels based on departure ratio from reference condition.

Depositional patterns adjustment

The categories representing depositional patterns B3, B5, B6 and B7 in **Figure 4-28** represent a higher risk of excess sediment deposition. A reach with any one or combination of B3, B5, B6 or B7 raises the overall aggradation risk one full category (see Column (13) in **Worksheet 4-11**).

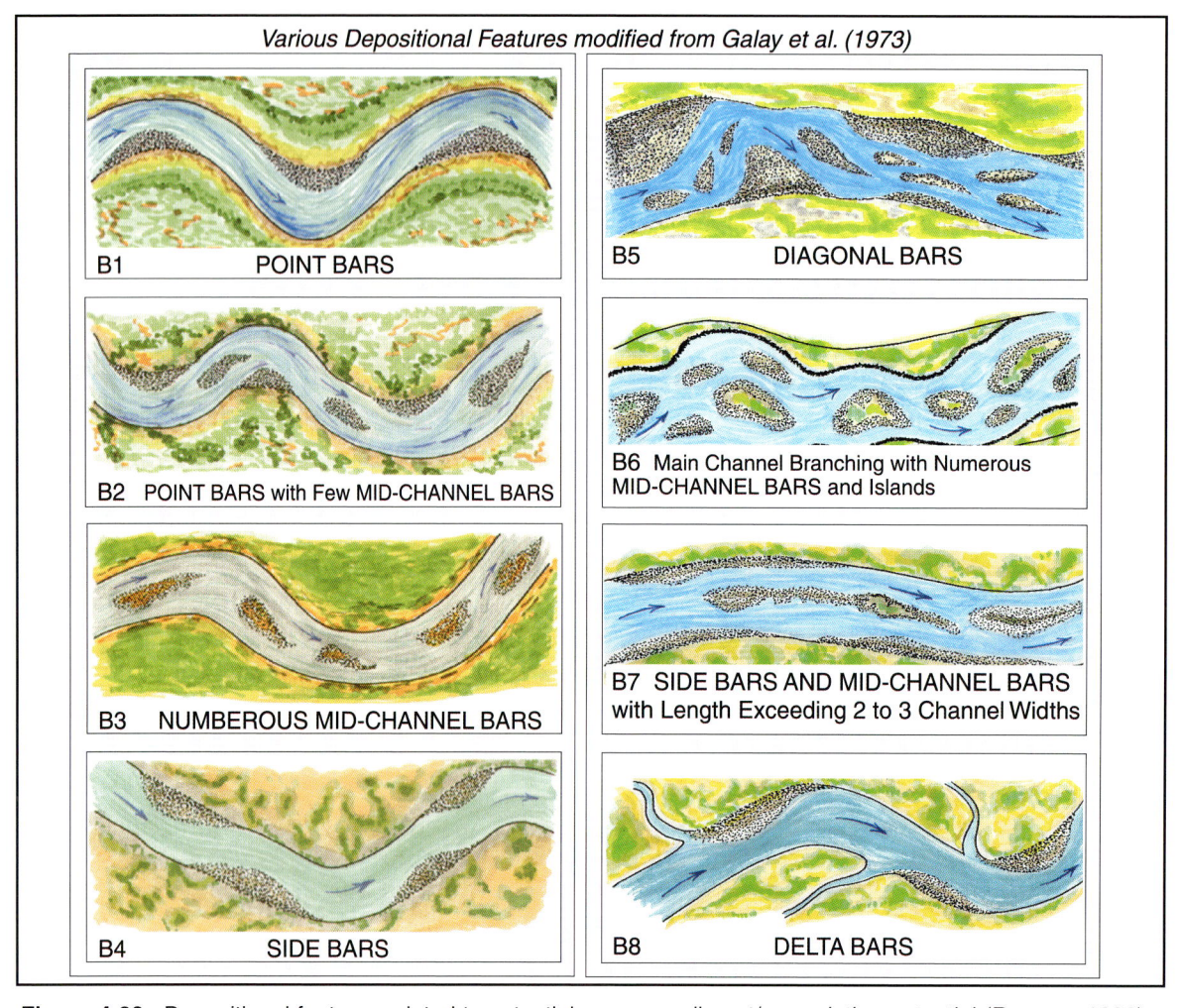

Figure 4-28. Depositional features related to potential excess sediment/aggradation potential (Rosgen, 1996).

Decrease in streamflow due to diversions and/or flow regulation adjustment

The user is advised to review **Figure 4-17** in the streamflow changes **Step 10** and Column (12) in **Worksheet 4-6**. Any diversion and/or flow regulation that *decreases* streamflow in a regulated reach needs to be compared to unregulated tributaries. Streamflow reduction decreases the shear stress and stream power required to route the sediment provided by its watershed. Unregulated tributaries will transport their normal sediment load to the regulated main-shear, potentially causing aggradation/excess deposition. In this situation, the rating is adjusted upward by one full risk category in Column (14) in **Worksheet 4-11**.

The final adjective risk rating for aggradation is recorded in the *RRISSC* summary **Worksheet 4-2** for the reaches assessed. The multiple criteria for aggradation require mitigation based on the specific criteria that result in any *moderate* risk ratings.

Step 18. Determine channel evolution potential

Stream channels change over time due to geologic influences, such as climate change and anthropogenic influences. The tendency of rivers is to seek their own stability within a certain climatic regime. Following disturbance, streams will try to re-establish a dimension, pattern and profile of the pre-disturbance morphology. The existing stream type must be compared to the potential stable form. Reference reaches that represent the stable form of stream type within a similar valley type are used to verify stable morphological relations. The information required to complete this risk assessment includes the following:

- Stream types (Level II);
- Reference condition; and
- Scenarios of successional stages of stream channel evolution (**Figure 2-38**).

The various successional stage departures and their corresponding risk ratings are shown in **Table 4-5**. The first letter in each row refers to the stable condition stream type, while the second letter refers to the current reach stream type. The risk rating for channel evolution is recorded in the *RRISSC* summary **Worksheet 4-2**.

Table 4-5. Risk rating for various stream channel successional state scenarios.

Channel successional states of stream type evolution*	Risk rating
E to C	Moderate (3)
C to D	Very High (5)
B, C, E or D to G	Very High (5)
G to F	High (4)
G to B	Very Low (1)
F to B	Very Low (1)
F to C	Low (2)
F to D	Moderate (3)
All others (e.g., C to E)	Low (2)
* See **Figure 2-38**: Stream type succession scenarios	

The interpretation of channel successional states relates to the nature of adjustment processes associated with a given scenario. For example, a risk rating of *very high* for a C to D shift would be associated with channel enlargement, excess sediment deposition/aggradation, increase in width/depth ratio and increased bank erosion. These changes correspond to a *very high* risk of instability, sediment supply and impairment of various uses. Filling of pools, increases in surface water temperatures and substrate composition shifting to bi-modal (invasion of fine sediments) can create loss of depth and instream cover, food chain shifts and other adverse biological changes.

A change from either a B, C, E or D to a G stream type is also associated with a *very high* risk. This risk rating relates to floodplain abandonment due to channel incision and accelerated bed and bank erosion rates. The G channels are poor habitat due to their hydraulic and sedimentologic characteristics. The very high shear stress and stream power decrease and/or eliminate holding cover for all age classes of fish.

The *very low* and *low* risk ratings are associated with stream reaches that are presently recovering or have recovered to their physical potential stable state.

For *moderate*-risk areas, revised land use management recommendations should be directed at encouraging the processes necessary for improving recovery rates. This would include riparian vegetation management, riparian grazing strategies, specifying riparian management zones that identify criteria for allowable site disturbance, etc.

High to *very high* risk ratings require advancement to *PLA*. This level is necessary for actual river restoration/natural channel design.

Step 19. Calculate degradation risk

Lowering of local baseflow level through channel incision becomes a major adverse impact to stream stability, sediment supply and water resource uses. It also limits riparian vegetation's usefulness in stream recovery, as the normal high flows are often still below the rooting depth, generating less root mass and leading to accelerated streambank erosion. Water tables are lowered, land productivity decreases, land loss increases due to streambank erosion and tributaries are over-steepened, causing headward advancement of the incision. Excess sediment is produced downstream of degrading channels, leading to further channel instability and loss of fish habitat. Degradation can be created by contraction scour below culverts or narrow bridges; thus, road crossings may contribute to degradation. The information required to complete this step includes the following:

- Stream types;
- Stream channel evolution risk (**Step 18** in **Worksheet 4-2**; **Table 4-5**);
- Streamflow changes risk (**Step 10** in **Worksheet 4-2**; **Worksheet 4-6**);
- Roads, drainageway crossing designs (potential base-level shifts) (**Worksheet 4-13**);
- In-channel mining associated with base-level shifts (**Step 14** in **Worksheet 4-2**; **Worksheet 4-8**); and
- Direct channel impact risk (straightening, channelization, dredging) (**Step 15** in **Worksheet 4-2**; **Worksheet 4-9**).

The overall risk rating for degradation potential for each reach is determined by using the highest individual risk rating in **Worksheet 4-12** for any specific category rather than summarizing total points. The risk rating for road drainage design (Column (5) in **Worksheet 4-12**) is determined by using **Worksheet 4-13** and **Figure 4-29**. A decrease in width/depth ratio from reference (**Figure 4-29**) is completed that compares departure from the reference condition. This risk assessment is implemented with the following assumptions:

- That a decrease in width/depth ratio increases shear stress for the same discharge, which increases the likelihood for channel incision; and
- That the stream channel shows evidence of incision (by increasing bank-height ratio of lowest bank).

Note: If no bridges or culverts in any specific reach, Worksheet 4-13 does not need to be completed for that reach; a risk rating of "Very Low (1)" is automatically given for road drainage density to be inserted in Column (5) in Worksheet 4-12.

All other risk ratings necessary to determine overall degradation risk have previously been determined.

Insert the overall risk rating for degradation in the *RRISSC* summary **Worksheet 4-2**.

For *moderate* risk ratings of degradation, the user needs to identify the various causes, locations and land uses responsible for the risk rating. General mitigation may be very effective at reversing some of the problems encountered by specific activities/criteria. For *high* or *very high* risk ratings, the *PLA* level provides a more quantitative basis for mitigation, restoration and/or grade control stabilization design. Culvert and/or bridge retro-fits must be sufficient to correct the identified problems.

Worksheet 4-12. Risk rating worksheet for degradation.

(1)	(2)	(3)	(4)	(5)	(6)	(7)
Location code/ river reach I.D.	Risk rating: streamflow changes **(Step 10 in Worksheet 4-2; Worksheet 4-6)**	Risk rating: in-channel mining associated with base-level shifts **(Step 14 in Worksheet 4-2; Worksheet 4-8)**	Risk rating: channel evolution **(Step 18 in Worksheet 4-2; Table 4-5)**	Risk rating: road drainage designs, "shot gun" culverts (base-level shifts) **(Worksheet 4-13)**	Risk rating: direct channel impacts **(Step 15 in Worksheet 4-2; Worksheet 4-9)**	Overall risk rating for degradation (Insert highest adjective rating from Columns 2–6)
1.						
2.						
3.						
4.						
5.						
6.						
7.						
8.						
9.						
10.						
11.						
12.						
13.						
14.						
15.						

Chapter 4   The Rapid Resource Inventory for Sediment and Stability Consequence (RRISSC)

Worksheet 4-13. Risk rating worksheet for potential contraction scour/degradation/channel incision due to culverts or bridges.

(1)	(2)	(3)	(4)	(5)	(6)	(7)	(8)	(9)	(10)
Location code/ river reach I.D.	Percent reduction of sinuosity (insert numeric rating)	Stream crossing structure (insert numeric rating)	Subtotal Σ[(2)+(3)]	Increase in energy slope (use (4) points and insert numeric rating)	Ratio of a decrease in w/d ratio to existing reference w/d ratio (**Figure 4-29**) (insert numeric rating)	Backwater potential above structure (insert numeric rating)	Presence of floodplain drains (through fills) (insert numeric rating)	Subtotal Σ[(5)+(6)+(7)+(8)]	Overall risk rating: culverts or bridges
	(1) = No change	(1) = Bridge		VL (1) = 2	VL (1) > 8.0	VL (1) = None	VL (1) = All floods greater than bankfull drain through fill		VL (1) = 4
	(2) = Sinuosity reduced up to 50%	(2) = Arch culvert		L (2) = 3	L (2) = 0.61–0.80	L (2) = Slight only for floods > 50 yr recurrence interval	L (2) = Accommodates 90% of floods		L (2) = 5–8
	(3) = Sinuosity reduced 50–80%	(3) = Culvert		M (3) = 4	M (3) = 0.41–0.60	M (3) = Some for floods 11–50 yr recurrence interval	M (3) = Accommodates 50–89% of floods		M (3) = 9–12
	(4) = Sinuosity reduced more than 80%	(4) = Over-steepened culvert		H (4) = 5–6	H (4) = 0.21–0.40	H (4) = Evident for floods 2–10 yr recurrence interval	H (4) = Evident for floods 2–10 yr recurrence interval		H (4) = 13–16
				VH (5) = 7–8	VH (5) ≤ 0.20	VH (5) = Backwater at bankfull discharge	VH (5) = Backwater at bankfull discharge		VH (5) = 17–20
1.									
2.									
3.									
4.									
5.									
6.									
7.									
8.									
9.									
10.									
11.									
12.									
13.									
14.									
15.									

Channel incision leading to degradation and vertical containment of river channels is determined by dividing the lowest bank height by the bankfull stage maximum depth, termed bank-height ratio. The relation in **Figure 4-29** of width/depth ratio decrease from reference is used only for incising channels or for bank-height ratios greater than 1. The risk rating using **Figure 4-29** is inserted in Column (6) in **Worksheet 4-13**.

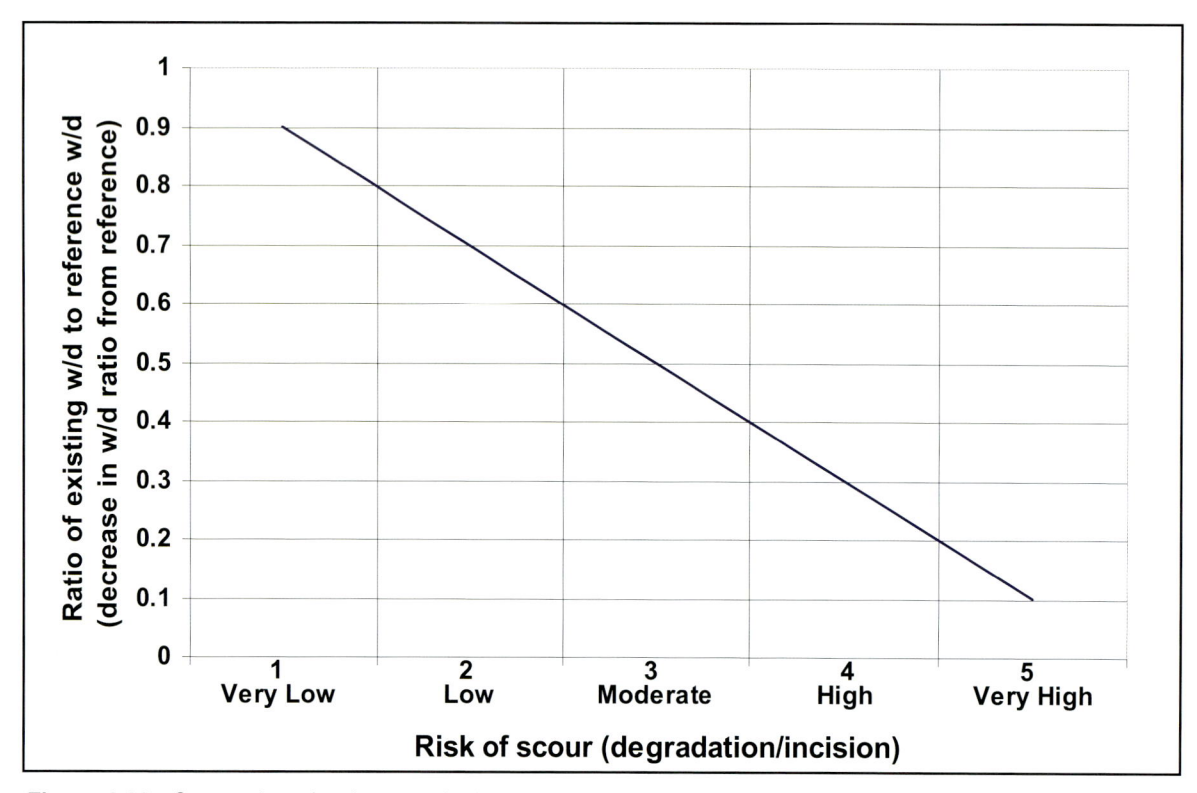

Figure 4-29. Conversion of a decrease in the existing width/depth ratio compared to reference width/depth ratio for potential degradation (incision due to excess energy). This relation is used only if the lowest bank height is greater than the maximum bankfull depth (bank-height ratio > 1).

Step 20. Summarize total potential sediment and stream channel stability risk

This summary identifies specific areas and reaches that, based on the severity and potential adverse consequence of the particular rating, require either mitigation and/or a more detailed risk prediction assessment. This allows the user to identify which locations are affected by specific processes and land use activities. The *RRISSC* summary **Worksheet 4-2** is used to match risk, processes and practices to direct this phase in the *WARSSS* assessment.

Step 21. Create overall risk rating summary

This step allows the user to interpret the potential severity of change, departure from reference condition, recovery potential, long- versus short-term potential impacts, sources and causes of impairment and the consequences of potential impairments to beneficial uses. The results document which processes and associated areas warrant further assessment at the *PLA* level. This information is summarized in **Worksheet 4-2** and directly affects subsequent actions taken in **Steps 22**, **23** and **24**. This information should be used to prepare a narrative summary that indicates the major contributing factors and rationale for the ratings assigned to each location. For areas of *high* and *very high* risk, the summary should identify the processes influenced and the land use practices associated with the potential risk impairment. This information will assist the user in justifying the need to advance these processes to the *PLA* phase. Changes in land use management, stabilization, enhancement and restoration proposals will result from the more detailed and site-specific *PLA* analysis. For the *moderate* risk ratings, the summary should also formulate recommendations for mitigation and monitoring, as discussed in **Step 23**.

Step 22. Discard processes with low overall risk ratings

Recommendations for areas associated with *very low* and *low* risk ratings are generally to proceed with no change in management practices and no mitigation design. Because of low likelihood of disproportionate sediment supply, river instability or associated adverse consequences, no monitoring is recommended unless it will serve to document a stable reference condition within the low-risk reaches.

Step 23. Create management change recommendations for processes with moderate overall risk ratings

The pressing question for areas associated with *moderate* risk ratings is whether to continue with a *PLA* assessment or whether to develop site- and process-specific mitigation and schedule monitoring. Part of the answer to this question is associated with the potential severity and/or type of adverse consequences that would affect these locations. The uncertainty of the relations used in the *RRISSC* analysis may lead the user (or critics) to suggest that a more rigorous, quantitative method be implemented. The question of whether to proceed to *PLA* can also be answered by evaluating the potential magnitude and adverse consequences of the impairment associated with the water resource's uses and values.

If mitigation is indicated, users should design mitigation and management strategies that relate specifically to the activity that caused the risk rating and the processes affected. Results may also include recommendations for stabilization, enhancement and/or restoration. Monitoring is necessary to ensure that the mitigation is effective.

Mitigation

Management prescriptions can address these *moderate*-risk areas/processes. Prescribing specific management practices for the large variety of settings one might encounter is well beyond the scope of *WARSSS*. As an example, a season-long grazing system in poor condition on a C4 stream type with willow riparian vegetation may result in a recommended mitigation of a deferred or rotation grazing practice with spring riparian pasture grazing not to exceed 10–15% utilization of new willow growth or hoof shear on streambanks. This mitigation practice would be a better management choice than excluding grazing or the expensive option of mechanical stabilization of streambanks, unless subsequent monitoring indicated poor results.

Monitoring

Effectiveness monitoring evaluates the stream's response to mitigation. Monitoring streambank erosion resulting from the changed grazing management practices mentioned above would involve the installation and annual measurement of bank and/or toe pins. Bank pins installed in stable reference streambanks in the same vicinity or valley type would indicate the natural, acceptable rate of bank erosion compared to departure rates of the *moderate*-risk reach. This type of effectiveness monitoring is not very expensive or time-consuming, and adds a wealth of data that will either support an effective management program or identify additional stabilization or restoration needs. Additional information is included in Chapter 6.

Step 24. Advance high-risk and/or high-consequence processes to the PLA assessment phase

Locations associated with *high* or *very high* risk ratings have a high risk of impairment in the form of disproportionate sediment supply, channel instability and associated severe adverse consequences. In many cases, potential mitigation is so extensive that its implementation is questioned. These sub-watersheds, slopes and/or river reaches may require the more detailed assessment performed in the *PLA* phase of the *WARSSS* methodology. The *PLA* assessment will isolate and quantify sediment sources by specific processes and locations. This information can then be used to determine mitigation design and priorities.

"Inductive reasoning from facts observed in nature is verified only by the persistence and patience of consistent field measurements over time."

CHAPTER 5

The Prediction Level Assessment (PLA)

The *Prediction Level Assessment* (*PLA*), the most detailed level of the *WARSSS* methodology, is reserved for sub-watersheds and river reaches previously identified as being at high risk for sediment and/or river stability problems. These selected areas require this level of analysis due to their sensitive nature, the value of the resource and/or the severity of the adverse consequences of impairment. *PLA* compares direction, rate, nature and extent of departure of existing sediment and channel stability to a reference reach condition typical of stable, natural land and stream systems. The same assessment methodology is used for the reference condition and the previously identified high-risk areas. The following are seven objectives of *PLA*:

1. Locate and quantify sediment sources;
2. Link sediment sources to various land uses;
3. Identify disproportionate sediment supply;
4. Evaluate sediment impacts on river channels;
5. Integrate hydrology, river morphology and river stability with land use impacts by specific location;
6. Determine departure and degree of impairment due to sediment sources, watershed hydrology and riparian impacts; and
7. Provide sufficient detail to design site- and process-specific mitigation.

The general sequence of analysis is shown in **Flowcharts 5-1** and **5-2**. These comparisons of reference sediment and stability conditions are used to identify potential departure and to document "acceptable" erosion and sedimentation rates.

The results of *PLA* link quantitative evaluations of sediment sources and river stability problems to an individual source at a specific location that affects a particular process or a combination of processes. This allows users to identify proportional distributions of sediment yield, consequences of sediment on river stability and the influences of river instability on sediment yields. This information enables the design of well-targeted, site-specific and process-specific management prescriptions. To address possible prediction uncertainty, validation monitoring compares predicted values to observed values. The same monitoring design can also determine mitigation effectiveness. Chapter 6 presents monitoring concepts and methodologies.

WATERSHED ASSESSMENT OF RIVER STABILITY AND SEDIMENT SUPPLY

The time required for one person to assess a third-order watershed using the *PLA* methodology is approximately one month (rather than days or weeks), depending on the availability of basic data, maps, photographs, soils mapping, size of watershed, miles of stream length and the observer's experience and access to the site. The timeframe to complete the assessment can be shortened once an individual becomes familiar with the methods and builds a local database. The methods rely heavily on field measurements relating to hillslope, hydrologic and channel processes.

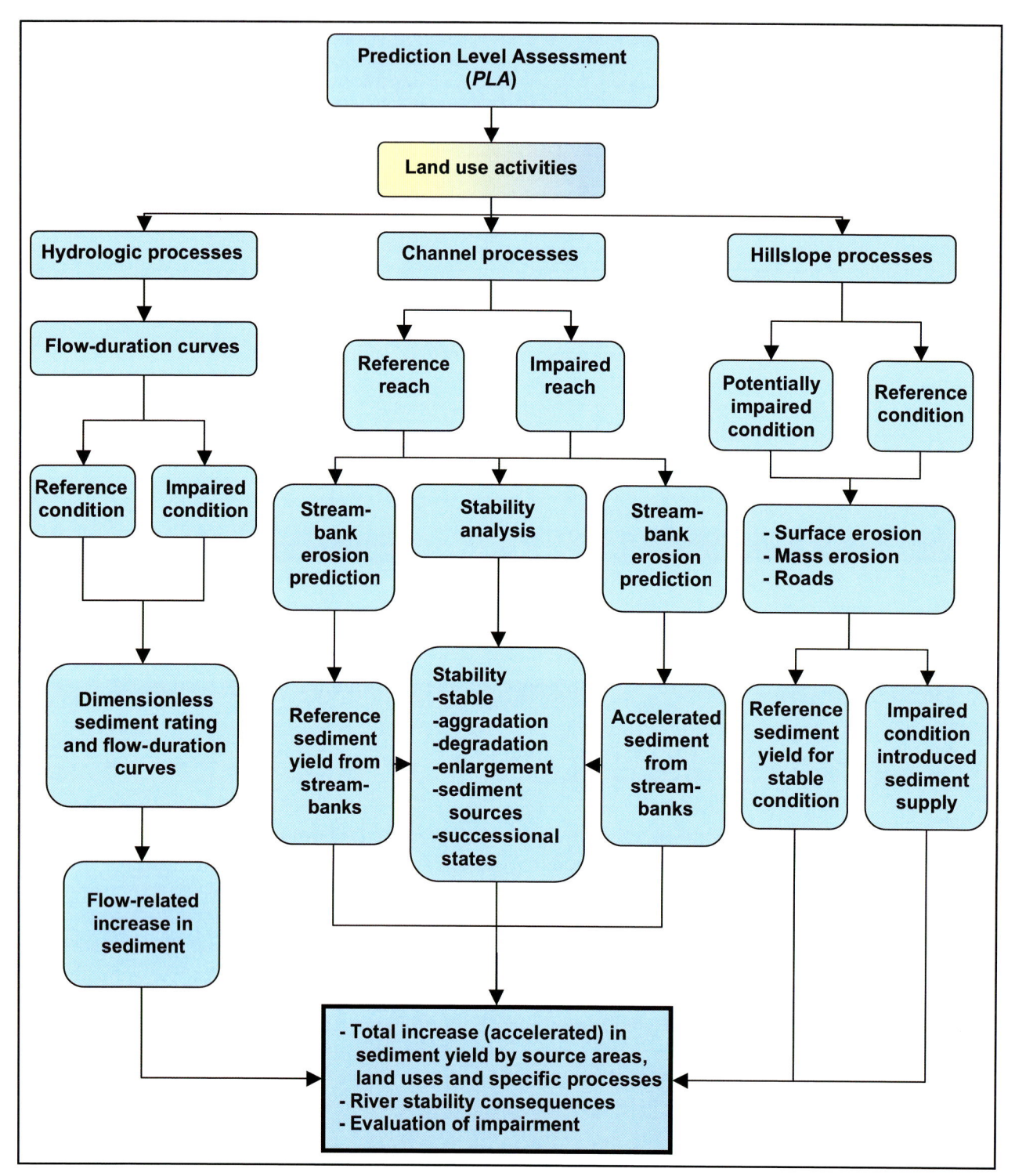

Flowchart 5-1. *PLA* comparative analysis of reference condition and impaired condition in parallel.

5-2

Chapter 5 — The Prediction Level Assessment (PLA)

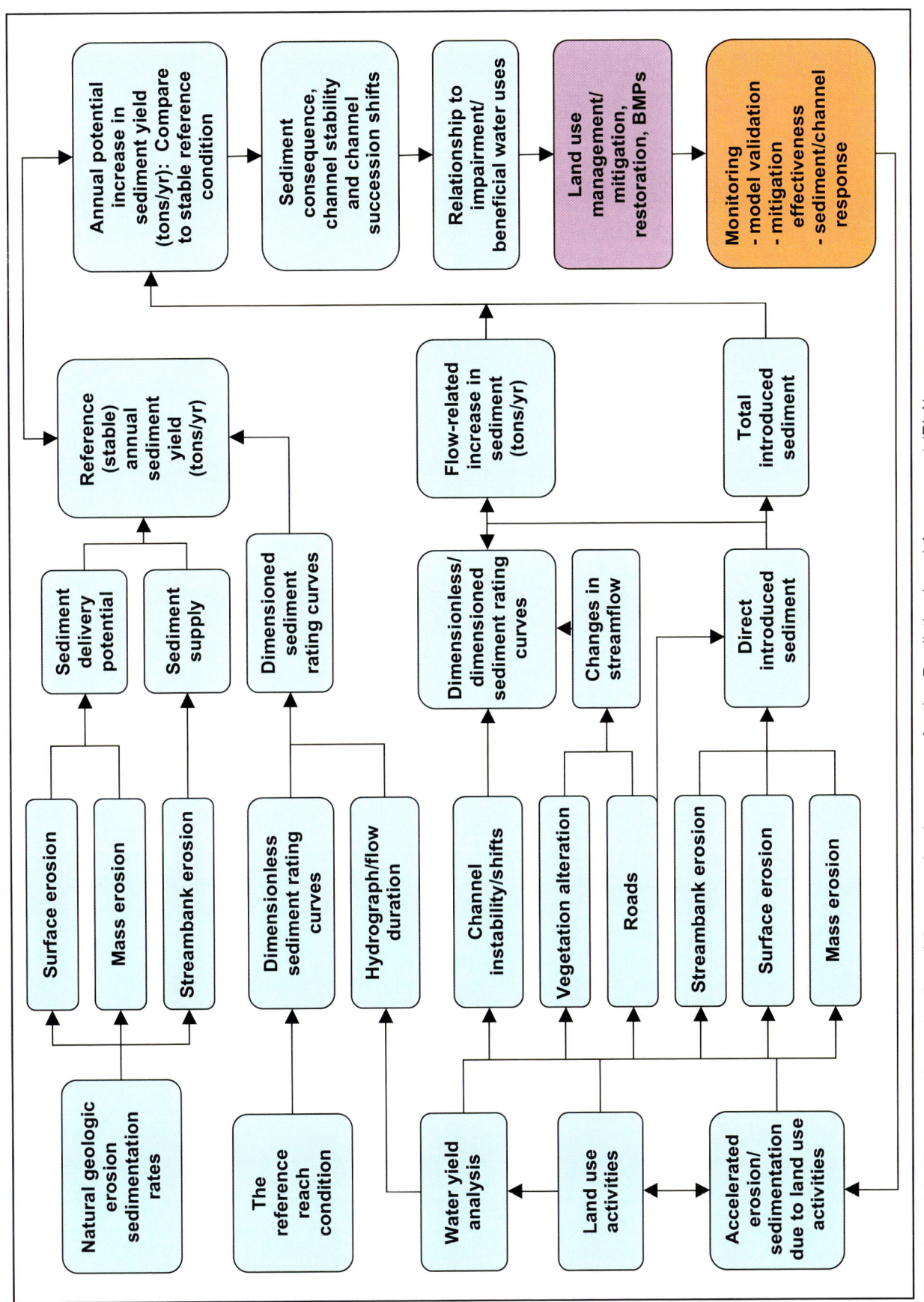

Flowchart 5-2. The general organization of the procedural sequence for the *Prediction Level Assessment (PLA)*.

PLA analytical concepts

The *PLA* assessment completes and compares the same field measurements and analysis for both stable hillslopes and reference stream reaches and impaired reaches. The analysis is applied first to stable hillslopes and reference reaches to calibrate the method and then to impaired hillslopes and reaches to quantify departure. Users will record data for reference and individual reaches and will transfer final data to summary worksheets. This summary allows decision makers to look comparatively at multiple locations as they prioritize management and mitigation needs. This assessment is designed for individuals trained and experienced in geomorphology, hydrology, engineering, geology, soils science and related fields. ***All of the prediction methods included in this approach can be validated by process-specific monitoring procedures.*** If concerns exist over the uncertainty of prediction, validation monitoring will adaptively improve the prediction.

The *PLA* methodology involves detecting sediment sources, measuring sediment sizes and determining the ability of the stream to transport sediment without adverse channel adjustment. Sediment storage and routing dealing with seasonal and annual distribution and redistribution of sediment would require a complex, three-dimensional model. Rather than calculate sediment storage, potential impairment is assessed by determining increases in sediment supply for the receiving stream channel. How well the stream channel can accommodate sediment supply increases and/or flow changes is determined by competence and capacity predictions. Measurements of *competence*, the ability to move the largest available particle, and sediment transport *capacity*, the ability to accommodate sediment volume, are calculated. Potential aggradation and/or degradation, based on changes in the sediment regime, channel characteristics and streamflow, are key portions of the assessment.

The *PLA* methodology provides recommended models and analytical procedures to predict sediment and stability conditions. The framework allows the option to substitute a more familiar model to complete specific process predictions. It is important, however, that the output variables of the substitute model be consistent with the numeric values and methods presented for consistency of interpretation.

The sequential steps in **Flowchart 5-3** route the assessment through a logical sequence to determine stream stability and to quantify sediment source by process (hillslope, channel and hydrologic). The analysis provides sufficient detail to determine stream instability and disproportionate sediment supply related to individual land use and specific erosional/depositional processes (such as streambank erosion rate).

Flowchart 5-3 also assists as a user's guide to follow the multiple steps throughout Chapter 5. Each step is described separately and may be further organized into many sub-steps within individual procedures. Specific flowcharts, worksheets, tables and figures are provided to complete the analysis.

Furthermore, a software program, RIVERMorph™, has been developed and extensively tested to assist the user in accomplishing a majority of the analysis in Chapter 5. RIVERMorph™ speeds up the process and builds a database to be shared with others.

Initial preparation

The information in **Table 5-1** is recommended for this approach.

Chapter 5 — The Prediction Level Assessment (PLA)

Stability

- Steps 1–4: Bankfull discharge and hydraulic relations
- Steps 5–6: Level II stream classification and dimensionless ratios of channel features
- Step 7: Identify stream stability indices
- Steps 18–19: Sediment transport capacity model (POWERSED)
- Step 22: Calculate sediment entrainment/competence
- Step 23: Predict channel response based on sediment competence and transport capacity
- Steps 24–27: Calculate channel stability ratings by various processes and source locations
- Step 28: Determine overall sediment supply rating based on individual and combined stability ratings

Sediment Supply

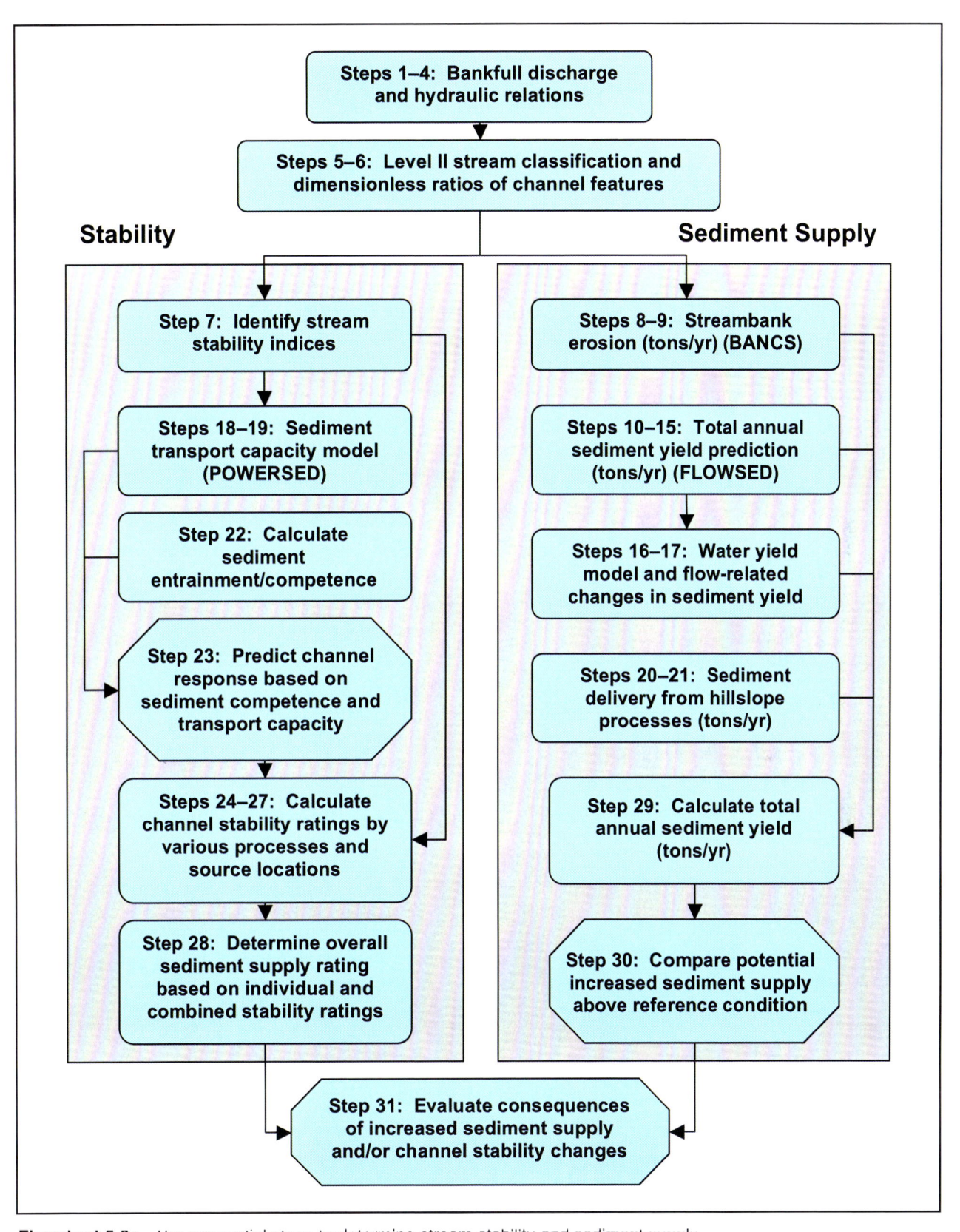

- Steps 8–9: Streambank erosion (tons/yr) (BANCS)
- Steps 10–15: Total annual sediment yield prediction (tons/yr) (FLOWSED)
- Steps 16–17: Water yield model and flow-related changes in sediment yield
- Steps 20–21: Sediment delivery from hillslope processes (tons/yr)
- Step 29: Calculate total annual sediment yield (tons/yr)
- Step 30: Compare potential increased sediment supply above reference condition

Step 31: Evaluate consequences of increased sediment supply and/or channel stability changes

Flowchart 5-3a. The sequential steps to determine stream stability and sediment supply.

Table 5-1. Recommended information sources for the *Prediction Level Assessment (PLA)*.

Information needed for *PLA*

- Soil type maps
- Topographic maps
- Aerial photography
    - current
    - historical
- Time-trend vegetation alteration maps
    - nature of change (% stand or area, composition change)
    - extent (acres)
    - location
    - dates of change
- Geologic maps
- Geologic hazard maps
- Precipitation isohyetal map
- Satellite imagery during run-off periods (color I-R)
- Level II stream classification and related channel measurements
    - bankfull width
    - max bankfull depth
    - mean depth
    - sinuosity
    - dominant bed material D_{50}
    - w/d ratio
    - flood-prone area width
    - entrenchment ratio
- River stability analysis
- Riparian vegetation inventory
- Flow regime type
- In-channel large wood inventory
- Stream order
- Meander pattern description
- Depositional pattern description
- Reference reach morphology and dimensionless ratios of dimension, pattern and profile
- Pfankuch channel stability rating
- Degree of channel incision: Bank-Height Ratio (BHR) = lowest bank height divided by max bankfull depth
- Stream channel successional state
- Sediment competence calculations
- Sediment capacity calculations
- Dimensionless flow-duration curves
- Dimensionless sediment rating curves
    - one representative data point obtained at bankfull stage for:
        a. bedload sediment
        b. suspended sediment
        c. streamflow (momentary maximum)
- Mean daily bankfull discharge
- Dimensioned flow-duration curves for pre- and post-treatment change (from water yield model)
- Stream channel: dimension, pattern, profile and materials
    - slope
    - mean bankfull velocity
    - mean bankfull depth
    - bankfull surface width
    - D_{50} riffle bed
    - D_{84} riffle bed
    - bar sample (D_{max}, D_{50})
- Unit stream power
- Friction factors
- Relative roughness
    - dimensionless hydraulic geometry (POWERSED)
- Streambank erosion prediction (BANCS)
    BEHI (Bank Erosion Hazard Index)
    - max depth at toe of bank
    - rooting depth
    - rooting density
    - study bank height
    - length of bank by condition
    - bank angle
    - bank material
    - bank stratigraphy
    NBS (Near-Bank Stress)
    - radius of curvature
    - bankfull width
    - max bankfull width
    - channel slope
    - near-bank slope
    - near-bank max depth
    - nature and extent of deposition
- Roads
    - number of stream crossings
    - acres of road
    - age of road
    - specific mitigation

Chapter 5 · The Prediction Level Assessment (PLA)

Steps 1–4: Bankfull discharge and hydraulic relations (See Flowchart 5-4)

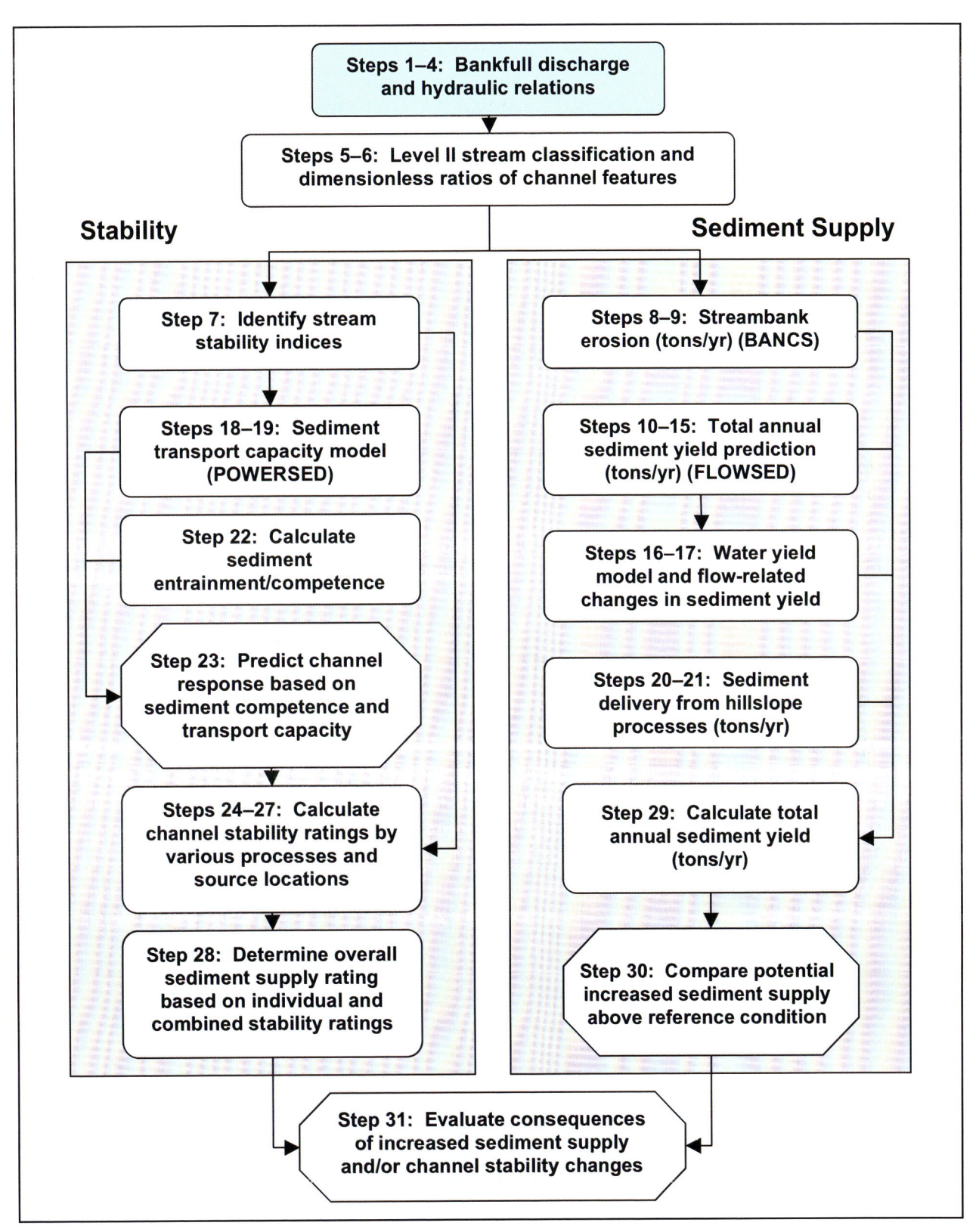

Steps 1–4: Bankfull discharge and hydraulic relations

↓

Steps 5–6: Level II stream classification and dimensionless ratios of channel features

Stability

Step 7: Identify stream stability indices

↓

Steps 18–19: Sediment transport capacity model (POWERSED)

↓

Step 22: Calculate sediment entrainment/competence

↓

Step 23: Predict channel response based on sediment competence and transport capacity

↓

Steps 24–27: Calculate channel stability ratings by various processes and source locations

↓

Step 28: Determine overall sediment supply rating based on individual and combined stability ratings

Sediment Supply

Steps 8–9: Streambank erosion (tons/yr) (BANCS)

↓

Steps 10–15: Total annual sediment yield prediction (tons/yr) (FLOWSED)

↓

Steps 16–17: Water yield model and flow-related changes in sediment yield

↓

Steps 20–21: Sediment delivery from hillslope processes (tons/yr)

↓

Step 29: Calculate total annual sediment yield (tons/yr)

↓

Step 30: Compare potential increased sediment supply above reference condition

↓

Step 31: Evaluate consequences of increased sediment supply and/or channel stability changes

Flowchart 5-3b. The sequential steps to determine stream stability and sediment supply, highlighting **Steps 1–4**.

5-7

Steps 1–4: Bankfull discharge and hydraulic relations

Step 1: Develop and/or obtain regional curves of bankfull discharge versus drainage area and bankfull dimensions versus drainage area (Figure 5-2)

- Cross-sectional area (ft^2) vs. drainage area
- Width (ft) vs. drainage area
- Mean depth (ft) vs. drainage area
- Bankfull discharge (cfs) vs. drainage area
- Bankfull velocity (ft/s) vs. drainage area

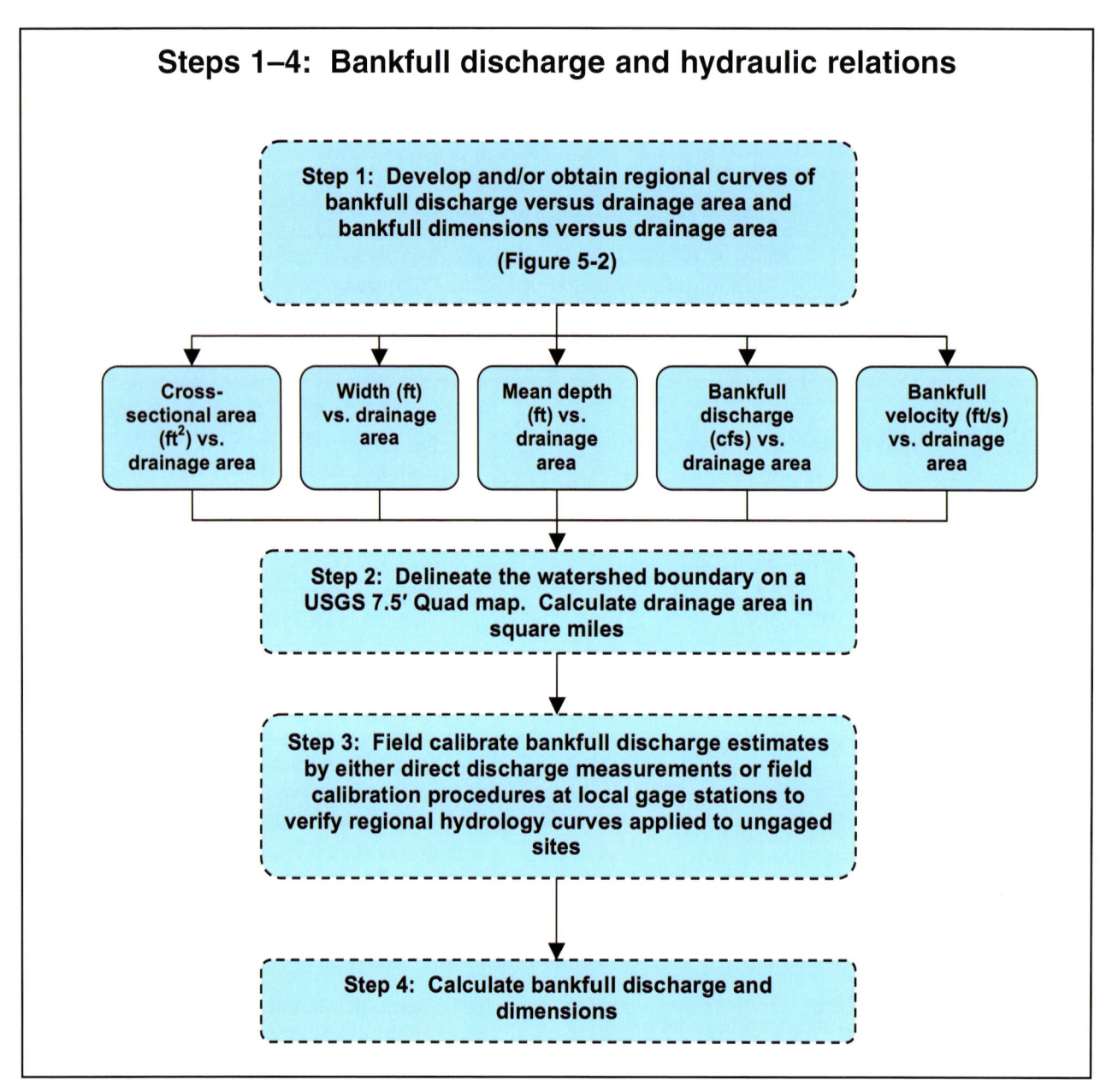

Step 2: Delineate the watershed boundary on a USGS 7.5' Quad map. Calculate drainage area in square miles

Step 3: Field calibrate bankfull discharge estimates by either direct discharge measurements or field calibration procedures at local gage stations to verify regional hydrology curves applied to ungaged sites

Step 4: Calculate bankfull discharge and dimensions

Flowchart 5-4. Sequential steps and variables to obtain bankfull discharge and bankfull channel dimensions.

Step 1. Develop and/or obtain regional curves of bankfull discharge versus drainage area and bankfull channel dimensions versus drainage area

Bankfull channel dimensions of cross-sectional area, width, mean depth and the related streamflow velocities tend to increase proportionately with increases in drainage area (Leopold et al., 1964). When stratified by stream type, plots of bankfull channel dimensions versus drainage area prove even more useful for estimating similar channel dimensions for ungaged areas. Selected average bankfull channel dimensions for four hydro-physiographic regions are shown in **Figure 5-1**. The regional curves are most valuable when bankfull indicators are not evident at ungaged sites. Actively incising channels are a good example of a difficult location to estimate bankfull stage. The curves, however, must be validated locally and separated by hydro-physiographic provinces (changes in precipitation, lithology and/or major land use, such as percent of watershed in impervious condition).

The majority of USGS streams are alluvial channels. The bankfull channel dimensions that have been collected and field-calibrated at streamgages should be plotted to build a supporting database that can be used to refine estimates of bankfull channel dimensions for ungaged areas. As one constructs local relationships and curves from the gage data, the plot of bankfull channel dimensions by drainage area should be integrated or stratified by hydro-physiographic province and by stream type.

As changes in stream type reflect changes in width/depth ratio, caution should be used in extrapolating width and depth measurements from regional curves. Cross-sectional area, however, is not as sensitive to various stream types compared to width and depth. When constructing bankfull dimensions by drainage area relations, it is valuable to denote individual stream types with their corresponding width/depth ratio. Follow the progression in **Figure 5-2** to field calibrate the bankfull discharge and to develop regional curves (Rosgen and Silvey, 2005).

Watershed Assessment of River Stability and Sediment Supply

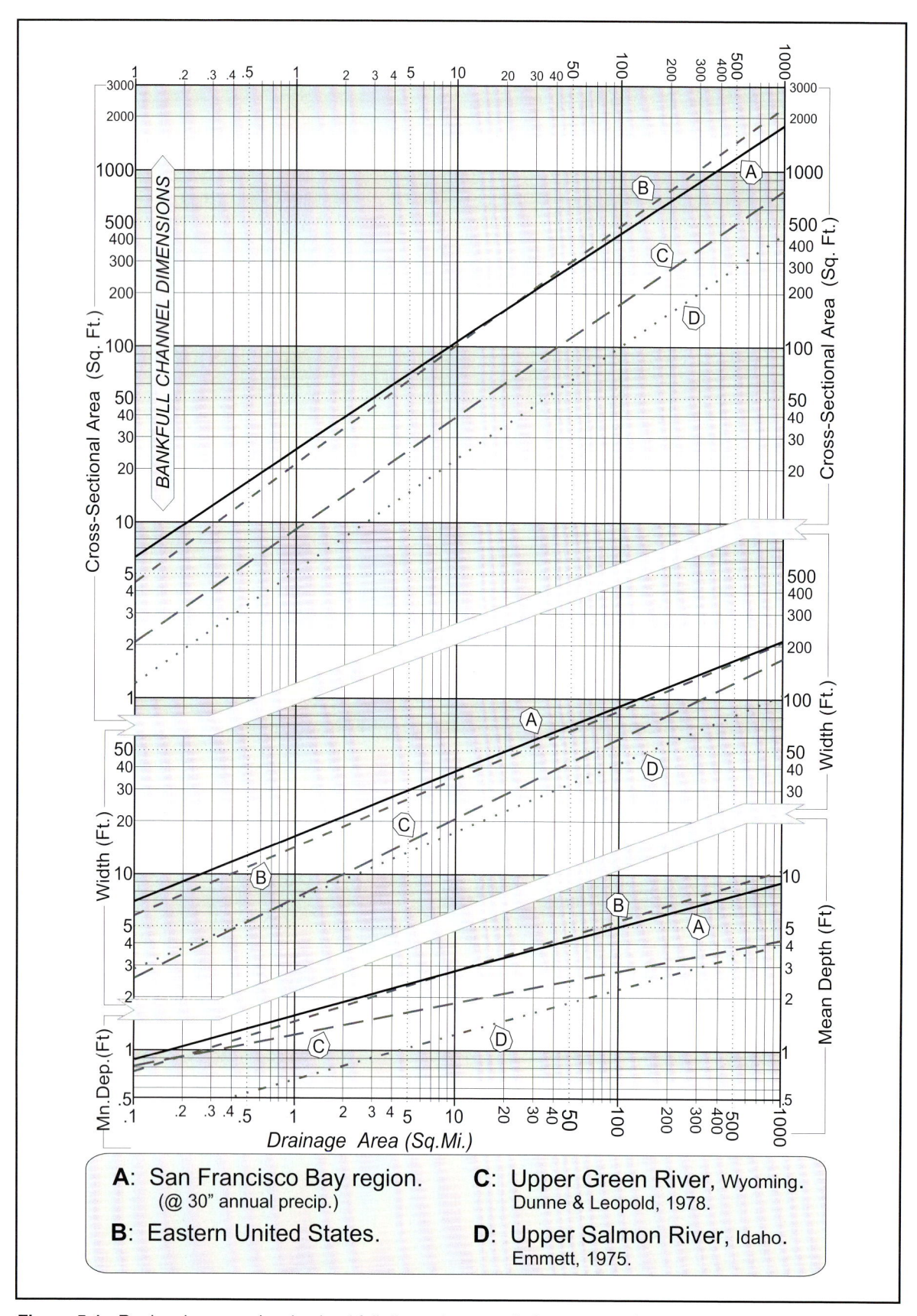

Figure 5-1. Regional curves showing bankfull dimensions vs. drainage areas for various hydro-physiographic provinces (adapted from Dunne and Leopold, 1978; Emmett, 1975; Rosgen and Silvey, 2005).

A: San Francisco Bay region. (@ 30" annual precip.)
B: Eastern United States.
C: Upper Green River, Wyoming. Dunne & Leopold, 1978.
D: Upper Salmon River, Idaho. Emmett, 1975.

Chapter 5 — The Prediction Level Assessment (PLA)

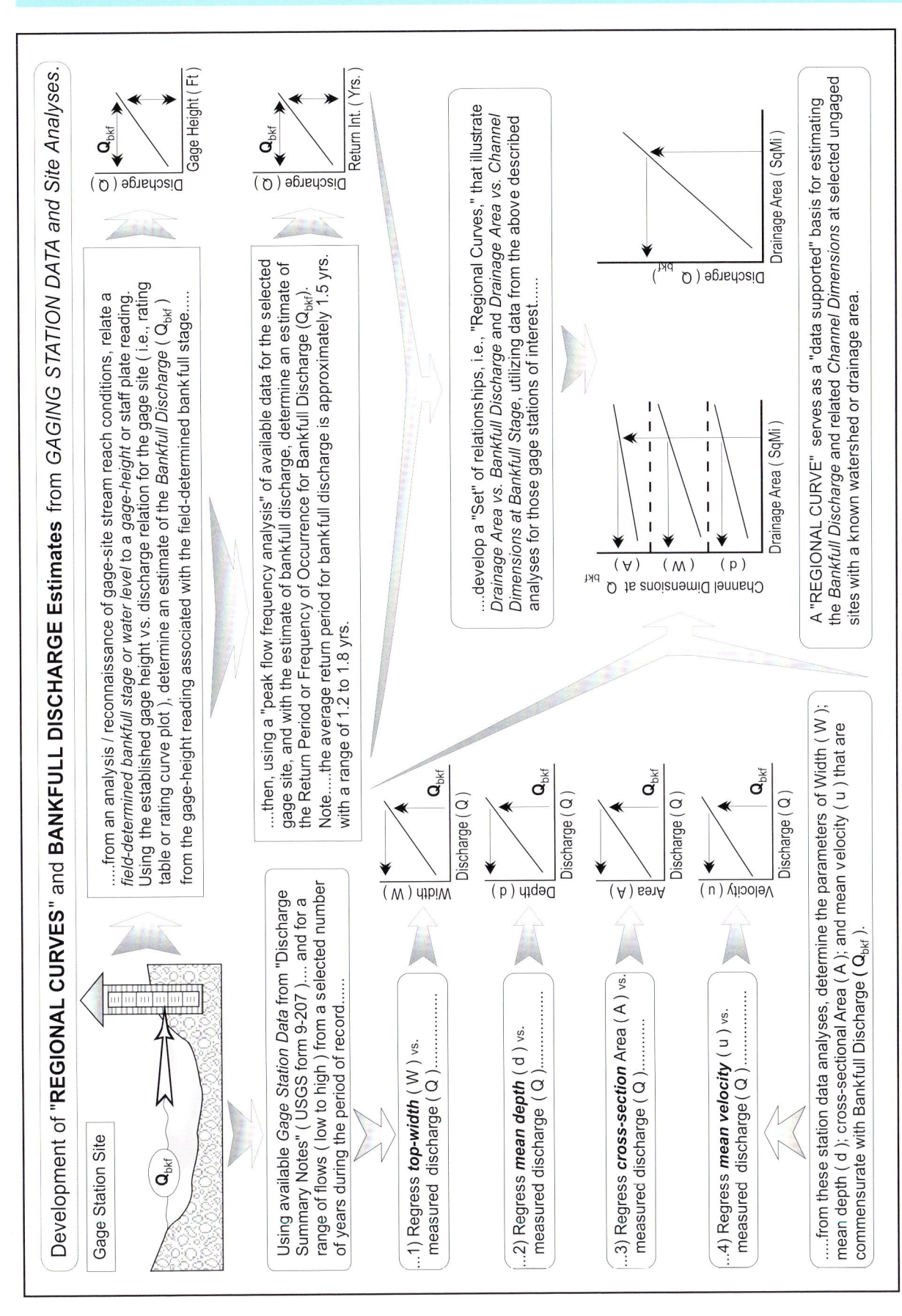

Figure 5-2. Development of regional curves and bankfull discharge estimates from gaging station data and site analyses (Rosgen and Silvey, 2005).

5-11

Step 2. Delineate the watershed boundary on a USGS 7.5' Quad map. Calculate drainage area in square miles

Delineate the watershed boundary on a U.S. Geological Survey (USGS) 7.5' Quad map or use previous delineations from the *RLA* or *RRISSC* phases. Calculate the drainage area in square miles using GIS (preferably) or a manual area grid- or dot-sample estimation technique. Drainage area is entered in **Worksheet 5-1** for gage station locations and **Worksheet 5-2** (**Step 4**) and **Worksheet 5-3** (**Step 5**) for assessment watersheds.

Also note the location in **Worksheet 5-1** to record the Hydrologic Unit Code (HUC). This code is a national standard, unique numbering system of rivers, consisting of two to eight digits, as described in the *USGS Water-Supply Paper* (Seaber et al., 1987). **Worksheet 5-1** provides additional space to allow the user the option of further delineating the river based on the user's coding scheme.

Chapter 5   The Prediction Level Assessment (PLA)

Worksheet 5-1. Sample form for recording gage station and field data from *The Reference Reach Field Book* (Rosgen and Silvey, 2005).

Summary....**USGS GAGE STATION** Data / Records...for...
STREAM CHANNEL CLASSIFICATION

Station NAME: Station Number:

LOCATION:

Period of RECORD: _____ Yrs.   Mean Annual DISCHARGE: _____ CFS

Drainage AREA: _____ Ac. _____ SqMi.   Drainage Area Mn. ELEV: _____ Ft.

Reference REACH SLOPE: _____ Ft/Ft.   **VALLEY TYPE:** _____

STREAM TYPE: _____   HUC: _ _ _ _ _ _ _ _ _ _ _ _ _ _

"BANKFULL" CHARACTERISTICS

Determined by FIELD MEASUREMENT	*Determined from GAGE DATA Analyses*
Bankfull WIDTH (W_{bkf}) _____ Ft	Bankfull WIDTH: (W_{bkf}) _____ Ft
Bankfull MEAN DEPTH (d_{bkf}) _____ Ft	Bankfull MEAN DEPTH (d_{bkf}) _____ Ft
Bankfull Xsec AREA (A_{bkf}) _____ SqFt	Bankfull Xsec AREA (A_{bkf}) _____ SqFt
Wetted PERIMETER (WP) _____ Ft	Wetted PERIMETER (WP) _____ Ft
Bankfull STAGE (Gage Ht.) _____ Ft	Bankfull STAGE (Gage Ht.) _____ Ft
Est. Mean VELOCITY (u) _____ Ft/sec	Mean VELOCITY (u) _____ Ft/sec
Est. Bnkfl. DISCHARGE (Q_{bkf}) _____ CFS	Bankfull DISCHARGE (Q_{bkf}) _____ CFS

Bankfull DISCHARGE associated with *"field-determined"* Bankfull Stage _____ CFS (Q_{bkf})
(From Gage Height reading at Staff Plate and tabular Stage-Discharge curve data.)

Recurrence **I**nterval (Log-Pearson) associated with *"field-determined"* Bankfull Discharge.
R.I. = _____ Years

From the *Annual Peak Flow Frequency Analysis* data for the Gage Station, determine:

1.5 Year R.I. Discharge = _____ Cfs.   **10** Year R.I. Discharge = _____ Cfs.
2.0 Year R.I. Discharge = _____ Cfs.   **25** Year R.I. Discharge = _____ Cfs.
5.0 Year R.I. Discharge = _____ Cfs.   **50** Year R.I. Discharge = _____ Cfs.

MEANDER GEOMETRY

Meander Length (L_M) = _____ Ft.   Radius of Curvature (R_C) = _____ Ft
Belt Width (W_B) = _____ Ft.   Meander Width Ratio = (W_B / W_{BKF}) = _____

HYDRAULIC GEOMETRY

Based on: *USGS Discharge Summary Notes* data (Form 9-207) and regression analyses of measured discharge (**Q**) with the hydraulic parameters of Width (**W**), Area (**A**), Mean Depth (**d**), & Mean Velocity (**u**); determine the *intercept coefficient* (**a**) and the *slope exponent* (**b**) values for a power function of the form **Y** = a**X**^{b}, when **Y** is one of the selected hydraulic parameters, and **X** is a given discharge value (**Q**).

	Width (**W**)	Depth (**d**)	Area (**A**)	Velocity (**u**)
Coefficient: (**a**)				
Slope Expn: (**b**)				

Hydraulic **R**adius: (aka...Hydraulic Mean Depth) ($R = A / W_P$) _____ Ft.

Manning's "n" (Roughness Coefficient) at Bankfull Stage _____

"n" = 1.4895 [(**A**rea) (**H**ydraulic **R**adius$^{2/3}$) (**S**lope$^{1/2}$)] / Q_{BKF}

5-13

Step 3. Field calibrate bankfull discharge estimates by either direct discharge measurements or field calibration procedures at local gage stations to verify regional hydrology curves applied to ungaged sites

The most consistent bankfull determination is associated with the top of the floodplain. This is the elevation where incipient flooding begins at a flow stage above the bankfull discharge. Many floodplains are constructed when rivers abandon their point bars through lateral migration. The common elevation of the highest tops of point bars in high bedload streams is often the bankfull stage. The lateral abandonment of point bars can often be related to floodplain development within the valley under the current climate regime. Alluvial channels with well-developed floodplains often have river terraces (abandoned floodplains) adjacent to the channel. Thus, for correct bankfull stage determination, it is important for the field observer to distinguish between a terrace and a floodplain. For stream types with a well-developed floodplain (C, D, DA and E stream types), the bankfull stage is easily and reliably identified as the elevation of the floodplain, coincident with the incipient point of flooding.

Where floodplains are not well developed, the bankfull stage must be determined by field indicators that reveal corroborating evidence of a common elevation. The appropriate use of any bankfull indicator requires adherence to four basic principles:

1. Seek indicators in the locations appropriate for specific stream types;

2. Know the recent flood and/or drought history of the area to avoid being misled by spurious indicators (e.g., colonization of riparian species within the bankfull channel during drought, scour lines or flood debris accumulations caught in willows that have rebounded after flood flows have receded);

3. Use multiple indicators where possible for reinforcement of a common elevation; and

4. Where possible, calibrate field-determined elevations and corresponding bankfull dimensions to known recurrence interval discharges at gaged stations. This verifies the difference between the active floodplain and a low terrace.

There are several visual indicators of the bankfull stage that enable field determination of this important parameter for areas where streamflow records are not available. These indicators vary in their importance and discriminating power for different stream types. A partial listing of these indicators follows:

- Presence of a floodplain at the elevation of incipient flooding;

- The elevation associated with the top of the highest active channel depositional features in high bedload streams (e.g., point bars and central bars within the bankfull channel). These are especially good indicators for channels in the presence of terraces or channels adjacent to colluvial slopes;

- A break in slope of the banks and/or a change in the particle size distribution (because finer material is associated with deposition by overflow rather than deposition of coarser material within the active channel);

- Evidence of an inundation feature, such as defined benches inside of incised rivers;

- Staining of rocks;

- Exposed root hairs below an intact soil layer, indicating exposure to frequent erosive flow; and
- Vegetation species specific to particular locations and/or stream types.

Before selecting a reference reach for bankfull cross-section data, it is essential to study the longitudinal profile for at least 20–30 bankfull widths both upstream and downstream to determine representative bankfull indicators for that reach. Consistent measured values for bankfull width can be determined when corroborating, representative indicators have been identified. Establishing corroborating indicators greatly improves the resulting field estimates. Again, it is important to first measure bankfull cross-sections at gaged reaches to calibrate the interpretation of geomorphic features to known streamflow quantities and corresponding return periods.

Appropriate locations to measure bankfull widths are summarized for riffle/pool and step/pool channels in **Figures 5-3** and **5-4**, respectively.

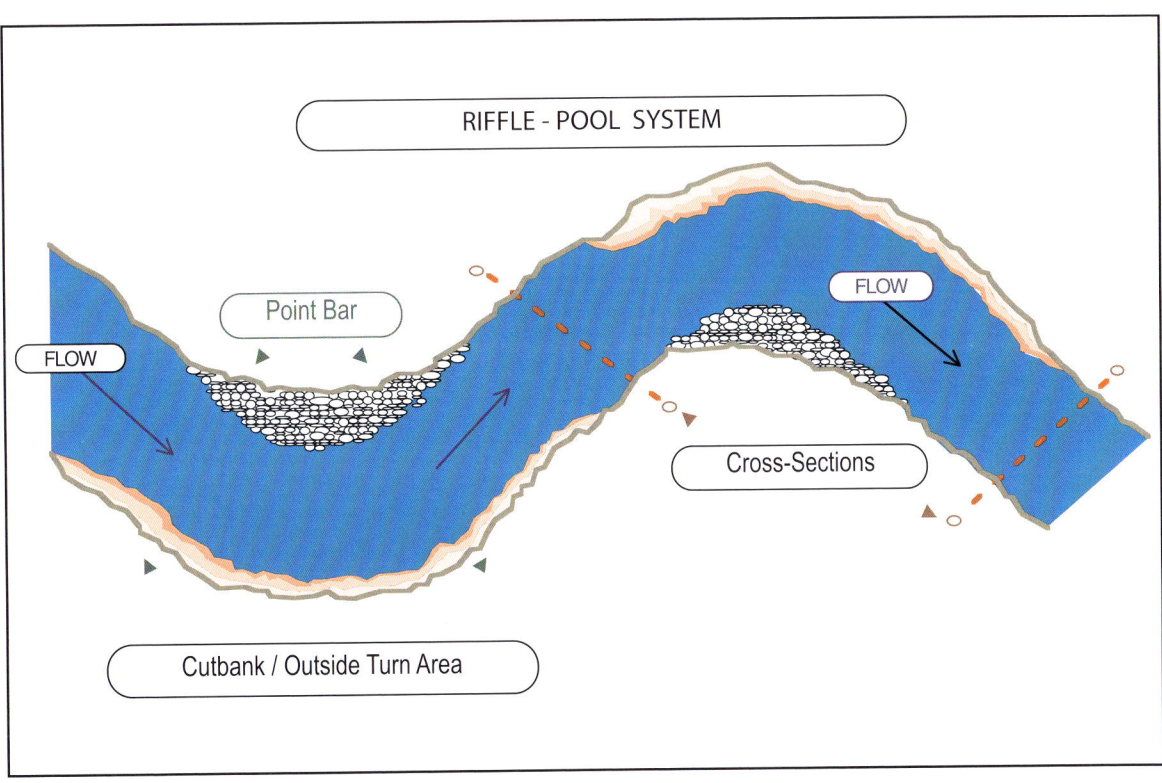

Figure 5-3. Recommended cross-section locations for bankfull stage measurements in riffle/pool systems (Rosgen, 1996).

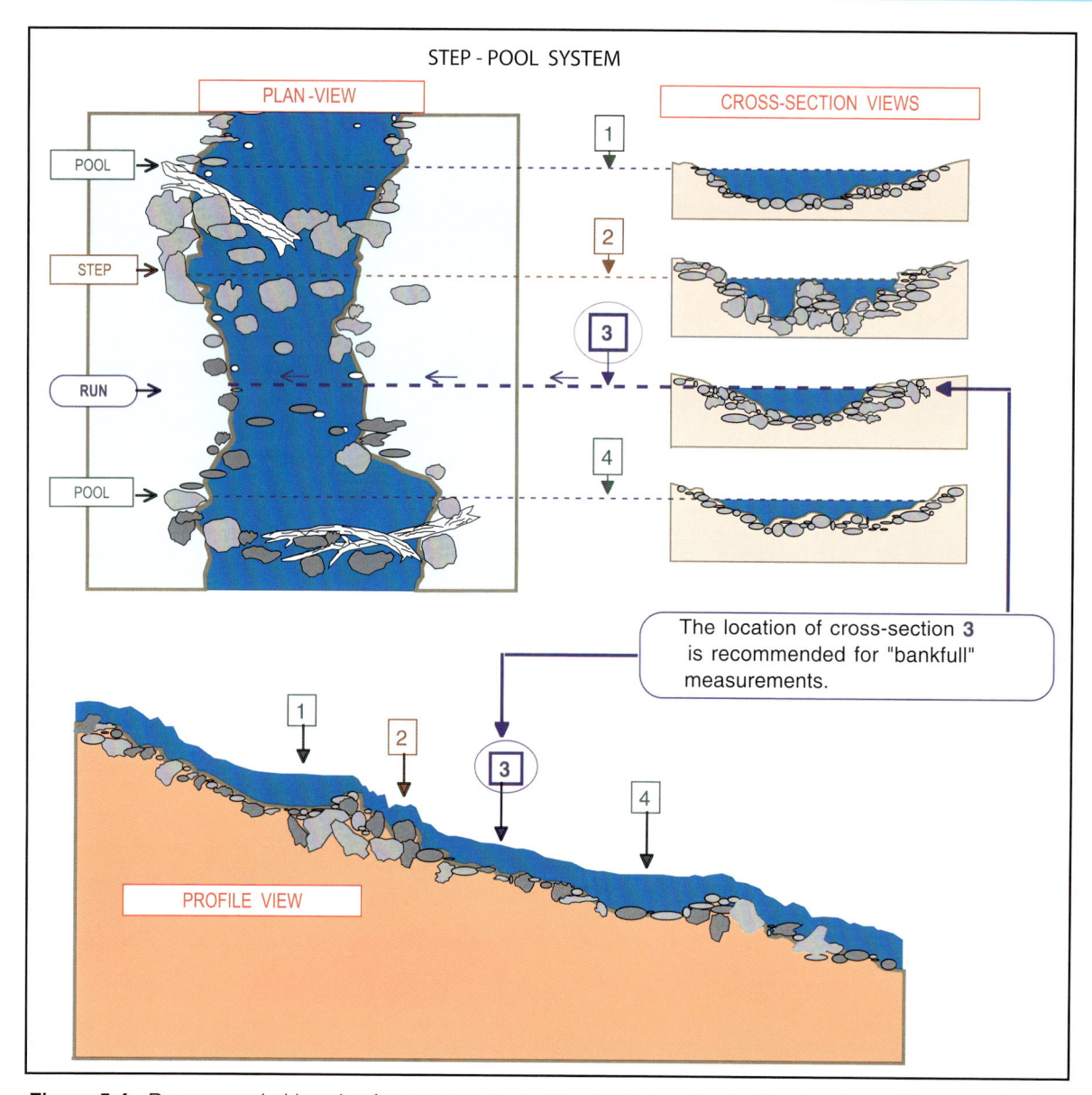

Watershed Assessment of River Stability and Sediment Supply

Figure 5-4. Recommended location for measurement of bankfull stage in step/pool systems (Rosgen, 1996).

In general, for all stream types, the best location to measure bankfull width is the narrowest portion of the cross-section where the channel can freely adjust its boundaries under existing streamflow conditions. The position of the narrowest portion varies by stream type. For example, the best locations for determining bankfull dimensions are at the riffle or "cross-over" reach for C, E and F stream types; the middle of the "rapid" for B stream types; and the narrow width of the transition from the "step" into the "head" of the pool for Aa+, A and G stream types. Avoid measuring pools to determine bankfull dimensions for regional curves. Deflectors, such as rocks, logs, nickpoints or constrictions that make the stream unusually narrow or that create unusually wide backwater conditions, must also be avoided. Additional examples of field methods for locating measurements of bankfull stage for riffle/pool channels are summarized by Lowham (1976) and described for step/pool channels by Osterkamp (1994).

Many species of riparian plants are widely distributed, occurring across a variety of hydro-physiographic provinces. Using vegetation to identify bankfull stage must be done cautiously because some species and age classes can often establish themselves within the bankfull channel. Bankfull stage is frequently underestimated when determined solely on the basis of vegetation, and such unilateral determinations should be avoided. Nonetheless, some common riparian species can be used as indicators of bankfull stage, such as certain mature species of birch (*Betula* spp.), dogwood (*Cornus* spp.), cottonwood (*Populus* spp.) and alder (*Alnus* spp.), which can colonize from seed and become established at levels very close to bankfull stage. Mature alders are not generally found within the bankfull channel, unless a steep, erodible bank was undercut and "slumped in." Smaller woody plants, grasses and forbs can colonize within the bankfull channel, especially during drought, as can certain species and age classes of willows (*Salix* spp.). These species should not be used as indicators. In cases where there is little choice but to use vegetation as an indicator, it is best to seek the advice of a riparian ecologist familiar with the study area and to verify the relations of these species to stream stage at gaged sites within the same hydro-physiographic province.

Calibrating bankfull stage to known streamflows

A common error of field observers is failing to calibrate the appropriate field indicators of bankfull stage to known streamflows. *Such calibration is essential to validate the proper interpretation of the bankfull stage until sufficient field experience in a given locale is gained.* The recommended procedure for calibrating field-identified bankfull stage with known streamflows and return period is as follows:

1. Locate all current and discontinued streamgaging stations within the study basin or in nearby similar basins; and

2. Collect supplemental data needed to interpret existing hydrologic records at each station. These field visits are not an unnecessary extravagance nor are they likely to be a major encumbrance. It is fortunate to find a half-dozen gaging stations within a study area. It may be necessary to travel outside the study area to obtain representative data for extrapolation.

Table 5-2 serves as a list of procedures performed at gaged sites and other measurement locations. This data is used for stream classification and subsequent sediment and hydraulic analyses and should be recorded as appropriate on maps or in reference documentation. **Figure 5-5** shows where to measure meander geometry. Record gaged and field measurement data in **Worksheet 5-1**.

Caution: The user should not use an arbitrary recurrence interval discharge to obtain bankfull discharge. Recurrence interval discharge is best determined only after field calibration at gage stations. Once established, however, recurrence interval discharge values can be applied to discontinued stations. For example, in high precipitation zones and high percent impervious areas due to urban development, return periods of bankfull discharge (using annual series flood frequency analysis) are closely grouped between the 1.05 to 1.20 year return period discharge. For low precipitation areas, the return period is generally between 1.50 and 1.80. There are exceptions, however, to these general statements.

Table 5-2. Recommended procedure at USGS gage or other streamflow measurement locations.

Recommended procedure at USGS gage station or other measurement locations

1. **Describe site**
   - Geomorphic setting – valley types I through XI (**Table 2-3**). (see Rosgen, 1996, Chapter 4 for detailed descriptions)
   - Channel materials (pebble count) (D_{16}, D_{35}, D_{50}, D_{84}, D_{95})
     - bed material: pebble count
     - bank material: pebble count and core sample
     - bar material: core sample
   - Locate on topographic map
   - Photo document: up/downstream
   - Compute percentage of watershed hydraulically impacted

2. **Create longitudinal profile**
   - Measure average water surface slope
     - riffle slope
     - pool slope
   - Measure valley slope
   - Sequence of riffle/pool or step/pool as a function of bankfull width
   - Locate bankfull stage along longitudinal profile

3. **Describe plan-view**
   - Measure sinuosity (SL/VL) or (VS/CS), where SL=stream length, VL=valley length, VS=valley slope and CS=channel slope
   - Measure meander geometry
     - meander wavelength (L_M)
     - belt width (W_{blt})
     - radius of curvature (R_C)
     - meander arc angle (A_a)
     - Meander Width Ratio (MWR)

4. **Measure cross-section (dimension)**
   - Measure cross-section of channel (valley features such as the terrace/floodplains to be identified on cross-section plot)
     - bankfull width (W_{bkf})
     - bankfull mean depth (d_{bkf})
     - bankfull maximum depth (d_{mbkf})
     - flood-prone area width (W_{fpa})
     - entrenchment ratio (W_{fpa}/W_{bkf})
     - bankfull cross-sectional area (A_{bkf})
     - bankfull mean velocity (u_{bkf})*
     - estimated bankfull discharge (Q_{bkf})

   Estimated from various sources

   - Calibrate bankfull estimates
     - survey estimated bankfull stage
     - from staff gage plate, extrapolate stage reading associated with estimated "Bankfull"
     - read discharge from rating curve at gage (stage/discharge relation)
     - determine recurrence interval in years from flood frequency curves at station (should be 1–2 years or average of 1.5 year Q)
     - analyze hydraulic geometry data from 9-207 forms (discharge notes) for width, depth, velocity and cross-sectional area vs. stream discharge. Plot data on log-log paper, and run a regression to obtain slope and intercept values for each variable
     - develop dimensionless hydraulic geometry relations. This is to be applied for extrapolation purposes to rivers of the same stream type but for various sizes (W/W_{bkf} vs. Q/Q_{bkf}) (complete for depth, velocity and cross-sectional area)

Chapter 5   The Prediction Level Assessment (PLA)

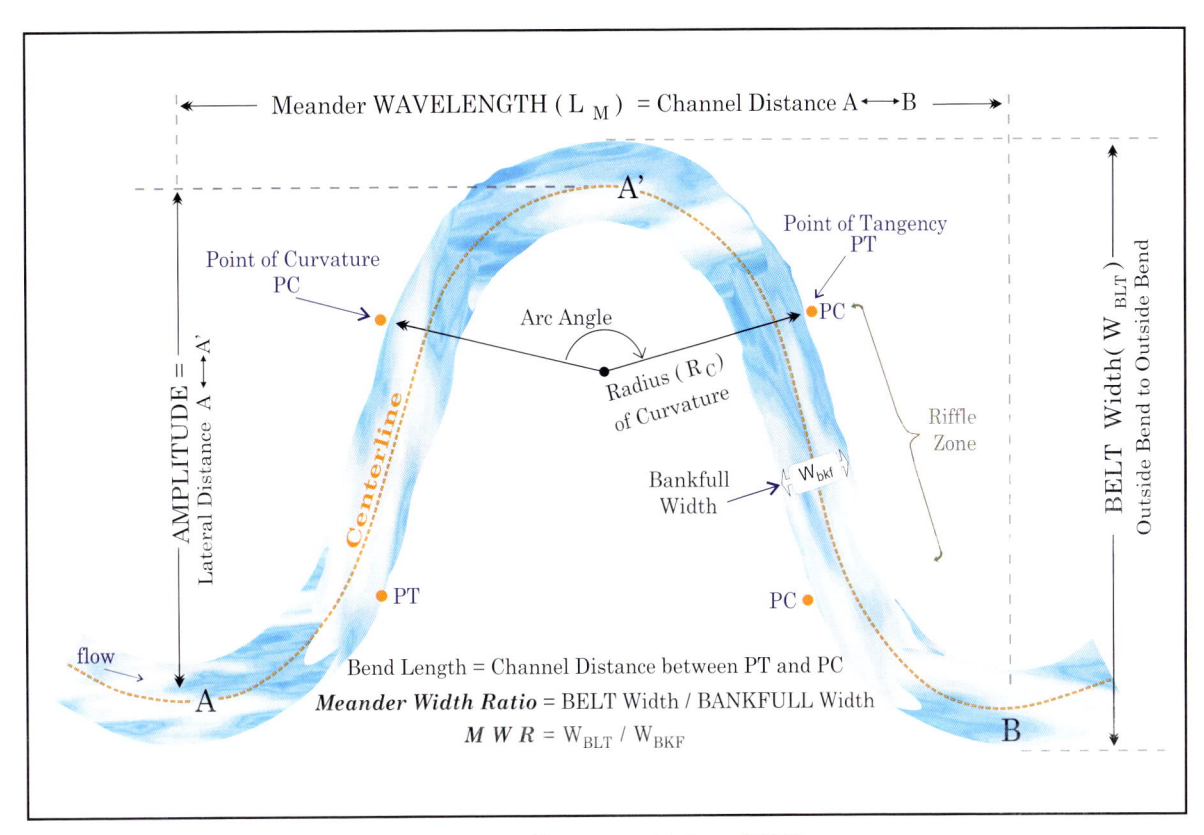

Figure 5-5. Meander geometry measurements (Rosgen and Silvey, 2005).

This gage station information is used to calibrate field-estimated bankfull stage with a corresponding measured discharge and associated return period. The return period of bankfull discharge is most frequently 1–2 years, with an average of 1.5 years. If field-determined bankfull discharge yields a return period greater than this, the field indicators selected may indicate a low terrace or other features not typical of the bankfull stage. **Worksheet 5-1** is completed by using flood frequency data and rating curve information (stage-discharge relationships) published by the USGS in its annual water resources summaries and by using stream discharge notes (available from the USGS on Form 9-207; for some states, information is available in electronic format). Field data is collected to supplement published data that describes the reference reach slope, particle size distribution, bankfull characteristics and stream type.

All gaging stations have a permanent benchmark installed for elevation control at corresponding cross-sections. These cross-sections should be resurveyed and expanded to include the active floodplain, low terraces and other valley features.

A helpful guide by the USDA Forest Service describes bankfull stage determination and stream channel surveys (Harrelson et al., 1994). This field guide includes stream survey methods, bankfull stage surveys, pebble counts and other channel inventory methods. A bankfull video produced by the USDA Forest Service (2003) is also helpful in ascertaining bankfull stage for both the Eastern and Western United States.

Step 4. Calculate bankfull discharge and dimensions

At the selected assessment site (reference and impaired reaches), identify the bankfull stage and measure a cross-section of a riffle where the stream is free to adjust its boundaries to the frequent flows. Measure the bankfull width, bankfull mean depth and bankfull cross-sectional area. Obtain the drainage area of the reach (**Step 2**), and then check the bankfull cross-sectional area against the regional curve for the hydro-physiographic region. Estimate field-determined bankfull discharge by multiplying estimated bankfull velocity by the cross-sectional area. Record this information in **Worksheet 5-2**.

Various methods are appropriate for estimating bankfull mean velocity as long as the assumptions of the computational method are met. RIVERMorph™ is a computer program that allows a variety of velocity equations to be applied to the prediction. Calibrating several velocity equations at gage stations, where velocity is measured for a wide range of flows, will provide the familiarity and confidence to select the appropriate formula for a similar stream type at an ungaged site. For gravel- and cobble-bed streams, an approach that uses a resistance (friction) factor and a relative roughness factor (**Figure 5-6**) is recommended.

USGS streamgage station data in conjunction with morphological field data at the site can also be used to verify various resistance equations and other prediction methods for velocity. The 9-207 data contains measured mean velocity data for a wide range of flows. This data will determine what velocity prediction models are most appropriate for various stream types, sizes, etc., to be implemented at ungaged sites.

Chapter 5 — The Prediction Level Assessment (PLA)

Worksheet 5-2. Computations of velocity and bankfull discharge using various methods (Rosgen and Silvey, 2005).

Bankfull VELOCITY / DISCHARGE Estimates

Site		Location	
Date		Valley Type	
	Stream Type		
Observers		Hydrologic Unit Code	

INPUT VARIABLES

Variable	Value	Symbol	Units
Bankfull Cross-section AREA		A_{bkf}	(SqFt)
Bankfull WIDTH		W_{bkf}	(Ft)
D84 @ Riffle		Dia.	(mm)
Bankfull SLOPE		S	(Ft/Ft)
Gravitational Acceleration	32.2	g	(Ft/Sec²)
Drainage AREA		DA	(SqMi)

OUTPUT VARIABLES

Computation	Value	Symbol	Units
Bankfull Mean DEPTH $D_{bkf} = A_{bkf} / W_{bkf}$		D_{bkf}	(Ft)
Wetted Perimeter $\sim 2 * D_{bkf} + W_{bkf}$		WP_{bkf}	(Ft)
D84 mm / 304.8 =		D84	(Ft)
Hydraulic Radius A_{bkf} / WP_{bkf}		R	(Ft)
$R_{(Ft)} / D84_{(Ft)}$			
Shear Velocity: $u^* = \sqrt{gRS}$		U^*	(Ft/Sec)

ESTIMATION METHODS

1. Friction Factor / Relative Roughness

$$u = \left[2.83 + 5.66 \, \text{Log} \left(\frac{R}{D84} \right) \right] u^*$$

Bankfull VELOCITY (Ft/Sec) | Bankfull DISCHARGE (CFS)

2. Roughness Coefficient: a) Manning's "n" from friction factor and relative roughness. (Figs. 5-6, 5-7) n = _____

$$u = 1.4895 * R^{2/3} * S^{1/2} / n$$

2. Roughness Coefficient:
b) Manning's "n" from Jarrett (USGS): $n = 0.39 * S^{.38} * R^{-.16}$
n = _____
$$u = 1.4895 * R^{2/3} * S^{1/2} / n$$

Note: This equation is for applications involving **steep, step-pool, high boundary roughness, cobble- boulder-dominated** stream systems; i.e., for stream types A1, A2, A3 B1, B2, B3, C2 and E3.

2. Roughness Coefficient: c) Manning's "n" from Stream Type
n = _____
$$u = 1.4895 * R^{2/3} * S^{1/2} / n$$

3. Other Methods, i.e. Hydraulic Geometry; (Hey, Darcy-Weisbach, Chezy C, etc.)

4. Continuity Equations: b) USGS Gage: $u = Q / A$
Return Period for Bankfull Q = _____ Yr.

4. Continuity Equations: a) Regional Curves: $u = Q / A$

Options for using the D84 term in the relative roughness relation (R / D84), when using estimation method 1.

Option 1. For **sand-bed** channels: measure the "**protrusion height**" (h_{sd}) of sand dunes above channel bed elevations. Substitute an average sand dune protrusion height (h_{sd} in feet) for the D84 term in estimation method 1.

Option 2. For **boulder-dominated** channels: measure several "**protrusion heights**" (h_{bo}) of boulders above channel bed elevations. Substitute an average boulder protrusion height (h_{bo} in ft) for the D84 term in est. Method 1.

Option 3. For **bedrock-dominated** channels: measure "**protrusion heights**" (h_{br}) of rock separations/steps/joints/uplifted surfaces above channel bed elevations. Substitute an average bed-rock protrusion height (h_{br} in ft) for the D84 term in estimation method 1.

$$u = \left[2.83 + 5.66 \log\left(\frac{R}{D_{84}}\right)\right] u^*$$

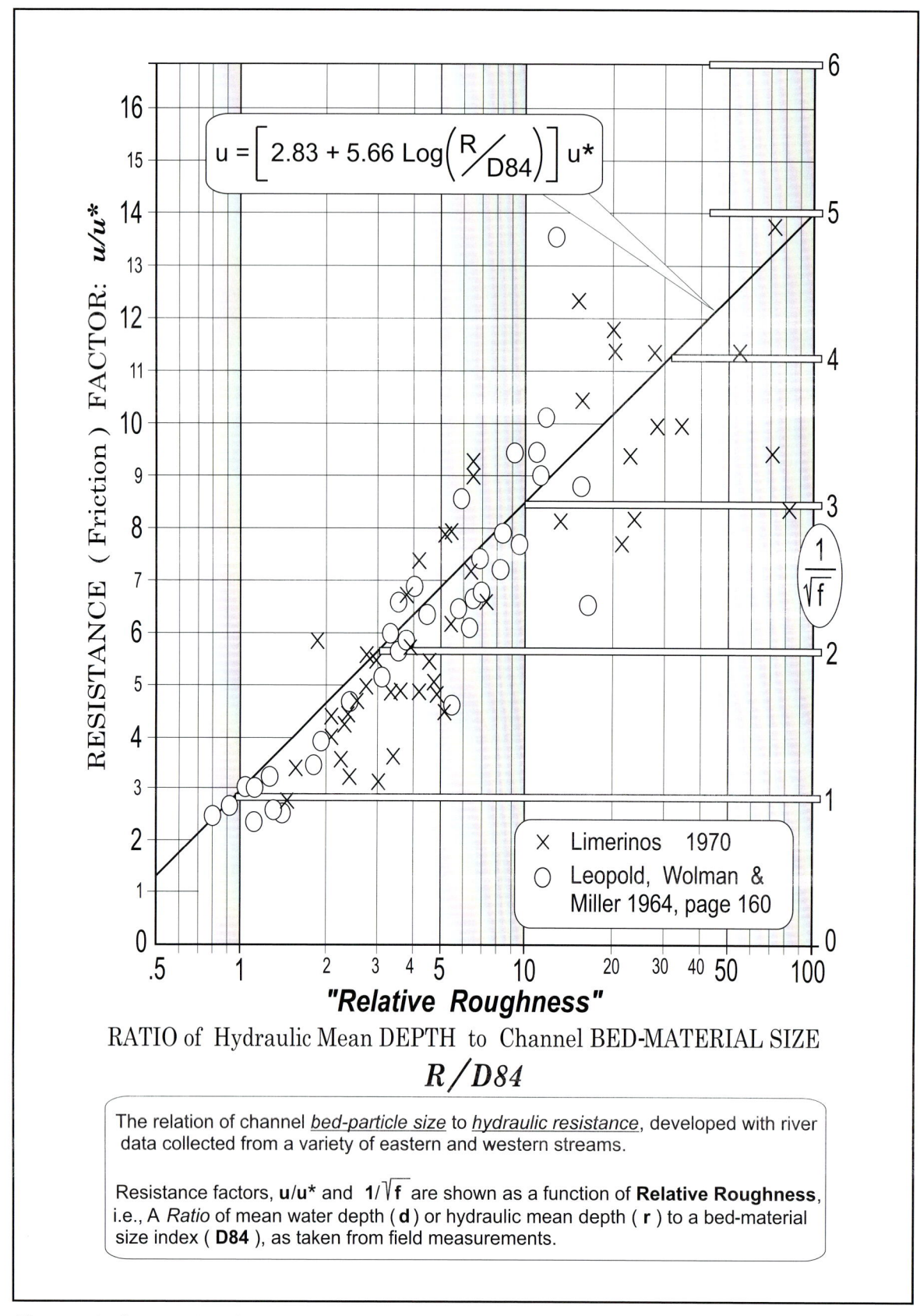

The relation of channel *bed-particle size* to *hydraulic resistance*, developed with river data collected from a variety of eastern and western streams.

Resistance factors, **u/u*** and **1/√f** are shown as a function of **Relative Roughness**, i.e., A *Ratio* of mean water depth (**d**) or hydraulic mean depth (**r**) to a bed-material size index (**D84**), as taken from field measurements.

Figure 5-6. Computation of velocity from a resistance factor and relative roughness (Rosgen and Silvey, 2005).

The relation in **Figure 5-6** is presented as:

$$u = [2.83 + 5.66 \log(R/D_{84})] u^* \qquad \textbf{(Equation 5-1)}$$

where:
- u = mean velocity in ft/sec
- u^* = shear velocity ($\sqrt{gRS}$)
- g = gravitational acceleration (32.2 ft/sec^2)
- R = hydraulic radius (a.k.a. hydraulic mean depth)
- S = water surface slope
- D_{84} = particle size of bed material of the 84th percentile

Another option uses the Darcy-Weisbach friction factor (f), where u/u^* can approximate values of ($1/\sqrt{f}$) (**Figure 5-6**). The use of a friction factor to predict a Manning's "n" roughness factor is shown in **Figure 5-7** where:

$$f = \frac{8gRS}{u^2}$$

R/D_{84} values are used to obtain u/u^* values in **Figure 5-6**. The u/u^* values are then transferred to **Figure 5-7** to read Manning's "n" values. The equation for mean velocity prediction using Manning's procedure is shown in **Figure 5-7**. Record the selected velocity equation and computation in **Worksheet 5-2**.

In field practice, the author often substitutes an average *protrusion height* (height of the roughness element as measured from the bed surface to the top of the feature) for the D_{84} size. This substitution is applied to boulder-bed, sand-bed (height of dunes), bedrock and very large, shelf-like, overlapping-but-flat particles (**Worksheet 5-2**).

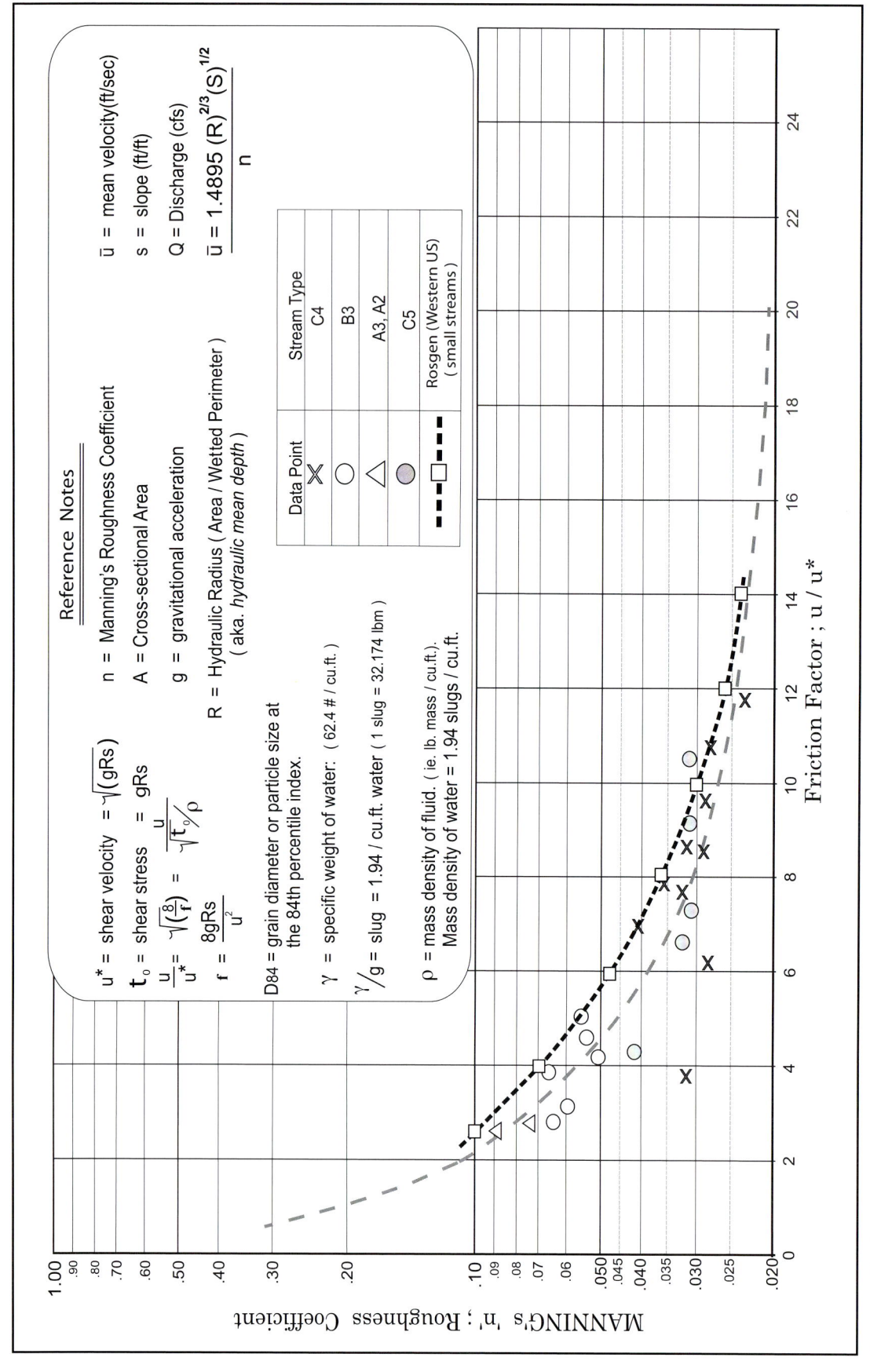

Figure 5-7. Conversion of a resistance (friction) factor to Manning's "n" roughness coefficient (Rosgen and Silvey, 2005).

Chapter 5   *The Prediction Level Assessment (PLA)*

The use of hydraulic geometry relations is also recommended for this application. Hydraulic geometry relations are associated with the plotting of actual measured velocity, width and depth against discharge. This data is obtained from USGS gage stations or other current meter measurement sources. An example of hydraulic geometry is shown in **Figure 5-8** (Leopold and Maddock, 1953).

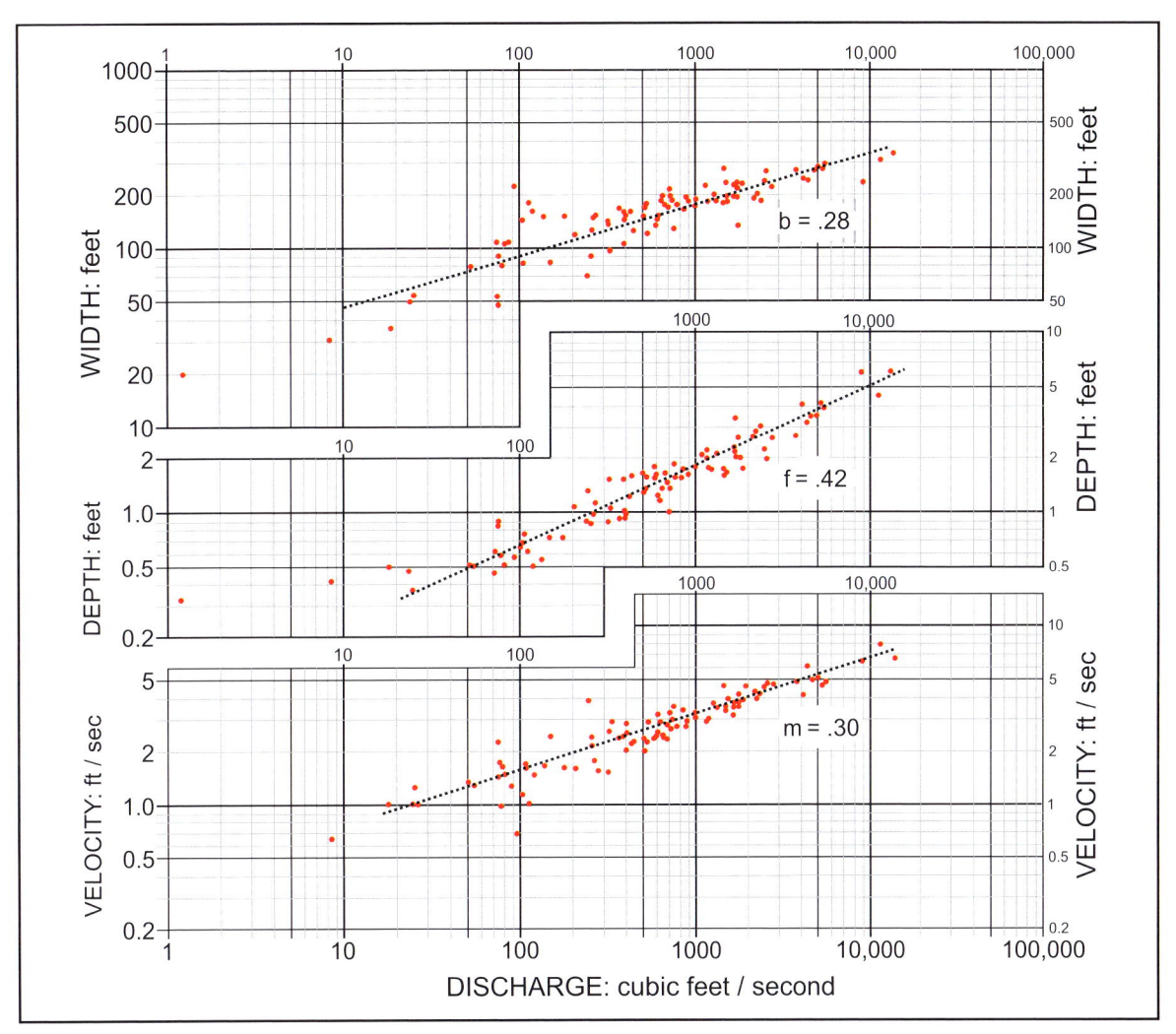

Figure 5-8. Hydraulic geometry for the Powder River, Montana (Leopold and Maddock, 1953).

Examples of hydraulic geometry relations by stream type are shown in Chapter 2 in **Figure 2-22**. Hydraulic geometry relations can also be predicted by using RIVERMorph™ software in conjunction with the POWERSED sediment transport capacity model described later in this chapter.

It is generally good practice to obtain the drainage area first (**Step 2**), and then obtain the bankfull cross-sectional area from the regional curve. Check this value against the field-determined cross-sectional area. Divide the field-determined cross-sectional area into the bankfull discharge from the regional curve to see if the mean velocity is reasonable compared to the velocity predictions (**Worksheet 5-2**).

5-25

WATERSHED ASSESSMENT OF RIVER STABILITY AND SEDIMENT SUPPLY

Steps 5–6: Level II stream classification and dimensionless ratios of channel features (See Flowchart 5-5)

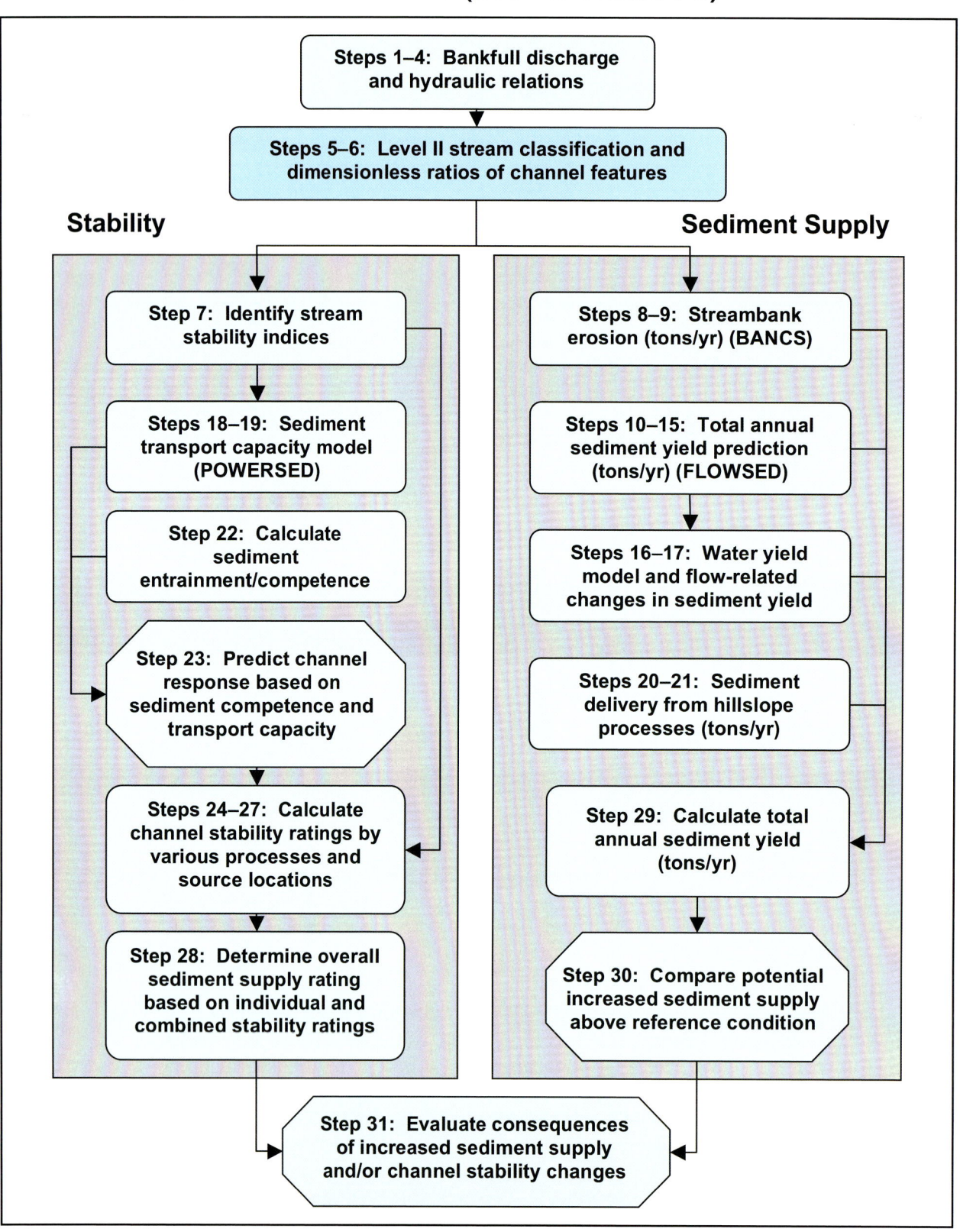

```
Steps 1–4: Bankfull discharge and hydraulic relations
        ↓
Steps 5–6: Level II stream classification and dimensionless ratios of channel features
```

Stability

- Step 7: Identify stream stability indices
- Steps 18–19: Sediment transport capacity model (POWERSED)
- Step 22: Calculate sediment entrainment/competence
- Step 23: Predict channel response based on sediment competence and transport capacity
- Steps 24–27: Calculate channel stability ratings by various processes and source locations
- Step 28: Determine overall sediment supply rating based on individual and combined stability ratings

Sediment Supply

- Steps 8–9: Streambank erosion (tons/yr) (BANCS)
- Steps 10–15: Total annual sediment yield prediction (tons/yr) (FLOWSED)
- Steps 16–17: Water yield model and flow-related changes in sediment yield
- Steps 20–21: Sediment delivery from hillslope processes (tons/yr)
- Step 29: Calculate total annual sediment yield (tons/yr)
- Step 30: Compare potential increased sediment supply above reference condition

Step 31: Evaluate consequences of increased sediment supply and/or channel stability changes

Flowchart 5-3c. The sequential steps to determine stream stability and sediment supply, highlighting **Steps 5–6**.

Steps 5–6: Level II stream classification and dimensionless ratios of channel features

Step 5: Classify entire length of the reference and impaired stream reaches (Worksheet 5-3)

- Valley type
  - Bankfull width (ft)
  - Bankfull depth (ft)
  - Bankfull cross-sectional area (ft^2)
  - Max depth (ft)
  - Width of floodprone area (ft)
  - Particle size D$_{50}$ (mm)
  - Width/depth ratio
  - Check with regional curves
  - Water surface slope
  - Sinuosity (k)
  - Entrenchment ratio

- Stream classification

Step 6: Calculate detailed morphological descriptions (including dimensionless ratios) (Worksheet 5-4)

- Valley type
- Stream type
  - Channel dimensions
  - Channel pattern
  - Channel profile
  - Channel materials

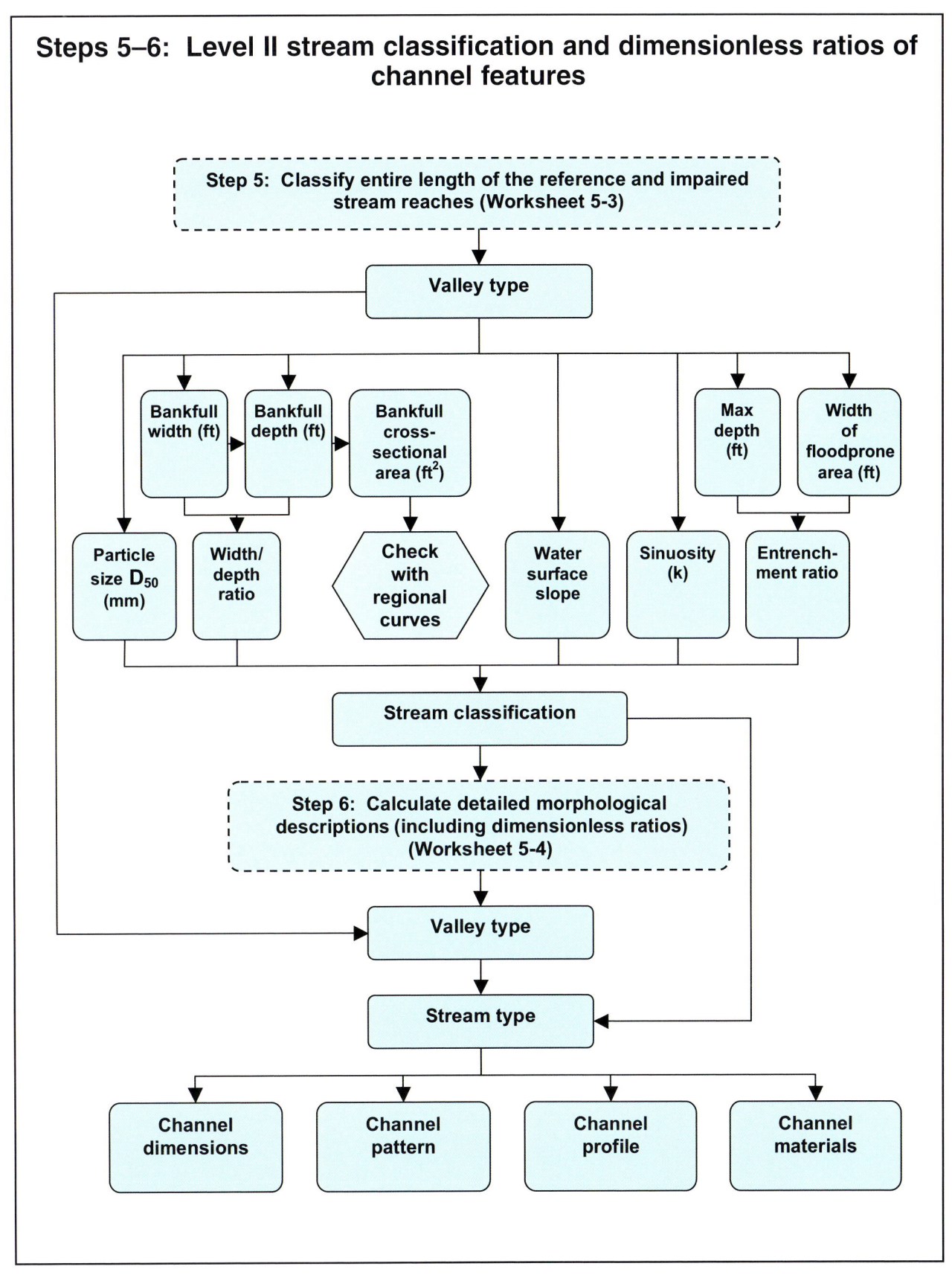

Flowchart 5-5. Sequential steps and variables to classify streams and develop dimensionless ratios.

Level II stream classification and dimensionless ratios of channel features: Steps 5–6

Stream type classification and an understanding of morphological relations assist in determining departure from a stable reference condition of the same stream type in the same valley type. Detailed descriptions of the field methods and examples for stream classification are shown in Chapters 4 and 5 of *Applied River Morphology* (Rosgen, 1996), and field forms are included in *The Reference Reach Field Book* (Rosgen and Silvey, 2005).

Step 5. Classify streams

Classify the entire length of the reference and impaired stream reaches using the descriptions and key shown in **Figures 2-13** and **2-14**. In Chapter 4 *RRISSC*, great distances are rapidly classified through extrapolation from individual measurements. In *PLA*, however, the reference and impaired reaches require field verification with site-specific measurements of the Level II delineative criteria. The recorded morphological data appears in **Worksheet 5-3**.

Worksheet 5-3. Field form for Level II stream classification (Rosgen, 1996; Rosgen and Silvey, 2005).

Stream:			
Basin:	Drainage Area:	acres	mi²
Location:			
Twp.&Rge:	Sec.&Qtr.:		
Cross-Section Monuments (Lat./Long.):		Date:	
Observers:		Valley Type:	

Bankfull WIDTH (W_{bkf})
WIDTH of the stream channel at bankfull stage elevation, in a riffle section. — ft

Bankfull DEPTH (d_{bkf})
Mean DEPTH of the stream channel cross-section, at bankfull stage elevation, in a riffle section ($d_{bkf} = A / W_{bkf}$). — ft

Bankfull X-Section AREA (A_{bkf})
AREA of the stream channel cross-section, at bankfull stage elevation, in a riffle section. — ft²

Width/Depth Ratio (W_{bkf} / d_{bkf})
Bankfull WIDTH divided by bankfull mean DEPTH, in a riffle section. — ft/ft

Maximum DEPTH (d_{mbkf})
Maximum depth of the bankfull channel cross-section, or distance between the bankfull stage and Thalweg elevations, in a riffle section. — ft

WIDTH of Flood-Prone Area (W_{fpa})
Twice maximum DEPTH, or ($2 \times d_{mbkf}$) = the stage/elevation at which flood-prone area WIDTH is determined in a riffle section. — ft

Entrenchment Ratio (ER)
The ratio of flood-prone area WIDTH divided by bankfull channel WIDTH (W_{fpa} / W_{bkf}) (riffle section). — ft/ft

Channel Materials (Particle Size Index) D_{50}
The D_{50} particle size index represents the mean diameter of channel materials, as sampled from the channel surface, between the bankfull stage and Thalweg elevations. — mm

Water Surface SLOPE (S)
Channel slope = "rise over run" for a reach approximately 20–30 bankfull channel widths in length, with the "riffle-to-riffle" water surface slope representing the gradient at bankfull stage. — ft/ft

Channel SINUOSITY (k)
Sinuosity is an index of channel pattern, determined from a ratio of stream length divided by valley length (SL / VL); or estimated from a ratio of valley slope divided by channel slope (VS / S).

Stream Type ⟩ ☐ ⟨ (See **Figure 2-14**)

Step 6. Calculate detailed morphological descriptions (including dimensionless ratios)

Obtain detailed morphological relations of stream dimension, pattern, profile and materials for the reference and impaired reaches. These quantitative morphological relations assist in determining departure from a stable reference condition of the same stream type in the same valley type. Valley types are described in detail in Chapter 4 of *Applied River Morphology* (Rosgen, 1996). Enter river reach data into **Worksheet 5-4** and calculate dimensionless ratios. These ratios are used to compare morphological data representing dimension, pattern and profile for river reaches of different sizes, including the reference and impaired reaches.

Chapter 5   *The Prediction Level Assessment (PLA)*

Worksheet 5-4. Morphological relations, including dimensionless ratios of river reach sites (Rosgen and Silvey, 2005).

Stream: _____ Location: _____
Observers: _____ Date: _____ Valley Type: _____ Stream Type: _____

River Reach Summary Data

Channel Dimension

Mean Riffle Depth (d_{bkf})		ft	Riffle Width (W_{bkf})		ft	Riffle Area (A_{bkf})		ft²
Mean Pool Depth (d_{bkfp})		ft	Pool Width (W_{bkfp})		ft	Pool Area (A_{bkfp})		ft²
Mean Pool Depth/Mean Riffle Depth		d_{bkfp}/d_{bkf}	Pool Width/Riffle Width		W_{bkfp}/W_{bkf}	Pool Area / Riffle Area		A_{bkfp}/A_{bkf}
Max Riffle Depth (d_{mbkf})		ft	Max Pool Depth (d_{mbkfp})		ft	Max riffle depth/Mean riffle depth		
Max pool depth/Mean riffle depth						Point Bar Slope		
Streamflow: Estimated Mean Velocity at Bankfull Stage (u_{bkf})		ft/s				Estimation Method		
Streamflow: Estimated Discharge at Bankfull Stage (Q_{bkf})		cfs				Drainage Area		mi²

Channel Pattern

Geometry	Mean	Min	Max		Dimensionless Geometry Ratios	Mean	Min	Max
Meander Length (Lm)				ft	Meander Length Ratio (L_m/W_{bkf})			
Radius of Curvature (Rc)				ft	Radius of Curvature/Riffle Width (R_c/W_{bkf})			
Belt Width (W_{blt})				ft	Meander Width Ratio (W_{blt}/W_{bkf})			
Individual Pool Length				ft	Pool Length/Riffle Width			
Pool to Pool Spacing				ft	Pool to Pool Spacing/Riffle Width			
Riffle Length				ft	Riffle Length/Riffle Width			

Channel Profile

Valley Slope (VS)		ft/ft	Average Water Surface Slope (S)		ft/ft	Sinuosity (VS/S)		
Stream Length (SL)		ft	Valley Length (VL)		ft	Sinuosity (SL/VL)		
Low Bank Height (LBH)	start / end	ft	Max Riffle Depth	start / end	ft	Bank-Height Ratio (BHR) (LBH/Max Riffle Depth)	start / end	

Facet Slopes	Mean	Min	Max	Dimensionless Slope Ratios	Mean	Min	Max
Riffle Slope (S_{rif})			ft/ft	Riffle Slope/Average Water Surface Slope (S_{rif}/S)			
Run Slope (S_{run})			ft/ft	Run Slope/Average Water Surface Slope (S_{run}/S)			
Pool Slope (S_p)			ft/ft	Pool Slope/Average Water Surface Slope (S_p/S)			
Glide Slope (S_g)			ft/ft	Glide Slope/Average Water Surface Slope (S_g/S)			

Feature Midpoint [a]	Mean	Min	Max	Dimensionless Depth Ratios	Mean	Min	Max
Riffle Depth (d_{rif})			ft	Riffle Depth/Mean Riffle Depth (d_{rif}/d_{bkf})			
Run Depth (d_{run})			ft	Run Depth/Mean Riffle Depth (d_{run}/d_{bkf})			
Pool Depth (d_p)			ft	Pool Depth/Mean Riffle Depth (d_p/d_{bkf})			
Glide Depth (d_g)			ft	Glide Depth/Mean Riffle Depth (d_g/d_{bkf})			

Channel Materials

	Reach[b]	Riffle[c]	Bar		Reach[b]	Riffle[c]	Bar	Protrusion Height[d]
% Silt/Clay				D_{16}				mm
% Sand				D_{35}				mm
% Gravel				D_{50}				mm
% Cobble				D_{84}				mm
% Boulder				D_{95}				mm
% Bedrock				D_{100}				mm

a Min, max, mean depths are the average mid-point values except pools, which are taken at deepest part of pool.
b Composite sample of riffles and pools within the designated reach.
c Active bed of a riffle.
d Height of roughness feature above bed.

Step 7: Stream channel stability indices (See Flowchart 5-6)

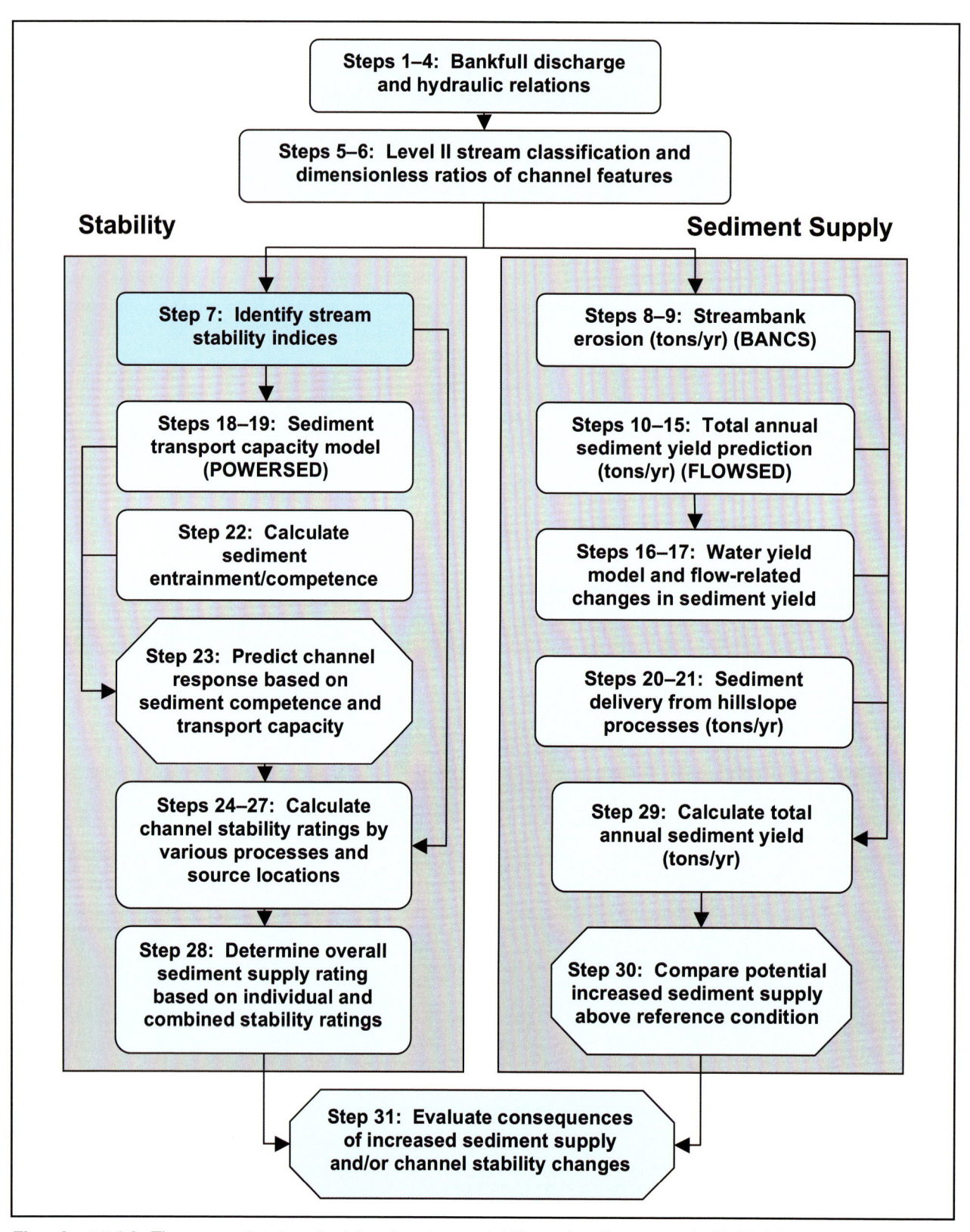

```
                    Steps 1–4: Bankfull discharge
                        and hydraulic relations
                                  ↓
                Steps 5–6: Level II stream classification and
                    dimensionless ratios of channel features
```

Stability — **Sediment Supply**

Stability column:
- Step 7: Identify stream stability indices
- Steps 18–19: Sediment transport capacity model (POWERSED)
- Step 22: Calculate sediment entrainment/competence
- Step 23: Predict channel response based on sediment competence and transport capacity
- Steps 24–27: Calculate channel stability ratings by various processes and source locations
- Step 28: Determine overall sediment supply rating based on individual and combined stability ratings

Sediment Supply column:
- Steps 8–9: Streambank erosion (tons/yr) (BANCS)
- Steps 10–15: Total annual sediment yield prediction (tons/yr) (FLOWSED)
- Steps 16–17: Water yield model and flow-related changes in sediment yield
- Steps 20–21: Sediment delivery from hillslope processes (tons/yr)
- Step 29: Calculate total annual sediment yield (tons/yr)
- Step 30: Compare potential increased sediment supply above reference condition

Step 31: Evaluate consequences of increased sediment supply and/or channel stability changes

Flowchart 5-3d. The sequential steps to determine stream stability and sediment supply, highlighting **Step 7**.

Chapter 5 The Prediction Level Assessment (PLA)

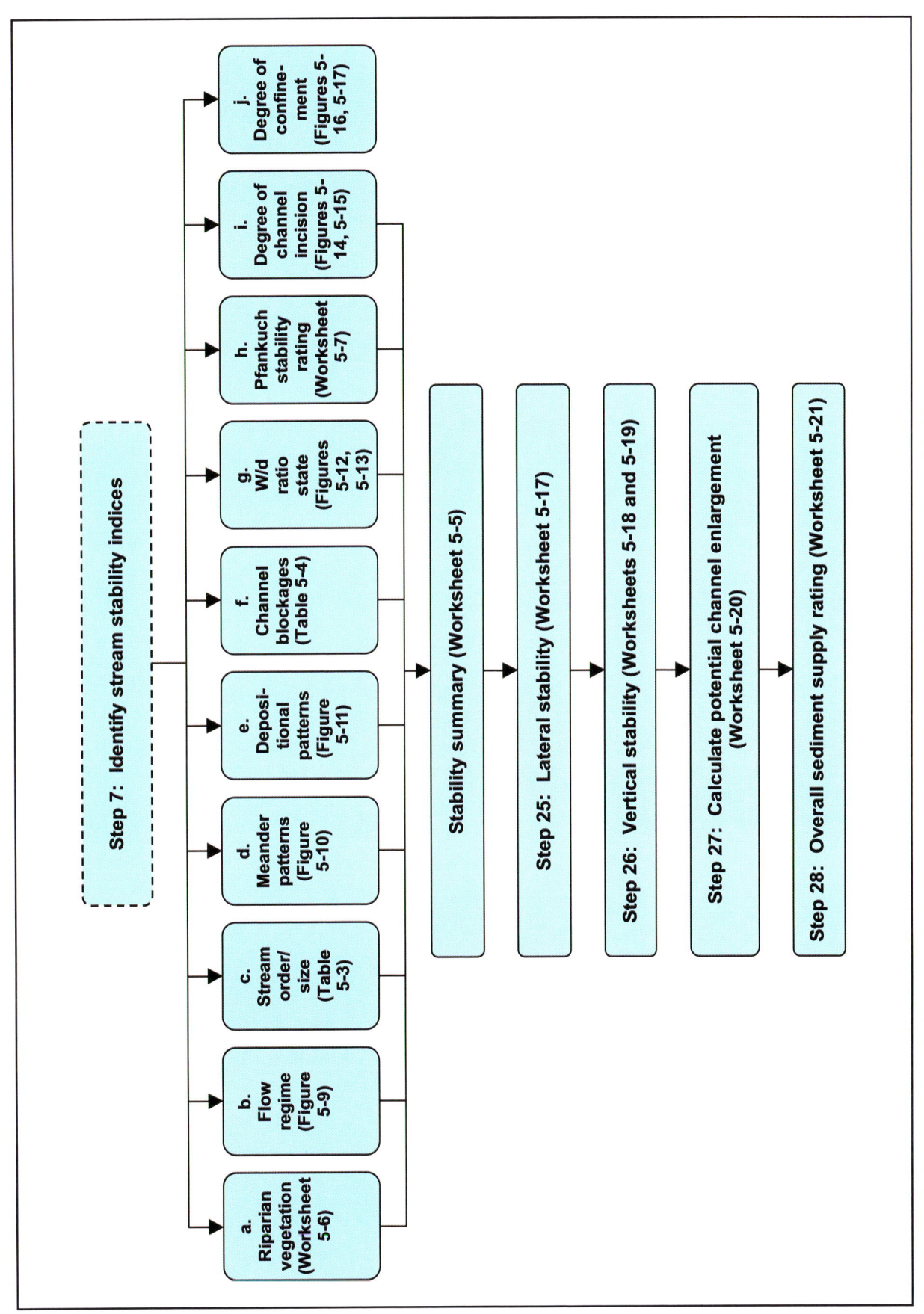

Step 7: Identify stream stability indices

- a. Riparian vegetation (Worksheet 5-6)
- b. Flow regime (Figure 5-9)
- c. Stream order/size (Table 5-3)
- d. Meander patterns (Figure 5-10)
- e. Depositional patterns (Figure 5-11)
- f. Channel blockages (Table 5-4)
- g. W/d ratio state (Figures 5-12, 5-13)
- h. Pfankuch stability rating (Worksheet 5-7)
- i. Degree of channel incision (Figures 5-14, 5-15)
- j. Degree of confinement (Figures 5-16, 5-17)

Stability summary (Worksheet 5-5)

Step 25: Lateral stability (Worksheet 5-17)

Step 26: Vertical stability (Worksheets 5-18 and 5-19)

Step 27: Calculate potential channel enlargement (Worksheet 5-20)

Step 28: Overall sediment supply rating (Worksheet 5-21)

Flowchart 5-6. The multiple variables and procedures (a.–j.) used for stream stability prediction in Steps 25–28.

Step 7. Identify stream stability indices

This step requires a stream channel stability analysis, using departure analysis summaries of morphological variables and collection and analysis of specific channel stability variables. The overall assessment for each variable is recorded in the stream stability summary **Worksheet 5-5**. This information is used in subsequent **Steps 25–28** for process-based channel stability determinations and is helpful in formulating site- and process-specific mitigation. The stability variables are assessed by reach using specific worksheets, figures and tables and include the following:

a. Riparian vegetation (**Worksheet 5-6**);

b. Flow regime (**Figure 5-9**);

c. Stream order/size (**Table 5-3**);

d. Meander patterns (**Figure 5-10**);

e. Depositional patterns (**Figure 5-11**);

f. Channel blockages (**Table 5-4**);

g. Width/depth ratio state (**Figures 5-12** and **5-13**);

h. Pfankuch channel stability rating (**Worksheet 5-7**);

i. Degree of channel incision (**Figures 5-14** and **5-15**); and

j. Degree of channel confinement (lateral containment) (**Figures 5-16** and **5-17**).

Detailed interpretations and examples of these stability variables are presented in Chapter 6 of *Applied River Morphology* (Rosgen, 1996). Some natural, stable rivers have few anthropogenic influences yet are very active. Reference reach evaluations are conducted to determine differences between natural, stable channel adjustment and genuine departure. General descriptions and assessments for these stability variables follow.

Chapter 5 — The Prediction Level Assessment (PLA)

Worksheet 5-5. Summary of stream stability ratings for multiple reaches.

Stream: Location: Observers: Date:

(1) Reach location	(2) Stream type (Worksheet 5-3)	(3) a. Riparian vegetation (Worksheet 5-6) — Existing species composition / Potential species composition	(4) b. Flow regime (Fig. 5-9)	(5) c. Stream order/size (Table 5-3)	(6) d. Meander patterns (Fig. 5-10)	(7) e. Depositional patterns (Fig. 5-11)	(8) f. Channel blockages (Table 5-4)	(9) g. Width/depth ratio state — W/d ratio (Worksheet 5-3) / W/d stability rating (Fig. 5-13)	(10) h. Pfankuch channel stability rating (Worksheet 5-7)	(11) i. Degree of channel incision — Bank-Height Ratio (BHR) (Worksheet 5-4) / Stability rating (Fig. 5-15)	(12) j. Degree of channel confinement — MWR divided by MWR$_{ref}$ / Degree of confinement (Fig. 5-17)
1.											
2.											
3.											
4.											
5.											
6.											
7.											
8.											
9.											
10.											
11.											
12.											
13.											
14.											
15.											

a. Riparian vegetation

The riparian vegetation alteration assessment determines existing composition and density of a river reach. The riparian area assessed includes:

1. Overstory,
2. Understory, and
3. Ground level.

Departure of the current vegetative community from the potential vegetative community is an important interpretation. Riparian vegetation is documented in **Worksheet 5-6**. If the composition in alluvial channel types changes from a woody riparian community to a grass/forb, it generally increases the risk of enlargement by streambank erosion due to the change in rooting depth and is associated with a high sediment supply.

The riparian vegetation **Worksheet 5-6** separates ground cover, understory (shrub layer) and overstory for species composition, density and/or bare soil. This information assists the assessment by comparing the reference condition to existing. This is helpful when calculating bank erosion rates and recommending mitigation for altered, unstable stream types where riparian vegetative changes have occurred. Potential riparian vegetative communities vary by soil type, soil moisture, elevation, geographic location and stream type (slope, particle size, entrenchment ratio, etc.). As a result, reference riparian communities must be stratified at least by stream type within a region.

Using **Worksheet 5-6**, input the existing and potential species composition in Column (3) in the stream stability summary **Worksheet 5-5**. If the user has a more detailed riparian vegetation scorecard or assessment methodology, it can be incorporated at this step. The procedure in **Worksheet 5-6** represents a minimum, general riparian assessment and does not preclude the use of a more detailed site inventory.

Chapter 5

The Prediction Level Assessment (PLA)

Worksheet 5-6. Riparian vegetation composition/density used for channel stability assessment.

	Stream:			Location:		
	Observers:	Reference reach	Disturbed (impacted reach)		Date:	
	Existing species composition:			Potential species composition:		

Riparian cover categories	Percent aerial cover*	Percent of site coverage**	Species composition	Percent of total species composition
1. Overstory — Canopy layer				100%
2. Understory — Shrub layer				100%
3. Ground level — Herbaceous				100%
3. Ground level — Leaf or needle litter			**Remarks:** Condition, vigor and/or usage of existing reach:	
3. Ground level — Bare ground				

*Based on crown closure.
** Based on basal area to surface area.

Column total = 100%

5-37

b. Flow regime

Flow regime categories are shown in **Figure 5-9**. Interpretations for sediment supply and stability relations are based on flow regime and vary with stream type. For example, a spring-fed E4 stream type has little bedload transport and very low suspended sediment concentration. The riparian vegetation is saturated yearlong and, therefore, is very dense with little evidence of bank erosion unless direct alterations occur. Stormflow-dominated streams, in contrast to snowmelt, ephemeral and spring-fed systems, have a certain influence for their respective flow category. Both flow increases, from reservoirs and/or diversions, and flow decreases, from diversions or reservoir regulations, influence channel and hydrologic processes. Altered flow regime is evaluated in more detail later in this chapter. Rain-on-snow events, which contribute to mid-winter floods, are identified in **Figure 5-9**.

Using **Figure 5-9**, insert all combinations from the general category and specific category that apply in Column (4) in the stream stability summary **Worksheet 5-5**.

General Category	
E	Ephemeral stream channels: flows only in response to precipitation. Often used in conjunction with intermittent.
S	Subterranean stream channel: flows parallel to and near the surface for various seasons - a sub-surface flow that follows the stream bed.
I	Intermittent stream channel: one that flows only seasonally or sporadically. Surface sources involve springs, snowmelt, artificial controls, etc. Often this term is associated with flows that reappear along various locations of a reach then run subterranean.
P	Perennial stream channels: surface water persists yearlong.

Specific Category	
1	Seasonal variation in streamflow dominated primarily by snowmelt runoff.
2	Seasonal variation in streamflow dominated primarily by stormflow runoff.
3	Uniform stage and associated streamflow due to spring-fed condition, backwater, etc.
4	Streamflow regulated by glacial melt.
5	Ice flows/ice torrents from ice dam breaches.
6	Alternating flow/backwater due to tidal influence.
7	Regulated streamflow due to diversions, dam release, de-watering, etc.
8	Altered due to development, such as urban streams, cut-over watersheds or vegetation conversions (forested to grassland) that change flow response to precipitation events.
9	Rain-on-snow generated runoff.

Figure 5-9. Flow regime variables that influence channel characteristics, sediment regime and biological interpretations.

c. Stream order and stream size

Strahler stream order gives an approximate reference for source area of contribution. For example, interpretations for an ephemeral, first-order A3 stream type would indicate a much lower risk of sediment supply, based on watershed size alone, than a third-order A3 stream type. Stream sizes expressed as bankfull width ranges also help to summarize data by indicating the approximate channel width. **Table 5-3** shows stream size and stream order categories for stratification by stream type. Use **Table 5-3** to input the category that applies for a given reach in Column (5) in the stability summary **Worksheet 5-5**. Add the specific stream order of the reach in parenthesis after the listed category. For example, a third-order stream with a bankfull width of 6.1 meters (20 ft) would be indexed as: S-4(3). Dimensionless ratios by stream type/valley type are stratified further by these size categories. Reference reach data can vary by river size; thus, sorting by stream size stratification allows for comparison of ratios between similar stream types.

Table 5-3. Stream order/stream size categories for stratification by stream type (Rosgen, 1996).

Category	STREAM SIZE: Bankfull width	
	meters	feet
S-1	<0.3	<1
S-2	0.3 – 1.5	1 – 5
S-3	1.5 – 4.6	5 – 15
S-4	4.6 – 9	15 – 30
S-5	9 – 15	30 – 50
S-6	15 – 22.8	50 – 75
S-7	22.8 – 30.5	75 – 100
S-8	30.5 – 46	100 – 150
S-9	46 – 76	150 – 250
S-10	76 – 107	250 – 350
S-11	107 – 150	350 – 500
S-12	150 – 305	500 – 1000
S-13	>305	>1000

d. Meander patterns

Meander patterns express adjustments to past events or impacts. They also indicate the direction and mode of lateral adjustment. The patterns indicate a central tendency for the "wandering river," lateral migration and other related processes using time-trend aerial photography. The various categories are summarized in **Figure 5-10**. Use this figure to list individual or combined categories that apply in Column (6) in the stability summary **Worksheet 5-5**. These meander pattern ratings will be used in the assessment of lateral stability in **Step 25**.

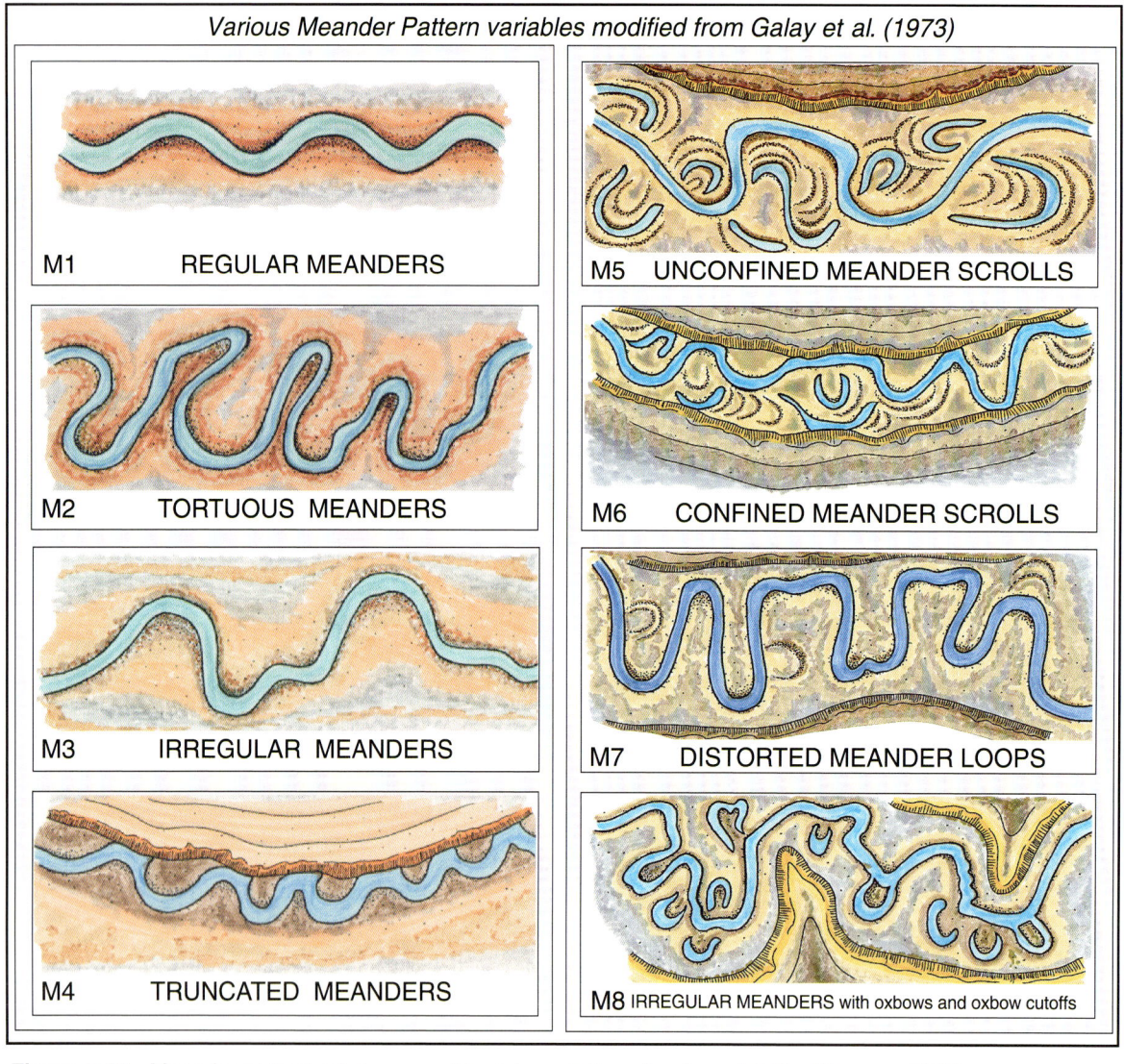

Various Meander Pattern variables modified from Galay et al. (1973)

M1	REGULAR MEANDERS
M2	TORTUOUS MEANDERS
M3	IRREGULAR MEANDERS
M4	TRUNCATED MEANDERS
M5	UNCONFINED MEANDER SCROLLS
M6	CONFINED MEANDER SCROLLS
M7	DISTORTED MEANDER LOOPS
M8	IRREGULAR MEANDERS with oxbows and oxbow cutoffs

Figure 5-10. Meander pattern relations used for interpretations for river stability (Rosgen, 1996).

Chapter 5   *The Prediction Level Assessment (PLA)*

e. Depositional patterns

These ratings describe the nature and extent of bar features in rivers. Many of these features are very active yet stable. Point bars, for example, on C4 stream types are a stable, depositional feature. When point bars start to occur on E4 stream types, however, they are generally associated with accelerated sediment supply and/or increased width/depth ratio. Furthermore, alternating side bars, as shown in category B4 in **Figure 5-11**, are often a depositional feature indicating the process of stabilization. Rivers that have been straightened often develop alternating bars in an effort to reduce width/depth ratio, increase sinuosity, reduce slope and eventually develop morphological features representing the stable stream type form.

Excess deposition leading to channel enlargement and/or aggradation/excess sediment supply is often associated with multiple mid-channel bars and chute-cutoffs, identified by the categories in **Figure 5-11**. Use this figure to list individual or combined categories in Column (7) in the stability summary **Worksheet 5-5**. These depositional pattern ratings will be used in the assessment of lateral and vertical stability (**Steps 25** and **26**).

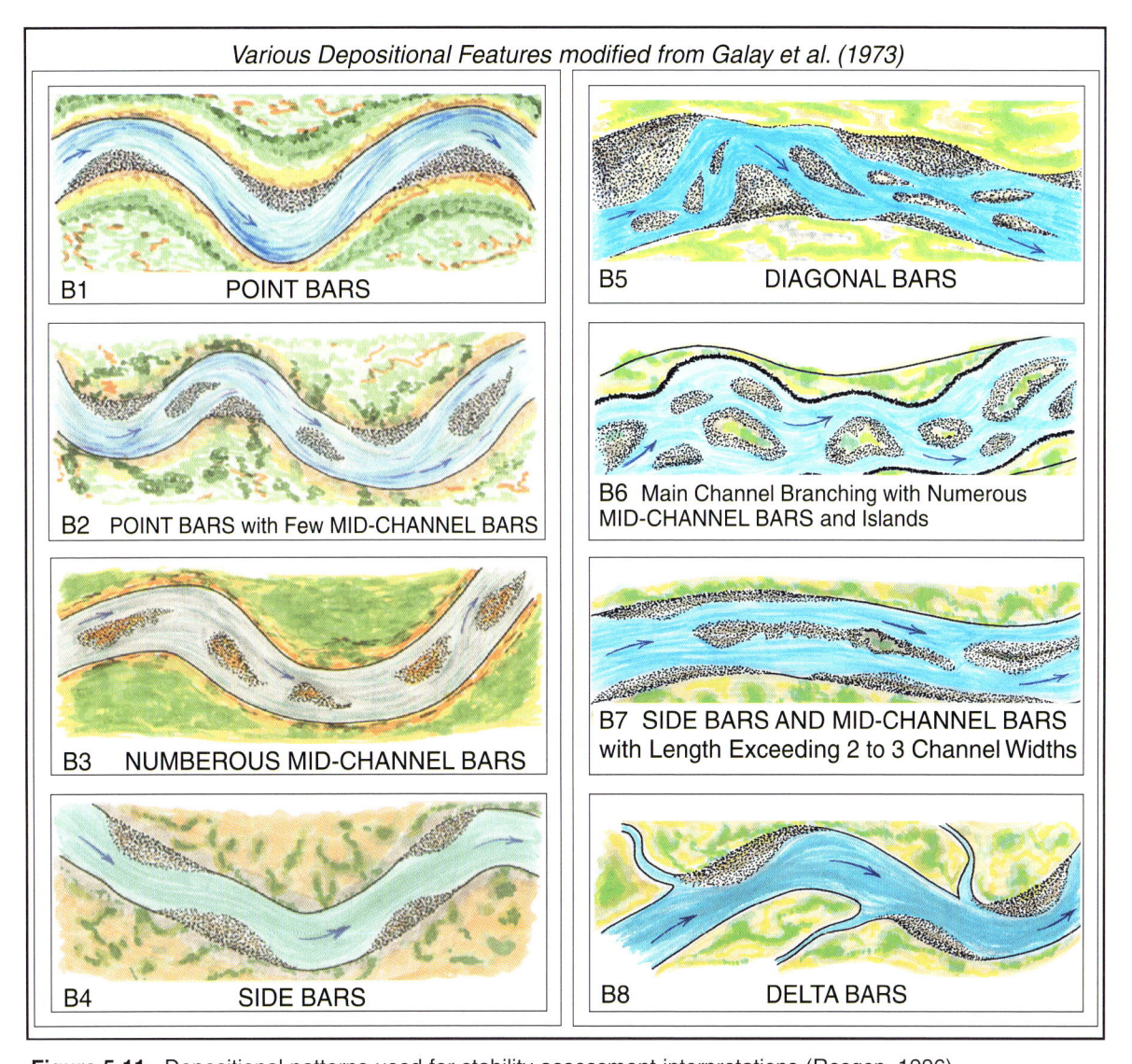

Figure 5-11. Depositional patterns used for stability assessment interpretations (Rosgen, 1996).

f. Channel blockages

A certain amount of large, woody debris is desirable for both physical and biological channel functions. However, when the magnitude and frequency of debris is such that the stream aggrades, loses sediment transport capacity and provides fish migration barriers, then the debris is likely adding to sediment supply and instability. Debris-driven sediment supply increases can often result in avulsions, lateral migration or streambank or side-slope rejuvenation, which often accelerate mass failures. Categories of channel debris also include channel blockage by in-channel structures, check dams, diversion structures or similar installations. The stability rating also considers different categories of active and abandoned beaver dams. **Table 5-4** summarizes these various categories. List all the categories that apply in Column (8) in the stability summary **Worksheet 5-5**. These channel blockage ratings will be used in the vertical stability assessment in **Step 26**.

Table 5-4. Various categories of in-channel debris, dams and/or channel blockages used to evaluate channel stability (Rosgen, 1996).

	Description/extent	Materials, which upon placement into the active channel or flood-prone area, may cause adjustments in channel dimensions or conditions due to influences on the existing flow regime.
D1	None	Minor amounts of small, floatable material.
D2	Infrequent	Debris consists of small, easily moved, floatable material, e.g., leaves, needles, small limbs and twigs.
D3	Moderate	Increasing frequency of small- to medium-sized material, such as large limbs, branches and small logs, that when accumulated, affect 10% or less of the active channel cross-section area.
D4	Numerous	Significant build-up of medium- to large-sized materials, e.g., large limbs, branches, small logs or portions of trees that may occupy 10–30% of the active channel cross-section area.
D5	Extensive	Debris "dams" of predominantly larger materials, e.g., branches, logs and trees, occupying 30–50% of the active channel cross-section area, often extending across the width of the active channel.
D6	Dominating	Large, somewhat continuous debris "dams," extensive in nature and occupying over 50% of the active channel cross-section area. Such accumulations may divert water into the flood-prone areas and form fish migration barriers, even when flows are at less than bankfull.
D7	Beaver dams: Few	An infrequent number of dams spaced such that normal streamflow and expected channel conditions exist in the reaches between dams.
D8	Beaver dams: Frequent	Frequency of dams is such that backwater conditions exist for channel reaches between structures where streamflow velocities are reduced and channel dimensions or conditions are influenced.
D9	Beaver dams: Abandoned	Numerous abandoned dams, many of which have filled with sediment and/or breached, initiating a series of channel adjustments, such as bank erosion, lateral migration, avulsion, aggradation and degradation.
D10	Human influences	Structures, facilities or materials related to land uses or development located within the flood-prone area, such as diversions or low-head dams, controlled by-pass channels, velocity control structures and various transportation encroachments that have an influence on the existing flow regime, such that significant channel adjustments occur.

Chapter 5   The Prediction Level Assessment (PLA)

g. Width/depth ratio state

Width/depth ratio is a key variable for assessing departure from a stable reference condition. Increases in width/depth ratio are generally associated with accelerated streambank erosion rates, excess deposition/aggradation processes and over-widening due to direct mechanical impacts. Width/depth ratio state is rated as a comparison or departure from a reference condition. Examples of width/depth ratio increases and decreases from a reference width/depth ratio are shown in **Figure 5-12**.

Width/Depth Ratio Increase

- Reference: C4 Stream Type — W/D Ratio: 13
- C4 Stream Type — W/D Ratio: 15 — 15 / 13 = 1.15 (**Stable**)
- C4 Stream Type — W/D Ratio: 17 — 17 / 13 = 1.3 (**Mod Unstable**)
- C4 Stream Type — W/D Ratio: 20 — 20 / 13 = 1.5 (**Unstable**)

Width/Depth Ratio Decrease

- Reference: B4 Stream Type — W/D Ratio: 16
- B4 Stream Type — W/D Ratio: 12 — 12 / 16 = 0.75 (**Stable**)
- G4 Stream Type (Incising) — W/D Ratio: 8 — 8 / 16 = 0.50 (**Unstable**)
- G4 Stream Type (Incising) — W/D Ratio: 6 — 6 / 16 = 0.37 (**Highly Unstable**)

Width/Depth Ratio Decrease (Increasing Stability)

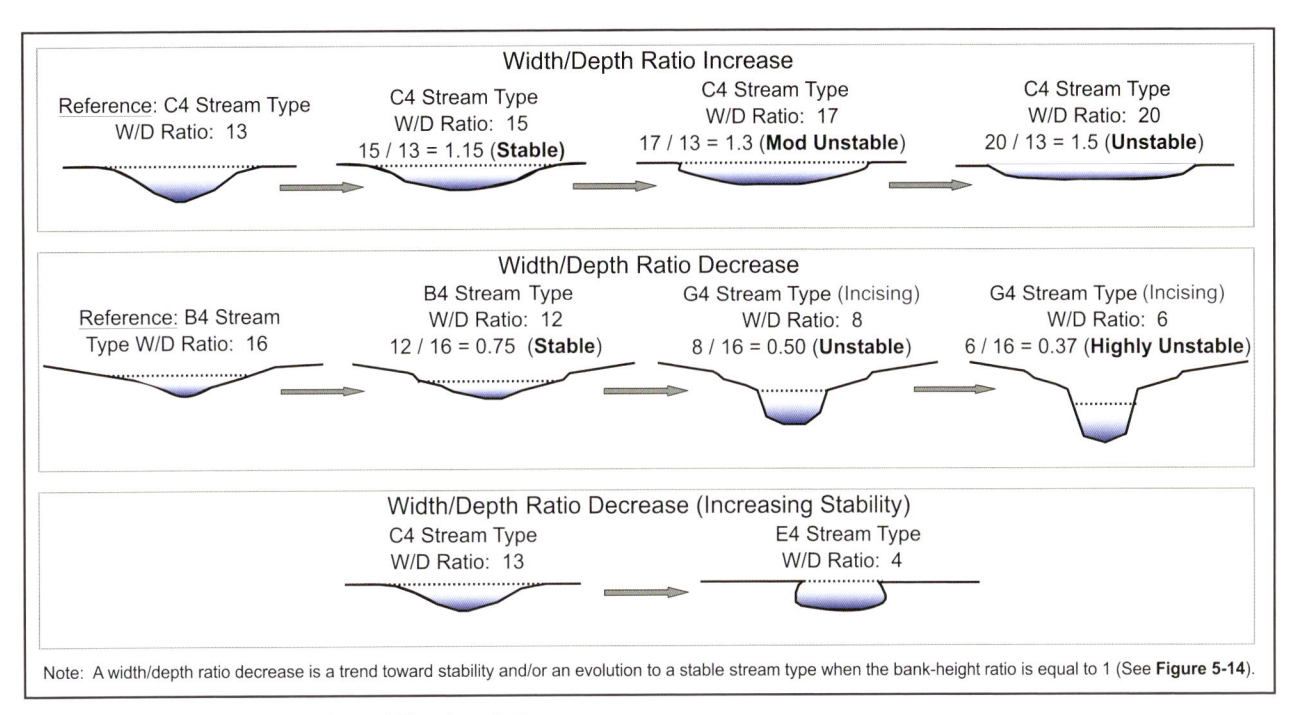

- C4 Stream Type — W/D Ratio: 13
- E4 Stream Type — W/D Ratio: 4

Note: A width/depth ratio decrease is a trend toward stability and/or an evolution to a stable stream type when the bank-height ratio is equal to 1 (See **Figure 5-14**).

Figure 5-12. Width/depth ratio stability descriptions.

WATERSHED ASSESSMENT OF RIVER STABILITY AND SEDIMENT SUPPLY

It is important to distinguish between width/depth ratio increases and decreases. The decrease category is rated as high risk only when accompanied by a low Bank-Height Ratio (BHR). *Bank-height ratio* is the lowest bank height divided by maximum bankfull depth that determines the degree of channel incision (see river stability indices, Step 7i, for channel incision). As the width/depth ratio decreases, there is an associated increase in shear stress and unit stream power potentially leading to accelerated incision. BHR values larger than 1.0 indicate incision. Examples of width/depth ratio departures are shown in Chapter 2 of this book and in Chapter 6 of *Applied River Morphology* (Rosgen, 1996). **Figure 5-13** shows width/depth ratio stability ratings.

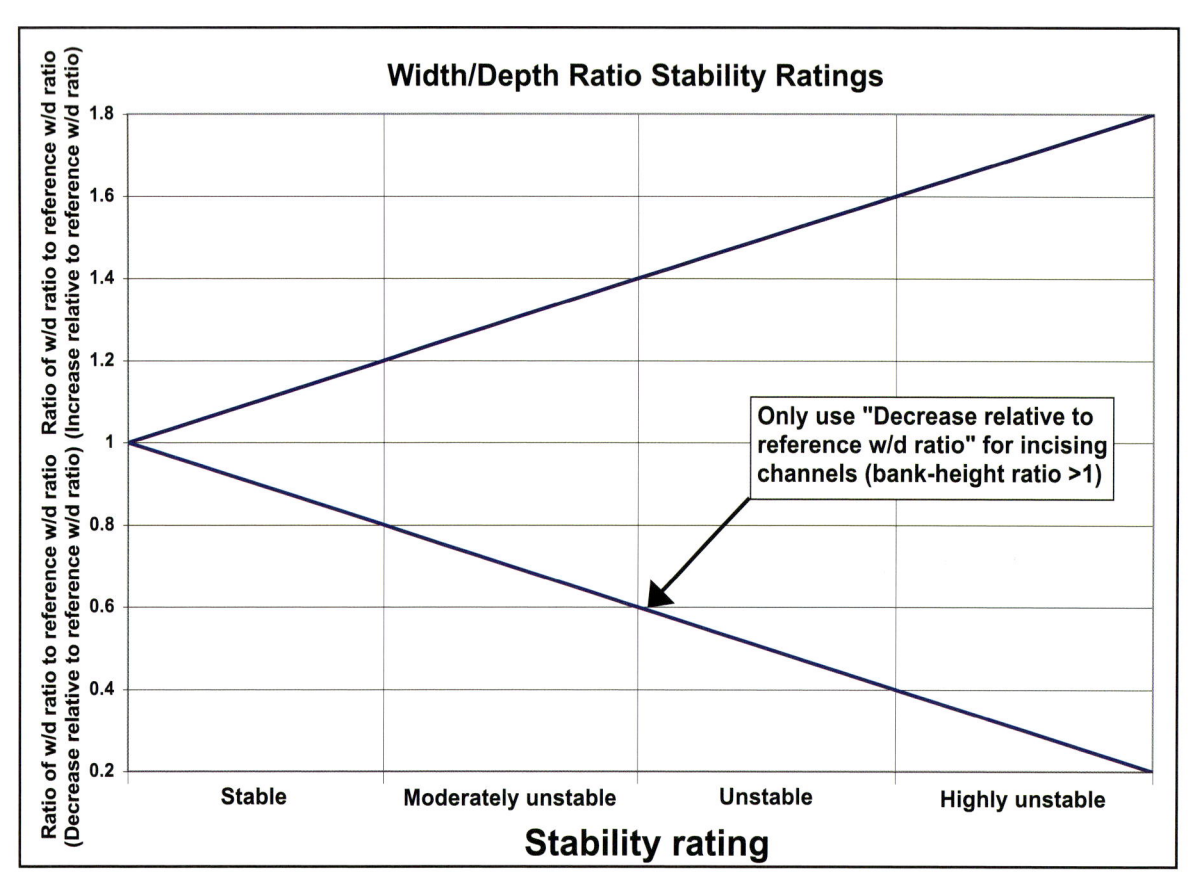

Figure 5-13. Stability ratings based on departure of width/depth ratio from reference condition (Rosgen, 2001b).

Use data from **Worksheet 5-3** to express bankfull surface width/mean bankfull depth as a ratio. Use **Figure 5-13** to determine the stability rating. Record this information in Column (9) in the stability summary **Worksheet 5-5**. Width/depth ratio stability ratings are used in **Steps 25** and **26** to assess lateral and vertical stability.

h. Modified Pfankuch stability rating

Use the variables assessed above and **Worksheet 5-7** to determine a Pfankuch (1975) stability rating, which is then adjusted according to stream type (Rosgen, 1996, 2001b). A stability rating of 88 for a C4 would be *good* or *stable* due to the morphological characteristics associated with the rating. However, the same rating for a B4 stream type would be *poor* or *unstable* and atypical of the stable B4 stream type. Preliminary interpretations are identified for potential vertical stability, width/depth ratio condition and sediment supply. The higher the numerical score, the higher the sediment supply. Record the Pfankuch stability rating in Column (10) in the stability summary **Worksheet 5-5**.

Pfankuch stability ratings by stream type are used to select appropriate dimensionless bedload and suspended sediment ratings curves in **Step 12** of the FLOWSED model. The ratings are also used to determine overall sediment supply ratings in **Step 28**.

WATERSHED ASSESSMENT OF RIVER STABILITY AND SEDIMENT SUPPLY

Worksheet 5-7. Pfankuch (1975) stream channel stability rating procedure, as modified by Rosgen (1996, 2001b).

Stream: _____ Location: _____ Valley Type: _____ Observers: _____ Date: _____

Location	Key	Category	Excellent Description	Rating	Good Description	Rating	Fair Description	Rating	Poor Description	Rating
Upper banks	1	Landform slope	Bank slope gradient <30%.	2	Bank slope gradient 30–40%.	4	Bank slope gradient 40–60%.	6	Bank slope gradient > 60%.	8
	2	Mass erosion	No evidence of past or future mass erosion.	3	Infrequent. Mostly healed over. Low future potential.	6	Frequent or large, causing sediment nearly yearlong.	9	Frequent or large, causing sediment nearly yearlong OR imminent danger of same.	12
	3	Debris jam potential	Essentially absent from immediate channel area.	2	Present, but mostly small twigs and limbs.	4	Moderate to heavy amounts, mostly larger sizes.	6	Moderate to heavy amounts, predominantly larger sizes.	8
	4	Vegetative bank protection	> 90% plant density. Vigor and variety suggest a deep, dense soil-binding root mass.	3	70–90% density. Fewer species or less vigor suggest less dense or deep root mass.	6	50–70% density. Lower vigor and fewer species from a shallow, discontinuous root mass.	9	<50% density plus fewer species & less vigor indicating poor, discontinuous and shallow root mass.	12
Lower banks	5	Channel capacity	Bank heights sufficient to contain the bankfull stage. Width/depth ratio departure from reference width/depth ratio = 1.0. Bank-Height Ratio (BHR) = 1.0.	1	Bankfull stage is contained within banks. Width/depth ratio departure from reference width/depth ratio = 1.0–1.2. Bank-Height Ratio (BHR) = 1.0–1.1	2	Bankfull stage is not contained. Width/depth ratio departure from reference width/depth ratio = 1.2–1.4. Bank-Height Ratio (BHR) = 1.1–1.3.	3	Bankfull stage is not contained; over-bank flows are common with flows less than bankfull. Width/depth ratio departure from reference width/depth ratio > 1.4. Bank-Height Ratio (BHR) >1.3.	4
	6	Bank rock content	> 65% with large angular boulders. 12"+ common.	2	40–65%. Mostly boulders and small cobbles 6–12".	4	20–40%. Most in the 3–6" diameter class.	6	<20% rock fragments of gravel sizes, 1–3" or less.	8
	7	Obstructions to flow	Rocks and logs firmly imbedded. Flow pattern w/o cutting or deposition. Stable bed.	2	Some present causing erosive cross currents and minor pool filling. Obstructions fewer and less firm.	4	Moderately frequent, unstable obstructions move with high flows causing bank cutting and pool filling.	6	Frequent obstructions and deflectors cause bank erosion yearlong. Sediment traps full, channel migration occurring.	8
	8	Cutting	Little or none. Infrequent raw banks <6".	2	Some, intermittently at outcurves and constrictions. Raw banks may be up to 12".	4	Significant. Cuts 12–24" high. Root mat overhangs and sloughing evident.	6	Almost continuous cuts, some over 24" high. Failure of overhangs frequent.	16
	9	Deposition	Little or no enlargement of channel or point bars.	4	Some new bar increase, mostly from coarse gravel.	4	Moderate deposition of new gravel and coarse sand on old and some new bars.	12	Extensive deposit of predominantly fine particles. Accelerated bar development.	16
Bottom	10	Rock angularity	Sharp edges and corners. Plane surfaces rough.	1	Rounded corners and edges. Surfaces smooth and flat.	2	Corners and edges well rounded in 2 dimensions.	3	Well rounded in all dimensions, surfaces smooth.	4
	11	Brightness	Surfaces dull, dark or stained. Generally not bright.	1	Mostly dull, but may have <35% bright surfaces.	2	Mixture dull and bright, i.e., 35–65% mixture range.	3	Predominantly bright, > 65% exposed or scoured surfaces.	4
	12	Consolidation of particles	Assorted sizes tightly packed or overlapping.	2	Moderately packed with some overlapping.	4	Mostly loose assortment with no apparent overlap.	6	No packing evident. Loose assortment, easily moved.	8
	13	Bottom size distribution	No size change evident. Stable materials 80–100%.	4	Distribution shift light. Stable material 50–80%.	8	Moderate change in sizes. Stable materials 20–50%.	12	Marked distribution change. Stable materials 0–20%.	16
	14	Scouring and deposition	<5% of bottom affected by scour or deposition.	6	5–30% affected. Scour at constrictions and where grades steepen. Some deposition in pools.	12	30–50% affected. Deposits and scour at obstructions, constrictions and bends. Some filling of pools.	18	More than 50% of the bottom in a state of flux or change nearly yearlong.	24
	15	Aquatic vegetation	Abundant growth moss-like, dark green perennial. In swift water, too.	1	Common. Algae forms in low velocity and pool areas. Moss here, too.	2	Present but spotty, mostly in backwater. Seasonal algae growth makes rocks slick.	3	Perennial types scarce or absent. Yellow-green, short-term bloom may be present.	4

Excellent total = _____   Good total = _____   Fair total = _____   Poor total = _____

Stream type	A1	A2	A3	A4	A5	A6	B1	B2	B3	B4	B5	B6	C1	C2	C3	C4	C5	C6	D3	D4	D5	D6
Good (Stable)	38-43	38-43	54-90	60-95	60-95	50-80	38-45	38-45	40-60	40-64	48-68	40-60	38-50	38-50	60-85	70-90	70-90	60-85	85-107	85-107	85-107	67-98
Fair (Mod. unstable)	44-47	44-47	91-129	96-132	96-142	81-110	46-58	46-58	61-78	65-84	69-88	61-78	51-61	51-61	86-105	91-110	91-110	86-105	108-132	108-132	108-132	99-125
Poor (Unstable)	48+	48+	130+	133+	143+	111+	59+	59+	79+	85+	89+	79+	62+	62+	106+	111+	111+	106+	133+	133+	133+	126+

Stream type	DA3	DA4	DA5	DA6	E3	E4	E5	E6	F1	F2	F3	F4	F5	F6	G1	G2	G3	G4	G5	G6
Good (Stable)	40-63	40-63	40-63	40-63	40-63	50-75	50-75	40-63	60-85	60-85	85-110	85-110	90-115	80-95	40-60	40-60	40-60	85-107	90-112	85-107
Fair (Mod. unstable)	64-86	64-86	64-86	64-86	64-86	76-96	76-96	64-86	86-105	86-105	111-125	111-125	116-130	96-110	61-78	61-78	61-78	108-120	113-125	108-120
Poor (Unstable)	87+	87+	87+	87+	87+	97+	97+	87+	106+	106+	126+	126+	131+	111+	79+	79+	79+	121+	126+	121+

*Rating should be adjusted to potential stream type, not existing.

Grand total = _____
Existing stream type = _____
*Potential stream type = _____
Modified channel stability rating = _____

5-46

i. Degree of channel incision

This field measurement uses the Bank-Height Ratio (BHR) to determine the degree of channel incision. BHR is calculated by dividing the lowest bank height by the maximum bankfull depth. Some rivers may be entrenched (vertical containment of floods); others may have a BHR indicative of a stream that is lowering its local base level but is not yet entrenched. **Figure 5-14** indicates degree of incision according to BHR ranges. Streams with high bank-height ratios generally contribute disproportionate amounts of sediment from streambanks and channel beds due to high shear stress. The relation of BHR values to stability risk is shown in **Figure 5-15**. For actively incising channels, field-tested regional hydrology curves are often used to obtain bankfull discharge and cross-sectional area by drainage area. This assists in obtaining maximum bankfull stage depth in such channels where depositional surfaces are not present. Obtain the BHR from **Worksheet 5-4**, and use **Figure 5-15** to determine the stability rating. Insert this information in Column (11) in the stream stability summary **Worksheet 5-5**. BHR values are used to assess vertical stability in **Step 26**.

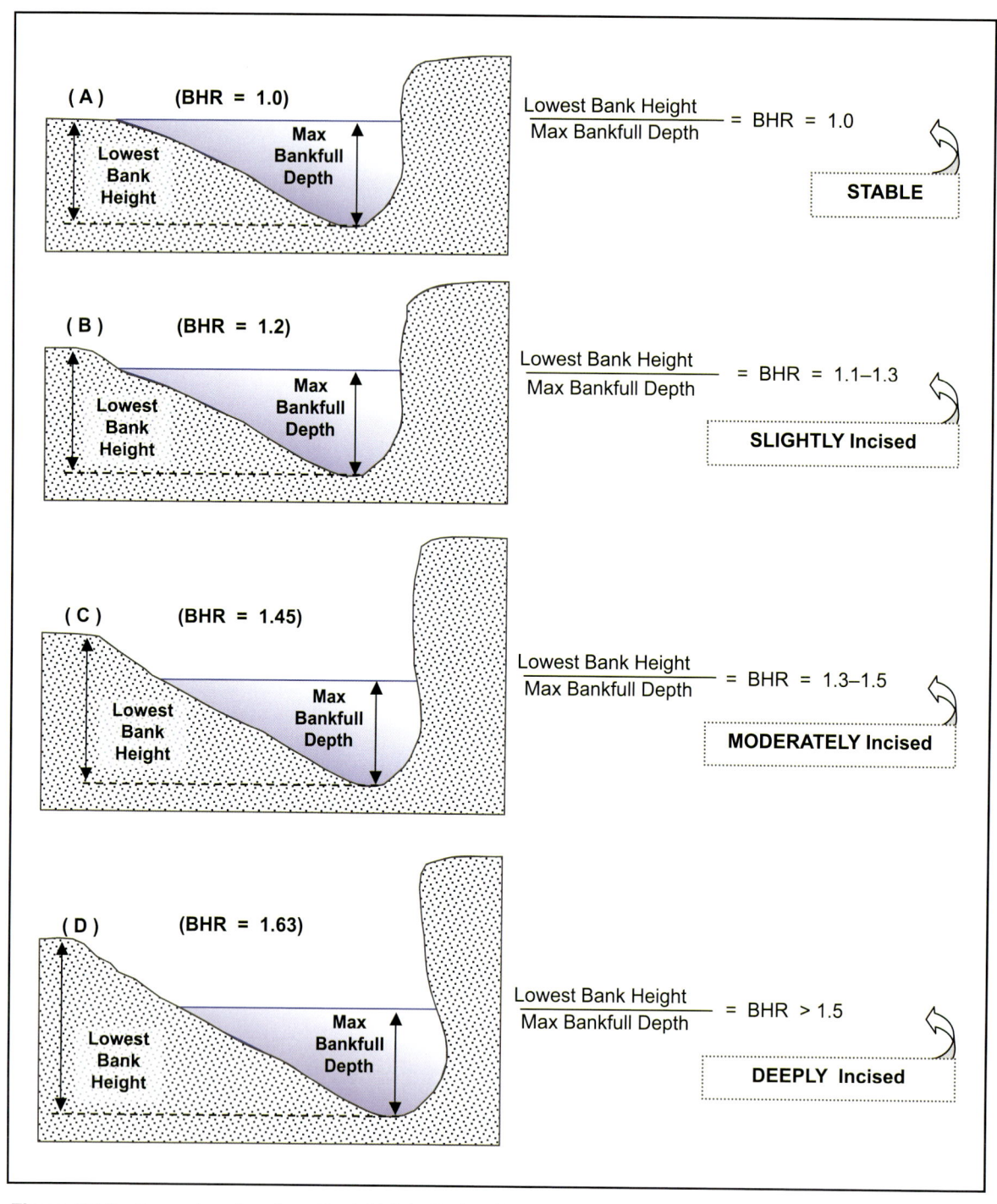

Figure 5-14. Examples of various Bank-Height Ratios (BHR) indicating degree of channel incision (Rosgen, 2001b).

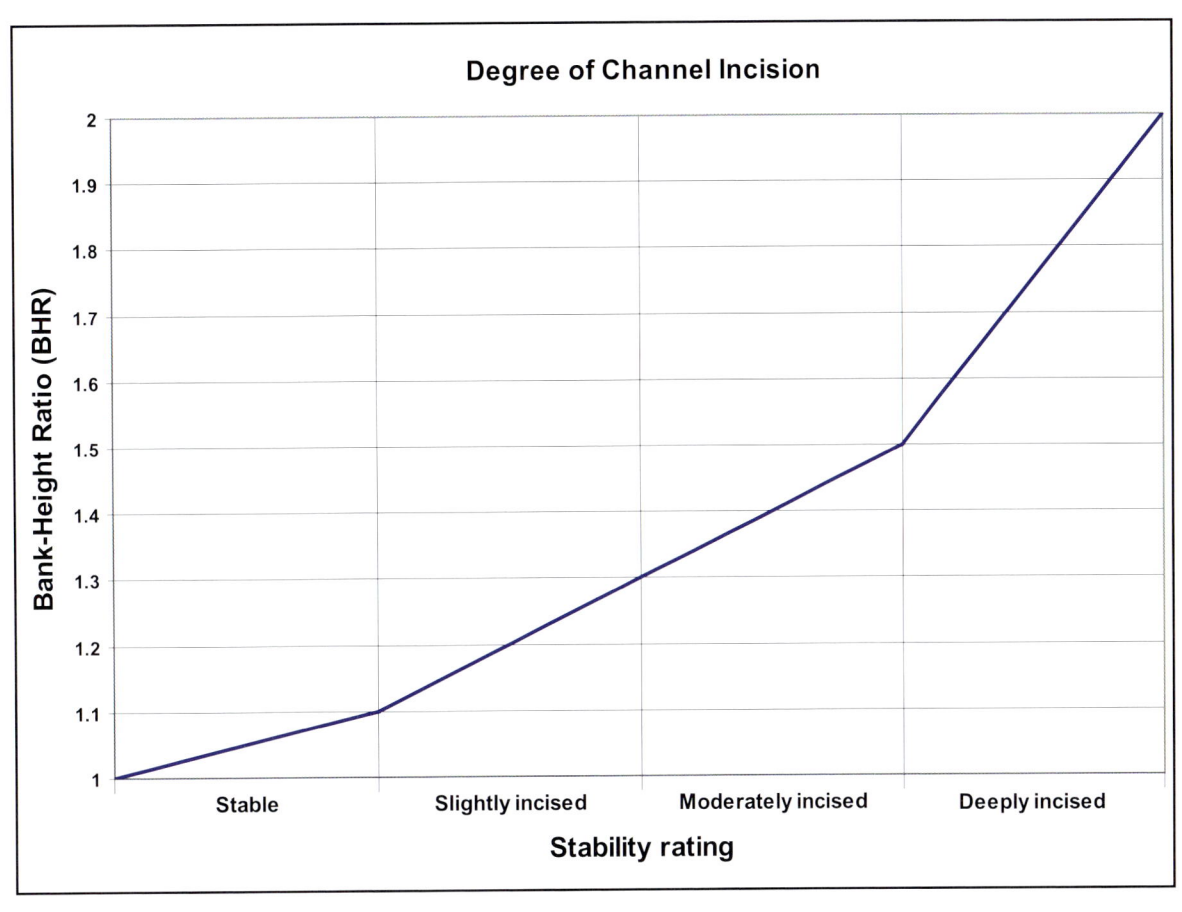

Figure 5-15. Relationship of BHR ranges to corresponding stream stability ratings (Rosgen, 2001b).

j. Degree of channel confinement (lateral containment)

Instability can often be caused due to encroachment on natural channels. The extent of encroachment can limit the lateral containment of the stream reach. The containment can change the dimension, pattern and profile of the river. The measure used to determine confinement (lateral containment) is *Meander Width Ratio* (MWR), which equals belt width divided by bankfull width. *Belt width* is the farthest lateral extent of the stream in its valley measured from the outside to outside of opposing stream bends (see **Figure 5-5**). **Figure 5-16** depicts the range and average values of confinement using MWR by stream type. Streams that are confined (laterally contained) often are associated with channel enlargement, lateral accretion, high bank erosion rates and sediment transport problems.

Meander Width Ratio (MWR) by Stream Type Categories

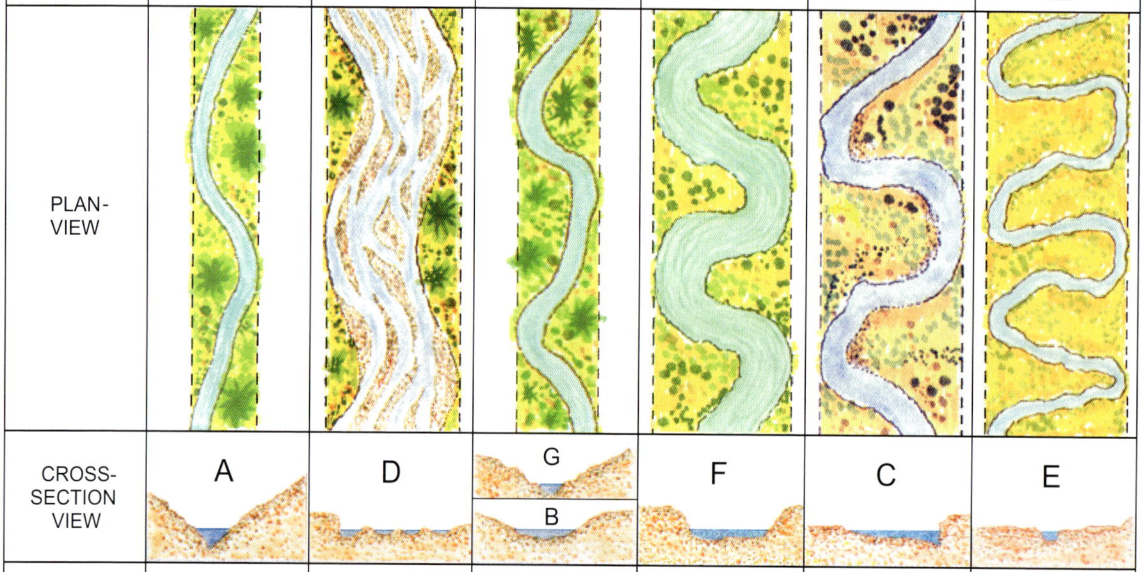

STREAM TYPE	A	D	B & G	F	C	E
AVERAGE VALUES	1.5	1.1	3.7	5.3	11.4	24.2
RANGE	1 - 3	1 - 2	2 - 8	2 - 10	4 - 20	20 - 40

Figure 5-16. The range and average values of confinement using MWR by stream type (Rosgen, 1996).

To determine the degree of channel confinement, calculate the meander width ratio. Divide the meander width ratio by the reference condition meander width ratio (MWR$_{ref}$). Use **Figure 5-17** to determine the degree of confinement. Record this information in Column (12) in the stream stability summary **Worksheet 5-5**. MWR/MWR$_{ref}$ values are used to assess lateral and vertical stability in **Steps 25 and 26**.

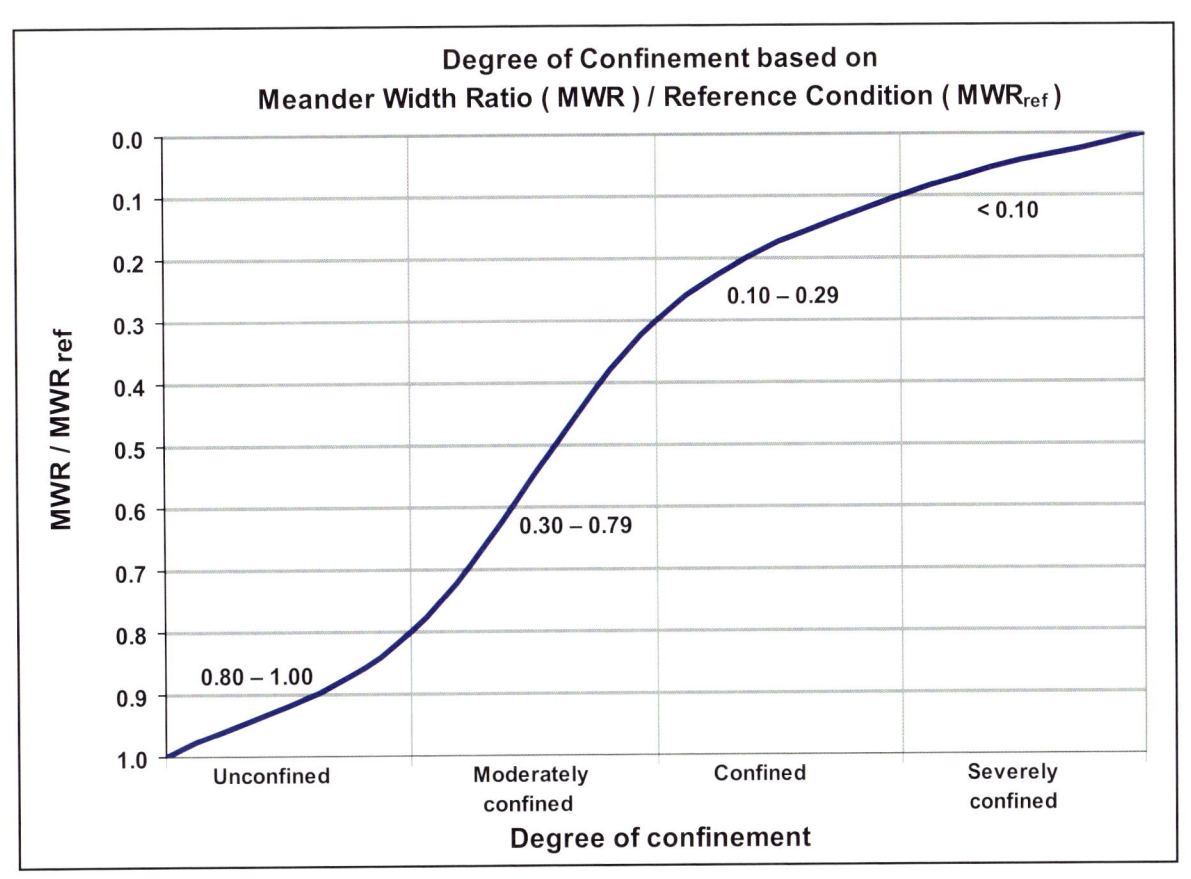

Figure 5-17. Degree of confinement based on meander width ratio divided by reference condition meander width ratio.

The use of MWR is primarily intended for single-thread channels. For braided, D stream types, MWR is not applicable to infer vertical instability (**Step 26**). MWR, however, can be used to predict lateral stability in **Step 25** for braided, D stream types.

Steps 8–9: Streambank erosion (BANCS model) (See Flowchart 5-7)

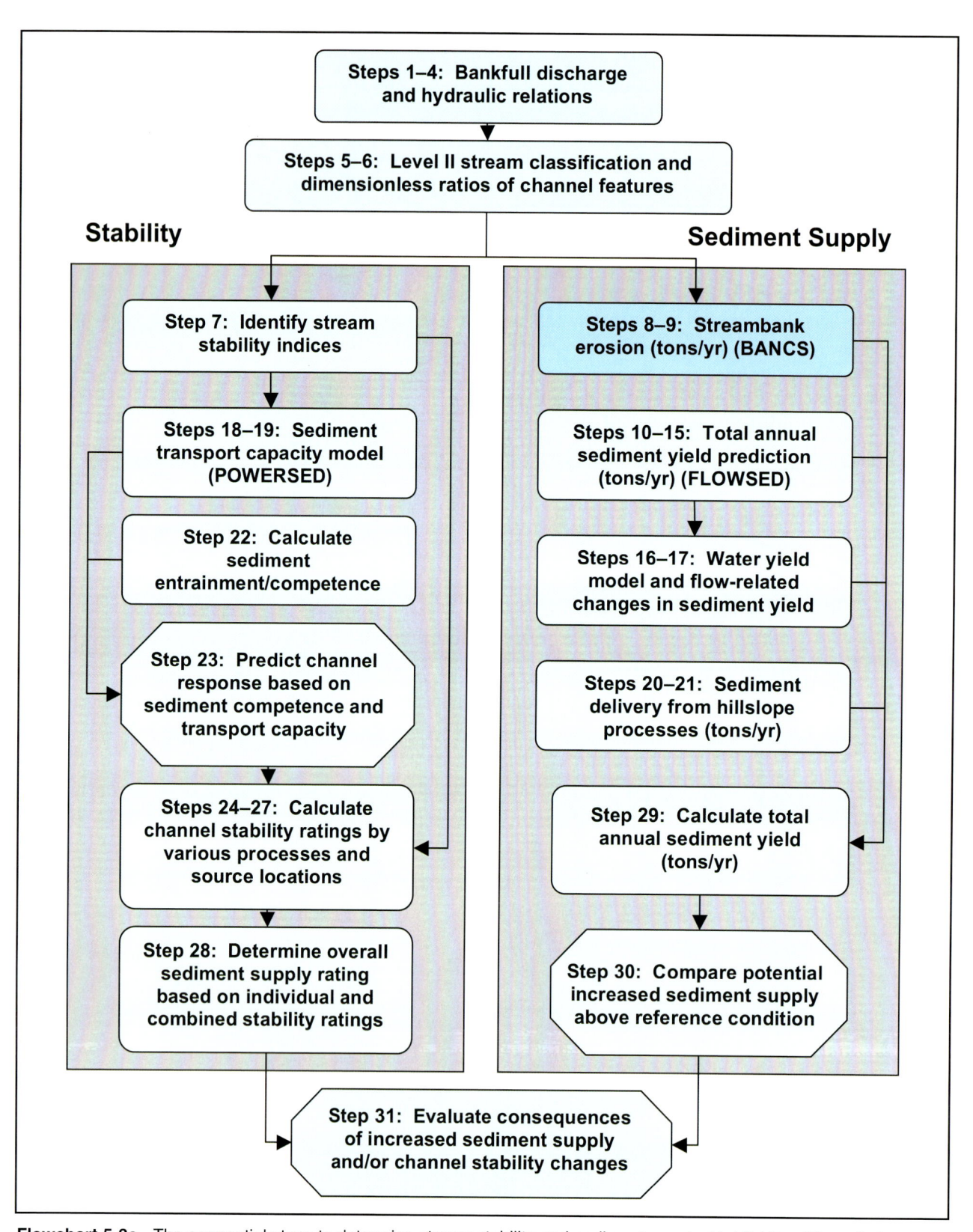

Steps 1–4: Bankfull discharge and hydraulic relations

Steps 5–6: Level II stream classification and dimensionless ratios of channel features

Stability

Step 7: Identify stream stability indices

Steps 18–19: Sediment transport capacity model (POWERSED)

Step 22: Calculate sediment entrainment/competence

Step 23: Predict channel response based on sediment competence and transport capacity

Steps 24–27: Calculate channel stability ratings by various processes and source locations

Step 28: Determine overall sediment supply rating based on individual and combined stability ratings

Sediment Supply

Steps 8–9: Streambank erosion (tons/yr) (BANCS)

Steps 10–15: Total annual sediment yield prediction (tons/yr) (FLOWSED)

Steps 16–17: Water yield model and flow-related changes in sediment yield

Steps 20–21: Sediment delivery from hillslope processes (tons/yr)

Step 29: Calculate total annual sediment yield (tons/yr)

Step 30: Compare potential increased sediment supply above reference condition

Step 31: Evaluate consequences of increased sediment supply and/or channel stability changes

Flowchart 5-3e. The sequential steps to determine stream stability and sediment supply, highlighting **Steps 8–9**.

Chapter 5 — The Prediction Level Assessment (PLA)

Steps 8–9: Streambank erosion (BANCS model)

Step 8: Calculate Bank Erosion Hazard Index (BEHI) and Near-Bank Stress (NBS) ratings using the BANCS model

BEHI

Bank Erosion Hazard Index (BEHI) (Figure 5-19 and Worksheet 5-8)
- Bank height/bankfull height
- Rooting depth/bank height
- Weighted root density
- Bank angle (slope steepness)
- Surface protection
- Bank material
- Material stratification

BEHI rating (Worksheet 5-8)

NBS

Near-Bank Stress (NBS) (Methods 1–7 in Worksheet 5-9)
1. Transverse bar or split channel creating NBS
2. Ratio of radius of curvature to bankfull width
3. Ratio of pool slope to average water surface slope
4. Ratio of pool slope to riffle slope
5. Ratio of near-bank max depth to bankfull mean depth
6. Ratio of near-bank shear stress to bankfull shear stress
7. Velocity gradient

NBS rating (Worksheet 5-9)

Step 9: Predict annual streambank erosion rate (ft/yr) using BEHI and NBS ratings (use Figure 5-38 or 5-39 and Worksheet 5-10)

Step 25: Lateral stability (Worksheet 5-17)

Summary of sediment sources in tons/yr (*PLA* summary Worksheet 5-22)

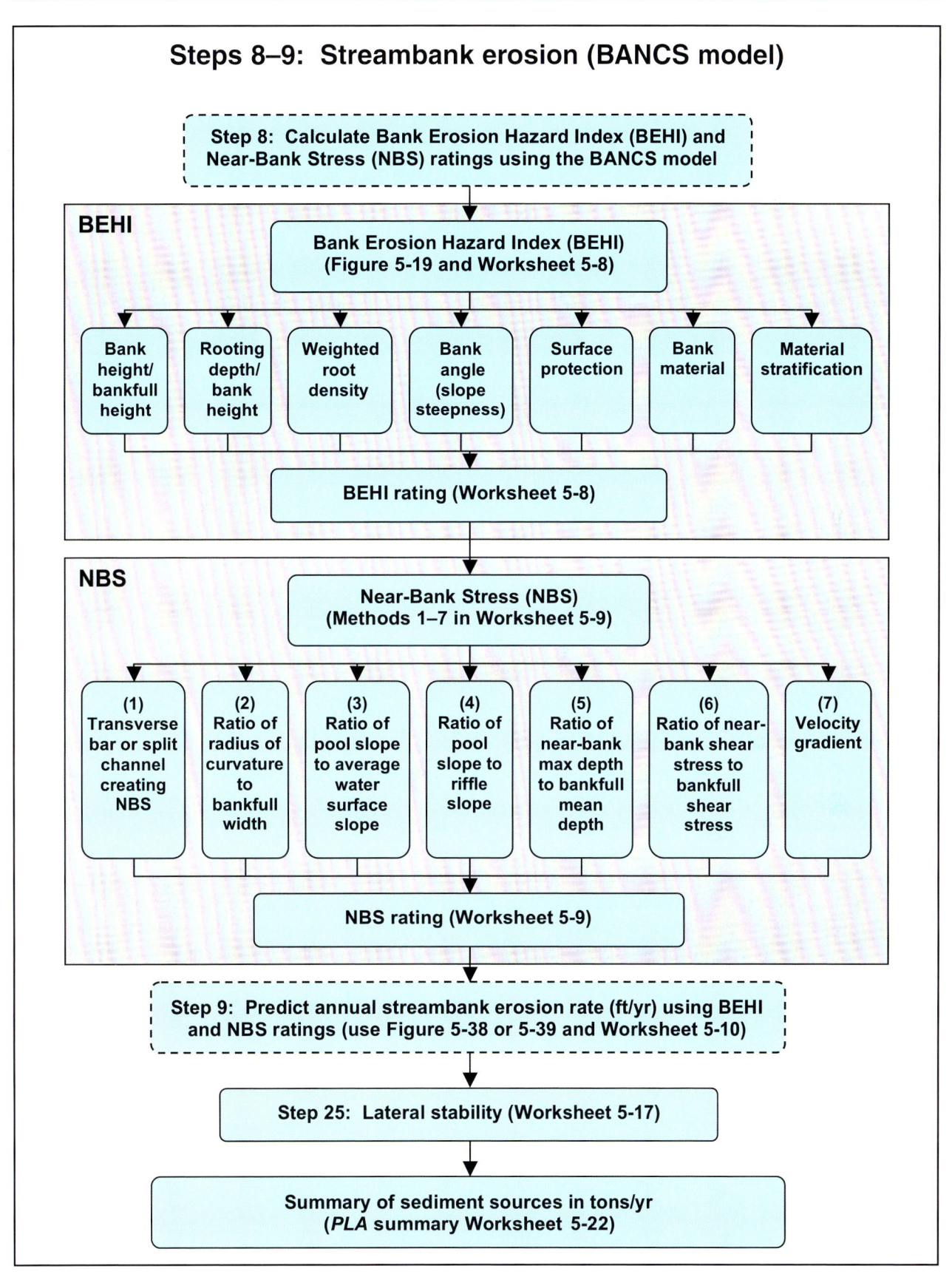

Flowchart 5-7. The BANCS model variables, ratios and procedures associated with the Bank Erosion Hazard Index (BEHI) and Near-Bank Stress (NBS) to predict annual streambank erosion.

Step 8. Calculate Bank Erosion Hazard Index (BEHI) and Near-Bank Stress (NBS) ratings

The prediction of streambank erosion rates uses the Bank Assessment for Non-point source Consequences of Sediment (BANCS) model. This model, as published by Rosgen (1996, 2001a), uses two bank erosion estimation tools:

1. The Bank Erosion Hazard Index (BEHI); and

2. Near-Bank Stress (NBS).

The application evaluates the bank characteristics and flow distribution along river reaches and maps BEHI and NBS risk ratings commensurate with streambank and channel changes. Annual erosion rates are estimated and then multiplied by the bank height and by a corresponding bank length of a similar condition, providing an estimate of cubic yards and tons of sediment per year. This information can be compared to the annual sediment yield data to apportion the amount of sediment potentially contributed by streambanks.

For great distances of stream reaches, multiple combinations of BEHI and NBS ratings may be necessary. The field observer can rapidly determine multiple ratings over great distances by calibrating ocular estimates to detailed site measurements. The following field procedures are recommended:

1. Prior to obtaining detailed measurements for a particular river reach, estimate the BEHI rating for the bank most susceptible to erosion using the methods presented in the BEHI section;

2. Measure each of the variables to obtain a BEHI rating. Compare estimated versus measured values. Calibrate field estimates by repeating the procedure until the field observer's estimations closely match measured values; and

3. Repeat steps 1 and 2 for NBS using individual ratings or a combination of the seven methods presented in the NBS section. Compare a rapid field estimate of NBS with the more detailed measurements. Repeat procedure until rapid estimates closely match the detailed measured values.

In such a manner, the "field calibration" of the observer eliminates the need to complete the BEHI and NBS worksheets for each BEHI and NBS evaluation along several miles of stream length. The field practice involves noting the BEHI and NBS values as one walks along a stream reach and measures distance (stationing) and study bank height. These values can be recorded using the worksheet in **Step 9** to predict annual streambank erosion. This allows the user to rapidly predict bank erosion rates for entire lengths of stream. It is important, however, to occasionally check the "field calibration." The Wolf Creek Watershed was quantified using this procedure, with examples of the detailed BEHI and NBS predictions for the C4 and D4 stream types shown in the *PLA* section of Chapter 7.

Note: It is imperative to plot the BEHI and NBS ratings and/or annual erosion rates in tons/yr/ft on a photo or map to identify specific high erosion rate locations in need of mitigation, restoration or changed riparian management.

Chapter 5 The Prediction Level Assessment (PLA)

The Bank Erosion Hazard Index (BEHI)

The Bank Erosion Hazard Index (BEHI) evaluates the susceptibility to erosion for multiple erosional processes. The BEHI model is a process-integration approach that does not isolate individual processes of erosion (such as surface erosion, fluvial entrainment or mass erosion). The process integrates multiple variables that relate to "combined" erosional processes leading to annual erosion rates. Erosion risk is then established for a variety of BEHI variables and is eventually used to establish corresponding streambank erosion rates.

The individual BEHI variables used for the erosion prediction model are broken into seven categories:

1. Study bank height/bankfull height (study bank-height ratio);
2. Root depth/bank height (root depth ratio);
3. Weighted root density;
4. Bank angle;
5. Surface protection;
6. Bank material; and
7. Stratification of bank material.

Figure 5-18 depicts study bank height, bankfull height, root depth, bank angle and surface protection.

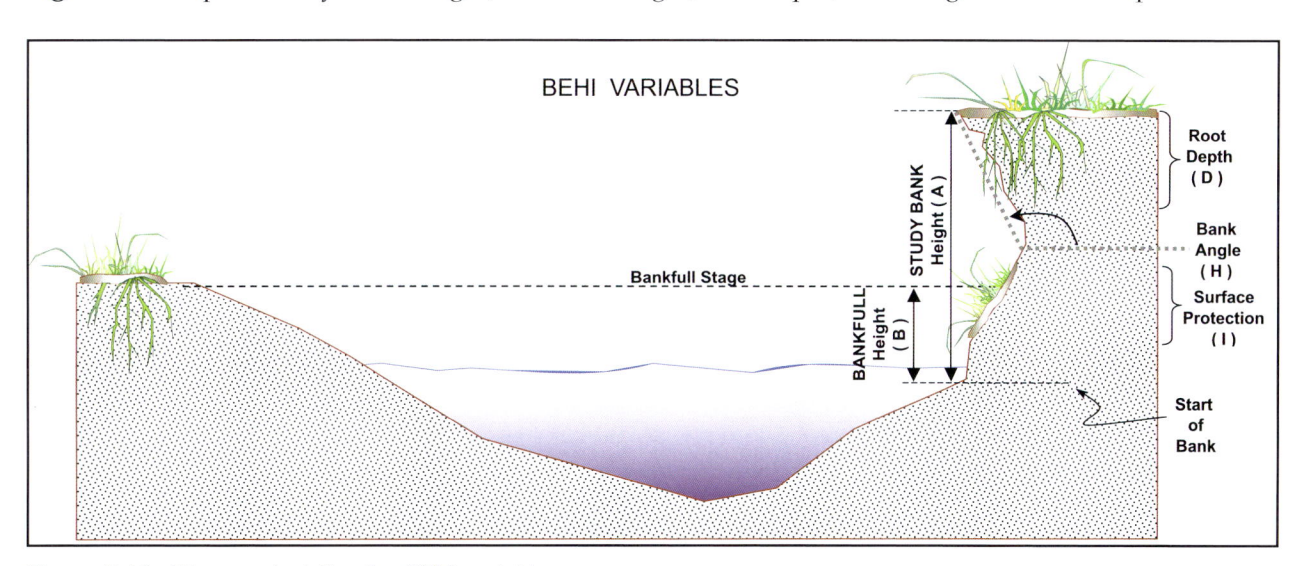

Figure 5-18. Diagram depicting the BEHI variables.

Study bank height/bankfull height, root depth/bankfull height, weighted root density, bank angle and surface protection are each calculated and/or recorded in **Worksheet 5-8**. These variables are converted to a BEHI rating using the relations in **Figure 5-19**. These ratings vary between Very Low to Very High and have values between 0 and 10. The numerical BEHI ratings are then totaled and adjusted according to bank materials and stratification of the bank materials in **Worksheet 5-8** to obtain an overall BEHI risk rating. The overall BEHI risk ratings are used with Near-Bank Stress (NBS) ratings to obtain annual erosion rate values in **Step 9** (Rosgen 1996, 2001a). Descriptions of the BEHI variables and examples are given in detail in the following pages.

Watershed Assessment of River Stability and Sediment Supply

Worksheet 5-8. Form to calculate Bank Erosion Hazard Index (BEHI) variables and an overall BEHI rating (Rosgen, 1996, 2001a). Use **Figure 5-19** with BEHI variables to determine BEHI score.

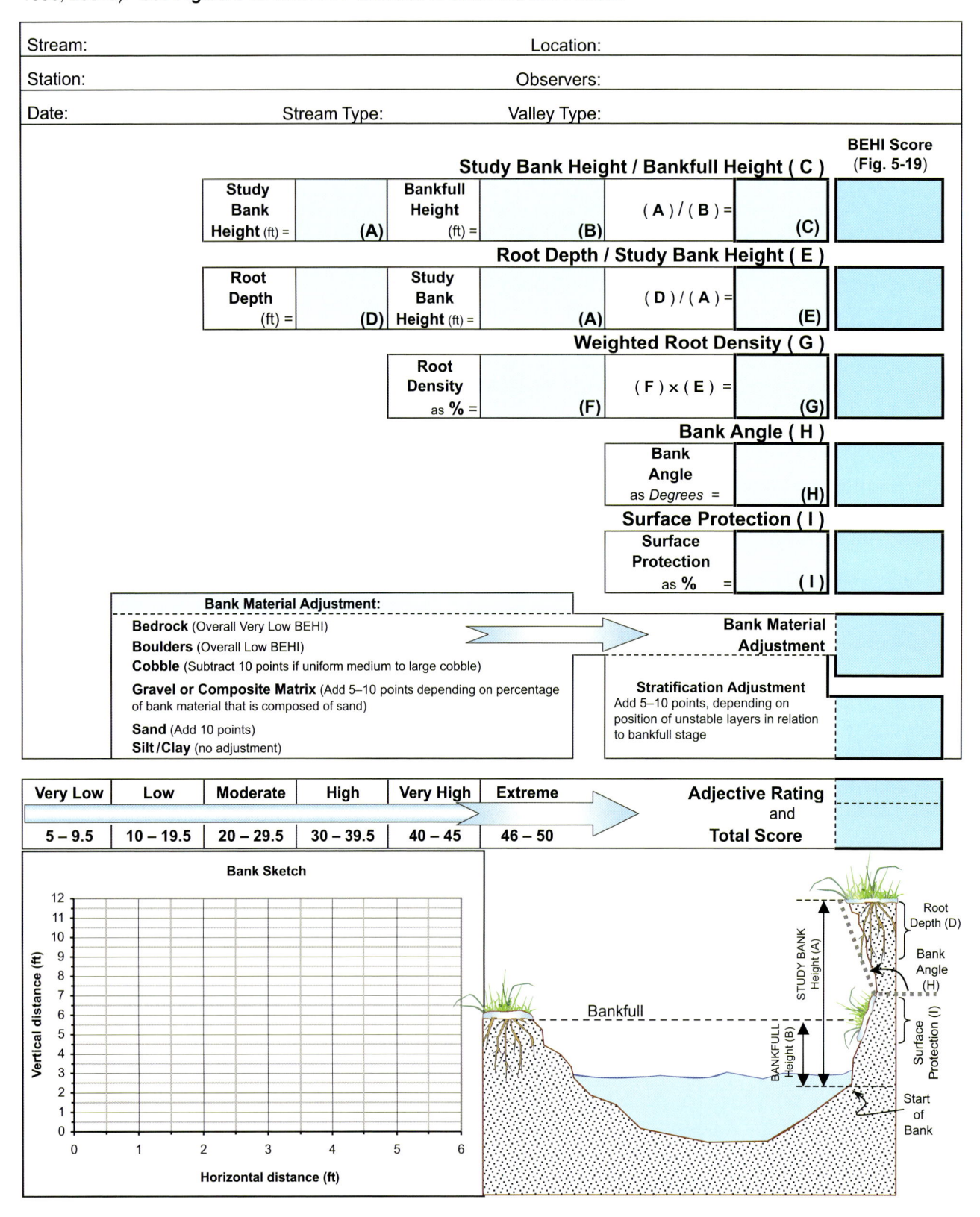

Stream:		Location:	
Station:		Observers:	
Date:	Stream Type:	Valley Type:	

BEHI Score (Fig. 5-19)

Study Bank Height / Bankfull Height (C)

| Study Bank Height (ft) = | (A) | Bankfull Height (ft) = | (B) | (A) / (B) = | (C) |

Root Depth / Study Bank Height (E)

| Root Depth (ft) = | (D) | Study Bank Height (ft) = | (A) | (D) / (A) = | (E) |

Weighted Root Density (G)

| | | Root Density as % = | (F) | (F) × (E) = | (G) |

Bank Angle (H)

| | | | | Bank Angle as *Degrees* = | (H) |

Surface Protection (I)

| | | | | Surface Protection as % = | (I) |

Bank Material Adjustment:
- **Bedrock** (Overall Very Low BEHI)
- **Boulders** (Overall Low BEHI)
- **Cobble** (Subtract 10 points if uniform medium to large cobble)
- **Gravel or Composite Matrix** (Add 5–10 points depending on percentage of bank material that is composed of sand)
- **Sand** (Add 10 points)
- **Silt/Clay** (no adjustment)

→ **Bank Material Adjustment**

Stratification Adjustment Add 5–10 points, depending on position of unstable layers in relation to bankfull stage

Very Low	Low	Moderate	High	Very High	Extreme
5 – 9.5	10 – 19.5	20 – 29.5	30 – 39.5	40 – 45	46 – 50

Adjective Rating and Total Score

Bank Sketch

5-56

Chapter 5   The Prediction Level Assessment (PLA)

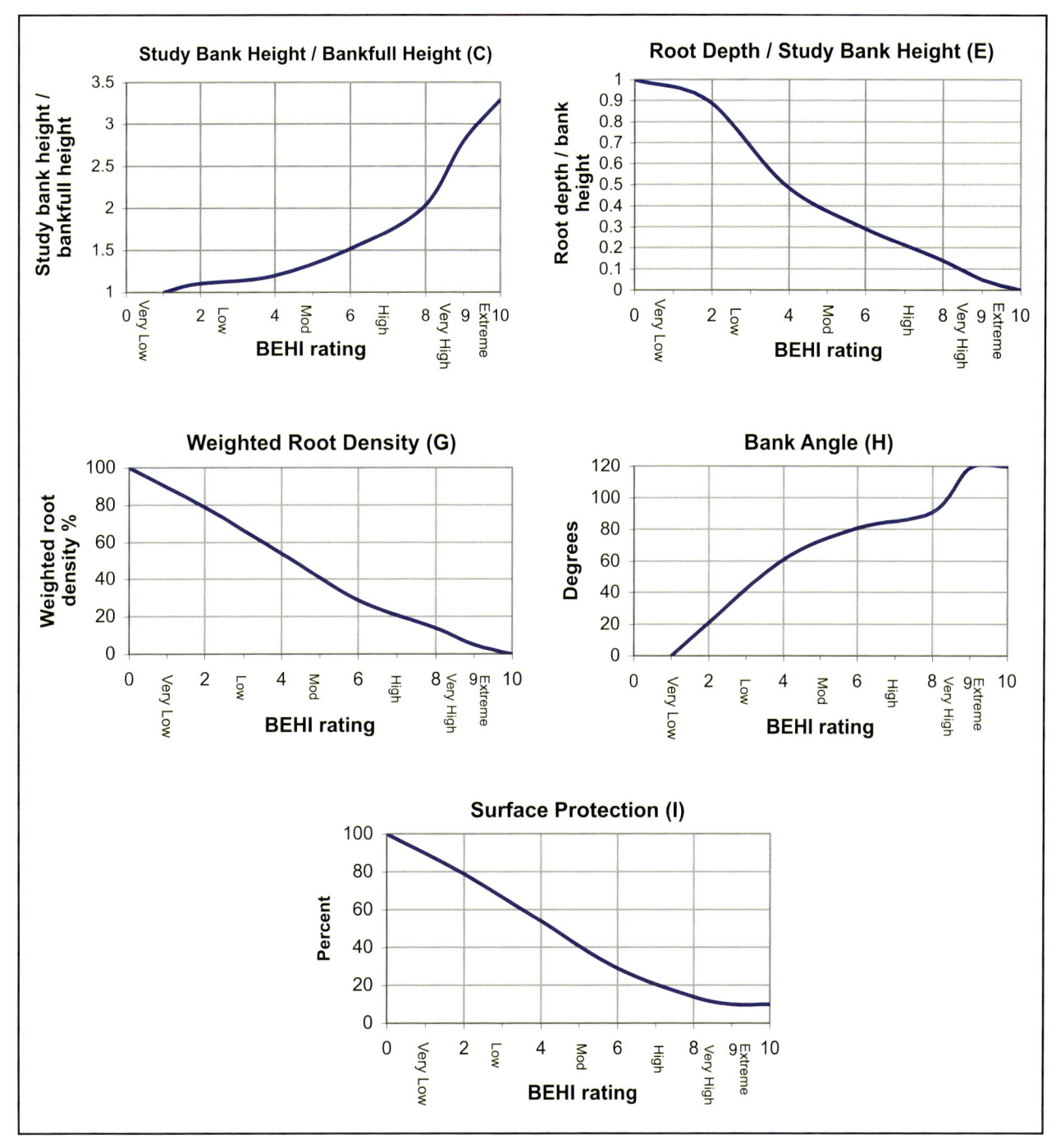

Figure 5-19. Streambank erodibility criteria showing conversion of measured ratios and bank variables to a BEHI rating (Rosgen, 1996, 2001a). Use **Worksheet 5-8** variables to determine BEHI score.

1. Study bank height (A) / bankfull height (B) (study bank-height ratio)

This ratio is used instead of total bank height to adjust for scale in comparing different-sized streams. The study bank height is measured from bank toe to bank top (**Figure 5-18**). The bankfull stage is then used to calculate the bankfull height from the reference elevation at the toe of the bank to the bankfull stage (**Figure 5-18**). The calculation of the study bank-height ratio is obtained by dividing the study bank height (A) by the bankfull height (B) as calculated in **Worksheet 5-8**. The ratio value A/B is converted to a BEHI rating based on the relationship in **Figure 5-19**. Examples of study bank-height ratio calculations for different field conditions are shown in **Figure 5-20**.

The relationship in **Figure 5-19** for study bank height (A) / bankfull height (B) indicates that the higher the study bank-height ratio is above 1.0, the higher the risk of erosion. For example, if more bank surface is exposed above the normal high flow level (bankfull), the more likely the erosional processes of surface erosion, dry/ravel, freeze-thaw, cantilever and shear block failures and other mass erosion processes are to occur.

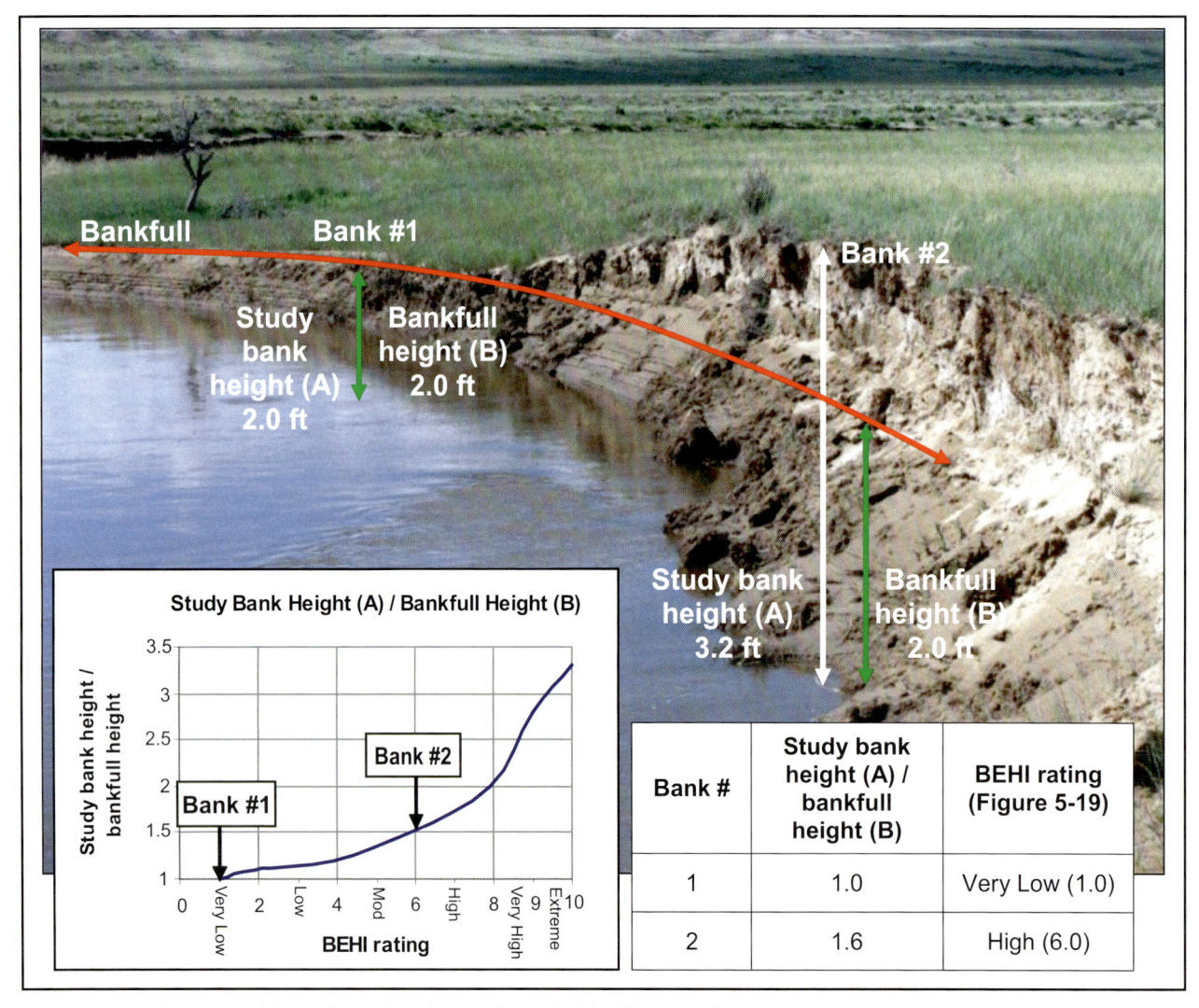

Figure 5-20. Example of the calculation of study bank height (A) / bankfull height (B) for two different bank locations, lower Little Snake River, Wyoming.

Chapter 5   *The Prediction Level Assessment (PLA)*

2. Root depth (D) / study bank height (A) (root depth ratio)

The predominant rooting depth (**Figure 5-18**) is divided by the study bank height (A) as calculated in **Worksheet 5-8**. The rooting depth ratio is converted to a BEHI rating using **Figure 5-19**. The greater the ratio, the lower the risk of erosion due to the fact that roots can have a major stabilizing influence on many erosional processes. Roots help maintain internal strength of the soil and can reduce fluvial entrainment due to individual particle detachment. If the root mass does not extend to the full bank height, undercutting of banks causing bank collapse can occur. An example of converting the root depth ratio to a BEHI rating is shown in **Figure 5-21**.

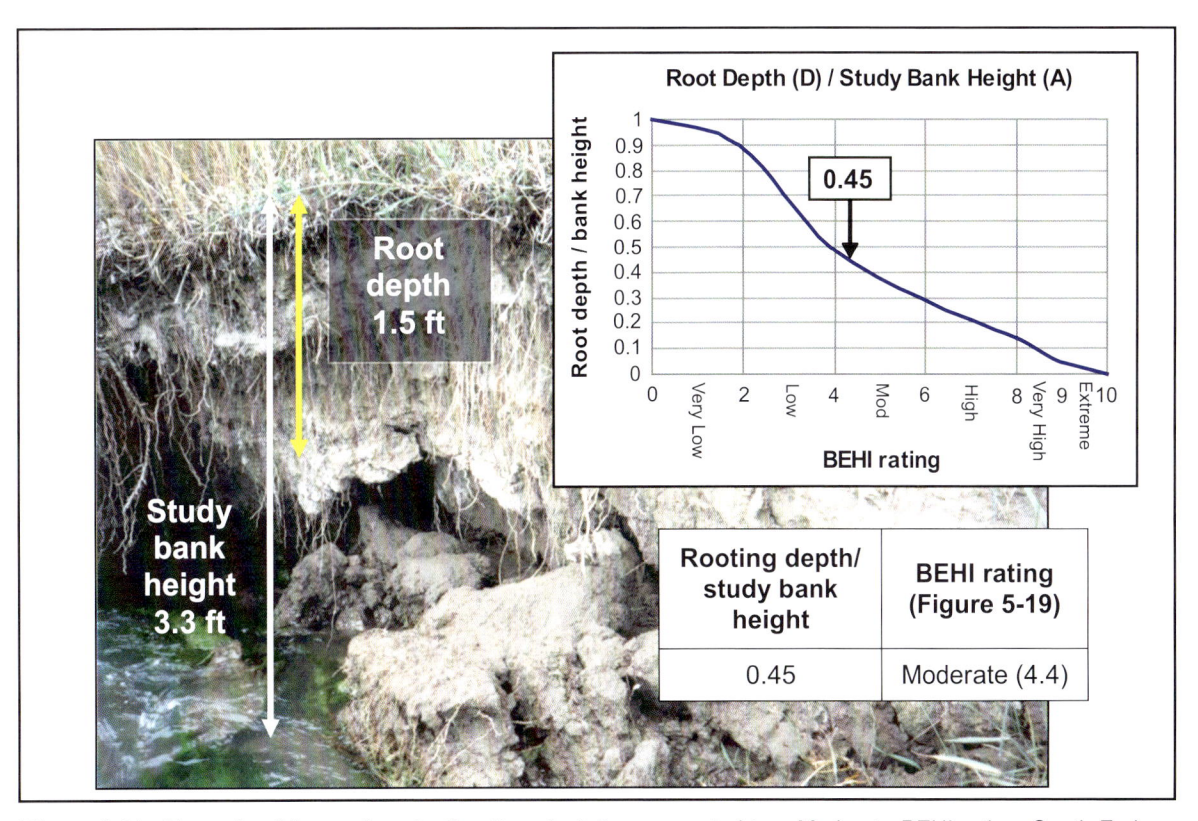

Figure 5-21. Example of the rooting depth ratio calculation converted to a Moderate BEHI rating, South Fork Mitchell River, North Carolina.

3. Weighted root density

To determine weighted root density, the user must first determine the density of root mass (*mass* is the amount of roots in the soil, whereas *density* is the amount of root mass per unit volume of soil). This is an ocular estimate rather than a bank sample taken to a laboratory. The density of root mass is often evaluated as the percent of the soil composed of roots where the roots have extended. The root density is then multiplied by the rooting depth ratio (root depth (D) / study bank height (A)) to obtain the weighted root density of the entire bank height (**Worksheet 5-8**). The weighted root density relationship in **Figure 5-19** is used to determine a BEHI rating. The greater the weighted density of roots within the soil column, the lower the risk of erosional processes. An example of a weighted root density calculation to determine a BEHI rating is shown in **Figure 5-22**.

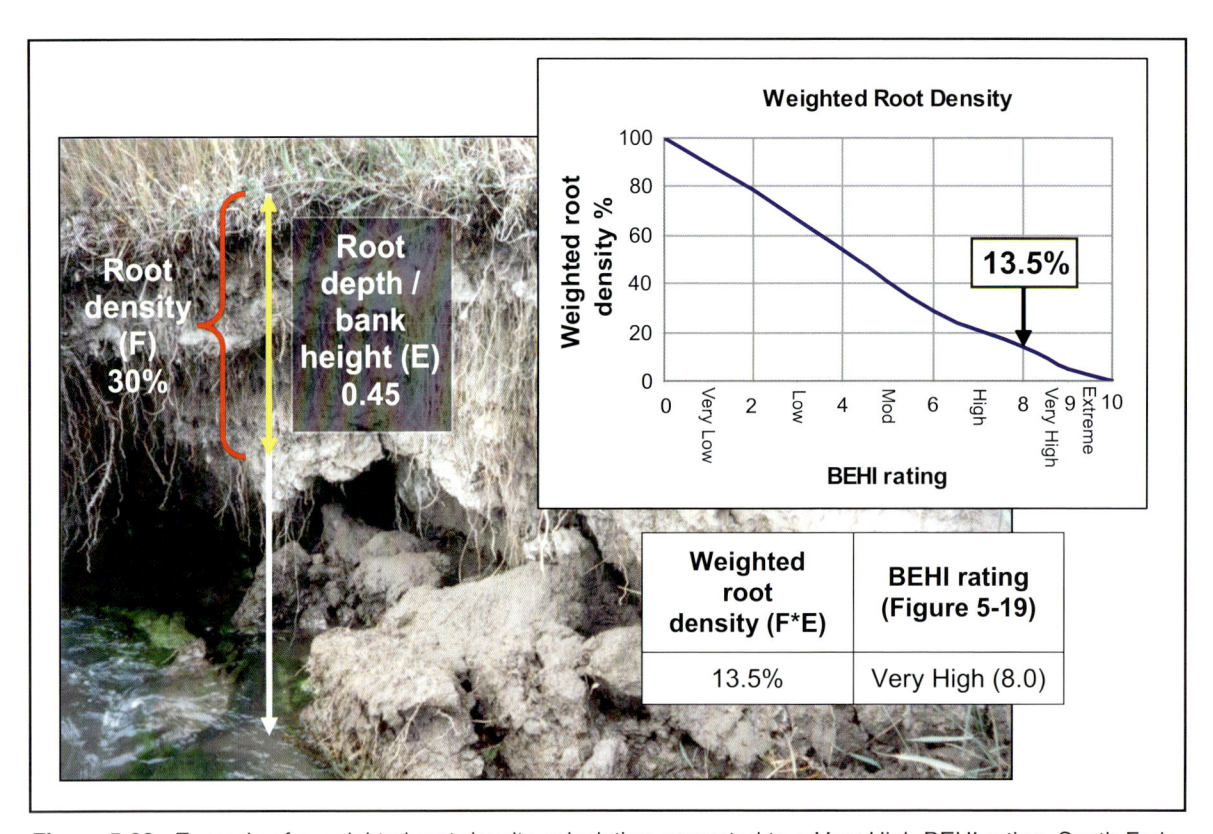

Figure 5-22. Example of a weighted root density calculation converted to a Very High BEHI rating, South Fork Mitchell River, North Carolina.

Chapter 5   *The Prediction Level Assessment (PLA)*

4. Bank angle

Bank angle is used to determine risk to bank failure mechanisms. The steeper the bank (as measured from a horizontal plane in degrees, 90° being vertical), the more susceptible the streambanks are to mass erosion processes (planar failures, cantilever collapse, etc.). Examples of different bank angles that are most sensitive for this evaluation are shown in **Figure 5-23**.

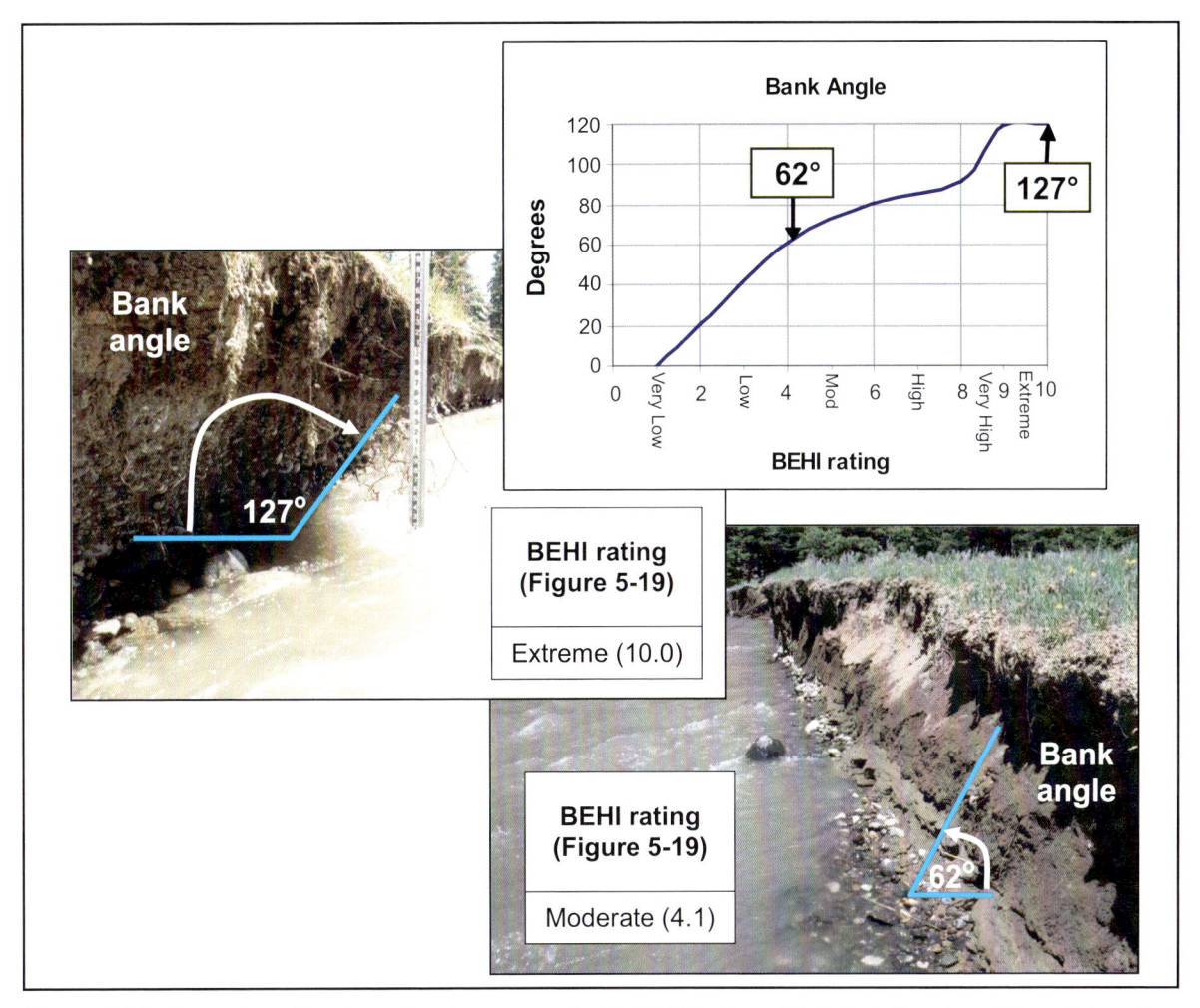

Figure 5-23. Examples of bank angle values converted to BEHI ratings, West Fork San Juan River, Colorado.

5. Surface protection

Surface protection is measured in the field by determining how much of the streambank is exposed to the agents of erosion. It is measured as the surface area protected by sod mats, large woody debris and/or existing revetments (**Figure 5-18**). The conversion of surface protection values to corresponding BEHI ratings is shown in **Figure 5-19** and is depicted in the examples shown in **Figure 5-24**.

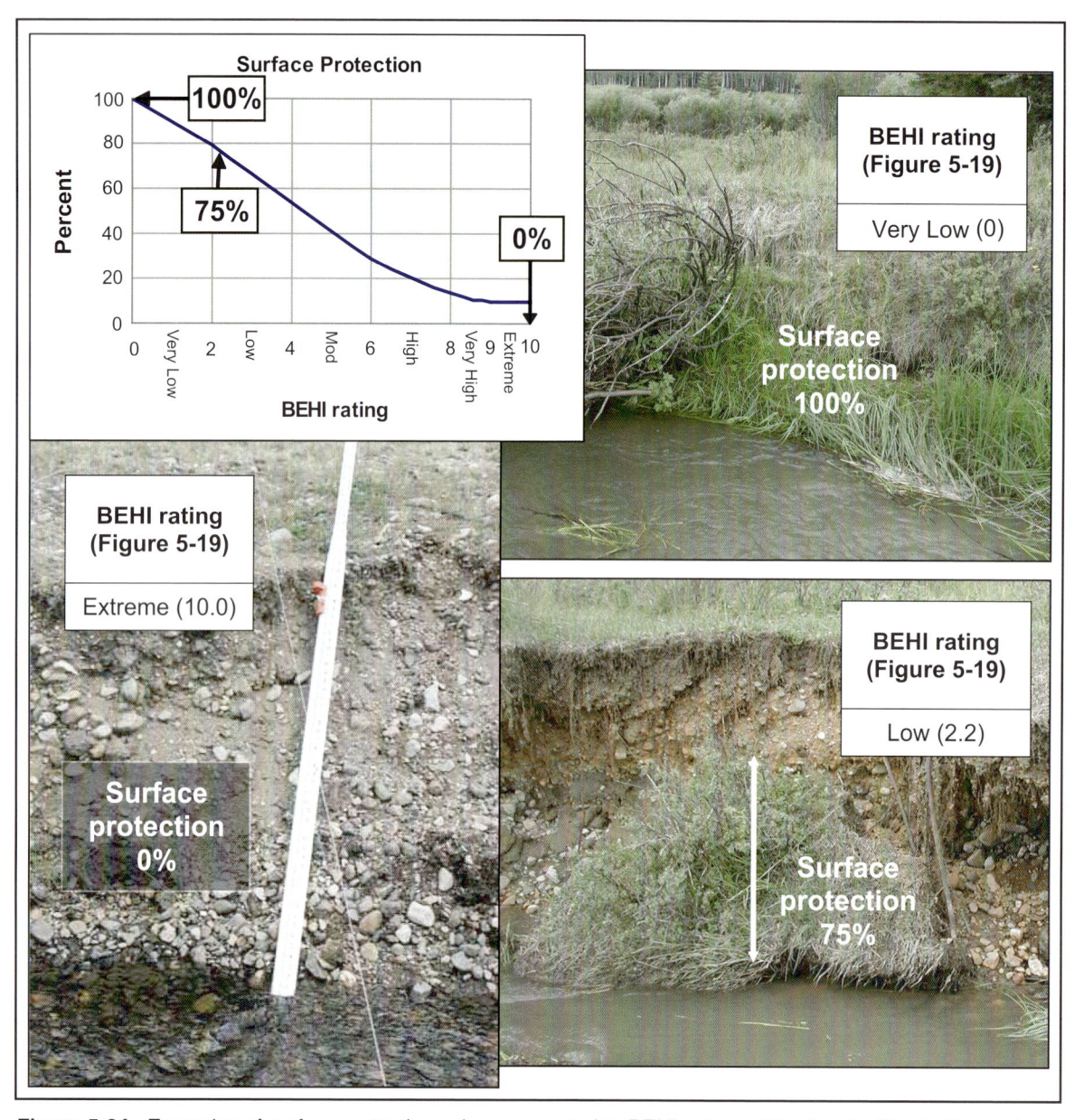

Figure 5-24. Examples of surface protection values converted to BEHI ratings, Weminuche Creek, Colorado.

Chapter 5 *The Prediction Level Assessment (PLA)*

6. Bank material adjustment

The more erodible the soil type, the higher the susceptibility to erosion; thus, composition of bank material is obtained to adjust overall BEHI ratings accordingly. Obviously, for bedrock and large boulders, the BEHI values are always Very Low. **Worksheet 5-8** is used to adjust the BEHI rating up or down based on the nature of the bank material. If the matrix is composed of uniform medium to large cobble, then 10 points are subtracted from the BEHI score. If the matrix is gravel or a composite of small cobble, gravel and/or sand, 5 to 10 points are added depending on the percent of the bank material composed of sand. For a sand matrix, 10 points are added to the BEHI score (**Worksheet 5-8**). No adjustments are made for cohesive (silt/clay) bank materials. Examples of point adjustments are shown in **Figures 5-25** and **5-26**.

Figure 5-25. Bank materials: Composite matrix of small cobble, gravel and sand (add 5 points), Weminuche Creek, Colorado.

Figure 5-26. Bank materials: Sand (add 10 points), Florida.

7. Stratification adjustment

Layers in the soil matrix can concentrate water, create weak layers, provide weak zones associated with slip failures, pop-out failures and piping (collapse of the bank due to seepage pressures and flow rates) and can cause bank disintegration at various layers by sapping (particles entrained from seepage flows). **Worksheet 5-8** is used to adjust the BEHI score depending on the number and position of layers. Bank stratification adjustments of 5 and 10 points are shown in **Figures 5-27** and **5-28**.

Figure 5-27. Stratification: Presence of layers (add 5 points), Weminuche Creek, Colorado.

Figure 5-28. Stratification: Presence of multiple layers (add 10 points). Note the differential erosion at contact zones and exposed erosion pins from one year of erosion. Pins were installed flush to bank one year earlier, yielding approximately 2.0 ft of erosion, Weminuche Creek, Colorado.

Chapter 5 | The Prediction Level Assessment (PLA)

Near-Bank Stress (NBS)

Annual streambank erosion rate prediction must include the Near-Bank Stress (NBS) assessment associated with energy distribution against streambanks. NBS determinations are broken into seven different options. The NBS variables used in the prediction methodology indicate potential disproportionate energy distribution in the near-bank region (the third of the channel cross-section associated with the bank being evaluated), which can accelerate streambank erosion. For example, in some instances, the presence of a gravel bar directs the velocity vectors directly into a bank and increases the local energy slope; erosion rates are then higher. *The user must select one or more of the methods that best represent the onsite conditions. The average of all methods is not recommended; in practice, the resultant highest near-bank stress consequence method is selected.* **Worksheet 5-9** is used to determine an NBS rating using one or more of the seven methods:

1. Channel pattern, transverse bar or split channel/central bar creating NBS/high velocity gradient;
2. Ratio of radius of curvature to bankfull width (R_c / W_{bkf});
3. Ratio of pool slope to average water surface slope (S_p / S);
4. Ratio of pool slope to riffle slope (S_p / S_{rif});
5. Ratio of near-bank maximum depth to bankfull mean depth (d_{nb} / d_{bkf});
6. Ratio of near-bank shear stress to bankfull shear stress (τ_{nb} / τ_{bkf}); and
7. Velocity profiles/isovels/velocity gradient.

The various levels (I–IV) in **Worksheet 5-9** are associated with the level of detail of the assessment (Level I being the most broad and rapid and Level IV being the most complex and time-consuming). *The levels are not necessarily synonymous with reliability of prediction.*

Each of the seven methods is presented below. Data for the user-selected method is entered in **Worksheet 5-9**.

Watershed Assessment of River Stability and Sediment Supply

Worksheet 5-9. Various field methods of estimating Near-Bank Stress (NBS) risk ratings to calculate erosion rate.

Estimating Near-Bank Stress (NBS)		
Stream:	Location:	
Station:	Stream Type:	Valley Type:
Observers:		Date:

Methods for estimating Near-Bank Stress (NBS)

(1)	Channel pattern, transverse bar or split channel/central bar creating NBS...............	Level I	Reconaissance
(2)	Ratio of radius of curvature to bankfull width (R_c / W_{bkf})...........................	Level II	General prediction
(3)	Ratio of pool slope to average water surface slope (S_p / S)...........................	Level II	General prediction
(4)	Ratio of pool slope to riffle slope (S_p / S_{rif})..	Level II	General prediction
(5)	Ratio of near-bank maximum depth to bankfull mean depth (d_{nb} / d_{bkf})...........	Level III	Detailed prediction
(6)	Ratio of near-bank shear stress to bankfull shear stress (τ_{nb} / τ_{bkf}).............	Level III	Detailed prediction
(7)	Velocity profiles / Isovels / Velocity gradient..	Level IV	Validation

Level I (1)
- Transverse and/or central bars-short and/or discontinuous... NBS = High / Very High
- Extensive deposition (continuous, cross-channel)... NBS = Extreme
- Chute cutoffs, down-valley meander migration, converging flow...................................... NBS = Extreme

Level II

(2)
Radius of Curvature R_c (ft)	Bankfull Width W_{bkf} (ft)	Ratio R_c / W_{bkf}	Near-Bank Stress (NBS)

(3)
Pool Slope S_p	Average Slope S	Ratio S_p / S	Near-Bank Stress (NBS)

(4)
Pool Slope S_p	Riffle Slope S_{rif}	Ratio S_p / S_{rif}	Near-Bank Stress (NBS)

Dominant Near-Bank Stress

Level III

(5)
Near-Bank Max Depth d_{nb} (ft)	Mean Depth d_{bkf} (ft)	Ratio d_{nb} / d_{bkf}	Near-Bank Stress (NBS)

(6)
Near-Bank Max Depth d_{nb} (ft)	Near-Bank Slope S_{nb}	Near-Bank Shear Stress τ_{nb} (lb/ft²)	Mean Depth d_{bkf} (ft)	Average Slope S	Bankfull Shear Stress τ_{bkf} (lb/ft²)	Ratio τ_{nb} / τ_{bkf}	Near-Bank Stress (NBS)

Level IV

(7)
Velocity Gradient (ft / sec / ft)	Near-Bank Stress (NBS)

Converting values to a Near-Bank Stress (NBS) rating

Near-Bank Stress (NBS) ratings	Method number						
	(1)	(2)	(3)	(4)	(5)	(6)	(7)
Very Low	N / A	> 3.00	< 0.20	< 0.40	< 1.00	< 0.80	< 0.50
Low	N / A	2.21 – 3.00	0.20 – 0.40	0.41 – 0.60	1.00 – 1.50	0.80 – 1.05	0.50 – 1.00
Moderate	N / A	2.01 – 2.20	0.41 – 0.60	0.61 – 0.80	1.51 – 1.80	1.06 – 1.14	1.01 – 1.60
High	See	1.81 – 2.00	0.61 – 0.80	0.81 – 1.00	1.81 – 2.50	1.15 – 1.19	1.61 – 2.00
Very High	(1)	1.50 – 1.80	0.81 – 1.00	1.01 – 1.20	2.51 – 3.00	1.20 – 1.60	2.01 – 2.40
Extreme	Above	< 1.50	> 1.00	> 1.20	> 3.00	> 1.60	> 2.40

Overall Near-Bank Stress (NBS) rating

Chapter 5   The Prediction Level Assessment (PLA)

NBS Method #1. Channel pattern, transverse bar or split channel/central bar creating NBS/high velocity gradient

This method is the most rapid due to lack of required measurements but, nevertheless, can be an accurate, appropriate method. Channel pattern features that create a disproportionate energy distribution in the near-bank region include chute cutoffs and split channels that converge against study banks; near-bank stress ratings in these areas range from High to Extreme. Depositional features including transverse bars and central bars that create a disproportionate distribution of energy in the near-bank region also have NBS ratings of High to Extreme. In these areas, changes in slope and velocity against banks are evident. Examples of Extreme NBS due to channel pattern and/or deposition are depicted in **Figures 5-29** and **5-30**. Use **Worksheet 5-9** to determine the appropriate NBS rating using method #1.

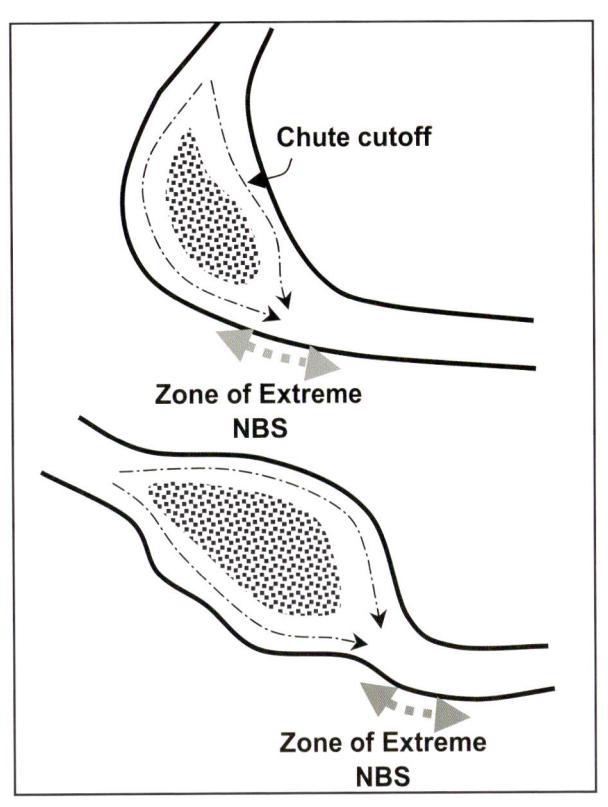

Figure 5-29. Examples of Extreme Near-Bank Stress (NBS) due to depositional features.

Figure 5-30. Use of NBS method #1 indicating transverse bar, chute cutoff and steep slope against a near-bank region with an Extreme NBS rating, lower Blanco River, Colorado.

NBS Method #2. Ratio of radius of curvature to bankfull width (R_c / W_{bkf})

This method can be rapidly completed in the field or on an aerial photograph. To compute the R_c/W_{bkf} ratio, measure radius of curvature and divide that value by the bankfull width of the channel at riffle reach. **Worksheet 5-9** is used to record R_c/W_{bkf} values and to determine NBS ratings. **Table 5-5** shows the conversion of R_c/W_{bkf} values to NBS ratings, which was developed from actual measured bank erosion rates. The relationship of R_c/W_{bkf} is associated with High to Extreme boundary shear stress with R_c/W_{bkf} values less than 2.0. **Figure 5-31** shows examples of High and Extreme R_c/W_{bkf} NBS ratings.

Table 5-5. Conversion table of R_c/W_{bkf} values to NBS ratings.

NBS ratings based on R_c/W_{bkf}	
R_c/W_{bkf} ratio	NBS rating
> 3.00	Very Low
2.21 – 3.00	Low
2.01 – 2.20	Moderate
1.81 – 2.00	High
1.50 – 1.80	Very High
< 1.50	Extreme

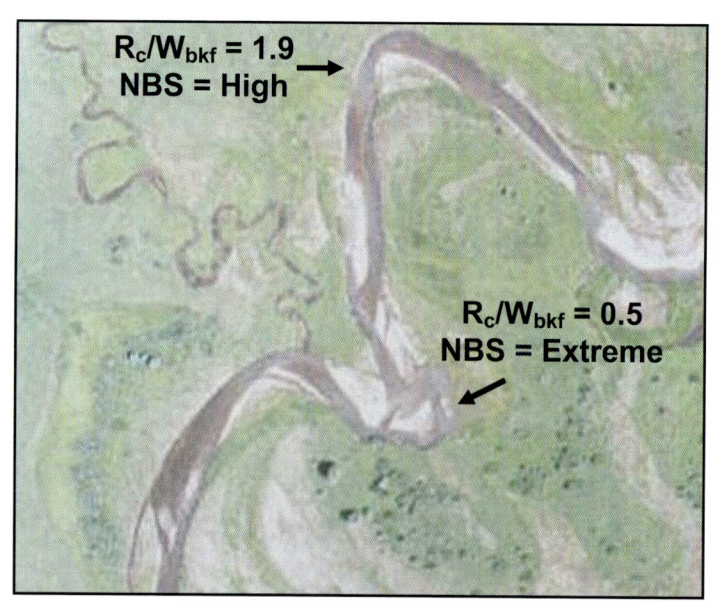

Figure 5-31. Example of High and Extreme NBS ratings due to R_c/W_{bkf} values of 1.9 and 0.5 using NBS method #2, Weminuche Creek, Colorado.

NBS Method #3. Ratio of pool slope to average slope (S_p / S)

Steep pool slopes accelerate streambank erosion; if the pool slope is steeper than the average slope of the river, Near-Bank Stress (NBS) is increased with centrifugal force on lateral scour pools. The ratio of a stable pool slope to average water surface slope is generally 0.1 to 0.2 for meandering C4 stream types. As this ratio increases toward and above 1.0, the risk of accelerated erosion is due to NBS. Values between 0.61 and 1.00 are rated as High or Very High NBS, while values above 1.00 rate as Extreme NBS (**Table 5-6**). The photograph in **Figure 5-32** depicts a pool slope ratio greater than 1.0 indicating Extreme NBS.

Table 5-6. Conversion table of S_p/S values to NBS ratings.

NBS ratings based on S_p/S	
S_p/S ratio	NBS rating
< 0.20	Very Low
0.20 – 0.40	Low
0.41 – 0.60	Moderate
0.61 – 0.80	High
0.81 – 1.00	Very High
> 1.00	Extreme

The longitudinal profile example in **Figure 5-33** shows the gentle facet slope of a pool compared to average water surface slope. Pool slope is measured as the facet slope of the feature at low flow, whereas average water surface slope is obtained from head of riffle to head of riffle elevations over stream length.

Figure 5-32. Example of Extreme NBS due to an S_p/S value greater than 1.0 using NBS method #3, South Fork Little Snake River, Colorado.

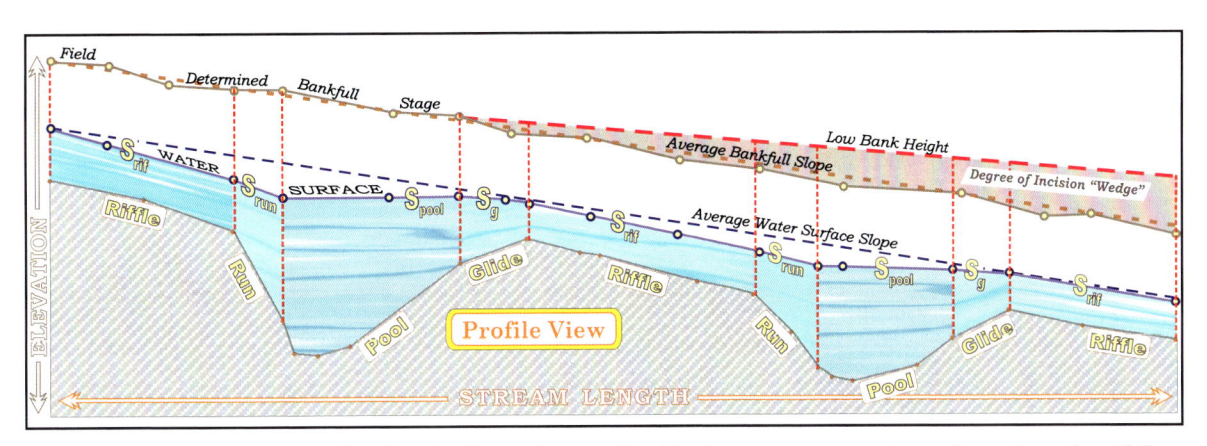

Figure 5-33. Longitudinal profile showing facet slopes of bed features, average water surface slope, bankfull stage and average bankfull slope.

NBS Method #4. Ratio of pool slope to riffle slope (S_p / S_{rif})

This procedure does not require the measurement of an average water surface slope but, instead, is a measure of the local facet slopes of both a pool and a riffle. The riffle and pool slopes need to be measured from the same riffle/pool sequence as shown in **Figure 5-33**. Conversion of the slope ratio to an NBS rating is shown for method #4 in **Table 5-7** and **Worksheet 5-9**. Riffle slopes are characteristically much steeper than pool slopes. When "slope reversals" occur (i.e., steep pools and flat riffles), streambank erosion is significantly affected. The field observer can often see the slope reversal without measurement once the observer is "calibrated" from multiple measured slope values. **Figure 5-34** depicts a pool/riffle slope reversal. This photo also indicates that the user could use method #1 associated with a transverse bar to determine an Extreme NBS rating.

Table 5-7. Conversion table of S_p/S_{rif} values to NBS ratings.

NBS ratings based on S_p/S_{rif}	
S_p/S_{rif} ratio	NBS rating
< 0.40	Very Low
0.41 – 0.60	Low
0.61 – 0.80	Moderate
0.81 – 1.00	High
1.01 – 1.20	Very High
> 1.20	Extreme

Figure 5-34. Example of a slope reversal of a steep pool and a flat riffle where NBS is Extreme using NBS method #4, Nevada Creek, Montana.

Chapter 5 The Prediction Level Assessment (PLA)

NBS Method #5. Ratio of near-bank maximum depth to bankfull mean depth (d_{nb} / d_{bkf})

This method calculates the ratio of the near-bank maximum bankfull depth at a study site to mean depth from a riffle cross-section. The disadvantage of method #5 is the time required to measure a cross-section to obtain bankfull mean depth (d_{bkf}). To avoid measuring a cross-section, the user can calibrate his/her "eye" to estimate d_{bkf}. This involves the user to first estimate width/depth ratio and the ratio of max depth to mean depth (d_{max}/d_{bkf}) prior to measuring these variables in the field. Once enough max depth/mean depth ratios are measured to calibrate the observer, the user can apply the estimated ratio (d_{max}/d_{bkf}) to calculate bankfull mean depth (d_{bkf}), where $d_{bkf} = d_{nb} / (d_{max}/d_{bkf})$. Once bankfull mean depth is obtained, divide the measured near-bank maximum depth (d_{nb}) value by the estimated bankfull mean depth value (d_{bkf}) to obtain the NBS rating.

Table 5-8. Conversion table of d_{nb}/d_{bkf} values to NBS ratings.

d_{nb}/d_{bkf} ratio	NBS rating
< 1.00	Very Low
1.00 – 1.50	Low
1.51 – 1.80	Moderate
1.81 – 2.50	High
2.51 – 3.00	Very High
> 3.00	Extreme

Table 5-8 converts d_{nb}/d_{bkf} ratio values to NBS ratings; values greater than 1.80 have High to Extreme NBS. **Figure 5-35** depicts the measurement of maximum bankfull depth at the study site relating to a Very High NBS rating.

Figure 5-35. Example of a maximum depth ratio to determine Very High NBS using method #5, East Fork San Juan River, Colorado.

NBS Method #6. Ratio of near-bank shear stress to bankfull shear stress (τ_{nb}/τ_{bkf})

This method requires the observer to measure near-bank maximum depth (d_{nb}), near-bank water surface slope (S_{nb}), bankfull mean depth (d_{bkf}) determined from a permanent cross-section and average water surface slope (S). The method obtains the bankfull shear stress (τ_{bkf}) and the near-bank shear stress (τ_{nb}). The ratio of near-bank bankfull shear stress to bankfull shear stress for the entire cross-section and average slope of the river is used to obtain an NBS rating (**Table 5-9**). **Figure 5-36** depicts an Extreme NBS rating. This method is shown in **Flowchart 5-8** and is related to the alternative of measured velocity gradients to obtain NBS ratings. Due to the level of detail, this approach is often used to establish empirical relations of NBS, BEHI and measured streambank erosion.

Table 5-9. Conversion table of τ_{nb}/τ_{bkf} values to NBS ratings.

NBS ratings based on τ_{nb}/τ_{bkf}	
τ_{nb}/τ_{bkf} ratio	NBS rating
< 0.80	Very Low
0.80 – 1.05	Low
1.06 – 1.14	Moderate
1.15 – 1.19	High
1.20 – 1.60	Very High
> 1.60	Extreme

Figure 5-36. Example of an Extreme NBS rating based on the near-bank shear stress/bankfull shear stress ratio (NBS method #6), upper Little Snake River, Colorado.

Chapter 5 — *The Prediction Level Assessment (PLA)*

NBS method #6: Ratio of near-bank shear stress to bankfull shear stress (τ_{nb} / τ_{bkf})

```
Select permanent cross-section
           ↓
Select study bank and the channel that
influences the study bank
           ↓
Measure bankfull area (A_bkf)  →  Divide bankfull channel
and width (W_bkf)                  width (W_bkf) into thirds
     ↓                                       ↓
Calculate mean depth:              Measure near-bank*
d_bkf = A_bkf / W_bkf              maximum depth (d_nb)
     ↓                                       ↓
Measure water surface              Measure near-bank water
slope (S)                          surface slope (S_nb)
     ↓                                       ↓
Calculate total bankfull           Calculate near-bank shear
shear stress                       stress
τ_bkf = γ d_bkf S                  τ_nb = γ d_nb S_nb
              ↓           ↓
         Divide near-bank shear stress
         by total shear stress
         (τ_nb / τ_bkf)
                  ↓
         Convert value to a Near-Bank Stress
         (NBS) rating using Worksheet 5-9
```

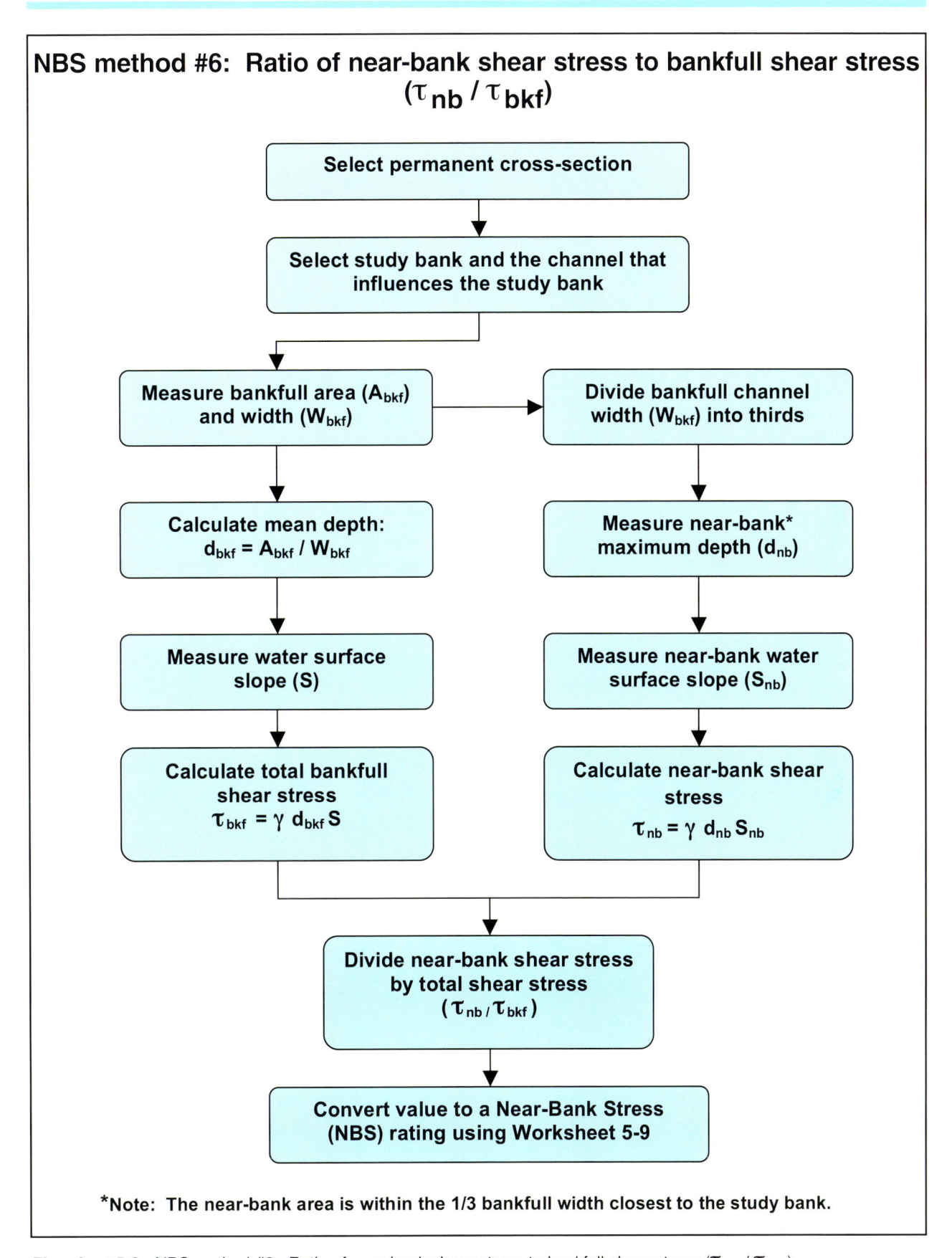

*Note: The near-bank area is within the 1/3 bankfull width closest to the study bank.

Flowchart 5-8. NBS method #6: Ratio of near-bank shear stress to bankfull shear stress (τ_{nb} / τ_{bkf}).

NBS Method #7. Velocity profiles/isovels/velocity gradient

This procedure involves the most detailed data that includes measured isovels (vertical velocity profile data using current meters across the entire channel at high flow). This method is often used to develop empirically derived values of NBS and BEHI correlations with measured annual streambank erosion rates. NBS ratings can be calculated using one of the following method #7 options.

Option A. For this option, velocity gradient is obtained from an orthogonal line (perpendicular to the velocity isovels). The units are ft/sec/ft, derived from velocity data in ft/sec from the isovel core to the bank per unit length of orthogonal distance in feet. **Table 5-10** converts these velocity gradient values to NBS ratings, as also shown in **Worksheet 5-9**. The initial research correlating measured annual streambank erosion rates versus BEHI and NBS used velocity gradient data and near-bank shear stress (NBS method #6). **Figure 5-37** shows two examples of velocity gradients obtained from orthogonal lines to determine an NBS rating.

Option B. This option uses measured vertical velocity profile data to back-calculate near-bank stress. This is similar to NBS method #6, except actual vertical velocity profile data is measured to obtain near-bank shear stress.

The procedure to determine NBS ratings using vertical velocity data and isovels is shown in **Flowchart 5-9** and **Table 5-11**.

Table 5-10. Conversion table of velocity profiles/isovels/velocity gradient values to NBS ratings.

Velocity profiles/isovels/velocity gradient values	NBS rating
< 0.50	Very Low
0.50 – 1.00	Low
1.01 – 1.60	Moderate
1.61 – 2.00	High
2.01 – 2.40	Very High
> 2.40	Extreme

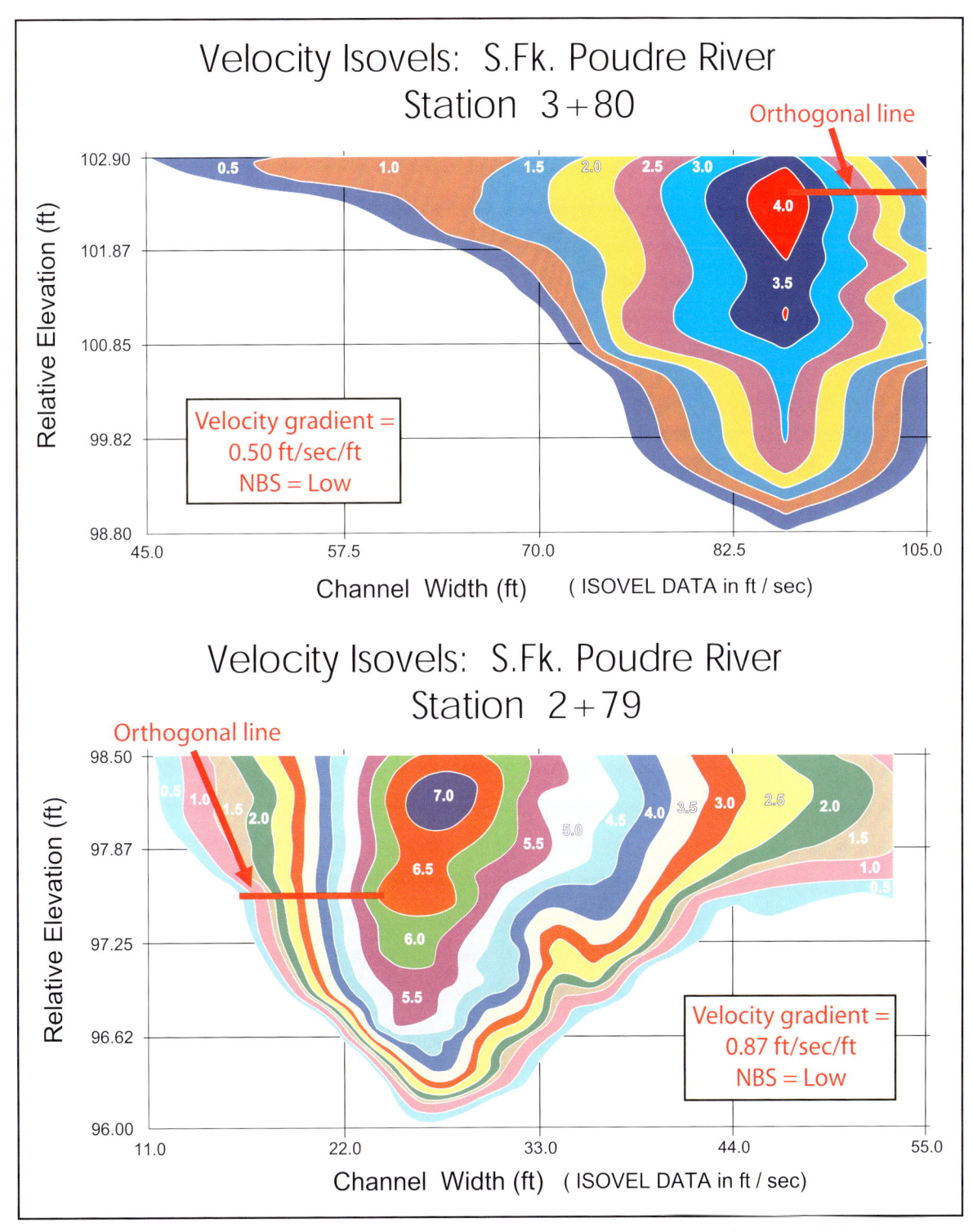

Figure 5-37. Velocity isovels at two separate locations on a C3 stream type reach showing variation in stress velocity distribution (Rosgen, 1996).

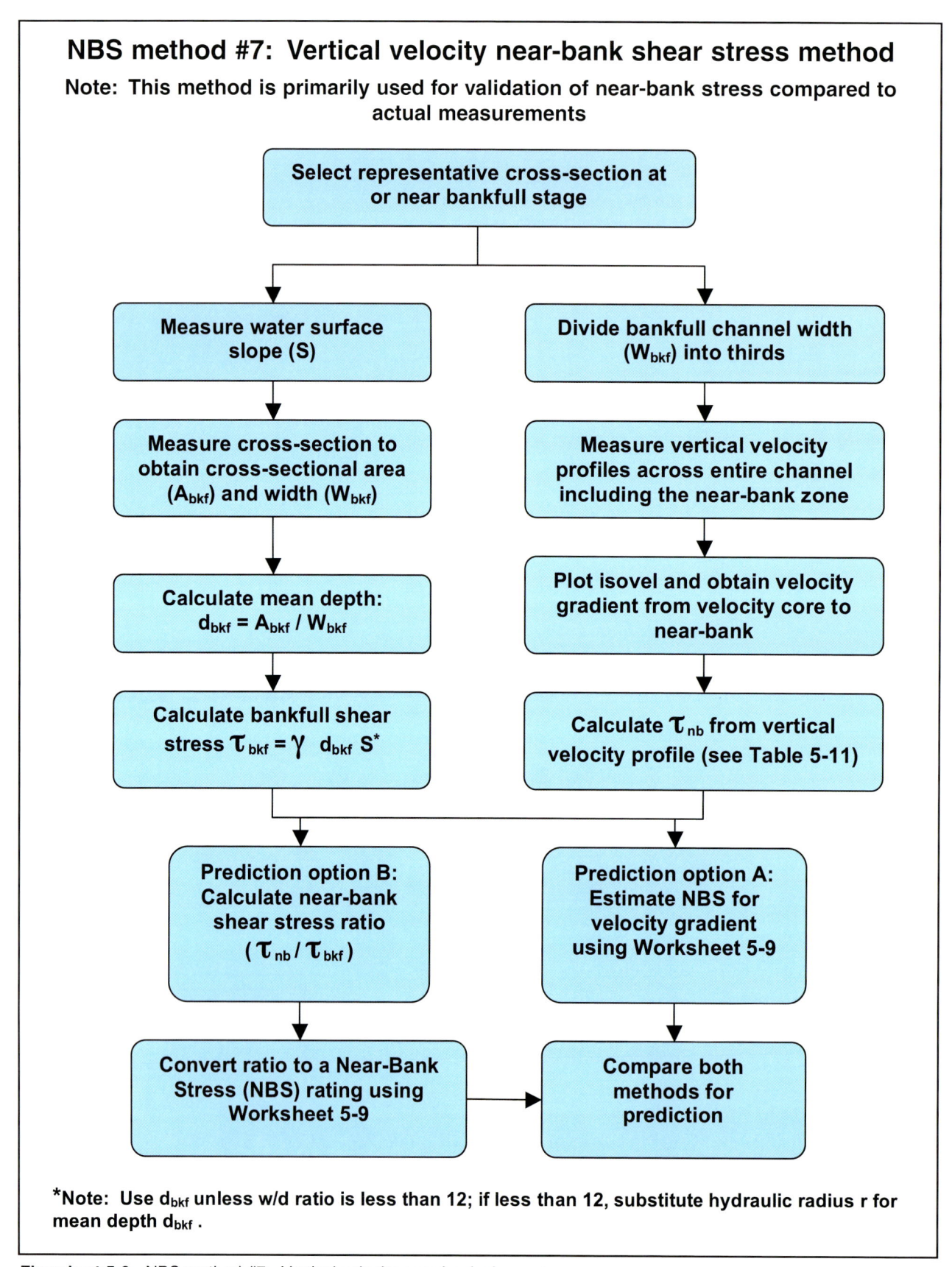

WATERSHED ASSESSMENT OF RIVER STABILITY AND SEDIMENT SUPPLY

NBS method #7: Vertical velocity near-bank shear stress method

Note: This method is primarily used for validation of near-bank stress compared to actual measurements

```
                    Select representative cross-section at
                              or near bankfull stage
                    ┌─────────────────────┴─────────────────────┐
                    ▼                                           ▼
        Measure water surface                      Divide bankfull channel width
              slope (S)                              (W_bkf) into thirds
                    │                                           │
                    ▼                                           ▼
        Measure cross-section to                    Measure vertical velocity
         obtain cross-sectional area                profiles across entire channel
         (A_bkf) and width (W_bkf)                  including the near-bank zone
                    │                                           │
                    ▼                                           ▼
         Calculate mean depth:                      Plot isovel and obtain velocity
           d_bkf = A_bkf / W_bkf                    gradient from velocity core to
                                                           near-bank
                    │                                           │
                    ▼                                           ▼
         Calculate bankfull shear                   Calculate τ_nb from vertical
         stress τ_bkf = γ d_bkf S*                  velocity profile (see Table 5-11)
                    │                                           │
                    ▼                                           ▼
         Prediction option B:                       Prediction option A:
         Calculate near-bank                        Estimate NBS for
         shear stress ratio                         velocity gradient
         ( τ_nb / τ_bkf )                           using Worksheet 5-9
                    │                                           │
                    ▼                                           ▼
         Convert ratio to a Near-Bank               Compare both
         Stress (NBS) rating using  ──────────►     methods for
         Worksheet 5-9                              prediction
```

*Note: Use d_{bkf} unless w/d ratio is less than 12; if less than 12, substitute hydraulic radius r for mean depth d_{bkf}.

Flowchart 5-9. NBS method #7: Vertical velocity near-bank shear stress.

Table 5-11. Procedure to predict near-bank shear stress using velocity isovels/velocity gradient.

Procedure to predict near-bank stress using velocity isovels/velocity gradient

1. At near-bankfull stage, conduct vertical velocity profiles across entire channel.
   a. Measure point averaged velocities from near-bed to water surface.
   b. Collect a minimum of 5–7 velocity measurements per vertical profile.
   c. Collect at least 10–15 verticals per cross-section at bank study site.
   d. Be sure to place vertical velocity profile near both banks.

2. Plot lines of equal velocity (isovel) for cross-section.

3. Construct orthogonal line perpendicular to isovels from core of isovel to study bank.

4. Measure velocity gradient: Velocity in ft/sec per length of orthogonal line.

5. Optional: Calculate shear stress from vertical velocity profile.
   a. Divide river into thirds.
   b. Calculate near-bank shear stress (τ_{nb}) from vertical velocity profile from velocity profile closest to study bank or representing 1/3 width in near-bank.
   c. Calculate average shear stress (τ_{bkf}) for entire channel.
   d. Calculate near-bank shear stress ratio: τ_{nb}/τ_{bkf}.
   e. Compare with measured values in **Worksheet 5-9**.

To calculate shear stress and near-bank shear stress ratios from vertical velocity profiles use:

$$\tau = 1.94 \left(\frac{u_2 - u_1}{y} \right)^2$$

Where: τ = shear stress (lbs/ft^2)

u_2 = point velocity near bed (ft/sec)

u_1 = point velocity near surface in the vertical (ft/sec)

y = depth or distance between the 2 velocity points

The relation assumes a logarithmic profile; plot velocity data points on semi-log paper. Draw line between the two points to establish a velocity difference/depth.

Take cross-sectioned averaged velocity profile data to establish bankfull shear stress (lbs/ft^2). Compare with $\tau = \gamma RS$ for same stage.

Calculate ratio τ_{nb}/τ_{bkf} and compare velocity gradient and near-bank shear stress ratios with values in **Worksheet 5-9**.

Step 9. Predict annual streambank erosion rate using BEHI and NBS ratings

The combination of the Bank Erosion Hazard Index (BEHI) and the Near-Bank Stress (NBS) ratings (**Step 8**) are used in the BANCS model to derive annual streambank erosion rates. Annual streambank erosion rates are predicted using the relations shown in **Figures 5-38** (derived from Colorado data) for streams found in sedimentary and/or metamorphic geology and **Figure 5-39** (derived from Yellowstone National Park data) for streams found in alpine glaciation and/or volcanism areas. These relations have been proven to be statistically valid for these locations (Rosgen, 1996, 2001a).

However, the relations in **Figure 5-38** for Colorado and in **Figure 5-39** for Yellowstone are not intended to be universal for alluvial streams. Rather, they provide a framework for others to develop similar empirical relations or to validate these relations locally. Regional validation for variations in soil-bulk density and streambank failure mechanics would necessitate validation of the curves presented in **Figure 5-38** or **5-39**. These relations allow the field practitioner to obtain BEHI and NBS ratings along river reaches of varying conditions.

Use **Figure 5-38** or **5-39** (dependent upon condition), or similar empirical relations, to convert combined NBS and BEHI ratings to total annual lateral streambank erosion rates. By measuring bank heights and stream lengths for associated BEHI and NBS values, these total lateral erosion rates can be converted to annual sediment supply in tons/year. For a Very Low BEHI in **Figure 5-38**, the user is recommended to use a trend line representing the lowest observed values in the Low BEHI data set or to use the Very Low BEHI values from the Yellowstone relation (**Figure 5-39**). Use **Worksheet 5-10** to record summary erosion rates for multiple locations from **Step 8** and to convert erosion rates to tons/yr/ft. This conversion allows a unit erosion comparison to reference and other reach rates.

Worksheet 5-10 is also used to determine the dominant BEHI/NBS combination to use in **Step 25** to assess lateral stability. It is often the case that streambank erosion rates are not uniform along a stream reach. The dominant NBS and BEHI ratings as required in the lateral stability analysis are selected based on the highest combination of erosion rate that was observed. Generally the user would obtain the highest unit length of erosion (tons/yr/ft) and select the corresponding NBS and BEHI ratings from **Worksheet 5-10**.

Streambank erosion rates are recorded in the *PLA* summary worksheet located in **Steps 29–31**.

Chapter 5   *The Prediction Level Assessment (PLA)*

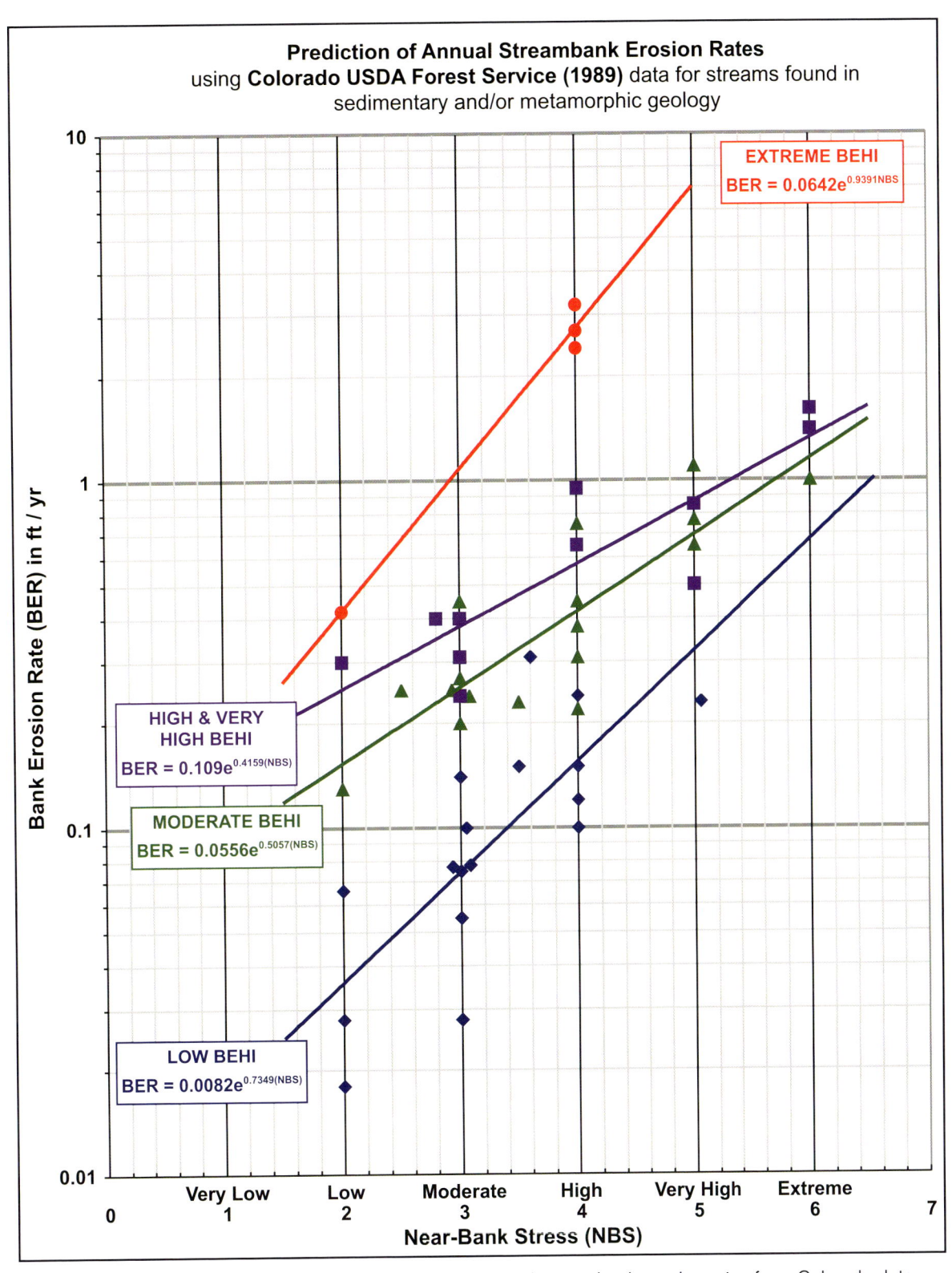

Figure 5-38. Relationship of BEHI and NBS to predict annual streambank erosion rates from Colorado data (1989) for streams found in sedimentary and/or metamorphic geology (Rosgen, 1996, 2001a).

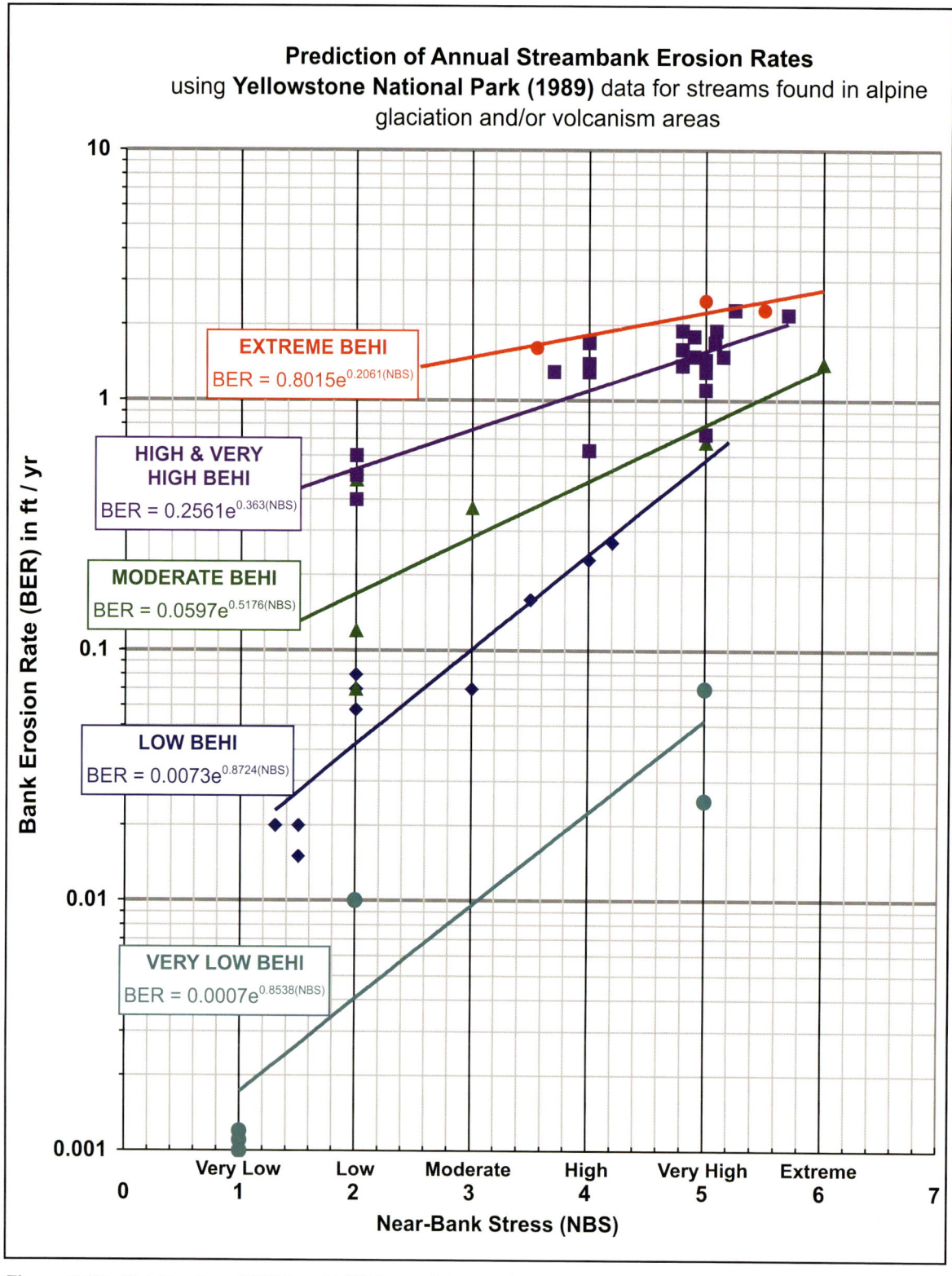

Figure 5-39. Relationship of BEHI and NBS to predict annual streambank erosion rates from Yellowstone National Park data (1989) for streams found in alpine glaciation and/or volcanism geology (Rosgen, 1996, 2001a).

Chapter 5

The Prediction Level Assessment (PLA)

Worksheet 5-10. Summary form of annual streambank erosion estimates for various study reaches.

Stream:				Location:		
Graph Used:		Stream Type:		Total Bank Length (ft):		
Observers:			Valley Type:		Date:	
(1) Station (ft)	(2) BEHI rating (Worksheet 5-8) (adjective)	(3) NBS rating (Worksheet 5-9) (adjective)	(4) Bank erosion rate (Figure 5-38 or 5-39) (ft/yr)	(5) Length of bank (ft)	(6) Study bank height (ft)	(7) Erosion subtotal [(4)X(5)X(6)] (ft^3/yr)
1.						
2.						
3.						
4.						
5.						
6.						
7.						
8.						
9.						
10.						
11.						
12.						
13.						
14.						
15.						
Sum erosion subtotals in Column (7) for each BEHI/NBS combination					Total erosion (ft^3/yr)	
Convert erosion in ft^3/yr to yds^3/yr {divide Total erosion (ft^3/yr) by 27}					Total erosion (yds^3/yr)	
Convert erosion in yds^3/yr to tons/yr {multiply Total erosion (yds^3/yr) by 1.3}					Total erosion (tons/yr)	
Calculate erosion per unit length of channel {divide Total erosion (tons/yr) by total length of stream (ft) surveyed}					Total erosion (tons/yr/ft)	

Steps 10–15: Total annual sediment yield prediction using the FLOWSED model (See Flowcharts 5-10 and 5-11)

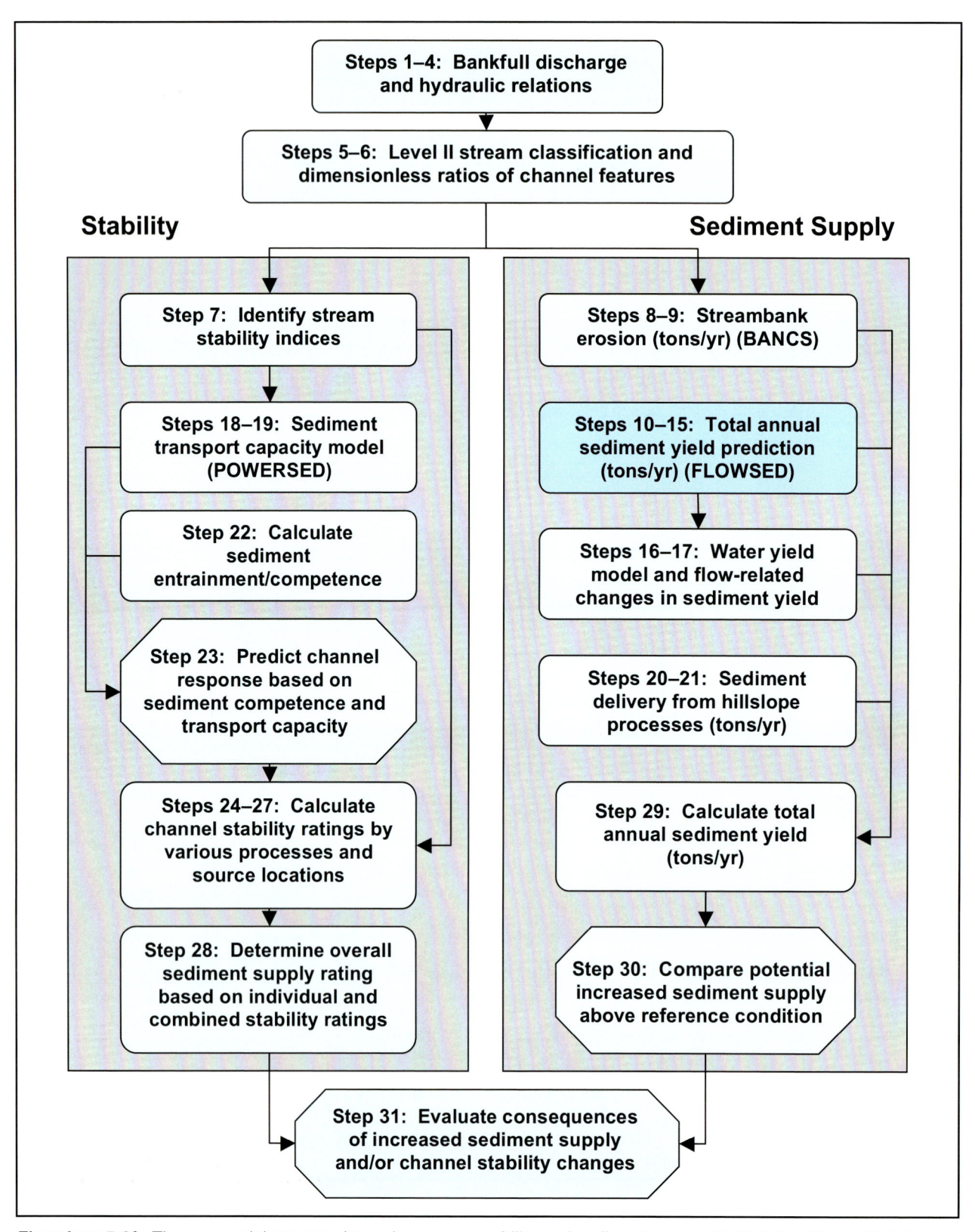

Steps 1–4: Bankfull discharge and hydraulic relations

Steps 5–6: Level II stream classification and dimensionless ratios of channel features

Stability

Step 7: Identify stream stability indices

Steps 18–19: Sediment transport capacity model (POWERSED)

Step 22: Calculate sediment entrainment/competence

Step 23: Predict channel response based on sediment competence and transport capacity

Steps 24–27: Calculate channel stability ratings by various processes and source locations

Step 28: Determine overall sediment supply rating based on individual and combined stability ratings

Sediment Supply

Steps 8–9: Streambank erosion (tons/yr) (BANCS)

Steps 10–15: Total annual sediment yield prediction (tons/yr) (FLOWSED)

Steps 16–17: Water yield model and flow-related changes in sediment yield

Steps 20–21: Sediment delivery from hillslope processes (tons/yr)

Step 29: Calculate total annual sediment yield (tons/yr)

Step 30: Compare potential increased sediment supply above reference condition

Step 31: Evaluate consequences of increased sediment supply and/or channel stability changes

Flowchart 5-3f. The sequential steps to determine stream stability and sediment supply, highlighting **Steps 10–15**.

Chapter 5 — The Prediction Level Assessment (PLA)

Steps 10–15: Total annual sediment yield prediction using the FLOWSED model

FLOWSED model

Step 10: Develop dimensionless flow-duration curve (e.g., Figures 5-40 and 5-41) from USGS data:
- Flow-duration curve
- Bankfull discharge (Q_{bkf})
- Mean daily discharge record

Step 11: Collect field data by stream type and valley type:
- Bankfull discharge (Q_{bkf}) (cfs)
- Bankfull bedload sediment (kg/s)
- Bankfull suspended sediment (mg/l)
- Bankfull suspended sand (mg/l) if used with POWERSED

Step 12: Obtain or establish dimensionless bedload and suspended sediment rating curves by stream type/stability (e.g., Figures 2-52 and 2-53)

Step 13: Convert dimensionless bedload and suspended sediment rating curves to dimensioned sediment rating curves (Figures 5-42 and 5-43)

Step 14: Convert dimensionless flow-duration curve to dimensioned flow-duration curve (Figure 5-44)

Step 15: Calculate total annual sediment yield for bedload and suspended sediment (tons/yr) (Worksheet 5-11)

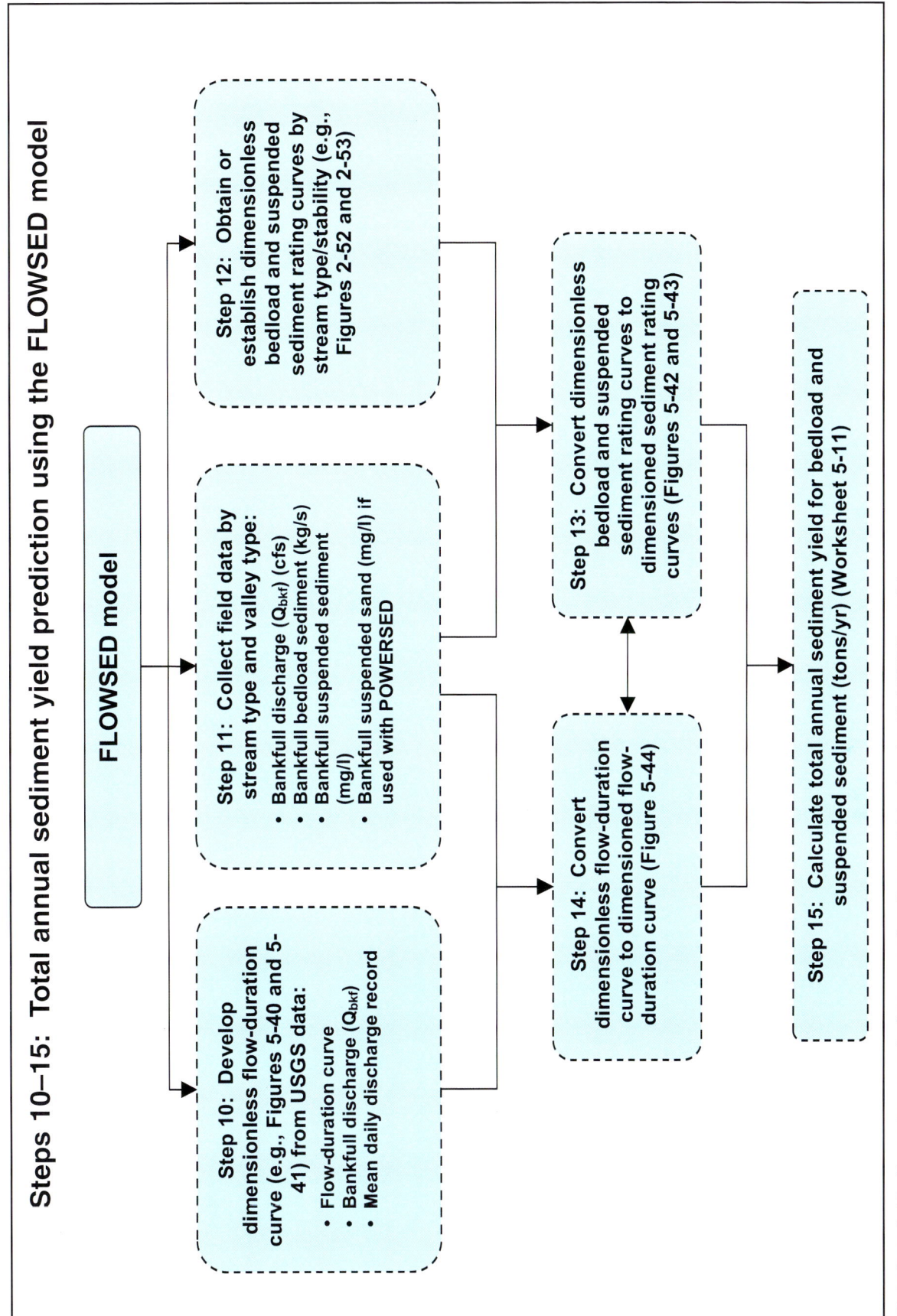

Flowchart 5-10. Procedure to calculate annual sediment yield using the FLOWSED model.

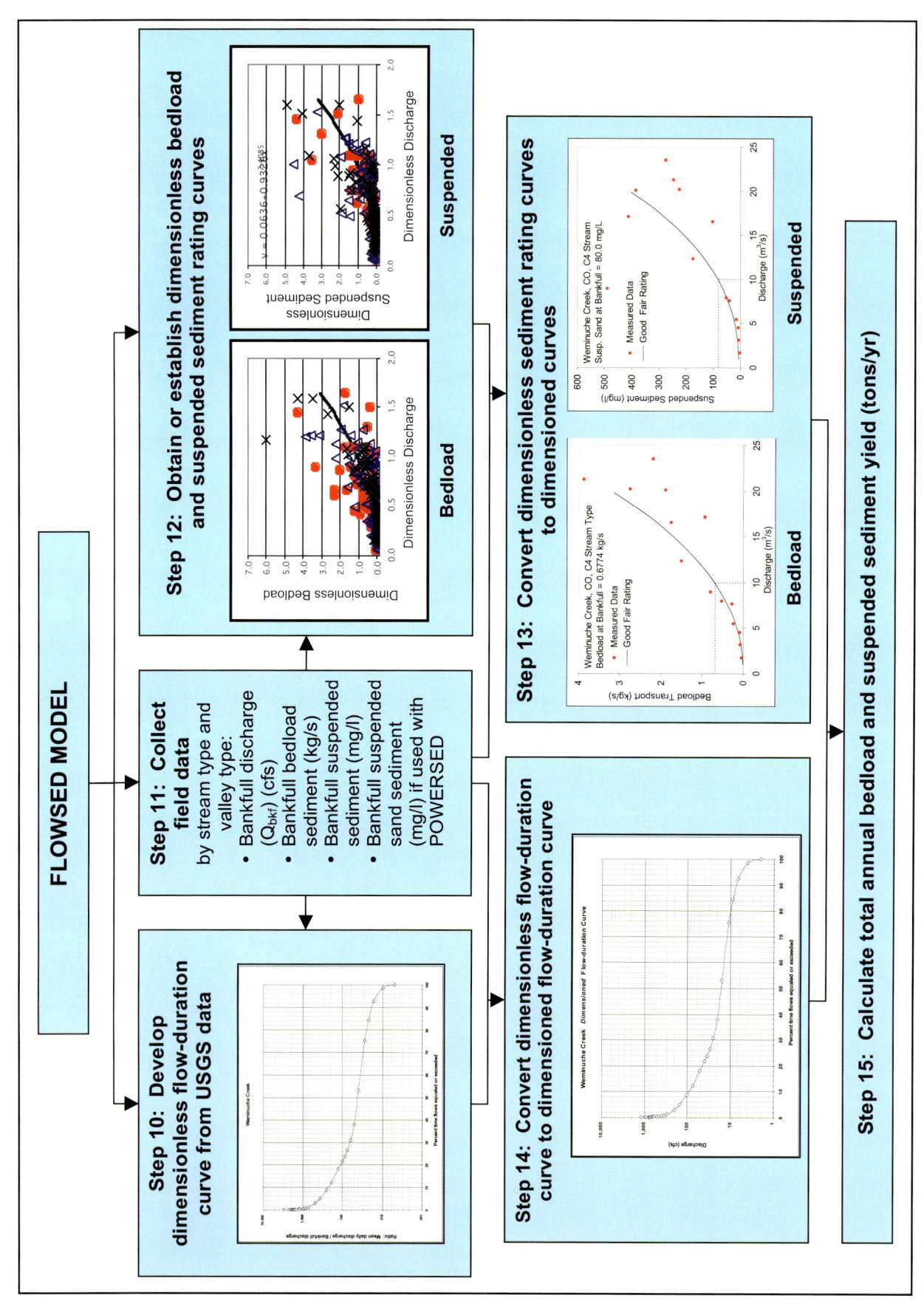

WATERSHED ASSESSMENT OF RIVER STABILITY AND SEDIMENT SUPPLY

FLOWSED MODEL

Step 10: Develop dimensionless flow-duration curve from USGS data

Step 11: Collect field data by stream type and valley type:
- Bankfull discharge (Q_{bkf}) (cfs)
- Bankfull bedload sediment (kg/s)
- Bankfull suspended sediment (mg/l)
- Bankfull suspended sand sediment (mg/l) if used with POWERSED

Step 12: Obtain or establish dimensionless bedload and suspended sediment rating curves

Step 13: Convert dimensionless sediment rating curves to dimensioned curves

Step 14: Convert dimensionless flow-duration curve to dimensioned flow-duration curve

Step 15: Calculate total annual bedload and suspended sediment yield (tons/yr)

Flowchart 5-11. Visual representation of the FLOWSED model.

5-84

Chapter 5   *The Prediction Level Assessment (PLA)*

Total annual sediment yield prediction using the FLOWSED model: Steps 10–15

A key to predicting river instability and associated loss of physical and biological river function rests with the ability of a stream to move the sediment size and load made available from its watershed. The first aspect of assessing sediment capacity or load is described in **Steps 10–15** using the FLOWSED model (**Flowcharts 5-10** and **5-11**).

FLOWSED is a model that is used to determine disproportional contributions of annual sediment supply or load at specific locations related to various land uses and erosional or depositional processes. FLOWSED was developed for *WARSSS* to predict:

- Total annual bedload sediment yield;
- Total annual suspended sediment yield;
- Total annual suspended sand sediment yield; and
- Total annual sediment yield increases or decreases from a reference condition due to changes in streamflow magnitude, timing and duration.

The FLOWSED model can be used in different ways based on user objectives. The main objective for the *WARSSS* methodology is to compare the upstream sediment supply for an existing or reference condition to the downstream condition for a specified flow regime. The prediction allows comparisons of sediment delivery (tons/yr) from hillslope and channel processes. The procedure to meet this objective is outlined in **Steps 10–15** and **Flowcharts 5-10** and **5-11**. The procedure is first completed for the reference condition followed by the impaired condition.

Furthermore, if the user objective is to determine increases or decreases in sediment supply due to streamflow change, the FLOWSED model can be used in conjunction with a water yield model. In this case, FLOWSED does not use the dimensionless flow-duration curve in **Step 10,** as shown in **Flowcharts 5-10** and **5-11**. Instead, flow-duration curves for pre- and post-treatment conditions produced from a water yield model in **Step 16** are used directly in **Step 14**. The water yield procedure in conjunction with the FLOWSED model to determine flow-related changes in sediment supply is discussed in **Steps 16** and **17**.

Another objective is to use the water yield model in conjunction with the FLOWSED model to predict proposed flow-related sediment yields based on future land use projections (discussed in **Steps 16** and **17**).

Once sediment supply is determined from the FLOWSED model, changes in sediment disposition are calculated using the POWERSED model in **Steps 18** and **19**. The POWERSED model tracks changes in the hydraulic character of river reaches. The hydraulic changes govern whether a stream will be stable, aggrade or degrade based on changes in channel shape and slope for a given sediment supply.

FLOWSED is also used for sediment supply in the POWERSED model. In this case, suspended sand sediment is separated from the washload sand (<0.062 mm) of the suspended sediment. The suspended sand data is used with the bedload data in the FLOWSED model to determine the capacity for hydraulically controlled sediment in POWERSED. The FLOWSED and POWERSED steps start to apportion the causes of impairment to identify potential logical management changes and specific mitigation needs to be addressed at the completion of the *PLA* phase.

The FLOWSED and POWERSED models are programmed in RIVERMorph™, which significantly simplifies the sediment analysis procedure. RIVERMorph™ has the capability to link to the USGS database to obtain the necessary data and compute all the necessary model conversions and calculations. Thus, the models can be implemented rapidly and easily following the basic understanding and mechanics of the models and the software instructions.

The FLOWSED model requirements

The FLOWSED model as discussed in **Steps 10–15** is based on the use of dimensionless flow-duration and sediment rating curves. The normalization parameters include:

- Bankfull discharge;
- Bankfull stage bedload;
- Bankfull stage suspended sediment; and
- Bankfull stage suspended sand sediment (for use with POWERSED).

The dimensionless flow-duration curves are developed from USGS streamgage data that represents a similar hydro-physiographic province as the study reach (**Step 10**). The dimensionless sediment rating curves are developed or are selected based on representative stream types and stability ratings (**Step 12**). **Steps 13** and **14** use the field-measured bankfull values (**Step 11**) to convert the dimensionless flow-duration and sediment curves to dimensioned curves. These dimensioned curves are then used to calculate total annual bedload and suspended sediment yield (tons/yr) in **Step 15**.

Table 5-12 lists the data required to run the FLOWSED model.

Table 5-12. Data required for the FLOWSED model.

Data requirements for FLOWSED (at ungaged sites)

- Background reference data (flow and sediment)
  - dimensionless flow-duration curve (from a local or representative hydro-physiographic province)
  - dimensionless bedload rating curve by stream type/stability rating
  - dimensionless suspended sediment rating curve by stream type/stability rating
  - momentary maximum bankfull discharge
  - mean daily bankfull discharge (the mean daily discharge the day bankfull occurs at a gage station)
  - flow-duration curve from gage
- Field measured values (for both reference and impaired reaches)
  - stream classification (Level II)
  - Pfankuch channel stability rating
  - measured bankfull discharge (cfs)
  - measured bedload sediment (kg/s) (Helley-Smith bedload sampler)
  - measured suspended sediment (mg/l) (depth-integrated sediment sampler)
  - measured suspended sand sediment (mg/l) (separated from suspended sediment sample in lab)

The computations for total annual bedload and suspended sediment yield (FLOWSED model) are summarized in **Worksheet 5-11**. Detailed descriptions of **Steps 10–15** follow to assist the user in the application of the FLOWSED model.

WATERSHED ASSESSMENT OF RIVER STABILITY AND SEDIMENT SUPPLY

Worksheet 5-11. FLOWSED calculation of total annual sediment yield.

Stream: _____ Location: _____ Date: _____

Observers: _____ Gage Station #: _____ Stream Type: _____ Valley Type: _____

Equation type	Intercept	Coefficient	Exponent	Form (e.g., linear, non-linear, etc.)	Equation name	Bankfull discharge (cfs)	Bankfull bedload (kg/s)	Bankfull suspended (mg/l)
1. Bedload (dimensionless)								
2. Suspended sediment (dimensionless)								
3. User-defined relations (bedload)					Notes:			
4. User-defined relations (suspended sediment)								

From dimensioned flow-duration curve					Dimensionless streamflow (7)	From sediment rating curves				Calculate (12)	Calculate sediment yield		
(1) Flow exceedence	(2) Daily mean discharge	(3) Mid-ordinate	(4) Time increment (percent)	(5) Time increment (days)	(6) Mid-ordinate streamflow	(7) Dimensionless streamflow	(8) Dimensionless suspended sediment discharge	(9) Suspended sediment discharge	(10) Dimensionless bedload discharge	(11) Bedload	(12) Time adjusted streamflow	(13) Suspended sediment [(5)×(9)]	(14) Bedload sediment [(5)×(11)]	(15) Suspended + bedload [(13)+(14)]
(%)	(cfs)	(%)	(%)	(days)	(cfs)	(Q/Q$_{bkf}$)	(S/S$_{bkf}$)	(tons/day)	(b$_s$/b$_{bkf}$)	(tons/day)	(cfs)	(tons)	(tons)	(tons)

Annual totals: _____ (tons/yr) _____ (tons/yr) _____ (tons/yr)

5-88

Step 10. Develop dimensionless flow-duration curve

To develop a dimensionless flow-duration curve, the user must first obtain data from the USGS or a related streamflow record that represents a hydro-physiographic region similar to the reach being assessed. The following data is required:

- Flow-duration curve from the gage (mean daily discharge (Q_{mnd}) versus percent of time flows equaled or exceeded);

- Mean daily discharge record; and

- The field-calibrated bankfull discharge (Q_{bkf}) determined at the reach location of the gage as in **Steps 3** and **4** of *PLA*.

Bankfull discharge (Q_{bkf}) is a momentary maximum discharge value, as determined in the field and obtained from flood-frequency data. If the bankfull discharge (Q_{bkf}) is the same as the mean daily discharge (Q_{mnd}) the day bankfull occurs, then the method for snowmelt-dominated hydrographs is used. If Q_{bkf} is greater than Q_{mnd} at the time bankfull occurs, then the procedure to develop dimensionless flow-duration curves for stormflow-dominated hydrographs will apply to convert the momentary maximum discharge value (Q_{bkf}) to a new mean daily bankfull discharge value (Q_{mndbkf}). The following are the procedures for both snowmelt- and stormflow-dominated hydrographs.

Snowmelt-dominated hydrograph

For reaches in these areas where the field-calibrated bankfull discharge (Q_{bkf}) is similar to the 24-hour average discharge (mean daily discharge, Q_{mnd}), the following is the procedure to develop a dimensionless flow-duration curve:

1. Divide the field-calibrated bankfull discharge (Q_{bkf}) into all of the mean daily discharge (Q_{mnd}) values of the flow-duration curve to make a dimensionless flow-duration curve.

Stormflow-dominated hydrograph

If the gage has a stormflow-dominated hydrograph, then complete the following:

1. Determine if the bankfull discharge (Q_{bkf}) is greater than the mean daily discharge (Q_{mnd}) the day bankfull occurs from the mean daily record:

    a. If Q_{bkf} is equal to Q_{mnd}, the procedure described for the snowmelt-dominated hydrograph above can be used.

    b. If Q_{bkf} is greater than Q_{mnd}, then continue with the following procedures:

2. Establish a new mean daily bankfull discharge value (Q_{mndbkf}), which equals the mean daily discharge the day bankfull occurs from the mean daily discharge record;

3. Divide the Q_{mnd} values from the flow-duration curve by Q_{mndbkf} to develop a dimensionless flow-duration curve; and

4. Establish a conversion ratio for this gage to be used to convert bankfull discharge (Q_{bkf}) at an ungaged site to mean daily bankfull discharge (Q_{mndbkf}):

    a. Divide Q_{mndbkf} (the mean daily discharge the day bankfull occurs from the mean daily record) by Q_{bkf}. This ratio (Q_{mndbkf}/Q_{bkf}) will be used in **Step 14** by multiplying the field-determined Q_{bkf} by this ratio prior to converting dimensionless flow-duration curves to dimensioned values.

Dimensionless flow-duration curves can be developed rapidly using the RIVERMorph™ program that links directly to the USGS database to obtain the required information for **Step 10**.

Figures 5-40 and **5-41** show examples of the application for Weminuche Creek and the upper Salmon River.

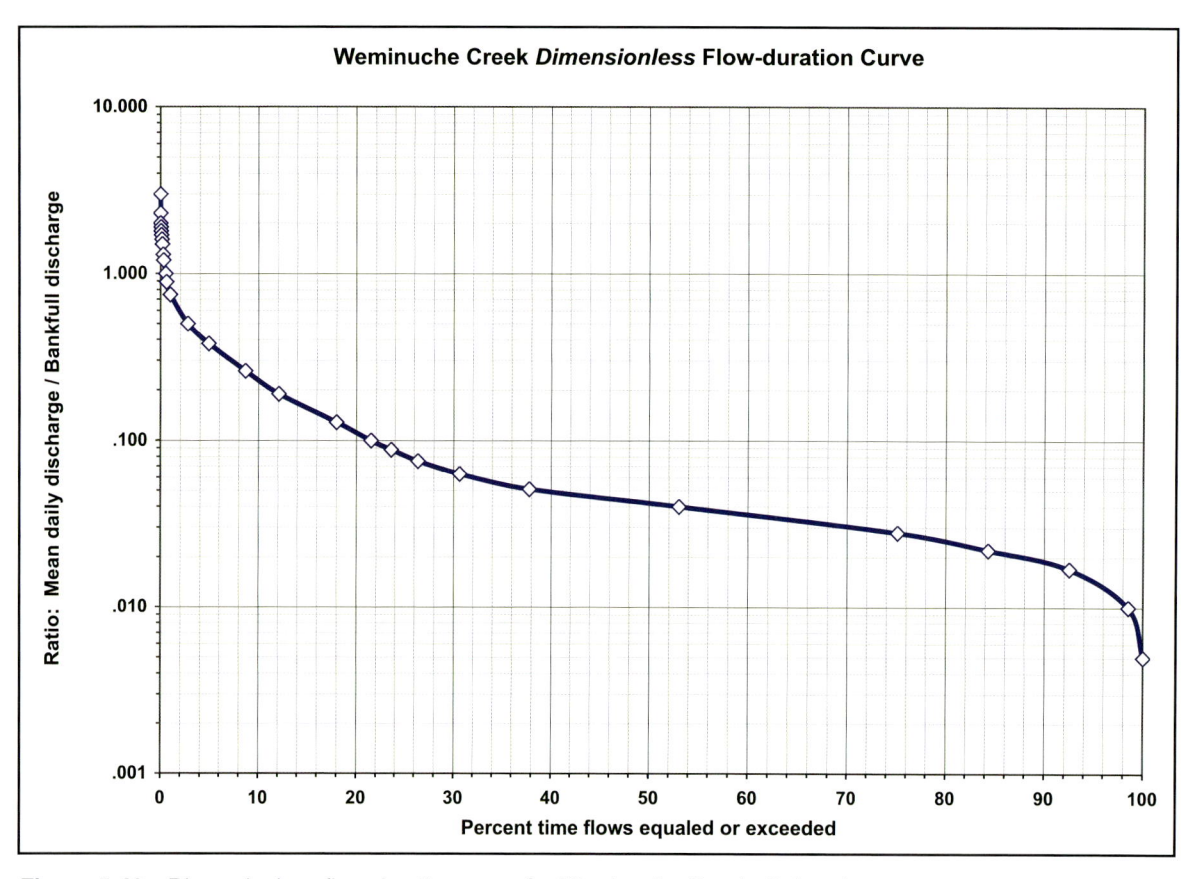

Figure 5-40. Dimensionless flow-duration curve for Weminuche Creek, Colorado.

Chapter 5

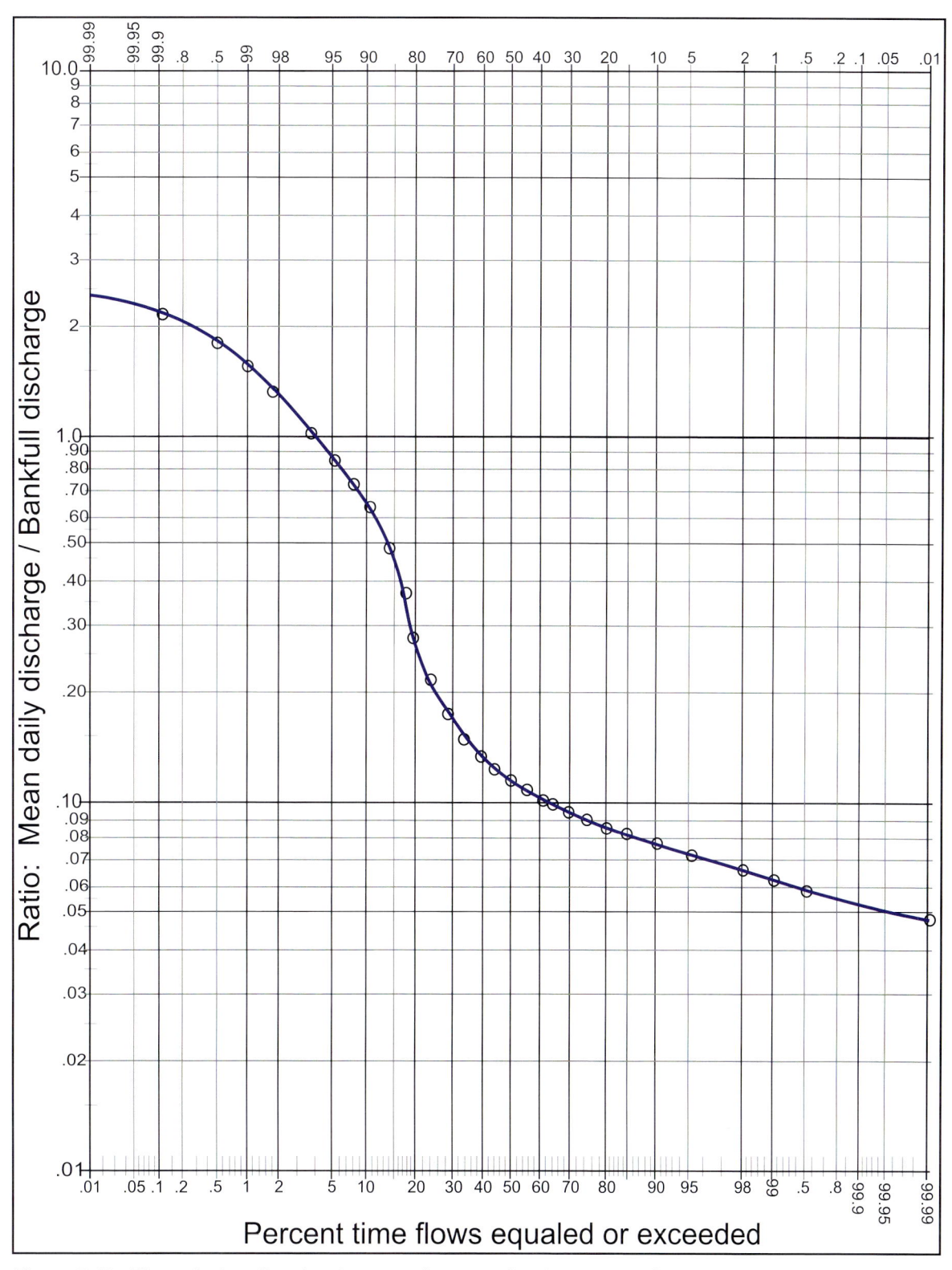

The Prediction Level Assessment (PLA)

Figure 5-41. Dimensionless flow-duration curve for streamflow in the upper Salmon River area (adapted from Emmett, 1975).

Step 11. Collect bankfull discharge, bedload sediment and suspended sediment

The user should use the field methods and equipment outlined in Book 3, Chapter C2 of *Field Methods for Measurement of Fluvial Sediment* (USGS, 1999) to collect a range of bedload and suspended sediment data at or near the bankfull stage. Bankfull discharge and the average bankfull values of bedload and suspended sediment collected in this step will be used in **Steps 13** and **14** to convert dimensionless flow-duration and sediment rating curves to dimensioned curves. It is important to capture the bankfull discharge and to measure several values of bedload and suspended sediment at or near the bankfull stage. *Due to the inherent variability in measured values (variable source of sediment supply), the user is advised to obtain an average value over several observations representing a wide range of sediment data at or near the bankfull stage.*

For use in the POWERSED model, suspended sand sediment is separated in a lab analysis from the depth-integrated suspended sediment data. The suspended sand data in mg/l is used in place of the suspended sediment data in the FLOWSED model to obtain dimensioned sediment rating curves for capacity calculations in POWERSED (**Steps 18** and **19**).

Step 12. Obtain or establish dimensionless bedload and suspended sediment rating curves

This step requires the user to obtain or establish dimensionless bedload and suspended sediment rating curves. The field-collected data in **Step 11** will be used to convert the dimensionless sediment rating curves to dimensioned bedload and suspended sediment curves in **Step 13**.

The field evidence of stream type and Pfankuch stability ratings are used to select the appropriate reference dimensionless sediment rating curves for both bedload and suspended sediment. The reference condition generally represents "good/fair" stability stream types and uses the associated dimensionless bedload and suspended sediment rating curves shown in Chapter 2 in **Figures 2-52** and **2-53**. The "poor" stability/high supply stream types will generate a higher sediment yield for the same streamflows. If, due to mitigation or restoration, the stream stability changes (including the BEHI and NBS ratings), the sediment yield prediction will reflect this condition due to a *before* vs. *after* scenario of sediment response. The change is due to higher sediment supply reflective of the stability ratings. Examples of the dimensionless bedload and suspended sediment rating curves for "poor" stability stream types are shown in **Figures 2-58** and **2-59**.

An extensive data set, involving measured bedload and suspended sediment rating curves, was used to test significance among "poor," "good" and "fair" stream types (Troendle et al., 2001). The dimensionless sediment rating curves for the "poor" stream types were statistically different than the dimensionless curves for the "good" and "fair" stream types, reflecting the change in sediment supply. Following the results of this study, for streams in Southwestern Colorado, a similar approach was made by grouping streams as either "good" or "fair" stability. Dimensionless curves were made by dividing all of the values in the measured sediment rating curves by their respective bankfull value (**Figures 2-52** and **2-53**). The same procedure was established for the "poor" stability streams (**Figures 2-59** and **2-59**). As a result, dimensionless sediment rating curves were developed for both bedload and suspended sediment relations. The rationale for this approach allows for extrapolation of the dimensionless sediment rating curves by using measured bankfull sediment and discharge values to locally convert to dimensioned sediment rating curves.

Step 13. Convert dimensionless bedload and suspended sediment rating curves to dimensioned sediment rating curves

Convert dimensionless bedload and suspended sediment rating curves to dimensioned sediment rating curves by multiplying the bankfull discharge and sediment values obtained in **Step 11** by each of the ratios appropriate for the relation selected in **Step 12**. Examples of converting a dimensionless relation (**Figures 2-52** and **2-53**) using bankfull values of sediment and discharge to a dimensioned relation are shown in **Figures 5-42** and **5-43** for Weminuche Creek, Colorado (Rosgen, 2006). The predicted sediment rating curves from the FLOWSED model for Weminuche Creek are plotted comparing the observed values. The close agreement of predicted-to-observed values provides reasonable annual sediment yield values.

Validation tests were completed using the dimensionless sediment rating curves from Pagosa Springs, Colorado (**Figures 2-52** and **2-53**). These sediment rating curves were developed using an individual data point representing the bankfull values of discharge, bedload sediment and suspended sediment. The predicted sediment rating curves using this method for independent data sets from a variety of sources, including the USGS data in Alaska, Wyoming, Idaho, Oregon, Nevada and Colorado, are shown in Chapter 2, **Figures 2-54** through **2-57**. The Pagosa curves were not intended to be universal relations; however, the application of the methodology of dimensionless sediment rating curves obtains reasonable predictions.

If it is not possible to obtain measured bankfull discharge, bedload sediment data and suspended sediment data to convert dimensionless sediment rating curves to dimensioned sediment rating curves, regional curves may be temporarily substituted. The user must obtain drainage area (mi^2) to calculate bankfull discharge from a similar hydro-physiographic province (**Figure 5-2**). The bankfull flow is also used to convert the dimensionless flow-duration to dimensioned flow-duration. The bankfull discharge is used to convert the dimensionless discharge portion of the dimensionless bedload and suspended rating curve to dimensioned values. The sediment data obtained from a known drainage area (mi^2) may be used to also develop regional sediment rating curves. The existing measured bankfull bedload and suspended sediment data is then converted to unit area sediment values and is plotted with the corresponding drainage area. This data needs to be stratified by the corresponding lithology, stream type and stability condition.

Examples of various unit area suspended sediment data at approximated bankfull discharges from USGS sites throughout the United States are shown in Simon et al. (2004). These measured sediment values were separated by evolutionary stages. The evolutionary stages can be used to infer sediment supply and approximations of stability. This data can be used to convert a dimensionless sediment rating curve to dimensioned values. Additional stability or stream type data may help to identify specific relations for extrapolation. This drainage area extrapolation procedure represents only an interim procedure until measured bankfull values can be obtained using standard measurement equipment and field methods.

Chapter 5   The Prediction Level Assessment (PLA)

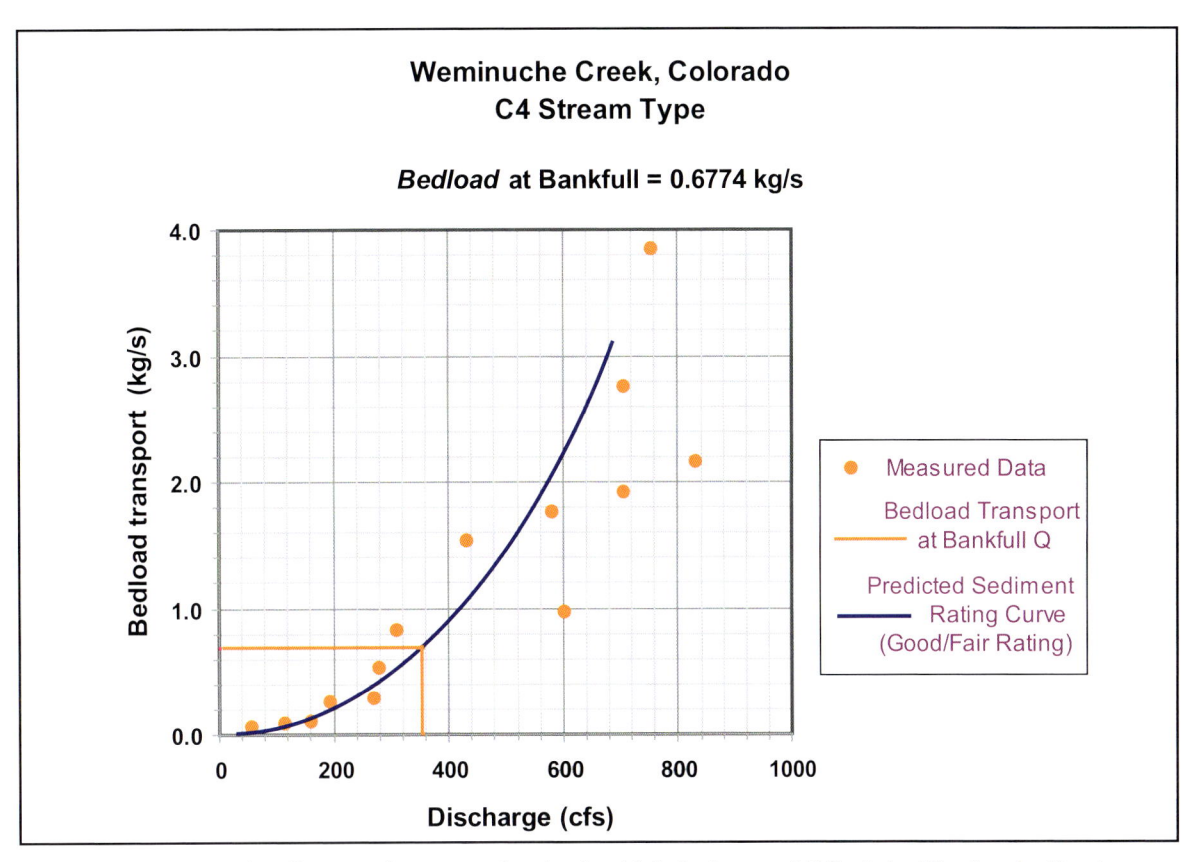

Figure 5-42. Bedload sediment rating curve showing bankfull discharge of 350 cfs for Weminuche Creek, Colorado.

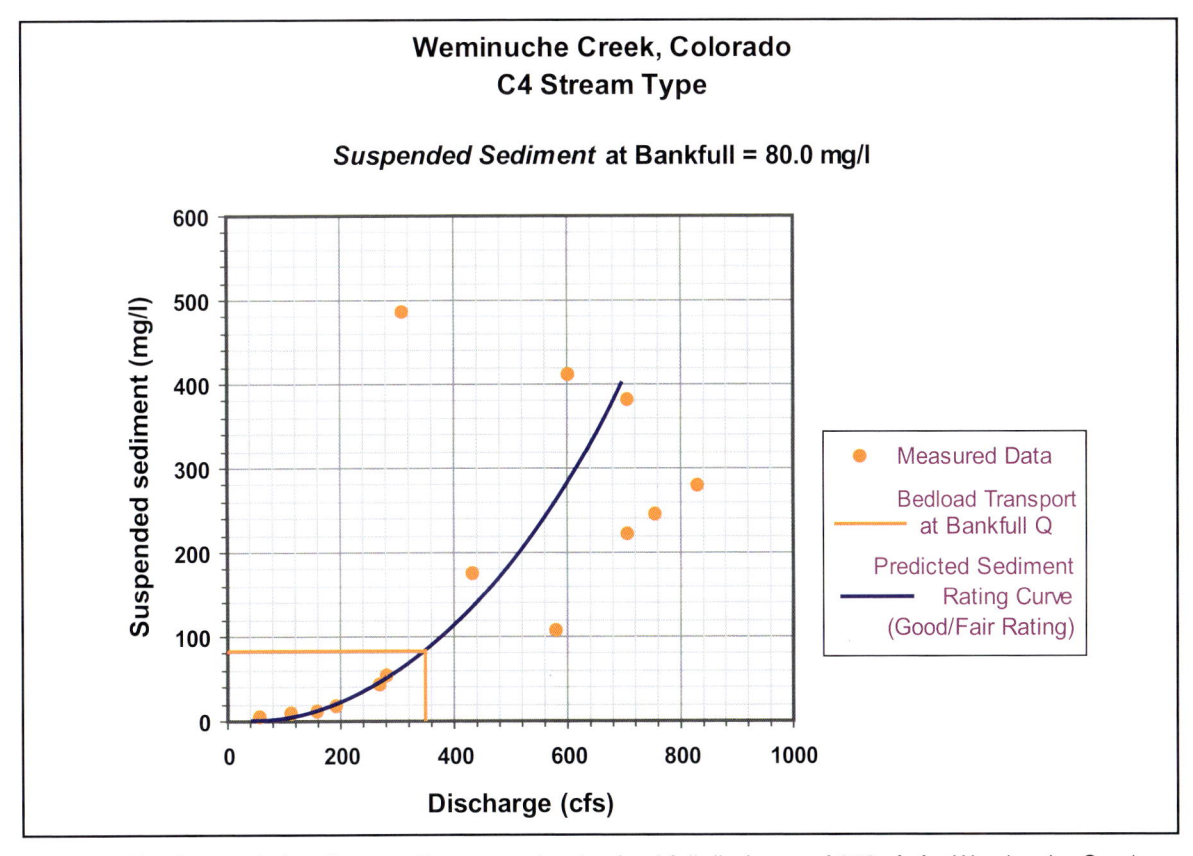

Figure 5-43. Suspended sediment rating curve showing bankfull discharge of 350 cfs for Weminuche Creek, Colorado.

Step 14. Convert dimensionless flow-duration curve to dimensioned flow-duration curve

This step creates a dimensioned flow-duration curve by multiplying the bankfull discharge collected in **Step 11** by the Q_{mnd}/Q_{bkf} ratios from the dimensionless flow-duration curve developed in **Step 10**. The flow-duration curve represents mean daily discharge for each percentile.

However, if the reach is in a stormflow-dominated area and the stormflow procedure in **Step 10** was used to create the dimensionless flow-duration curve, the following procedure is used to convert the dimensionless flow-duration curve to a dimensioned curve:

1. Multiply the field-determined bankfull discharge (Q_{bkf}) obtained in **Step 11** for the ungaged site by the ratio calculated in **Step 10** (Q_{mndbkf}/Q_{bkf}); and

2. Multiply this converted bankfull discharge by the dimensionless flow-duration curve values (Q_{mnd}/Q_{mndbkf}).

An example of a dimensioned flow-duration curve using bankfull discharge to convert from the dimensionless relation (**Figure 5-40**) is shown in **Figure 5-44**.

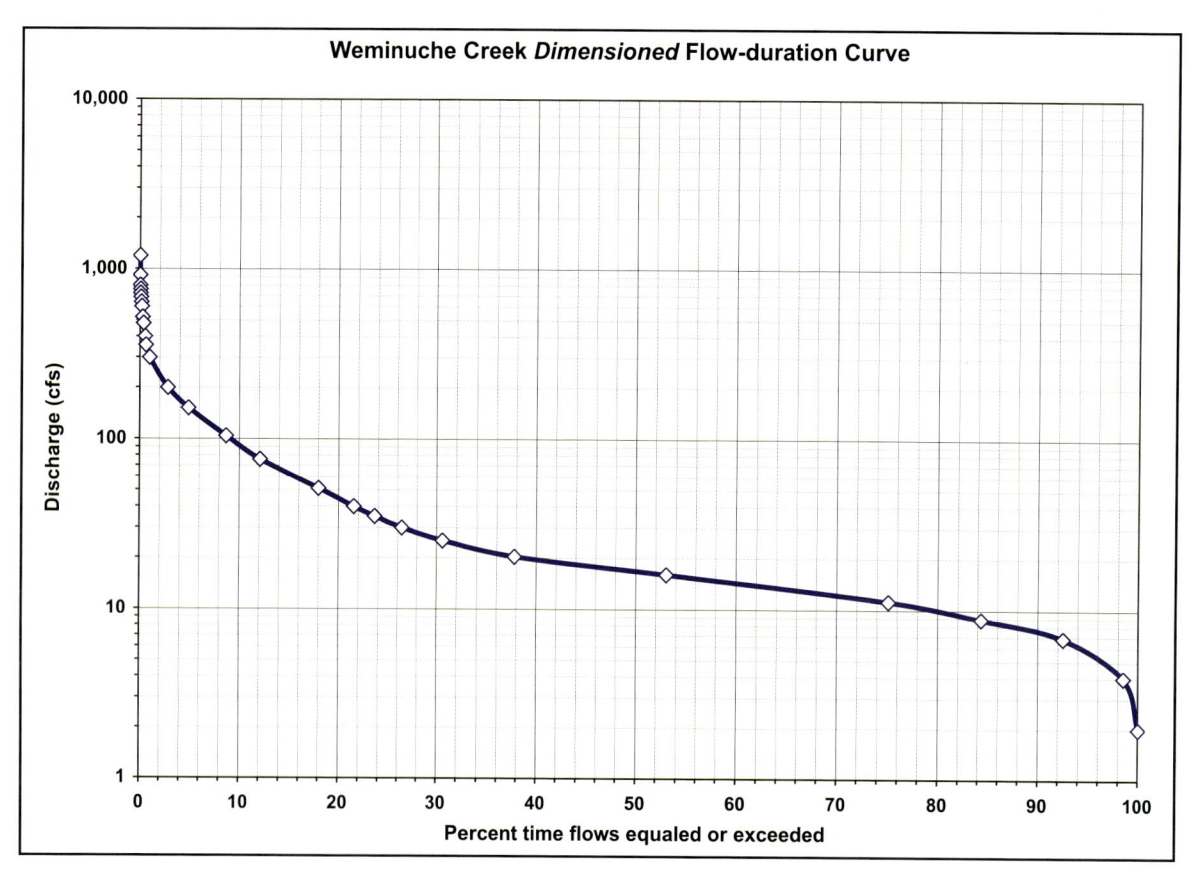

Figure 5-44. Example of dimensioned flow-duration curve for Weminuche Creek, Colorado.

Step 15. Calculate total annual sediment yield for both bedload and suspended sediment

This step calculates total annual bedload and suspended sediment yield for both the reference and existing conditions. This is accomplished by using the dimensioned bedload and suspended sediment rating curves in **Step 13** and the dimensioned flow-duration curve in **Step 14**. The flow increments for duration of time in days (from the dimensioned flow-duration curve) are multiplied by the sediment yield in tons/day associated with that flow (from the dimensioned bedload and suspended sediment rating curve). Enter these calculations in **Worksheet 5-11**.

The total annual sediment yield (tons/yr) for bedload and suspended sediment is recorded in the *PLA* summary worksheet to use in **Steps 29–31** to analyze disproportionate contributions of sediment from various sources. This assists the user to identify or design sediment reduction and river mitigation/restoration measures. Also, the user can prioritize mitigation once magnitude, location, nature and consequence of the sediment problems are quantified.

Steps 16–17: Calculate flow-related changes in annual sediment yield using water yield and FLOWSED models (See Flowchart 5-12)

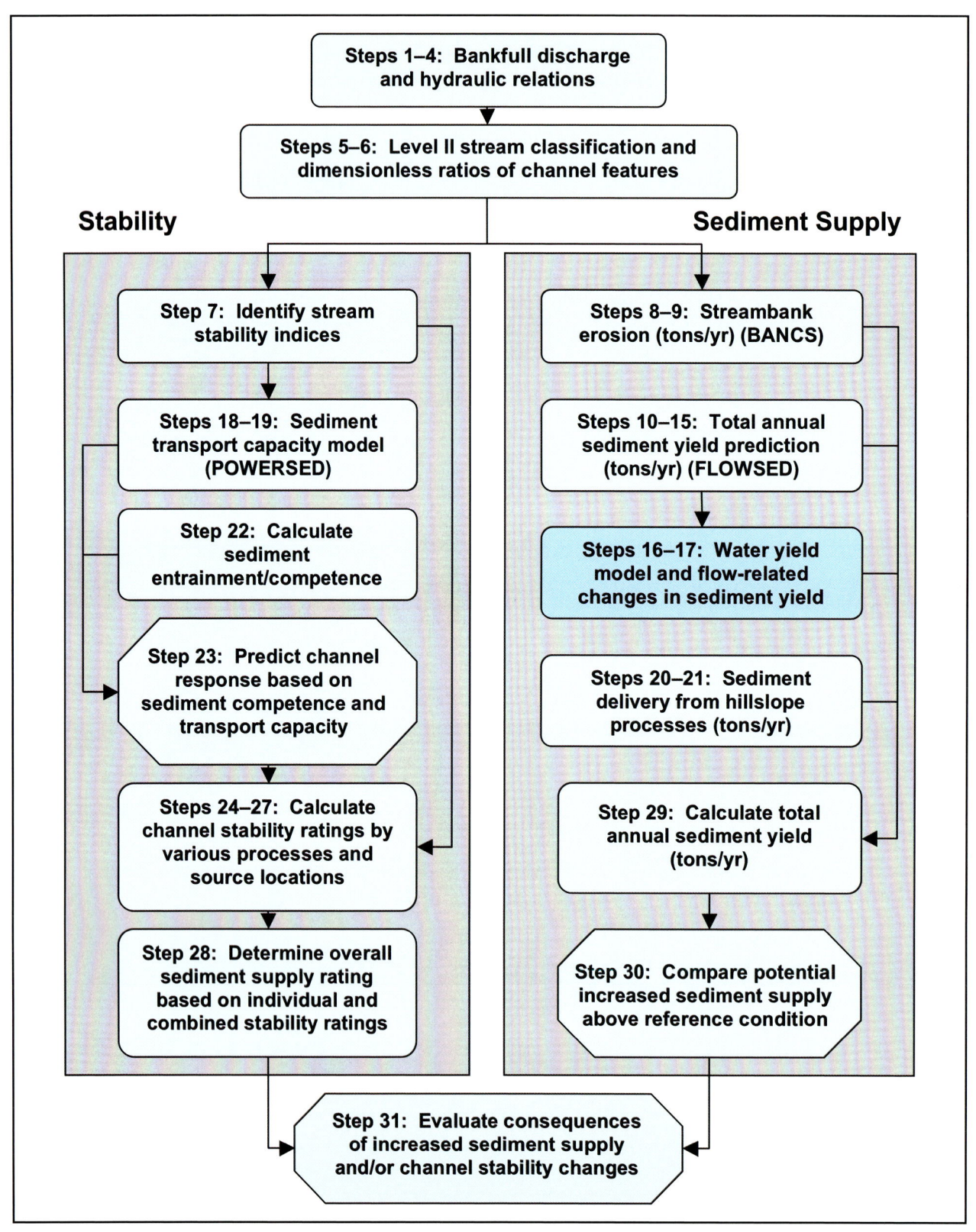

Flowchart 5-3g. The sequential steps to determine stream stability and sediment supply, highlighting **Steps 16–17**.

Chapter 5 — The Prediction Level Assessment (PLA)

Steps 16–17: Calculate flow-related changes in annual sediment yield using water yield and FLOWSED models

- Step 16: Water yield model
- Calculate sediment yield using FLOWSED Steps 11–15
  - Pre-treatment condition
  - Post-treatment condition

- Determine pre-treatment bankfull discharge (Q_{bkf}) from water yield model

- Step 12: Obtain or establish dimensionless sediment rating curves by stream type/stability

- Step 11: Collect field data by stream type/valley type
  - Bankfull discharge (Q_{bkf})
  - Bankfull bedload sediment (kg/s)
  - Bankfull suspended sediment (mg/l)

- Step 14 (modified): Flow-duration curves from water yield model
  - Pre-treatment condition
  - Post-treatment condition

- Step 13: Convert dimensionless sediment rating curves to dimensioned curves
  - Pre-treatment condition
  - Post-treatment condition

- Step 15: Calculate total annual bedload and suspended sediment yield
  - Pre-treatment condition
  - Post-treatment condition

- Step 17: Calculate flow-related changes in annual sediment yield

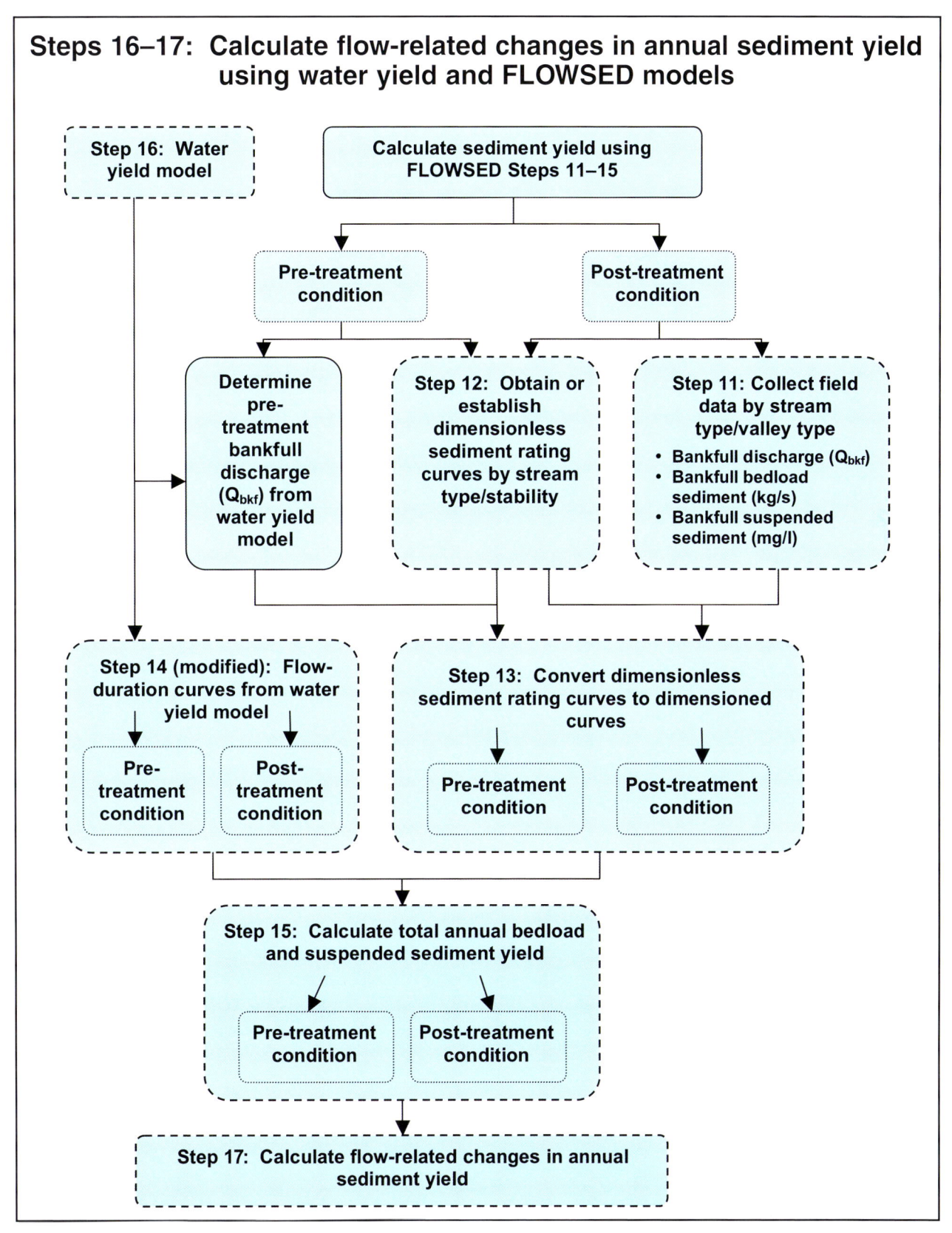

Flowchart 5-12. Procedure to calculate total annual sediment yield increase by comparing pre-treatment with post-treatment water yield conditions (combining water yield and FLOWSED models).

Calculate flow-related changes in annual sediment yield using water yield and FLOWSED models: Steps 16–17

The overall objective of **Steps 16–17** is to predict the response of sediment yield due to an increase or change in streamflow. The discharge routed through different stream types and/or stability conditions, which may vary for pre- and post-treatment conditions, will yield different sediment yield values.

Step 16. Select and run a water yield model

A water yield model is selected and run at this step to determine flow regime for both reference and existing conditions. The model selected should best represent the flow regime, e.g., snowmelt-dominated, stormflow-dominated, rain-on-snow, spring-fed, urban hydrology, etc. The selected water yield model also needs to reflect evapo-transpiration time-trend recovery, road acres, change in vegetative cover and percent impervious surfaces.

This step also considers operational hydrology from reservoirs, diversions and other flow modifications that influence the magnitude, duration and timing of streamflow. The input variables for most models are precipitation data, a vegetation alteration map by aspect and elevation, drainage area computations, percent of drainage area in impervious condition and similar data specified based on the specific model being selected.

The output for these models needs to be altered bankfull discharge and flow-duration curves for pre-treatment versus post-treatment condition for the reference and impaired reaches. The model can also be used to predict a proposed condition. This data is used in the FLOWSED model in **Steps 11–15**.

Several water yield models exist. Users should select a model that they are most familiar with and one that is locally calibrated. The following discusses snowmelt and stormflow models and pre-treatment, post-treatment and proposed conditions in more detail.

Snowmelt models

A snowmelt model that has been calibrated nationally is WRENSS (USEPA, 1980). This model has been applied successfully for water resource management projects for over 25 years. The change in evapo-transpiration and snowpack hydrology due to altered timber stands is related to changes in flow-duration curves and bankfull discharge. An updated version of WRENSS (Troendle and Swanson, in press) is supported by the U.S. Forest Service in Fort Collins, Colorado. This model was used in conjunction with the FLOWSED model in the test case example for Wolf Creek in Chapter 7.

Stormflow models

Several stormflow models exist including TR-55 (available at http://www.scisoftware.com and in RIVERMorph™) and the unit hydrograph approach (U.S. Army Corps of Engineers, 1998). Whatever model is selected to be used with FLOWSED, the output should generate an altered bankfull discharge and flow-duration curves for pre- and post-treatment conditions on reference and impaired reaches.

In the use of stormflow models, it is important to calibrate a given return period storm under specified antecedent moisture conditions, runoff curve number, etc. to the bankfull condition. This is often best applied at a gage station where precipitation data and flood frequency data is available.

Pre-treatment condition

Water yield determinations must indicate full hydrologic utilization. This means that the watershed is fully recovered from any vegetative alteration, roads, imported water diversions, etc. The flow-duration curves associated with the pre-treatment condition are associated with no changes in flow from the natural potential runoff.

Post-treatment condition: Changes in runoff

The post-treatment curves must represent existing hydrologic conditions. The hydrologic response of vegetative alteration, roads, imported water, operational hydrology below dams, diversions, depletions, etc. is assessed for the existing condition.

Proposed condition

Proposed alteration of vegetation, timber harvest, roads, flow modification due to diversion, reservoir release, etc. can also be evaluated at this step. The change in bankfull discharge and flow-duration curves due to proposed flow changes are treated the same as the post-treatment computations as shown in **Flowchart 5-12**. The proposed annual sediment yield consequence for both bedload and suspended sediment load are compared to both pre-treatment and post-treatment (existing) conditions. The predicted flow-related sediment yield consequence for the proposed condition can also be entered in **Worksheet 5-11**, similar to the post-treatment computations.

Flow-related sediment yield predictions offer an opportunity to understand potential adverse consequences. Preventing potential impairment and adverse impacts is often less expensive than treating impairment "after the fact." Land management alternatives can modify streamflow, often in a positive way, to avoid obvious adverse impacts. Stabilizing streambanks, reconnecting floodplains and installing grade-control structures can all reduce channel source sediment increases due to flow changes. If managers are made aware prior to management actions, alternatives can often be prescribed to offset such potential adverse impacts to the water resources.

Step 17. Calculate flow-related changes in annual sediment yield

The procedure to calculate flow-related changes in sediment yield is shown in **Flowchart 5-12**. The procedure is applied to both the reference and impaired reaches. The flow-related sediment procedure requires the user to run the water yield model selected in **Step 16** to develop dimensioned flow-duration curves for the pre- and post-treatment conditions for the reference and impaired reaches. The flow-duration curve for the pre-treatment condition is first routed through the FLOWSED **Steps 11–15**; the procedure is then repeated for the post-treatment (existing) condition. Because the water yield model output is a dimensioned flow-duration curve, the use of a dimensionless flow-duration curve is negated as in previous FLOWSED steps. Flow-related changes in sediment yield for proposed conditions can also be simulated based on projected changes in vegetative cover and or flow regulation as discussed in **Step 16**.

For pre-treatment bankfull discharge used in the FLOWSED model, the value is derived from the flow model based on water yield increases shown on the flow-duration curve. The pre-treatment sediment supply is obtained by selecting a stream type/stability from aerial photo interpretations of stream conditions that may have occurred prior to the land use change; for example, a "good/fair" dimensionless sediment rating curve may be selected. The post-treatment sediment measurements are used in the FLOWSED model to convert a dimensionless sediment rating value to a dimensioned value. The same measured sediment values are used for the pre-treatment conversion; however, the dimensionless sediment rating curve selected may be different than the post-treatment curve. Often, a "poor" stability rating curve represents the post-treatment or impaired condition.

The pre-treatment and post-treatment flow-duration curves from the selected water yield model are then routed through both the dimensioned bedload and suspended sediment rating curves from **Step 13** (**Figures 5-42** and **5-43**). For each time increment of flow, a value of sediment transport is obtained. The tons/day values are multiplied by the total days in each time/flow/sediment category as in **Worksheet 5-11**. The flow-related annual sediment yield is the difference between the pre-treatment annual sediment yield versus the post-treatment annual sediment yield.

The FLOWSED model in RIVERMorph™ version 4.1 allows the user to make the proper calculations to adjust the "pre-treatment bankfull" discharge and directly import the pre-treatment and post-treatment condition flow-duration curves. The output of this step provides an interpretation of the potential flow-related sediment increase. Bank erosion, bed sediment scour and redistribution of stored sediment processes are integrated into such flow-related sediment increases.

The summary of the calculations for flow-related sediment are included in **Worksheet 5-11**, and the overall flow-related increase values are transferred to the *PLA* summary worksheet in **Steps 29–31**.

Chapter 5 — The Prediction Level Assessment (PLA)

Steps 18–19: Sediment transport capacity model (POWERSED) (See Flowcharts 5-13 and 5-14)

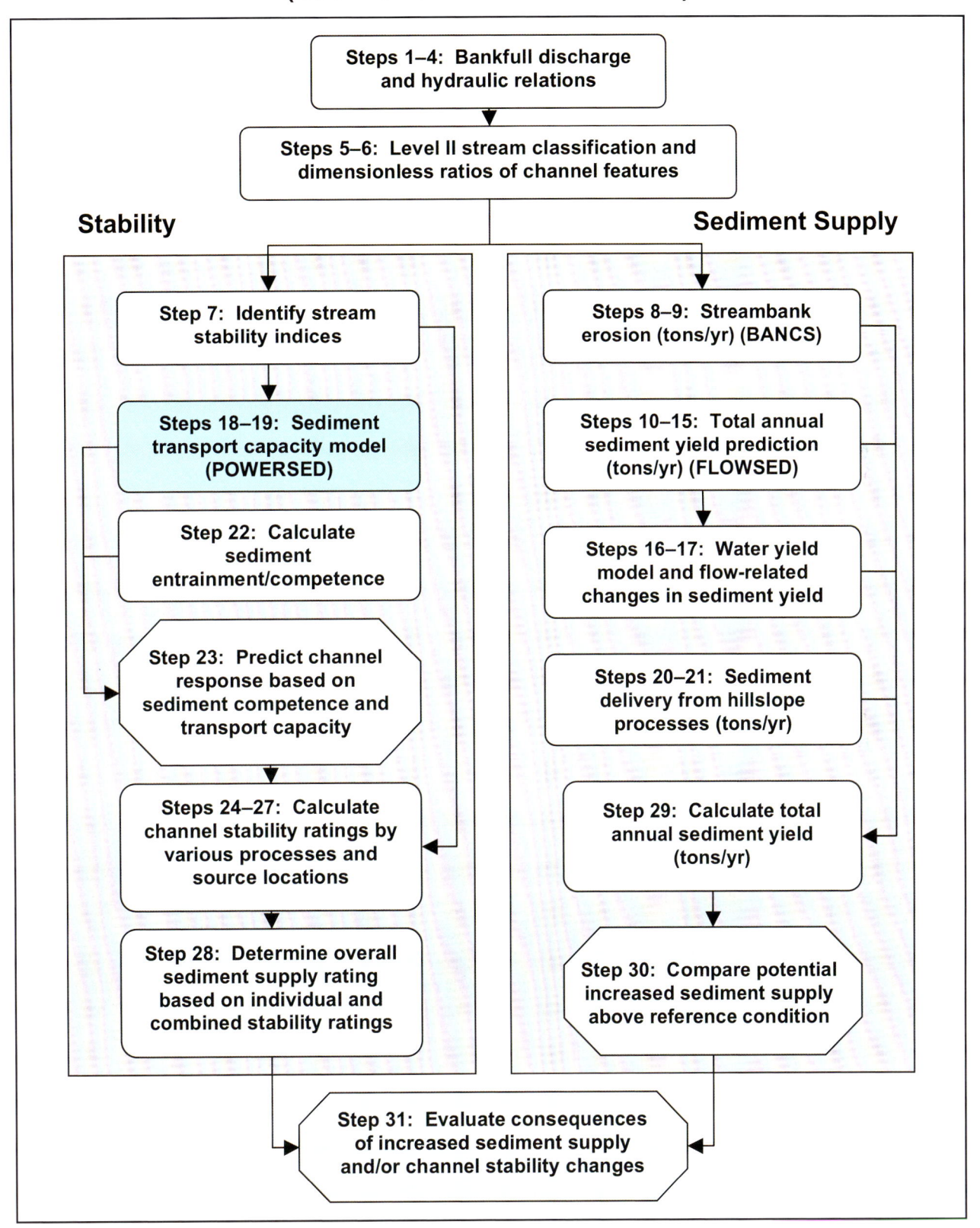

```
        Steps 1–4: Bankfull discharge
            and hydraulic relations
                    │
                    ▼
      Steps 5–6: Level II stream classification and
        dimensionless ratios of channel features
```

Stability | **Sediment Supply**

Stability branch:
- Step 7: Identify stream stability indices
- **Steps 18–19: Sediment transport capacity model (POWERSED)**
- Step 22: Calculate sediment entrainment/competence
- Step 23: Predict channel response based on sediment competence and transport capacity
- Steps 24–27: Calculate channel stability ratings by various processes and source locations
- Step 28: Determine overall sediment supply rating based on individual and combined stability ratings

Sediment Supply branch:
- Steps 8–9: Streambank erosion (tons/yr) (BANCS)
- Steps 10–15: Total annual sediment yield prediction (tons/yr) (FLOWSED)
- Steps 16–17: Water yield model and flow-related changes in sediment yield
- Steps 20–21: Sediment delivery from hillslope processes (tons/yr)
- Step 29: Calculate total annual sediment yield (tons/yr)
- Step 30: Compare potential increased sediment supply above reference condition

Step 31: Evaluate consequences of increased sediment supply and/or channel stability changes

Flowchart 5-3h. The sequential steps to determine stream stability and sediment supply, highlighting **Steps 18–19**.

5-103

Steps 18–19: Sediment transport capacity model (POWERSED)

Step 18: Evaluate channel characteristics that change hydraulic and morphological variables and develop hydraulic geometry relations for a wide range of flows for the upstream and impaired reaches

- Width
- Depth
- Slope
- Velocity
- Discharge

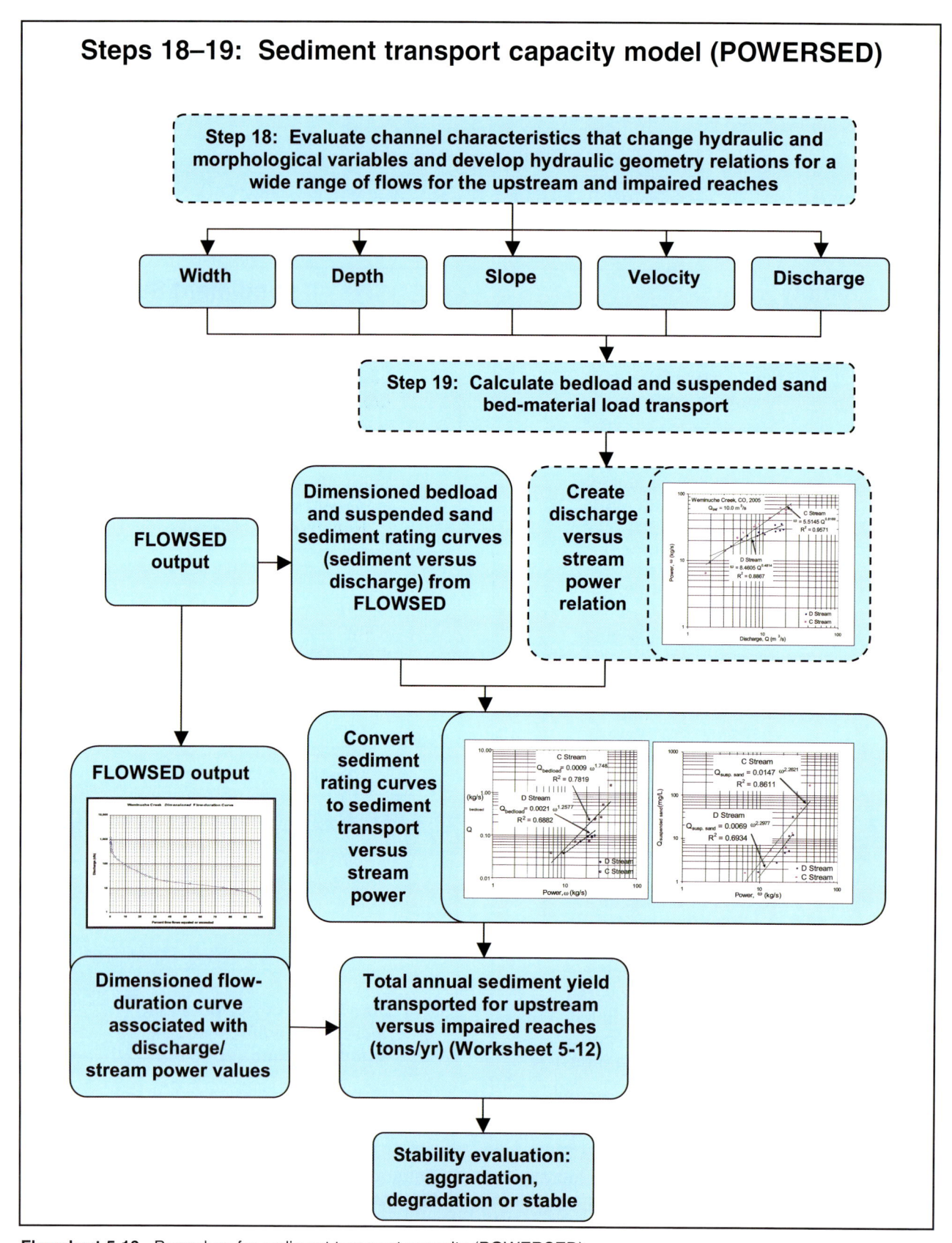

Step 19: Calculate bedload and suspended sand bed-material load transport

- FLOWSED output
- Dimensioned bedload and suspended sand sediment rating curves (sediment versus discharge) from FLOWSED
- Create discharge versus stream power relation

- FLOWSED output
- Convert sediment rating curves to sediment transport versus stream power
- Dimensioned flow-duration curve associated with discharge/stream power values
- Total annual sediment yield transported for upstream versus impaired reaches (tons/yr) (Worksheet 5-12)
- Stability evaluation: aggradation, degradation or stable

Flowchart 5-13. Procedure for sediment transport capacity (POWERSED).

Chapter 5 — The Prediction Level Assessment (PLA)

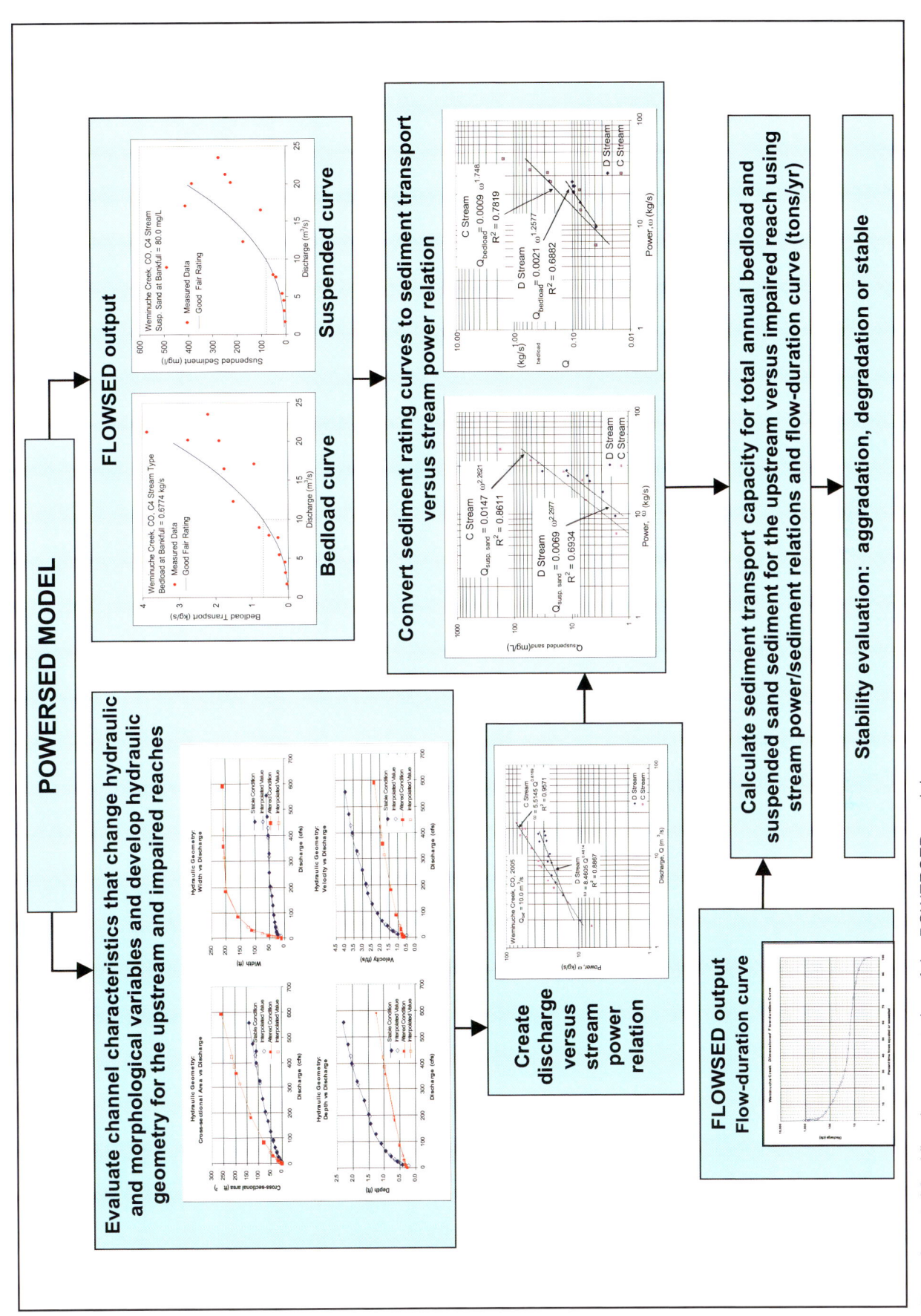

Flowchart 5-14. Visual representation of the POWERSED model.

The POWERSED sediment transport capacity model: Steps 18–19

The POWERSED model is used in conjunction with the FLOWSED model to determine channel stability. Reaches may be stable (sediment in versus sediment out), aggrading or degrading. The POWERSED model compares sediment transport capacity from an upstream, adjacent sediment supply reach by predicting transport rate change due to channel hydraulics. The hydraulics reflect potential changes in morphological variables, such as channel width, depth, slope and/or channel materials. The corresponding changes in flow resistance are used to predict velocity, shear stress and unit stream power (velocity multiplied by shear stress). Sediment rating curves from the FLOWSED model are converted from discharge to unit stream power for a wide range of flows. Revised values of annual sediment transport can then be compared to the upstream sediment supply from the subsequent change in the hydraulic geometry of the stream channel and corresponding response in sediment transport.

The stable, reference condition is used in *PLA* to compare with existing or potentially impaired conditions for departure analysis. For sediment transport capacity and evaluation of potential aggradation, degradation and/or stable channel inference using POWERSED, only the upstream, adjacent sediment supply reach is used. If the upstream reach is a stable, reference reach, an advantageous opportunity for comparison between the reference and impaired reaches exists. It is not imperative, however, that the upstream reach be a reference condition to determine how the downstream, impaired reach may or may not accommodate the sediment being made available from the watershed.

The POWERSED model for sediment transport capacity was developed to predict the following:

- Total annual bedload sediment transport;
- Total annual suspended sand sediment transport;
- Total annual sediment transport; and
- Channel stability evaluation (stable, aggrading or degrading) based on departure from the upstream sediment supply reach.

For application in the POWERSED model, the suspended sediment transport data must represent the hydraulically controlled suspended sand sediment transport. This adjustment is made in the FLOWSED model by substituting the mg/l for suspended sediment with the mg/l for suspended sand sediment concentration. The conversion ratio is obtained from laboratory analysis by separating the silt/clay (washload) fraction from the sand size particles from the depth-integrated suspended sediment sample. However, this would not be the case if there were concerns over accelerated fine sediment deposition into extremely low-gradient streams, deltas, reservoirs, lakes, marshes, tidal streams or saltwater estuaries. Colloidal sediments can present problems for impaired waters; thus, washload may need to be retained in the suspended sediment analysis in the FLOWSED model (**Steps 10–15**).

The bankfull discharge, bedload and suspended sand sediment values are then used with the appropriate dimensionless sediment rating curves to create dimensioned sediment rating curves in FLOWSED. Any flow modifications can also be simulated by revised flow-duration curves.

Chapter 5 — The Prediction Level Assessment (PLA)

The analytical procedures for POWERSED are shown in **Flowcharts 5-13** and **5-14**. The data requirements to run POWERSED are listed in **Table 5-13**. Bedload transport and suspended sand bed-material load is calculated using **Worksheet 5-12**. Two versions of **Worksheet 5-12** exist. **Worksheet 5-12a** is used for the upstream, adjacent reach, and **Worksheet 5-12b** is used for the impaired reach to determine channel stability.

Table 5-13. Data required for the POWERSED model.

Data requirements for POWERSED
• The output data from FLOWSED 　- bedload and suspended sand sediment rating curves for the upstream and impaired reaches using the appropriate dimensionless sediment rating curves 　- flow-duration curve (curve can also be from water yield model) • Field measured values: 　- cross-section data 　- longitudinal profile data 　- pebble count data 　- stream type 　- Pfankuch channel stability rating

RIVERMorph™ greatly minimizes the time complexity required to run FLOWSED and POWERSED. The program has been tested successfully on many data sets throughout the United States.

Other bedload and/or suspended sand bed-material load transport models can be employed by the user based on familiarity with and calibration/validation of the model for application to the particular stream type being analyzed. If necessary, the user should review the model assumptions, calibration, detailed discussion and associated references presented in Chapter 2.

Watershed Assessment of River Stability and Sediment Supply

Worksheet 5-12a. Bedload and suspended sand bed-material load transport prediction for the upstream reach, using the POWERSED model.

Stream: _____ Location: _____ Valley Type: _____ Gage Station #: _____ Date: _____

Observers: _____ Stream Type: _____

Flow-duration curve		Calculate	Hydraulic geometry				Measure				Calculate						
(1)	(2)	(3)	(4)	(5)	(6)	(7)	(8)	(9)	(10)	(11)	(12)	(13)	(14)	(15)	(16)	(17)	(18)
Exceedance probability	Daily mean discharge	Mid-ordinate stream-flow	Area	Width	Depth	Velocity	Slope	Shear stress	Stream power	Unit power	Time increment	Time increment	Daily mean bedload transport	Daily mean suspended sand transport	Time adjusted bedload transport [(13)×(14)]	Time adjusted suspended sand transport [(13)×(15)]	Time adjusted total transport [(16)+(17)]
(%)	(cfs)	(cfs)	(ft²)	(ft)	(ft)	(ft/s)	(ft/ft)	(lb/ft²)	(lb/s)	(lb/ft/s)	(%)	(days)	(tons/day)	(tons/day)	(tons)	(tons)	(tons)

Total annual sediment yield (bedload and suspended sand bed-material load) (tons/yr): _____

Chapter 5
The Prediction Level Assessment (PLA)

Worksheet 5-12b. Bedload and suspended sand bed-material load transport prediction for the potentially impaired reach, using the POWERSED model.

Stream: _____ Location: _____ Date: _____
Observers: _____ Stream Type: _____ Gage Station #: _____
  Valley Type: _____

Flow-duration curve		Calculate	Measure							Calculate						
			Hydraulic geometry													
(1)	(2)	(3)	(4)	(5)	(6)	(7)	(8)	(9)	(10)	(11)	(12)	(13)	(14)	(15)	(16)	(17)	(18)
Exceedance probability	Daily mean discharge	Mid-ordinate stream-flow	Area	Width	Depth	Velocity	Slope	Shear stress	Stream power	Unit power	Time increment	Time increment	Daily mean bedload transport	Daily mean suspended sand transport	Time adjusted bedload transport [(13)×(14)]	Time adjusted suspended sand transport [(13)×(15)]	Time adjusted total transport [(16)+(17)]
(%)	(cfs)	(cfs)	(ft²)	(ft)	(ft)	(ft/s)	(ft/ft)	(lb/ft²)	(lb/s)	(lb/ft/s)	(%)	(days)	(tons/day)	(tons/day)	(tons)	(tons)	(tons)

Notes:

Total annual sediment yield (bedload and suspended sand bed-material load) (tons/yr): _____
Upstream total annual sediment supply (tons/yr) (**Worksheet 5-12a**): _____
Difference in sediment transport capacity (tons/yr) (+ or −): _____
Stability evaluation: Aggradation, Degradation or Stable: _____

Step 18. Evaluate channel characteristics that change hydraulic and morphological variables and develop hydraulic geometry relations for a wide range of flows

This step involves collecting cross-section data, slope and channel materials for both the upstream and impaired conditions to generate hydraulic geometry by stage for each condition. The RIVERMorph™ POWERSED program simulates hydraulic geometry (width, depth, slope, velocity and discharge) for a wide range of stages for the upstream and impaired reach hydraulic evaluations. POWERSED also computes changes in hydraulic character due to modified channel dimension, pattern, profile and/or materials. Examples of typical hydraulic geometry plots comparing a stable condition to an altered condition are shown in **Figure 5-45**.

The hydraulic geometry relations are used to determine changes in unit stream power in **Step 19** for the same discharge or for increased or decreased discharge.

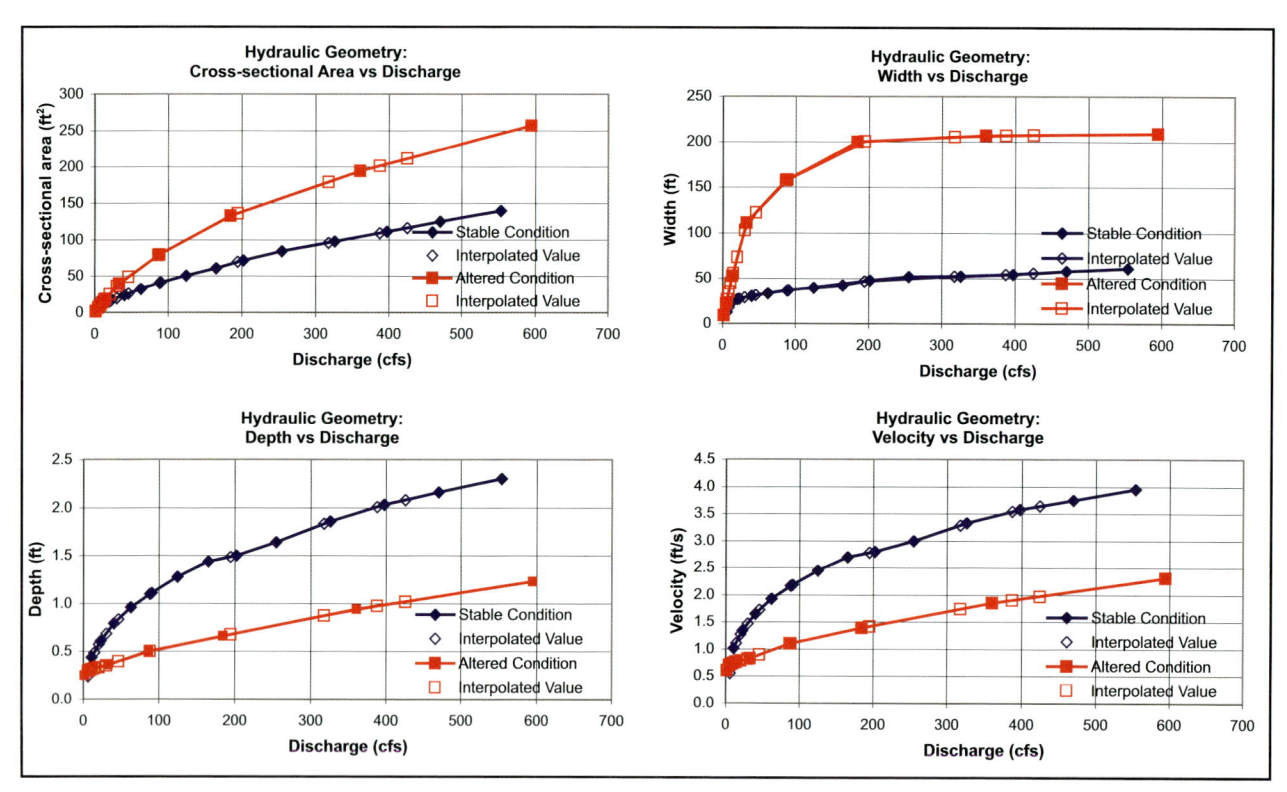

Figure 5-45. Examples of hydraulic geometry for a stable, meandering channel versus a braided (altered condition) channel, Weminuche Creek, Colorado (Rosgen, 2006).

Step 19. Calculate bedload and suspended sand bed-material load transport capacity (stream power)

This step applies the hydraulic geometry from **Step 18** and the flow-duration and bedload and suspended sand sediment rating curves from FLOWSED to generate changes in coarse bedload and suspended sand bed-material transport that are influenced by changes in channel cross-section, channel materials and/or slope. Total annual sediment yield in tons/yr and channel stability are predicted using **Worksheet 5-12**.

Bedload and suspended sand bed-material load transport calculations may use various equations, such as the Bagnold equation. The following equations from Chapter 2 are used in POWERSED as part of the bed-material load transport capacity prediction:

Shear stress and unit stream power are calculated using **Equation 2-2** and **Equation 2-11** from Chapter 2:

$$\tau = \gamma RS \qquad \text{(Equation 2-2)}$$

where:
γ = specific weight of the fluid

R = hydraulic radius (hydraulic mean depth)

S = water surface slope

Unit stream power or power per unit of streambed area (ω_a) is defined as:

$$\omega_a = \tau u \qquad \text{(Equation 2-11)}$$

where:
τ is defined in **Equation 2-2**

u = mean velocity

The integration of FLOWSED and the hydraulic geometry relations are depicted specifically in **Flowchart 5-14**. The following summarizes the POWERSED procedure:

1. Create a discharge (Q) versus stream power (ω_a) relation from the hydraulic geometry relations in **Step 18** (any changes in width, depth, boundary roughness (channel materials), velocity and/or slope will be reflected in this relation);

2. Convert the sediment rating curves from FLOWSED to sediment transport versus stream power (ω_a) using the discharge (Q) versus stream power relation (ω_a); and

3. Calculate total annual bedload and suspended sand sediment transport capacity using the sediment transport versus stream power (ω_a) relation and the flow-duration curve from FLOWSED.

The total annual bedload and suspended sand sediment yield for the upstream condition (**Worksheet 5-12a**) is then compared to the sediment yield for the impaired reach in **Worksheet 5-12b** to evaluate channel stability. A rating of aggradation, degradation or stable is recorded in **Worksheet 5-12b** and is transferred to the *PLA* summary worksheet in **Steps 29–31**.

WATERSHED ASSESSMENT OF RIVER STABILITY AND SEDIMENT SUPPLY

Steps 20–21: Sediment delivery from hillslope processes
(See Flowchart 5-15)

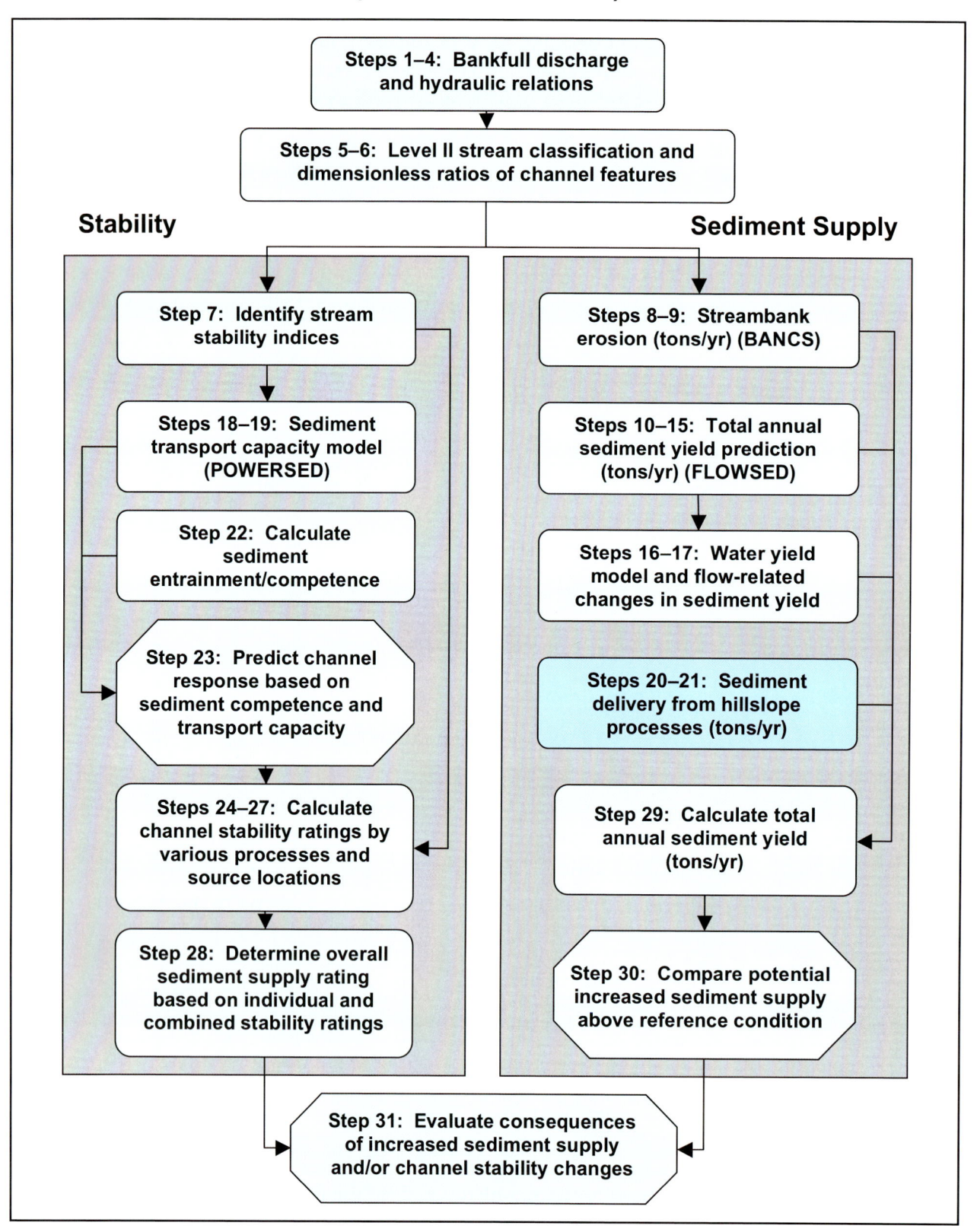

```
Steps 1–4: Bankfull discharge and hydraulic relations
        ↓
Steps 5–6: Level II stream classification and dimensionless ratios of channel features
```

Stability | **Sediment Supply**

Stability branch:
- Step 7: Identify stream stability indices
- Steps 18–19: Sediment transport capacity model (POWERSED)
- Step 22: Calculate sediment entrainment/competence
- Step 23: Predict channel response based on sediment competence and transport capacity
- Steps 24–27: Calculate channel stability ratings by various processes and source locations
- Step 28: Determine overall sediment supply rating based on individual and combined stability ratings

Sediment Supply branch:
- Steps 8–9: Streambank erosion (tons/yr) (BANCS)
- Steps 10–15: Total annual sediment yield prediction (tons/yr) (FLOWSED)
- Steps 16–17: Water yield model and flow-related changes in sediment yield
- **Steps 20–21: Sediment delivery from hillslope processes (tons/yr)**
- Step 29: Calculate total annual sediment yield (tons/yr)
- Step 30: Compare potential increased sediment supply above reference condition

Step 31: Evaluate consequences of increased sediment supply and/or channel stability changes

Flowchart 5-3i. The sequential steps to determine stream stability and sediment supply, highlighting **Steps 20–21**.

Chapter 5 — The Prediction Level Assessment (PLA)

Steps 20–21: Sediment delivery from hillslope processes (tons/yr)

Step 20: Determine potential sediment delivery from roads, surface erosion and mass erosion

- Sediment delivery from mass erosion → Soil mass movement (tons/yr) (Chapter V, USEPA, 1980) → Sediment delivery
- Sediment delivery from surface erosion → User's choice in model → e.g. WEPP / e.g. WRENSS procedure / e.g. RUSLE → Sediment delivery index (USEPA, 1980) (annual sediment yield in tons/yr)
- Sediment delivery from roads → Road Impact Index (RII) (Table 5-14, Figure 5-46) → Preliminary annual sediment yield (Worksheet 5-13) → Total annual sediment yield due to roads (tons/yr)
  - Mitigation adjustment (Table 5-14) → Preliminary annual sediment yield
  - Recovery adjustment (Figure 5-47) → Preliminary annual sediment yield

Step 21: Summarize total annual sediment yield (tons/yr) from hillslope processes

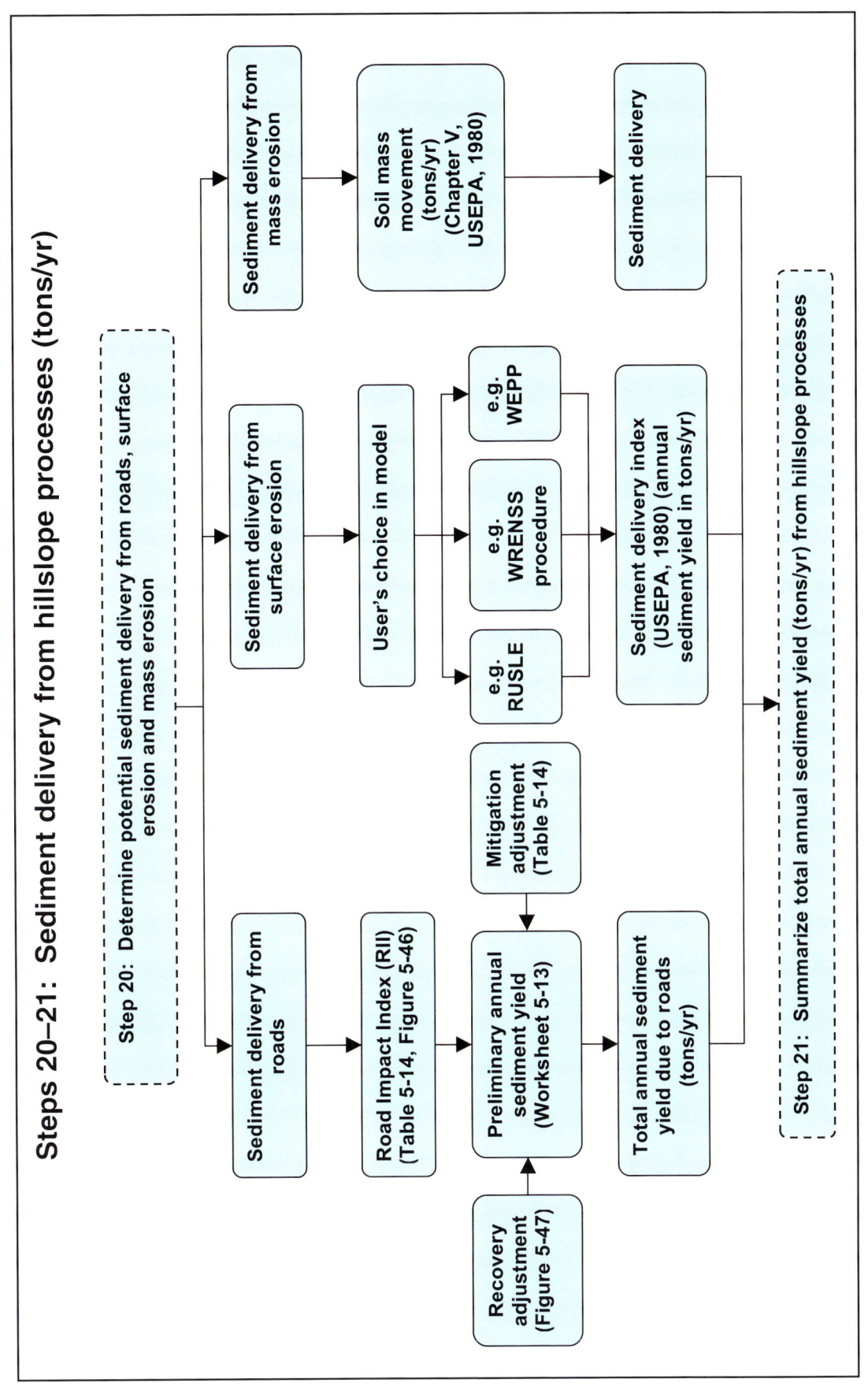

Flowchart 5-15. Procedural sequence of hillslope erosional processes contributing to total annual sediment yield.

5-113

Sediment delivery from hillslope processes: Steps 20–21

The output from the hillslope processes of roads, surface erosion and mass erosion will be used in several upcoming *PLA* steps. The tons/yr of potential sediment supply from hillslope processes will be used to compare total annual sediment yield and to determine changes from reference and baseline conditions (**Step 30**) and consequences of sediment yield (**Step 31**). The procedural sequence for sediment delivery involving roads, surface erosion and mass erosion is shown in **Flowchart 5-15**. The entrainment calculations in **Step 22** address sediment size or river competence. The POWERSED, Bagnold or similar bedload transport models simulate the consequences of increased coarse-grained sediment supply for annual load and potential stream channel aggradation (**Worksheet 5-12**).

Step 20. Determine potential sediment delivery from roads, surface erosion and mass erosion

Hillslope processes may contribute significant amounts of sediment to streams. Aggradation or excess deposition can occur based on the size and magnitude of introduced sediment. The following variables will be combined to assess the interrelated effects of these processes.

Sediment delivery from roads

Follow the procedure outlined in **Table 5-14**, **Figures 5-46** and **5-47** and **Worksheet 5-13** to determine potential sediment delivery from roads.

Note: The relations developed to predict potential sediment delivery from roads were derived from watersheds with a low mass-erosion hazard. If there appears to be a high risk of mass erosion, the user should follow the soil mass movement procedure outlined in WRENSS (USEPA, 1980, Chapter V).

The annual sediment yield from roads is entered and summarized in the *PLA* summary worksheet in **Steps 29–31**.

Chapter 5   *The Prediction Level Assessment (PLA)*

Table 5-14. Procedure to calculate Road Impact Index (RII) and corresponding annual sediment yield from roads (record data in **Worksheet 5-13**).

Annual sediment yield calculation based on Road Impact Index (RII)

1. Divide watershed into 200–5,000 acre sub-watersheds (Column (2) in **Worksheet 5-13**).
2. Locate roads on topographic map and/or aerial photos.
3. Calculate total acres of road (Column (3) in **Worksheet 5-13**).
4. Count number of stream crossings (Column (4) in **Worksheet 5-13**).
5. Calculate sediment delivered by Road Impact Index (RII) (Column (5) in **Worksheet 5-13**):

    $$RII = \frac{(acres\ of\ road)}{(acres\ of\ sub\text{-}watershed)} \times (number\ of\ stream\ crossings)^*$$

    *if no stream crossings, multiply by 1.

6. Determine dominant (most representative) slope position: lower, mid or upper 1/3 slope (Column (6) in **Worksheet 5-13**).
7. Obtain sediment yield (tons/acre of road) by using **Equations 5-2** or **5-3** and comparing with **Figure 5-46** (Column (7) in **Worksheet 5-13**).

    $y = 1.7 + 40x$ (lower 1/3 slope position)   **(Equation 5-2)**

    $y = -0.1595 + 3.0913x$ (mid or upper 1/3 slope position)   **(Equation 5-3)**

    where:
    y = Sediment yield (tons/acre of road)
    x = Road Impact Index

8. Obtain potential sediment yield, assuming road construction is new. Use the recovery potential or other adjustments, below, to reduce sediment yield based on road age and/or quality of construction.

9. Calculate recovery potential using Megahan's (1974) negative exponential recovery relation, **Equation 5-4**:

    $$\varepsilon_t = 3.52 + 523.9 e^{-0.00956 t}$$ **(Equation 5-4)**

    where:
    ε_t = Erosion rate recovery (tons/mi^2/day)
    e = Natural logarithm
    t = Elapsed time since disturbance (days)

    See **Figure 5-47**. Because this empirical relation calculates erosion rate recovery for an Idaho site, to use this equation for recovery, a conversion of the elapsed time versus erosion rate reduction needs to be converted to *percent reduction* (**Figure 5-47**). This percent reduction can then be applied to the initial sediment yield as calculated from the Road Impact Index (RII). Insert erosion rate recovery in Column (9) in **Worksheet 5-13**.

    For recovery rate for durations greater than 1,200 days, a reduction of 95% of the initial sediment delivery is recommended for use. Rate of recovery beyond 1,200 days is relatively low.

10. Mitigation adjustments. Use 95% reduction in delivered sediment from calculated rates if roads are surfaced, with stable ditch and/or out-sloped road grade (Column (11) in **Worksheet 5-13**).

    a. If road fills are less than 200 ft from stream, treat as surface erosion. Decrease sediment yields if mitigation by surfacing, out-sloping, etc.; use with sediment delivery ratio relations.
    b. If road is involved in mass erosion or debris torrent, use mass erosion procedure from WRENSS (USEPA, 1980, Chapter V).
    c. If a road encroaches on the stream and subsequently alters dimension, pattern and/or profile, use channel process analysis to quantify impacts.

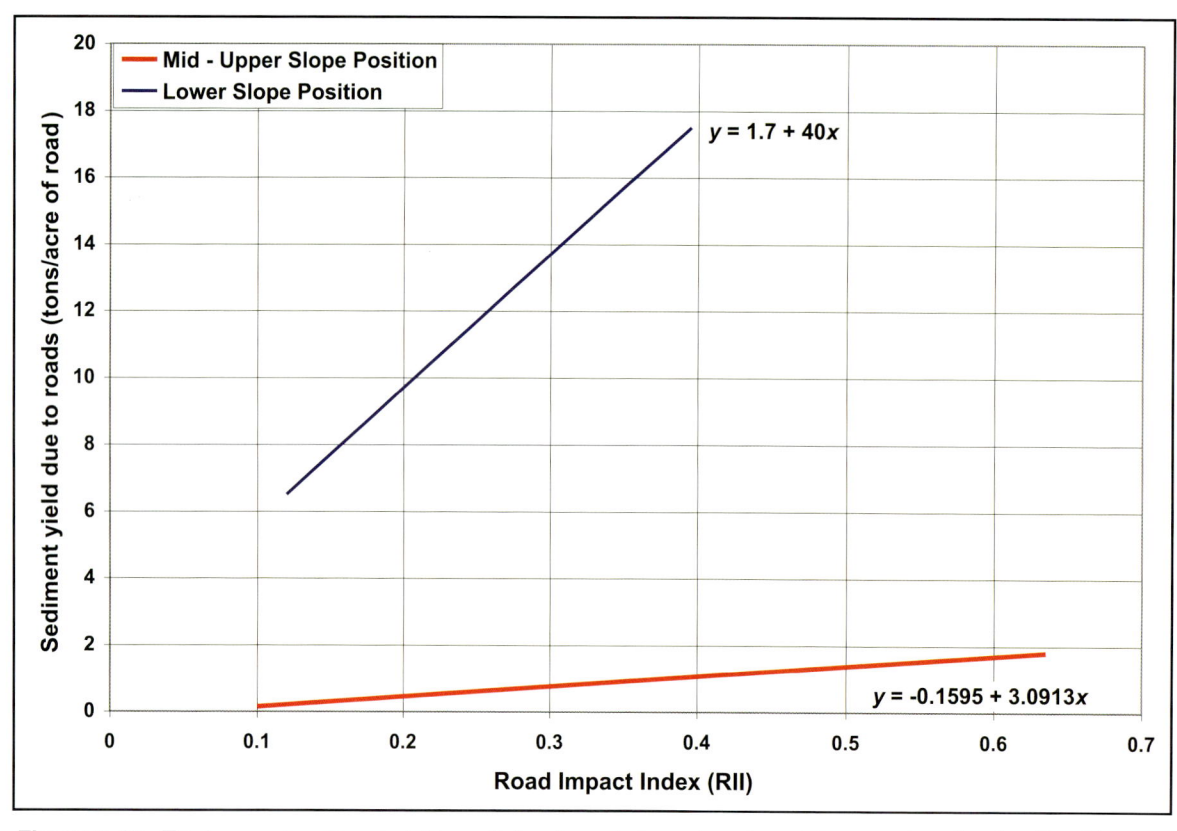

Figure 5-46. Total annual sediment yield prediction from delivered sediment from roads using the Road Impact Index (RII) model (data from Horse Creek experimental watershed, Idaho and Fool Creek experimental forest, Colorado, from annual debris basin accumulation related to road source sediment).

Lower Slope Position: $y = 1.7 + 40x$

Mid - Upper Slope Position: $y = -0.1595 + 3.0913x$

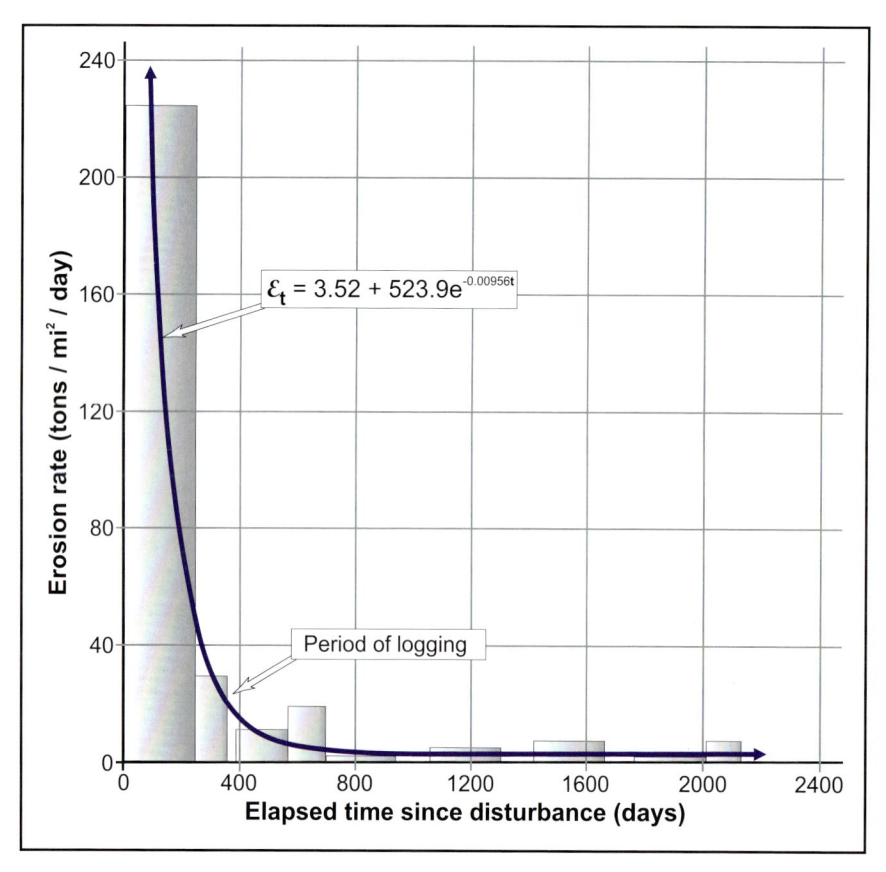

$$\varepsilon_t = 3.52 + 523.9e^{-0.00956t}$$

Figure 5-47. Erosion rate recovery over time (Megahan, 1974). Use the relations to calculate percent reduction from initial erosion based on elapsed time.

Chapter 5   The Prediction Level Assessment (PLA)

Worksheet 5-13. Road Impact Index (RII) and corresponding annual sediment yield from roads (use the steps in **Table 5-14**).

Stream: _____ Location: _____
Observers: _____ Date: _____

(1) Sub-watershed location ID#	(2) Total acres of sub-watershed (Step 1)	(3) Total acres of road (Step 3)	(4) Number of stream crossings (Step 4)	(5) Road impact index [(3)/(2)X(4)] (Step 5)	(6) Dominant slope position (lower, mid or upper slope) (Step 6)	(7) Sediment yield (tons/acre of road) (**Fig. 5-46**, Step 7)	(8) Total tons [(3)X(7)]	(9) Erosion rate recovery (% from **Fig. 5-47**, Step 9) (convert to decimal)	(10) Total tons/yr [(8) - (8)×(9)]	(11) Mitigation adjustments (Step 10)
1.										
2.										
3.										
4.										
5.										
6.										
7.										
8.										
9.										
10.										
11.										
12.										
13.										
14.										
15.										

Total road sediment yield (tons/year): _____

Sediment delivery from surface erosion

Every surface erosion model assumes that the site has the potential for overland flow. Prediction of surface erosion using the Revised Universal Soil Loss Equation (RUSLE) procedure, or the Watershed Erosion Prediction Project (WEPP) for forested watersheds, is recommended. The USDA manual of soil loss prediction uses RUSLE from the Agricultural Handbook No. 703 (Renard et al., 1997), which has been adapted in the RUSLE computer program. Because these procedures are well documented and readily available, the user is directed to these sources for the surface erosion rate computations.

Sediment delivery ratios, which convert surface erosion rates into potential sediment supply, can be determined using the procedure in WRENSS (USEPA, 1980) (**Figures 5-48, 5-49** and **5-50**, and **Table 5-15**) or adjusted by Williams (1975). WEPP (Laflen et al., 1997) is also used for roads and forested environments. The information from this portion of **Step 20** is used to determine potential disproportionate sediment sources from surface erosion processes. This helps in assigning mitigation for various erosional processes contributing to river instability and impairment. An example of the procedural steps to convert surface erosion per year to annual sediment yield is included in the sediment delivery index section of Chapter IV in WRENSS (USEPA, 1980), reprinted below. Note that the figure numbers in this section have changed from the original document.

A big challenge is to convert the soil erosion prediction to annual sediment yield in tons. The following procedure is recommended to convert erosion rate to sediment delivery. Once this analysis is completed, the sediment yield data from surface erosion is entered into the *PLA* summary worksheet in **Steps 29–31**.

The sediment delivery index (USEPA, 1980)

An index approach is recommended to help bridge the gap between the need to estimate how much sediment reaches a stream channel and the lack of a working sediment delivery model to provide such estimates. This approach provides a relative evaluation of seven generally accepted environmental factors and one site specific factor that are considered important in the sediment delivery process. These eight factors are not necessarily the only ones that may be needed in all situations. This indexing procedure has not been validated by research. Therefore, the computed quantities may be different from measured quantities of sediment delivered to a stream channel. Use of the index is only an aid in evaluating the relative effects of different management practices on sediment delivery from a given forest area.

Evaluation factors

For this discussion, each of the following eight factors is considered as though it acts independently of any other factor. In reality, these factors interact with each other in complex ways.

1. **Transport agent (e.g., water availability).** —Surface runoff from rainfall and snowmelt is an important factor in the movement of eroded material. It is estimated that overland flow rates from sheet and rill erosion rarely exceed 1 cfs on agricultural land and generally are less than 0.1 cfs on forest lands in the United States.

2. **Texture of eroded material.** —Assuming that aggregates do not form, individual particles of fine-textured soil material require less energy for delivery than particles of coarse-textured material. Sediment delivery efficiencies are higher on an area dominated by fine-textured material than on an area dominated by coarse-textured materials if the other factors influencing sediment delivery are equal.

3. **Ground cover.** —Ground cover (forest floor litter, vegetation, and rocks) creates a tortuous pathway for eroded particles to travel which allows time for eroded material to settle from surface runoff water (Tollner et al., 1976). Protective ground cover may also prevent raindrop impact energy from creating increased flow turbulence which would increase the carrying capacity of the runoff flow.

4. **Slope shape.** —Concave slopes between the source area and the stream channel promote deposition of the larger size fraction of the transported material (Neibling and Foster, 1977). Concave slopes create more favorable conditions for increasing the material carrying capacity of the transporting agent. Slope shape is a difficult factor to quantify, but it seems to play an important role in sediment delivery.

5. **Slope gradient.** —Slope gradient, along with the volume of water available for sediment delivery, provides the necessary energy to deliver the eroded material. The efficiency of the sediment delivery process increases with increasing slope gradient.

6. **Delivery distance.** — Increasing the distance from a sediment source to a stream channel or diversion ditch increases the effect that other factors have on the amount of sediment actually delivered. On the other hand, if a sediment source is very close to a stream channel, the other factors affecting sediment delivery have proportionally less opportunity to reduce the amount of sediment delivered.

7. **Surface roughness.** —Roughness of the soil surface affects sediment delivery similarly to that of ground cover. Rougher surfaces create more tortuous pathways for eroded particles to pass over and more surface area for water infiltration than smooth surfaces for a given area (Meeuwig, 1970).

8. **Site specific factors.** —In many parts of the United States, unique forest environments and/or soil factors influence the sediment delivery efficiency. For example, soil non-wettability (DeBano and Rice, 1975), mineralogy such as the Idaho batholith described by Megahan (1974), biological activity, or fire can change the sediment delivery efficiency of some forest lands. Within forested areas of the southeast United States, microrelief adjacent to stream channels may cause concentrated water flows, thus having a large effect on sediment delivery efficiency. Some soils have a greater tendency than others to form stable aggregates, hence reducing the sediment delivery efficiency.

Determining the sediment delivery index

The stiff diagram shown in **Figure 5-48** uses vectors to display the magnitude and scale of each major factor identified as influencing sediment delivery. The area of the polygon created by connecting the observed, anticipated, or measured value for each factor is determined and related to the total possible area (the polygon formed by connecting the outer limits of each vector) of the graph. The percentage of area inside the polygon is coupled to the delivery index through the use of skewed probit transformations (Bliss, 1935). Small polygonal areas surrounding the midpoint indicate a low probability of efficient sediment delivery, or, in other words, a very low sediment delivery index. Sediment delivery indexes will be low in most forest ecosystems managed by the best forest practices. Polygons approaching the outer limits of the stiff diagram indicate a high probability of efficient sediment delivery. The fraction of the total stiff diagram area formed by a given polygon is adjusted using **Figure 5-49**, to give the sediment delivery index.

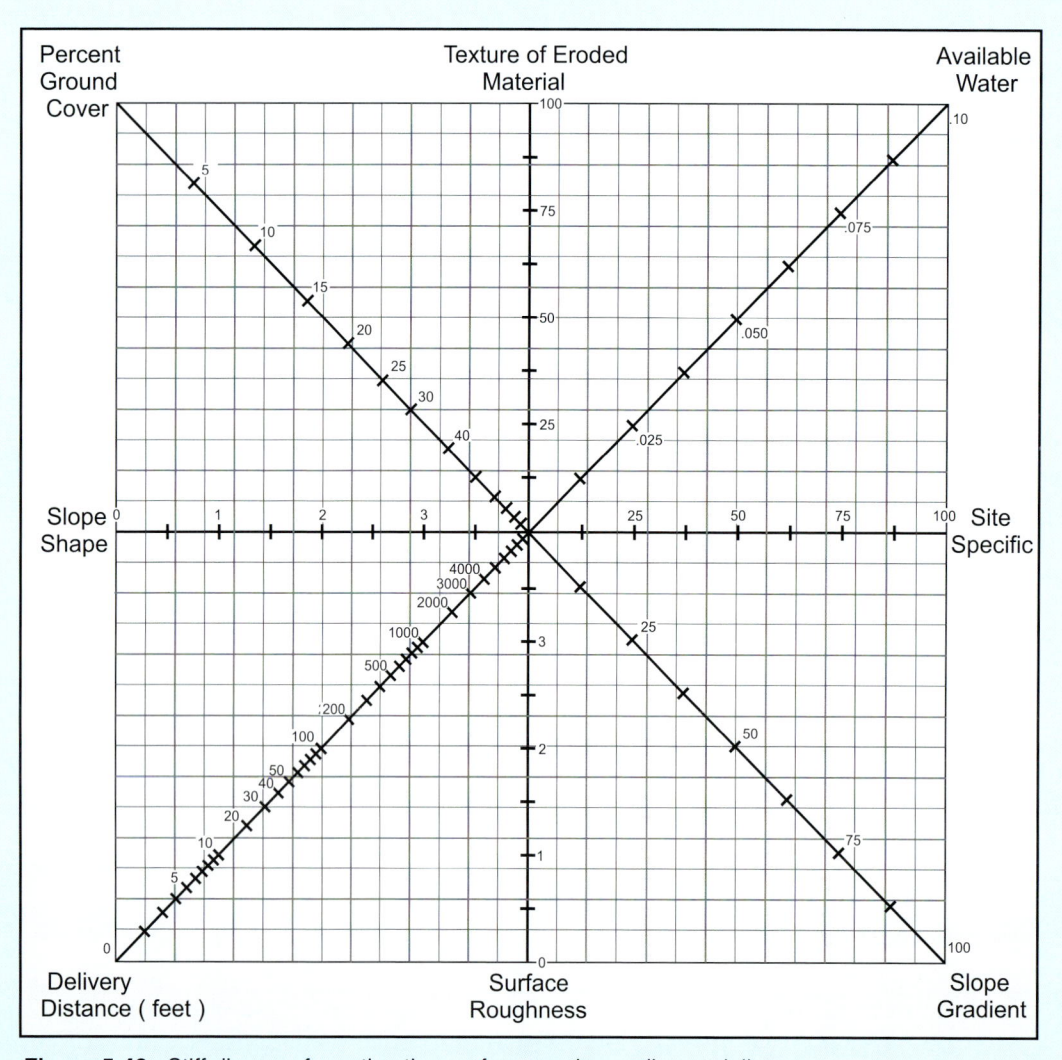

Figure 5-48. Stiff diagram for estimating surface erosion sediment delivery.

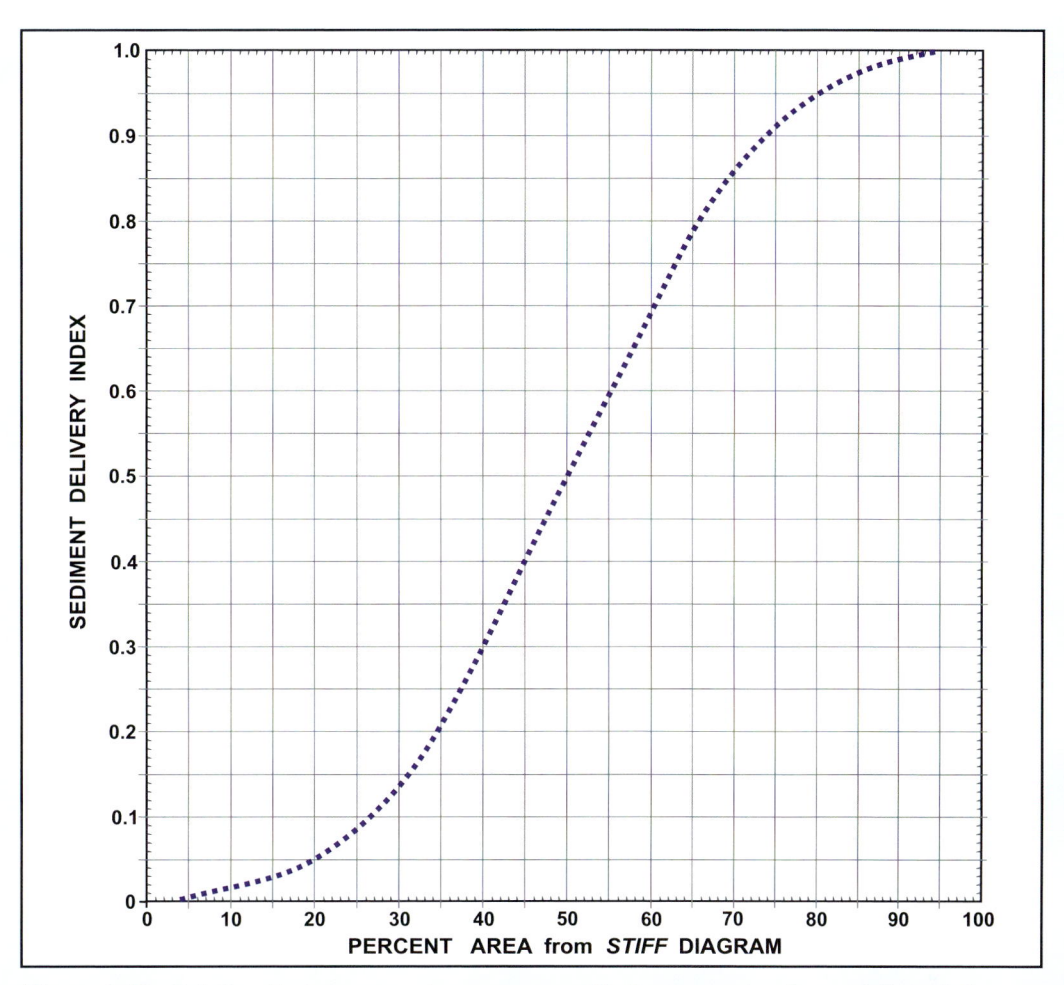

Figure 5-49. Relationship between polygon area on stiff diagram and sediment delivery index.

The scale and magnitude of the vectors in **Figure 5-48** have been defined as follows:

1. The magnitude of the transport agent is determined by the equation:

    F = CRL

    where:

    F = water availability,

    C = 2.31 x 10^{-5} ft^2 hr/in sec (a conversion constant),

    R = maximum anticipated precipitation and/or snowmelt rate minus infiltration in units of in/hr from local records, and

    L = slope length in feet of the sediment source area (perpendicular to contours).

Values of F for given values R and L are in **Table 5-15**.
The maximum scale value in **Figure 5-48** is 0.1 cfs. If the flow is calculated to exceed 0.1 cfs, use the scale factor of 0.1 for water availability. This model assumes that the precipitation input exceeds the site infiltration capacity causing overland flow conditions at the lower boundary of the eroded material source area. If no water is available, then the sediment delivery index is zero (0.0).

Table 5-15. Water availability values for given source area slope length (ft) and runoff (in/hr)[1].

Surface slope length	\multicolumn{17}{c}{Runoff}																
	.025	.05	.75	1.0	1.25	1.5	1.75	2.0	2.25	2.5	2.75	3.0	3.25	3.5	3.75	4.0
10	.00006	.00012	.00017	.00023	.00029	.00035	.0004	.00046	.00052	.00058	.00064	.00069	.00075	.00081	.00087	.00092
20	.00012	.00023	.00035	.00046	.00058	.00069	.00081	.00092	.001	.0012	.0013	.0014	.0015	.0016	.0017	.0018
30	.00017	.00035	.00052	.00069	.00087	.001	.0012	.0014	.0016	.0017	.0019	.0021	.0023	.0024	.0026	.0028
40	.00023	.00046	.00069	.00092	.0012	.0014	.0016	.0018	.0021	.0023	.0025	.0028	.003	.0032	.0035	.0037
50	.00029	.00058	.00087	.0012	.0014	.0017	.002	.0023	.0026	.0029	.0032	.0035	.0038	.004	.0043	.0046
75	.00043	.00087	.0013	.0017	.0022	.0026	.003	.0035	.0039	.0043	.0048	.0052	.0056	.0061	.0065	.0069
100	.00058	.0012	.0017	.0023	.0029	.0035	.004	.0046	.0052	.0058	.0064	.0069	.0075	.0081	.0087	.0092
150	.00087	.0017	.0026	.0035	.0043	.0052	.0061	.0069	.0078	.0087	.0095	.01	.011	.012	.013	.014
200	.0012	.0023	.0035	.0046	.0058	.0069	.0081	.0092	.01	.012	.013	.014	.015	.016	.017	.018
250	.0014	.0029	.0043	.0058	.0072	.0087	.01	.012	.013	.014	.016	.017	.019	.02	.022	.023
300	.0017	.0035	.0052	.0069	.0087	.01	.012	.014	.016	.017	.019	.021	.023	.024	.026	.028
350	.002	.004	.0061	.0081	.01	.012	.014	.016	.018	.02	.022	.024	.026	.028	.03	.032
400	.0023	.0046	.0069	.0092	.012	.014	.016	.018	.021	.023	.025	.028	.03	.032	.035	.037
450	.0026	.0052	.0078	.01	.013	.016	.018	.021	.023	.026	.029	.031	.034	.036	.039	.042
500	.0029	.0058	.0087	.012	.014	.017	.02	.023	.026	.029	.032	.035	.038	.04	.043	.046
1000	.0058	.012	.017	.023	.029	.035	.04	.046	.052	.058	.064	.069	.075	.081	.087	.092

[1] The table values were obtained by the formula: $F = (2.31 \times 10^{-5} \text{ ft}^2 \text{ hr/in sec})(\text{Runoff in/hr})(\text{slope length ft.})$

2. Texture of eroded material is expressed as percent of eroded material that is finer than 0.05 mm (silt size). A particle diameter less than 0.05 mm was shown to be highly transportable for sediment movement (Neibling and Foster, 1977). A scale factor of zero indicates that the eroded material contains no material less than 0.05 mm diameter, and a factor of 100 percent indicates that all of the eroded material is 0.05 mm or less in diameter.

3. Ground cover that is in actual contact with the soil surface, is expressed in percent cover between 0 (bare soil surface) and 100 (mineral soil surface completely covered). This factor is scaled based on unpublished data by Dissmeyer, which relates relative ground cover density influence to overland water flow.

4. Slope shape is scaled in magnitude between 0 and 4, with 4 being a slope that is convex from the boundary of the source area to the stream channel. A scale factor of 0 describes a slope concave from the boundary of the source area to the stream channel, while a factor of 2 shows that one-half of the slope is concave and the other half is convex or that the entire slope is uniformly straight. A factor of 3 indicates that a larger percentage of the slope is convex in shape.

5. The slope gradient is the vertical elevation difference between the lower boundary of the source area and the stream channel divided by the horizontal distance and expressed as a percent between 0 and 100.

6. The distance factor is the $\log_{10}$ of the distance in feet from the boundary of the source area to a stream channel or ditch. Distances greater than 10,000 feet (3,050 m) are considered infinite. The distance vector is marked using a $\log_{10}$ scale so that distances are entered directly onto the vector in **Figure 5-48**.

7. The roughness factor is scaled in magnitude between 0 and 4 with 0 being an extremely smooth forest floor surface condition and 4 being a very rough surface. This is a subjective evaluation of soil surface conditions.

8. The site specific factor influencing delivery ratios is scaled between 0 and 100 and must be **assigned** its effective magnitude by a user familiar with the unique condition of the site.

Appropriate factor values are plotted on each vector of the graphic sediment delivery model (**Figure 5-50**). Lines are drawn to connect all plotted points to form an enclosed, irregular polygon. If a site specific factor is not used, draw a line directly between plotted points on the slope gradient and available water vectors. Determine the area inside the polygon by: measuring with a planimeter, estimating with a dot grid, or calculating and summing the areas of the individual triangles. Determine the percent of the total graph area that is within the polygon. Using the S-shaped probit curve in **Figure 5-49**, determine the sediment delivery index by using the percent area of the polygon from **Figure 5-50**.

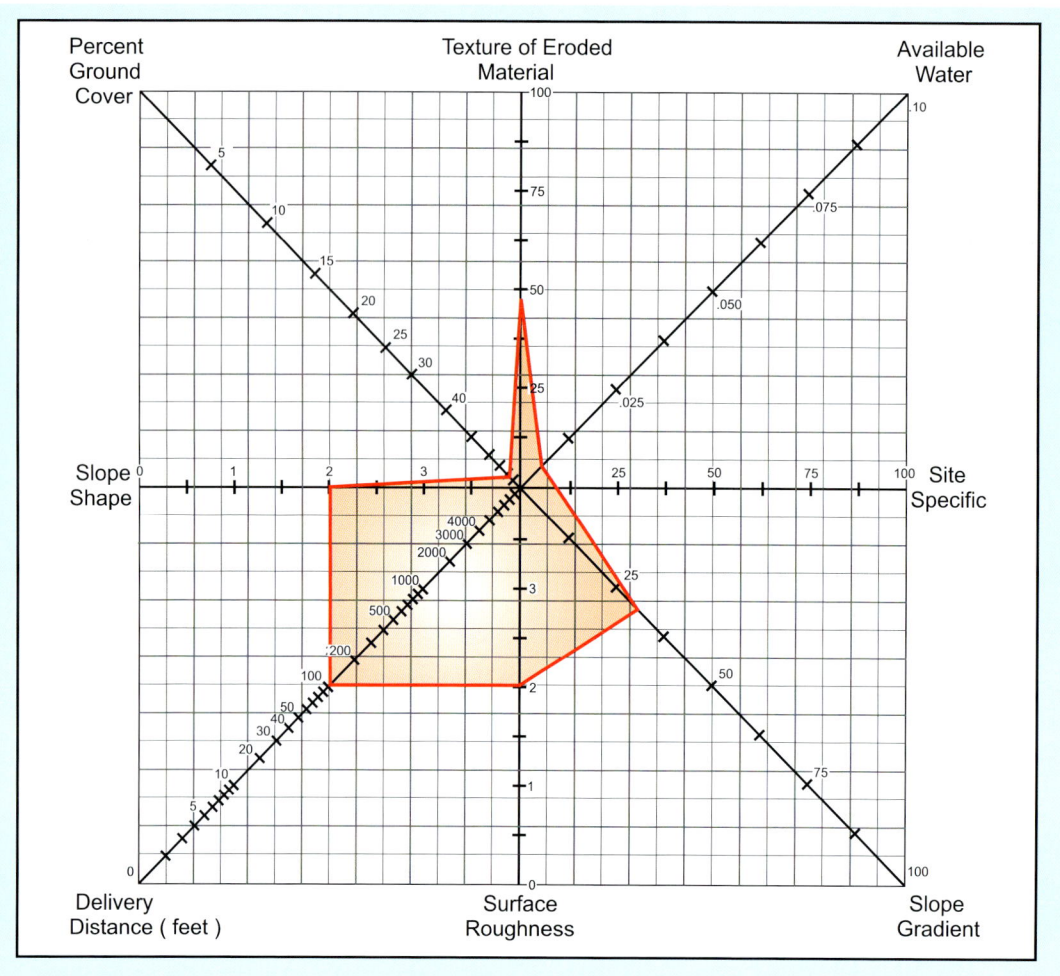

Figure 5-50. Example of graphic sediment delivery model for road R3.1.

Chapter 5 The Prediction Level Assessment (PLA)

Sediment delivery from mass erosion

Follow the soil mass movement procedure in Chapter V of WRENSS (USEPA, 1980) to determine the potential sediment delivery from mass erosion. Useful information for mass erosion assessments includes:

- Soil type maps;
- Topographic maps;
- Aerial photography, during runoff periods if possible
  - current photographs
  - historical photographs
- Geologic maps;
- Geologic hazard maps; and
- Slope shape and gradient.

The delivery to convert erosion to sediment yield from WRENSS (USEPA, 1980) is shown for debris flows in **Figure 5-51** and for slump/earthflow processes in **Figure 5-52**.

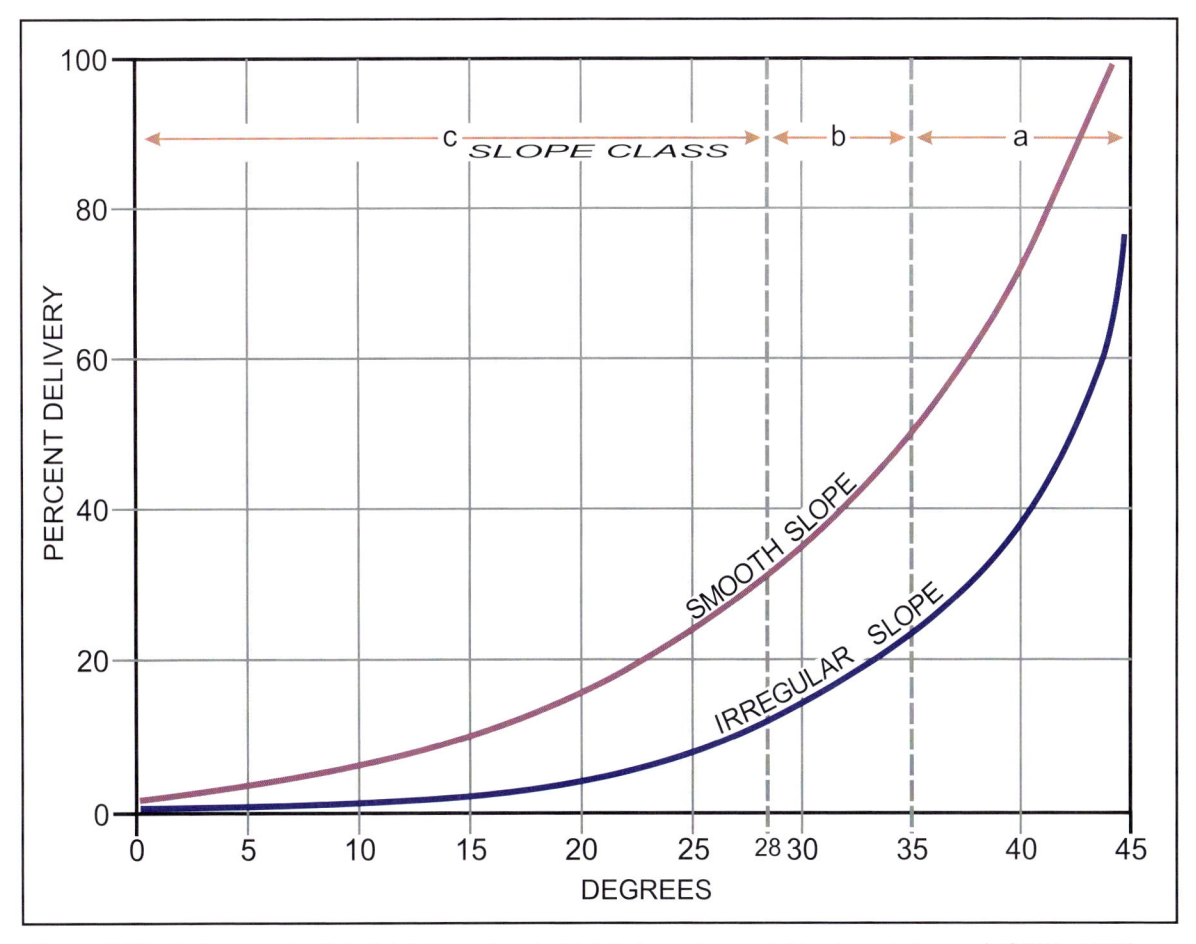

Figure 5-51. Delivery potential of debris avalanche/debris torrent material to closest stream (USEPA, 1980).

5-125

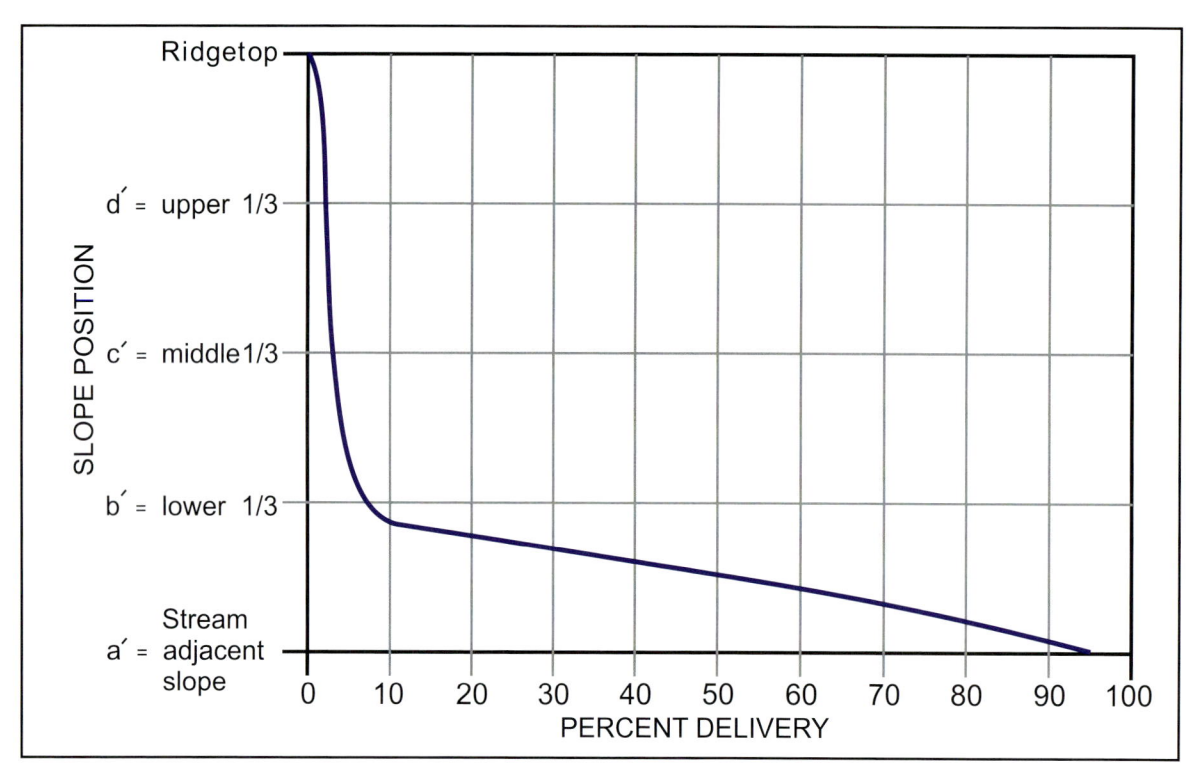

Figure 5-52. Delivery potential of slump/earthflow material to closest stream (USEPA, 1980).

The output of the mass erosion prediction converted to an annual sediment yield is entered in the *PLA* summary worksheet in **Steps 29–31**. This information is used to determine potential disproportionate sediment sources from mass erosion processes.

Step 21. Summarize total annual sediment yield (tons/year) from hillslope processes

This step summarizes the sediment load and sizes for all sources of introduced sediment. Sediment delivery relations are used to assess the conversion of mass soil erosion to sediment supply in **Step 20**. The overall annual sediment sources for hillslope processes are recorded in the *PLA* summary worksheet in **Steps 29–31**. The nature of the analysis required to predict annual sediment yield from mass erosion allows the user to specify exact locations and processes that may be contributing to sediment stability problems. This information can be used to create and prioritize subsequent mitigation strategies that are potentially more cost-effective and efficient in reducing sediment source problems.

Chapter 5   The Prediction Level Assessment (PLA)

Step 22: Sediment competence prediction (See Flowchart 5-16)

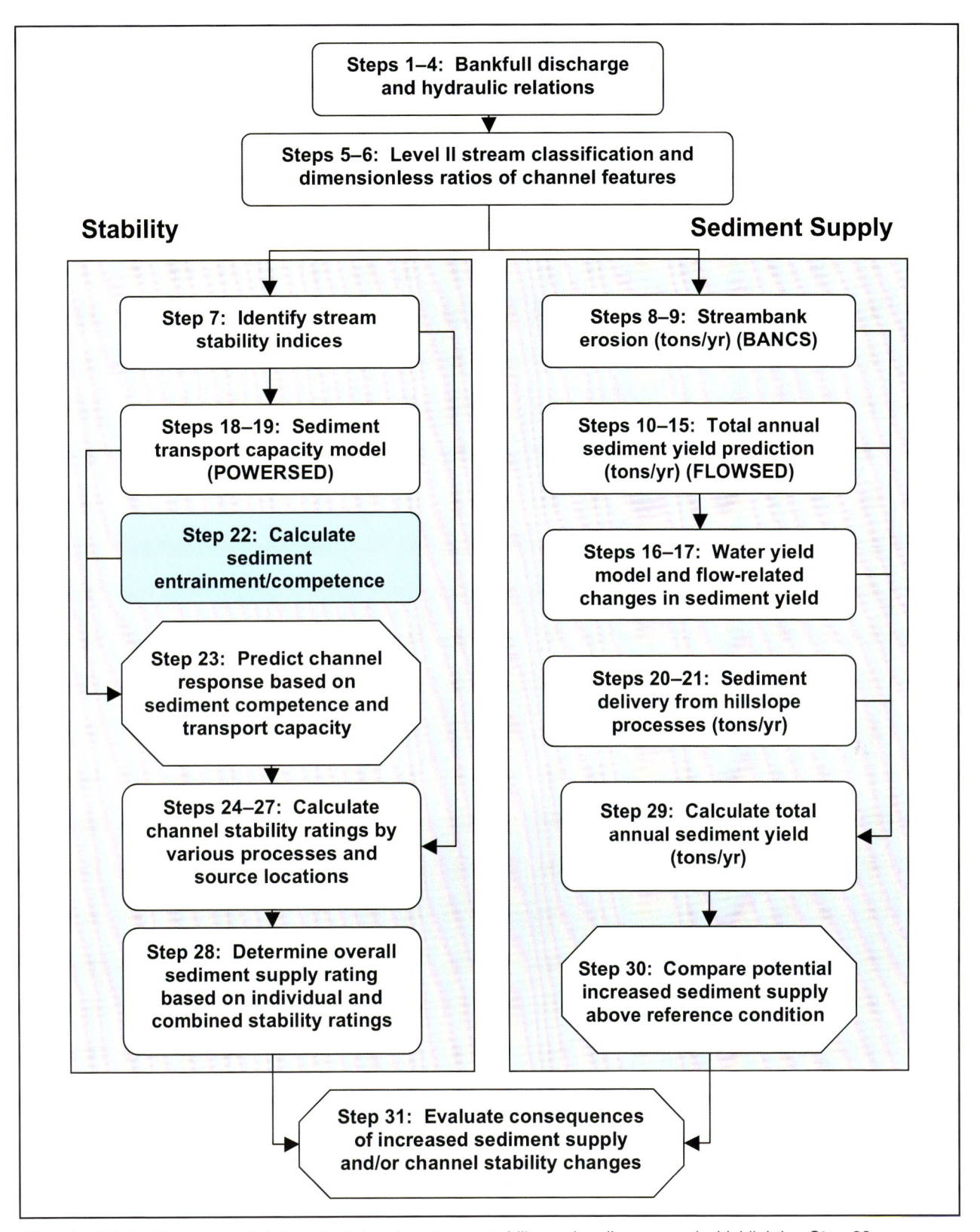

```
Steps 1–4: Bankfull discharge and hydraulic relations
            ↓
Steps 5–6: Level II stream classification and dimensionless ratios of channel features
```

Stability | **Sediment Supply**

Stability branch:
- Step 7: Identify stream stability indices
- Steps 18–19: Sediment transport capacity model (POWERSED)
- **Step 22: Calculate sediment entrainment/competence**
- Step 23: Predict channel response based on sediment competence and transport capacity
- Steps 24–27: Calculate channel stability ratings by various processes and source locations
- Step 28: Determine overall sediment supply rating based on individual and combined stability ratings

Sediment Supply branch:
- Steps 8–9: Streambank erosion (tons/yr) (BANCS)
- Steps 10–15: Total annual sediment yield prediction (tons/yr) (FLOWSED)
- Steps 16–17: Water yield model and flow-related changes in sediment yield
- Steps 20–21: Sediment delivery from hillslope processes (tons/yr)
- Step 29: Calculate total annual sediment yield (tons/yr)
- Step 30: Compare potential increased sediment supply above reference condition

Step 31: Evaluate consequences of increased sediment supply and/or channel stability changes

Flowchart 5-3j. The sequential steps to determine stream stability and sediment supply, highlighting **Step 22**.

5-127

WATERSHED ASSESSMENT OF RIVER STABILITY AND SEDIMENT SUPPLY

Step 22: Calculate sediment entrainment/competence

Collect field data:
- Bed material, riffle bed (D_{50})
- Bar samples (D_{max}, $D_{50}^{\wedge}$) (Figure 5-53)
- Average water surface slope (bankfull)
- Cross-section (mean bankfull depth)

↓

Obtain ratio of D_{max}/D_{50} (Worksheet 5-15)

Ratio outside range of 1.3–3.0

Calculate ratio $D_{50}/D_{50}^{\wedge}$

Ratio within range of 1.3–3.0

Calculate dimensionless shear stress:
$\tau^* = 0.0384 (D_{max}/D_{50})^{-0.887}$

Ratio outside range of 3.0–7.0

Calculate dimensional shear stress:
$\tau = \gamma RS$
(Figure 5-54, Worksheet 5-15)

Ratio within range of 3.0–7.0

Calculate dimensionless shear stress:
$\tau^* = 0.0834 (D_{50}/D_{50}^{\wedge})^{-0.872}$

Determine slope and depth requirements to transport D_{max}: $d = \dfrac{\tau}{\gamma S}$, $S = \dfrac{\tau}{\gamma d}$

Calculate the depth and slope necessary to transport D_{max} (Worksheet 5-15):
$d = \dfrac{\tau^* \gamma_s D_{max}}{S}$,
$S = \dfrac{\tau^* \gamma_s D_{max}}{d}$

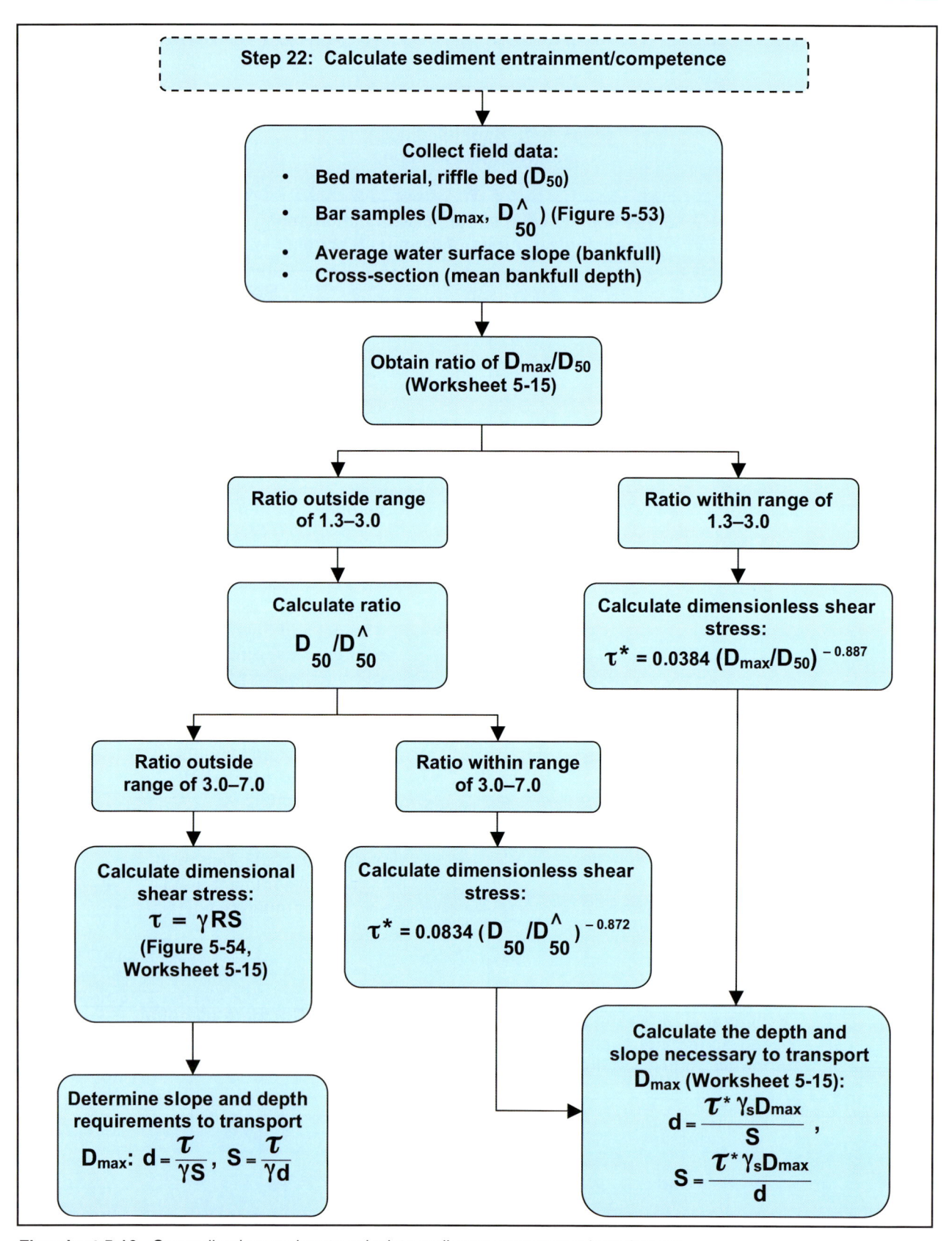

Flowchart 5-16. Generalized procedure to calculate sediment competence/entrainment.

Step 22. Calculate sediment entrainment/competence

The objective of this step is to determine the required depth and/or slope of a stream channel necessary to move the largest particle made available from its upstream reach at the bankfull stage. If the channel dimensions and slope are not sufficient to transport the largest size, aggradation potential exists. If the depth and/or slope create excess shear stress that potentially transports sizes greater than the D_{100} of the bed, then potential degradation is predicted. The method to evaluate sediment entrainment/competence is shown in **Flowchart 5-16**.

This procedure is appropriate for gravel-, cobble- and boulder-bed stream types, generally of a heterogeneous mixture. Sand-bed streams are not evaluated for competence, as the entire bed is assumed to be mobile at the bankfull discharge; i.e., the D_{100} of the bed material is entrained at flows of bankfull and less. Sediment transport capacity for sand, however, is predicted (**Steps 18** and **19**).

The calculations have been described in Chapter 2; however, field procedures for bar sampling, pavement/sub-pavement sampling and wet-sieving onsite are presented in **Tables 5-16** and **5-17**. The user is advised to review additional details of particle size sampling by Bunte and Abt (2001). Bar samples are field-sieved using the procedure shown in **Figure 5-53** and recorded in **Worksheet 5-14**.

Table 5-16. Field procedure for bar samples.

Bar sample field procedure

1. Collect sediment core samples from point bars along the project and reference reaches. At least one sample should be collected from each reach associated with a change in stream type. Conduct a bankfull shear stress analysis using the following procedures.

2. Locate a sampling point on the downstream one-third of a meander bend. The sample location on the point bar is halfway between the Thalweg elevation (the point of maximum depth) and the bankfull stage elevation. Scan the point bar in this area to determine the sampling location by observing the maximum particles on the surface of the bar.

3. Place a 5-gallon bottomless bucket at the sampling location over one of the representative larger particles that are observed on the lower one-third of the point bar. Remove the two largest particles from the surface covered by the bottomless bucket. Measure the intermediate axis for each particle and individually weigh the particles. Record these values. The largest particle obtained from the bar sample is D_{max}. Push the bottomless bucket into the bar material. Excavate the materials from the bottomless bucket to a depth that is equal to **twice** the intermediate axis width of the largest surface particle. Place these materials in a bucket or bag for sieving and weighing.

4. For fine bar materials: Follow the directions above, except that when the bottomless bucket is pushed into the bar material, excavate materials from the bucket to a depth of 4 to 6 inches. Place these materials in a bucket or bag for sieving and weighing.

5. Wet-sieve the collected bar materials using water and a standard sieve set with a 2-millimeter screen size for the bottom sieve. Weigh the bucket with sand after draining off as much water as possible. Subtract the tare weight of the bucket to obtain the net weight of the sand.

6. Weigh the sieved materials and record weights (less tare weight) by size class. Be sure to include the intermediate axis measurements and individual weights of the two largest particles that were collected.

7. Determine a material size-class distribution for all of the collected materials. The data represents the range of channel materials subject to movement or transport as bedload sediment materials at bankfull discharge.

8. Plot data; determine size-class indices, i.e., D_{16}, D_{35}, D_{50}, D_{84}, D_{95}. The D_{100} should represent the actual intermediate axis width and weight (not the tray size) when plotted. The largest size measured will be plotted at the D_{100} point. (Note: $D_{100} = D_{max}$). The intermediate axis measurement of the second largest particle will be the top end of the catch range for the last sieve that retains material (use the procedure in **Figure 5-53** and record data in **Worksheet 5-14**).

9. Survey a typical cross-section of a riffle reach at a location where the stream is free to adjust its boundaries. Plot the survey data. Determine the hydraulic radius of the cross-section.

10. Conduct a Wolman Pebble Count (100 count in riffle) of the bed material in the coarsest portion of the wetted riffle area (active channel). The pebble count should be conducted at multiple transects that represent the riffle. Plot data and determine the size-class indices.

Chapter 5 | The Prediction Level Assessment (PLA)

Table 5-17. Field procedure for pavement/sub-pavement samples.

Pavement/sub-pavement sample (Alternate procedures for obtaining a pavement/sub-pavement sample if you are unable to collect a bar sample)

1. Locate a sampling point in the same riffle where cross-section survey was conducted. The sampling point should be to the left or right of the Thalweg, not in the Thalweg, in a coarse-grained size portion of the riffle.

2. Push a 5-gallon bottomless bucket into the riffle at the sampling location to cut off the streamflow. The diameter of the bucket (sample size) should be at least twice the diameter of the largest rock on the bed of the riffle.

3. Remove the pavement material (surface layer only) by removing the smallest to the coarsest particles. Measure the intermediate axis and weight of the largest and second largest particles. Record these values. Place the remaining pavement materials into a bucket or bag for sieving and weighing.

4. Remove the sub-pavement material to a depth that is equal to twice the intermediate axis width of the largest particle in the pavement layer, or at least 150 mm depth. Caution: If a coarser bed material persists under the sub-pavement, it generally is material remnant of the previous bed. Stop at this condition and do not excavate deeper, even if the depth is not at twice the maximum pavement particle diameter. This residual layer is generally not associated with the size distribution of bedload transported at the bankfull stage. Collect the sub-pavement materials into a separate bucket or a bag. Measure the intermediate axis and weight of the two largest particles in the sub-pavement sample. Record these values. Sieve and weigh the remaining sub-pavement materials. The sub-pavement sample is the equivalent of the bar sample; therefore, use the largest particle from the sub-pavement sample in lieu of the largest particle from a bar sample in the entrainment calculations. Note: If the largest particle collected from the sub-pavement is larger than the pavement layer, the largest rock should be discarded from the sub-pavement layer. Drop back to the next largest particle size to determine the largest particle size to be used in the entrainment calculation.

5. Wet-sieve the collected pavement materials and then the sub-pavement materials, using water and a standard sieve set with a 2-millimeter screen size for the bottom sieve. Weigh the bucket with sand after draining off as much water as possible. Subtract the tare weight of the bucket to obtain the net weight of the sand.

6. Weigh the sieved materials and record weights (less tare weight) by size class for both the pavement and sub-pavement samples. Be sure to include the mean intermediate axis width and individual net weights of the two largest particles that were collected (**Worksheet 5-14**).

7. Determine a material size-class distribution for the materials. The sub-pavement data represents the range of channel materials subject to movement or transport as bedload sediment materials at bankfull discharge.

8. Plot data; determine size-class indices, i.e., D_{16}, D_{35}, D_{50}, D_{84}, D_{95}. The D_{100} should represent the actual intermediate axis width and weight (not the tray size) when plotted. The largest size measured will be plotted at the D_{100} point. (Note: $D_{100} = D_{max}$). The intermediate axis measurement of the second largest particle will be the top end of the catch range for the last sieve that retains material.

9. The pavement material size-class distribution may be used to determine the D_{50} of the riffle bed instead of doing the 100 count in the riffle bed.

10. Determine the average bankfull slope (approximated by the average water surface slope) for the study reach from the longitudinal profile.

11. Calculate the bankfull dimensionless shear stress required to mobilize and transport the largest particle from the bar sample (or sub-pavement sample). Use the equations and record the data in **Worksheet 5-15**.

Watershed Assessment of River Stability and Sediment Supply

BAR / BULK SAMPLE of *Representative Channel Materials* Subject to Movement as "Bedload". Samples collected at selected cross-section and bar locations within a reference reach.

1. Locate sampling point within the downstream 1/3 of the lateral or point-bar area and approx. 1/2 the distance (elev.) between the **Thalweg** and the **Bankfull** stage elevations.

2. For "**coarse material**" systems: Remove the 2 largest particles from "bottomless" bucket. Measure mean diameters and individually weigh particles.

3. Excavate materials from "bottomless bucket" to a depth equal to twice the diameter of the largest surface particle. Place materials in second bucket or bag for sieving / weighing.

4. For "**fine material**" systems: Remove the 2 largest particles and set aside. Excavate materials from bucket to a depth of 4 to 6 inches. Place materials in a second bucket or bag for sieving / weighing.

5. "Wet-sieve" the collected channel materials, with water and standard sieve set, using a 2-millimeter screen size for the bottom sieve. Materials 2mm and less will be collected in the sieving bucket and weighed after coarse material sieving is completed.

6. Weigh materials sieved and record weights (less tare wt.) by size class. *Include weights and mean diameters of the two largest particles collected.*

7. Determine a *material size-class distribution* for all of the collected materials. These data represent the range of channel materials subject to movement or transport as "bedload" sediment materials.

8. Plot data; determine size-class indices, i.e., D16, D35, D50, D84, etc.

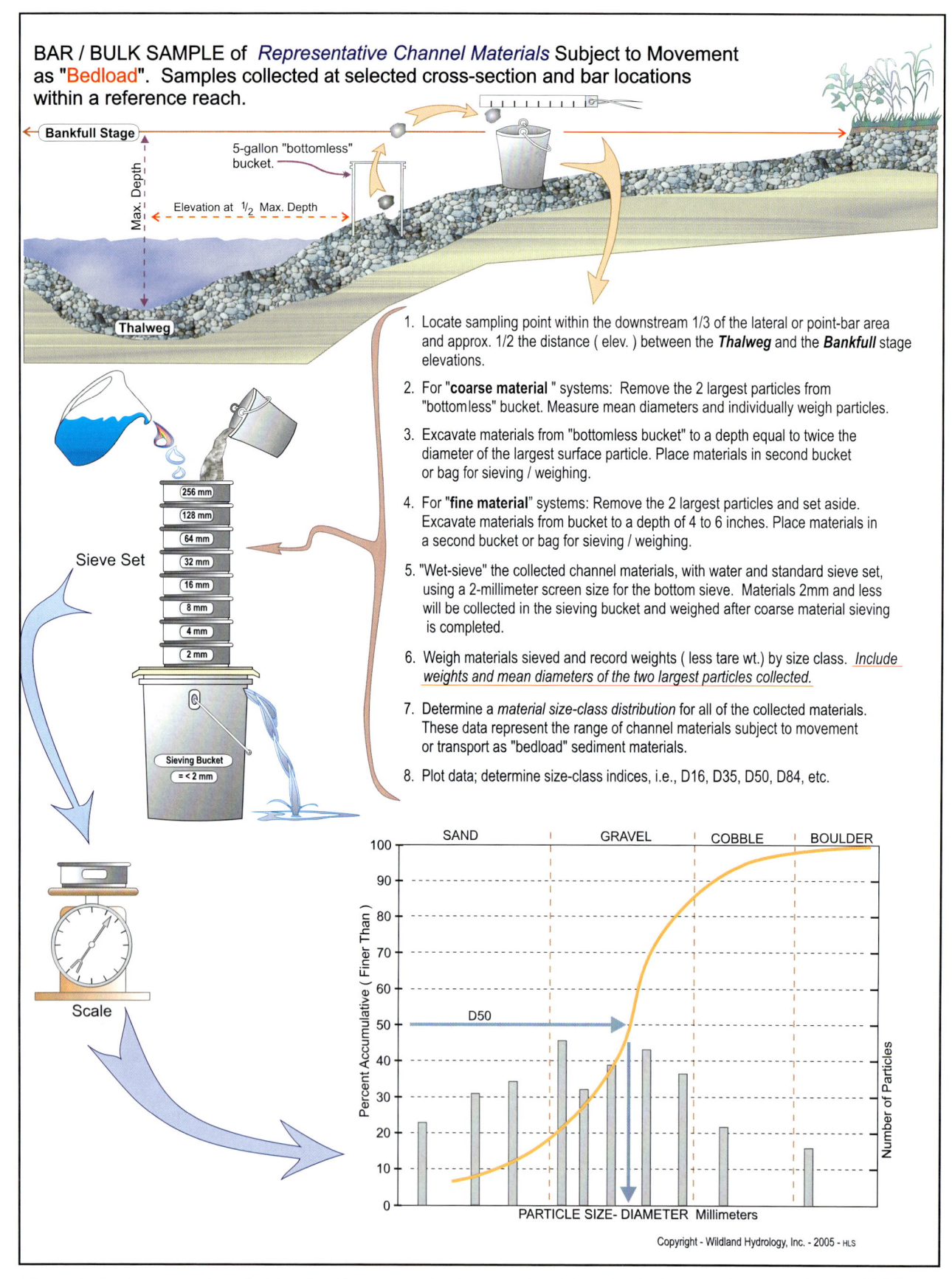

Copyright - Wildland Hydrology, Inc. - 2005 - HLS

Figure 5-53. Diagram of the field method for wet-sieving bar samples (drawn by H.S. Silvey in Leopold et al., 2000).

Chapter 5 — The Prediction Level Assessment (PLA)

Worksheet 5-14. Bar sample data collection and sieve analysis form.

WATERSHED ASSESSMENT OF RIVER STABILITY AND SEDIMENT SUPPLY

The sediment competence computations that determine bed stability (aggradation/degradation) are completed and summarized in **Worksheet 5-15**. This method has shown consistency when actual bedload/scour chain data is compared to predicted values. Use the value of the largest particle in the bar sample (or sub-pavement sample), D_{max} in millimeters, and the revised Shields Diagram to predict the shear stress required to initiate movement of the largest particle in the bar and/or sub-pavement (**Figure 5-54**).

If the protrusion ratios described in **Equations 2-7** or **2-8** are outside the ranges indicated in **Worksheet 5-15**, the user should use the dimensional shear stress equation **(Equation 2-2)** and apply it with a revised Shields relation using Colorado data or local data if available (**Figure 5-54**). A grain size corresponding with shear stress is selected to determine what sizes the river can potentially move. Based on measured bedload sizes, in a heterogeneous mixture of bed material comprised of a mixture of sand to gravel and cobble, the previously published Shields relation generally underestimates particle sizes of heterogeneous bed material in the shear stress range of 0.05 lbs/ft^2 to 1.5 lbs/ft^2. The Shields relation is appropriately used for entrainment sizes below or above this value range. Without this adjustment, most computations underestimate the largest sizes of heterogeneous bed material moved during bankfull discharge. The measured data in **Figure 5-54** indicates the magnitude of the underestimate of particle size entrainment from comparing published relations to measured values.

To determine the ability of the existing stream reach to transport the largest clast size of the bedload sediment, it is necessary to calculate the bankfull dimensionless shear stress (τ^*). This calculation determines the depth and slope necessary to mobilize and transport the largest particle made available to the channel. The dimensionless shear stress at bankfull stage is used in the entrainment analysis for both the reference reach and project reach. This analysis of the reference, stable condition is compared to the potentially disturbed reach. To maintain stability, a stream must be competent to transport the largest size of sediment and have the capacity to transport the load (volume) on an annual basis. These calculations provide a prediction of sediment competence.

Worksheet 5-15 is also used to determine an overall bed stability rating of stable, aggrading or degrading, which is recorded in the *PLA* summary worksheet in **Steps 29–31** to use in the final analysis.

Chapter 5 — The Prediction Level Assessment (PLA)

Worksheet 5-15. Sediment competence calculation form to assess bed stability.

Stream:		Stream Type:
Location:		Valley Type:
Observers:		Date:

Enter required information

	D_{50}	Riffle bed material D_{50} (mm)
	$\hat{D}_{50}$	Bar sample D_{50} (mm)
	D_{max}	Largest particle from bar sample (ft) (mm) 304.8 mm/ft
	S	Existing bankfull water surface slope (ft/ft)
	d	Existing bankfull mean depth (ft)
	γ_s	Submerged specific weight of sediment

Select the appropriate equation and calculate critical dimensionless shear stress

$D_{50}/\hat{D}_{50}$	Range: 3 – 7	Use EQUATION 1: $\tau^* = 0.0834\,(D_{50}/\hat{D}_{50})^{-0.872}$
D_{max}/D_{50}	Range: 1.3 – 3.0	Use EQUATION 2: $\tau^* = 0.0384\,(D_{max}/D_{50})^{-0.887}$
τ^*	Bankfull Dimensionless Shear Stress	EQUATION USED:

Calculate bankfull mean depth required for entrainment of largest particle in bar sample

d	Required bankfull mean depth (ft)	$d = \dfrac{\tau^* \gamma_s D_{max}}{S}$

Check ✓: ☐ Stable   ☐ Aggrading   ☐ Degrading

Calculate bankfull water surface slope required for entrainment of largest particle in bar sample

S	Required bankfull water surface slope (ft/ft)	$S = \dfrac{\tau^* \gamma_s D_{max}}{d}$

Check ✓: ☐ Stable   ☐ Aggrading   ☐ Degrading

Sediment competence using dimensional shear stress

	Bankfull shear stress $\tau = \gamma d S$ (lbs/ft²) (substitute hydraulic radius, R, with mean depth, d)
	Moveable particle size (mm) at bankfull shear stress **(Figure 5-54)**
	Predicted shear stress required to initiate movement of D_{max} (mm) **(Figure 5-54)**
	Predicted mean depth required to initiate movement of D_{max} (mm) $d = \dfrac{\tau}{\gamma S}$
	Predicted slope required to initiate movement of D_{max} (mm) $S = \dfrac{\tau}{\gamma d}$

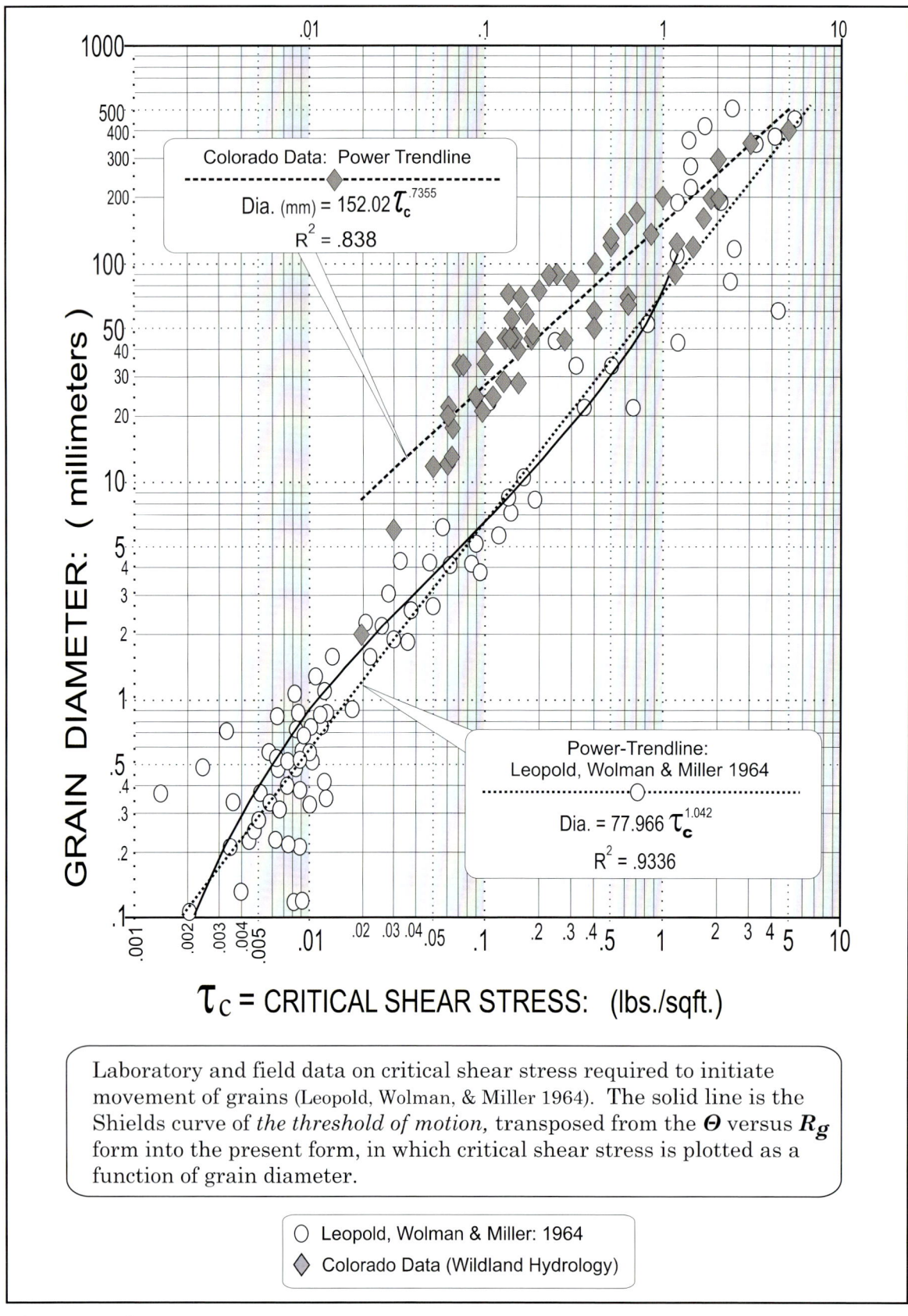

Figure 5-54. Critical shear stress required to initiate movement of bed-material grains following the Shields relation (Leopold et al., 1964), as modified by field data from Colorado (Rosgen and Silvey, 2005).

Colorado Data: Power Trendline
Dia. (mm) = $152.02 \tau_c^{.7355}$
$R^2 = .838$

Power-Trendline: Leopold, Wolman & Miller 1964
Dia. = $77.966 \tau_c^{1.042}$
$R^2 = .9336$

τ_c = CRITICAL SHEAR STRESS: (lbs./sqft.)

GRAIN DIAMETER: (millimeters)

Laboratory and field data on critical shear stress required to initiate movement of grains (Leopold, Wolman, & Miller 1964). The solid line is the Shields curve of *the threshold of motion,* transposed from the Θ versus R_g form into the present form, in which critical shear stress is plotted as a function of grain diameter.

○ Leopold, Wolman & Miller: 1964
◆ Colorado Data (Wildland Hydrology)

Chapter 5 — The Prediction Level Assessment (PLA)

Step 23: Channel response based on sediment competence and/or transport capacity

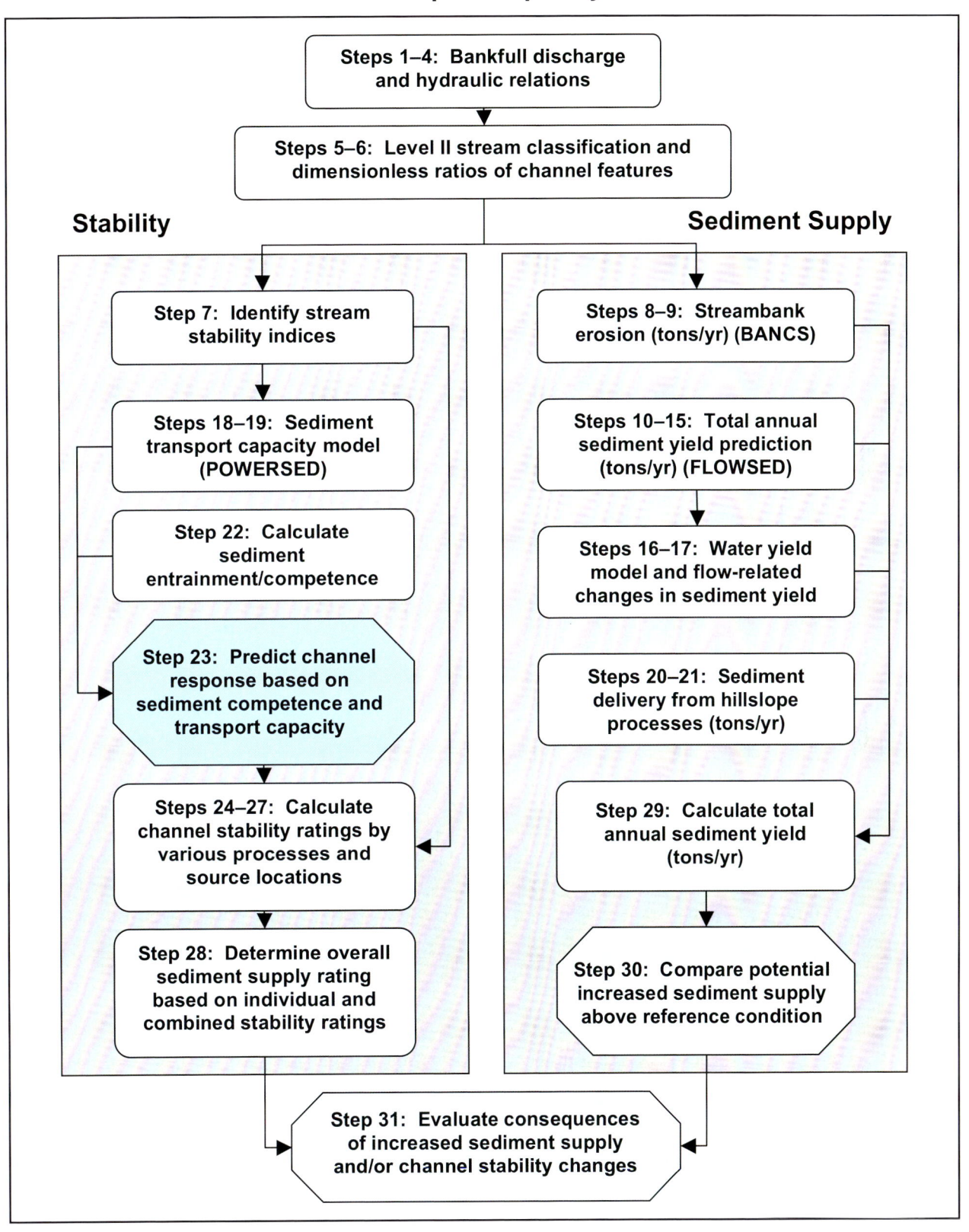

Stability

- Steps 1–4: Bankfull discharge and hydraulic relations
- Steps 5–6: Level II stream classification and dimensionless ratios of channel features
- Step 7: Identify stream stability indices
- Steps 18–19: Sediment transport capacity model (POWERSED)
- Step 22: Calculate sediment entrainment/competence
- **Step 23: Predict channel response based on sediment competence and transport capacity**
- Steps 24–27: Calculate channel stability ratings by various processes and source locations
- Step 28: Determine overall sediment supply rating based on individual and combined stability ratings

Sediment Supply

- Steps 8–9: Streambank erosion (tons/yr) (BANCS)
- Steps 10–15: Total annual sediment yield prediction (tons/yr) (FLOWSED)
- Steps 16–17: Water yield model and flow-related changes in sediment yield
- Steps 20–21: Sediment delivery from hillslope processes (tons/yr)
- Step 29: Calculate total annual sediment yield (tons/yr)
- Step 30: Compare potential increased sediment supply above reference condition

- Step 31: Evaluate consequences of increased sediment supply and/or channel stability changes

Flowchart 5-3k. The sequential steps to determine stream stability and sediment supply, highlighting **Step 23**.

5-137

Step 23. Predict channel response based on sediment competence and transport capacity

This step uses both capacity (**Steps 18** and **19**) and competence (**Step 22**) calculations to determine bed stability of the study reach by comparing the reference and/or upstream reach to the impaired reach. This is a critical step in the analytical process as it determines potential aggradation or degradation using both competence and capacity calculations. The procedure is similar to previous steps but instead uses the output of the changes in introduced sediment and changes in energy (flow and hydraulics due to channel change) for a given sediment supply.

Sediment transport capacity (Steps 18 and 19)

Sediment capacity evaluates the ability of the stream to transport the annual yield (tons/yr) for a given sediment supply and channel hydraulics. Changes in width/depth ratio and slope can change the ability of the stream to transport the annual sediment yield due to changes in velocity, shear stress and stream power (POWERSED, **Step 19**).

Even if the largest size can be moved at the bankfull stage for a given depth and slope, the sediment transport capacity (load or total volume) must also be calculated. Aggradation has occurred in rivers that have had sufficient competence to move the largest particle but have insufficient capacity to move the load. Deposition of sand and small gravel can occur even though the competence may indicate potential to move large cobble (Rosgen, 2006).

The prediction of sediment transport capacity involves models, such as FLOWSED, POWERSED, Bagnold, etc., for both bedload and suspended sand sediment.

Sediment competence (Step 22)

The sediment competence calculation involves the ability of the stream to entrain transport and/or initiate motion of channel bed-material particles for a range of flows primarily at the bankfull stage.

The bed-material size required to be transported associated with the competence calculation is the largest size made available as bedload at the bankfull stage. This size is obtained by measuring the largest grain size contained in the lower third of a point bar or in the sub-pavement region of the bed. The required depth and/or slope at the bankfull stage necessary to transport the largest size, previously described, is predicted to determine channel stability (**Worksheet 5-15**). A comparison between sediment competence for the reference and the potentially impaired condition is made in this step.

Channel Response Summary

Analyze the stability evaluation made for the impaired reach in **Steps 18** and **19** along with the stability evaluations made in **Step 22** for sediment competence to predict if the channel response will be stable or unstable, indicating aggradation or degradation. The predicted channel response is transferred to the *PLA* summary worksheet in **Steps 29–31** to use for the final summary analysis. This step is also used for restoration to help design the dimension, pattern and profile of the river to have the competence and capacity to transport the sediment made available by the watershed.

Chapter 5   The Prediction Level Assessment (PLA)

Steps 24–27: Channel stability ratings by various processes and source locations (See Flowchart 5-17)

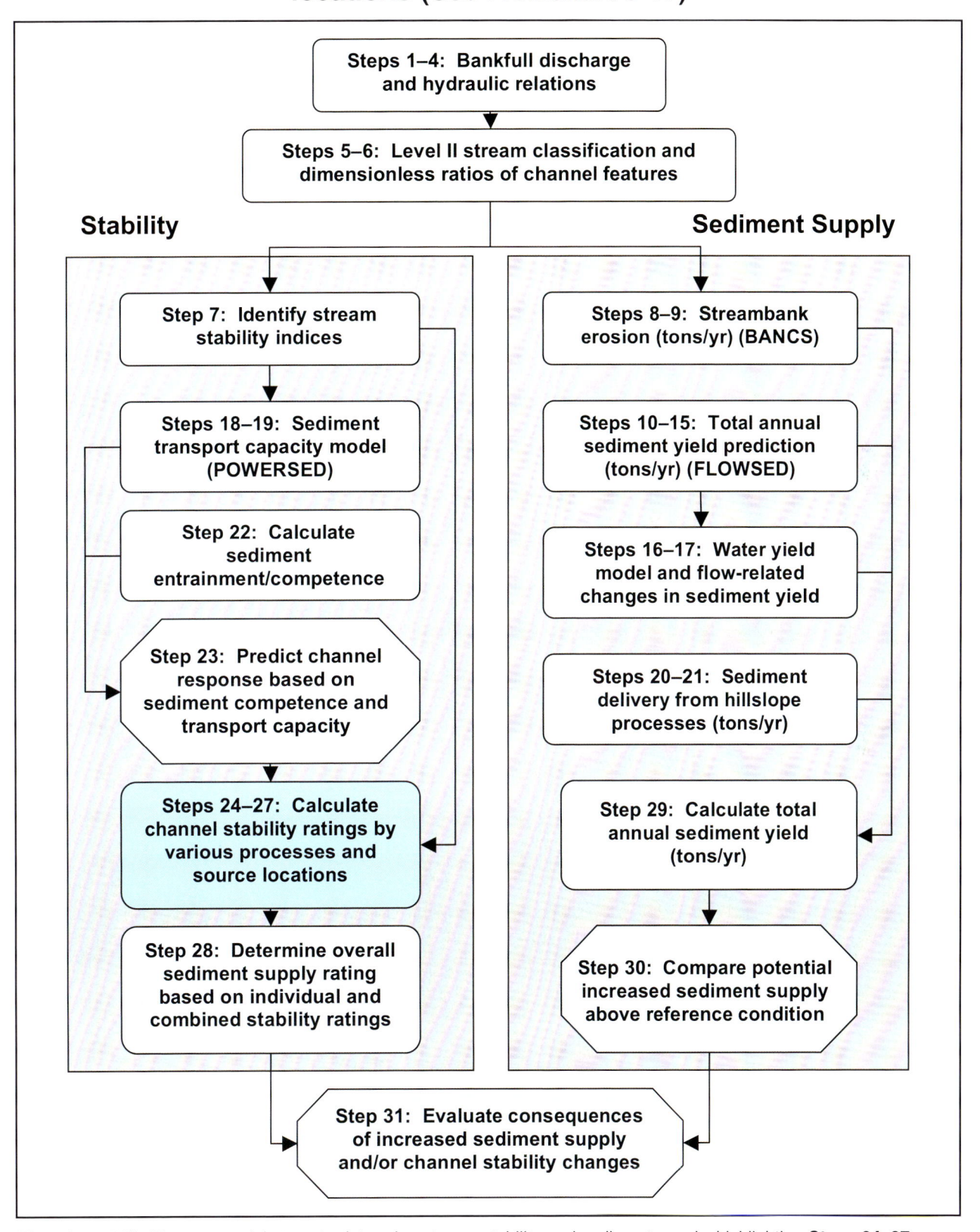

```
                  Steps 1–4:  Bankfull discharge
                        and hydraulic relations
                                  │
                                  ▼
              Steps 5–6:  Level II stream classification and
                  dimensionless ratios of channel features
```

Stability   **Sediment Supply**

- Step 7: Identify stream stability indices
- Steps 18–19: Sediment transport capacity model (POWERSED)
- Step 22: Calculate sediment entrainment/competence
- Step 23: Predict channel response based on sediment competence and transport capacity
- **Steps 24–27: Calculate channel stability ratings by various processes and source locations**
- Step 28: Determine overall sediment supply rating based on individual and combined stability ratings

- Steps 8–9: Streambank erosion (tons/yr) (BANCS)
- Steps 10–15: Total annual sediment yield prediction (tons/yr) (FLOWSED)
- Steps 16–17: Water yield model and flow-related changes in sediment yield
- Steps 20–21: Sediment delivery from hillslope processes (tons/yr)
- Step 29: Calculate total annual sediment yield (tons/yr)
- Step 30: Compare potential increased sediment supply above reference condition

- Step 31: Evaluate consequences of increased sediment supply and/or channel stability changes

Flowchart 5-3l. The sequential steps to determine stream stability and sediment supply, highlighting **Steps 24–27**.

5-139

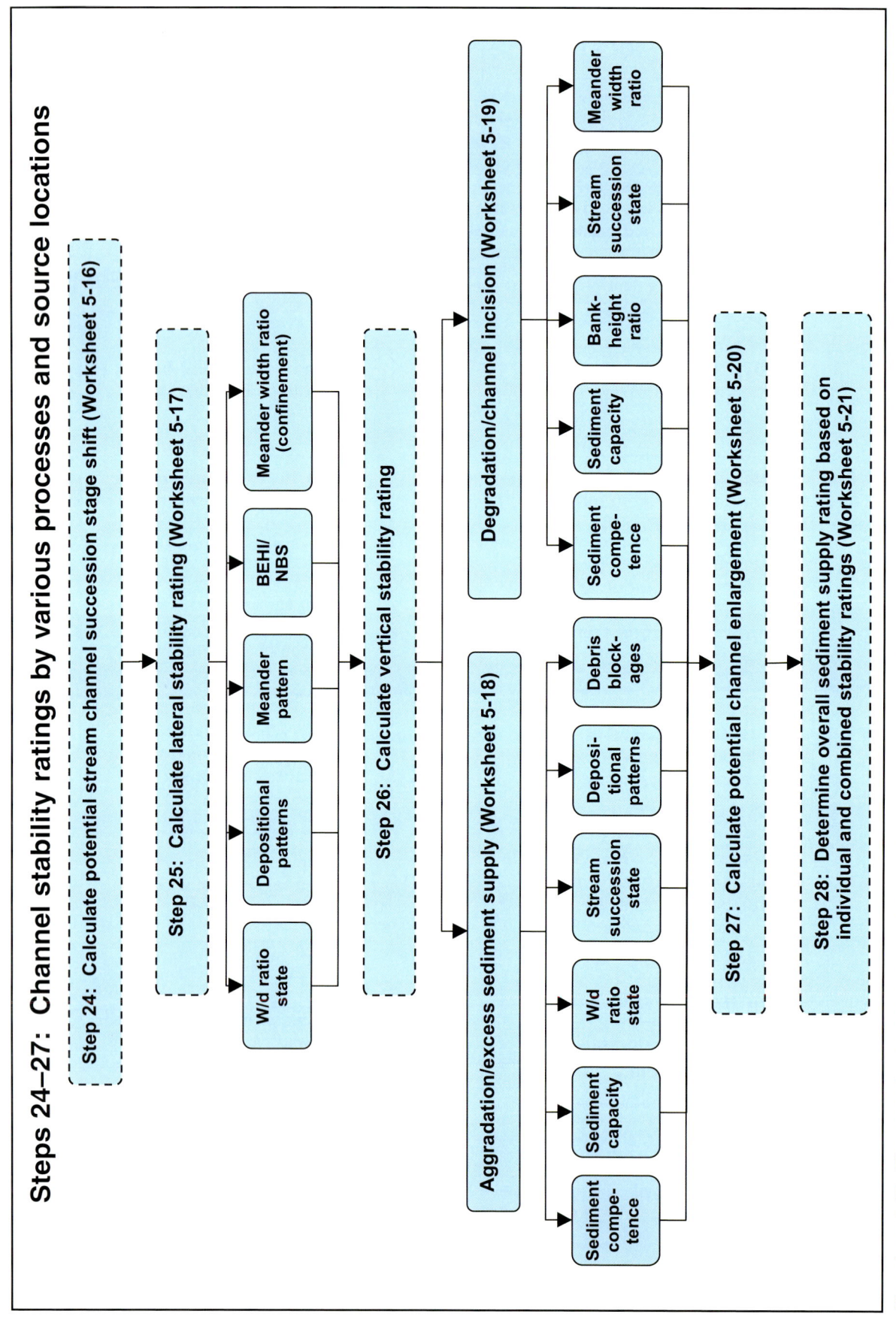

WATERSHED ASSESSMENT OF RIVER STABILITY AND SEDIMENT SUPPLY

Flowchart 5-17. The procedural steps and variables used in the assessment of channel stability.

5-140

Channel stability ratings by various processes and source locations: Steps 24–27

Steps 24–27 assess various categories and processes related to channel stability by integrating multiple prediction methods. The various stability indices previously evaluated for individual reaches in **Step 7** (**Worksheet 5-5**) and the stability predictions in previous steps are evaluated and used to identify specific processes, locations and the nature of potential instability.

Overall channel stability predictions and interpretations of aggradation, degradation, enlargement, channel succession stage shifts and sediment supply are documented for river reach locations within sub-watersheds. The various category and process relations are not only important as evidence for predicting stability change but are also important as documentation of the variables that may need to be addressed for mitigation purposes to reduce sediment supply and/or river impairment. These variables and processes are diagrammed in **Flowchart 5-17**. **Steps 24–27** calculate individual stability ratings by reach location and are then used to determine the overall sediment supply rating in **Step 28**.

Step 24. Calculate potential stream channel successional stage shift

Increased sediment size, supply, aggradation, degradation, vertical and/or lateral stability and channel enlargement can lead to long-term significant channel adjustments, changing stream type and related beneficial uses. **Worksheet 5-16** is used to categorize the existing stream type's stability using successional stage shift categories. The various successional scenarios are depicted in **Figure 5-55**. For example, a potential shift from stream type C to G would create long-term serious adverse effects on sediment supply and beneficial uses. The user is advised to review the discussion of the physical and biological consequences of stream channel succession in Chapter 2. Certain ratings, however, could indicate a positive response or a direction toward natural recovery or stabilization.

The stream channel successional stage shift stability ratings are used in **Worksheets 5-18** and **5-19** to predict vertical stability and in **Worksheet 5-20** to determine potential channel enlargement.

Worksheet 5-16. Stability ratings for corresponding successional stage shifts of stream types. Check (✓) the appropriate stability rating.

Stream:	Stream Type:
Location:	Valley Type:
Observers:	Date:

Stream type changes due to successional stage shifts (Figure 5-55)	Stability rating (check (✓) appropriate rating)
Stream type at potential, (C→E), (F$_b$→B), (G→B), (F→B$_c$), (F→C), (D→C)	☐ Stable
(E→C)	☐ Moderately unstable
(G→F), (F→D)	☐ Unstable
(C→D), (B→G), (D→G), (C→G), (E→G)	☐ Highly unstable

Chapter 5 — The Prediction Level Assessment (PLA)

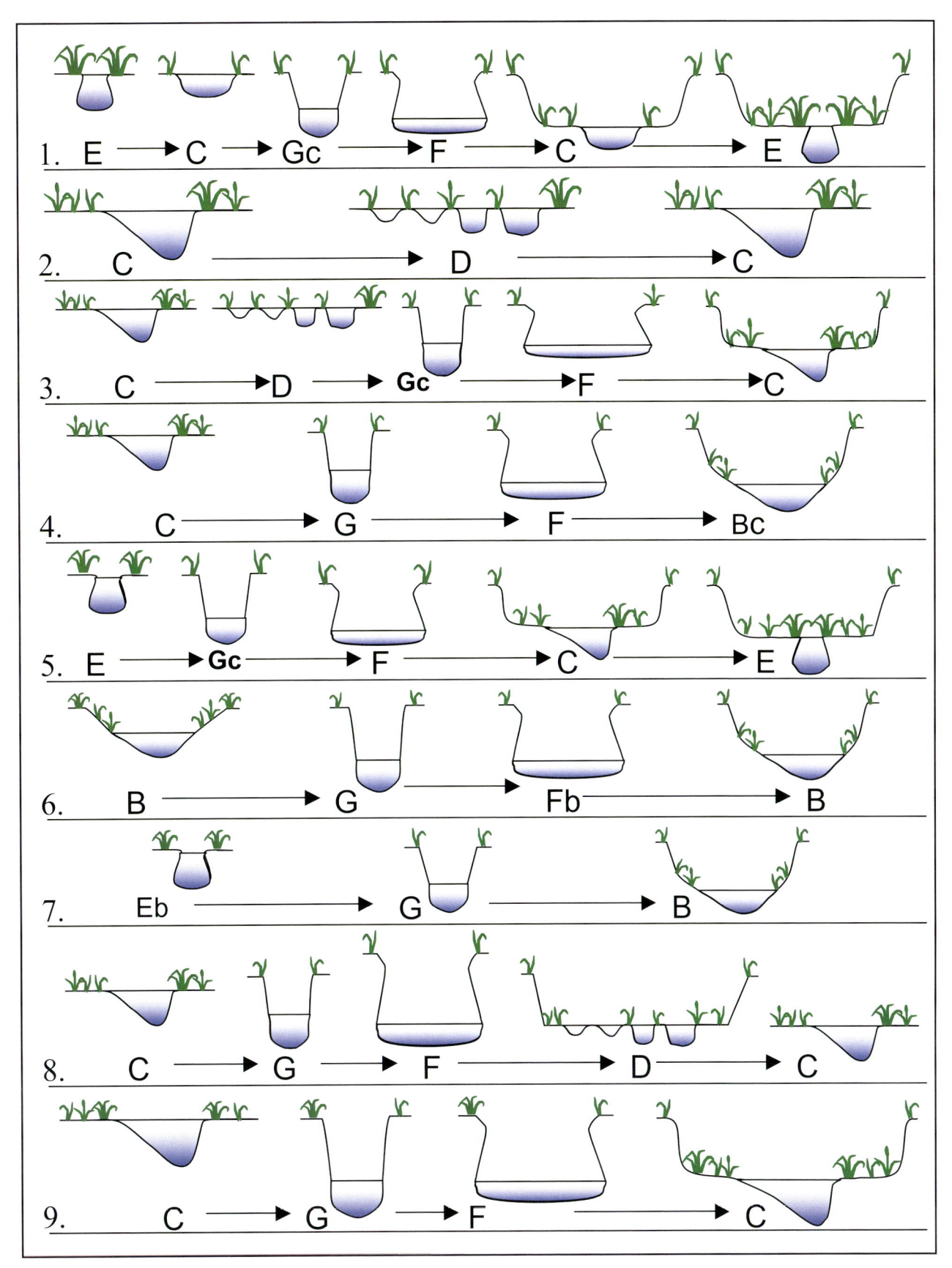

1. E → C → Gc → F → C → E
2. C → D → C
3. C → D → Gc → F → C
4. C → G → F → Bc
5. E → Gc → F → C → E
6. B → G → Fb → B
7. Eb → G → B
8. C → G → F → D → C
9. C → G → F → C

Figure 5-55 (Figure 2-38). Various channel evolution scenarios involving stream type succession (Rosgen, 1999, 2001b).

5-143

Step 25. Calculate lateral stability rating

In this step, five field stability indicators are used to calculate lateral stability ratings:

1. W/d ratio state: W/d ratio divided by reference w/d ratio (**Step 7**, **Worksheet 5-5**);
2. Depositional pattern (**Step 7**, **Worksheet 5-5**);
3. Meander pattern (**Step 7**, **Worksheet 5-5**);
4. Dominant BEHI/NBS (**Step 9**, **Worksheet 5-10**); and
5. Degree of confinement: Meander width ratio divided by reference meander width ratio (**Step 7**, **Worksheet 5-5**).

Use **Worksheet 5-17** to calculate lateral stability for impaired and reference reaches. The adjective stability rating is used in **Worksheet 5-20** to determine potential channel enlargement in **Step 27** and to calculate the overall sediment supply rating in **Worksheet 5-21** in **Step 28**.

Worksheet 5-17. Lateral stability prediction summary.

Stream:	Stream Type:
Location:	Valley Type:
Observers:	Date:

Lateral stability criteria (choose one stability category for each criterion 1–5)	Lateral stability categories					Selected points (from each row)
	Stable	**Moderately unstable**	**Unstable**	**Highly unstable**	
1 W/d ratio state (Worksheet 5-5)	< 1.2	1.2 – 1.4	1.4 – 1.6	> 1.6	
	(2)	(4)	(6)	(8)	
2 Depositional pattern (Worksheet 5-5)	B1, B2	B4, B8	B3	B5, B6, B7	
	(1)	(2)	(3)	(4)	
3 Meander pattern (Worksheet 5-5)	M1, M3, M4		M2, M5, M6, M7, M8		
	(1)		(3)		
4 Dominant BEHI / NBS (Worksheet 5-10)	L/VL, L/L, L/M, L/H, L/VH, M/VL	M/L, M/M, M/H, L/Ex, H/L	M/VH, M/Ex, H/L, H/M, H/H, VH/VL, Ex/VL	H/H, H/Ex, Ex/M, Ex/H, Ex/VH, VH/VH, Ex/Ex	
	(2)	(4)	(6)	(8)	
5 Degree of confinement (MWR / MWR$_{ref}$) (Worksheet 5-5)	0.8 – 1.0	0.3 – 0.79	0.1 – 0.29	< 0.1	
	(1)	(2)	(3)	(4)	
				Total points	

	Lateral stability category point range			
Overall lateral stability category (use total points and check (✓) stability rating)	**Stable** 7 – 9 ☐	**Moderately unstable** 10 – 12 ☐	**Unstable** 13 – 21 ☐	**Highly unstable** > 21 ☐

WATERSHED ASSESSMENT OF RIVER STABILITY AND SEDIMENT SUPPLY

Step 26. Calculate vertical stability ratings

The objectives in **Step 26** are similar to those in **Step 25**, except the variables and stability categories selected are those that influence vertical stability (change in local base level). These ratings are separated by aggradation or excess sediment deposition (**Worksheet 5-18**) and degradation/channel incision (**Worksheet 5-19**).

Aggradation or excess deposition

The variables and individual stability ratings from various steps selected for vertical stability for aggradation/excess deposition include the following six categories:

1. Sediment competence (**Step 22, Worksheet 5-15**);

2. Sediment capacity (**Steps 18 and 19, Worksheet 5-12**);

3. W/d ratio state (**Step 7, Worksheet 5-5**);

4. Stream succession states (**Step 24, Worksheet 5-16**);

5. Depositional patterns (**Step 7, Worksheet 5-5**); and

6. Debris blockages (**Step 7, Worksheet 5-5**).

Degradation/channel incision

The following are the five categories selected to evaluate the overall vertical stability for degradation/channel incision:

1. Sediment competence (**Step 22, Worksheet 5-15**);

2. Sediment capacity (**Steps 18 and 19, Worksheet 5-12**);

3. Bank-Height Ratio (BHR) (degree of channel incision) (**Step 7, Worksheet 5-5**);

4. Stream succession states (**Step 24, Worksheet 5-16**); and

5. Channel confinement (Meander Width Ratio (MWR)) (**Step 7, Worksheet 5-5**).

Use **Worksheets 5-18** and **5-19** to calculate vertical stability for impaired and reference reaches. The adjective stability ratings are used in **Worksheet 5-20** to determine potential channel enlargement in **Step 27** and to calculate the overall sediment supply rating in **Worksheet 5-21** in **Step 28**.

Worksheet 5-18. Vertical stability prediction for excess deposition/aggradation.

Stream:			Stream Type:		
Location:			Valley Type:		
Observers:			Date:		

	Vertical stability criteria (choose one stability category for each criterion 1–6)	\multicolumn{4}{c	}{Vertical stability categories for excess deposition / aggradation}	Selected points (from each row)
		No deposition	**Moderate deposition**	**Excess deposition**	**Aggradation**	
1	Sediment competence (Worksheet 5-15)	Sufficient depth and/or slope to transport largest size available	Trend toward insufficient depth and/or slope-slightly incompetent	Cannot move D_{35} of bed material and/or D_{100} of bar material	Cannot move D_{16} of bed material and/or D_{100} of bar or subpavement size	
		(2)	(4)	(6)	(8)	
2	Sediment capacity (POWERSED) (Worksheet 5-12)	Sufficient capacity to transport annual load	Trend toward insufficient sediment capacity	Reduction up to 25% of annual sediment yield of bedload and/or suspended	Reduction over 25% of annual sediment yield for bedload and/or suspended	
		(2)	(4)	(6)	(8)	
3	W/d ratio state (Worksheet 5-5)	1.0 – 1.2	1.2 – 1.4	1.4 – 1.6	>1.6	
		(2)	(4)	(6)	(8)	
4	Stream sucession states (Worksheet 5-16)	Current stream type at potential	(E→C)	(C→High w/d C), (B→High w/d B)	(C→D), (F→D)	
		(2)	(4)	(6)	(8)	
5	Depositional patterns (Worksheet 5-5)	B1	B2, B4	B3, B5	B6, B7, B8	
		(1)	(2)	(3)	(4)	
6	Debris / blockages (Worksheet 5-5)	D1, D2, D3	D4, D7	D5, D8	D6, D9, D10	
		(1)	(2)	(3)	(4)	
					Total points	

	\multicolumn{4}{c	}{Vertical stability category point range for excess deposition / aggradation}	
Vertical stability for excess deposition / aggradation (use total points and check (✓) stability rating)	**No deposition** 10 – 14 ☐	**Moderate deposition** 15 – 20 ☐	**Excess deposition** 21 – 30 ☐	**Aggradation** > 30 ☐	

Worksheet 5-19. Vertical stability prediction for channel incision/degradation.

Stream:			Stream Type:		
Location:			Valley Type:		
Observers:			Date:		

Vertical stability criteria (choose one stability category for each criterion 1–5)	Vertical stability categories for channel incision / degradation				Selected points (from each row)
	Not incised	Slightly incised	Moderately incised	Degradation	
1 Sediment competence (Worksheet 5-15)	Does not indicate excess competence	Trend to move larger sizes than D_{100} of bar or > D_{84} of bed	D_{100} of bed moved	Particles much larger than D_{100} of bed moved	
	(2)	(4)	(6)	(8)	
2 Sediment capacity (POWERSED) (Worksheet 5-12)	Does not indicate excess capacity	Slight excess energy: up to 10% increase above reference	Excess energy sufficient to increase load up to 50% of annual load	Excess energy transporting more than 50% of annual load	
	(2)	(4)	(6)	(8)	
3 Degree of channel incision (BHR) (Worksheet 5-5)	1.00 – 1.10	1.11 – 1.30	1.31 – 1.50	> 1.50	
	(2)	(4)	(6)	(8)	
4 Stream sucession states (Worksheets 5-5 and 5-16)	Does not indicate incision or degradation	If BHR > 1.1 and stream type has w/d between 5–10	If BHR > 1.1 and stream type has w/d less than 5	(B→G), (C→G), (E→G), (D→G)	
	(2)	(4)	(6)	(8)	
5 Confinement (MWR / MWR$_{ref}$) (Worksheet 5-5)	0.80 – 1.00	0.30 – 0.79	0.10 – 0.29	< 0.10	
	(1)	(2)	(3)	(4)	
				Total points	
	Vertical stability category point range for channel incision / degradation				
Vertical stability for channel incision/ degradation (use total points and check (✓) stability rating)	Not incised 9 – 11 ☐	Slightly incised 12 – 18 ☐	Moderately incised 19 – 27 ☐	Degradation > 27 ☐	

Chapter 5 — The Prediction Level Assessment (PLA)

Step 27. Calculate potential channel enlargement

The ratings for this category are based on stream channel successional stage shift (**Step 24**) and potential lateral and vertical instability (**Steps 24** and **25**). Use **Worksheet 5-20** to determine a final channel enlargement category, and use the adjective potential channel enlargement rating in the overall sediment supply rating (**Worksheet 5-21**).

Worksheet 5-20. Channel enlargement prediction summary.

Stream:			Stream Type:		
Location:			Valley Type:		
Observers:			Date:		
Channel enlargement prediction criteria (choose one stability category for each criterion 1–4)	\multicolumn{4}{c}{**Channel enlargement prediction categories**}	**Selected points (from each row)**
	No increase	**Slight increase**	**Moderate increase**	**Extensive**	
1 Successional stage shift (Worksheet 5-16)	Stream type at potential, (C→E), (F_b→B), (G→B), (F→B_c), (F→C), (D→C) (2)	(E→C) (4)	(G→F), (F→D) (6)	(C→D), (B→G), (D→G), (C→G), (E→G) (8)	
2 Lateral stability (Worksheet 5-17)	Stable (2)	Moderately unstable (4)	Unstable (6)	Highly unstable (8)	
3 Vertical stability excess deposition/ aggradation (Worksheet 5-18)	No deposition (2)	Moderate deposition (4)	Excess deposition (6)	Aggradation (8)	
4 Vertical stability incision/ degradation (Worksheet 5-19)	Not incised (2)	Slightly incised (4)	Moderately incised (6)	Degradation (8)	
				Total points	
	\multicolumn{4}{c}{**Category point range**}	
Channel enlargement prediction (use total points and check (✓) stability rating)	No increase 8 – 10 ☐	Slight increase 11 – 16 ☐	Moderate increase 17 – 24 ☐	Extensive > 24 ☐	

5-149

WATERSHED ASSESSMENT OF RIVER STABILITY AND SEDIMENT SUPPLY

Step 28: Overall sediment supply rating by individual and combined stability ratings (See Flowchart 5-18)

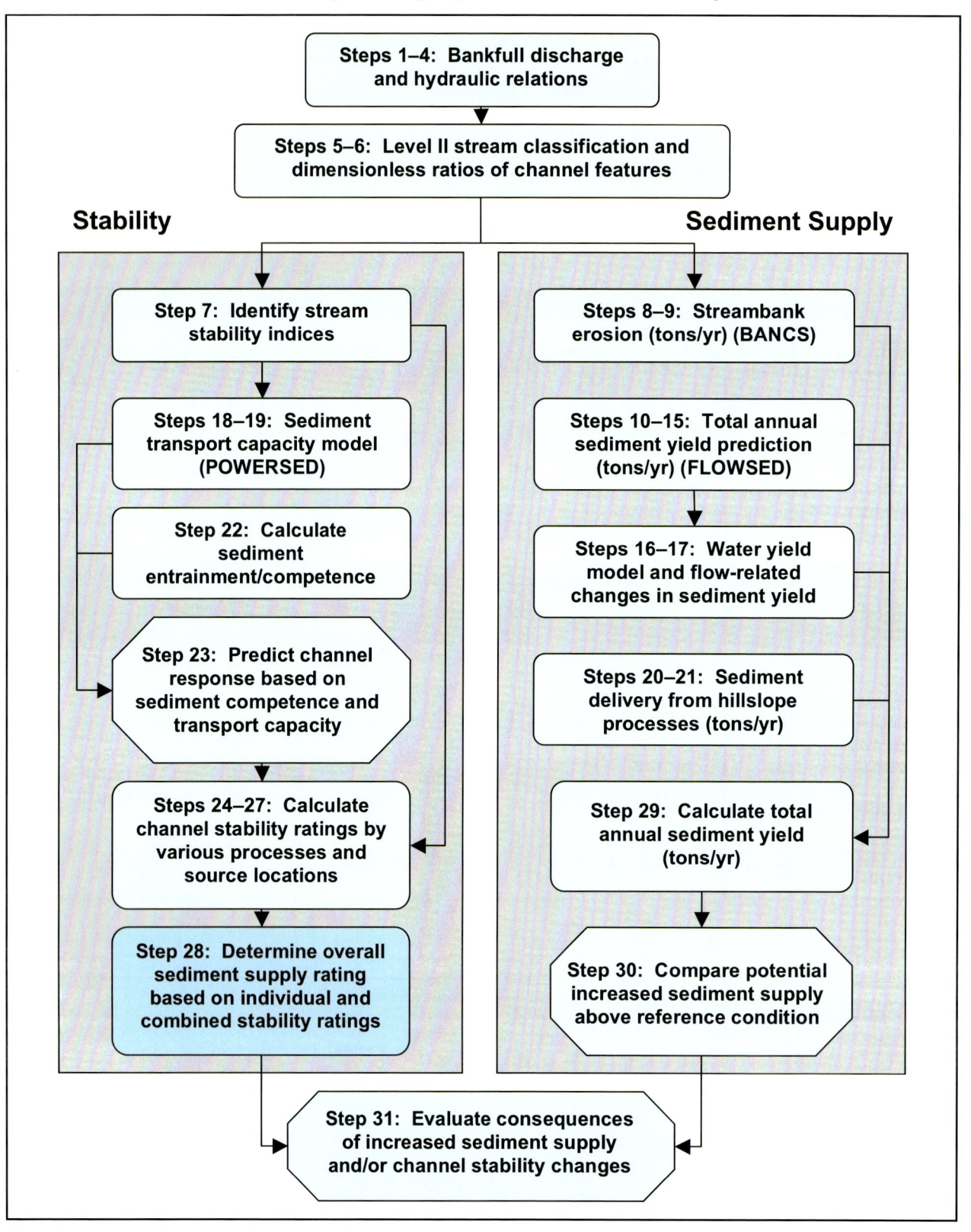

```
Steps 1–4: Bankfull discharge and hydraulic relations
      │
      ▼
Steps 5–6: Level II stream classification and dimensionless ratios of channel features
```

Stability

- Step 7: Identify stream stability indices
- Steps 18–19: Sediment transport capacity model (POWERSED)
- Step 22: Calculate sediment entrainment/competence
- Step 23: Predict channel response based on sediment competence and transport capacity
- Steps 24–27: Calculate channel stability ratings by various processes and source locations
- **Step 28: Determine overall sediment supply rating based on individual and combined stability ratings**

Sediment Supply

- Steps 8–9: Streambank erosion (tons/yr) (BANCS)
- Steps 10–15: Total annual sediment yield prediction (tons/yr) (FLOWSED)
- Steps 16–17: Water yield model and flow-related changes in sediment yield
- Steps 20–21: Sediment delivery from hillslope processes (tons/yr)
- Step 29: Calculate total annual sediment yield (tons/yr)
- Step 30: Compare potential increased sediment supply above reference condition

Step 31: Evaluate consequences of increased sediment supply and/or channel stability changes

Flowchart 5-3m. The sequential steps to determine stream stability and sediment supply, highlighting **Step 28**.

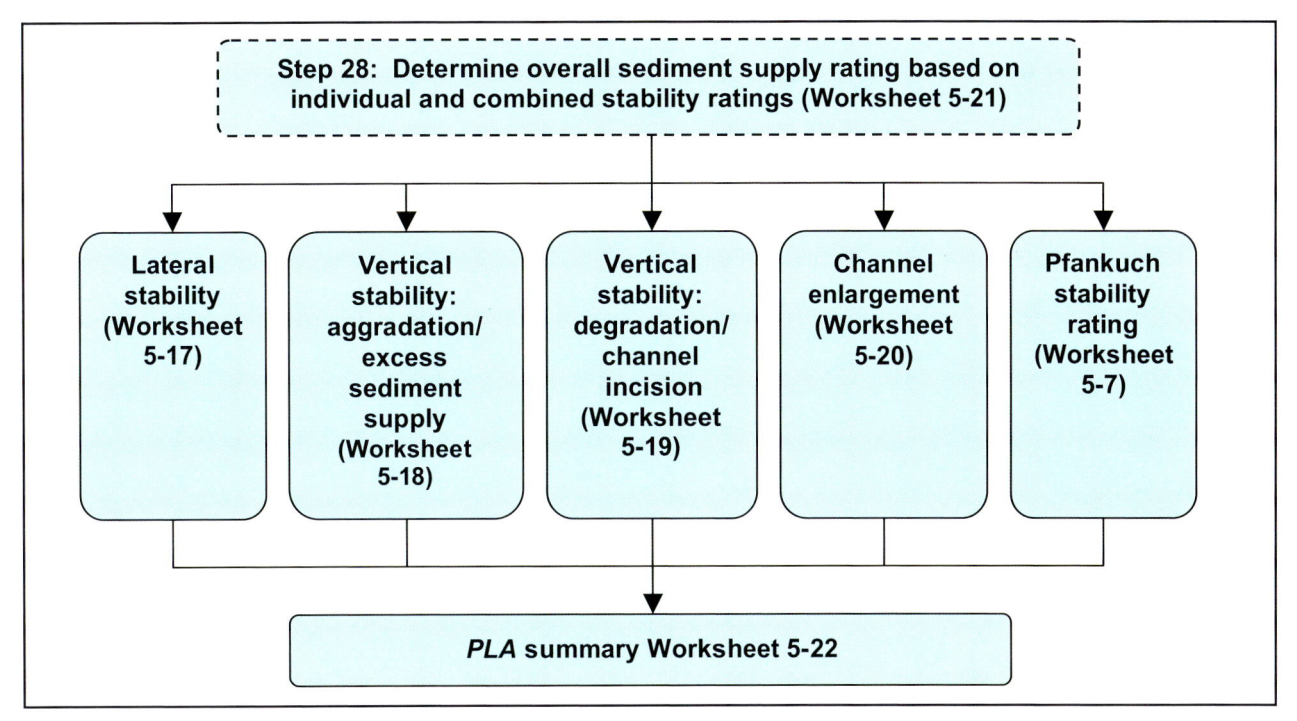

Flowchart 5-18. The combined influence of individual stability ratings on sediment supply.

Step 28. Determine overall sediment supply rating from individual and combined stability ratings

The stability categories from **Steps 24–27** are used to summarize the overall stability summary, potential impairment and the associated channel source sediment supply in **Worksheet 5-21**. **Flowchart 5-18** shows the individual and combined stability ratings used to evaluate overall sediment supply. Disproportionate or high sediment sources can be related to specific processes and locations. This information is crucial for mitigation proposals.

The overall analysis helps the user focus on specific processes, locations, causes and consequences of instability. In designing stabilization, restoration and/or management changes, the variables influencing the resultant instability and/or impairment are identified. These designs can be effective at targeting the specific problem areas to mitigate the impairment.

Each worksheet used to complete **Step 28** can be reviewed to ascertain the variables, categories, processes, conditions, etc., by location that result in a high or very high sediment supply. This assessment is completed in **Step 31**. By targeting the sources, nature and consequences of impact, process-specific mitigation for sediment supply and channel stability can be directed. The overall sediment supply rating is transferred to the *PLA* summary worksheet in **Steps 29–31**.

WATERSHED ASSESSMENT OF RIVER STABILITY AND SEDIMENT SUPPLY

Worksheet 5-21. Overall sediment supply ratings for multiple reaches determined from individual stability rating categories.

Stream: _____ Location: _____ Observers: _____ Date: _____

(1) Reach location	(2) Successional stage shift stability rating (Worksheet 5-16)*		(3) Lateral stability rating (Worksheet 5-17)		(4) Vertical stability for excess deposition/aggradation (Worksheet 5-18)		(5) Vertical stability for channel incision/degradation (Worksheet 5-19)		(6) Channel enlargement prediction (Worksheet 5-20)		(7) Pfankuch channel stability (Worksheet 5-7)		(8) Add points from Columns (3) to (7)	(9) Overall sediment supply rating; use column (8) points to determine adjective rating
	Stability rating:		Stability rating:	Points:	Stability rating:	Points:	Stability rating:	Points:	Stability rating:	Points:	Stability rating:	Points:		Sediment supply rating:
	Stable		Stable	1	No deposition	1	Not incised	1	No increase	1	Good: stable	1		5 = Low
	Mod.		Mod. unstable	2	Mod. deposition	2	Slightly incised	2	Slight increase	2	Fair: mod. unstable	2		6–10 = Moderate
	Unstable		Unstable	3	Excess deposition	3	Mod. incised	3	Mod. increase	3				11–15 = High
	Highly unstable		Highly unstable	4	Aggradation	4	Degradation	4	Extensive	4	Poor: unstable	4		16–20 = Very high
1.														
2.														
3.														
4.														
5.														
6.														
7.														
8.														
9.														
10.														
11.														
12.														
13.														
14.														
15.														

*Successional stage shift stability rating is not used to determine overall sediment supply rating.

Steps 29–31: Summary evaluations

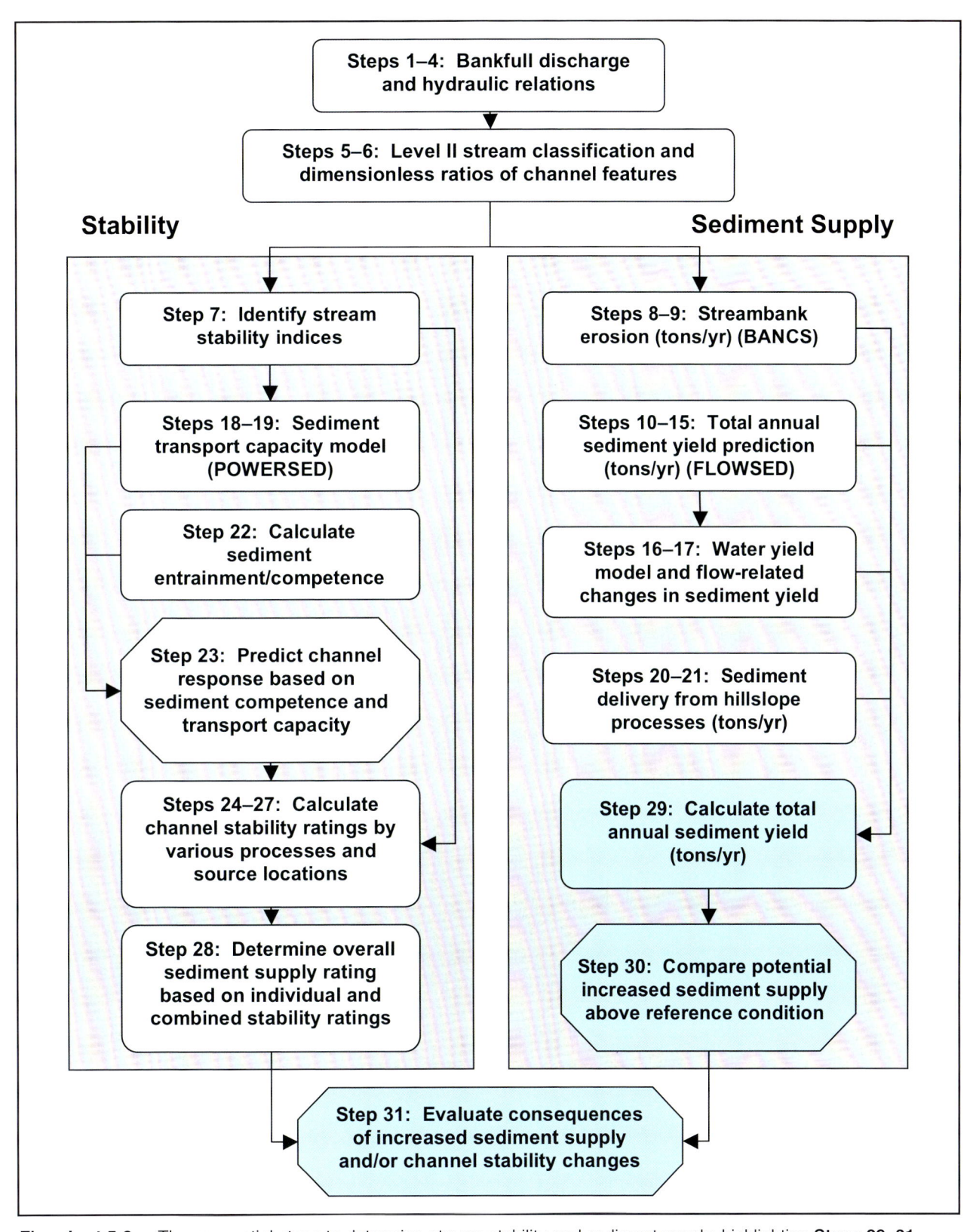

Stability

- Steps 1–4: Bankfull discharge and hydraulic relations
- Steps 5–6: Level II stream classification and dimensionless ratios of channel features
- Step 7: Identify stream stability indices
- Steps 18–19: Sediment transport capacity model (POWERSED)
- Step 22: Calculate sediment entrainment/competence
- Step 23: Predict channel response based on sediment competence and transport capacity
- Steps 24–27: Calculate channel stability ratings by various processes and source locations
- Step 28: Determine overall sediment supply rating based on individual and combined stability ratings

Sediment Supply

- Steps 8–9: Streambank erosion (tons/yr) (BANCS)
- Steps 10–15: Total annual sediment yield prediction (tons/yr) (FLOWSED)
- Steps 16–17: Water yield model and flow-related changes in sediment yield
- Steps 20–21: Sediment delivery from hillslope processes (tons/yr)
- Step 29: Calculate total annual sediment yield (tons/yr)
- Step 30: Compare potential increased sediment supply above reference condition

Step 31: Evaluate consequences of increased sediment supply and/or channel stability changes

Flowchart 5-3n. The sequential steps to determine stream stability and sediment supply, highlighting **Steps 29–31**.

Summary evaluations: Steps 29–31

Many erosional/depositional processes and stream channel stability relations have been identified in *PLA*. The objectives of these sequential steps are to identify, quantify and understand the river and sediment consequences of land use impacts on water resources and the river ecosystem. The key in the summary of the final three steps is to:

1. Calculate the total annual sediment supply in tons/yr (**Step 29**), including:

    a. baseline condition

    b. reference condition

    c. existing or potentially impaired condition

    d. proposed condition (optional)

    This includes total bedload and suspended sediment and the total sediment supply from the combined hillslope and channel processes (**Worksheet 5-22**).

2. Compare potential increased sediment supply above reference and baseline conditions (**Step 30**). This step analyzes the sediment yield in tons/yr for:

    a. hillslope processes (roads, surface erosion and mass erosion)

    b. hydrologic change (pre- and post-treatment flow-related sediment yields)

    c. channel processes sediment supply (streambank erosion)

    Comparisons of existing and/or proposed conditions are made for the various sediment sources to baseline and reference condition (**Worksheet 5-23**).

3. Evaluate the consequences (**Step 31**) of:

    a. sediment supply (**Step 30**)

    b. channel stability (**Step 28**, **Worksheet 5-21**)

    Step 31 also interprets the impairment due to departure analysis.

Step 31 is undoubtedly the most important step in *PLA*. It allows the user the ability to apply a consistent, comparative, quantitative procedure to specifically isolate the ***causes, location*** and ***magnitude*** of instability, impairment and sediment supply consequences. Disproportional contributions of sediment can be quantified by specific locations and land uses responsible for the sediment supply problems and/or impairment. This is the basis for site- and process-specific mitigation to offset the adverse consequences of past land use. The results also help set priorities for sediment reduction and improved stream stability and function.

Worksheet 5-22 is completed for multiple locations in **Steps 29–31** to assist in the sequential step analyses.

Chapter 5 — The Prediction Level Assessment (PLA)

Worksheet 5-22. PLA summary of sediment sources and stability ratings for multiple locations.

Stream: _____ Location: _____ Observers: _____ Date: _____

(1) Sub-watershed or reach location	(2) Step 9: Stream-bank erosion (Worksheet 5-10)	(3) Step 15: Total annual sediment yield (Worksheet 5-11)	(4) Step 17: Flow-related sediment (Worksheet 5-11)	(5) Step 19: Sediment transport capacity stability rating (Worksheet 5-12b)	(6) Step 20: Roads (total sediment yield) (Worksheet 5-13)	(6) Step 20: Surface erosion (total sediment yield)	(6) Step 20: Mass erosion (total sediment yield)	(7) Step 21: Hillslope (total sediment yield)	(8) Step 22: Sediment competence/ entrainment (Worksheet 5-15)	(9) Step 23: Overall channel response due to sediment competence and capacity	(10) Step 28: Overall channel source sediment supply rating (Worksheet 5-21)	(11) Step 29: Total sediment yield (Worksheet 5-11) Bedload	(11) Suspended	(11) Total	(12) Step 30: Difference in sediment from baseline (Worksheet 5-23) Bedload	(12) Suspended	(12) Total	(13) Step 31: Potential consequence for overall stability
	(tons/yr)	(tons/yr)	(tons/yr)	stable/ aggrading/ degrading	(tons/yr)	(tons/yr)	(tons/yr)	(tons/yr)	stable/ aggrading/ degrading	stable/ aggrading/ degrading	low/ moderate/ high/ very high	(tons/yr)	(tons/yr)	(tons/yr)	(tons/yr)	(tons/yr)	(tons/yr)	stable/ aggradation/ degradation
1. Baseline condition			n/a	n/a					n/a	n/a	n/a				n/a	n/a	n/a	
2.																		
3.																		
4.																		
5.																		
6.																		
7.																		
8.																		
9.																		
10.																		
11.																		
12.																		
13.																		
14.																		
15.																		

Step 29. Calculate total sediment yield (all sources)

This step identifies the sources, extent and locations of sediment-related impairments. The *PLA* summary **Worksheet 5-22** is used in this step to summarize all of the sediment sources in tons/yr and the corresponding channel stability evaluations, including:

- Streambank erosion (**Steps 8** and **9**, **Worksheet 5-10**);
- Total annual sediment yield (**Steps 10** and **15**, **Worksheet 5-11**);
- Flow-related sediment changes (**Steps 16** and **17**);
- Sediment transport capacity rating (**Steps 18** and **9**, **Worksheet 5-12**);
- Total annual sediment yield from hillslope processes:
    - roads (**Step 20**, **Worksheet 5-13**)
    - surface erosion (**Step 20**)
    - mass erosion (**Step 20**)
- Sediment competence summary (**Step 22**, **Worksheet 5-15**);
- Channel stability process summary (**Step 23**); and
- Overall sediment supply risk rating (**Step 28**).

The comparison of existing sediment yields to baseline (pristine condition and/or full recovery from disturbance) provides the user with the magnitude of departure from a pristine or undisturbed condition. **Worksheet 5-22** also documents sediment yield and/or stability conditions at various assessment locations for the baseline and reference conditions.

Baseline sediment yield represents a no-impact or relatively pristine, natural stream. The baseline sediment yield is calculated in row one in the *PLA* summary **Worksheet 5-22** to compare with the annual sediment yields for the reference condition and the impaired condition. The sum of the hillslope and channel erosional processes and their combined total annual sediment yield results in the baseline yield. The baseline bedload is adjusted by the same pre-treatment ratio of bedload to total load. Suspended sediment is adjusted in the same manner.

The reference condition does not infer that there has been no disturbance; however, the assumption and field assessment validation is that the effects of past land uses have not created an unstable condition. The differences among baseline, reference (stable) and impaired conditions provide insight of change for the observer and are key in this analysis (**Step 30**).

The sediment supply data becomes valuable as reasonable mitigation priorities need to utilize such information. It is often not possible to return many watersheds to a pristine or "no disturbance" state. This is based on the fact that hydrologic recovery can take up to 100 years to return to pre-disturbance condition (full hydrologic utilization). However, stable stream types do not exhibit the same flow-related sediment increases as unstable reaches. Mitigation may address the channel stability problems in order to reduce sediment supply. Often, this involves stream restoration or stabilization to reduce accelerated streambank and/or streambed erosion problems. Identification of various locations of certain stable stream types, exposed to increases in flow yet still stable, can offer a "blue print" for mitigation.

Step 30. Compare potential accelerated or increased sediment supply above a reference or baseline condition

This step indicates the existing change or departure of the potentially impaired reach from the baseline and reference conditions. The increased sediment supply from **Step 29** (**Worksheet 5-22**) for an impaired reach is compared to baseline and reference rates and is documented in **Worksheet 5-23**. This step identifies the magnitude of the disproportionate contributions of sediment and indicates processes and places for which site-specific mitigation is needed.

The data in **Step 30** using **Worksheet 5-23** documents the location, processes (streambanks, roads, flow-related, etc.) and magnitude and indicates the percent and direction of departure from the baseline and reference conditions. The output from **Worksheet 5-22** provides the basis for these calculations in **Worksheet 5-23** and sets the stage for the discussion, interpretations and corresponding summary of the consequences in **Step 31**.

WATERSHED ASSESSMENT OF RIVER STABILITY AND SEDIMENT SUPPLY

Worksheet 5-23. Annual sediment yield by sources for an individual reach location, including hillslope, streambank erosion and flow-related processes.

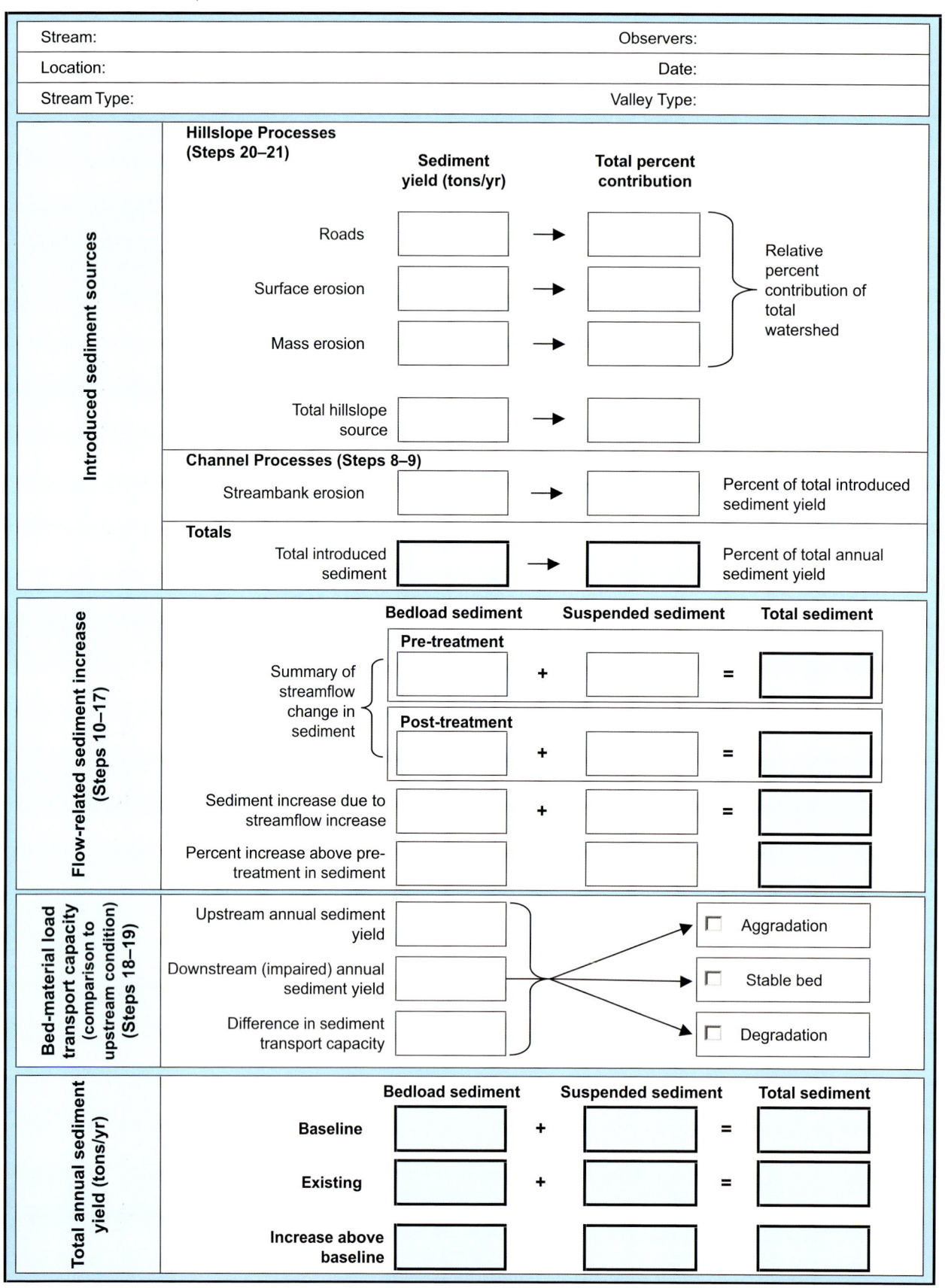

Stream: _____ Observers: _____
Location: _____ Date: _____
Stream Type: _____ Valley Type: _____

Introduced sediment sources

Hillslope Processes (Steps 20–21)

	Sediment yield (tons/yr)	Total percent contribution
Roads	☐	☐
Surface erosion	☐	☐
Mass erosion	☐	☐

→ Relative percent contribution of total watershed

Total hillslope source	☐	☐

Channel Processes (Steps 8–9)

Streambank erosion	☐	☐	Percent of total introduced sediment yield

Totals

Total introduced sediment	☐	☐	Percent of total annual sediment yield

Flow-related sediment increase (Steps 10–17)

Summary of streamflow change in sediment

	Bedload sediment	Suspended sediment	Total sediment
Pre-treatment	☐ +	☐ =	☐
Post-treatment	☐ +	☐ =	☐
Sediment increase due to streamflow increase	☐ +	☐ =	☐
Percent increase above pre-treatment in sediment	☐	☐	☐

Bed-material load transport capacity (comparison to upstream condition) (Steps 18–19)

Upstream annual sediment yield	☐
Downstream (impaired) annual sediment yield	☐
Difference in sediment transport capacity	☐

☐ Aggradation
☐ Stable bed
☐ Degradation

Total annual sediment yield (tons/yr)

	Bedload sediment	Suspended sediment	Total sediment
Baseline	☐ +	☐ =	☐
Existing	☐ +	☐ =	☐
Increase above baseline	☐	☐	☐

Step 31. Evaluate potential consequences of increased sediment supply and/or channel stability changes

Step 31 is the *key* step in summarizing and interpreting the previous analysis results. Adverse potential channel adjustments that cause channel instability, morphological shifts and disproportionate sediment supply have long-term adverse effects on beneficial uses. Understanding the causes of impairment is the first step toward mitigation; thus, **Step 31** deserves a thorough evaluation.

Worksheets 5-21, **5-22** and **5-23** document the magnitude of sediment changes and river stability consequence. The sources or causes of these changes are critical to isolate the responsible specific activities in various sub-watersheds. In doing so, mitigation including management changes and/or restoration can be recommended to offset problem areas or disproportionate sediment sources. Prior to developing the correction by specifying general guidelines in management direction, the following evaluations are suggested.

A. Channel stability considerations and summary interpretations

Review in detail the stability categories in **Worksheet 5-21**. Where ratings of *unstable, highly unstable, excess deposition, aggradation, moderately incised, degradation, moderate increase* or *extensive* are entered, obtain the worksheets that summarize the variables or conditions responsible for the rating. For example, in **Worksheet 5-21**, if the vertical stability rating for excess deposition/aggradation is *unstable* or *highly unstable*, then the following steps are recommended:

1. Review **Worksheet 5-18** and locate the reasons and locations that contributed to the unstable rating. The following variables were used to obtain the rating:

    a. sediment competence

    b. sediment capacity

    c. width/depth ratio state

    d. stream succession states

    e. depositional patterns

    f. debris/blockages

2. Review the variables above in detail to provide insight into prescriptive management and mitigation to offset the contributing variables or criteria leading to the excess sediment deposition/aggradation condition. Solutions to reverse the impairment and secure stability may be to change the grazing strategy; to conduct stream restoration that decreases width/depth ratio and increases slope to improve sediment competence and capacity; or to design a conversion from an F4 to a C4 stream type due to successional stage ratings.

Another example is if the review of **Worksheet 5-21** indicates incision or degradation (**Worksheet 5-19**), then one of the variables that influences the rating is the degree of channel incision (bank-height ratio: lowest bank height divided by max bankfull depth). This data that indicates the magnitude, location and direction of incision leads the user to provide site-specific solutions. Typical solutions may be grade-control structures, energy dissipaters below culverts, floodplain drains or a shift in successional states based on stream restoration designs.

B. Sediment supply considerations and summary interpretations

The quantitative summary of sediment yield as presented in both **Worksheets 5-22** and **5-23** provides insight into the primary processes contributing to the total annual sediment yield.

The review of **Worksheet 5-23** indicates percent departure of total annual sediment yield from baseline and/or reference condition. This review directs the user to those percentages that indicate the greatest sediment sources and departure. The cause of such disproportionate sediment supply contributions can be identified by reviewing the worksheets representing the processes and/or land uses. The worksheets in the previous steps of *PLA* are of such detail to help formulate management prescriptions for mitigation. The previous steps and worksheets provide the basis for understanding the causes of sediment problems. Without such interpretations, management prescriptions may be ineffective or misguided.

The overall summaries (**Worksheets 5-22** and **5-23**) offer a rapid overview to identify disproportionate sediment sources. The following interpretations are presented to offset the adverse effects of the processes contributing to excessive or disproportionate sediment supply.

B1. Sediment competence

The sediment summary of competence (**Step 22**, **Worksheet 5-15**) leads the user to back-calculate the slope and or depth necessary to transport the largest sediment clast. Thus, a river restoration recommendation would include these sediment competence criteria.

B2. Sediment capacity

The sediment transport capacity (**Steps 18** and **19**, **Worksheet 5-12**) is key to maintaining river stability. The POWERSED model can be used to predict the stable dimension, pattern and profile of river channels to secure the sediment transport capacity.

B3. Sediment supply from roads

The Road Impact Index (RII) data summary (**Step 20**, **Worksheet 5-13**) helps locate those roads disproportionately contributing sediment (high density with multiple stream crossings on lower slope position). Such roads would be prioritized for obliteration, stabilization, retro-fits, etc. Existing roads that are also contributing sediment due to mass erosion, rill erosion and ditch-line erosion need to be mitigated as mapped for specific location of impact in *PLA*.

B4. Surface erosion

Mitigation priorities for excess sediment from surface erosion (**Step 20**) are related to:

- Unstable soils;
- Poor plant cover density;
- Steep, concave slopes; and
- High sediment delivery potential.

B5. Mass erosion

Evidence of mass erosion disproportionately contributing to sediment supply (**Step 20**) needs to be summarized and priorities need to be set for mitigation or stabilization based on:

- Proximity to stream (high sediment delivery potential); and
- Relationship to causes, such as roads, that can be prevented or altered by rehabilitation or stabilization designs.

B6. Direct channel impacts

Identify streams that have had a history of straightening, levee construction, widening and other influences on their dimension, pattern and profile. These reaches need to be prioritized and restored to a stable form.

B7. Streambank erosion

The streambank erosion summary using the BANCS model (**Steps 8** and **9**, **Worksheet 5-10**) identifies the following information:

- A map that indicates the streambank erosion rate that shows locations of high to extreme erosion rates;
- Bank Erosion Hazard Index (BEHI) variables associated with the disproportionate high streambank erosion rates. The use of these BEHI variables, such as study bank-height ratio, rooting depth and density, bank angle and surface protection, provide insight into restoration criteria to reduce BEHI variables; and
- Near-Bank Stress (NBS) variables, such as radius of curvature to width ratio, transverse and central bars, ratio of slopes, etc. The NBS variables can be used to set restoration criteria and in-channel structure design to reduce streambank erosion.

B8. Flow-related sediment yield

The flow-related sediment yield (**Worksheet 5-11**) indicates increases in total annual sediment yield from changes in streamflow due to vegetative change and roads. The analysis is conducted by sub-watershed. The overlay of streambank erosion rates (tons/yr/ft) on a map indicates the sensitivity of various stream reaches to increases in flow. The map also identifies locations to restore or stabilize streambanks to potentially reduce flow-related increases. Road decommissioning on high road density areas can also be effective at reducing flow-related sediment yield.

B9. Total annual sediment yield (tons/yr)

The results are summarized using the FLOWSED model to predict total annual sediment yield for the following conditions:

- Baseline;
- Reference (existing);
- Impaired (existing); and
- Proposed (optional).

The results often include the integration with a flow model to simulate flow-related sediment. Comparisons of various sources of supply from hillslope and channel processes are made at this step. The purpose of the summaries and discussions are to identify the locations and the significant sediment contributors and associated land uses.

Discussion and application of *PLA*

The causes and consequences of the resultant processes associated with channel instability, including aggradation, degradation and accelerated streambank erosion, are shown in **Table 5-18**. An evaluation of such adverse consequences matched with watershed management objectives provides the impetus for mitigation.

By tracking the processes through the worksheet summaries, it is possible to:

- Identify specific locations and land uses contributing to the impacts;
- Understand the specific variables contributing to the impairment; and
- Design site- and process-specific mitigation that directly relates to the causes and consequences of impairment.

The advantage of having reasonable sediment yield values by activity/location/process is to:

- Set priorities for mitigation based on the severity, nature and extent of the sediment yield or river impairment;
- Predict potential effectiveness for sediment reduction or river improvement;
- Obtain specific information as to the variables responsible for the sediment or river impairment; and
- Understand the cumulative effects of a wide range of land uses on a watershed-based assessment.

Table 5-18. The causes and consequences of changes in erosional and depositional processes.

Process	Causes	Consequences
Aggradation	• Increased size and/or load of sediment • Change in channel hydraulics and morphology	• Land loss • Increase in flood hazard • Loss of aquatic habitat • Increase in streambank erosional lateral accretion • Change in water temperature, dissolved oxygen, etc. • Change in stream type
Degradation	• Clear water discharge • Change in channel hydraulics and morphology • Change in slope from straightening • Storm drains • Convergence scour due to culverts or bridges • Streamflow change (magnitude, duration and timing)	• Lowering of local base level • Channel below rooting depth and unstable banks • Lowering of water table causing loss of productivity and health of riparian vegetation • Land loss • Damage to infrastructure • Change in stream type • Increased downstream sediment supply • Loss of reservoir storage

Accelerated streambank erosion	• Change in riparian vegetation composition and density
	• Direct disturbance to channel
	• Aggradation leading to deposition patterns changing near-bank stress
	• Degradation, undercutting streambanks below rooting depth and excess shear sress
	• Enlargement causing channel widening
	• Streamflow alteration (magnitude, duration and timing)
		• Land loss
		• Increase in downstream sediment supply
		• Filling of reservoirs
		• Loss of aquatic habitat
		• Change in width/depth ratio leading to:
		  - excess deposition
		  - flood hazard increase
		  - water quality degradation (temperature, dissolved oxygen and phosphorus loading)
		• Damage to river infrastructure

Conclusion

These assessment steps and corresponding worksheets conclude *PLA*. Interpretations of this data are instrumental for formulating future watershed plans and restoration or stabilization efforts. *PLA* represents a strong foundation for understanding the causes and consequences of watershed problems that cause adverse river outcomes. The *PLA* worksheets help document exact locations and sources for site-specific mitigation.

This assessment leads the analyst to recommend process-specific mitigation and management changes that will potentially balance sediment supply and channel stability to improve beneficial water uses. It is critical that the final steps of this assessment provide a consistent and clear linkage to sources and erosional/depositional processes, for only in this manner can a proactive approach be taken for stabilization, remediation, restoration and/or enhancement. Best Management Practices (BMPs) are most effective when specifically directed to the processes and sources of adverse change.

Overall, this procedure can develop the accountability for those studying watersheds to document the problems then design the solutions. Monitoring is essential to keep the predictions in perspective and the mitigation directed correctly. An overview of related monitoring practices is presented in Chapter 6. Monitoring is critical to:

- Verify the relations used to predict sediment and stability change;
- Determine the effectiveness of process-specific mitigation, stabilization and restoration enhancement measures;
- Evaluate implementation qualtiy and correctness; and
- Document short-term and long-term sediment stability (and beneficial use) responses.

*"To monitor is to correct what you thought you knew
and is the essence of responsible management."*

CHAPTER 6

Monitoring

Watershed and river assessments involve complex process interactions, making accurate predictions precarious. Monitoring specific stream, watershed and biotic processes continually improves our understanding of those processes and their relationships. Well-planned monitoring can also demonstrate any reduced sediment or improved river stability that results from changes in management and/or mitigation practices.

Monitoring is generally recommended to:

- Measure the response of a *system* from combined process interactions due to imposed change;
- Document or observe the response of a *specific process* and compare it with the predicted response for a prescribed treatment;
- Define short-term versus long-term changes;
- Document spatial variability of process and system response;
- Reduce prediction uncertainty levels;
- Provide confidence in specific management practice modifications or mitigation recommendations;
- Determine if mitigation is implemented correctly;
- Evaluate effectiveness of stabilization or restoration approaches; and
- Build a database to extrapolate for similar applications.

Because monitoring techniques must be case-specific due to differences in regional climate, land use, stream type and many other factors, *WARSSS* provides only an overview of those elements that transcend these differences. Numerous specialized sources of monitoring guidance already exist. An excellent summary of the nature and design of monitoring, statistical guidance, methods, parameters, various strategies and biological monitoring appears in *Monitoring Guidelines to Evaluate Effects of Forestry Activities on Streams in the Pacific Northwest and Alaska* (MacDonald et al., 1991). *Monitoring Guidance for Determining the Effectiveness of Nonpoint Source Controls* (USEPA, 1997b) is also an exceptional reference. Riparian vegetation monitoring is described in detail in *Monitoring the Vegetation Resources in Riparian Areas*, a USDA Forest Service document (Winward, 2000).

Because of the complexity and variability of hillslope, hydrologic and channel processes, some uncertainty is inevitable in the *WARSSS* assessments. Iterative use of monitoring in connection with the *RRISSC* and *PLA* phases plays a vital role in validating calculations and setting the stage for future feedback on management practices. In particular, monitoring in conjunction with the *RRISSC* phase is critical to:

- Move the reaches or sub-watersheds that were given a rating of *moderate* risk directly into improved management practices with monitoring to provide feedback for adaptive management, where feasible;
- Verify the relations used to predict sediment and stability change;
- Determine the future effectiveness of process-specific mitigation measures, stabilization and restoration enhancement; and
- Document short- and long-term sediment stability and related effects on beneficial uses.

This chapter addresses the use of monitoring in *WARSSS* in an overview discussion that is divided into three parts:

1. Effectiveness monitoring (response of a process or system to imposed change);
2. Calibration and validation monitoring (matching predicted-to-observed response, including model calibration and validation); and
3. Field methods and procedures (techniques specific to the *WARSSS* assessment steps).

Effectiveness monitoring

For watersheds, slopes and river reaches that rated *moderate* risk for the *RRISSC* procedure, any implementation of process-specific mitigation or management prescriptions should be monitored. Monitoring determines the effectiveness of management prescriptions. Monitoring requires site-specific measurements (including temporal, spatial, scale, streamflow variation, site or reach variation and implementation variation measurements) to properly represent variability and extrapolate process and/or system responses to imposed change. These measurements are discussed in more detail.

Temporal measurements

To isolate the variability of season and/or annual change, effectiveness monitoring designs should include time scales. For example, measurements of lateral erosion rates should include annual measurements obtained at the same time of year. If the objective is to identify seasons where disproportionate erosion occurs, measurements might be obtained during snowmelt runoff, post-stormflow runoff, ice-off or other periods associated with a given erosional process. Annual replicate surveys of particle-size gradation of bed material under a permanent cross-section of a glide provide valuable information about the magnitude, direction and consequences of annual shifts.

Bedload sampling must also include a range of flows and temporal measurements throughout the runoff season, as discontinuous surges or "slugs" of bedload often occur. Bedload sampling over recommended time periods for a given flow, generally a minimum of 20 minutes per site visit, helps capture this variability. Short-term versus long-term monitoring must also be considered based on the probability of change, the severity and consequence of effects and the likelihood of variation. Sampling over many years, although costly, may be warranted.

Spatial measurements

Spatial changes or responses can be identified by measuring the same process at more than one site location (cross-section). For example, a longitudinal profile measured over several meander wavelengths will indicate changes in the maximum depth and/or slope of pools more accurately than monitoring only one pool. Identifying more than one reach of the same morphological type can also be used to understand response trends. Sampling the spatial variability, both vertically and laterally, within a cross-section for velocity and sediment measurements, helps identify or at least integrate this variability into a documented observation.

Scale measurements

Monitoring streams of various sizes and/or stream orders, but of the same morphological type and condition, helps identify variability in system response for proper interpretation of results. For example, stability measurements should be made on river reaches of the same condition and type but at locations that reflect various widths (sizes) and stream orders.

Streamflow variation measurements

Measurements of channel process relations should be stratified over a range of seasonal and annual flows. For example, both suspended and bedload sediment should be measured over a wide range of flows during the freshet, low-elevation snowmelt, high-elevation snowmelt, rising and recession, stormflow runoff and baseflow stages. These measurements allow the field observer to plot a sediment rating curve that represents the widest range of seasonal flows.

Site or reach variation measurements

Monitoring a site for soil loss should include a soil-type designation for potential extrapolation to similar soil types and conditions. The same is true for stream types. Sediment, hydraulic and stability monitoring needs to be stratified by stream type. This information helps observers to detect changes that suggest departure from a reference condition, as opposed to data that simply reflects normal differences among stream types.

Implementation variation measurements

Often, the "best-laid plans" are not implemented in the best possible manner. Effectiveness monitoring designs must assess whether the mitigation plan was implemented correctly. Riparian vegetation responses to livestock exclusion using fencing would be misleading if fences were broken or gates were left open. If the restoration designs were correct, but the contractor installed structures at the wrong angle, slope or position on the bank, then near-bank stress reduction or erosion rates would not be a fair observation of the effectiveness of the mitigation design.

Without these measurements, effectiveness monitoring designs cannot accurately determine a departure from stable conditions and cannot separate geologic and anthropogenic influences. A single sediment data point collected at the mouth of a large-order drainage contributes little value toward specifying its source, possible influences or specific process or land use contributions. Monitoring efforts have often been ineffective due to inconsistent or inadequate measurements of natural system variability or mitigation implementation.

Design concepts for effectiveness monitoring

The process to design an effectiveness monitoring plan includes the following steps:

1. Summarize the causes of land use impacts responsible for the impairment;
2. Understand the processes affected;
3. Identify specific locations and reaches associated with adverse impacts;
4. Predict the time trends of impacts (potential recovery periods);
5. Assess the specific nature of impairment (direction, magnitude and trend of change);
6. Predict the consequences of change; and
7. Summarize the nature, location, extent and quality of mitigation (implementation).

Once this information is analyzed, the monitoring design can proceed. The next step is to identify a monitoring strategy. Effectiveness monitoring should always be conducted near the activity responsible for the initial impairment. The following four design strategies are often used:

1. Obtain measurements *before* and *after* the initiation of a management change at the same location in the land use activity, mitigation, restoration, enhancement, etc. This can be very effective, as it identifies the pre-mitigation variability of the measured parameters. Following mitigation, departure can be readily determined, assuming that measurements take into consideration the aforementioned variability factors.

2. Obtain measurements or observations taken *above* and *below* impact areas related to specific land uses and mitigation. If two different grazing strategies are implemented, measurements of effectiveness can be observed above and below fence-line contrasts. This can also be implemented where a mitigation influences only the lower reach of a river, as compared to the upper reach (assuming the same stream type).

3. Obtain measurements that determine departure from a *paired watershed*. The "pairing" contrasts a watershed that has had extensive mitigation or land management change with an adjacent one that has not been changed. Again, this assumes that scale, temporal, spatial, landscape and stream type variations have been identified.

4. Obtain measurements that determine *departure* from a reference condition. This type of monitoring can occur at locations far removed from the reference reach. The reference condition, however, must be of the same soil, stream, valley, lithology and vegetative type.

The information supplied in the preceding summary will allow the observer to identify the locations, the nature of processes affected, the extent of the impact and the quality of the mitigation implementation. For example, if the dominant process impacted by a land use is causing disproportionate sediment supply, land loss and river instability, and is determined to be *accelerated streambank erosion*, then the accompanying lateral stability monitoring design would incorporate the following strategies:

- Locate reaches of the same stream type that represent an *unstable* bank;
- Locate reaches of the same stream type that represent a *stable* bank;
- Install permanent cross-sections on each set of reaches;
- Install bank pins and/or toe pins (if conditions warrant);
- Inventory vegetation and bank material, slope, etc., for each site;
- Resurvey both streambanks at least once per year to measure soil loss (lateral erosion) and total volume (in cubic feet and tons/year); and
- Compare annual lateral erosion rates over time to the stable reach and document the rate of recovery based on the mitigation.

Vertical stability, enlargement rates and direction can also be monitored using permanent cross-sections in a similar stratification procedure. Although this monitoring seems extensive, it can be accomplished for a stable/unstable reach comparison in one day of field inventory per year.

Relating physical to biological monitoring

The sediment and river stability changes assessed by the *RLA*, *RRISSC* and *PLA* methodologies are primarily physical changes. However, such physical changes are directly related to aquatic biological communities and functions. Changes in river stability, such as aggradation, degradation, enlargement or stream type alteration, may reduce pool volume and frequency and eliminate food chains by reducing intolerant fish or invertebrate taxa.

Limiting factor analyses are used to assess habitat loss due to river instability and/or excess sediment, such as relations of holding cover, instream/overhead cover, water temperature, dissolved oxygen and benthic invertebrates. Biological monitoring information associated with stream condition, such as diversity indices, population dynamics, age class distribution and spawning and rearing habitat, can be stratified by stream type and stream stability to reduce within-class variability. Biological monitoring should follow similar approaches for stratification based on the natural variability of streams. Good ambient monitoring designs and the use of reference condition concepts that are physical (geomorphic) and biological set the stage for well-informed monitoring at the single-stream scale. Many emerging methods for stream biomonitoring contrast observed species assemblages (O) with expected species assemblages (E) (developed with the reference data) in an O/E ratio that approaches a value of 1 in the best cases of stream recovery.

If a biologist is studying only the biological parameters within a specific eco-region and the stratification does not include stream type, then many natural biotic differences attributable to channel form differences cannot be properly identified. A stable C4 stream type does not have the attributes of a stable E4 or B4 stream type, even though they are all gravel-bed streams. Reference conditions that reflect biological potential must be stratified by stream type and stream stability at a minimum for adequate departure analysis to help explain the degree, direction and magnitude of impairment. This data can in turn be related to the limiting factor analysis of the aquatic ecosystem.

Calibration and validation monitoring

Calibration monitoring

Models are often used to predict potential impairment. Model calibration involves testing a model and adjusting the parameters to a set of field data. Field data guides the modeler in choosing the empirical coefficients used to predict the effect of management techniques. An example is to test the data set of measured suspended sediment and bedload sediment by stream type and stability to convert dimensionless sediment rating curves to dimensioned values used in the FLOWSED and POWERSED models. If sediment data was not collected in all areas where the model is potentially applied, then calibration monitoring helps to determine whether the model is appropriate for extrapolation to a particular region, stream type and/or stability condition.

Validation monitoring

Model validation tests a model with a data set representing "local" field data. This data set represents an independent source that is different from the data used to develop the relation. Often, this independent data is used to extend the range of conditions under which the model may be applied. This step is very important prior to widespread application of model output. Models can be extremely helpful for comparative analysis, even if observed values depart from measured. Often, validation monitoring outcomes result in tighter relations or sub-sets of the initial relation, improving the understanding of the processes being predicted. For example, using the previous example of streambank erosion, validation monitoring might require the observer to measure the parameters used to predict streambank erosion, the Bank Erosion Hazard Index (BEHI) and Near-Bank Stress (NBS). Subsequent analysis would plot the observed values with the predicted values for these variables. With sufficient numbers of observations, validation monitoring can improve local and/or regional models by adapting them to unique soil and vegetation types. Validation monitoring documents model performance, as well as mitigation performance.

Validation monitoring is designed to answer specific questions at specific sites and/or reaches. Designs must be matched with a strong understanding of the prediction model. Validation monitoring for the dimensionless sediment rating curves in Chapter 5 would involve sampling sediment over the full range of streamflows to compare predicted-to-observed values. The measurements would then be stratified according to the stream type and stability ratings used for the prediction. Monitoring involving sediment measurements should follow the protocols developed by the U.S. Geological Survey in *Field Methods for Measurement of Fluvial Sediment* (USGS, 1999, Book 3, Chapter 1) and *A Field Calibration of the Sediment-trapping Characteristics of the Helley-Smith Bedload Sampler* (Emmett, 1980).

Field methods and procedures

The prediction methods used in the *PLA* methodology include many models, empirical relations and stable reference channel descriptions. Even a basic prediction of mean bankfull velocity can be verified by direct measurements. This is accomplished by measuring a series of velocity readings with a flow meter across the stream channel at specific depths at the bankfull stage to calculate mean velocity. Procedures for velocity measurements and determination of streamflow are also necessary for sediment monitoring and analysis. The appropriate reference for these measurements is the U.S. Geological Survey manual, *Surface Water Investigations Measurement of Stream Discharge* (USGS, 1980, Book 3, Chapter A8).

Chapter 6 *Monitoring*

Table 6-1 shows typical monitoring procedures for various processes and sediment sources including hillslope, hydrologic and channel process relations. A number preceding the monitoring method indicates an additional process description discussed in the following pages.

Table 6-1. Monitoring procedures.

Process/source	Monitoring methods	Monitoring outputs
Surface erosion	• *(1) Frame and pin* • Precipitation gage	• Surface erosion rate in inches and tons
Sediment delivery from surface erosion	• Discontinuous contour trench parallel to stream	• Sediment yield
Mass erosion	• Total station survey of site to determine soil volume	• Erosion in yds^3 and tons
Sediment delivery	• Replicate annual survey to map "removed" or delivered sediment	• Sediment in yds^3 and tons
Roads: surfaces, ditch-line erosion, fill failures	• Measurement (physical survey) of surfaces, rill and ditch-line depth	• Erosion in yds^3 and tons
Sediment delivery from roads	• Installation of sediment basin **above** vs. **below** road or **before** vs. **after** • Install crest gage for peak flow	• Sediment delivered to channel in yds^3 and tons
Aggradation	• Install permanent cross-section • Size gradation at cross-section • Bed-material cross-section • *(2) Bar sample* • *(6) Replicate annual survey*	• Rate of change in channel dimension, pattern, profile and materials
Degradation/ sediment competence/ entrainment	• Install permanent cross-section • Size gradation at cross-section • *(2) Bar sample* • *(3) Scour chains* • *(6) Replicate annual survey*	• Rate of change in channel dimension, pattern, profile and materials • Bank-height ratio • Rate of incision • Depth of scour • Largest size particle moved
Lateral accretion/ streambank erosion	• Install permanent cross-section • *(4) Bank pins* • *(5) Bank profile* • *(6) Replicate annual survey*	• Lateral annual erosion rates and tons/yr from streambank erosion
Enlargement	• Install permanent cross-section • Size gradation at cross-section • *(6) Replicate annual survey*	• Rate of change in channel dimension, pattern, profile and materials
Sediment transport	• Helley-Smith bedload sampler • Depth-integrated suspended sediment sampler • Current meters • Sample over a wide range of flows, including bankfull discharge	• Bedload transport annual yield and bedload rating curve • Suspended sediment annual yield and suspended sediment rating curve • Velocity, discharge and change in bankfull sediment and discharge values

6-7

(1) Frame and pin (see Table 6-1)

The frame and pin method monitors soil erosion rate on a particular soil type, ground cover and slope gradient. For this method, a rectangular rigid frame is set on four, permanent benchmark corners (**Figure 6-1**). This frame is portable and can be made of aluminum for easy transport. Interior cross rods with fixed conduit pipe allow a calibrated pin (**Figure 6-2**) to be lowered vertically through the conduit to the ground surface. The pin depths within the fixed frame determine soil loss between runoff periods (following storms or annual resurvey).

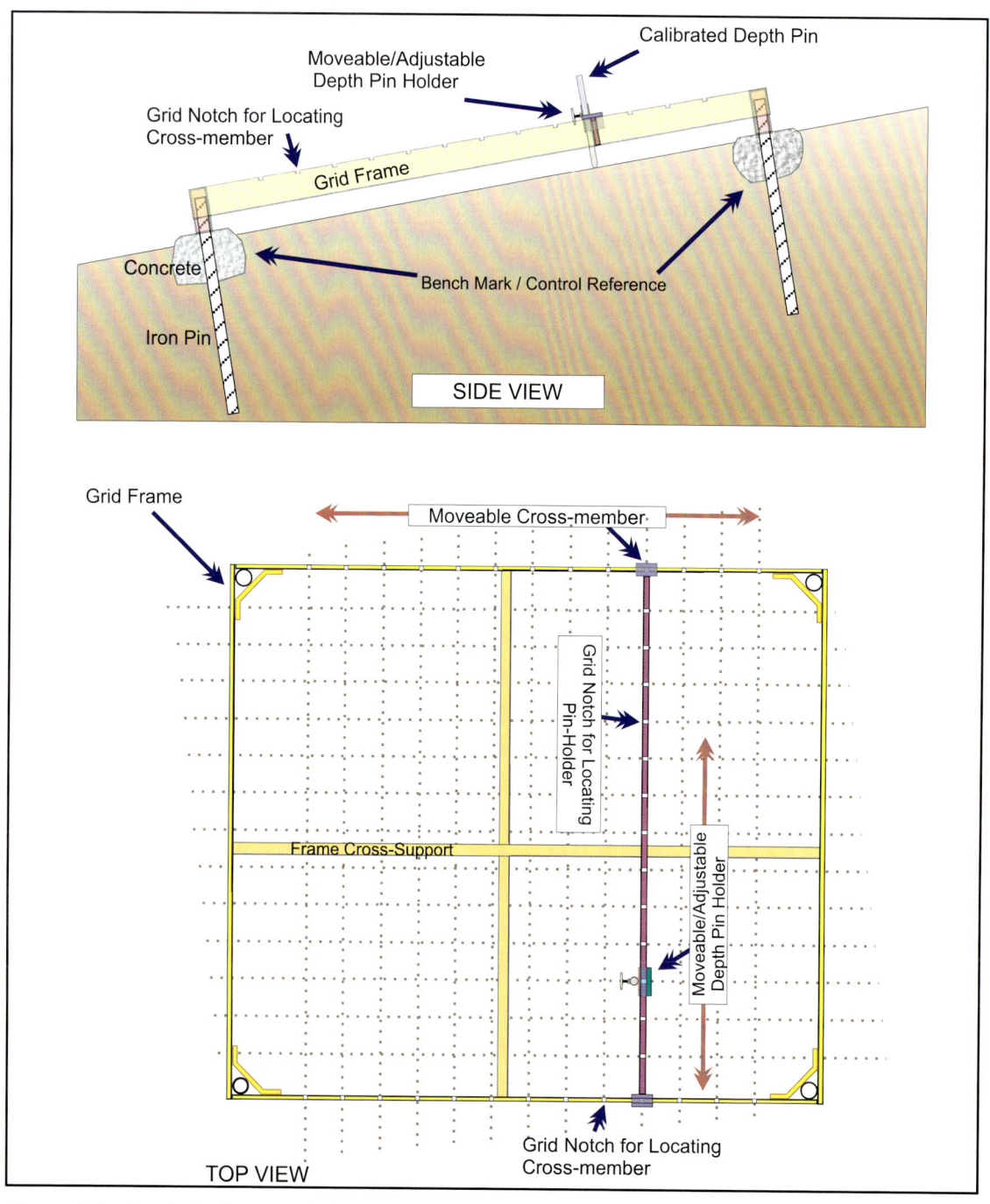

Figure 6-1. A portable frame and pin monitoring device to measure surface erosion following storms or on an annual resurvey basis.

Chapter 6 *Monitoring*

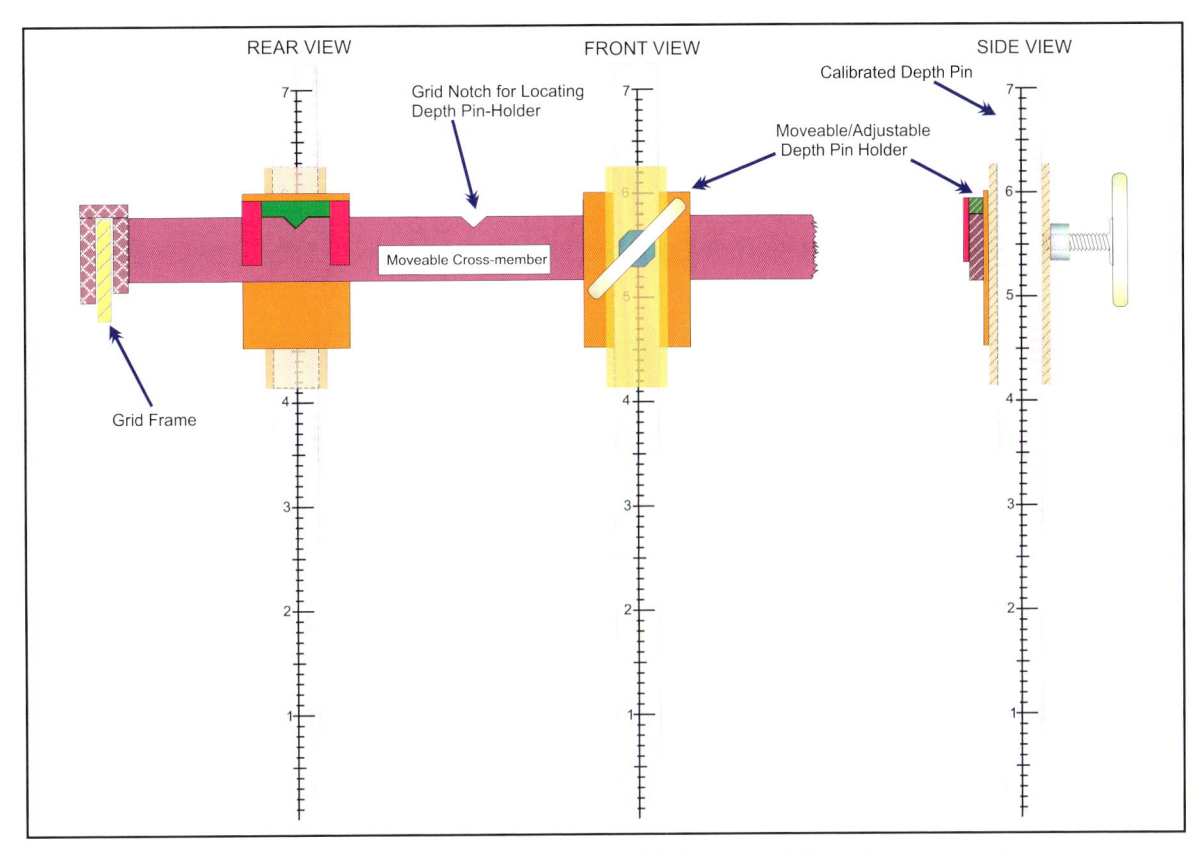

Figure 6-2. Detailed view of erosion pins used in the portable frame and pin surface erosion device.

(2) Bar sample (see Table 6-1)

A bar sample is taken on the downstream third of a bend on a point bar at an elevation halfway between the Thalweg and the bankfull stage. The core sample that is field-sieved (**Figure 6-3**) generally represents the size gradation of bedload at the bankfull stage. This data is also used in the sediment competence calculation. Descriptions of the procedure are discussed in Chapter 5 in **Tables 5-16** and **5-17**, **Figure 5-53** and **Worksheet 5-14**. Resurvey of the same location on an annual basis can also indicate if there are changes in size distribution of bedload.

Figure 6-3. Collecting and field-sieving a bar sample.

6-9

(3) Scour chains (see Table 6-1)

Scour chains are small-diameter chains attached to a duck-bill anchor and are driven vertically into the bed at a specific location under the permanent cross-section tape. The chain is left flush with the surface and is resurveyed annually or after a large flow event to determine scour depth, entrainment sizes of bed material and/or deposition and size of particle in the deposition. **Figure 6-4** shows a field observer installing a scour chain.

Table 6-2 provides instructions to install scour chains. **Worksheet 6-1** is used for documenting scour chain data. Scenario #1 in **Worksheet 6-1** indicates no change in bed elevation and no scour or deposition. Scenario #2 indicates net scour without deposition and also indicates the largest particle that was moved to scour the chain. Scenario #3 indicates scour and subsequent deposition back to the pre-runoff elevation. This indicates a dynamic but stable bed and also the size of particles that are entrained at bankfull and subsequently deposited. Scenario #4 demonstrates that the chain did not scour (still vertical) but the channel bed aggraded at that location. The last scenario (#5) indicates that an error occurred in the installation as the chain collapsed into its own hole. To prevent this, flagging is tied to the end of the chain, and once driven into the bed vertically, the bed material is "tamped" with constant upward pressure being applied with the flagging. A chain can be installed in this manner in less than five minutes. The validation of the entrainment models and prediction of bed stability can be verified rapidly by measuring before vs. after a bankfull event and/or once per year following the runoff season. It is also recommended to install a crest gage to obtain the highest stage. This stage data is used to determine the depth of flow for the highest flow stage for the time period the scour chains are installed. If two PVC pipe/crest gages are installed upstream and downstream of the scour chain, energy slope can also be determined. This allows a check on the entrainment relations by an accurate calculation of shear stress related to the size of particles that moved and were subsequently deposited over the chains.

Scour chains installed in glides and riffles indicate the depth of scour and particle sizes entrained on an annual basis for various stream types and bed-material gradations. It is not uncommon for these bed features to scour and fill back to their initial elevations without changing particle size distribution. These active bed features can be stable, and their function is better understood with the scour chain data. Those who replace riffles with angular large rock to ensure the riffles don't scour or to maintain a stable bed are advised to obtain scour chain data on their stream reaches to determine the actual streambed activity of the stable reach on an annual basis.

Figure 6-4. Installing a scour chain with a driving bar and duck-bill anchor.

Table 6-2. Instructions for scour chain installation.

Instructions to install scour chains

1. Install permanent cross-section at location where vertical stability data is needed (generally on a riffle reach).

2. Stretch tape across channel making sure that the zero end is over the left pin (if not at zero, make a note in field documentation of the actual distance).

3. Install a minimum of 2 scour chains per cross-section.

4. Drive rod with either a "T-post driver" (heavy gage pipe with top end welded plate with handles) or use a sledgehammer.

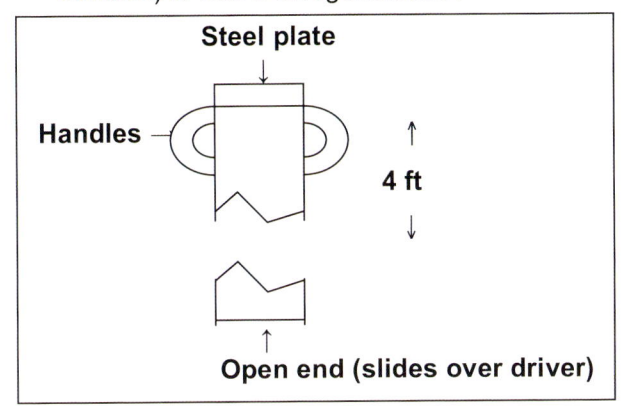

Steel plate
Handles
4 ft
Open end (slides over driver)

Move driving rod 5–10° to keep from "sticking."

5. Use plumb–bob to get accurate location of scour chain below tape. Record location (distance from left bank).

6. Measure largest 2 particles immediately upstream and downstream of chain (that would be entrained should bed scour occur).

7. Survey cross-section noting location of scour chain.

8. Obtain pebble count (100) under cross-section.

9. Resurvey cross-section, locate scour chain and resurvey pebble count to determine potential shifts. The amount of chain exposed horizontally is associated with the depth of scour (bed may refill over chain; thus, you will need to excavate material to locate chain). A metal detector may be used to help locate chain.

10. Record largest 2 particles over chain following excavation (this represents largest size entrained for measured shear stress).

11. Measure longitudinal profile identifying high water lines; calculate energy slope and depth (use hydraulic radius from cross-section data over chain).

12. Compare entrainment calculation with actual sizes.

13. Compare bed scour/aggradation prediction with actual measurements. Determine depth of scour with cross-section data and subsequent refilling.

WATERSHED ASSESSMENT OF RIVER STABILITY AND SEDIMENT SUPPLY

Worksheet 6-1. Field form for documenting scour chain results and corresponding bed-elevation changes.

Stream Name: _____ Location: _____

Observers: _____ Stream Type: _____ Valley Type: _____ Date: _____

		Installation Data (1st Year)			Recovery Data (2nd Year)				
		From cross-section	Particles near chain			Chain recovery		Particles near chain	
		Station (ft)	Elevation (ft)	Largest (mm)	2nd Largest (mm)	Scenario # (1–5)	Scour depth[a] (ft)	Elevation[b] (ft)	Net change[c] (ft)	Largest (mm)	2nd Largest (mm)
Riffle	Chain #1										
	Chain #2										
Glide	Chain #3										
	Chain #4										

Scenario #1.   Scenario #2.   Scenario #3.   Scenario #4.   Scenario #5. (Oops)

[a] Scenario 2 or 3. Scenario 2: Enter length of chain exposed. Scenario 3: Enter length of chain exposed then subsequently buried.
[b] Scenario 3 or 4. Scenario 3: Enter elevation of bed at same station @ 2nd year. Scenario 4: Enter depth of material over chain.
[c] Scenario 3: Subtract 1st and 2nd year elevations to calculate net change in bed.

6-12

Chapter 6

Monitoring

(4) Bank pins (see Table 6-1)

Bank pins are smooth steel rods, four feet long, which are driven horizontally and flush to the bank surface at various positions in the streambank. The amount of exposed pin upon resurvey following runoff events or annually is measured as the amount of lateral erosion at that site. Sod mats cannot be "pinned," as the pins hold the mats in place. Pins also cannot be installed in cobble/gravel matrix bank material due to the direct disturbance to the bank. An example of installed pins in a streambank is shown in the following bank profile section.

(5) Bank profile (see Table 6-1)

The most detailed measure of bank erosion rate is implementing a bank profile. This is also used when sod mats overhang the site or in cobble/gravel matrix bank material where installations of bank pins cause disturbance to the bank material.

The profile is accomplished in conjunction with a permanent cross-section. A toe pin is installed off-set from the bank. An elevation rod (plumb) is set on the toe pin with corresponding horizontal measurements taken to intercept the bank (**Figure 6-5**). A resurvey at the toe pin location allows a comparison of a detailed computation of a change in bank profiles, yielding lateral erosion rate. The bank profile is completed at least once per year to obtain annual erosion and contribution to the sediment supply. **Figure 6-6** depicts the procedure to monitor bank erosion using erosion pins and bank profiling methods.

Bank erosion monitoring also validates or develops empirically-based relations for specific regions using the BANCS model for corresponding BEHI and NBS ratings. The bank erosion data also shows the effectiveness of mitigation efforts and contribution of streambank erosion to annual sediment yield.

Bank profile data is documented in the bank profile **Worksheet 6-2**. The RIVERMorph™ software program also has a bank profiling tool that computes the lateral erosion rate and sediment yield from the streambank.

Figure 6-5. Measuring the bank profile at a toe pin.

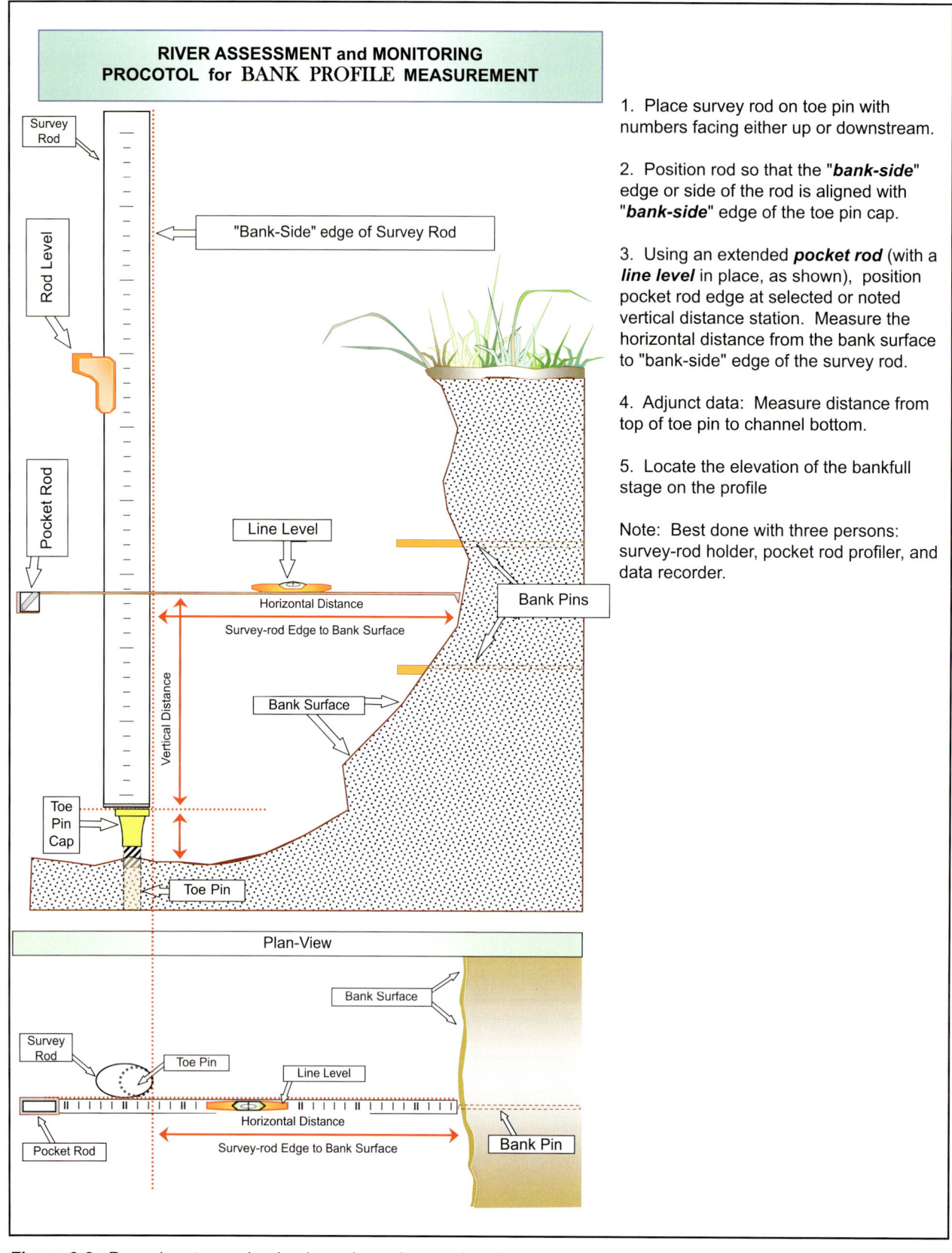

WATERSHED ASSESSMENT OF RIVER STABILITY AND SEDIMENT SUPPLY

Figure 6-6. Procedure to monitor bank erosion using erosion pins and bank profiling methods.

6-14

Worksheet 6-2. Bank profile and bank erosion summary data form.

BANK PROFILE WORKSHEET

Stream:	Observers:
Location:	Date:
Cross-section:	BEHI adjective:
Bank:	NBS adjective:
Toe pin station (ft):	Predicted erosion (ft):
Toe pin elevation (ft):	Measured erosion (ft):

Date:

Horizontal	Vertical	Notes

Date:

Horizontal	Vertical	Notes

Vertical Bank Profile

(Graph: Vertical Distance (ft) 0–8 vs. Horizontal Distance (ft) 0–6)

(6) Replicate annual survey (see Table 6-1)

Permanent measured sites are resurveyed annually (at a minimum), including cross-sections, pebble count, bar sample, Pfankuch stability rating, scour chains and bank pins/profiles. Longitudinal profiles are resurveyed and tied into benchmarks for the cross-sections and toe pins. The overlay comparing annual resurveys provides data on change in depth and slope of bed features, including pools, runs, glides and riffles. Generally, a resurvey of two full meander lengths or 20–30 channel widths provides ample data to evaluate channel response. **Worksheet 6-3** is used to compare data for annual resurveys (with the exception of the scour chain data, which is recorded in **Worksheet 6-1**).

Worksheet 6-3. Summary of annual data comparisons.

Data Comparison Form							
Stream:			Reach:			Observers:	
Date - Year 1:				Date - Year 2:			
Revised Pfankuch channel stability rating							
Year 1:				Year 2:			
	Riffle XS:		Pool XS:		Glide XS:		Notes
	Year 1	Year 2	Year 1	Year 2	Year 1	Year 2	
Cross-section dimensions							
Width$_{BKF}$							
Mean depth$_{BKF}$							
Width/depth ratio							
XS Area$_{BKF}$							
Max depth$_{BKF}$							
Pebble count							
D_{35} (mm)							
D_{50} (mm)							
D_{84} (mm)							
Bar sample							
D_{35} (mm)							
D_{50} (mm)							
D_{84} (mm)							
Largest particle (mm)							
Bank erosion							
BEHI rating							
NBS rating							
Predicted erosion (ft/yr)							
Measured erosion (ft/yr)							

Conclusion

For every prediction methodology, there is a procedure to validate the model. Some methods are more difficult and time-consuming to validate than others. Monitoring improves the predictive capability and eases the uncertainty of prediction.

If monitoring indicates that specific management practices are creating obvious impairment, revised practices or specific, process-based mitigation, such as restoration, may be recommended. The user will gain the most confidence in any monitoring procedure by incorporating field measurements that validate predictions and determine whether the initial assessment and resource management objectives have been met.

Assessment and monitoring are important and responsible tasks. The efforts and costs of mitigation to meet specific objectives can only be justified by effectiveness monitoring to demonstrate watershed management success or failure. If the short-term system response does not meet expectations, then a basis for change sets a positive, long-term direction for the future.

*"Land abuse creates impacts;
impairment generates concern;
assessment isolates cause;
understanding implements correction."*

CHAPTER 7

The WARSSS Methodology: A Case Study of Wolf Creek, Colorado

Chapter 7 is an application of the *WARSSS* assessment methodology to the Wolf Creek Watershed in Colorado. The three-phase assessment was conducted to provide a procedural example of the method using historic and current available data. Each section of the chapter provides a step-by-step explanation of that level of the *WARSSS* assessment and includes completed computations and data entry worksheets. Location- and process-specific mitigation and monitoring strategies are offered based on these findings. Figures, tables, flowcharts and worksheets that have been shown in prior chapters are referred to in the captions throughout Chapter 7.

The *Reconnaissance Level Assessment* (*RLA*): Wolf Creek Case Study

The *Reconnaissance Level Assessment* (*RLA*) provides the user with a review of the land use history overlaying the watershed landscape. Primary goals include the identification of sensitive soils, river types and general locations of erosional/depositional features influenced by land use. The step-wise sequence for this assessment is shown in **Flowchart 7-1**. For a detailed discussion of the steps and theories used in this process, see Chapter 3.

WATERSHED ASSESSMENT OF RIVER STABILITY AND SEDIMENT SUPPLY

Reconnaissance Level Assessment (RLA)

Start with problem identification

1) Compile existing data

What sources and effects?
2) Review the landscape history
3) Summarize activities that potentially affect sediment supply and channel stability
4) Identify specific process relations

5) Review the landscape overview and map the watershed

Where do they occur?
6) Identify hillslope processes
7) Document surface erosion
8) Document mass erosion
9) Assess hydrologic processes
10) Identify streamflow changes
11) Analyze channel processes
12) Detect direct impacts to streambanks and channels

13) Summarize problem verification: recognition of places, processes and sources

What places should be targeted?
14) Eliminate areas, sub-watersheds and reaches that are not contributors to impairment
15) Select areas, sub-watersheds and reaches for further assessment

Proceed to RRISSC

Refine problem identification if necessary

Flowchart 7-1 (Flowchart 3-1). The *Reconnaissance Level Assessment (RLA)* step-wise sequence.

7-2

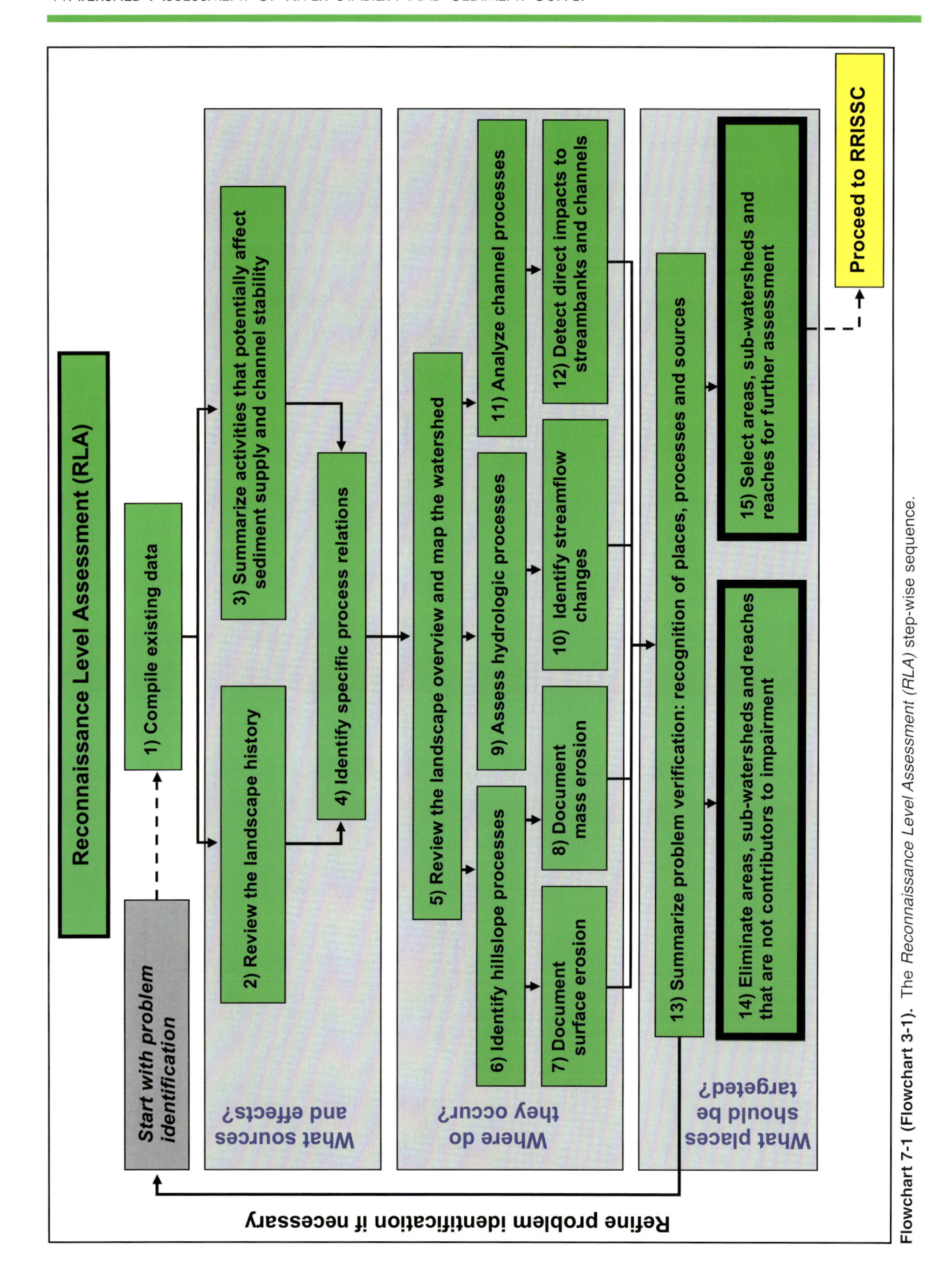

Problem identification

The Wolf Creek Watershed is located in Southwestern Colorado. Driving over Wolf Creek Pass on Highway 160, thousands of clearcut timber acres come into view, as does evidence of landslide activity adjacent to the highway. Heavy sediment deposits are visible at the mouth of Wolf Creek, just above the West Fork of the San Juan River.

The *WARSSS* assessment method is used to determine whether any disproportionate sediment sources due to land management activities occur in the Wolf Creek Watershed. Timber harvest, road construction and grazing have been, and are, the primary land uses. The change in water yield due to timber harvest needs to be determined, as does its influence (if any) on channel source sediment. The sediment supply from older timber harvest roads, as well as from State Highway 160 (Wolf Creek Pass highway), needs to be assessed. Riparian vegetation impacts due to herbicide spraying and heavy grazing pressure require evaluation, as vegetation changes may have affected streambank erosion rates. These land uses may be contributing to sediment and channel impairment problems. This hypothesis will be assessed using all three phases of the *WARSSS* assessment methodology, beginning at the *RLA* level.

Step 1. Compile existing data

To compile existing data, maps and photographs were obtained. A general vicinity map of the Wolf Creek Watershed in Southwestern Colorado is shown in **Figure 7-1**. A 1986 aerial photograph for the Wolf Creek Watershed indicating historic land use is shown in **Figure 7-2**. A USGS 7.5' topographic map is shown in **Figure 7-3**.

Physical Characteristics. The Wolf Creek Watershed has a drainage area of 22 square miles. The elevation at the mouth is 7,900 ft, rising to over 12,000 ft at the Continental Divide. Mean annual precipitation rates vary from approximately 40 inches in the lower watershed to over 60 inches in the upper watershed. Spring snowmelt dominates the annual hydrograph; however, the floods of record are associated with late summer/fall convectional storms.

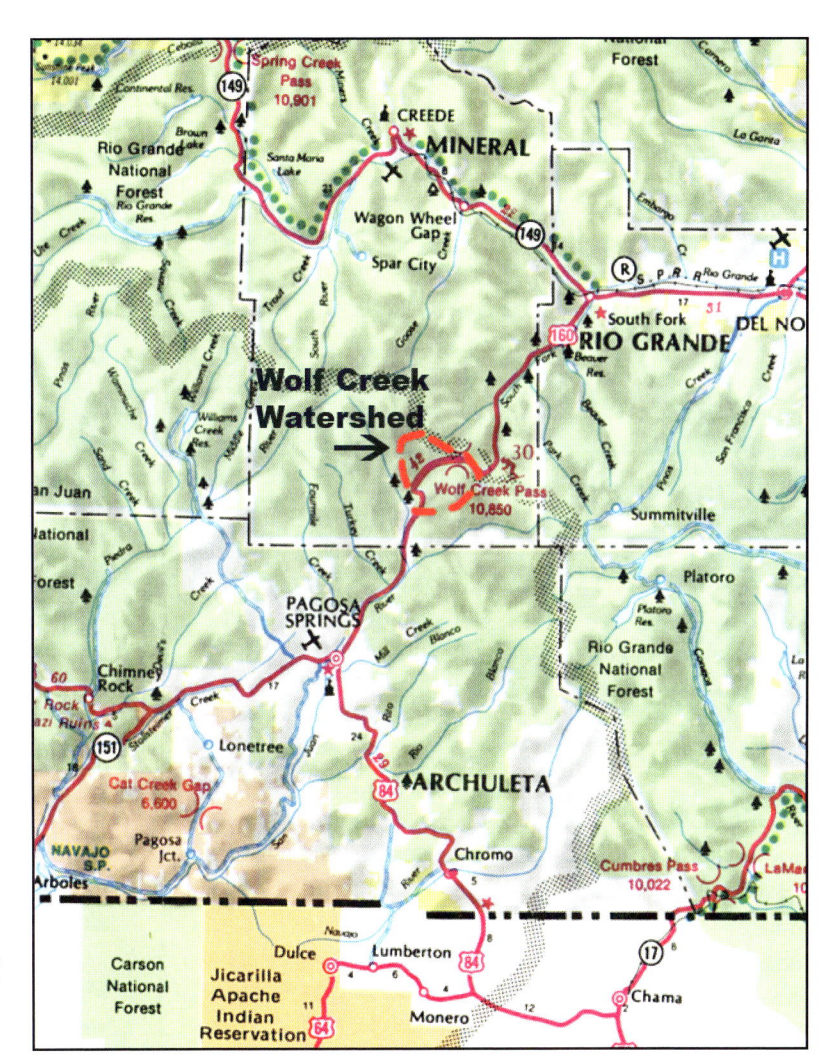

Figure 7-1. Vicinity map of the Wolf Creek Watershed, Southwestern Colorado.

WATERSHED ASSESSMENT OF RIVER STABILITY AND SEDIMENT SUPPLY

The area's geology is associated with volcanism, reworked by alpine glaciation and river alluviation. The soils are associated with glacial tills, colluvium, alluvium and pyroclastic soils that are often subject to mass erosion. The drainage density is very low with many of the tributaries associated with structural control (bedrock). Bedrock-controlled cascades and waterfalls are prevalent in the watershed. Approximately 30% of the watershed on the north and northwest portion are in the Weminuche wilderness and/or in areas without roads.

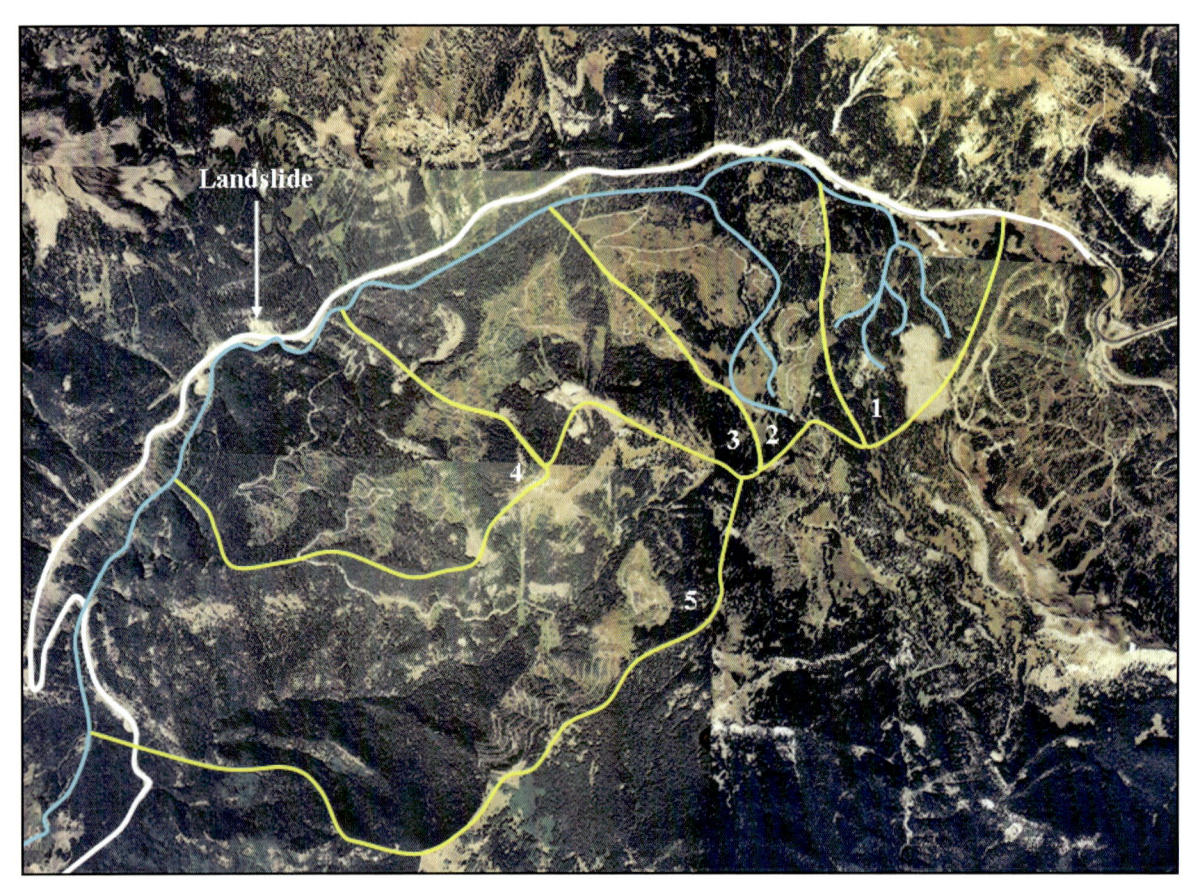

Figure 7-2. Aerial photograph of the Wolf Creek Watershed, showing sub-watersheds, roads and timber harvest.

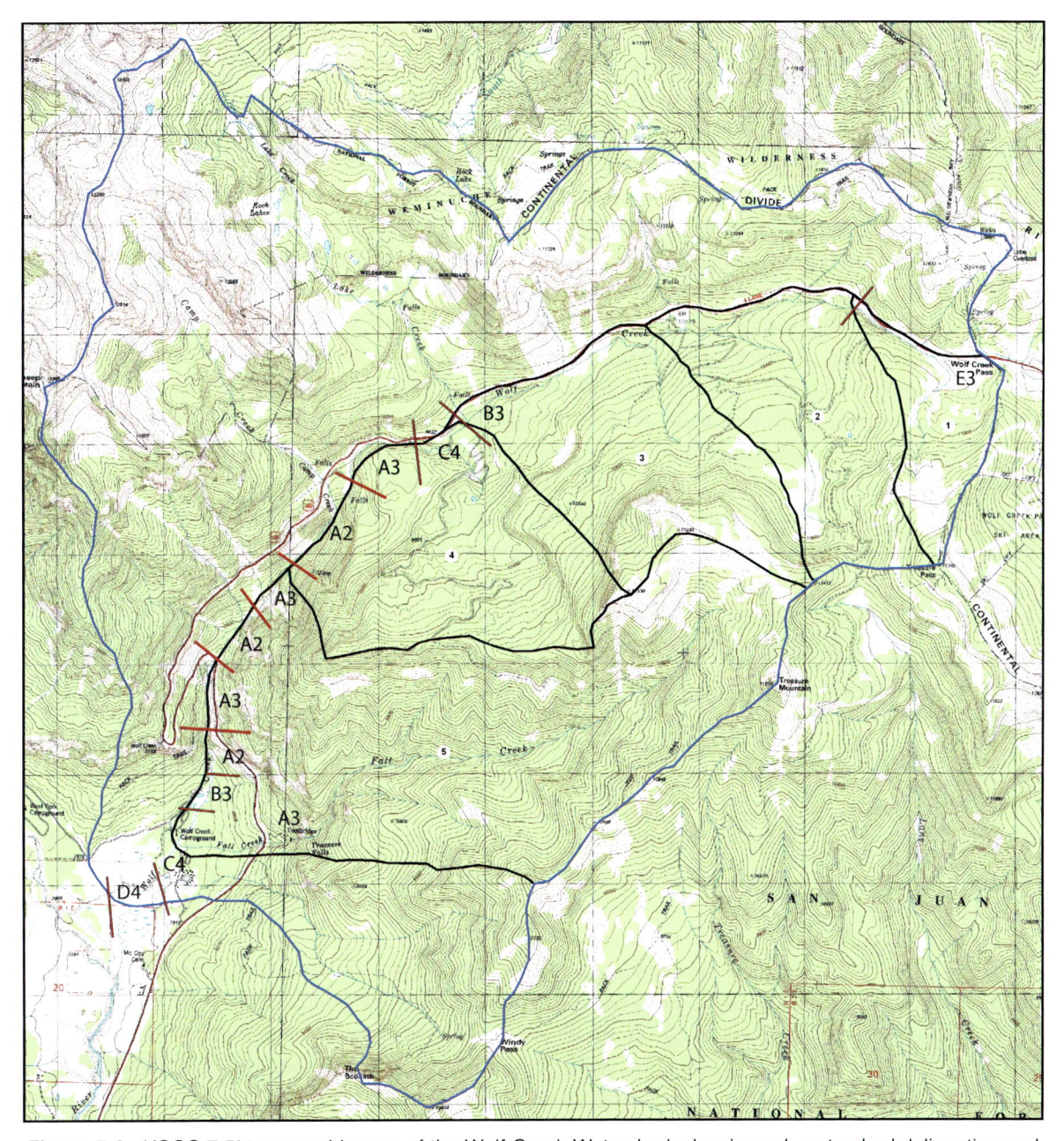

Figure 7-3. USGS 7.5′ topographic map of the Wolf Creek Watershed, showing sub-watershed delineation and stream classification.

Step 2. Review the landscape history

The watershed has had a long history of road construction, including State Highway 160 from Wolf Creek Pass over the Continental Divide to Pagosa Springs, Colorado. The road has been closed several times in the past due to major mudslides, and two large landslides occurred during road reconstruction activities in the 1970s. The soils adjacent to the highway are very susceptible to mass erosion. The road alignment is adjacent to Wolf Creek for the majority of its length.

The San Juan National Forest of the U.S. Department of Agriculture has conducted extensive timber harvest, primarily by clearcutting that included secondary road construction, timber skidding, decking, site scarification and other related silvicultural activities. The stands of removed timber were primarily pine, fir and spruce forests at elevations between 9,000 and 10,000 ft. The majority of the skidding was done by tractor. The timber harvest occurred over a 25-year period and ended in the early 1980s. Aerial photographs depict the extent of the roads and clearcuts in the basin. Most of the roads constructed for timber harvest are still hydraulically active, as they have not been adequately closed or back-sloped to grade. Thus, sub-surface flow interception, ditch-line erosion, snow accumulation in the road prism and lack of vegetative cover are characteristic of the miles of existing haul and skid roads.

Grazing impacts in the upper watershed have been low due to unavailability of suitable range. Private lands in the lower Wolf Creek Watershed, however, have been extensively grazed. Willows were chemically sprayed in the 1980s on private lands in the watershed, in cooperation with the federal government. Adverse riparian impacts appear to have resulted due to vegetative conversion from willows to grass/forb communities. Mining impacts are negligible. Off-road vehicle use occurs on timber roads for hunting and recreation.

Step 3. Summarize activities that potentially affect sediment supply and channel stability

Silvicultural activities, road construction and livestock grazing have been the predominant land use activities in the watershed. Large clearcuts were made in the late 1960s and 1970s, along with road and skid trails to support the logging activities. Evidence of the clearcut and roads are shown in the aerial photo in **Figure 7-2** (see **Step 1**). Road reconstruction of State Highway 160 has been on-going since the 1970s. Two major landslides were caused by the highway widening projects. Encroachment of road fills on Wolf Creek has occurred in numerous areas, along with concentrated runoff from cut bank flow interception and ditch-line drainage.

In the early 1980s, willows were sprayed with 2,4-D and 2,4,5-T along the riparian corridor of lower Wolf Creek. The spraying was done to convert vegetation from willows to a grass/forb community to improve grazing for domestic livestock. The results are shown in the time-trend aerial photographs in **Figure 7-4** (1976) and **Figure 7-5** (1991). The riparian conversion caused a shift from a single-thread, meandering C stream type to an over-wide, braided, D stream type.

The existing land use activities that have potential direct impacts on the flow, sediment and channel stability of Wolf Creek are highlighted in the appropriate columns in **Table 7-1**. The circled cells indicate high potential impact for the areas, reaches and sub-watersheds identified and associated with past land use.

A summary of the potential impacts due to land use activities that can alter stream channel stability and sediment supply for the Wolf Creek Watershed includes:

- Streamflow changes (silviculture and roads);
- Riparian vegetation change: composition and density (agriculture);
- Surface and sub-surface slope hydrology (silviculture and roads);
- Loss of stream buffers (agriculture);
- Altered channel dimension, pattern and profile (agriculture and roads);
- Excess sediment supply (silviculture, agriculture and roads);
- Stream power change: energy redistribution (silviculture, agriculture and roads);
- Excess sediment supply (agriculture and roads);
- Grazing impacts (agriculture); and
- Direct channel impacts (silviculture, agriculture, grazing and roads).

Figure 7-4. Aerial photograph (1976) of Wolf Creek, Colorado showing a C4 stream type prior to spraying.

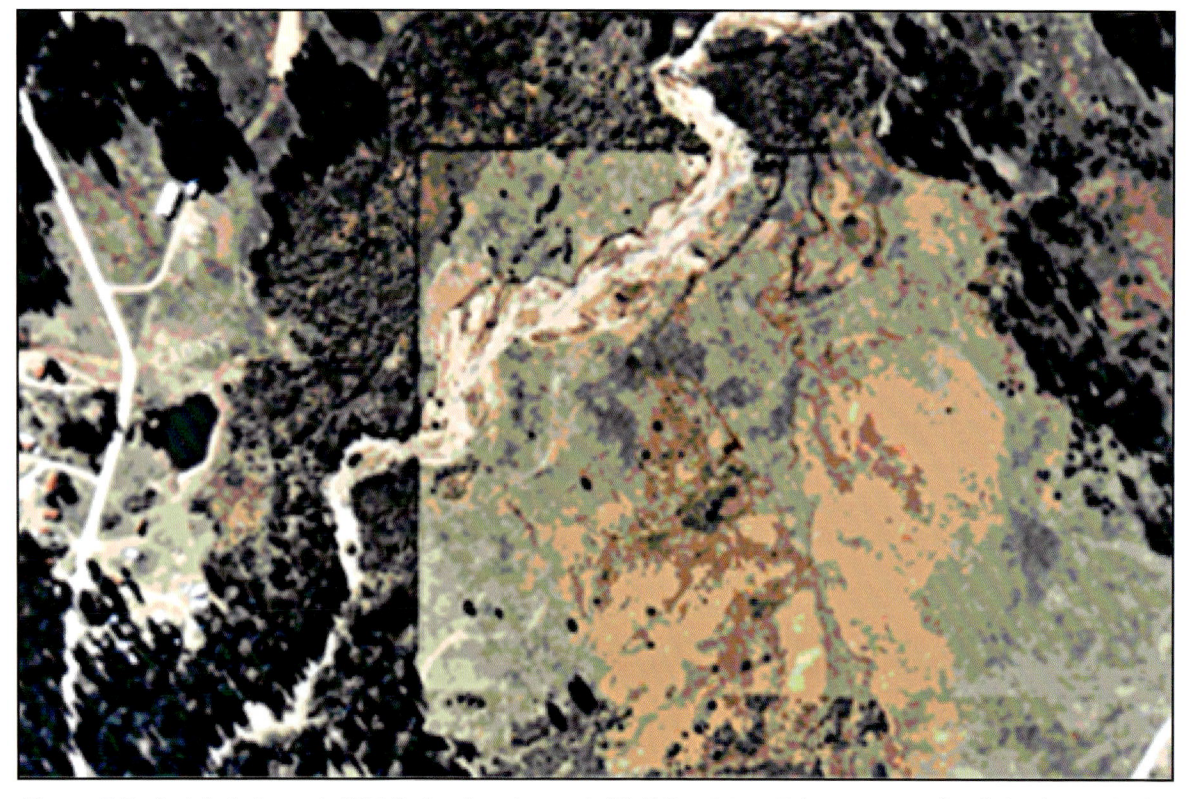

Figure 7-5. Aerial photograph (1991) showing change in Wolf Creek to a D4 stream type but little change in the upstream, untreated, (above fence line) C4 stream type. Flow is left to right.

Chapter 7 — RLA   The WARSSS Methodology: A Case Study of Wolf Creek, Colorado

Table 7-1 (Table 3-1). Land use variables potentially altering stream channels and sediment supply for the Wolf Creek Watershed.

Land Uses	(1) Streamflow changes (magnitude/ timing/duration)	(2) Riparian vegetation change (composition/ density)	(3) Surface disturbance (% bare ground/ compaction)	(4) Surface/ sub-surface slope hydrology	(5) Direct channel impacts that destabilize channel	(6) Clear water discharge	(7) Loss of stream buffers, surface filters, ground cover	(8) Altered dimension, pattern and profile	(9) Excess sediment deposition/ supply (all sources)	(10) Large woody debris in channel	(11) Stream power change (energy distribution)	(12) Floodplain encroachment channel confinement (lateral containment)
Urban development	D	D	D	D	D	D	D	D	—	D	D	D
Silvicultural	(D)	D	D	(D)	D	D	D	—	(D)	D	—	D
Agricultural	D	(D)	D	D	(D)		(D)	(D)	(D)	D	(D)	D
Channelization	D	D	D	D	D	D	D	D	D	D	D	D
Fires	D	D	D	D	—	—	D	D	D	D	D	
Flood control, clearing, veg. removal, dredging, levees	—	—		—	D	D	D	D	—	—	D	D
Reservoir storage, hydropower	D	—		—	D	D	—	—	I/D	—	D	
Diversions, depletions (-) Imported (+)	D	—		D	D	D			I/D			
Grazing	—	(D)	D	D	(D)		(D)	(D)	(D)	D	(D)	D
Roads	(D)	D	(D)	(D)	(D)		—	D	(D)	D	(D)	D
Mining	D	D	D	D	D		D	D	D	D	D	D
In-channel mining	D	D	D	D	D		D	D	D	D	D	D

D = Direct potential impact
I = Indirect potential impact
Blank = Little to no impact

Step 4. Identify specific process relations

The variables influenced, including streamflow change, surface/sub-surface hydrology, riparian vegetation, etc., associated with the potential processes of mass erosion, streambank erosion, channel enlargement and aggradation are summarized and circled in **Table 7-2**. This evaluation was completed rapidly, based on the evidence of land use by location. The major processes related to adverse erosional impacts within the Wolf Creek Watershed include:

- Mass erosion;
- Streambank erosion;
- Channel enlargement;
- Aggradation;
- Degradation;
- Channel successional stage shift; and
- Sediment delivery efficiency changes.

These processes will be evaluated at the *RLA* level. Depending on the outcome of this initial assessment, further analysis may be required at the *RRISSC* and *PLA* levels.

Chapter 7 — RLA   The WARSSS Methodology: A Case Study of Wolf Creek, Colorado

Table 7-2 (Table 3-2). Variables influenced by land management activities and associated potential erosional process impacts for the Wolf Creek Watershed.

Variables Influenced	Surface erosion	Mass erosion	Gully erosion	Streambank erosion	Channel enlargement	Aggradation	Degradation	Channel succession state	Sediment delivery efficiency
(1) Streamflow changes (magnitude/ timing/ duration)		I	D	(D)	(D)	(D)	D	(D)	I
(2) Riparian vegetation change (composition/ density)		(D)	D	(D)	(D)	(D)	D	(D)	I
(3) Surface disturbance (% bare ground/ compaction)	D		D (rills-gully)	I	I	I	I	I	D
(4) Surface/ sub-surface slope hydrology	D		D	I	I	I	I	I	D
(5) Direct channel impacts that destabilize channel			D	(D)	(D)	(D)	D	D	I
(6) Clear water discharge			D	D	D	I	D	D	I
(7) Loss of stream buffers, surface filters, ground cover	D		I	D	D	D	D	D	(D)
(8) Altered dimension, pattern and profile				(D)	(D)	(D)	(D)	(D)	
(9) Excess sediment deposition/ supply			D	(D)	(D)	(D)	(D)	(D)	
(10) Large woody debris in-channel		D		D	D	D	D	D	
(11) Stream power change (energy redistribution)			D	(D)	(D)	(D)	(D)	(D)	
(12) Floodplain encroachment channel confinement (lateral containment)		I	I	D	D	D		I	D

D = Direct potential contribution
I = Indirect potential contribution
Blank = Little to no influence

Potential Erosional Process Impacts

7-11

Step 5. Review the landscape overview and map the watershed

The locations of sub-watersheds that need a more detailed assessment within the Wolf Creek Watershed are delineated on the topographic map shown in **Figure 7-3** (see **Step 1**). The reason for this delineation is the obvious concentration of roads and clearcuts in these sub-watersheds that may potentially change the flow and sediment regimes in Wolf Creek. The north and west portions of the watershed are either in wilderness or have not experienced land use activities. As a result, *these north and west areas were not delineated for further study*.

Stream types (Level I) are also delineated within the Wolf Creek Watershed, as shown in **Figure 7-3**. The materials as indicated by the numerical identifiers were inferred by knowledge of the local geology and landforms. Channel materials were indicated in association with glacial tills, volcanism, landslide debris and river alluvium. Field validation (Level II classification) for specific locations will be accomplished during the *RRISSC* assessment level.

In addition to the sub-watershed and stream type delineation, river reaches are mapped that parallel steep, unstable road locations where eroded soil has a high transport (sediment delivery) efficiency relative to the channel (**Figure 7-6**). An obvious landslide, undercut by Highway 160, is also mapped and shows mass erosion sediment sources. No evidence of Hortonian overland flow exists, and no field evidence of rills or gullies associated with surface disturbance activities was noted. Thus, surface erosion potential is low. This is not the case, however, for cut and fills, un-surfaced roads and ditch lines.

Stream orders were also mapped by sub-watershed. Wolf Creek is a fourth-order stream on its lower reaches. The majority of land use occurs on the south and east side of Highway 160.

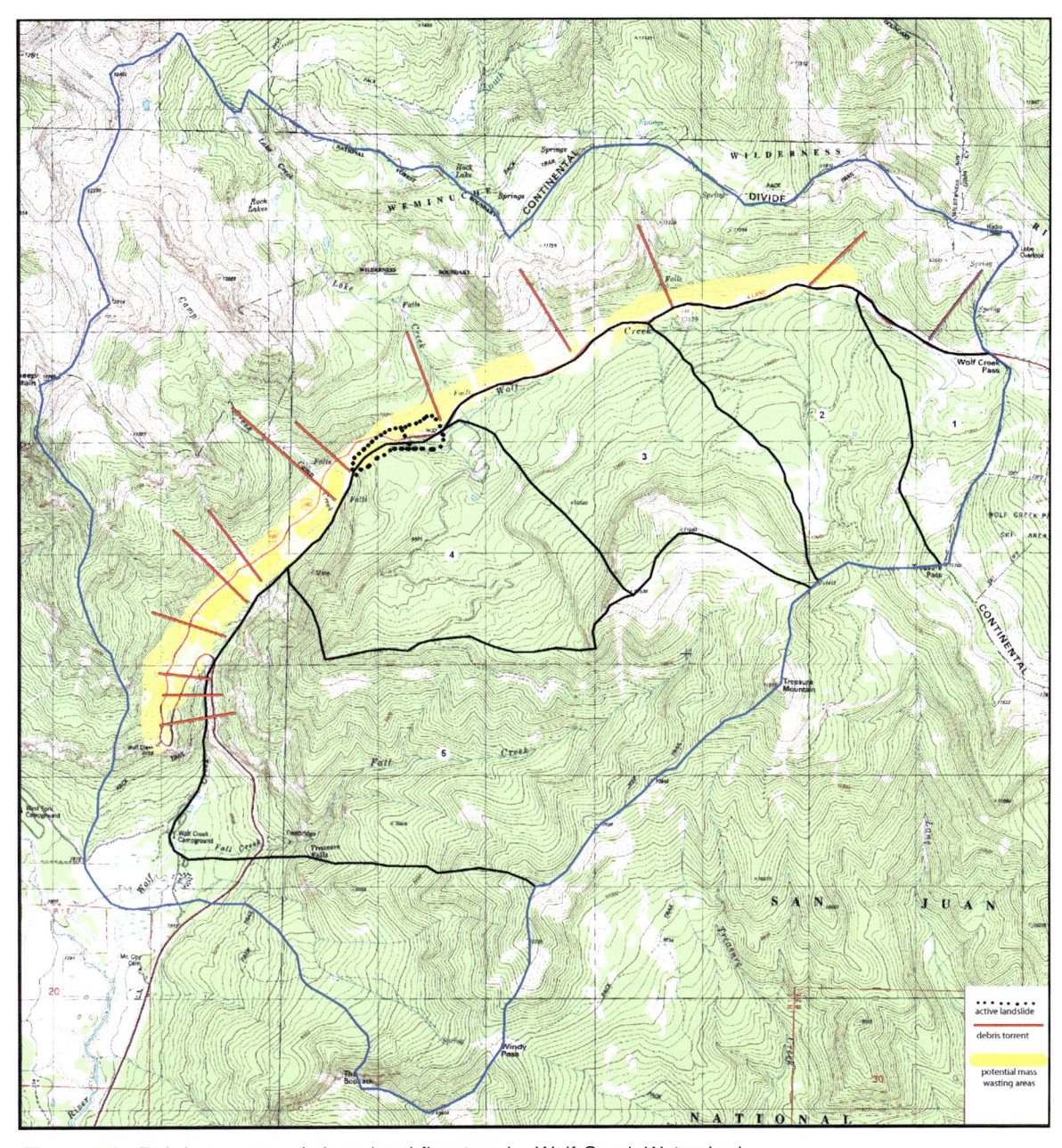

Figure 7-6. Debris torrent and slump/earthflow terrain, Wolf Creek Watershed.

Step 6. Identify hillslope processes

An obvious landslide caused by Highway 160 is located on the map in **Figure 7-6** (see **Step 5**). Landslide debris encroached on Wolf Creek and continues to provide an annual sediment source. Approximately 5,680 acres have been harvested in the basin, which represents approximately 38% in equivalent clearcut condition of a fourth-order watershed. The percentage clearcut for respective sub-watersheds would obviously be much higher for their corresponding second- or third-order streams. These sub-watersheds should advance to the *RRISSC* assessment phase.

Step 7. Document surface erosion

Due to low drainage density, good internal drainage and lack of evidence of rills and/or gullies associated with the various land uses, surface erosion and delivery has a low probability and is not considered further. The exception to this is road source erosion and potential delivery assessed in **Step 10**. None of the surface erosion guidance criteria 1–4 in **Table 7-3** justify advancement to *RRISSC* for the conditions in Wolf Creek. Guidance criteria ratings are recorded in **Worksheet 7-1** for both geographic locations and stream types. A completed version of this summary worksheet appears in **Step 15** at the end of this section.

Table 7-3 (Table 3-3). Guidance criteria for advancement to the *RRISSC* assessment based on surface erosion.

Surface erosion guidance criteria for advancement to *RRISSC*
1. If surface erosion is evident on steep, dissected slopes.
2. If surface erosion is evident on unstable soils at lower slope positions in close proximity to drainageways.
3. If activities such as skid trails are continuous down-slope indicating a high potential of surface erosion converted to sediment delivery to a drainageway.
4. If surface disturbance activities occur on rill-dominated slopes.

Step 8. Document mass erosion

The location of Highway 160 intersects numerous slump/earthflow and debris flow/debris torrent landforms. It is apparent that many of these erosional landforms are kept active due to the highway influence. The locations of a slump/earthflow landslide and debris flow/debris torrents were mapped in **Figure 7-6** (see **Step 5**). These slides met all five of the mass erosion criteria in **Table 7-4** required for advancement to *RRISSC*. Debris flows from road fills on the lower end of the switchback will require additional assessment due to very high risk. It is necessary to include Highway 160 for mass erosion prediction at the *RRISSC* assessment level.

Also, the A3 stream type reaches that are intercepted by State Highway 160 are kept active for debris torrent/debris avalanche processes. This is due to the increased flow/saturation from sub-surface flow interception and great distances between cross-drains from ditch-line water. Increased water routed down the ditch lines is contributed by reduced evapo-transpiration from the large acreages of cut bank and road surface as well as sub-surface flow interception. Guidance criteria 1, 4 and 5 are met; thus, the A3 stream types influenced by Highway 160 and associated with accelerated mass erosion need to advance to the *RRISSC* assessment level (**Worksheet 7-1**).

Table 7-4 (Table 3-4). Guidance criteria for advancement to the *RRISSC* assessment for mass erosion.

Mass erosion guidance criteria for advancement to *RRISSC*
1. If evidence exists of recent (within last 10 years) slump/earthflow and/or debris flow/debris avalanche activity.
2. If slide activity is located on steep, concave, continuous slopes.
3. If there is a high percentage of vegetation clearing in proximity to landslide prone terrain.
4. If the location of slide activity is in or adjacent to drainageways.
5. If evidence exists of slump/earthflow and/or debris flow/debris avalanche caused by road location.

Step 9. Assess hydrologic processes

The high percentage of equivalent clearcut areas and the evident A3, C and D stream types downstream of the timber harvest blocks within the fourth-order watershed are susceptible to channel adjustment due to flow-related change. This can be associated with either bed and/or streambank erosion. It is important to review stream types at this step to evaluate the potential flow response associated with the stream type of the receiving channel. Stream types in Wolf Creek are shown in **Figure 7-3** (see **Step 1**). Because some of the receiving stream types are A3, C and D, the criteria to advance to the *RRISSC* assessment level are met (**Step 10**).

Step 10. Identify streamflow changes

All of the sub-watersheds contributed to potential increases in streamflow and need to advance to *RRISSC*. The clearcut guidance criteria 1 and 3 (**Table 7-5**) were met for rural sub-watersheds 1–5 due to the large amount of clearcutting in a snowmelt-dominated hydrograph. Stream types A3, C4, D4 and E3 also need to advance to *RRISSC* due to guidance criterion 1. Hydrologic recovery for the clearcuts in this snow-dominated region would not be sufficiently long to preclude increases in flow. Road density and poor vegetative recovery of road surfaces also make the existing roads potential contributors to sediment supply and increases in streamflow; thus, Highway 160, sub-watersheds 1–5 and associated stream types A3, C4, D4 and E3 need further assessment at *RRISSC* (criterion 6).

Table 7-5 (Table 3-5). Guidance criteria for advancement to the *RRISSC* assessment for potential streamflow changes.

Streamflow change guidance criteria for advancement to *RRISSC*

1. If rural (non-urban) watersheds have a percentage of bare ground, hydrologic modification due to change in vegetative type and clearcutting timber stands that exceed 20% of first- to third-order watershed areas in the presence of A3–A6, C, D, E, F and G stream types.

2. If urban watersheds have impervious conditions that exceed 10% of second- to third-order watershed areas in the presence of A3–A6, C, D, E, F and G stream types. No hydrologic recovery is recognized.

3. *Time-trend of vegetation* (rural or non-urban). If the vegetative conversions occurred within the last 15–20 years for rain-dominated or temperate climates, or 80 years or less for snowmelt-dominated montane and/or sub-alpine climatic regions, there likely has not been sufficient time for full hydrologic recovery. These recovery times are based on revegetating sites and the time necessary to regain pre-treatment evapo-transpiration, snow deposition patterns and other similar processes reflecting consumptive water loss.

4. *Diversions, imported water, water depletion and/or return flows*. If the recipient or depleted stream types are alluvial and susceptible to degradation, aggradation, streambank erosion or enlargement (stream types A3–A6, C, D, E, F and G).

5. *Reservoirs*. All reservoirs located on alluvial channel types or those incised in landslide debris, glacial tills, etc. need to be assessed at the *RRISSC* or *PLA* level. This is due to the complexity of potential impacts, the nature of the stream type, the variation in the operational hydrology of the reservoir, potential ramping flows due to power generation (rapid raising and lowering of flow stage), timing of releases with downstream unregulated tributaries and clear water discharge effects. Temperature and other water quality parameters may also need to be assessed.

6. *Roads*. If roads are located in the lower one-third of slope position on moderate to steep slopes (sub-surface flow interception). Road densities over 10% of watershed area of first- and second-order watersheds. Roads traversing highly dissected slopes or with multiple stream crossings. Drainageway crossings associated with floodplain fill blockages, and base-level changes above and/or below culverts and/or bridges.

Step 11. Analyze channel processes

The guidance criteria for channel processes (**Table 7-6**) were met by:

- Potential increases in streamflow due to timber harvest/clearcut blocks (criterion 1);
- Having a D stream type in a valley type VIII (criterion 2);
- Successional change in stream stability from a C to D stream type at the lower end of the Wolf Creek valley (criterion 3);
- Aerial photo time-trend evidence of excess deposition, lateral channel migration and enlargement resulting in a change from a C to D stream type (see time-trend aerial photo evidence in **Figures 7-4** and **7-5**) (criteria 4 and 5); and
- Through fills around culverts and bridges that occur on Highway 160 through drainageway crossings without providing floodplain/flood-prone area drainage (criterion 6).

Based on these guidance criteria, it is necessary to proceed with additional assessments at the *RRISSC* level for the identified sub-watersheds and stream types (**Worksheet 7-1**).

Table 7-6 (Table 3-6). Guidance criteria for advancement to the *RRISSC* assessment for channel processes.

Channel process guidance criteria for advancement to *RRISSC*

1. If there are potential increases in streamflow within the sub-watershed associated with A3–A6, C, D, E, F and/or G stream types.

2. If there appear to be stream types that are of the unstable form for a given valley type, i.e., G and F types in valley types II, IX, and X, then proceeding to the *RRISSC* assessment level is recommended. The observer is reminded to compare reference to existing conditions to determine if the existing stream type is appropriate for the valley type being studied. For example, if a D stream type was mapped in a valley type IX (glacial outwash valley), it would be indicative of the stable form for that valley type. However, if a D stream type was mapped in valley types II, IV, VI, VIII or X, it would not represent the typical stable form and should be flagged to require the *RRISSC* assessment.

3. If the current stream type departs from the stable form as indicated in the potential channel evolution or successional stage of channel adjustment relations (**Figure 2-38**), then proceed to the *RRISSC* assessment level (see also *Succession* background information in Chapter 2).

4. If aerial photographs or site visits reveal the following channel-destabilizing processes:
    a. aggradation (excess deposition, wide/shallow)
    b. degradation (incision, floodplain abandonment)
    c. lateral accretion (excess bank erosion)
    d. avulsion (abandonment of previous channels)
    e. enlargement
    f. meandering to braided channels

5. If time-trend aerial photography analysis indicates little recovery of apparent channel condition associated with the magnitude, extent and/or obvious consequence of channel change.

6. If road drainage, stream crossings and/or lack of floodplain drains (through-fill crossings) cause adverse channel adjustment.

Step 12. Detect direct impacts to streambanks and channels

Willow spraying on the lower end of Wolf Creek changed not only the riparian vegetation composition from willows to grass/forb, but also changed the stream type from C to D (see time-trend aerial photo evidence in **Figures 7-4** and **7-5** (see **Step 3**), and **Figure 7-7**). The obvious change in channel dimension and pattern shown in the time-trend aerial photographs meets guidance criterion 1 in **Table 7-7** to advance to the *RRISSC* assessment level for the A3 and D4 stream types. The change from a willow to grass/forb composition on the D4 stream type from analysis of the same aerial photos also meets guidance criterion 2 (**Worksheet 7-1**).

Figure 7-7. Typical eroding bank on the D4 (braided) reach of Wolf Creek following willow removal due to herbicide spraying.

Also, the lower reach of the C4 stream type has been modified by the location of the bridge. There are no floodplain drains installed; thus, all flood flows must pass through the active channel, causing bed scour. The approach-section reach of the stream was straightened upstream and downstream from the bridge, and a levee was constructed from the excavated material. These direct impacts to the channel necessitate advancing to the *RRISSC* assessment level as criterion 1 was exceeded (**Worksheet 7-1**).

Table 7-7 (Table 3-7). Guidance criteria for advancement to the *RRISSC* assessment due to direct channel impacts.

Direct channel impact guidance criteria for advancement to *RRISSC*
1. If the stream's dimension, pattern and profile have been altered due to direct impacts from various sources, then the influence of time of disturbance on channel recovery must be determined at a more advanced level of assessment.
2. If evidence exists of riparian vegetation alteration from woody plants to a grass/forb community or annuals.

Step 13. Summarize problem verification: recognition of places, processes and sources

The maps and aerial photographs showing the delineation of five sub-watersheds, the mass erosion site, roads, clearcuts and stream types indicate the specific locations to be evaluated at the *RRISSC* level of assessment.

Step 14. Eliminate areas, sub-watersheds and/or river reaches that are not contributors to either sediment or stability problems

The areas of the Wolf Creek Watershed outside of the delineated sub-watersheds 1–5 and areas *not* adjacent to State Highway 160 are eliminated from further assessment. Elimination of these areas from the assessment is due to their wilderness and roadless designations and their overall undeveloped nature. The A2 and B3 stream types shown in **Figure 7-3** (see **Step 1**) are also eliminated from further assessment due to the analysis completed in **Worksheet 7-1** (see **Step 15**), where the guidance criteria to advance these stream types was not met.

Step 15. Select areas, sub-watersheds and river reaches for further assessment

The sub-watersheds 1–5, mass erosion sites, roads and stream type locations mapped in **Figures 7-2** and **7-3** (see **Step 1**) and **Figure 7-6** (see **Step 5**) will advance for assessment at the *RRISSC* level. As Hortonian overland flow is rare on the soils where timber harvest occurred, further surface erosion assessments in these areas are not recommended.

The guidance criteria selections provided for each process and/or land use assessment in *RLA* are summarized in **Worksheet 7-1**. Locations and stream types selected for advancement and their associated processes are as follows:

- State Highway 160, for mass erosion and streamflow change;
- Sub-watersheds 1–5, for streamflow change and forest access roads;
- Sub-watershed 5, for channel processes; and
- Stream types A3, upper C4, lower C4, D4 and E3, for mass erosion, channel processes, direct channel impacts, streamflow changes and/or roads.

The following section examines these areas at the *RRISSC* assessment level.

WATERSHED ASSESSMENT OF RIVER STABILITY AND SEDIMENT SUPPLY

Worksheet 7-1 (Worksheet 3-1). Evaluation and summary of guidance criteria for selection of sub-watersheds to proceed to *RRISSC* or to exclude from further assessment.

	Sub-watershed/ reach location ID	Step 7: Surface erosion — Circle selected guidance criteria number (Table 3-3)*	Reason for exclusion	Step 8: Mass erosion — Circle selected guidance criteria number (Table 3-4)*	Reason for exclusion	Step 10: Streamflow change — Circle selected guidance criteria number (Table 3-5)*	Roads	Reason for exclusion	Step 11: Channel processes — Circle selected guidance criteria number (Table 3-6)*	Reason for exclusion	Step 12: Direct channel impacts — Circle selected guidance criteria number (Table 3-7)*	Reason for exclusion	Step 15 — ✓location selected for advancement to *RRISSC***
1.	1	(1) (2) (3) (4)		(1) (2) (3) (4) (5)		(1) ②③(4) (5)	⑥		(1) (2) (3) (4) (5) (6)		(1) (2)		✓
2.	2	(1) (2) (3) (4)		(1) (2) (3) (4) (5)		(1)②③(4) (5)	⑥		(1) (2) (3) (4) (5) (6)		(1) (2)		✓
3.	3	(1) (2) (3) (4)		(1) (2) (3) (4) (5)		(1)②③(4) (5)	⑥		(1) (2) (3) (4) (5) (6)		(1) (2)		✓
4.	4	(1) (2) (3) (4)		(1) (2) (3) (4) (5)		(1)②③(4) (5)	⑥		(1) (2) (3) (4) (5) (6)		(1) (2)		✓
5.	5	(1) (2) (3) (4)		(1) (2) (3) (4) (5)		(1)②③(4) (5)	⑥		①②③(4) (5) (6)		(1) (2)		
6.	Wilderness areas	(1) (2) (3) (4)		(1) (2) (3) (4) (5)	wilderness	(1) (2) (3) (4) (5)	⑥	wilderness	(1) (2) (3) (4) (5) (6)		(1) (2)		
7.	State Highway 160	(1) (2) (3) (4)		①②③④⑤		(1) (2) (3) (4) (5)	⑥		①②(3) (4) (5)⑥		(1) (2)		✓
8.		(1) (2) (3) (4)		(1) (2) (3) (4) (5)		(1) (2) (3) (4) (5)	⑥		(1) (2) (3) (4) (5) (6)		(1) (2)		
9.	Stream type A2	(1) (2) (3) (4)		(1) (2) (3) (4) (5)		(1) (2) (3) (4) (5)	⑥		(1) (2) (3) (4) (5) (6)		(1) (2)		
10.	A3	(1) (2) (3) (4)		①(2) (3)④⑤		(1) (2) (3) (4) (5)	⑥		①(2) (3) (4) (5)⑥		①(2)		✓
11.	B3	(1) (2) (3) (4)		(1) (2) (3) (4) (5)		(1) (2) (3) (4) (5)	⑥		(1) (2) (3) (4) (5) (6)		(1) (2)		
12.	C4 upper	(1) (2) (3) (4)		(1) (2) (3) (4) (5)		①(2) (3) (4) (5)	⑥		①(2) (3) (4) (5)⑥		(1) (2)		✓
13.	C4 lower	(1) (2) (3) (4)		(1) (2) (3) (4) (5)		①(2) (3) (4) (5)	⑥		①(2) (3) (4) (5)⑥		(1) (2)		✓
14.	D4	(1) (2) (3) (4)		(1) (2) (3) (4) (5)		(1) (2) (3) (4) (5)	⑥		①(2)③④⑤(6)		①(2)		✓
15.	E3	(1) (2) (3) (4)		(1) (2) (3) (4) (5)		①(2) (3) (4) (5)	⑥		①(2) (3) (4) (5) (6)		①(2)		✓
16.		(1) (2) (3) (4)		(1) (2) (3) (4) (5)		(1) (2) (3) (4) (5)	⑥		(1) (2) (3) (4) (5) (6)		(1) (2)		

*Criteria based on overall review of the list in Tables 3-1 and 3-2 (Tables 7-1 and 7-2).

**Locations that meet one or more selection criteria should proceed to the *RRISSC* assessment level.

The *Rapid Resource Inventory for Sediment and Stability Consequence* (*RRISSC*) Assessment Level: Wolf Creek Case Study

The *Reconnaissance Level Assessment* (*RLA*) determined that selected sub-watersheds for approximately one-half of the watershed area of the Wolf Creek Watershed need to be assessed at a more detailed level due to combined hillslope, hydrologic and channel processes.

The riparian impacts on the lower reaches of Wolf Creek, evident accelerated streambank erosion, channel successional shift from a C to D stream type and the apparent stream instability all justify a more detailed channel assessment.

Due to the nature and extent of silvicultural treatments (clearcutting) and the potential changes in the hydrograph discovered in the *RLA* phase, the timber harvest impacts need to be evaluated to determine the potential for flow-related sediment sources.

The extent and nature of the roads in the watershed also show high potential for sediment delivery and should be assessed. Mass erosion processes were identified along Highway 160 due to the influence of highway reconstruction. Also, the road location parallels Wolf Creek and presents high potential for sediment introduction directly from ditch-line, fill slope and cut bank erosion.

The first steps of the *RRISSC* assessment involve a review of the landscape character and the nature and extent of land use activities. Applying assessment materials from *RLA* avoids duplication of effort. Conducting the *RLA* assessment first helps to organize data and familiarize the observer with the watershed and its land use history. Most importantly, *RLA* indicates what land uses are affecting the processes to evaluate and where to focus efforts spatially within the watershed.

Sequential assessment steps

The sequence for the *RRISSC* assessment is shown in **Flowchart 7-2**. The following steps in the Wolf Creek Watershed assessment are associated with this sequence. For a detailed explanation of the steps and calculations required at this level, refer to Chapter 4.

Watershed Assessment of River Stability and Sediment Supply

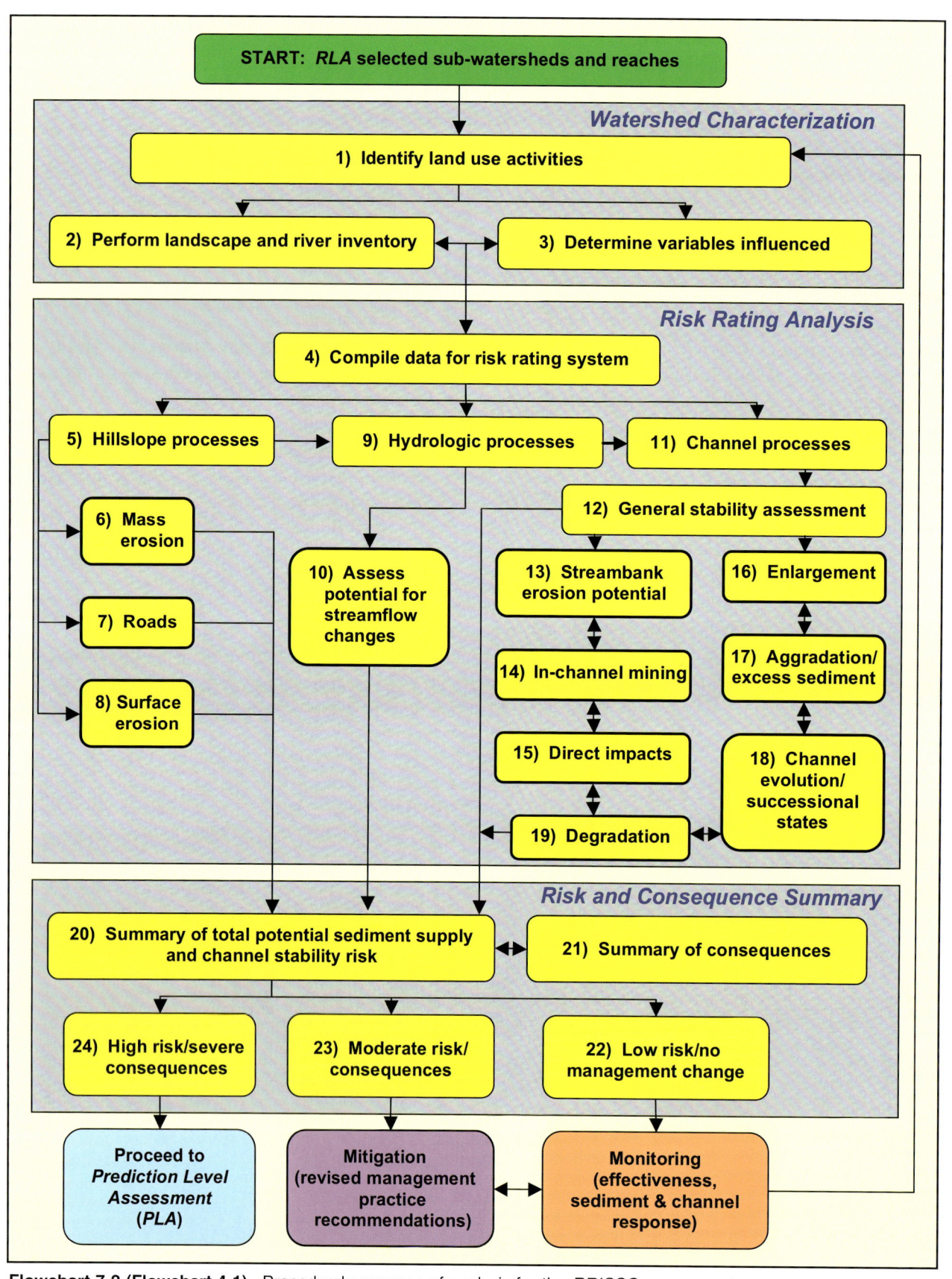

Flowchart 7-2 (Flowchart 4-1). Procedural sequence of analysis for the *RRISSC* assessment.

START: *RLA* selected sub-watersheds and reaches

Watershed Characterization
1) Identify land use activities
2) Perform landscape and river inventory
3) Determine variables influenced

Risk Rating Analysis
4) Compile data for risk rating system
5) Hillslope processes
6) Mass erosion
7) Roads
8) Surface erosion
9) Hydrologic processes
10) Assess potential for streamflow changes
11) Channel processes
12) General stability assessment
13) Streambank erosion potential
14) In-channel mining
15) Direct impacts
16) Enlargement
17) Aggradation/excess sediment
18) Channel evolution/successional states
19) Degradation

Risk and Consequence Summary
20) Summary of total potential sediment supply and channel stability risk
21) Summary of consequences
22) Low risk/no management change
23) Moderate risk/consequences
24) High risk/severe consequences

- Proceed to *Prediction Level Assessment* (*PLA*)
- Mitigation (revised management practice recommendations)
- Monitoring (effectiveness, sediment & channel response)

7-22

Step 1. Identify land use activities

The land uses that potentially influence sediment supply and stream channel stability have been assessed at the *RLA* level for Wolf Creek. If the user is starting at the *RRISSC* level, it is important to follow **Steps 1–5** of *RLA*.

The list of activities and categories within each activity shown in **Table 7-8** provides a highlighted list of management activities and their relative influences on sediment supply and river stability. The summary of the existing land use assessment highlights the variables and land use categories for hillslope, hydrologic and channel processes. This land use activity list identifies processes with potential to affect sediment supply and/or stream channel stability in the Wolf Creek Watershed.

Table 7-8 (Table 4-1). A list of various land and river management activities and inventory information by hillslope, hydrologic and channel processes.

Land and river management activities and inventory information

Hillslope processes
Roads
1. Road density
2. Location within the drainage network
3. Number and type of stream crossings (culverts, bridges, fords)
4. Type and quality of construction
5. Road surfacing
6. Type of drainage (culvert spacing, in-sloped, out-sloped, lined ditches)
7. Drainage area of second- or third-order watershed containing roads

Surface disturbance
1. Compaction potential
2. Percent bare ground
3. Vegetation composition
4. Vegetation density change
5. Location in regards to unstable soils

Mining
1. Altered vegetation (composition, density)
2. Altered surface and sub-surface soils
3. Altered surface and sub-surface hydrology
4. Spoil piles, tailings
5. Altered channel dimension, pattern and profile

Silvicultural activities
1. Silvicultural treatments (clearcutting, partial cutting)
2. Log landings
3. Log skidding
4. Site preparation (scarification, burning)
5. Fuel hazard reduction (burning, tractor bunching)
6. Haul roads (density, location, stream crossings)

Table 7-8 (Table 4-1). - *Continued*. A list of various land and river management activities and inventory information by hillslope, hydrologic and channel processes.

Land and river management activities and inventory information - *Continued*

Agricultural applications
1. Type of site preparation (plowing, no-till, summer fallow)
2. Concentration of animals
3. Vegetation type conversions

Urban development
1. Percent impervious of land surface changed due to development
2. Encroachment on drainageways/floodplains
3. Alteration of the drainage network (reducing infiltration, percolation and use of vegetation for reducing and dispersing urban runoff)
4. Stormwater management

Hydrologic processes
Roads
1. Affecting drainage density
2. Change in evapo-transpiration (excess runoff)
3. Change in time of concentration of flow
4. Increasing snowpack storage and water availability for runoff
5. Interception of sub-surface water, increased time of concentration
6. Drainageway crossings: provide direct sediment source from fill erosion and ditch-line source, as well as local channel instability

Surface disturbance
1. Compaction, change in time of concentration of flow, reduced infiltration, potential flow increase
2. Percent vegetative cover, type of vegetative cover and age class distribution affecting evapo-transpiration and water yield
3. Wildfires, controlled burns

Reservoirs, hydropower dams, diversions, imported water
1. Change in sediment regime (clear water discharge)
2. Change in flow magnitude, duration and timing
3. Change in water quality attributes affecting designated uses (water temperature)
4. Ramping flows affecting streambank stability

Many hillslope processes, including urban development, mining, agricultural and silvicultural applications, influence hydrologic processes.

Channel processes
Roads
1. Encroachment on active channels and floodplains, and/or straightening of channels

Table 7-8 (Table 4-1). - *Continued*. A list of various land and river management activities and inventory information by hillslope, hydrologic and channel processes.

Land and river management activities and inventory information - *Continued*

    2. Elimination of riparian buffers, and encroachment of road fills
    3. Drainage density increased sub-surface interception and surface runoff leading sediment directly into stream systems
    4. Time of concentration of flows potentially increased (nature of drainage)
    5. Source of direct sediment introduction
    6. Number and type of drainageway crossings
    7. Age of road

Surface and/or direct disturbance activities
1. Direct disturbance to stream channels by straightening, raising, lowering, lining and levee construction
2. Vegetation clearing and type conversion
3. Large woody debris clearing and tree removal
4. Dams for recreation, flood control and/or hydropower generation, in-channel structures (grade control, check dams)

Mining
1. Reworked channel materials
2. Altered channel dimension, pattern and profile
3. Vertical and lateral containment
4. Riparian vegetation change
5. Tailings; location and extent

Silvicultural activities
1. Timber harvest within riparian (stand removal/management)
2. Surface disturbance within riparian from skidding, site preparation or fuel hazard reduction
3. Roads or skid trails encroached on channel and multiple stream crossings

Agricultural applications
1. Dams and irrigation diversions
2. Poor grazing practices that change vegetation composition and density and add to hoof shear on streambanks
3. Direct disturbance to stream channels due to land drainage
4. Riparian vegetation change to cultivated crops
5. Feedlots and other animal concentration activities

Urban development
1. Encroachment on the drainage network
2. Elimination of floodplains
3. Concentration of flows (storm drain outfalls leading to stream degradation)
4. Direct alteration of channels (dimension, pattern, profile and materials)

Step 2. Perform landscape and river inventory

A Level II stream type classification is completed in **Worksheets 7-2a** through **7-2d** for the A3, C4, D4 and E3 stream types. The Level II classification used field observation to validate the interpretations from soils, geology and landforms made in *RLA*. See the *RLA* section for aerial photographs, topographic maps and Level I valley type and stream classification.

Worksheet 7-2a (Worksheet 4-1). Level II stream classification for the A3 stream type.

Stream Name: **Wolf Creek**	
Basin Name: **San Juan**	Drainage Area: **14080** acres **22** mi²
Location: **SW Colorado below Wolf Creek Pass**	
Twp.&Rge: **T37N, R33E**	Sec.&Qtr.:	Valley Type: **I**
Cross-Section Monuments (Lat./Long.):	
Observers: **Dave Rosgen**	Date: **10-Jul-04**

Bankfull WIDTH (W_{bkf}) — WIDTH of the stream channel at bankfull stage elevation, in a riffle section.	19	ft
Bankfull DEPTH (d_{bkf}) — Mean DEPTH of the stream channel cross-section, at bankfull stage elevation, in a riffle section ($d_{bkf} = A/W_{bkf}$).	2.7	ft
Bankfull X-Section AREA (A_{bkf}) — AREA of the stream channel cross-section, at bankfull stage elevation, in a riffle section.	51	ft²
Width/Depth Ratio (W_{bkf}/d_{bkf}) — Bankfull WIDTH divided by bankfull mean DEPTH, in a riffle section.	7	ft/ft
Maximum DEPTH (d_{mbkf}) — Maximum depth of the bankfull channel cross-section, or distance between the bankfull stage and Thalweg elevations, in a riffle section.	3.9	ft
WIDTH of Flood-Prone Area (W_{fpa}) — Twice maximum DEPTH, or ($2 \times d_{mbkf}$) = the stage/elevation at which flood-prone area WIDTH is determined in a riffle section.	25	ft
Entrenchment Ratio (ER) — The ratio of flood-prone area WIDTH divided by bankfull channel WIDTH (W_{fpa}/W_{bkf}) (riffle section).	1.3	ft/ft
Channel Materials (Particle Size Index) D_{50} — The D_{50} particle size index represents the mean diameter of channel materials, as sampled from the channel surface, between the bankfull stage and Thalweg elevations.	95	mm
Water Surface SLOPE (S) — Channel slope = "rise over run" for a reach approximately 20–30 bankfull channel widths in length, with the "riffle-to-riffle" water surface slope representing the gradient at bankfull stage.	0.088	ft/ft
Channel SINUOSITY (k) — Sinuosity is an index of channel pattern, determined from a ratio of stream length divided by valley length (SL / VL); or estimated from a ratio of valley slope divided by channel slope (VS / S).	1.03	

Stream Type: A3 (See **Figure 2-14**)

Worksheet 7-2b (Worksheet 4-1). Level II stream classification for the C4 stream type.

Stream Name: **Wolf Creek**	
Basin Name: **San Juan**	Drainage Area: **14080** acres **22** mi^2
Location: **SW Colorado below Wolf Creek Pass**	
Twp.&Rge: **T37N, R33E**	Sec.&Qtr.:	Valley Type: **VIII**
Cross-Section Monuments (Lat./Long.):	
Observers: **Dave Rosgen**	Date: **10-Jul-04**

Bankfull WIDTH (W_{bkf})
WIDTH of the stream channel at bankfull stage elevation, in a riffle section.
37 ft

Bankfull DEPTH (d_{bkf})
Mean DEPTH of the stream channel cross-section, at bankfull stage elevation, in a riffle section ($d_{bkf} = A/W_{bkf}$).
1.6 ft

Bankfull X-Section AREA (A_{bkf})
AREA of the stream channel cross-section, at bankfull stage elevation, in a riffle section.
59.2 ft^2

Width/Depth Ratio (W_{bkf} / d_{bkf})
Bankfull WIDTH divided by bankfull mean DEPTH, in a riffle section.
23 ft/ft

Maximum DEPTH (d_{mbkf})
Maximum depth of the bankfull channel cross-section, or distance between the bankfull stage and Thalweg elevations, in a riffle section.
2.6 ft

WIDTH of Flood-Prone Area (W_{fpa})
Twice maximum DEPTH, or (2 x d_{mbkf}) = the stage/elevation at which flood-prone area WIDTH is determined in a riffle section.
320 ft

Entrenchment Ratio (ER)
The ratio of flood-prone area WIDTH divided by bankfull channel WIDTH (W_{fpa} / W_{bkf}) (riffle section).
8.6 ft/ft

Channel Materials (Particle Size Index) D_{50}
The D_{50} particle size index represents the mean diameter of channel materials, as sampled from the channel surface, between the bankfull stage and Thalweg elevations.
48 mm

Water Surface SLOPE (S)
Channel slope = "rise over run" for a reach approximately 20–30 bankfull channel widths in length, with the "riffle-to-riffle" water surface slope representing the gradient at bankfull stage.
0.005 ft/ft

Channel SINUOSITY (k)
Sinuosity is an index of channel pattern, determined from a ratio of stream length divided by valley length (SL / VL); or estimated from a ratio of valley slope divided by channel slope (VS / S).
1.8

Stream Type ▸ **C4** ◂ (See **Figure 2-14**)

Worksheet 7-2c (Worksheet 4-1). Level II stream classification for the D4 stream type.

Stream Name: **Wolf Creek- Lower end**		
Basin Name: **San Juan**	Drainage Area: **14080** acres **22** mi²	
Location: **Near mouth of Wolf Creek**		
Twp.&Rge: **T37N, R33E**	Sec.&Qtr.:	Valley Type: **VIII**
Cross-Section Monuments (Lat./Long.):		
Observers: **Dave Rosgen**		Date: **10-Jul-04**

Bankfull WIDTH (W_{bkf})
WIDTH of the stream channel at bankfull stage elevation, in a riffle section.
203 ft

Bankfull DEPTH (d_{bkf})
Mean DEPTH of the stream channel cross-section, at bankfull stage elevation, in a riffle section ($d_{bkf} = A / W_{bkf}$).
0.8 ft

Bankfull X-Section AREA (A_{bkf})
AREA of the stream channel cross-section, at bankfull stage elevation, in a riffle section.
163 ft²

Width/Depth Ratio (W_{bkf} / d_{bkf})
Bankfull WIDTH divided by bankfull mean DEPTH, in a riffle section.
254 ft/ft

Maximum DEPTH (d_{mbkf})
Maximum depth of the bankfull channel cross-section, or distance between the bankfull stage and Thalweg elevations, in a riffle section.
3.5 ft

WIDTH of Flood-Prone Area (W_{fpa})
Twice maximum DEPTH, or (2 x d_{mbkf}) = the stage/elevation at which flood-prone area WIDTH is determined in a riffle section.
350 ft

Entrenchment Ratio (ER)
The ratio of flood-prone area WIDTH divided by bankfull channel WIDTH (W_{fpa} / W_{bkf}) (riffle section).
N/A ft/ft

Channel Materials (Particle Size Index) D_{50}
The D_{50} particle size index represents the mean diameter of channel materials, as sampled from the channel surface, between the bankfull stage and Thalweg elevations.
19 mm

Water Surface SLOPE (S)
Channel slope = "rise over run" for a reach approximately 20–30 bankfull channel widths in length, with the "riffle-to-riffle" water surface slope representing the gradient at bankfull stage.
0.009 ft/ft

Channel SINUOSITY (k)
Sinuosity is an index of channel pattern, determined from a ratio of stream length divided by valley length (SL / VL); or estimated from a ratio of valley slope divided by channel slope (VS / S).
1.1

Stream Type **D4** (See **Figure 2-14**)

Worksheet 7-2d (Worksheet 4-1). Level II stream classification for the E3 stream type.

Stream Name: **Wolf Creek**	
Basin Name: **San Juan**	Drainage Area: **2,560** acres **4** mi²
Location: **Near Wolf Creek pass, Colorado**	
Twp.&Rge: **T37N, R33E**	Sec.&Qtr.:	Valley Type: **VIII**
Cross-Section Monuments (Lat./Long.):	
Observers: **Dave Rosgen**	Date: **10-Jul-04**

Bankfull WIDTH (W_{bkf})
WIDTH of the stream channel at bankfull stage elevation, in a riffle section.
6.5 ft

Bankfull DEPTH (d_{bkf})
Mean DEPTH of the stream channel cross-section, at bankfull stage elevation, in a riffle section ($d_{bkf} = A / W_{bkf}$).
0.8 ft

Bankfull X-Section AREA (A_{bkf})
AREA of the stream channel cross-section, at bankfull stage elevation, in a riffle section.
28 ft²

Width/Depth Ratio (W_{bkf} / d_{bkf})
Bankfull WIDTH divided by bankfull mean DEPTH, in a riffle section.
1.5 ft/ft

Maximum DEPTH (d_{mbkf})
Maximum depth of the bankfull channel cross-section, or distance between the bankfull stage and Thalweg elevations, in a riffle section.
4.9 ft

WIDTH of Flood-Prone Area (W_{fpa})
Twice maximum DEPTH, or ($2 \times d_{mbkf}$) = the stage/elevation at which flood-prone area WIDTH is determined in a riffle section.
195 ft

Entrenchment Ratio (ER)
The ratio of flood-prone area WIDTH divided by bankfull channel WIDTH (W_{fpa} / W_{bkf}) (riffle section).
30 ft/ft

Channel Materials (Particle Size Index) D_{50}
The D_{50} particle size index represents the mean diameter of channel materials, as sampled from the channel surface, between the bankfull stage and Thalweg elevations.
75 mm

Water Surface SLOPE (S)
Channel slope = "rise over run" for a reach approximately 20–30 bankfull channel widths in length, with the "riffle-to-riffle" water surface slope representing the gradient at bankfull stage.
0.009 ft/ft

Channel SINUOSITY (k)
Sinuosity is an index of channel pattern, determined from a ratio of stream length divided by valley length (SL / VL); or estimated from a ratio of valley slope divided by channel slope (VS / S).
1.9

Stream Type → **E3** ← (See **Figure 2-14**)

Step 3. Determine variables influenced

A list of land use activities and their influences on process variables and potential consequences, along with the corresponding *RRISSC* prediction methods, appears in **Table 7-9**. This table indicates which screening variables must be used to determine specific risk ratings. The Wolf Creek example is highlighted to show the screening level relations used for risk assessment. Subsequent analysis is conducted using the worksheets that correspond to these activities.

Table 7-9 (Table 4-3). Relationship among land uses/activities, process influences, consequences and assessment methods.

Potential change from land uses/activities	Processes influenced	Potential consequences	RRISSC prediction method
Streamflow decrease in magnitude, duration and altered timing due to reservoirs or diversions	Shear stress ↓ Stream power ↓ Sediment transport competency and capacity ↓	Excess sediment deposition Aggradation Accelerated bank erosion Widening channel Successional state	Worksheet 4-11 Worksheet 4-11 Worksheet 4-7 Worksheet 4-10 Table 4-5
Streamflow discharge increase due to high % impervious and storm water drains from urban development. Clear water discharge "ramping flows" from reservoir releases	Shear stress ↑ Stream power ↑ Sediment transport capacity ↑	Degradation Channel enlargement Bank erosion Channel succession shift Increased sediment load (supply)	Worksheet 4-12 Worksheet 4-10 Worksheet 4-7 Table 4-5 Worksheet 4-11
Streamflow increase from vegetative alteration, clearcutting, land clearing and roads	Shear stress ↑ Stream power ↑ Magnitude of flow ↑ Duration of flows ↑	Channel enlargement Bank erosion Degradation Channel succession shift Increased sediment load (supply) Surface erosion	Worksheet 4-10 Worksheet 4-7 Worksheet 4-12 Table 4-5 Worksheet 4-11 Worksheet 4-5
Riparian vegetation alteration (% of channel length by stream type)	Bank erodibility ↑ Sediment transport capacity ↓ Stream power ↓ Shear stress ↓	Bank erosion Aggradation Enlargement Channel succession shift	Worksheet 4-7 Worksheet 4-11 Worksheet 4-10 Table 4-5
Surface disturbances (% of ground cover) and roads	Surface runoff ↑ Sub-surface flow interception (roads) ↑ Deposition ↑ Sediment transport capacity (aggradation) ↓ Excess scour (degradation) ↑	Surface erosion delivered to stream Road source sediment Gully erosion Aggradation Degradation Streambank erosion	Worksheet 4-5 Worksheet 4-4 Worksheets 4-7, 9, 10, 12 Worksheet 4-11 Worksheet 4-12 Worksheet 4-7
Water yield – harvest and roads – add to soil water influencing slope stability	Surface/sub-surface hydrology ↑ Soil saturation ↑ Internal strength by roots ↓ Slope equilibrium ↓	Mass erosion: - slump earthflow ↑ - debris torrent ↑ - sediment supply delivered to channel ↑ Aggradation ↑ Channel succession shift Enlargement ↑ Surface erosion ↑	Table 4-4 Worksheet 4-3 Worksheet 4-11 Table 4-5 Worksheet 4-10 Worksheet 4-5

Direct channel impacts
  Channelization
  Levees
  Straightening
  Dredging
	Shear stress ↑↓
	Stream power ↑↓
	Width ↑
	Confinement ↑
	Incision ↑
		Gully erosion ↑
		Bank erosion ↑
		Channel enlargement ↑
		Degradation ↑
		Aggradation ↑
		Channel succession shift
			Worksheets 4-7, 9, 10, 12
			Worksheet 4-7
			Worksheet 4-10
			Worksheet 4-12
			Worksheet 4-11
			Table 4-5

Channel clearing, cleaning, grubbing, large woody debris removal	Stream power ↑ Shear stress ↑ Sediment transport capacity ↓ Competence ↑ Degradation ↑ Energy dissipation ↓	Sediment deposition ↑ Degradation ↑ Bank erosion ↑ Channel enlargement ↑ Sediment supply ↑ Aggradation ↑	Worksheet 4-11 Worksheet 4-12 Worksheet 4-7 Worksheet 4-10 Worksheet 4-11 Worksheet 4-11

Note: Potential consequences column is directly related to RRISSC prediction method column; for example, potential excess sediment deposition is assessed in **Worksheet 4-11**.

Step 4. Compile data for risk rating system

Based on the *RLA* evaluation and **Table 7-9**, the following process/land use risk ratings are needed for selected sub-watersheds and reaches of Wolf Creek:

- Hydrologic changes due to timber harvest and roads;
- Riparian vegetation conversion/impacts;
- Sediment delivery potential from roads;
- Mass erosion sediment delivery potential; and
- Direct channel impacts.

During the risk rating steps (**Steps 6–19**), overall risk ratings are obtained for hillslope, hydrologic and channel processes by various land uses. These overall risk ratings for multiple sub-watersheds and river reaches are recorded in the *RRISSC* summary **Worksheet 7-3**. A completed version of this worksheet appears at the end of the *RRISSC* section in **Step 21** for discussion purposes.

WATERSHED ASSESSMENT OF RIVER STABILITY AND SEDIMENT SUPPLY

Worksheet 7-3 (Worksheet 4-2). *RRISSC* summary worksheet for multiple sites/river reaches within a study watershed.

Watershed Name: Date: Observer(s):

Location code/ river reach I.D.	Geographic Location					Stream Type Location							Processes identified by step for advancement to PLA	✓ Location selected for advancement to PLA
	Step 6: Mass erosion (Worksheet 4-3)	Step 7: Roads (Worksheet 4-4)	Step 8: Surface erosion (Worksheet 4-5)	Step 10: Streamflow change (Worksheet 4-6)	Step 13: Streambank erosion (Worksheet 4-7)	Step 14: In-channel mining (Worksheet 4-8)	Step 15: Direct channel impacts (Worksheet 4-9)	Step 16: Channel enlargement (Worksheet 4-10)	Step 17: Aggradation/ excess sediment (Worksheet 4-11)	Step 18: Channel evolution/ successional states (Table 4-5)	Step 19: Degradation (Worksheet 4-12)		
1.													
2.													
3.													
4.													
5.													
6.													
7.													
8.													
9.													
10.													
11.													
12.													
13.													
14.													
15.													

Hillslope processes: Steps 5–8

The *RRISSC* analysis of Wolf Creek Watershed hillslope processes revealed a wide range of risk levels. With the exception of some mass erosion risks around Highway 160, the risk of mass erosion related to existing land uses was generally low, as was the risk of surface erosion. The risks of sediment from roads, however, ranged from low to high.

Step 5. Select appropriate models and compile data for hillslope processes

The *RLA* results indicated the need for a more detailed mass erosion assessment for Highway 160. As documented in *RLA*, however, surface erosion processes are associated with very limited potential sediment contributions; thus, further surface erosion assessments were not conducted in *RRISSC*. *RLA* also identified roads to be assessed in *RRISSC*. The influences of flow regime and sediment supply due to road impacts are addressed in **Step 7**.

Step 6. Assess mass erosion risk

Sub-watersheds 1–5 did not meet any guidance criteria for mass erosion risk in the *RLA* assessment and will not be analyzed further; thus, an automatic rating of *very low* (1) is allotted to these areas in the *RRISSC* summary worksheet to complete the overall sediment summary. State Highway 160, however, was identified in *RLA* as a *high* risk and was analyzed for potential erosion using the variables in **Table 7-10**. Both debris avalanche/debris torrent, as well as slump/earthflow processes, were identified. The widening of the mid-portion of Highway 160 had triggered a massive slump/earthflow approximately 20 years previous. This slide has stabilized, although the toe of the slide debris is slowly being eroded by Wolf Creek. The predominant active process that is currently providing the most potential sediment delivery is from the very steep, continuous slopes on A3a+ stream types. These numerous, steep drainageways are intercepted by the highway, especially on the lower 40% of its length. Sub-surface water intercepted by the road cut and surface water concentrated from the road surface is routed by the ditch line into these debris flow/torrent drainageways. The risk rating for this process is highlighted in **Table 7-10**. The sediment delivery potential, as rated in **Figures 7-8** and **7-9** and summarized in **Tables 7-11** and **7-12**, is influenced more by the debris flow/debris torrent process than the slump/earthflow. Sediment delivery potential ratings using **Figures 7-8** and **7-9** reflect a *very high* risk for Highway 160.

The overall risk rating for potential sediment supply and delivery from mass erosion was *very high* as shown in **Worksheet 7-4**. The overall risk summary of the mass erosion risk for the Wolf Creek Watershed, compared to other processes for a variety of erosional processes, are documented in the *RRISSC* summary worksheet; a completed version of the summary worksheet appears in **Step 21** at the end of the *RRISSC* section.

Table 7-10 (Table 4-4). General guidelines for broad-level high risk of mass erosion potential (debris avalanche/debris torrent and slump/earthflow) (USEPA, 1980).

Guideline for various slopes	Debris avalanche/debris torrent	Slump/earthflow
Landform features • slope gradient • slope shape	Steep >34° • concave • continuous/uniform	2° → steep • hummocky • discontinuous/irregular
Soil characteristics • depth • type	Shallow • residual or colluvium • glacial till deposition	Deep • deeply weathered • fine-grained • presence of mica • high clay content (pyroclastic soils)
Hydrologic characteristics of site • concentration of ground water	• linear depressions parallel to the slope	• saturated depressions • bowls • springs • "elk wallows" (undermined by road at toe of slide activity)
Rock type/geology	• hard, resistant volcanic rock • granites • diorites • alpine glaciation • highly weathered volcanic rock	• volcanic ash • breccias • silt sandstones • mudstones • highly weathered rock • attitude (dip of beds, parallel to slope)
Vegetative cover	• rooting depth impacts	• indication of clearcuts/roads increase risk • old age vs. new stands (indication of activity) • leaning trees

Precipitation	• varies
	  - snowmelt
	  - convectional storms
		• high intensity
		• short-duration storms

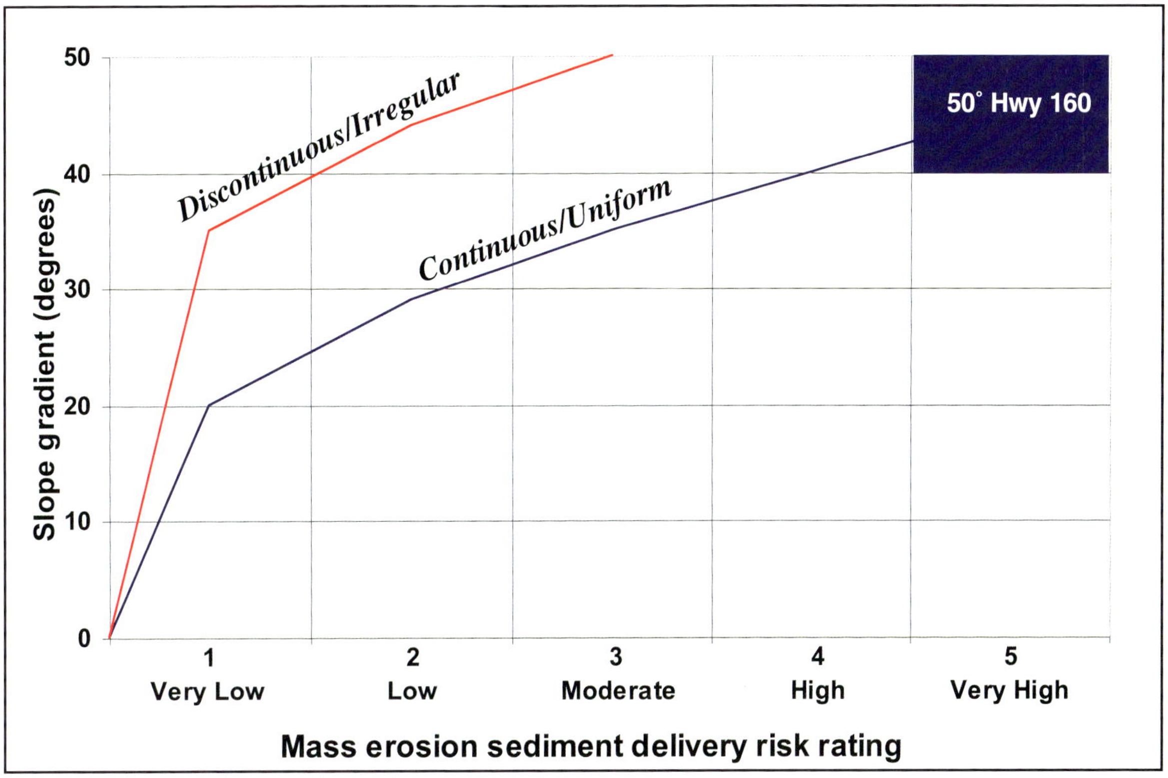

Figure 7-8 (Figure 4-1). Mass erosion sediment delivery risk for Highway 160 based on slope gradient (degrees) by slope shape.

Table 7-11. Slope gradient by slope shape risk rating summary for Wolf Creek Watershed.

Slope gradient by slope shape for sediment delivery risk rating using Figure 7-8 (Figure 4-1)			
Sub-watershed	**Slope gradient (degrees)**	**Slope shape (discontinuous or continuous)**	**Risk rating**
Hwy 160	50°, concave	Continuous/uniform	VH (5)

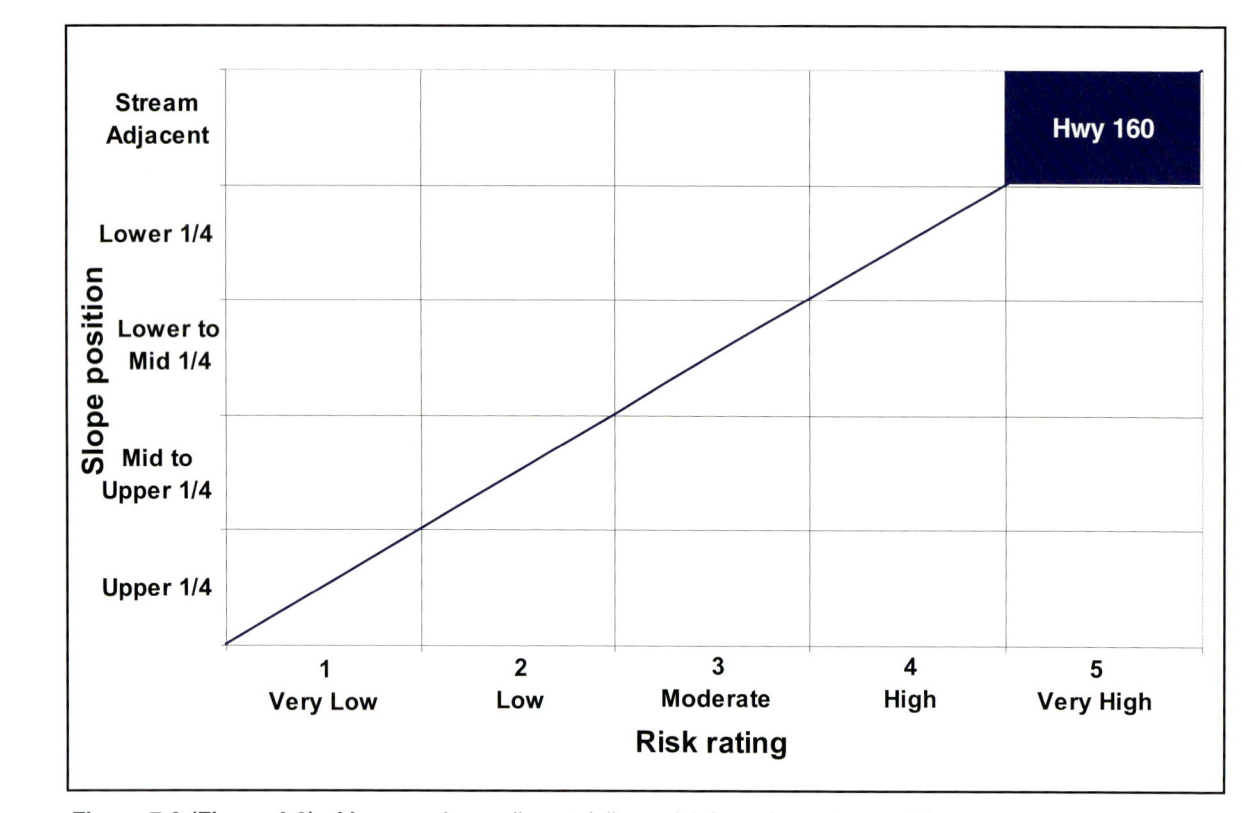

Figure 7-9 (Figure 4-2). Mass erosion sediment delivery risk based on slope position.

Table 7-12. Slope position risk rating summary for Wolf Creek Watershed.

Slope position risk rating using Figure 7-9 (Figure 4-2)		
Sub-watershed	Slope position	Risk rating
Hwy 160	Stream adjacent	VH (5)

Chapter 7 — RRISSC   The WARSSS Methodology: A Case Study of Wolf Creek, Colorado

Worksheet 7-4 (Worksheet 4-3). Risk rating worksheet for mass erosion sediment delivery.

(1) Sub-watershed location (I.D.)	(2) Slope gradient (degrees)	(3) Slope shape (discontinuous or continuous)	(4) Risk rating: slope gradient by slope shape (**Figure 4-1**)	(5) Slope position (lower 1/4, mid to lower 1/4, mid to upper 1/4, upper 1/4 or stream adjacent)	(6) Risk rating: slope position (**Figure 4-2**)	(7) Total risk rating points by sub-watershed ∑[(4)+(6)]	(8) Overall mass erosion risk rating (use column (7) points; insert adjective and numerical risk rating) VL(1) = 2–3 L(2) = 3–4 M(3) = 5–6 H(4) = 7–8 VH(5) = 9–10
1. Hwy 160	50°, concave	continuous	VH (5)	stream adjacent	VH (5)	(10)	VH (5)
2.							
3.							
4.							
5.							
6.							
7.							
8.							
9.							
10.							
11.							
12.							
13.							
14.							
15.							

Step 7. Evaluate road impact risk

The assessment of road impact (the acres of road divided by the acres in a sub-watershed multiplied by the number of stream crossings) and road position in relation to the drainage network was completed for Wolf Creek. The roads in the five sub-watersheds were assessed in addition to Highway 160. Because the roads servicing timber harvest in the sub-watersheds were constructed 25–30 years ago, there is considerable recovery represented by vegetative establishment on cut banks and fill slopes, although sub-surface flow interception processes and increases in snowpack depth/water storage still prevail.

The risk assessment conducted for the forest roads and State Highway 160, using the relations in **Figures 7-10** through **7-13**, is summarized in **Tables 7-13** through **7-16**.

This information is recorded in **Worksheet 7-5**. The overall risk of sediment delivery from roads is recorded in the *RRISSC* summary worksheet. The results indicate a high potential risk of delivered sediment due primarily to Highway 160 rather than the older, stabilized roads in the watershed. The older roads, however, may still be contributing to increased flow, which will be evaluated in the streamflow changes and channel enlargement steps, **Steps 10** and **16**, respectively. Before designing and proposing extensive and expensive mitigation and expecting the Colorado Department of Transportation to implement stabilization procedures on Highway 160, a more quantitative demonstration of the proportion of sediment contributed to the annual sediment yield is recommended. This effort may demonstrate the benefits of proposed stabilization procedures to river processes and downstream sediment and stability problems. Therefore, the areas along Highway 160 should advance to *PLA*.

Furthermore, all of the roads in each sub-watershed, as well as Highway 160, were carried forward to *PLA*. The rationale for this is to quantify the relative annual sediment yields from forest roads compared to other watershed sediment sources. The cumulative effects of the sediment contributions from all sources can thus be calculated.

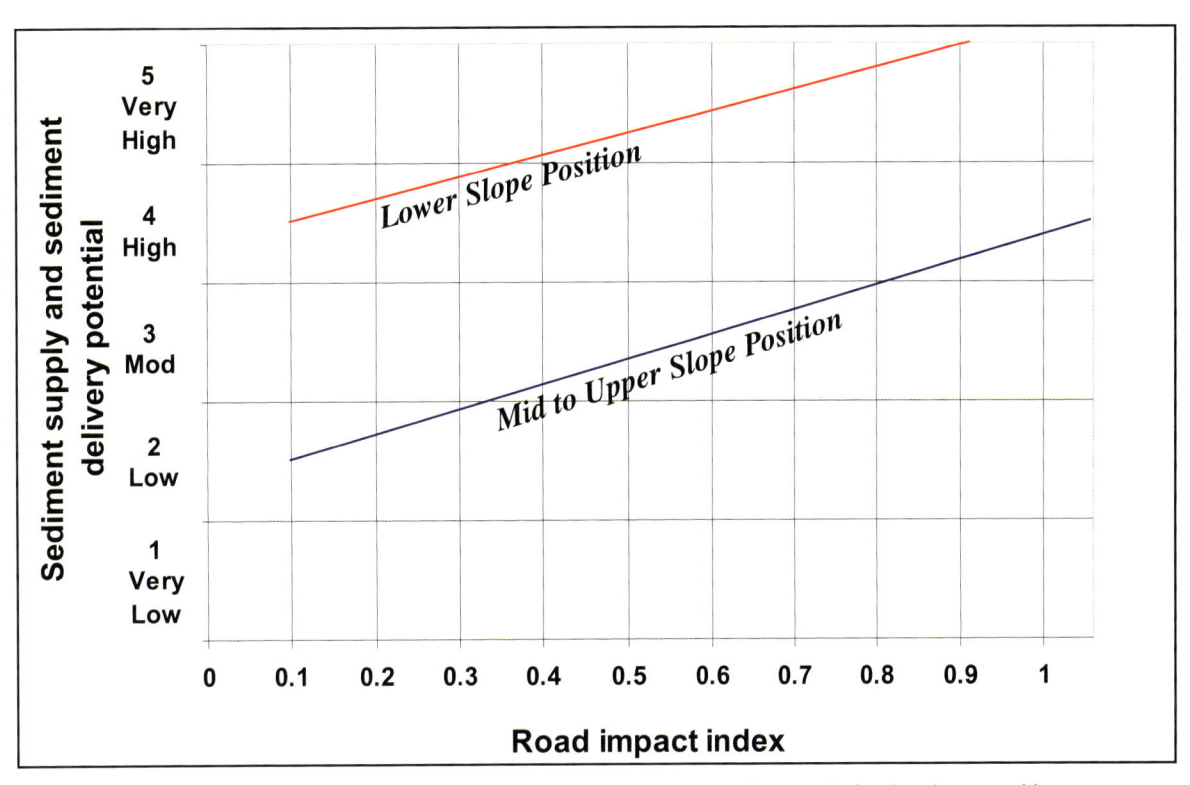

Chapter 7 — RRISSC   The WARSSS Methodology: A Case Study of Wolf Creek, Colorado

[Graph: Sediment supply and sediment delivery potential (y-axis, 1 Very Low to 5 Very High) vs. Road impact index (x-axis, 0 to 1), showing two lines: "Lower Slope Position" (upper red line) and "Mid to Upper Slope Position" (lower blue line).]

Figure 7-10 (Figure 4-3). Road sediment delivery risk based on road impact index by slope position. Figure modified from Rosgen (2001b) based on measured delivered road sediment to debris basins in Horse Creek Watershed, Idaho and Fool Creek, Colorado using experimental watershed data from USDA Forest Service.

Table 7-13. Sediment supply and delivery risk rating summary for forest roads and Highway 160.

Sediment supply risk ratings using Figure 7-10 (Figure 4-3)						
Sub-watershed	Acres of sub-watershed	Acres disturbance of road	Number of stream crossings	Road Impact Index (RII)	Position	Risk rating
1	355	34	4	0.38	lower slope	VH (5)
2	1024	144	3	0.42	mid to upper slope	M (3)
3	861	27	8	0.25	lower slope	H (4)
4	1013	88	3	0.26	lower slope	H (4)
5	2649	124	2	0.09	mid to upper slope	L (2)
Hwy 160	3975	157	27	1.07	mid to upper slope	H (4)

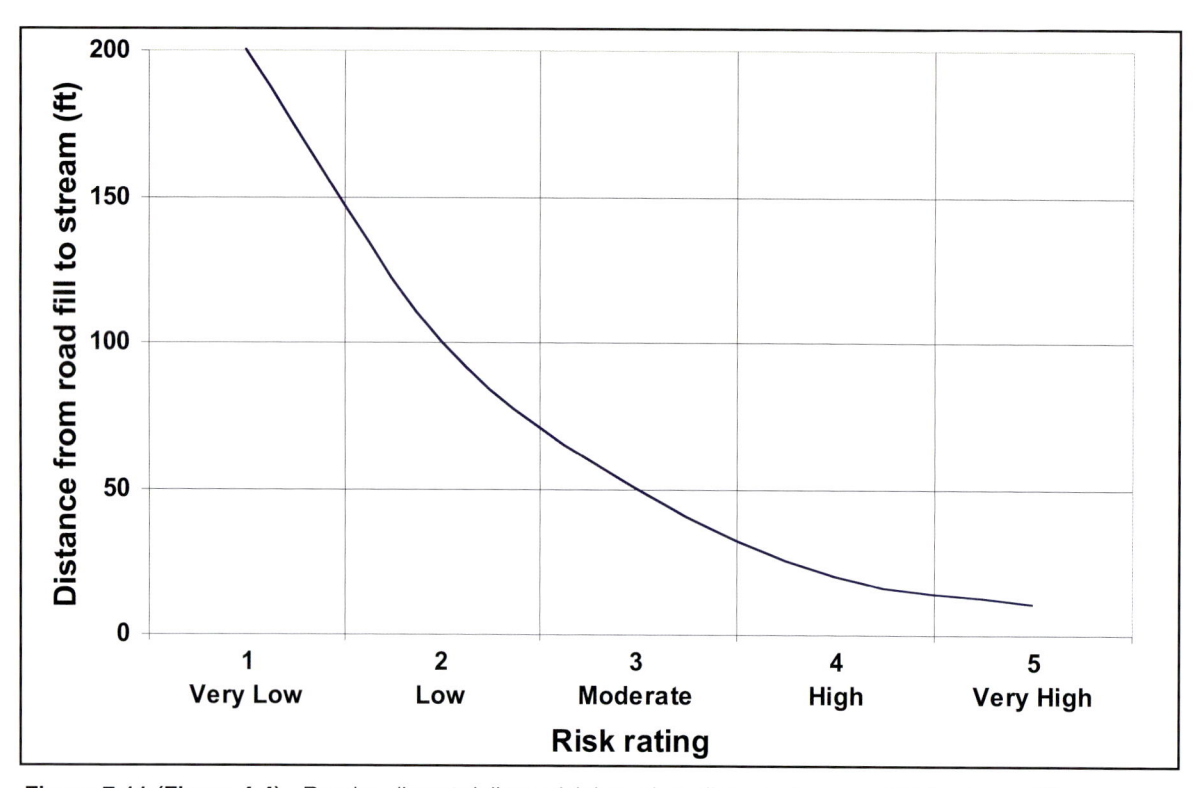

Figure 7-11 (Figure 4-4). Road sediment delivery risk based on distance from road fill to stream (ft).

Table 7-14. Sediment delivery risk rating summary based on proximity of road fill to stream for forest roads and Highway 160.

Road sediment delivery risk ratings using Figure 7-11 (Figure 4-4)		
Sub-watershed	Fill distance to stream (ft)	Risk rating
1	60	M (3)
2	60	M (3)
3	130	L (2)
4	50	M (3)
5	120	L (2)
Highway 160	70	M (3)

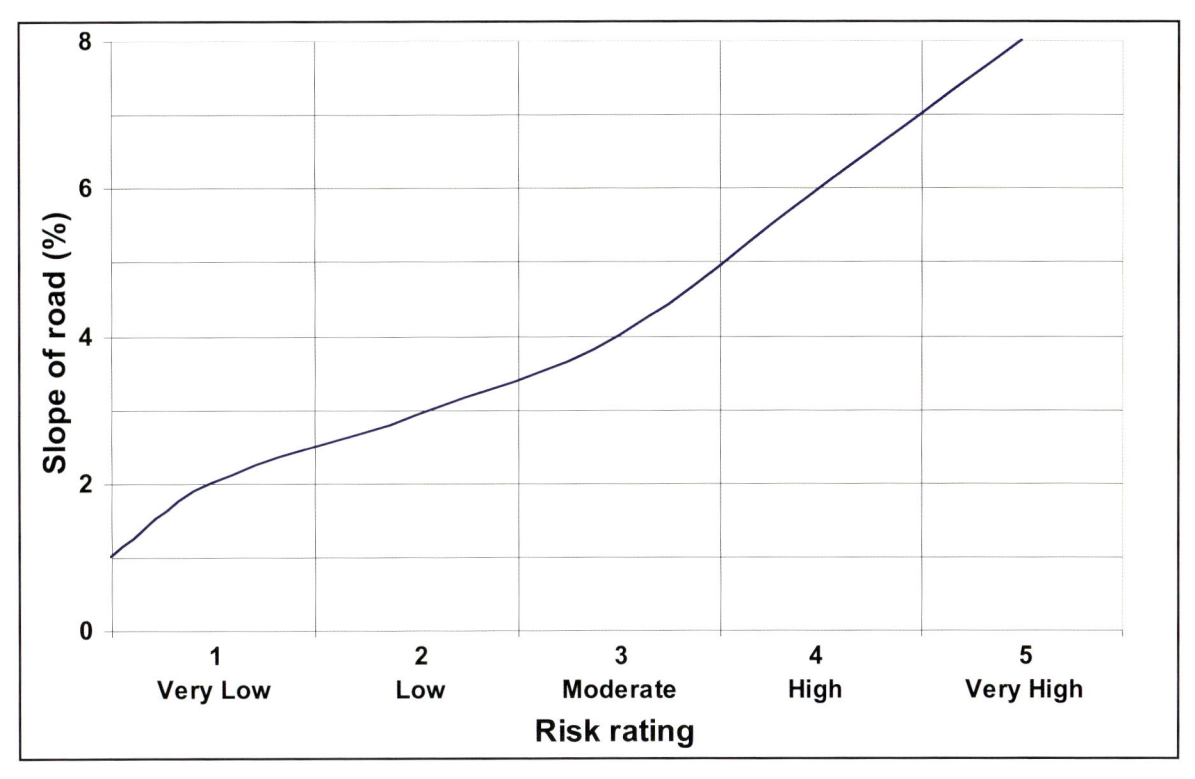

Chapter 7 — RRISSC   The WARSSS Methodology: A Case Study of Wolf Creek, Colorado

Figure 7-12 (Figure 4-5). Road sediment delivery risk based on slope of road (%) (curve derived from data from Reid and Dunne, 1984).

Table 7-15. Sediment supply risk rating summary based on road slopes for forest roads and Highway 160.

Road sediment supply risk ratings using Figure 7-12 (Figure 4-5)		
Sub-watershed	Fill distance to stream (ft)	Risk rating
1	4%	M (3)
2	3%	L (2)
3	4%	M (3)
4	5.5%	H (4)
5	4%	M (3)
Highway 160	7%	VH (5)

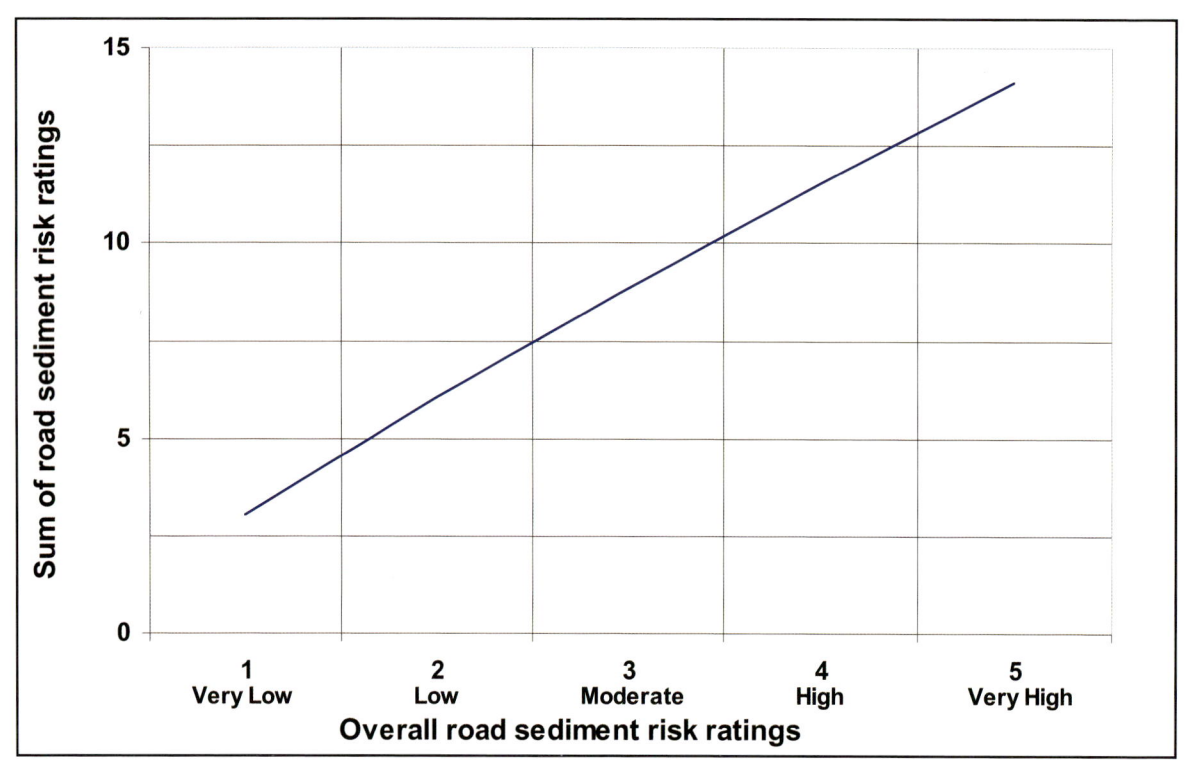

Figure 7-13 (Figure 4-6). Overall road sediment delivery risk based on the sum of individual sediment risk ratings.

Table 7-16. Overall road sediment risk rating summary for Wolf Creek Watershed.

Overall road sediment risk ratings using Figure 7-13 (Figure 4-6)					
Sub-watershed	Road sediment supply risk rating	Fill distance to stream risk rating	Road slope risk rating	Sum of road sediment risk ratings	Overall risk rating
1	VH (5)	M (3)	M (3)	11	H (4)
2	M (3)	M (3)	L (2)	8	M (3)
3	H (4)	L (2)	M (3)	9	M (3)
4	H (4)	M (3)	H (4)	11	H (4)
5	L (2)	L (2)	M (3)	7	L (2)
Highway 160	H (4)	M (3)	VH (5)	12	H (4)

Chapter 7 — RRISSC   The WARSSS Methodology: A Case Study of Wolf Creek, Colorado

Worksheet 7-5 (Worksheet 4-4). Risk rating worksheet for potential sediment delivery from roads.

(1) Location of sub-watershed (I.D.)	(2) Acres of sub-watershed (200–5000 acres)	(3) Acres disturbance of road (include cut bank, fill slope, road surface)	(4) Number of stream crossings	(5) Calculate road impact index [(3)/(2)X(4)] *If crossings = 0, multiply by 1.	(6) Slope position (lower or mid-upper)	(7) Risk rating: road impact index (5) by slope position (Fig. 4-3)	(8) Distance of road fill to stream (ft)	(9) Risk rating: distance of road fill to stream (ft) (Fig. 4-4)	(10) Slope of road (%)	(11) Risk rating: slope of road (%) (Fig. 4-5)	(12) Total individual risk rating points ∑[(7)+(9)+(11)]	(13) Overall risk rating for potential sediment from roads (Fig. 4-6)	(14) Adjustments for construction, design and age of road — Age of road: If > 7 yrs and sediment delivery potential = Low, reduce one risk category* / Road surfacing: If gravel/asphalt, then reduce one risk category** / Ditch line: If surfacing out-sloped, reduce one risk category / Vegetative condition of cut banks, road fills: If > 50% ground cover, reduce one risk category	(15) Risk rating adjustments for mass erosion potential slump/earthflow*** (Table 4-4, Figures 4-1, 4-2)	(16) Debris torrent/avalanche: If erosion risk and sediment delivery potential is High, raise final road risk rating to Very High (Table 4-4, Figures 4-1, 4-2)	(17) Final risk rating of potential sediment from roads
1. 1	355	34	4	0.38	lower	VH (5)	60	M (3)	4%	M (3)	11	H (4)				H (4)
2. 2	1024	144	3	0.42	mid-upper	M (3)	60	M (3)	3%	L (2)	8	M (3)				M (3)
3. 3	861	27	8	0.25	lower	H (4)	130	L (2)	4%	M (3)	9	M (3)				M (3)
4. 4	1013	88	3	0.26	lower	H (4)	50	M (3)	5.5%	H (4)	11	H (4)				H (4)
5. 5	2649	124	2	0.09	mid-upper	L (2)	120	L (2)	4%	M (3)	7	L (2)		✓		L (2)
6. Hwy 160	3975	157	27	1.07	upper	H (4)	70	M (3)	7%	VH (5)	12	H (4)			✓	VH (5)

*Unless: road has not recovered; poor maintenance; poor vegetative cover on cut bank and fill slopes - ditch line is still leading water into stream.
**Unless: road cut bank, fills and ditch line continue to provide sediment source to stream.
***If risk is *high* for potential sediment delivery of mass erosion (Worksheet 4-3), then adjust overall risk up one category.

7-45

Step 8. Determine surface erosion risk

Due to the nature of the soils in the location of land use disturbances, conditions to satisfy Hortonian overland flow are not prevalent. Infiltration rates are much higher than snowmelt ablation rates. Soil saturation conditions are rare during summer and late fall convectional storms; thus, infiltration rates are higher than storm intensity. The exception to this is associated with compacted running surfaces of roads, steep cut banks and road fill slopes. The sheet and rill erosion due to these sites, which are quite evident, are integrated into the evaluation in **Step 7** and associated with risk due to roads. An examination of the silvicultural treatment areas that potentially create the most surface disturbance due to logging, skidding and site preparation activities revealed exceptionally little evidence of extensive bare soil, rills or gullies from surface erosion processes.

Furthermore, the initial assessment in *RLA* determined that surface erosion potential was low; thus, no further analysis is required for sub-watersheds 1–5. These sub-watersheds automatically receive a rating of *very low* (1) to be inserted in the *RRISSC* summary worksheet.

Hydrologic processes: Steps 9–10

The Wolf Creek Watershed shows significant potential for increases in streamflow processes due to the large acreages of clearcuts and roads, as identified in *RLA*.

Step 9. Compile maps and data for hydrologic processes

This step determines the potential for water yield increase and identifies the nature of the receiving streams. There are approximately 5,860 clearcut acres in the upper Wolf Creek Watershed that have not recovered hydrologically due to their age, stand composition and high elevation (9,500–11,000 ft). The road densities in the watershed also show potential for increases in sub-surface flow interception and snowpack storage and changes in the magnitude and timing of flow.

The C4 and D4 stream types shown in the lower third of the basin are highly susceptible to sediment increases from flow-related activities (**Figure 7-14**). The willow removal on lower Wolf Creek has created accelerated streambank erosion and channel instability, as documented in the *RLA* section. Increases in streamflow magnitude and duration will potentially increase bank erosion rates and create subsequent increases in sediment supply and channel instability. Therefore, a risk assessment of potential streamflow change (**Step 10**) is necessary.

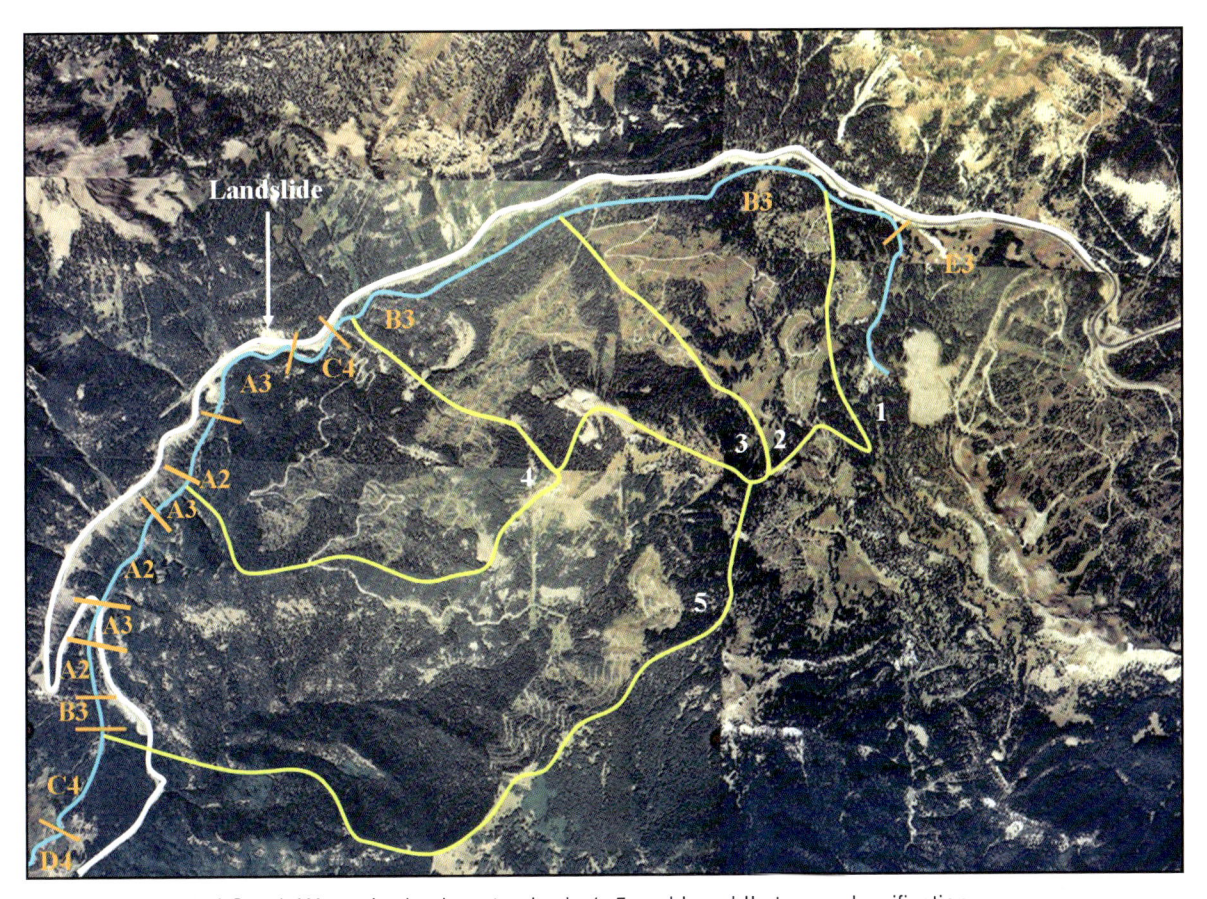

Figure 7-14. Wolf Creek Watershed sub-watersheds 1–5 and Level II stream classification.

Step 10. Assess potential for streamflow changes

The location of the rural sub-watersheds (**Figure 7-3**; see **Step 1**, *RLA*) and computed acres by sub-watershed obtained at the *RLA* assessment were accessed for this analysis. The Level II stream classification (**Worksheets 7-2a** through **7-2d**; see **Step 2**) is used in the risk rating for potential "channel source" flow-related sediment increase (**Figure 7-15**). The relations in **Figure 7-15** and summarized in **Table 7-17**, based on percent of watershed with altered vegetation, are used to obtain a risk rating by the dominant "weak link" stream type.

For snowmelt watersheds, it is common for watersheds that have 25% of a first-order watershed harvested to show an increase in peak flow (above bankfull) due to clearcutting (see **Figure 2-46**). The percent of the sub-watershed areas in equivalent clearcut condition (timber harvest and roads) varied from 20% (sub-watershed 5) to 74% (sub-watershed 2). This information is recorded in **Worksheet 7-6**.

The A3, C4 lower and D4 stream types are the "weak link" stream types used to assess risk because they are the receiving streams of the cumulative streamflow increase of all sub-watersheds and roads. The A3 stream type for Wolf Creek, located at the lower switchback about two miles above the confluence with the West Fork San Juan River, and the C4 and D4 stream types near the mouth of Wolf Creek, are also the "weak link" stream types selected for the total watershed. The high sediment supply from both bed and banks makes increases in streamflow highly sensitive for corresponding exponential increases in flow-related sediment increases. The overall risk ratings for Wolf Creek are *moderate* for sub-watershed 5 and Highway 160, *high* for sub-watersheds 1 and 3, and *very high* for sub-watersheds 2 and 4 (**Worksheet 7-6**). The overall risk rating for the total watershed is *very high*.

Furthermore, risk ratings are needed by individual stream type to be used in subsequent steps. Because the A3, C4 and D4 stream types are more susceptible to bed and bank erosional processes due to increased streamflow, they rated *very high* risk. The E3 stream type that is susceptible to flow-related sediment increases is located upstream of the cumulative streamflow increase potential; consequently, the risk rating for flow-related sediment for this stream type is *very low* (**Worksheet 7-6**).

These analyses indicate the need to proceed to the *PLA* phase to quantify the changes in flow and resulting sediment sources. Watershed restoration, including road reshaping and stream channel stabilization/bank stability work, is expensive and requires justification. A *PLA* analysis will provide site-specific data that isolates and computes the percent effectiveness (due to percent reduction in sediment) of proposed mitigation. *PLA* will also indicate the processes that are the largest contributors to impaired uses, as well as the specific locations where mitigation should be effectively implemented.

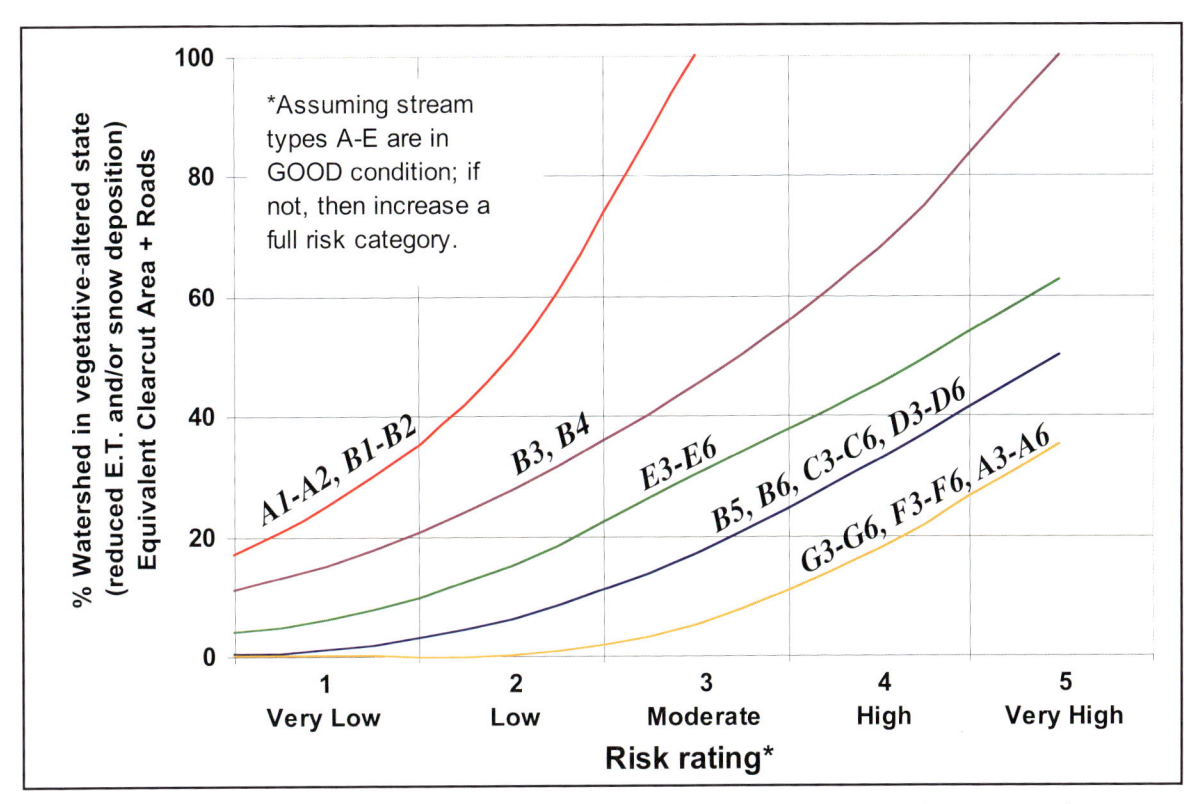

Chapter 7 — RRISSC   The WARSSS Methodology: A Case Study of Wolf Creek, Colorado

Figure 7-15 (Figure 4-14). Rural watershed flow-related sediment increase risk based on percent of watershed in vegetation-altered state by stream type.

Table 7-17. Flow-related sediment increase risk rating summary for Wolf Creek Watershed.

Rural watershed flow-related sediment increase risk ratings using Figure 7-15 (Figure 4-14)					
Sub-watershed	Total acres	Acres cleared/harvested (include roads) [roads + clearcut = total]	Percent cleared/harvested of total	Stream type most susceptible to change or "weak link"	Risk rating: rural sub-watershed risk (Fig. 4-14)
1	355	140	39%	C4	H (4)
2	1024	758	74%	C4	VH (5)
3	861	285	33%	C4	H (4)
4	1013	493	49%	C4	VH (5)
5	2679	521	20%	C4	M (3)
Highway 160	3975	157	4%	A3/C4/D4	M (3)
Total watershed	5902	2354	40%	A3/C4/D4	VH (5)

WATERSHED ASSESSMENT OF RIVER STABILITY AND SEDIMENT SUPPLY

Worksheet 7-6 (Worksheet 4-6). Risk rating worksheet for streamflow changes.

	(1) Sub-watershed location/river reach I.D. (include cumulative total watershed following sub-watershed I.D.s)	(2) Total acres	Rural Sub-watershed Risk (3) Acres cleared/ harvested (include roads) [roads + clearcut = total]	(4) % cleared/ harvested of total [(3)/(2)X100]	(5) Stream type most susceptible to change or "weak link"	(6) Risk rating: rural sub-watershed risk (Fig. 4-14) (4) by stream type (5)	Urban Sub-watershed Risk (7) Total impervious acres	(8) % impervious [(7)/(2)X100]	(9) Stream type most susceptible to change or "weak link"	(10) Risk rating: urban sub-watershed risk (Fig. 4-15) (8) by stream type (9)	Adjustments (11) Risk rating: % increase over bankfull discharge (Fig. 4-16)*	(12) Risk rating: % reduction in bankfull discharge (Fig. 4-17)*	(13) Overall risk rating: streamflow changes (insert adjective and numeric rating)
1.	1	355	140	39%	C4	H (4)							H (4)
2.	2	1024	758	74%	C4	VH (5)							VH (5)
3.	3	861	285	33%	C4	H (4)							H (4)
4.	4	1013	493	49%	C4	VH (5)							VH (5)
5.	5	2649	521	20%	C4	M (3)							M (3)
6.	Hwy 160	3975	157	4%	A3/C4/D4	M (3)							M (3)
7.	Total watershed	5902	2354	40%	A3/C4/D4	VH (5)							VH (5)
8.													
9.	A3					VH (5)							VH (5)
10.	C4 upper					VH (5)							VH (5)
11.	C4 lower					VH (5)							VH (5)
12.	D4					VH (5)							VH (5)
13.	E3					VL (1)							VL (1)
14.													
15.													

* Describe source of increased or decreased bankfull discharge adjustment, i.e., operational hydrology of reservoir.

Channel processes: Steps 11–15

Channel processes are not only influenced by increased streamflow but also by obvious changes in dimension, pattern and profile due to encroachment and direct disturbance. Changes in riparian vegetation in the lower portions of Wolf Creek have shown resultant time-trend changes in stream type as indicated in *RLA*. The cumulative effects on sediment supply, direct channel impacts, streamflow, streambank erosion, riparian vegetative changes and potential channel instability are assessed at the *RRISSC* level.

Step 11. Compile data for channel processes

Based on the work conducted in *RLA*, the Wolf Creek Watershed contains sensitive and impaired stream types. This data suggests that channel processes need to be looked at more carefully. The *high* and *very high* risk ratings for flow-related sediment in **Step 10** also support a closer look at channel processes.

Step 12. Assess broad-level channel stability

This step helps to isolate the stream types that are the most sensitive, previously impacted or perhaps stable. Their location and condition (state) help to determine source areas for impairment and the locations of sediment contributors. The majority of the channel problems in the Wolf Creek Watershed appear to be located along State Highway 160 and on the lower third of the drainage. Individual reaches are assessed by stream type at this level to help isolate specific, high-risk reaches. Evidence of channel change in lower Wolf Creek from a meandering, C4 stream type to a braided, D4 stream type warrants a more quantitative assessment of channel stability at the *PLA* level.

Step 13. Determine streambank erosion risk

The streambank erosion risk was assessed for the A3, C4 upper, C4 lower, D4 and E3 stream types, using the relations in **Figures 7-16** through **7-19** to determine risk ratings. These ratings are summarized in **Tables 7-18** through **7-21**. The information is recorded in **Worksheet 7-7**. The overall risk rating for streambank erosion is also recorded by stream type in the *RRISSC* summary worksheet. Streambank erosion risk was *low* for the E3 stream type and *moderate* for the upper C4 stream types. Streambank erosion risk was *high* for the lower C4 and D4 stream types that occur in the lower reaches of Wolf Creek and *very high* for the A3 stream types. The photograph of the D4 stream type on lower Wolf Creek, as shown in **Figure 7-7** in the *RLA* section, indicates a high sediment supply from accelerated streambank erosion. Such visual evidence and the corresponding risk ratings necessitate further evaluation. This will provide an estimate of streambank erosion contributions to total sediment. A more accurate prediction of these erosion and sedimentation rates will, in turn, clarify mitigation priorities.

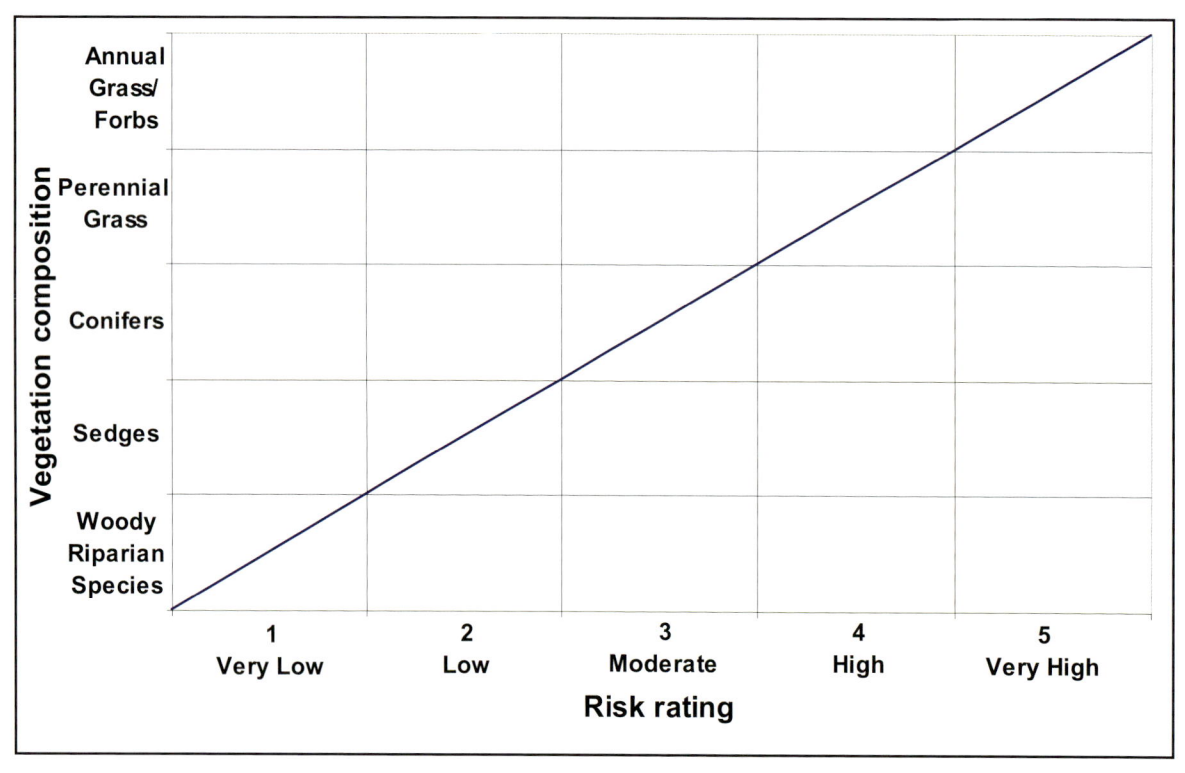

Figure 7-16 (Figure 4-18). Streambank erosion risk based on vegetation composition.

Table 7-18. Vegetation composition and corresponding streambank erosion risk rating summary for Wolf Creek Watershed.

Vegetation composition risk ratings using Figure 7-16 (Figure 4-18)		
Reach I.D.	Vegetation composition	Risk rating
A3	Conifers	M (3)
C4 upper	Conifers	M (3)
C4 lower	Perennial grass	H (4)
D4	Perennial grass	H (4)
E3	Sedges	L (2)

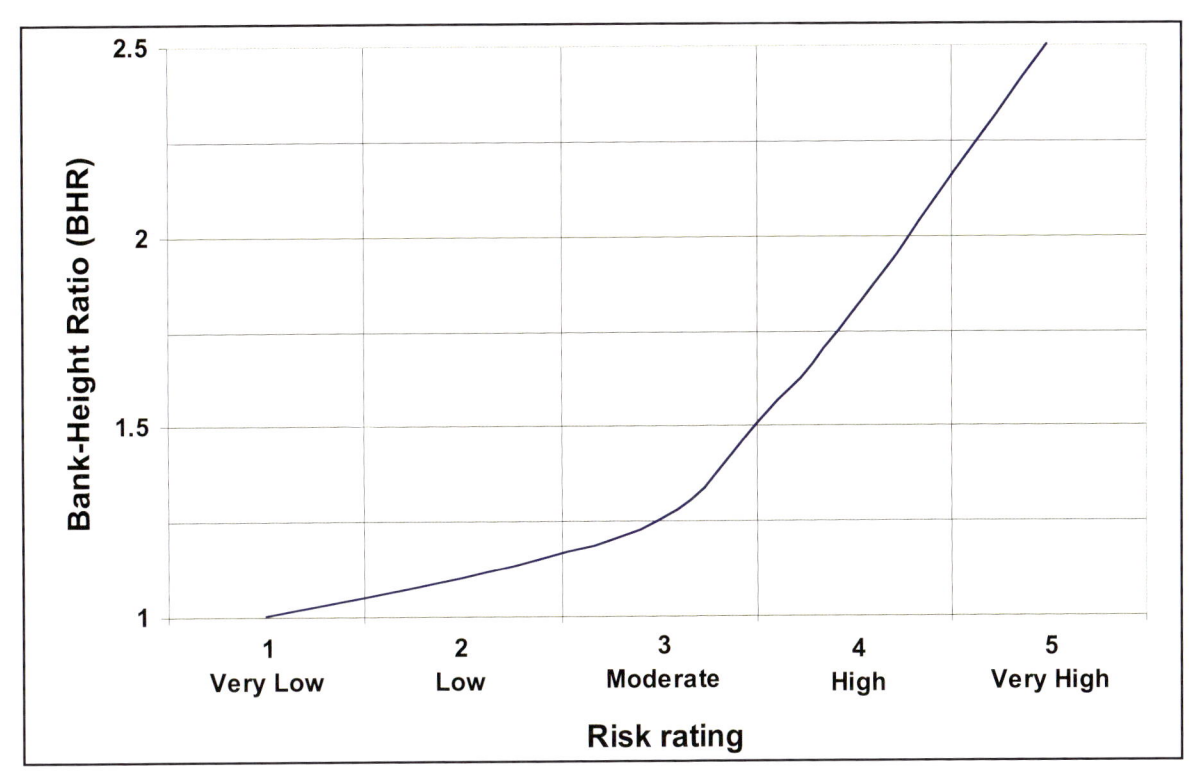

Chapter 7 — RRISSC   The WARSSS Methodology: A Case Study of Wolf Creek, Colorado

Figure 7-17 (Figure 4-19). Streambank erosion risk based on bank-height ratio.

Table 7-19. Bank-height ratio risk rating summary for Wolf Creek Watershed.

Bank-height ratio risk ratings using Figure 7-17 (Figure 4-19)		
Reach I.D.	Bank-height ratio	Risk rating
A3	2.3	VH (5)
C4 upper	1.5	H (4)
C4 lower	1.6	H (4)
D4	1.6	H (4)
E3	1.0	VL (1)

WATERSHED ASSESSMENT OF RIVER STABILITY AND SEDIMENT SUPPLY

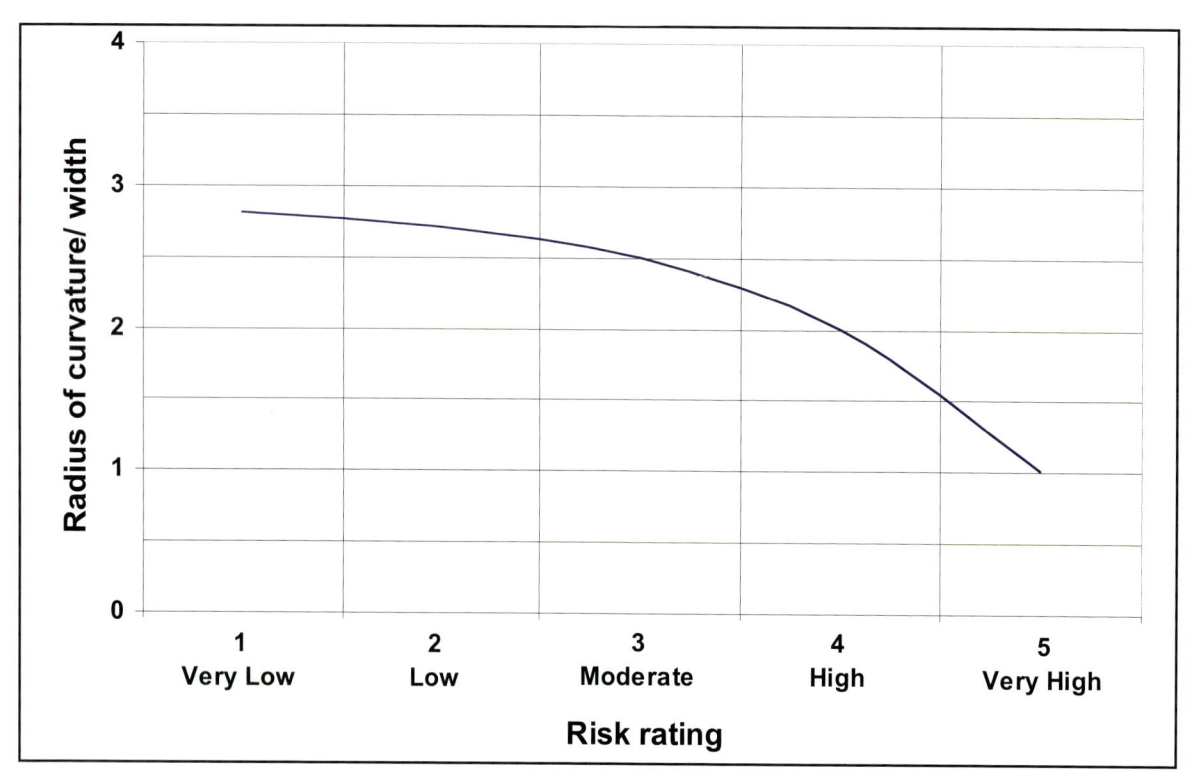

Figure 7-18 (Figure 4-20). Streambank erosion risk based on radius of curvature divided by width.

Table 7-20. Radius of curvature divided by width risk rating summary for Wolf Creek Watershed.

Radius of curvature divided by width risk rating using Figure 7-18 (Figure 4-20)		
Reach I.D.	Radius of curvature divided by bankfull width	Risk rating
A3	> 3.0	VL (1)
C4 upper	2.5	M (3)
C4 lower	1.8	H (4)
D4	1.3	VH (5)
E3	2.7	L (2)

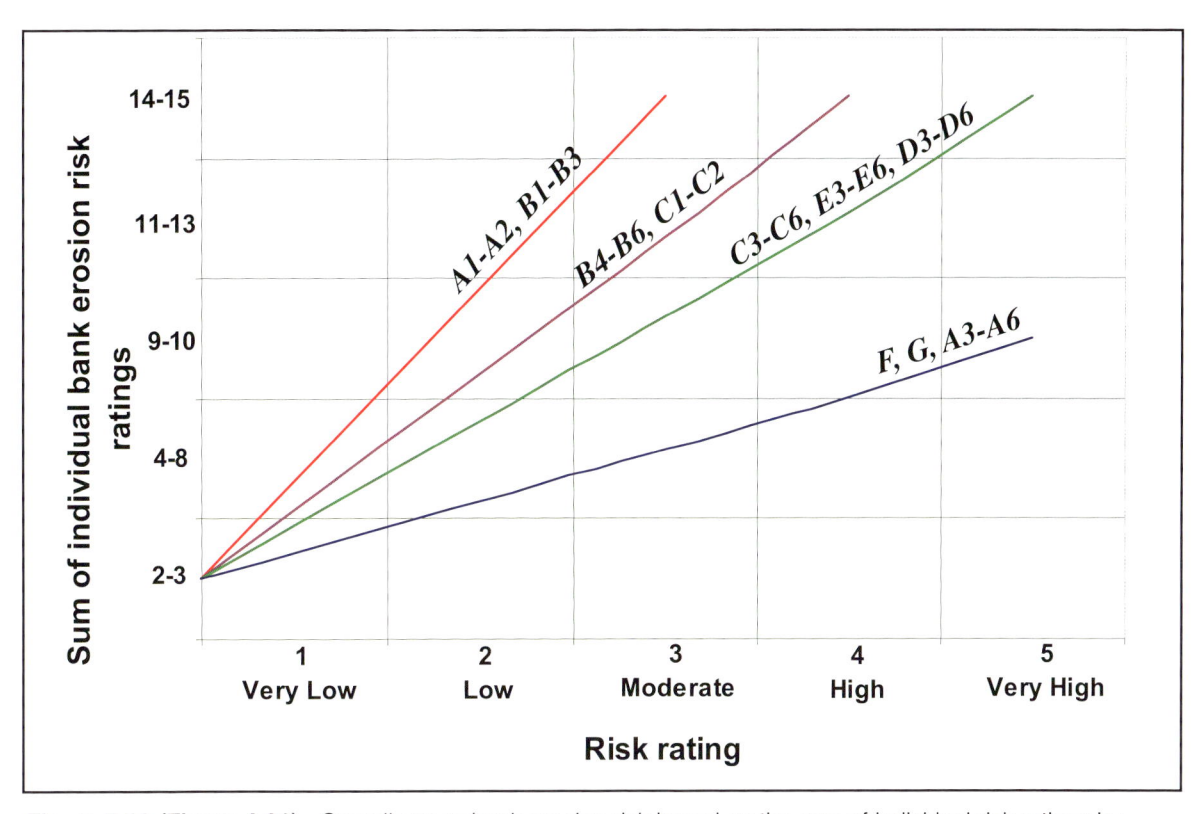

Figure 7-19 (Figure 4-21). Overall streambank erosion risk based on the sum of individual risk ratings by stream type.

Table 7-21. Overall streambank erosion risk rating summary for Wolf Creek Watershed.

Overall streambank erosion risk rating using Figure 7-19 (Figure 4-21)					
Reach I.D.	Risk rating: vegetation composition	Risk rating: Bank-height ratio	Risk rating: radius of curvature divided by bankfull width	Total individual risk rating points by reach	Overall risk rating by stream type
A3	M (3)	VH (5)	VL (1)	9	VH (5)
C4 upper	M (3)	H (4)	M (3)	10	M (3)
C4 lower	H (4)	H (4)	H (4)	12	H (4)
D4	H (4)	H (4)	VH (5)	13	H (4)
E3	L (2)	VL (1)	L (2)	5	L (2)

WATERSHED ASSESSMENT OF RIVER STABILITY AND SEDIMENT SUPPLY

Worksheet 7-7 (Worksheet 4-7). Risk rating worksheet for streambank erosion.

	(1) Location code/ river reach I.D.	(2) Vegetation composition	(3) Risk rating: vegetation composition (Fig. 4-18)	(4) Bank-height ratio	(5) Risk rating: bank-height ratio (Fig. 4-19)	(6) Radius of curvature divided by bankfull width	(7) Risk rating: radius of curvature divided by bankfull width (Fig. 4-20)	(8) Total individual risk rating points by reach $\sum[(3)+(5)+(7)]$	(9) Overall risk rating by stream type (Fig. 4-21)
1.	A3	Conifers	M (3)	2.3	VH (5)	> 3.0	VL (1)	9	VH (5)
2.	C4 upper	Conifers	M (3)	1.5	H (4)	2.5	M (3)	10	M (3)
3.	C4 lower	Perennial Grass	H (4)	1.6	H (4)	1.8	H (4)	12	H (4)
4.	D4	Perennial Grass	H (4)	1.6	H (4)	1.3	VH (5)	13	H (4)
5.	E3	Sedges	L (2)	1.0	VL (1)	2.7	L (2)	5	L (2)
6.									
7.									
8.									
9.									
10.									
11.									
12.									
13.									
14.									
15.									

Step 14. Assess in-channel mining

There are no known in-channel mining uses in Wolf Creek or its tributaries. Accordingly, no worksheet is included. However, in-channel mining was automatically given a rating of *very low* (1) for all stream types in the *RRISSC* summary worksheet to be used in subsequent risk ratings.

Step 15. Assess direct channel impacts

The A3, C4 upper, C4 lower, D4 and E3 stream types were assessed for direct channel impacts, implementing the relations outlined in **Figures 7-20**, **7-21** and **7-22**, and summarized in **Tables 7-22**, **7-23** and **7-24**. Most of the direct impacts came from road fill encroachment (Highway 160), riparian vegetation alteration and poor grazing practices. The risk summaries in **Worksheet 7-8** show ratings from *very low* (E3 stream type locations) to *very high* (A3 and D4 stream types). The overall risk ratings for direct impacts are also recorded in the *RRISSC* summary worksheet. As large woody debris was a low-risk factor in all locations, mitigation and/or management change is not recommended. The stream types associated with the higher densities of large woody debris can accommodate the debris load.

This risk analysis also indicates that there are potential increases in sediment and channel instability due to road and riparian impacts. It does not indicate how much sediment or what degree of instability is present; however, it does show that specific locations have a very high risk for these processes. As mentioned previously, requesting mitigation by the Colorado Department of Transportation requires a quantitative prediction of the annual sediment yields disproportionately contributed by Highway 160. Therefore, all roads and the A3, C4 and D4 reaches should be assessed at the *PLA* level.

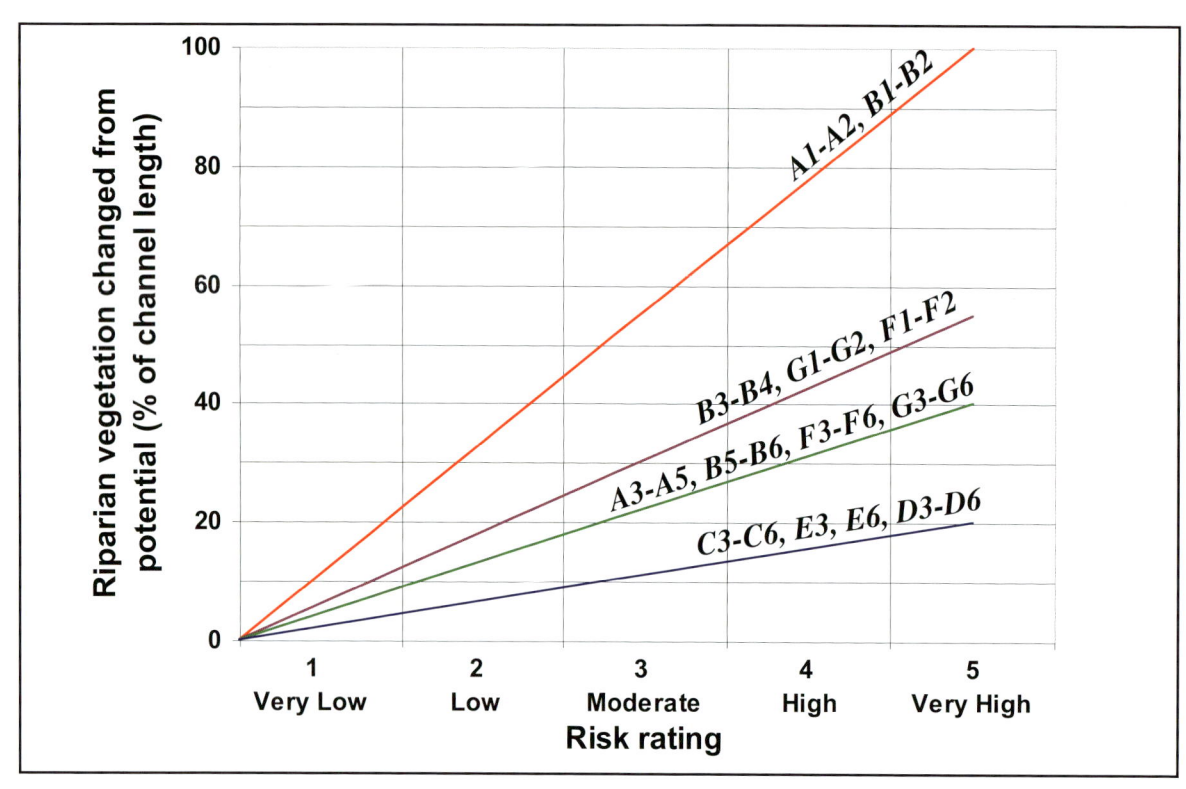

WATERSHED ASSESSMENT OF RIVER STABILITY AND SEDIMENT SUPPLY

Figure 7-20 (Figure 4-23). Risk rating for potential introduced sediment and channel instability by stream type based on percentage of channel length affected by vegetation change.

Table 7-22. Riparian vegetation change risk rating summary for Wolf Creek Watershed.

Riparian vegetation change risk rating using Figure 7-20 (Figure 4-23)				
Reach I.D.	Total channel length in feet	Riparian vegetation change in feet	Percent of total length impacted	Risk rating: percent riparian vegetation change
A3	3800	0	0%	VL (1)
C4 upper	3500	0	0%	VL (1)
C4 lower	1500	0	0%	VL (1)
D4	950	950	100%	VH (5)
E3	2000	0	0%	VL (1)

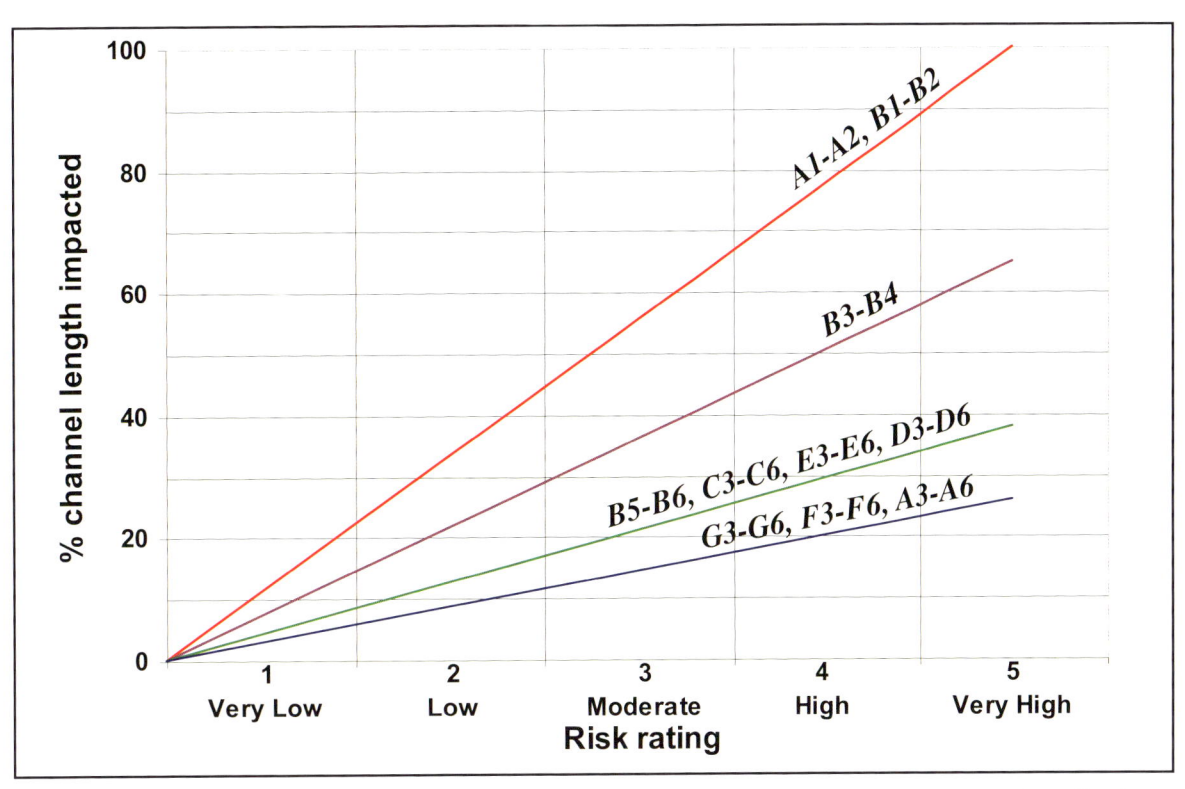

Figure 7-21 (Figure 4-24). Risk rating relation of percent of channel length impacted by vegetation utilization and bank impacts according to stream type.

Table 7-23. Channel length risk rating summary for Wolf Creek Watershed.

Percent channel length impacted risk rating using Figure 7-21 (Figure 4-24)				
Reach I.D.	Total channel length in feet	Riparian vegetation change in feet	Percent of total length impacted	Risk rating: percent riparian vegetation change
A3	3800	3000	79%	VH (5)
C4 upper	3500	200	6%	VL (1)
C4 lower	1500	500	33%	H (4)
D4	950	950	100%	VH (5)
E3	2000	0	0%	VL (1)

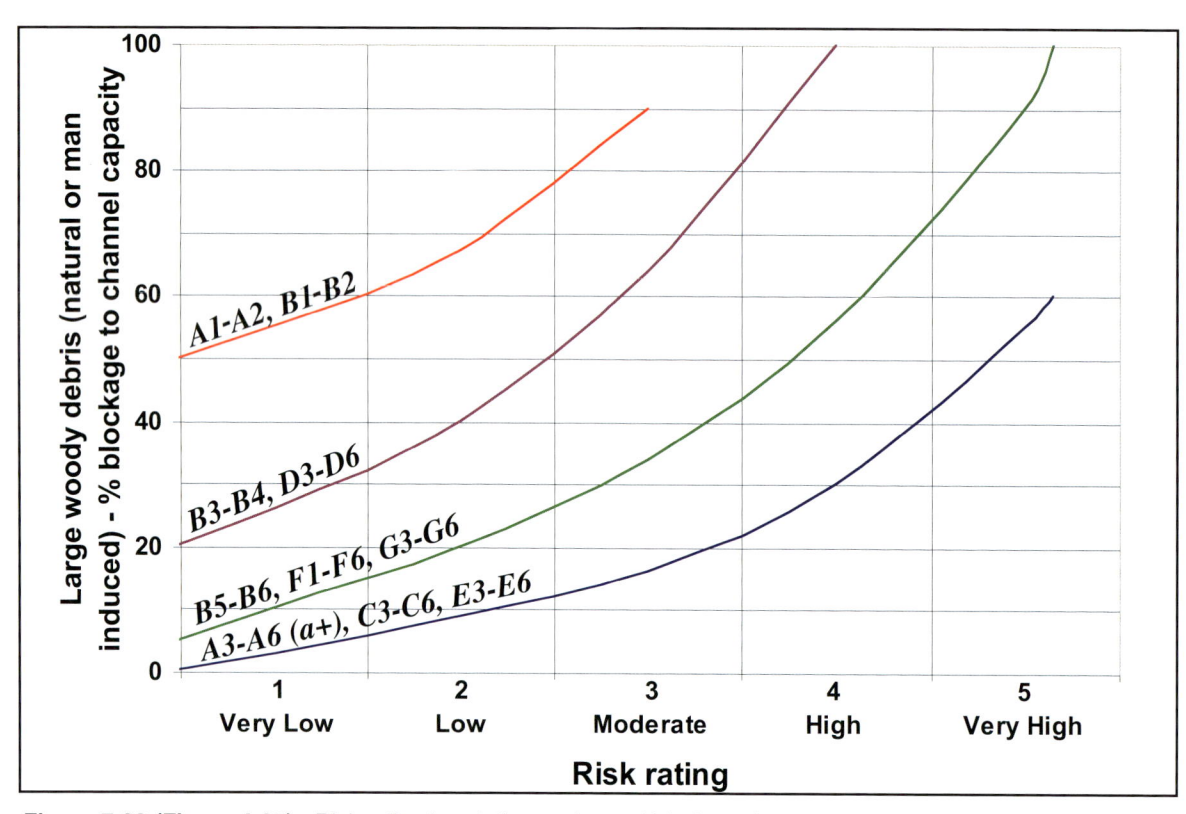

Figure 7-22 (Figure 4-25). Risk rating in relation to channel blockage from large woody debris by stream type.

Table 7-24. Debris blockage risk rating summary for Wolf Creek Watershed.

Debris blockage risk rating using Figure 7-22 (Figure 4-25)				
Reach I.D.	Total channel length in feet	Length impacted by large woody debris (ft)	Percent of total length impacted	Risk rating: debris blockage
A3	3800	300	8%	L (2)
C4 upper	3500	245	7%	L (2)
C4 lower	1500	150	10%	L (2)
D4	950	332	35%	L (2)
E3	2000	60	3%	VL (1)

Chapter 7 — RRISSC The WARSSS Methodology: A Case Study of Wolf Creek, Colorado

Worksheet 7-8 (Worksheet 4-9). Risk rating worksheet for direct channel impacts.

(1) Location code/ river reach I.D.	(2) Total channel length in feet	(3) Riparian vegetation change in feet	(4) % of total length impacted [(3)/(2)X100]	(5) Risk rating: % riparian vegetation change (Fig. 4-23) (4) by stream type	(6) Length impacted by direct channel disturbance in feet	(7) % of total length impacted [(6)/(2)X100]	(8) Risk rating: % channel length impacted (Fig. 4-24) (7) by stream type	(9) Length impacted by large woody debris in feet	(10) % of length of debris blockage [(9)/(2)X100]	(11) Risk rating: debris blockage (Fig. 4-25)	(12) Overall risk rating for direct channel impacts (Insert highest risk rating from Columns 5, 8 and 11)
1. A3*	3800	0	0%	VL (1)	3000	79%	VH (5)	300	8%	L (2)	VH (5)
2. C4 upper*	3500	0	0%	VL (1)	200	6%	VL (1)	245	7%	L (2)	L (2)
3. C4 lower*	1500	0	0%	VL (1)	500	33%	H (4)	150	10%	L (2)	H (4)
4. D4**	950	950	100%	VH (5)	950	100%	VH (5)	332	35%	L (2)	VH (5)
5. E3	2000	0	0%	VL (1)	0	0%	VL (1)	60	3%	VL (1)	VL (1)
6.											
7.											
8.											
9.											
10.											
11.											
12.											
13.											

* Road fill and debris torrent runoff due to road crossings. Encroachment of coarse sediment changing pattern, w/d ratio and increasing sediment storage (supply).
** Cause of impacted reach is related to spraying willows, converting to grass/forb riparian composition and heavy grazing pressure.

Risk potential assessment: Steps 16–20

Step 16. Calculate channel enlargement risk potential

The risk ratings for streamflow change (**Step 10**), streambank erosion (**Step 13**) and direct disturbance/riparian vegetation impacts (**Step 15**) are used to assess channel enlargement potential. **Figure 7-23** is used to determine overall risk for channel enlargement based on the individual risk ratings for streamflow change, streambank erosion and direct impacts; **Table 7-25** summarizes these risk ratings. All risk ratings are recorded in **Worksheet 7-9**. Channel enlargement risk results varied by stream type from *very low* (E3 stream type) to *high* (C4 lower stream type) and *very high* risk (A3 and D4 stream types). The overall risk rating for channel enlargement is recorded by stream type in the *RRISSC* summary worksheet.

These high risk ratings warrant obtaining more quantitative data before proceeding with mitigation/stabilization projects. Breaking this category into more specific reaches within stream types would help to prioritize specific locations and increase the effectiveness of treatment for specific objectives. These objectives are beyond the capability of the *RRISSC* procedure; thus, the channel processes associated with *high* or *very high* risk should advance to *PLA*.

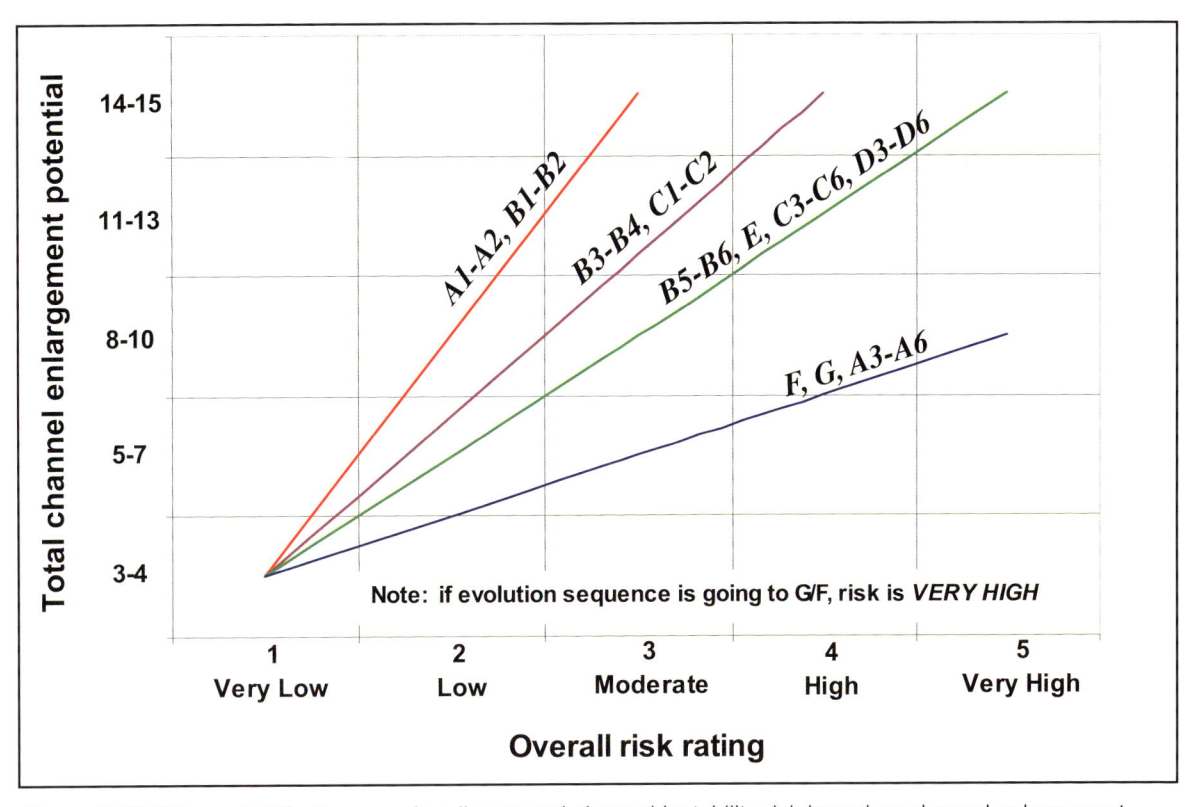

Chapter 7 — RRISSC   The WARSSS Methodology: A Case Study of Wolf Creek, Colorado

Figure 7-23 (Figure 4-26). Increased sediment and channel instability risk based on channel enlargement potential by stream type.

Table 7-25. Increased sediment and channel instability risk rating due to channel enlargement summary for Wolf Creek Watershed.

Channel enlargement risk rating using Figure 7-23 (Figure 4-26)					
Reach I.D.	Risk rating: streamflow changes	Risk rating: streambank erosion	Risk rating: direct channel impacts	Total individual risk rating points by reach	Overall risk rating by stream type
A3	VH (5)	VH (5)	VH (5)	15	VH (5)
C4 upper	VH (5)	M (3)	L (2)	10	M (3)
C4 lower	VH (5)	H (4)	H (4)	13	H (4)
D4	VH (5)	H (4)	VH (5)	14	VH (5)
E3	VL (1)	L (2)	VL (1)	4	VL (1)

Worksheet 7-9 (Worksheet 4-10). Risk rating worksheet for channel enlargement.

(1) Location code/ river reach I.D.	(2) Overall risk rating: streamflow changes **(Step 10 in Worksheet 4-2; Worksheet 4-6)**	(3) Overall risk rating: streambank erosion **(Step 13 in Worksheet 4-2; Worksheet 4-7)**	(4) Overall risk rating: direct channel impacts **(Step 15 in Worksheet 4-2; Worksheet 4-9)**	(5) Total numeric score $\sum[(2)+(3)+(4)]$	(6) Overall risk rating for channel enlargement **(Fig. 4-26)** (5) by stream type	(7) Adjustment due to in-channel mining*
1. A3	VH (5)	VH (5)	VH (5)	15	VH (5)	N/A
2. C4 upper	VH (5)	M (3)	L (2)	10	M (3)	N/A
3. C4 lower	VH (5)	H (4)	H (4)	13	H (4)	N/A
4. D4	VH (5)	H (4)	VH (5)	14	VH (5)	N/A
5. E3	VL (1)	L (2)	VL (1)	4	VL (1)	N/A
6.						
7.						
8.						
9.						
10.						
11.						
12.						
13.						
14.						
15.						

*Any in-channel mining automatically raises reach to *high* risk for enlargement and advances reach to *PLA*.

Step 17. Calculate aggradation/excess sediment risk

The introduced sediment from hillslope processes (roads and mass erosion), combined with channel processes, are used to assess potential aggradation/excess sediment deposition. In addition to the information collected in previous steps, the width/depth ratio relations in **Figure 7-24** (summarized in **Table 7-26**) were rated for each major stream type, as were the depositional feature relations shown in **Figure 7-25** (summarized in **Table 7-27**). This information is recorded in **Worksheet 7-10**.

Overall, the *very high* risk associated with mass erosion from Highway 160 (**Step 6**, **Worksheet 7-4**) indicates a high sediment supply and delivery potential. The receiving "weak link" stream types that are most susceptible to this sediment source are the A3, C4 and D4, which rated *very high* for aggradation/excess sediment risk (**Worksheet 7-10**). These ratings justify advancing to *PLA* to provide site- and process-specific mitigation to potentially reduce this accelerated sediment supply source. In addition to the hillslope sediment supply, excess sediment due to streamflow changes, streambank erodibility, width/depth ratio and excess sediment depositional indicators place the A3, C4 and D4 stream types at *high* or *very high* risk, requiring advancement to *PLA* (**Worksheet 7-10**). The E3 stream type rated *low* risk and does not require further assessment. The final risk ratings for aggradation/excess sediment supply are recorded in the *RRISSC* summary worksheet.

Width/Depth Ratio Stability Rating

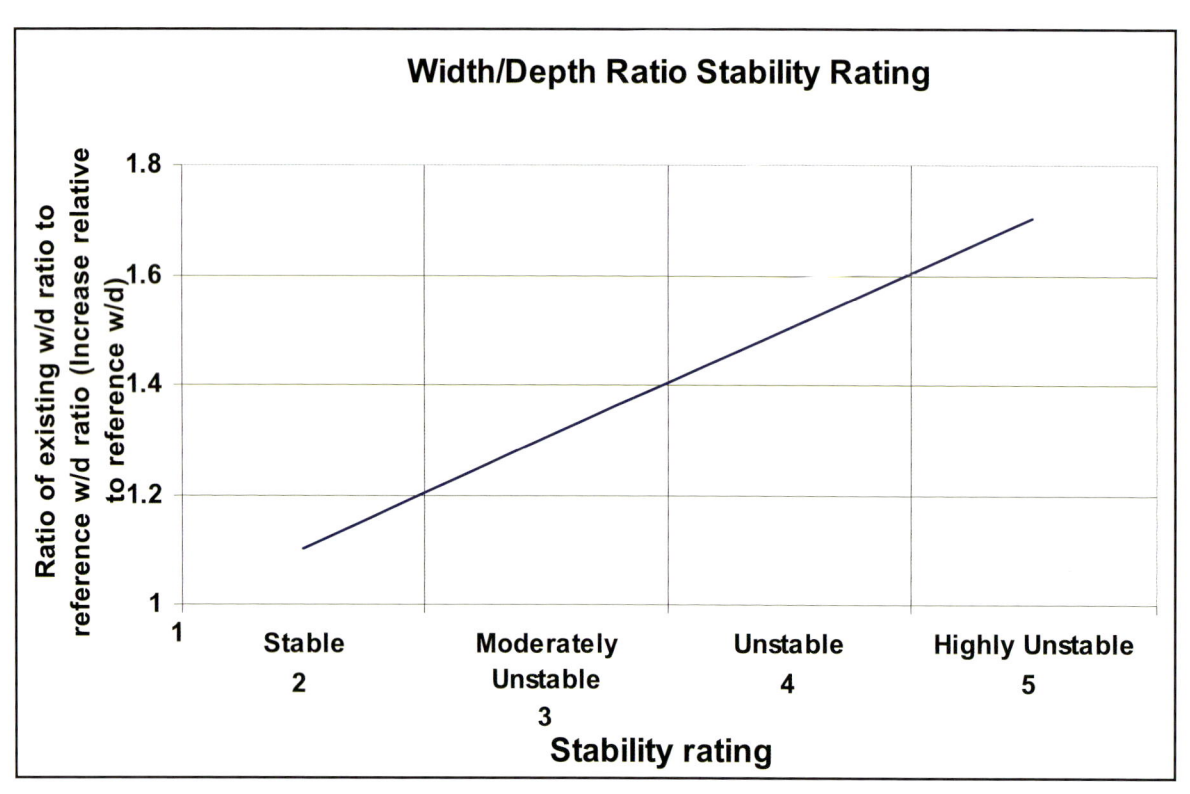

Figure 7-24 (Figure 4-27). Relation of risk rating for over-wide channels based on departure ratio from reference condition.

Table 7-26. Width/depth ratio stability risk rating summary for Wolf Creek Watershed.

Width/depth ratio departure risk rating using Figure 7-24 (Figure 4-27)				
Reach I.D.	Existing width/depth ratio	Reference width/depth ratio	Width/depth ratio impacted	Risk rating: width/depth ratio departure
A3	3	5	0.6	Highly Stable VL (1)
C4 upper	30	20	1.5	Unstable H (4)
C4 lower	23	20	1.2	Mod. Unstable M (3)
D4	165	20	8.3	Highly Unstable VH (5)
E3	1.5	2	0.8	Highly Stable VL (1)

Chapter 7 — RRISSC   The WARSSS Methodology: A Case Study of Wolf Creek, Colorado

The presence of numerous mid-channel and excess side bars leading to multiple channels and a conversion from meandering to braided channels indicates aggradation/excess deposition. The risk ratings for the A3, upper C4, lower C4 and D4 stream types were adjusted upward in **Worksheet 7-10** due to the related depositional features shown in **Figure 7-25**, summarized in **Table 7-27**.

Various Depositional Features modified from Galay et al. (1973)

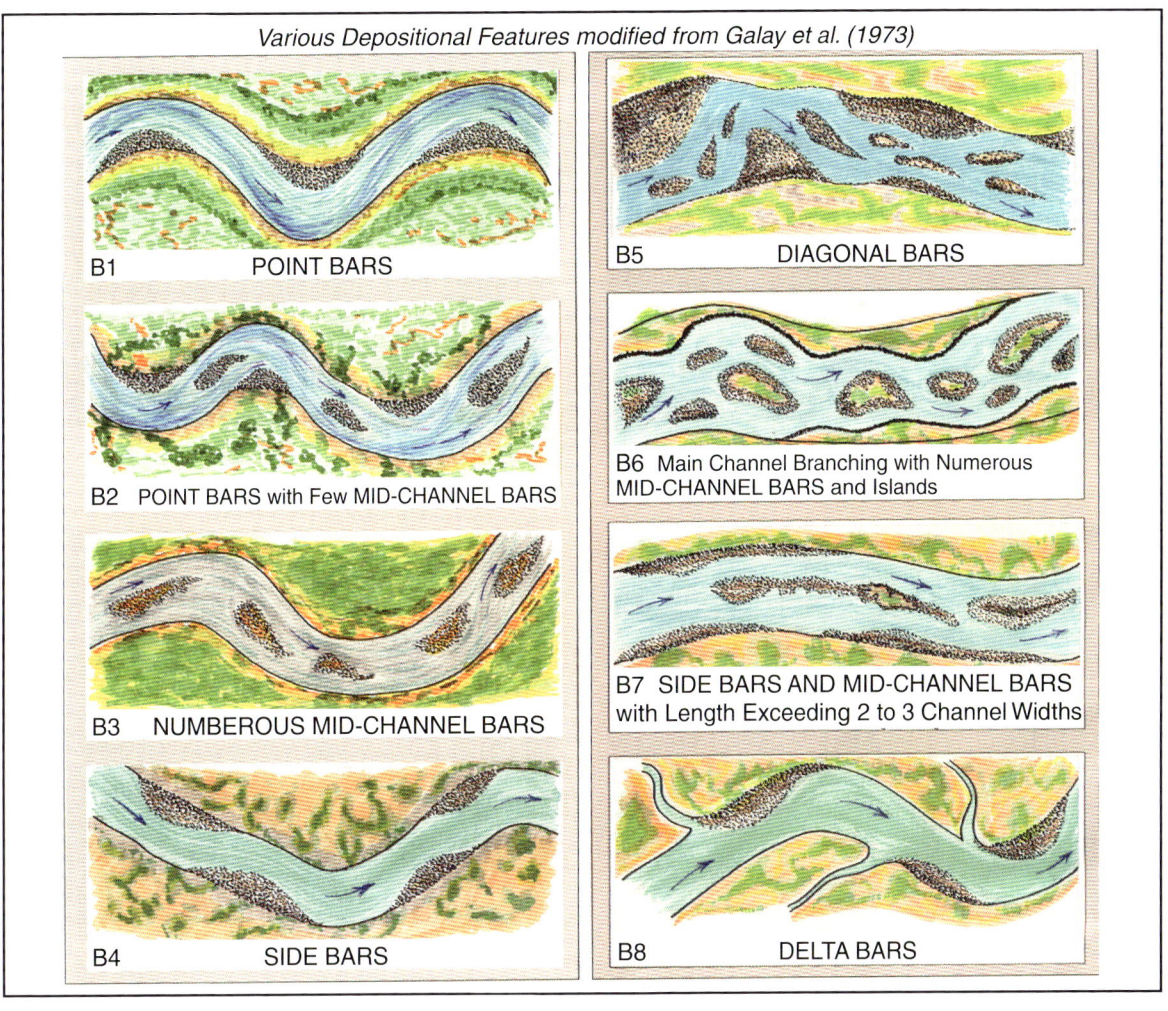

- B1 POINT BARS
- B2 POINT BARS with Few MID-CHANNEL BARS
- B3 NUMBEROUS MID-CHANNEL BARS
- B4 SIDE BARS
- B5 DIAGONAL BARS
- B6 Main Channel Branching with Numerous MID-CHANNEL BARS and Islands
- B7 SIDE BARS AND MID-CHANNEL BARS with Length Exceeding 2 to 3 Channel Widths
- B8 DELTA BARS

Figure 7-25 (Figure 4-28). Depositional features (bars) related to potential excess sediment/aggradation potential (Rosgen, 1996).

Table 7-27. Excess deposition/aggradation indicators for reaches in Wolf Creek Watershed.

Aggradation indicators for depositional patterns using Figure 7-25 (Figure 4-28)	
Reach I.D.	Depositional pattern
A3	B3
C4 upper	B7
C4 lower	B2, B5
D4	B5, B6, B7
E3	none

WATERSHED ASSESSMENT OF RIVER STABILITY AND SEDIMENT SUPPLY

Worksheet 7-10 (Worksheet 4-11). Summary of risk ratings for potential aggradation and/or excess sediment deposition.

(1)	(2)	(3)	(4)	(5)	(6)	(7)	(8)	(9)	(10)	(11)	(12)	(13)	(14)	(15)
	Hillslope Risk Ratings (Sediment Supply)					**Channel Process Response to Excess Sediment**								
Location code/ river reach I.D.	Risk rating: mass erosion (Step 6 in Worksheet 4-2; Worksheet 4-3)	Risk rating: roads (Step 7 in Worksheet 4-2; Worksheet 4-4)	Risk rating: surface erosion risk/ delivered sediment risk (Step 8 in Worksheet 4-2; Worksheet 4-5)	Point subtotal ∑[(2)+(3)+(4)]	Hillslope summary overall rating; use points from column (5) (Insert both numeric and adjective ratings) VL(1) = 3 L(2) = 4–7 M(3) = 8–10 H(4) = 11–14 VH(5) = >14	Representative location and associated rating points from column (6)*	Risk rating: width/depth ratio departure (Fig. 4-27) VL(1) = HS L(2) = S M(3) = MU H(4) = U VH(5) = HU	Risk rating: channel enlargement (Step 16 in Worksheet 4-2; Worksheet 4-10)	Risk rating: streambank erosion (Step 13 in Worksheet 4-2; Worksheet 4-7)	Point subtotal ∑[(7)+(8)+(9) +(10)]	Risk rating: use points from column (10) (Insert adjective risk rating) VL(1) < 5 L(2) = 5–8 M(3) = 9–12 H(4) = 13–16 VH(5) >16	Adjustments: aggradation/excess sediment indicators** **a.** obvious excess deposition **b.** filling of pools **c.** deposition of sand or larger material on floodplain **d.** bi-modal **e.** depositional patterns B3, B5–B7 (Fig. 4-28) (note categories that apply)	Adjustment: reduction in flow due to regulation**	Final aggradation/ excess sediment deposition risk rating (insert adjective risk rating)
1. **1**	VL(1)	H(4)	VL(1)	6	L(2)									
2. **2**	VL(1)	M(3)	VL(1)	5	L(2)									
3. **3**	VL(1)	M(3)	VL(1)	5	L(2)									
4. **4**	VL(1)	H(4)	VL(1)	6	L(2)									
5. **5**	VL(1)	L(2)	VL(1)	4	L(2)									
6. **Hwy 160**	VH(5)	VH(5)	VL(1)	11	H(4)									
7.														
8.														
9. **A3***						Hwy 160, H(4)	VL(1)	VH(5)	VH(5)	15	H	B3, n/a		VH
10. **C4 upper**						sub-water-shed 2, L(2)	H(4)	M(3)	M(3)	12	M	B7, a,b,c,d**		H
11. **C4 lower**						Hwy 160, H(4)	M(3)	H(4)	H(4)	15	H	B2, B5, d**		VH
12. **D4**						Hwy 160, H(4)	VH(5)	VH(5)	H(4)	18	VH	B5, B6, B7, a,b,c,d**		VH
13. **E3**						sub-water-shed 1, L(2)	VL(1)	VL(1)	L(2)	6	L	none, n/a		L
14.														
15.														

* To apply risk rating from Hillslope Processes for aggradation risk, it is important to identify the location of the sediment supply in relation to the most representative or "weak link" stream type.
** Adjust a full risk category upward if streamflow decrease and/or indicators provide evidence appropriate to the observed condition such as aggradation indicators on categories a,b,c,d and e.
*** Associated with Highway 160.

Step 18. Determine channel evolution potential

Aerial photography over a period of years is useful in ascertaining rate and direction of shifts in channel patterns, which are often readily visible. Aerial photographs taken of Wolf Creek from 1976 (**Figure 7-4**; see **Step 3**, *RLA*) to 1991 (**Figure 7-5**; see **Step 3**, *RLA*) show progressive stream channel changes. The channel changed from a single-thread, narrow C4 stream type to a wide, braided, D4 stream type during this period.

This successional change from a C4 to D4 stream type is associated with scenario #2 in **Figure 7-26**. The risk rating for the D4 stream type associated with this change is *very high*, as indicated in **Table 7-28**. Due to the significance of this adverse morphological change, a more quantitative analysis is necessary to indicate specific sediment and stability consequences. The C4 and E3 stream types are either recovering or have recovered to their physical potential stable states and have a *low* risk rating for channel evolution risk. The A3 stream type, although associated with high sediment supply, is a geologically active stream type. A3 stream types in valley type I in glacial tills rarely change to any other type; thus, converting to a stable stream type is inconsistent with natural evolution processes. The channel evolution risk ratings are recorded in the *RRISSC* summary worksheet by stream type.

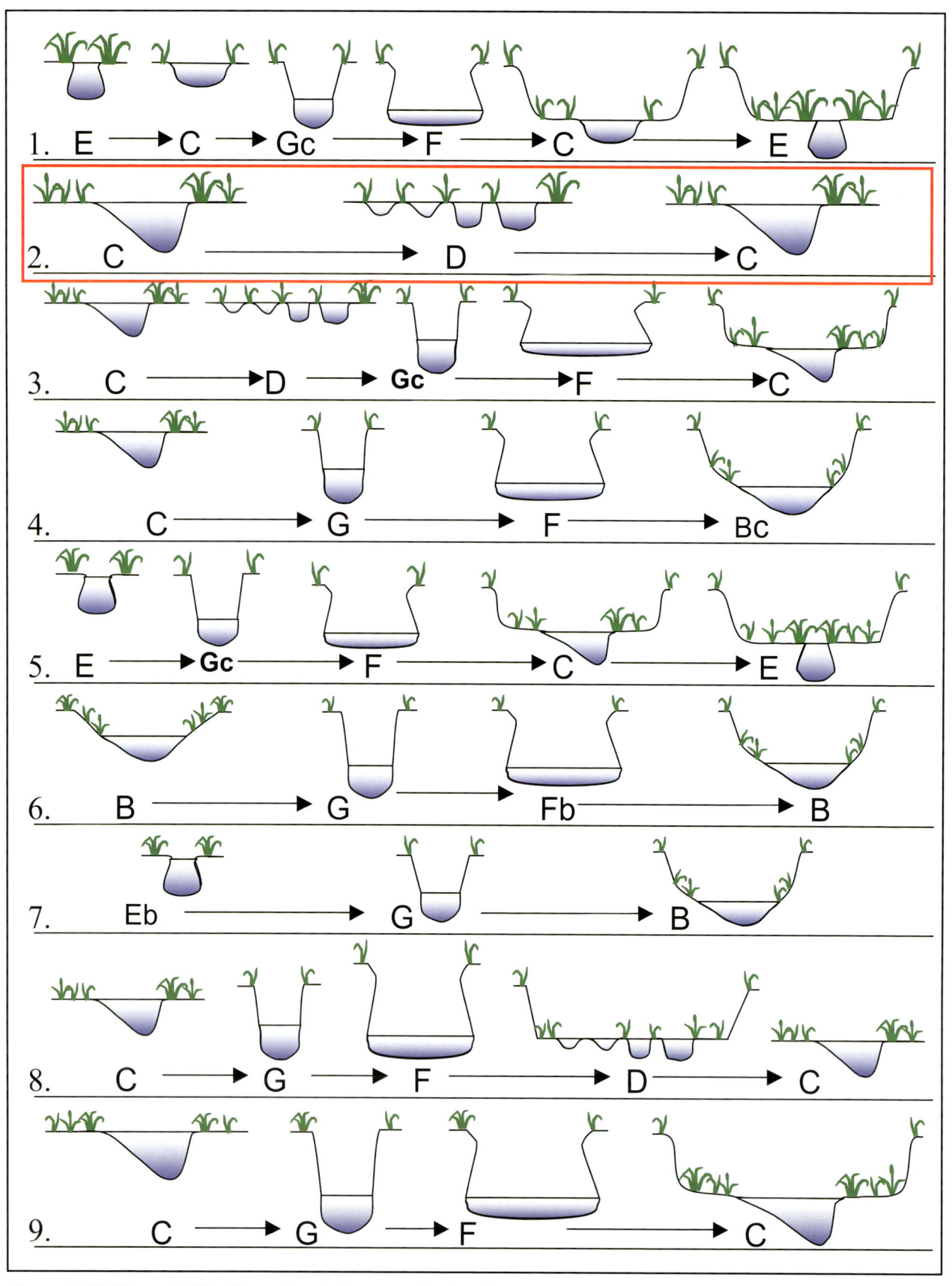

WATERSHED ASSESSMENT OF RIVER STABILITY AND SEDIMENT SUPPLY

1. E → C → Gc → F → C → E
2. C → D → C
3. C → D → Gc → F → C
4. C → G → F → Bc
5. E → Gc → F → C → E
6. B → G → Fb → B
7. Eb → G → B
8. C → G → F → D → C
9. C → G → F → C

Figure 7-26 (Figure 2-38). Channel evolution scenarios involving stream type succession scenario #2 for lower Wolf Creek (Rosgen, 1999, 2001b).

Table 7-28 (Table 4-5). Risk rating for various stream channel successional state scenarios.

Channel successional states of stream type evolution*	Risk rating
E to C	Moderate (3)
C to D	**Very High (5)**
B, C, E or D to G	Very High (5)
G to F	High (4)
G to B	Very Low (1)
F to B	Very Low (1)
F to C	Low (2)
F to D	Moderate (3)
All others (e.g., C to E)	Low (2)

* See **Figure 2-38**: Stream type succession scenarios

Step 19. Calculate degradation risk

One of the serious adverse consequences of watershed impacts is channel degradation, or a downward shift in local base level and abandonment of floodplains. The combined influences of streamflow change (**Step 10**), in-channel mining (**Step 14**), direct channel impacts (**Step 15**), channel evolution (**Step 18**) and bridges or culverts (**Figure 7-27** (summarized in **Table 7-29**) and **Worksheet 7-12**) are recorded in **Worksheet 7-12** by stream type location.

The D4 stream type is associated with excess sediment supply and aggradation and, therefore, was not analyzed for degradation risk. The E3 stream type had no culverts or bridges that could potentially cause degradation and was automatically given a *very low* risk rating in **Worksheet 7-11**, which is transferred to Column (5) in **Worksheet 7-12**. The E3 stream type rated *low* for overall degradation risk and will not require a more detailed *PLA* analysis.

The A3 stream type adjacent to the landslide caused by Highway 160 and the A3 stream type below the last culvert on the switchback on Highway 160 rated *very high* for degradation potential. There is a strong interrelation between the high mass erosion potential (debris flow/debris avalanche) along Highway 160 and degradation. The lowering of local base levels (degradation) causes streambank rejuvenation, or over-steepening of adjacent slopes and drainageways. This rejuvenation adds to slope disequilibrium, potentially accelerating mass erosion. Due to the serious consequences of degradation, grade control may need to be installed on these two A3 stream sites. This type of mitigation is expensive and complex; thus, a *PLA* analysis is recommended for these two reaches to specify exact locations and the nature of possible mitigation. The overall risk ratings for the C4 stream types also indicated *very high* risk due to potential increases in streamflow and will need to advance to *PLA*. The overall risk ratings for degradation from **Worksheet 7-12** are recorded in the *RRISSC* summary worksheet.

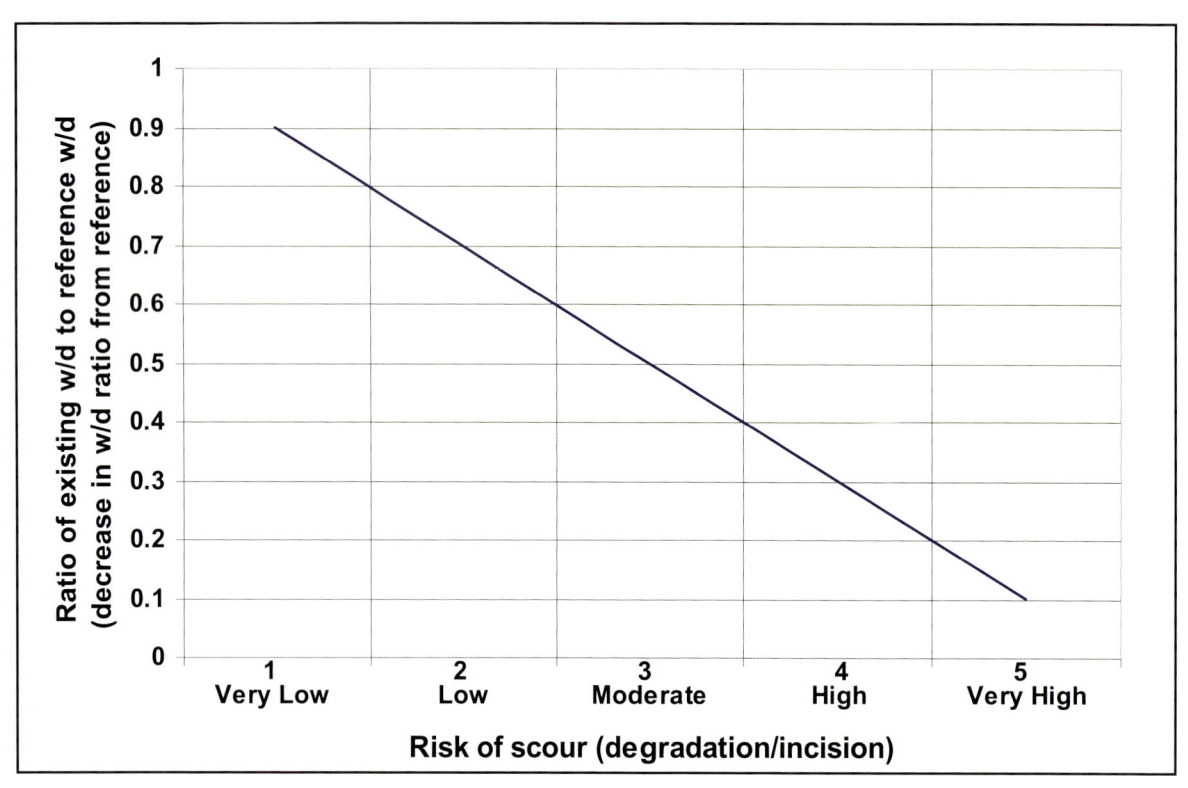

[Graph: Y-axis "Ratio of existing w/d to reference w/d (decrease in w/d ratio from reference)" from 0 to 1; X-axis "Risk of scour (degradation/incision)" with values 1 Very Low, 2 Low, 3 Moderate, 4 High, 5 Very High. A line descends from approximately (1, 0.9) to (5, 0.1).]

Figure 7-27 (Figure 4-29). Conversion of a decrease in the existing width/depth ratio compared to reference width/depth ratio for potential degradation (incision due to excess energy). This relation is used only if the lowest bank height is greater than the maximum bankfull depth (bank-height ratio > 1).

Table 7-29. Data for risk of scour by stream type.

Risk of scour by stream type using Figure 7-27 (Figure 4-29)				
Stream type	Reference w/d	Existing w/d	Existing w/d divided by reference w/d	Risk rating
A3	5	7	1.40	VL (1)
C4 upper	20	30	1.50	VL (1)
C4 lower	20	23	1.15	VL (1)
D4	20	165	8.25	VL (1)
E3	1.8	1.5	0.83	VL (1)

Chapter 7 — RRISSC   The WARSSS Methodology: A Case Study of Wolf Creek, Colorado

Worksheet 7-11 (Worksheet 4-13). Risk rating worksheet for potential contraction scour/degradation/channel incision due to culverts or bridges.

(1) Location code/ river reach I.D.	(2) Percent reduction of sinuosity (insert numeric rating)	(3) Stream crossing structure (insert numeric rating)	(4) Subtotal Σ[(2)+(3)]	(5) Increase in energy slope (use (4) points and insert numeric rating)	(6) Ratio of a decrease in w/d ratio to existing reference w/d ratio (Figure 4-29) (insert numeric rating)	(7) Backwater potential above structure (insert numeric rating)	(8) Presence of floodplain drains (through fills) (insert numeric rating)	(9) Subtotal Σ[(5)+(6)+(7)+(8)]	(10) Overall risk rating: culverts or bridges
	(1) = No change (2) = Sinuosity reduced up to 50% (3) = Sinuosity reduced 50–80% (4) = Sinuosity reduced more than 80%	(1) = Bridge (2) = Arch culvert (3) = Culvert (4) = Over-steepened culvert		VL (1) = 2 L (2) = 3 M (3) = 4 H (4) = 5–6 VH (5) = 7–8	VL (1) > 8.0 L (2) = 0.61–0.80 M (3) = 0.41–0.60 H (4) = 0.21–0.40 VH (5) ≤ 0.20	VL (1) = None L (2) = Slight only for floods > 50 yr recurrence interval M (3) = Some for floods 11–50 yr recurrence interval H (4) = Evident for floods 2–10 yr recurrence interval VH (5) = Backwater at bankfull discharge	VL (1) = All floods greater than bankfull drain through fill L (2) = Accomodates 90% of floods M (3) = Accomodates 50–89% of floods H (4) = Evident for floods 2–10 yr recurrence interval VH (5) = Backwater at bankfull discharge		VL (1) = 4 L (2) = 5–8 M (3) = 9–12 H (4) = 13–16 VH (5) = 17–20
1. A3*	(4)	(4)	8	(5)	(1)	(4)	(5)	15	**H (4)**
2. C4 upper	(2)	(2)	4	(3)	(1)	(2)	(3)	9	**M (3)**
3. C4 lower	(3)	(1)	4	(3)	(1)	(2)	(3)	9	**M (3)**
4. D4**									**VL (1)**
5. E3**									**VL (1)**
6.									
7.									
8.									
9.									
10.									
11.									
12.									
13.									
14.									
15.									

*Associated with Highway 160.

**No bridges or culverts on these stream types; therefore, risk rating is automatically *very low*.

Worksheet 7-12 (Worksheet 4-12). Risk rating worksheet for degradation.

(1) Location code/ river reach I.D.	(2) Risk rating: streamflow changes **(Step 10 in Worksheet 4-2; Worksheet 4-6)**	(3) Risk rating: in-channel mining associated with base-level shifts **(Step 14 in Worksheet 4-2; Worksheet 4-8)**	(4) Risk rating: channel evolution **(Step 18 in Worksheet 4-2; Table 4-5)**	(5) Risk rating: road drainage designs, "shot gun" culverts (base-level shifts) **(Worksheet 4-13)**	(6) Risk rating: direct channel impacts **(Step 15 in Worksheet 4-2; Worksheet 4-9)**	(7) Overall risk rating for degradation (Insert highest adjective rating from Columns 2–6)
1. A3*	VH (5)	VL (1)	L (2)	H (4)*	VH (5)	VH
2. C4 upper	VH (5)	VL (1)	L (2)	M (3)	L (2)	VH
3. C4 lower	VH (5)	VL (1)	L (2)	M (3)	H (4)	VH
4. D4**	N/A	N/A	N/A	N/A	N/A	N/A
5. E3	VL (1)	VL (1)	L (2)	VL (1)	VL (1)	L
6.						
7.						
8.						
9.						
10.						
11.						
12.						
13.						
14.						
15.						

*Associated with Highway 160.

**D4, braided reaches are not susceptible to degradation at this evolutionary state.

Step 20. Summarize total potential sediment and stream channel stability risk

The completed *RRISSC* summary **Worksheet 7-13** provides documentation of overall risk rating summaries by location. Not all the sites have the same ratings for all variables. This summary helps sort out the processes, land uses and locations where mitigation may be needed based on risk. It also identifies specific areas and/or reaches in need of a more detailed assessment depending on risk and associated land values.

Step 21. Create overall risk rating summary

This step allows the user to interpret the potential severity of change, departure from reference condition, recovery potential, long- versus short-term potential impacts, sources and causes of impairment, and the consequences of potential impairments to beneficial uses. The results document the processes and associated areas that warrant further assessment at the *PLA* level. This information is summarized in the *RRISSC* summary **Worksheet 7-13**. This information should be used to prepare a narrative summary that indicates the major contributing factors and rationale for the ratings assigned to each location. For areas of *high* and *very high* risk, the summary should identify the processes influenced and the land use practices associated with the potential risk impairment. This information will assist the user in justifying the need to advance these processes to *PLA*. Changes in land use management, stabilization, enhancement and restoration proposals will result from the more detailed and site-specific *PLA* phase. For the *moderate* risk ratings, the summary should also formulate recommendations for mitigation and monitoring, as discussed in **Step 23**.

For discussion purposes, the completed *RRISSC* summary **Worksheet 7-13** is shown. The accompanying narrative summary describes the ratings used to perform **Steps 22**, **23** and **24**.

WATERSHED ASSESSMENT OF RIVER STABILITY AND SEDIMENT SUPPLY

Worksheet 7-13 (Worksheet 4-2). Completed *RR/SSC* summary worksheet for the Wolf Creek Watershed.

Watershed Name: Wolf Creek Date: 7/04 Observer(s): Dave Rosgen

	Location code / river reach I.D.	Step 6: Mass erosion (Worksheet 4-3)	Step 7: Roads (Worksheet 4-4)	Step 8: Surface erosion (Worksheet 4-5)	Step 10: Streamflow change (Worksheet 4-6)	Step 13: Streambank erosion (Worksheet 4-7)	Step 14: In-channel mining (Worksheet 4-8)	Step 15: Direct channel impacts (Worksheet 4-9)	Step 16: Channel enlargement (Worksheet 4-10)	Step 17: Aggradation/excess sediment (Worksheet 4-11)	Step 18: Channel evolution/ successional states (Table 4-5)	Step 19: Degradation (Worksheet 4-12)	Processes identified by step for advancement to *PLA*	✓ Location selected for advancement to *PLA*
		Geographic Location						Stream Type Location						
1.	1	VL (1)	H (4)	VL (1)	H (4)								7, 10	✓
2.	2	VL (1)	M (3)	VL (1)	VH (5)								10	✓
3.	3	VL (1)	M (3)	VL (1)	H (4)								10	✓
4.	4	VL (1)	H (4)	VL (1)	VH (5)								7, 10	✓
5.	5	VL (1)	L (2)	VL (1)	M (3)									✓**
6.	Hwy 160	VH (5)	VH (5)	VL (1)	M (3)								6, 7, 10	✓
7.	Total Watershed				VH (5)									
8.														
9.	A3					VH (5)	VL (1)	VH (5)	VH (5)	VH	L (2)	VH	10, 13, 15, 16, 17, 19	✓
10.	C4 upper					M (3)	VL (1)	L (2)	M (3)	H	L (2)	VH	10, 17, 19	✓
11.	C4 lower					H (4)	VL (1)	H (4)	H (4)	VH	L (2)	VH	10, 13, 15, 16, 17, 19	✓
12.	D4					H (4)	VL (1)	VH (5)	VH (5)	VH	VH (5)	N/A*	10, 13, 15, 16, 17, 18	✓
13.	E3					L (2)	VL (1)	VL (1)	VL (1)	L	L (2)	L		
14.														
15.														

*Braided D4 channel does not have potential for degradation in present state.

**Although sub-watershed 5 is stable, the cumulative impacts of streamflow need to be assessed in *PLA* to determine total watershed impacts on the "weak link" stream types below sub-watershed 5.

7-76

Narrative Summary

The process-based risk analysis for sediment supply and stream channel stability (*RRISSC*) is summarized in **Worksheet 7-13**. Based on this assessment, the following recommendations are offered:

Mass erosion

All of the major sub-watersheds rated *very low*; no further analysis or mitigation is needed. The terrain adjacent to State Highway 160, however, rated *very high* and needs to advance to *PLA* to identify specific locations for mitigation rather than general treatments for the entire road segment within the watershed.

Roads

The older, secondary logging roads vary as to risk. The roads in sub-watershed 5 are *low* risk; due to their upper-slope position and low overall density, they do not contribute a disproportionate sediment supply. Mitigation, therefore, is not critical in this location. Sub-watersheds 2 and 3 rated *moderate*. Here, mitigation would require preventing vehicular off-road and four-wheeler use on most roads to allow revegetation of the road surface. Monitoring sediment sources and vegetal response would help to evaluate this strategy's effectiveness. Sub-watersheds 1 and 4 rated *high*. These ratings require advancement to *PLA* to identify specific, annual sediment supply contributions relative to all roads and to design restoration and/or rehabilitation plans at precise locations for specific processes contributing to the sediment supply.

Furthermore, all of the roads in each sub-watershed, as well as Highway160, were carried forward to *PLA*. The rationale for this is to quantify the relative annual sediment yields from forest roads compared to other watershed sediment sources. The cumulative effects of all sediment source contributions by land use and location can thus be calculated. The water yield increase must also use road acres and location at the *PLA* level to determine their contribution to streamflow increases.

Surface erosion

Due to the lack of overland flow processes other than roads, surface erosion rated as a *very low* risk with regard to annual sediment yield contribution. No mitigation or further analysis is recommended.

Streamflow change

Sub-watershed 5 had a *moderate* risk rating and requires some mitigation. Revegetation of clearcut areas that have not regenerated would be a first step. To reduce the streamflow runoff increase due to sub-surface flow interception, a road closure involving the replacement of fill back into the cut bank is recommended. This would be done only on side slope gradients greater than 10%. Revegetation of road surfaces would improve evapo-transpiration, further reducing water yield contributions. Monitoring of these practices would involve assessing vegetal response in older clearcuts and on road surfaces. Monitoring should also calculate reductions in water concentration due to ditch lines and/or sub-surface flow interception.

Streamflow change potential along Highway 160 also rated *moderate*. Here, the strategy would be to disperse runoff. Further mitigation plans should be deferred until the *PLA* assessment of sediment contribution potential is completed for this location.

Sub-watersheds 1, 2, 3 and 4 all rated *high* or *very high* and require assessment at the *PLA* level. All sub-watersheds should receive a detailed water yield analysis in order to quantify individual contributions. This analysis identifies priorities for mitigation by location and/or stream type to determine the cumulative sediment contribution from each watershed.

Streambank erosion

The E3 stream type rated *low*. Further assessment is not required for this stream type. The upper C4 stream type that rated *moderate*, however, requires mitigation specific to short distances of disturbance. Most of the disturbance is associated with road crossings and/or backwater on the upper C4. Placing large woody debris on streambank margins and dissipating energy on the streambank toes at road crossings would help mitigate these localized problems. These locations and the large woody debris management should be monitored to ensure effective mitigation.

The A3, lower C4 and D4 stream types rated *high* or *very high* with regard to streambank erosion risk. A detailed bank erosion prediction model (Bank Assessment for Non-point Source Consequences of Sediment) is required at the *PLA* level for these areas, located at the lower end of the watershed.

Direct channel impacts

The E3 and upper C4 stream types rated *very low* or *low* risk; no additional assessment is required. The A3, lower C4 and D4 stream types rated *high* or *very high* risk; therefore, a *PLA* analysis is recommended.

Channel enlargement

The E3 stream type rated *very low* risk; no additional assessment or mitigation is recommended. The upper C4 stream type rated *moderate* risk for channel enlargement due to the combined influences of *very high* streamflow increase risk and *moderate* streambank erosion risk. Mitigation to reduce flow increase, as previously discussed, and streambank erosion risk reduction (stabilization) on this reach would help offset the channel enlargement potential. The A3, lower C4 and D4 stream types rated *high* or *very high* risk. These stream types will be assessed at the *PLA* level.

Aggradation/excess sediment supply

The E3 stream type rated *low* risk; no further assessment or mitigation is recommended for this stream type. The A3 stream type rated *very high* risk. As this rating is caused primarily by roads, specific location and source-specific mitigation is recommended at the *PLA* level. The upper C4 stream type rated *high* and the lower C4 and D4 stream types rated *very high* and, therefore, require a more detailed evaluation at *PLA*. This analysis may result in restoration and/or stabilization plans designed to offset aggradation/excess sediment deposition and the resulting, obvious impairment to specific problem reaches.

Channel evolution/successional states

The only reach that indicated a *very high* risk for a time-trend shift, from a C4 to D4 stream type, was the lower section of the watershed located in the D4 stream reach. This location requires further assessment at the *PLA* level. The remaining stream types rated *low* risk and do not require additional assessment or mitigation.

Degradation

The E3 stream type rated *low* risk for degradation and does not require any additional assessment or mitigation. The D4, braided stream type was not assessed for degradation risk due to the low probability of the stream type to degrade or lower local base level; consequently, no further assessment is needed for this process. The A3, upper C4 and lower C4 stream types rated *very high* risk for degradation due to the influence of Highway 160 and streamflow increase risk; therefore, these locations require advancement to *PLA* to make site- and process-specific mitigation recommendations.

Overall review

As the overall *RRISSC* summary worksheet indicates, sub-watersheds 1–4 had at least one *high* or *very high* risk rating associated with streamflow increase and/or roads. A road and streamflow assessment at the *PLA* level will be conducted for each location associated with the A3, upper C4, lower C4 and D4 stream types.

Step 22. Discard processes with low overall risk ratings

Recommendations for areas associated with *low* or *very low* risk processes are generally to proceed with no change in management practices and/or mitigation design. Because of the low likelihood of disproportionate sediment supply, river instability or associated adverse consequences, no monitoring is recommended unless it will serve to document a stable reference condition within the low-risk reaches. In the Wolf Creek Watershed, the *RLA* assessment identified the A2 and B3 stream types as stable. Further assessment at the *RRISSC* level has identified the E3 stream type as stable and not in need of additional assessment.

Step 23. Create management change recommendations for processes with moderate overall risk ratings

The pressing question for areas associated with *moderate* risk ratings is whether to continue with a *PLA* assessment or whether to develop site- and process-specific mitigation and schedule monitoring. Part of the answer to this question is associated with the potential severity and/or adverse consequences of the sites and/or stream reaches. The uncertainty of the relations used in the *RRISSC* analysis may lead the user (or critics) to suggest that a more rigorous, quantitative method be implemented. The question of whether to proceed to *PLA* can also be answered by evaluating the potential magnitude and adverse consequences of the impairment associated with the water resource's uses and values.

If mitigation is deemed appropriate, users should design mitigation and management strategies that relate specifically to the activity that caused the risk rating and the processes affected. Otherwise, restoration or stabilization measures may "patch symptoms" rather than resolve the condition. Strategies may also include recommendations for stabilization, enhancement and/or restoration. One process associated with this rating is sediment delivery due to forest roads. As these roads contribute relatively low amounts of sediment when compared to Highway 160, areas in sub-watershed locations 1 and 2 that rated *moderate* risk are good candidates for mitigation and monitoring. Mitigation would include placing road fills back into the cut banks to reduce sub-surface interception. Road fills and culverts should be removed from stream crossings to eliminate water running down the ditch lines and minimize road surface contributions to sediment yields. This would reduce water yield increases that adversely affect the more sensitive C4 and D4 stream types near the mouth of the watershed. These mitigation plans place National Forest lands back into timber production, while still allowing for recreational and forest management access via existing roads.

Step 24. Advance high-risk and/or high-consequence processes to the *PLA* assessment phase

Locations associated with *high* or *very high* risk ratings have a high probability/likelihood of impairment in the form of disproportionate sediment supply, channel instability and associated severe adverse consequences. In many cases, the impacts are so extensive that effective implementation of general or broad-level mitigation is questioned. These sub-watersheds, slopes and/or river reaches require a more detailed assessment, performed in the *PLA* phase of the *WARSSS* methodology. This final risk assessment phase focuses on site-specific and process-based mitigation, as well as evaluation for model or prediction validation in addition to effectiveness monitoring.

Based on this *RRISSC* assessment, locations associated with the A3, C4 and D4 stream types will proceed to *PLA* due to the large number of *high* and *very high* risk ratings, the extensive potential adverse sediment supply and channel instability. These locations are found primarily along Highway 160. The *PLA* phase will isolate and quantify sediment sources by specific processes and locations.

The *Prediction Level Assessment (PLA)*: Wolf Creek Case Study

A review of the *RRISSC* assessment provides a detailed summary of the prediction of hillslope, hydrologic and channel processes. The high-risk sub-watersheds and river reaches recommended for a *PLA* assessment are documented in *RRISSC* in **Worksheet 7-13**. The *RRISSC* assessment of the Wolf Creek Watershed indicates that the following potential sediment source processes and potential impairment be assessed at the *PLA* level:

- Mass erosion related to Highway 160 (see **Step 20**, *PLA*);

- Roads: Highway 160 and forest access roads in sub-watersheds 1–5 (see **Step 20**, *PLA*);

- Streamflow change impacts of sediment supply for sub-watersheds 1–5 (see **Steps 10–17**, *PLA*);

- Streambank erosion for identifying accelerated and/or disproportionate sediment supply (see **Steps 8** and **9**, *PLA*);

- Direct channel impacts for the lower C4 and D4 stream types (see **Steps 7** and **24–28**, *PLA*);

- Channel enlargement for the A3, lower C4 and D4 stream types (see **Step 27**, *PLA*, for the lower C4 and D4 stream types; see **Step 31**, *PLA* summary, for the A3 stream type);

- Aggradation/excess sediment for the A3, upper C4, lower C4 and D4 stream types (see **Steps 7**, **10–19**, **22** and **24–28**, *PLA*, for the lower C4 and D4 stream types; see **Step 31**, *PLA* summary, for the A3 and upper C4 stream types);

- Degradation/channel incision for the A3, upper C4 and lower C4 stream types (see **Steps 7**, **10–19**, **22** and **24–28**, *PLA*, for the lower C4 stream type; see **Step 31**, *PLA* summary, for the A3 and upper C4 stream types); and

- Channel successional state for the D4 stream type (see **Step 24**, *PLA*).

Based on the channel successional state assessment in *RRISSC*, it is evident that the D4 stream type had recently evolved from a C4 stream type. If the reference, stable condition is a C4 stream type, located immediately upstream, then the *PLA* analysis needs to assess both the lower C4 and D4 stream types for a departure analysis (**Figure 7-28**). Furthermore, investigations of time-trend analysis from aerial photos from 1976 to the present show that the lower C4 stream type had not changed its dimension, pattern or profile; however, the D4 stream type below the lower C4 stream type had changed from a C4 to a D4 braided stream type. Due to this preliminary analysis, the *PLA* involves a comparison for departure of sediment supply and river stability between the lower C4 stream type (reference reach) and the lower D4 stream type (impaired reach). The locations of the detailed cross-sections to collect study reach data for the C4 reference and D4 impaired stream types are shown in **Figure 7-29**. In the following *PLA* channel assessment of Wolf Creek, these reach locations will be studied in more detail.

Furthermore, this assessment includes a quantification of streambank erosion rates and a sediment budget for the entire Wolf Creek Watershed. This information is designed to provide quantitative information about sediment contribution from various processes influenced by specific locations of land use practices. After such an analysis, specific mitigation can be planned according to a forecast of sediment reduction effectiveness.

WATERSHED ASSESSMENT OF RIVER STABILITY AND SEDIMENT SUPPLY

The following sequence of steps outlines the processes required for the *PLA* analysis. **Flowchart 7-3** illustrates the sequential steps necessary to complete *PLA* to determine stream stability and sediment supply. For a discussion of the variables and computational processes, see Chapter 5.

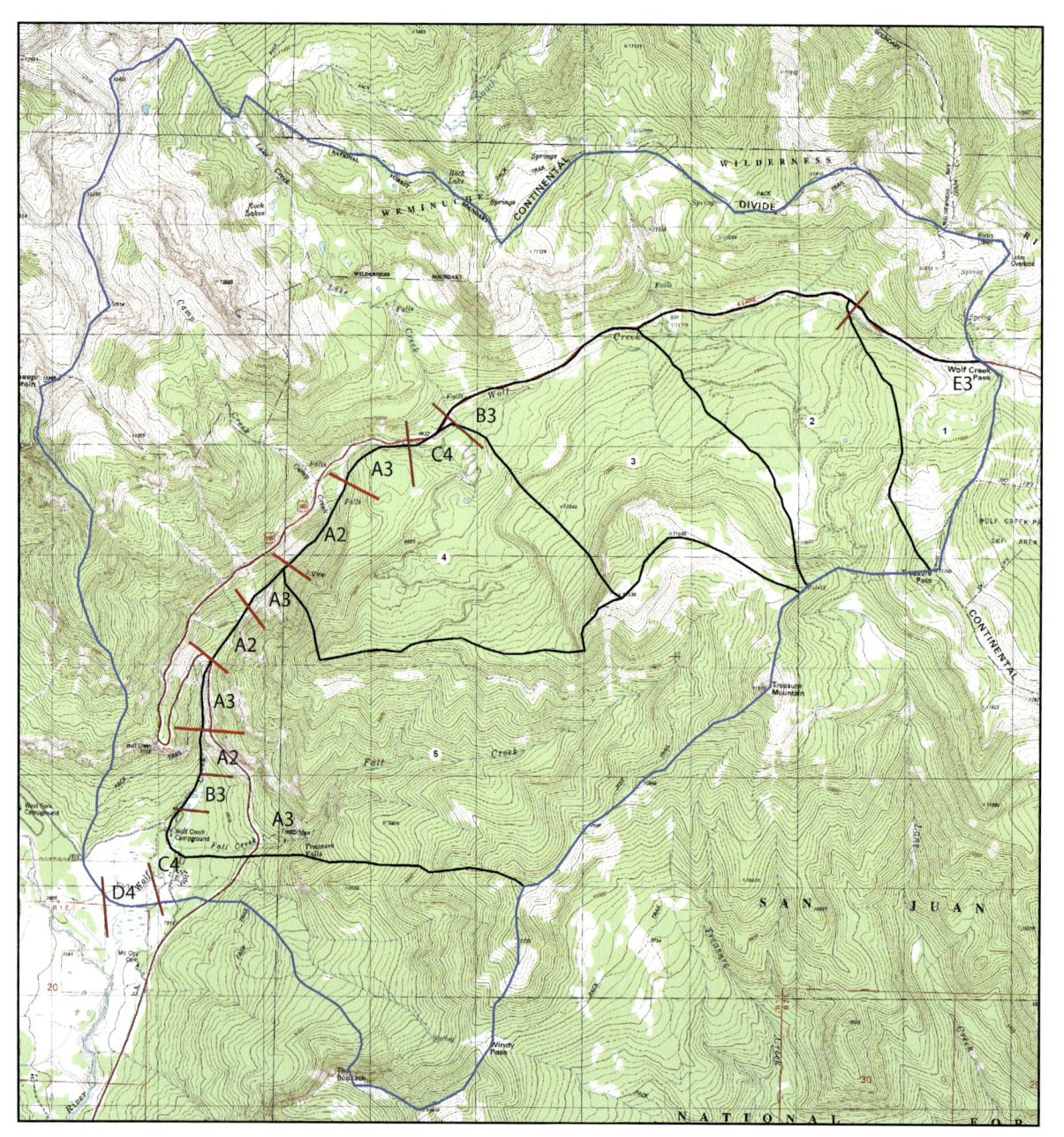

Figure 7-28. USGS 7.5′ topographic map of Wolf Creek, showing sub-watershed delineation and Level II stream classification.

7-84

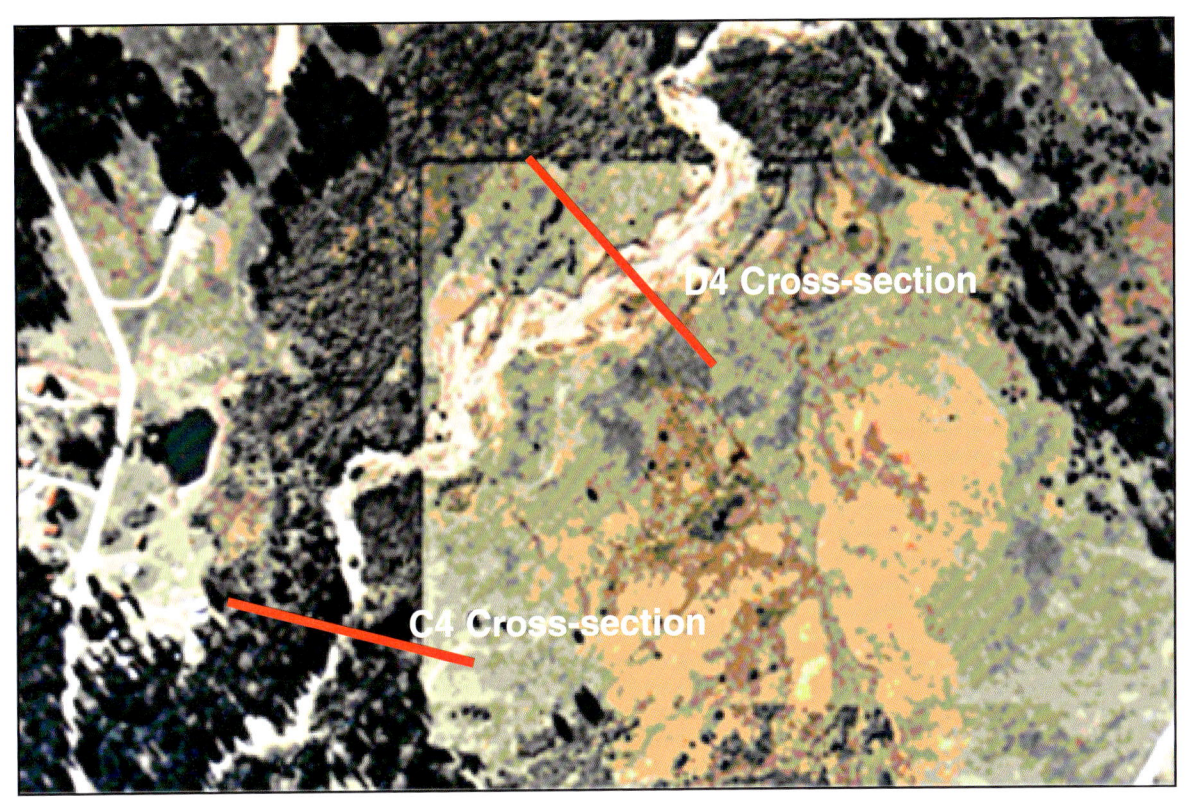

Chapter 7 — PLA   The WARSSS Methodology: A Case Study of Wolf Creek, Colorado

Figure 7-29. Aerial photograph (1991) showing downstream change in Wolf Creek to a D4 stream type but little change in the upstream, untreated, (above fence line) C4 stream type. Flow is left to right.

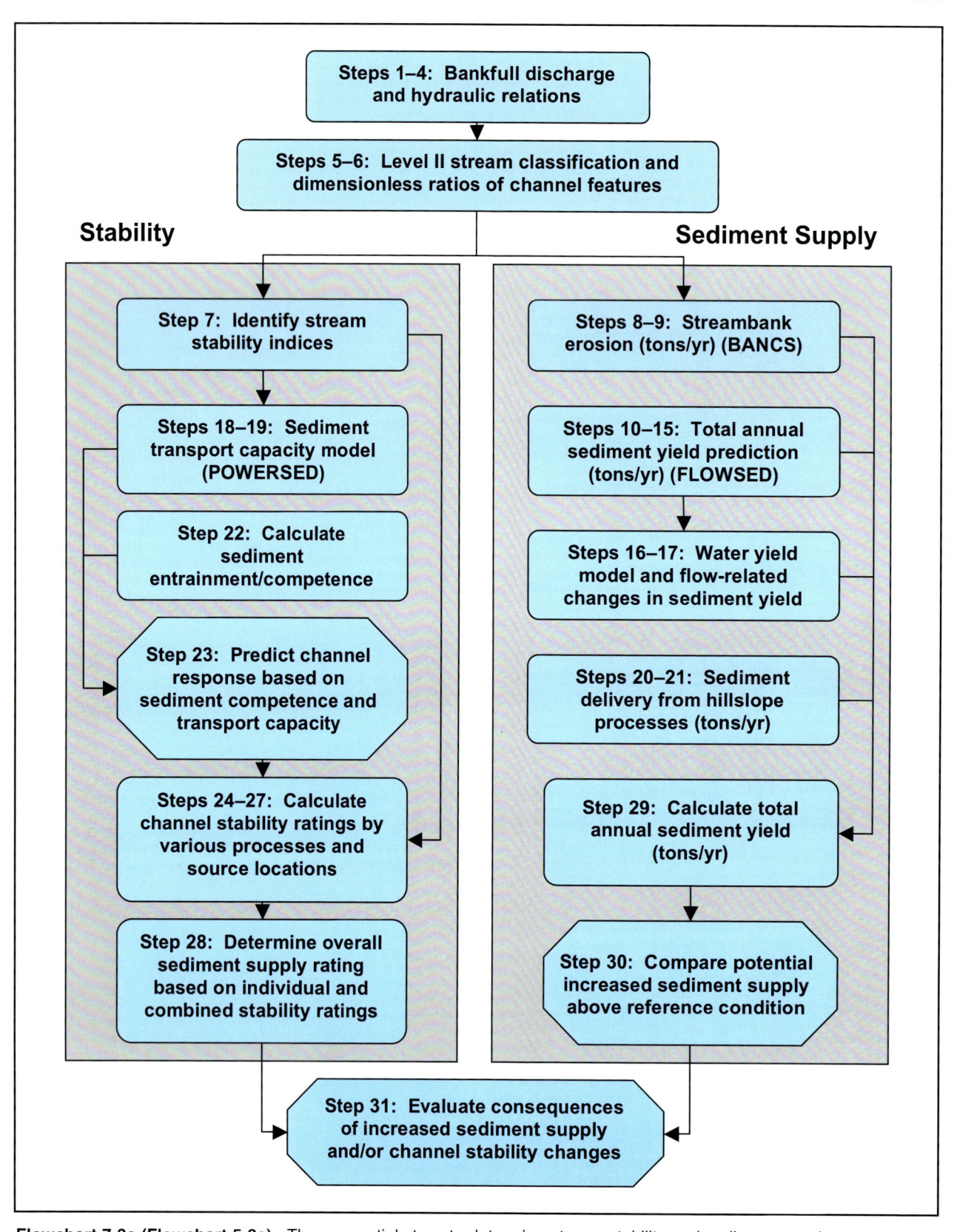

Watershed Assessment of River Stability and Sediment Supply

Stability

- Steps 1–4: Bankfull discharge and hydraulic relations
- Steps 5–6: Level II stream classification and dimensionless ratios of channel features
- Step 7: Identify stream stability indices
- Steps 18–19: Sediment transport capacity model (POWERSED)
- Step 22: Calculate sediment entrainment/competence
- Step 23: Predict channel response based on sediment competence and transport capacity
- Steps 24–27: Calculate channel stability ratings by various processes and source locations
- Step 28: Determine overall sediment supply rating based on individual and combined stability ratings

Sediment Supply

- Steps 8–9: Streambank erosion (tons/yr) (BANCS)
- Steps 10–15: Total annual sediment yield prediction (tons/yr) (FLOWSED)
- Steps 16–17: Water yield model and flow-related changes in sediment yield
- Steps 20–21: Sediment delivery from hillslope processes (tons/yr)
- Step 29: Calculate total annual sediment yield (tons/yr)
- Step 30: Compare potential increased sediment supply above reference condition

Step 31: Evaluate consequences of increased sediment supply and/or channel stability changes

Flowchart 7-3a (Flowchart 5-3a). The sequential steps to determine stream stability and sediment supply.

Steps 1–4: Bankfull discharge and hydraulic relations (See Flowchart 7-4)

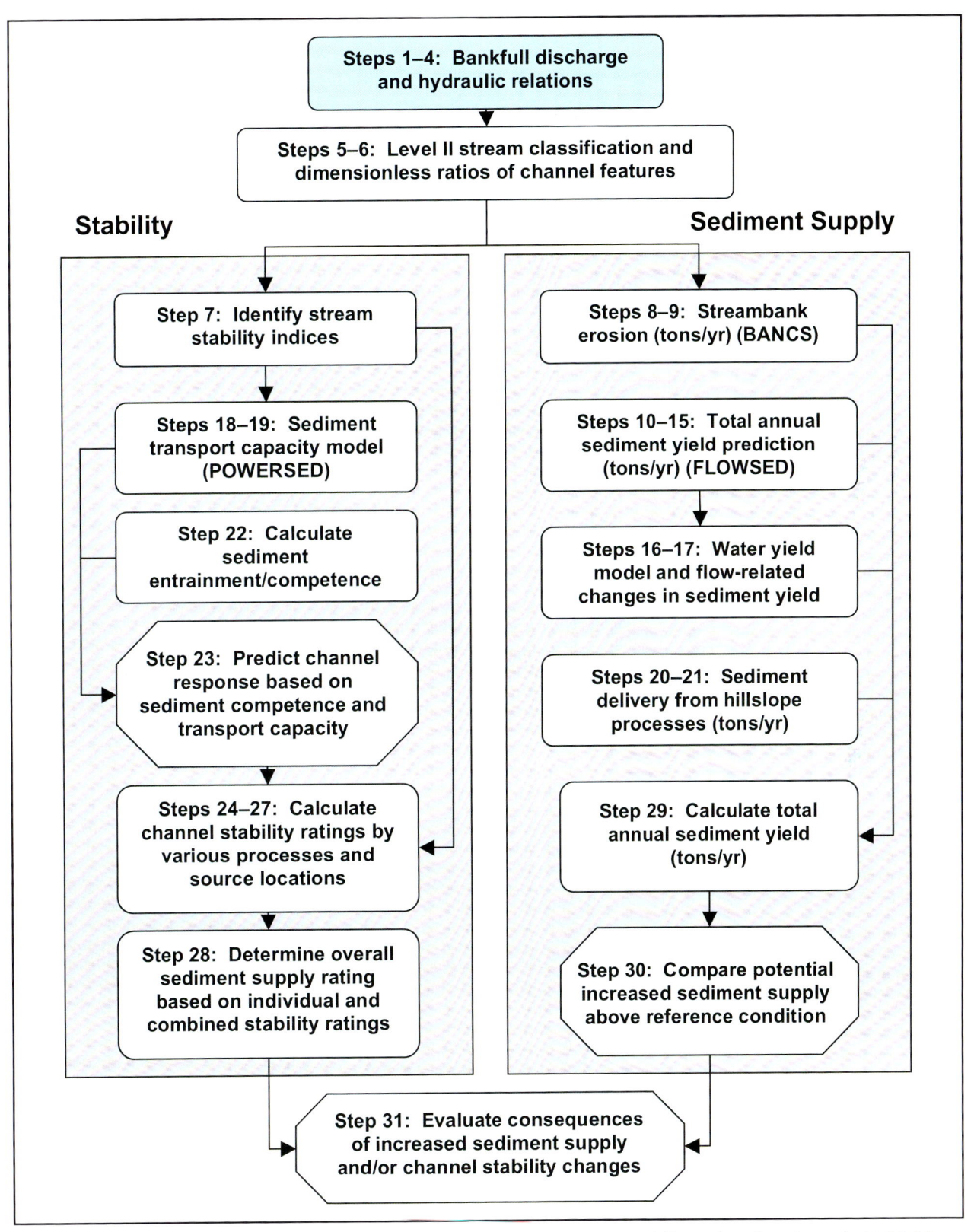

```
                    ┌─────────────────────────────────┐
                    │  Steps 1–4: Bankfull discharge   │
                    │      and hydraulic relations     │
                    └─────────────────┬───────────────┘
                                      ▼
                    ┌─────────────────────────────────┐
                    │ Steps 5–6: Level II stream      │
                    │ classification and dimensionless │
                    │    ratios of channel features    │
                    └─────────────────────────────────┘
```

Stability — **Sediment Supply**

- Step 7: Identify stream stability indices
- Steps 18–19: Sediment transport capacity model (POWERSED)
- Step 22: Calculate sediment entrainment/competence
- Step 23: Predict channel response based on sediment competence and transport capacity
- Steps 24–27: Calculate channel stability ratings by various processes and source locations
- Step 28: Determine overall sediment supply rating based on individual and combined stability ratings

- Steps 8–9: Streambank erosion (tons/yr) (BANCS)
- Steps 10–15: Total annual sediment yield prediction (tons/yr) (FLOWSED)
- Steps 16–17: Water yield model and flow-related changes in sediment yield
- Steps 20–21: Sediment delivery from hillslope processes (tons/yr)
- Step 29: Calculate total annual sediment yield (tons/yr)
- Step 30: Compare potential increased sediment supply above reference condition

Step 31: Evaluate consequences of increased sediment supply and/or channel stability changes

Flowchart 7-3b (Flowchart 5-3b). The sequential steps to determine stream stability and sediment supply, highlighting Steps 1–4.

Steps 1–4: Bankfull discharge and hydraulic relations

Step 1: Develop and/or obtain regional curves of bankfull discharge versus drainage area and bankfull dimensions versus drainage area
(Figure 5-2)

- Cross-sectional area (ft^2) vs. drainage area
- Width (ft) vs. drainage area
- Mean depth (ft) vs. drainage area
- Bankfull discharge (cfs) vs. drainage area
- Bankfull velocity (ft/s) vs. drainage area

Step 2: Delineate the watershed boundary on a USGS 7.5′ Quad map. Calculate drainage area in square miles

Step 3: Field calibrate bankfull discharge estimates by either direct discharge measurements or field calibration procedures at local gage stations to verify regional hydrology curves applied to ungaged sites

Step 4: Calculate bankfull discharge and dimensions

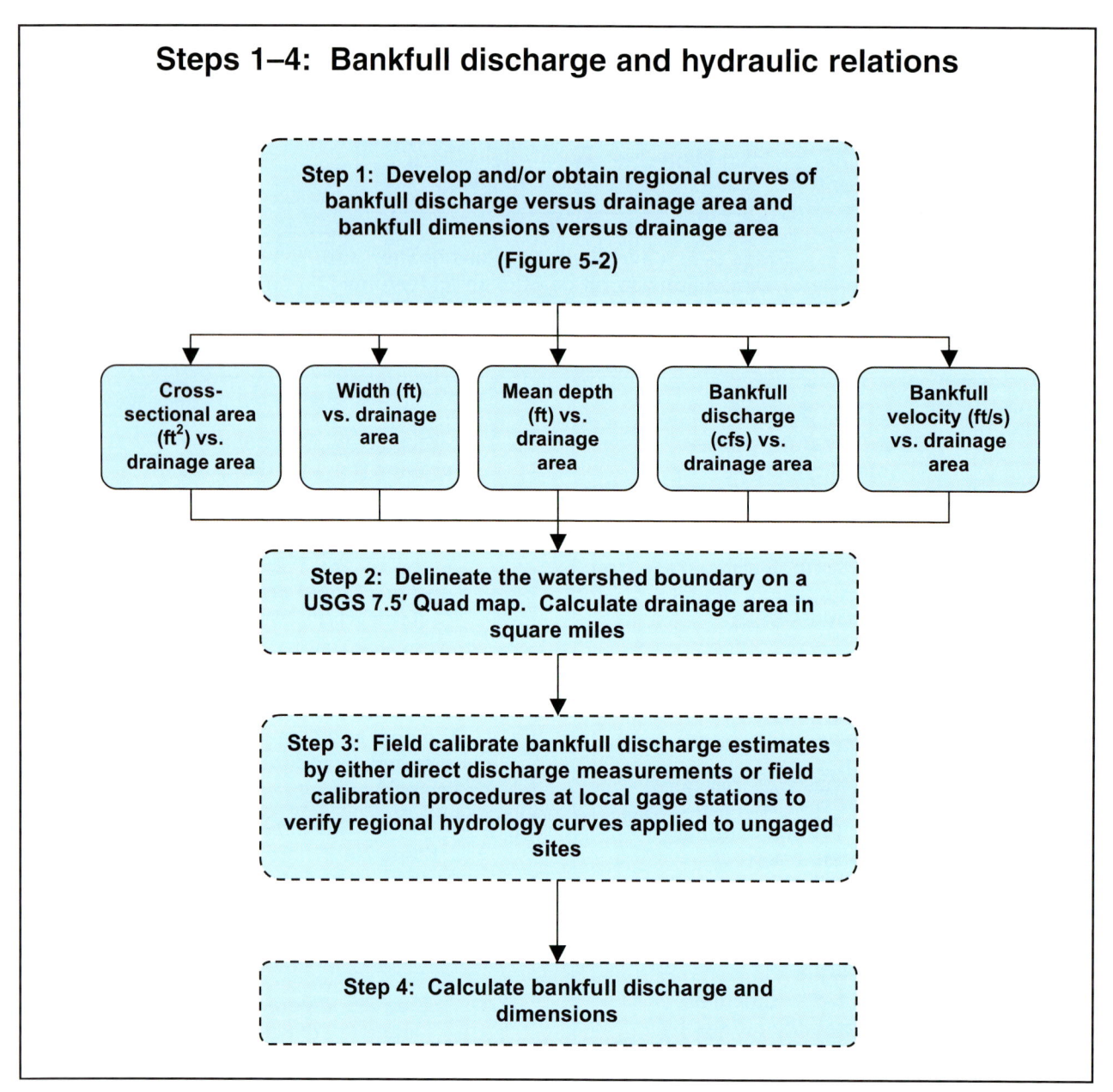

Flowchart 7-4 (Flowchart 5-4). Sequential steps and variables to obtain bankfull discharge and dimensions.

Step 1. Develop and/or obtain regional curves of bankfull discharge versus drainage area and bankfull channel dimensions versus drainage area

Regional curves developed by Wildland Hydrology for the upper San Juan River Basin are shown in **Figure 7-30**. These curves were used in conjunction with field-determined bankfull values predicted in **Steps 3** and **4**.

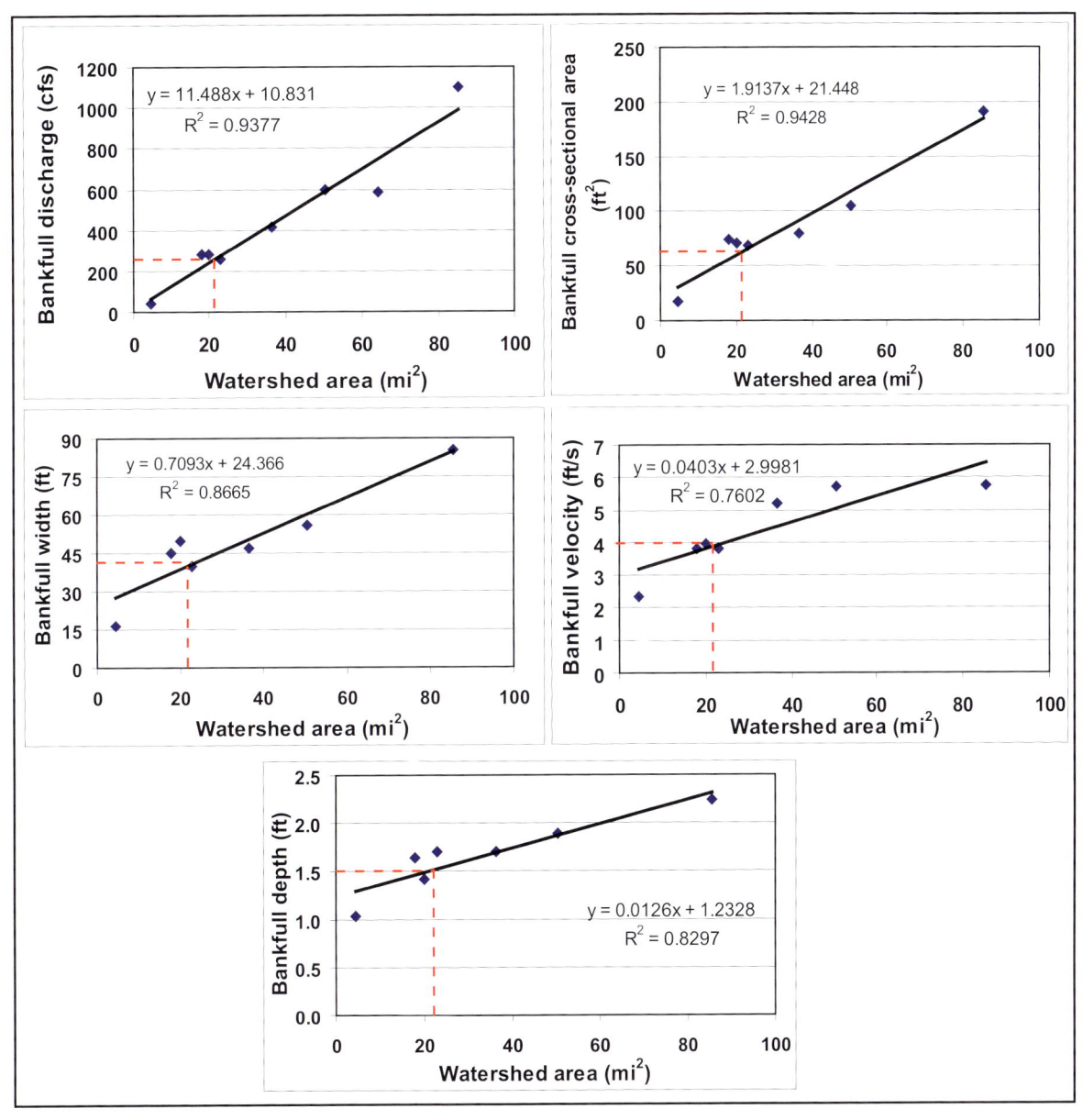

Figure 7-30. Regional bankfull curves, upper San Juan River, Colorado.

Step 2. Delineate the watershed boundary on a USGS 7.5' Quad map. Calculate drainage area in square miles

The watershed boundary is delineated on a USGS 7.5' map in **Figure 7-28**. The drainage area for the Wolf Creek Watershed is 22 mi^2.

Step 3. Field calibrate bankfull discharge estimates by either direct discharge measurements or field calibration procedures at local gage stations to verify regional hydrology curves applied to ungaged sites

The field-determined bankfull discharge is 280 ft^3/sec, and the cross-sectional area is 59.2 ft^2. Regional curve data was used for calibration because no long-term streamgage (established more than 10 years) exists on Wolf Creek to create a flood-frequency curve. The regional curve predicted approximately 264 cfs and 63.5 ft^2 (**Figure 7-30**). The 6% variation between the bankfull discharge regional curves and the field data is within the acceptable variability limits for regional curves.

Step 4. Calculate bankfull discharge and dimensions

To verify the bankfull discharge of 280 cfs, the field measurements of the bankfull channel dimensions compared to the regional curves (**Figure 7-31**) at the lower C4 reach of Wolf Creek are summarized below and are recorded in **Worksheet 7-14**:

- Bankfull width: *37 ft* (regional curve: *40 ft*)
- Bankfull mean depth: *1.6 ft* (regional curve: *1.5 ft*)
- Bankfull cross-sectional area: *59.2 square ft* (regional curve: *63.5 ft*)
- Slope: *0.005*
- Channel materials: Gravel-bed (D$_{84}$: *104 mm*)

Hydraulic calculations using resistance relations and other methods were conducted for the reference C4 stream type as shown in **Worksheet 7-14**. The use of the Manning's "n" from stream type (Rosgen and Silvey, 2005), where "n" for the C4 stream type is .031, resulted in a prediction of 4.5 ft/sec and a bankfull discharge of 266 cfs. This is close to the field-calibrated bankfull discharge of 280 cfs.

Continuity was used to calculate mean velocity for the D4 stream type immediately downstream. The field-calibrated bankfull discharge of 280 cfs was divided by the cross-sectional area of the D4 stream type (163 ft^2) to obtain a mean velocity of 1.7 ft/sec.

Chapter 7 — PLA The WARSSS Methodology: A Case Study of Wolf Creek, Colorado

Worksheet 7-14 (Worksheet 5-2). Computations of velocity and bankfull discharge using various methods (Rosgen and Silvey, 2005).

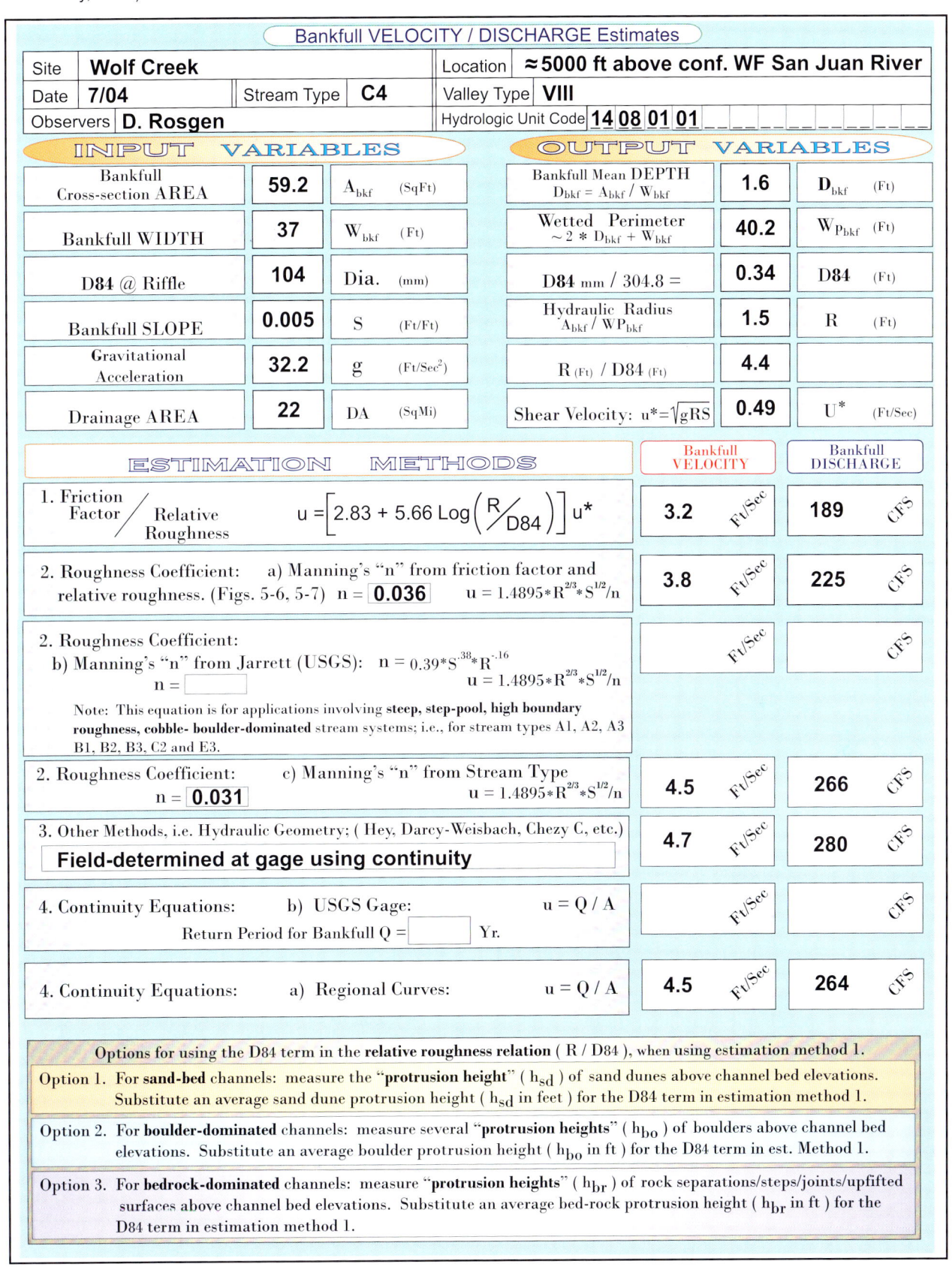

Bankfull VELOCITY / DISCHARGE Estimates

Site	Wolf Creek	Location	≈5000 ft above conf. WF San Juan River
Date	7/04	Stream Type	C4
		Valley Type	VIII
Observers	D. Rosgen	Hydrologic Unit Code	14 08 01 01

INPUT VARIABLES

Variable	Value	Symbol	Units
Bankfull Cross-section AREA	59.2	A_{bkf}	(SqFt)
Bankfull WIDTH	37	W_{bkf}	(Ft)
D84 @ Riffle	104	Dia.	(mm)
Bankfull SLOPE	0.005	S	(Ft/Ft)
Gravitational Acceleration	32.2	g	(Ft/Sec²)
Drainage AREA	22	DA	(SqMi)

OUTPUT VARIABLES

Variable	Value	Symbol	Units
Bankfull Mean DEPTH $D_{bkf} = A_{bkf} / W_{bkf}$	1.6	D_{bkf}	(Ft)
Wetted Perimeter ~ $2 * D_{bkf} + W_{bkf}$	40.2	Wp_{bkf}	(Ft)
D84 mm / 304.8 =	0.34	D84	(Ft)
Hydraulic Radius A_{bkf} / Wp_{bkf}	1.5	R	(Ft)
R (Ft) / D84 (Ft)	4.4		
Shear Velocity: $u^* = \sqrt{gRS}$	0.49	U^*	(Ft/Sec)

ESTIMATION METHODS

Method		Bankfull VELOCITY (Ft/Sec)	Bankfull DISCHARGE (CFS)
1. Friction Factor / Relative Roughness	$u = \left[2.83 + 5.66 \log\left(R/D84\right)\right] u^*$	3.2	189
2. Roughness Coefficient: a) Manning's "n" from friction factor and relative roughness. (Figs. 5-6, 5-7) n = **0.036**	$u = 1.4895 * R^{2/3} * S^{1/2} / n$	3.8	225
2. Roughness Coefficient: b) Manning's "n" from Jarrett (USGS): $n = 0.39 * S^{.38} * R^{-.16}$ n =	$u = 1.4895 * R^{2/3} * S^{1/2} / n$		

Note: This equation is for applications involving **steep, step-pool, high boundary roughness, cobble- boulder-dominated** stream systems; i.e., for stream types A1, A2, A3 B1, B2, B3, C2 and E3.

Method		Bankfull VELOCITY	Bankfull DISCHARGE
2. Roughness Coefficient: c) Manning's "n" from Stream Type n = **0.031**	$u = 1.4895 * R^{2/3} * S^{1/2} / n$	4.5	266
3. Other Methods, i.e. Hydraulic Geometry; (Hey, Darcy-Weisbach, Chezy C, etc.) **Field-determined at gage using continuity**		4.7	280
4. Continuity Equations: b) USGS Gage: Return Period for Bankfull Q = ___ Yr.	$u = Q/A$		
4. Continuity Equations: a) Regional Curves:	$u = Q/A$	4.5	264

Options for using the D84 term in the **relative roughness relation** (R / D84), when using estimation method 1.

Option 1. For **sand-bed** channels: measure the "protrusion height" (h_{sd}) of sand dunes above channel bed elevations. Substitute an average sand dune protrusion height (h_{sd} in feet) for the D84 term in estimation method 1.

Option 2. For **boulder-dominated** channels: measure several "protrusion heights" (h_{bo}) of boulders above channel bed elevations. Substitute an average boulder protrusion height (h_{bo} in ft) for the D84 term in est. Method 1.

Option 3. For **bedrock-dominated** channels: measure "protrusion heights" (h_{br}) of rock separations/steps/joints/upfifted surfaces above channel bed elevations. Substitute an average bed-rock protrusion height (h_{br} in ft) for the D84 term in estimation method 1.

WATERSHED ASSESSMENT OF RIVER STABILITY AND SEDIMENT SUPPLY

Steps 5–6: Level II stream classification and dimensionless ratios of channel features (See Flowchart 7-5)

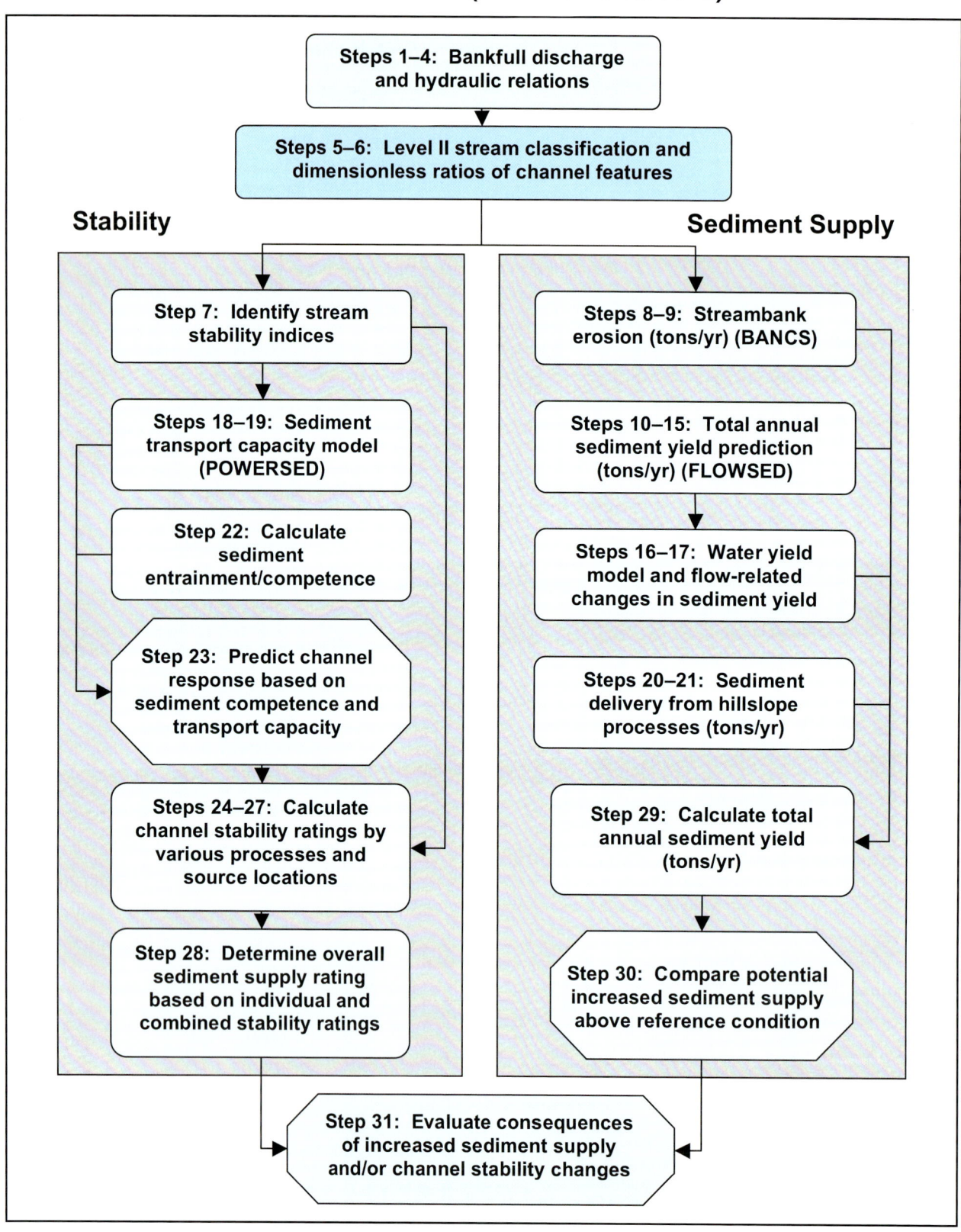

```
Steps 1–4: Bankfull discharge and hydraulic relations
        ↓
Steps 5–6: Level II stream classification and dimensionless ratios of channel features
```

Stability

- Step 7: Identify stream stability indices
- Steps 18–19: Sediment transport capacity model (POWERSED)
- Step 22: Calculate sediment entrainment/competence
- Step 23: Predict channel response based on sediment competence and transport capacity
- Steps 24–27: Calculate channel stability ratings by various processes and source locations
- Step 28: Determine overall sediment supply rating based on individual and combined stability ratings

Sediment Supply

- Steps 8–9: Streambank erosion (tons/yr) (BANCS)
- Steps 10–15: Total annual sediment yield prediction (tons/yr) (FLOWSED)
- Steps 16–17: Water yield model and flow-related changes in sediment yield
- Steps 20–21: Sediment delivery from hillslope processes (tons/yr)
- Step 29: Calculate total annual sediment yield (tons/yr)
- Step 30: Compare potential increased sediment supply above reference condition

Step 31: Evaluate consequences of increased sediment supply and/or channel stability changes

Flowchart 7-3c (Flowchart 5-3c). The sequential steps to determine stream stability and sediment supply, highlighting **Steps 5–6**.

7-92

Chapter 7 — PLA The WARSSS Methodology: A Case Study of Wolf Creek, Colorado

Steps 5–6: Level II stream classification and dimensionless ratios of channel features

Step 5: Classify entire length of the reference and impaired stream reaches (Worksheet 5-3)

↓

Valley type

↓

- Bankfull width (ft)
- Bankfull depth (ft)
- Bankfull cross-sectional area (ft²)
- Max depth (ft)
- Width of floodprone area (ft)

↓

- Particle size D_{50} (mm)
- Width/depth ratio
- Check with regional curves
- Water surface slope
- Sinuosity (k)
- Entrenchment ratio

↓

Stream classification

↓

Step 6: Calculate detailed morphological descriptions (including dimensionless ratios) (Worksheet 5-4)

↓

Valley type

↓

Stream type

↓

- Channel dimensions
- Channel pattern
- Channel profile
- Channel materials

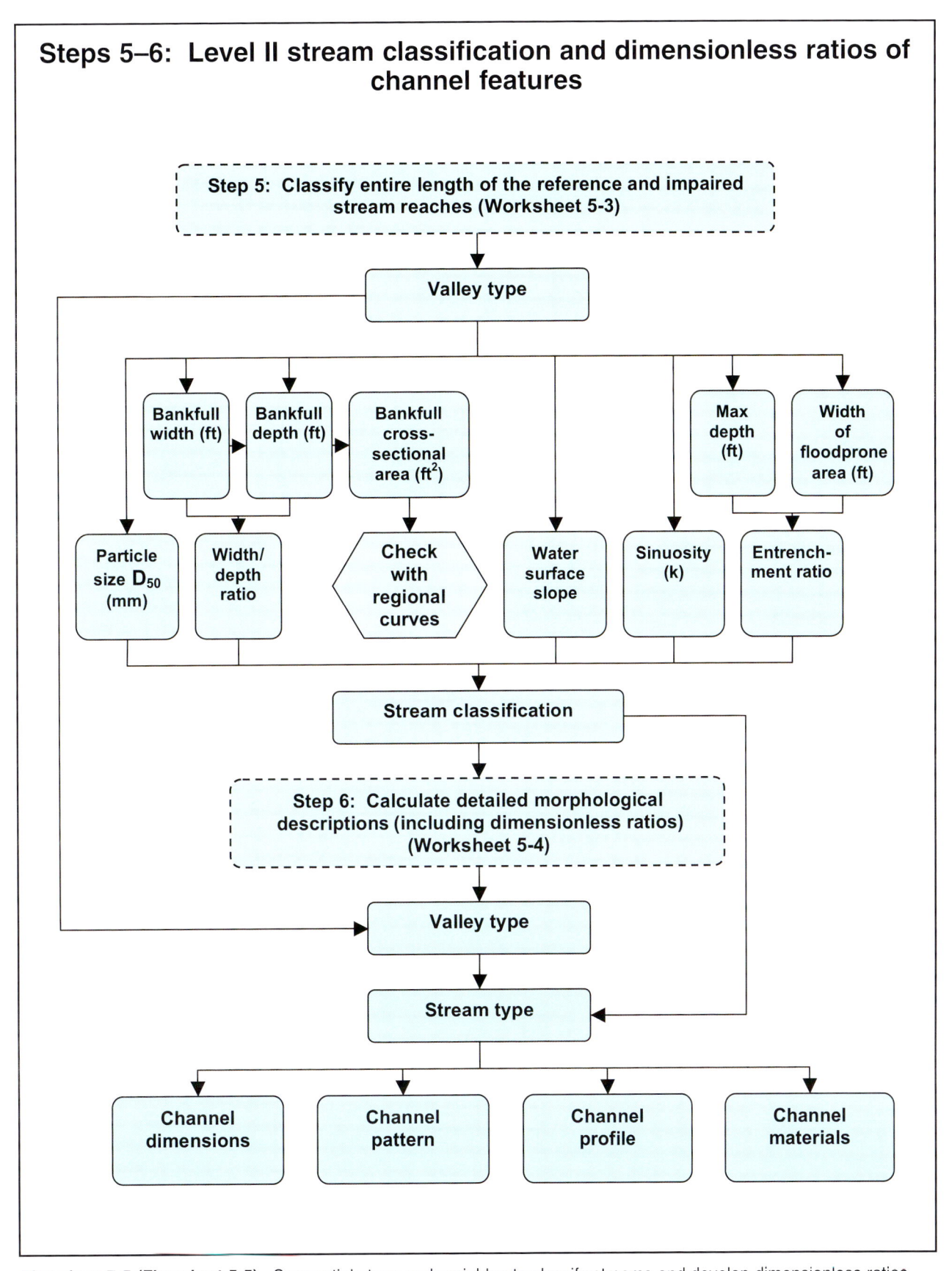

Flowchart 7-5 (Flowchart 5-5). Sequential steps and variables to classify streams and develop dimensionless ratios.

7-93

Step 5. Classify streams

The stream type dimensions and materials for the lower C4 and D4 stream types are summarized in **Worksheets 7-15a** and **7-15b**, as delineated at Level II. The locations of the stream types are the lower reaches shown in **Figure 7-28**.

Worksheet 7-15a (Worksheet 5-3). Level II stream classification for the lower C4 stream type (Rosgen, 1996; Rosgen and Silvey, 2005).

Stream:	**Wolf Creek**
Basin: **San Juan**	Drainage Area: **14,080** acres **22** mi^2
Location: **≈ 5000 ft above the confluence of the West Fork of the San Juan River**	
Twp.&Rge: **T37N, R33E**	Sec.&Qtr.:
Cross-Section Monuments (Lat./Long.):	Date: **7/04**
Observers: **D. Rosgen**	Valley Type: **VIII**

Bankfull WIDTH (W_{bkf})
WIDTH of the stream channel at bankfull stage elevation, in a riffle section. — **37** ft

Bankfull DEPTH (d_{bkf})
Mean DEPTH of the stream channel cross-section, at bankfull stage elevation, in a riffle section ($d_{bkf} = A / W_{bkf}$). — **1.6** ft

Bankfull X-Section AREA (A_{bkf})
AREA of the stream channel cross-section, at bankfull stage elevation, in a riffle section. — **59.2** ft^2

Width/Depth Ratio (W_{bkf} / d_{bkf})
Bankfull WIDTH divided by bankfull mean DEPTH, in a riffle section. — **23** ft/ft

Maximum DEPTH (d_{mbkf})
Maximum depth of the bankfull channel cross-section, or distance between the bankfull stage and Thalweg elevations, in a riffle section. — **2.6** ft

WIDTH of Flood-Prone Area (W_{fpa})
Twice maximum DEPTH, or (2 x d_{mbkf}) = the stage/elevation at which flood-prone area WIDTH is determined in a riffle section. — **320** ft

Entrenchment Ratio (ER)
The ratio of flood-prone area WIDTH divided by bankfull channel WIDTH (W_{fpa} / W_{bkf}) (riffle section). — **8.6** ft/ft

Channel Materials (Particle Size Index) D_{50}
The D_{50} particle size index represents the mean diameter of channel materials, as sampled from the channel surface, between the bankfull stage and Thalweg elevations. — **48** mm

Water Surface SLOPE (S)
Channel slope = "rise over run" for a reach approximately 20–30 bankfull channel widths in length, with the "riffle-to-riffle" water surface slope representing the gradient at bankfull stage. — **0.005** ft/ft

Channel SINUOSITY (k)
Sinuosity is an index of channel pattern, determined from a ratio of stream length divided by valley length (SL / VL); or estimated from a ratio of valley slope divided by channel slope (VS / S). — **1.8**

Stream Type **C4** (See **Figure 2-14**)

Worksheet 7-15b (Worksheet 5-3). Level II stream classification for the D4 stream type (Rosgen, 1996; Rosgen and Silvey, 2005).

Stream:	**Wolf Creek- Lower end near mouth of Wolf Creek**		
Basin:	**San Juan**	Drainage Area: **14,080** acres	**22** mi²
Location:	≈ **4000 ft above confluence of the West Fork of the San Juan River**		
Twp.&Rge: **T37N, R33E**		Sec.&Qtr.:	
Cross-Section Monuments (Lat./Long.):			Date: **7/04**
Observers: **D. Rosgen**		Valley Type: **VIII**	

Bankfull WIDTH (W_{bkf})
WIDTH of the stream channel at bankfull stage elevation, in a riffle section. — **203** ft

Bankfull DEPTH (d_{bkf})
Mean DEPTH of the stream channel cross-section, at bankfull stage elevation, in a riffle section ($d_{bkf} = A / W_{bkf}$). — **0.8** ft

Bankfull X-Section AREA (A_{bkf})
AREA of the stream channel cross-section, at bankfull stage elevation, in a riffle section. — **163** ft²

Width/Depth Ratio (W_{bkf} / d_{bkf})
Bankfull WIDTH divided by bankfull mean DEPTH, in a riffle section. — **254** ft/ft

Maximum DEPTH (d_{mbkf})
Maximum depth of the bankfull channel cross-section, or distance between the bankfull stage and Thalweg elevations, in a riffle section. — **3.5** ft

WIDTH of Flood-Prone Area (W_{fpa})
Twice maximum DEPTH, or (2 x d_{mbkf}) = the stage/elevation at which flood-prone area WIDTH is determined in a riffle section. — **350** ft

Entrenchment Ratio (ER)
The ratio of flood-prone area WIDTH divided by bankfull channel WIDTH (W_{fpa} / W_{bkf}) (riffle section). — **N/A** ft/ft

Channel Materials (Particle Size Index) D_{50}
The D_{50} particle size index represents the mean diameter of channel materials, as sampled from the channel surface, between the bankfull stage and Thalweg elevations. — **19** mm

Water Surface SLOPE (S)
Channel slope = "rise over run" for a reach approximately 20–30 bankfull channel widths in length, with the "riffle-to-riffle" water surface slope representing the gradient at bankfull stage. — **0.009** ft/ft

Channel SINUOSITY (k)
Sinuosity is an index of channel pattern, determined from a ratio of stream length divided by valley length (SL / VL); or estimated from a ratio of valley slope divided by channel slope (VS / S). — **1.1**

Stream Type **D4** (See **Figure 2-14**)

Step 6. Calculate detailed morphological descriptions (including dimensionless ratios)

Spraying willows caused accelerated streambank erosion, channel widening, aggradation, down-valley meander migration, downstream sedimentation, a change in stream type from C4 to D4 and severe land and fish habitat loss. The photographs in **Figure 7-31** depict the changes in the riparian vegetation and the corresponding shifts in morphological stream types.

A summary of dimension, pattern, profile, materials and dimensionless ratios for the C4 reference reach on Wolf Creek and the impaired D4 stream type are shown in **Worksheets 7-16a** and **7-16b**, respectively. The values for the D4 impaired reach show a major departure from the reference condition.

The reference condition C4 stream type is shown as station 45+56 in **Figure 7-32** and is compared to the impaired reach immediately downstream at station 3B-1. The impaired reach is a D4 (braided) stream type. The comparison of the cross-section data is shown for both the reference and impaired reaches of Wolf Creek in **Figure 7-33**. The bed-material size distribution of both the C4 and D4 stream types is shown in **Figure 7-34**. The D4 stream type bed-material size is much smaller than the reference material, with 14-mm (D_{16}) particles replaced by sand on the braided (D4) stream type.

Figure 7-31. Willow eradication and resulting stream type change, Wolf Creek, Colorado.

Worksheet 7-16a (Worksheet 5-4). Summary of dimension, pattern and profile data for the C4 stream type reference reach.

Stream: **Wolf Creek**   Location: **≈ 5000 ft above conf. of WF San Juan River**
Observers: **D. Rosgen**   Date: **7/04**   Valley Type: **VIII**   Stream Type: **C4**

River Reach Summary Data

Channel Dimension

Parameter	Value	Unit	Parameter	Value	Unit	Parameter	Value	Unit
Mean Riffle Depth (d_{bkf})	1.6	ft	Riffle Width (W_{bkf})	37	ft	Riffle Area (A_{bkf})	59.2	ft²
Mean Pool Depth (d_{bkfp})	2.0	ft	Pool Width (W_{bkfp})	36	ft	Pool Area (A_{bkfp})	72	ft²
Mean Pool Depth/Mean Riffle Depth	1.25	d_{bkfp}/d_{bkf}	Pool Width/Riffle Width	0.97	W_{bkfp}/W_{bkf}	Pool Area / Riffle Area	1.2	A_{bkfp}/A_{bkf}
Max Riffle Depth (d_{mbkf})	2.4	ft	Max Pool Depth (d_{mbkfp})	4.8	ft	Max riffle depth/Mean riffle depth	1.5	
Max pool depth/Mean riffle depth	3.0					Point Bar Slope	0.22	
Streamflow: Estimated Mean Velocity at Bankfull Stage (u_{bkf})	4.7	ft/s	Estimation Method	# 3 in W.S. 5-2				
Streamflow: Estimated Discharge at Bankfull Stage (Q_{bkf})	280	cfs	Drainage Area	22	mi²			

Channel Pattern

Geometry	Mean	Min	Max		Dimensionless Geometry Ratios	Mean	Min	Max
Meander Length (Lm)	463	407	518	ft	Meander Length Ratio (Lm/W_{bkf})	12.5	11	14
Radius of Curvature (Rc)	130	111	148	ft	Radius of Curvature/Riffle Width (Rc/W_{bkf})	3.5	3.0	4.0
Belt Width (W_{blt})	176	129	222	ft	Meander Width Ratio (W_{blt}/W_{bkf})	4.8	3.5	6.0
Individual Pool Length	34	28	40	ft	Pool Length/Riffle Width	0.9	0.76	1.1
Pool to Pool Spacing	232	204	259	ft	Pool to Pool Spacing/Riffle Width	6.3	5.5	7.0
Riffle Length	112	93	130	ft	Riffle Length/Riffle Width	3.0	2.5	3.5

Channel Profile

Valley Slope (VS)	0.009 ft/ft		Average Water Surface Slope (S)	0.005 ft/ft		Sinuosity (VS/S)	1.8
Stream Length (SL)	3240 ft		Valley Length (VL)	1800 ft		Sinuosity (SL/VL)	1.8
Low Bank Height (LBH)	start 2.4 ft / end 2.4 ft		Max Riffle Depth	start 2.4 ft / end 2.4 ft		Bank-Height Ratio (BHR) (LBH/Max Riffle Depth)	start 1.0 / end 1.0

Facet Slopes	Mean	Min	Max		Dimensionless Slope Ratios	Mean	Min	Max
Riffle Slope (S_{rif})	.015	.0125	.0175	ft/ft	Riffle Slope/Average Water Surface Slope (S_{rif}/S)	3.0	2.5	3.5
Run Slope (S_{run})	.009	.006	.011	ft/ft	Run Slope/Average Water Surface Slope (S_{run}/S)	1.7	1.2	2.2
Pool Slope (S_p)	.001	.0005	.0015	ft/ft	Pool Slope/Average Water Surface Slope (S_p/S)	0.2	0.1	0.3
Glide Slope (S_g)	.0015	.001	.002	ft/ft	Glide Slope/Average Water Surface Slope (S_g/S)	0.3	0.2	0.4

Feature Midpoint[a]	Mean	Min	Max		Dimensionless Depth Ratios	Mean	Min	Max
Riffle Depth (d_{rif})	2.4	2.1	2.6	ft	Riffle Depth/Mean Riffle Depth (d_{rif}/d_{bkf})	1.5	1.3	1.6
Run Depth (d_{run})	4.4	4.0	4.8	ft	Run Depth/Mean Riffle Depth (d_{run}/d_{bkf})	2.8	2.5	3.0
Pool Depth (d_p)	4.8	4.0	5.6	ft	Pool Depth/Mean Riffle Depth (d_p/d_{bkf})	3.0	2.5	3.5
Glide Depth (d_g)	1.6	1.3	1.8	ft	Glide Depth/Mean Riffle Depth (d_g/d_{bkf})	1.0	0.8	1.1

Channel Materials

	Reach[b]	Riffle[c]	Bar		Reach[b]	Riffle[c]	Bar	Protrusion Height[d]
% Silt/Clay				D_{16}	14	20	1	mm
% Sand	5	2	10	D_{35}	32	38	4	mm
% Gravel	65	60	80	D_{50}	48	48	8	mm
% Cobble	30	38	10	D_{84}	104	120	40	mm
% Boulder				D_{95}	139	150	60	mm
% Bedrock				D_{100}	160	170	78	mm

a Min, max, mean depths are the average mid-point values except pools, which are taken at deepest part of pool.
b Composite sample of riffles and pools within the designated reach.
c Active bed of a riffle.
d Height of roughness feature above bed.

Watershed Assessment of River Stability and Sediment Supply

Worksheet 7-16b (Worksheet 5-4). Summary of dimension, pattern and profile data for the potentially impaired D4 stream type reach.

Stream: **Wolf Creek** Location: **≈ 4000 ft above confluence of WF San Juan River**
Observers: **D. Rosgen** Date: **7/04** Valley Type: **VIII** Stream Type: **D4**

River Reach Summary Data

Channel Dimension

Mean Riffle Depth (d_{bkf})	0.8 ft	Riffle Width (W_{bkf})	203 ft	Riffle Area (A_{bkf})	163 ft²
Mean Pool Depth (d_{bkfp})	1.2 ft	Pool Width (W_{bkfp})	190 ft	Pool Area (A_{bkfp})	228 ft²
Mean Pool Depth/Mean Riffle Depth	1.5 d_{bkfp}/d_{bkf}	Pool Width/Riffle Width	0.9 W_{bkfp}/W_{bkf}	Pool Area / Riffle Area	1.4 A_{bkfp}/A_{bkf}
Max Riffle Depth (d_{mbkf})	1.1 ft	Max Pool Depth (d_{mbkfp})	1.6 ft	Max riffle depth/Mean riffle depth	1.4
Max pool depth/Mean riffle depth	2.0			Point Bar Slope	0.005
Streamflow: Estimated Mean Velocity at Bankfull Stage (u_{bkf})		1.7 ft/s	Estimation Method	Continuity	
Streamflow: Estimated Discharge at Bankfull Stage (Q_{bkf})		280 cfs	Drainage Area	22 mi²	

Channel Pattern

Geometry	Mean	Min	Max		Dimensionless Geometry Ratios	Mean	Min	Max
Meander Length (Lm)	N/A*			ft	Meander Length Ratio (Lm/W_{bkf})	N/A*		
Radius of Curvature (Rc)	N/A*			ft	Radius of Curvature/Riffle Width (Rc/W_{bkf})	N/A*		
Belt Width (W_{blt})	N/A*			ft	Meander Width Ratio (W_{blt}/W_{bkf})	N/A*		
Individual Pool Length	45	20	70	ft	Pool Length/Riffle Width	0.22	0.10	0.34
Pool to Pool Spacing	420			ft	Pool to Pool Spacing/Riffle Width	2.1		
Riffle Length	N/A			ft	Riffle Length/Riffle Width	N/A		

Channel Profile

Valley Slope (VS)	0.010 ft/ft			Average Water Surface Slope (S)		0.009 ft/ft	Sinuosity (VS/S)		1.1
Stream Length (SL)	880 ft			Valley Length (VL)		800 ft	Sinuosity (SL/VL)		1.1
Low Bank Height (LBH)	start 1.1 ft / end 1.1 ft			Max Riffle Depth	start 1.1 ft / end 1.1 ft		Bank-Height Ratio (BHR) (LBH/Max Riffle Depth)	start 1.0 / end 1.0		

Facet Slopes	Mean	Min	Max		Dimensionless Slope Ratios	Mean	Min	Max
Riffle Slope (S_{rif})	.012	0.011	0.013	ft/ft	Riffle Slope/Average Water Surface Slope (S_{rif}/S)	1.3	1.2	1.4
Run Slope (S_{run})	.010	.009	.010	ft/ft	Run Slope/Average Water Surface Slope (S_{run}/S)	1.1	1.0	1.1
Pool Slope (S_p)	.007	.005	.009	ft/ft	Pool Slope/Average Water Surface Slope (S_p/S)	0.8	0.6	1.0
Glide Slope (S_g)	.008	.007	.010	ft/ft	Glide Slope/Average Water Surface Slope (S_g/S)	0.9	0.8	1.1

Feature Midpoint [a]	Mean	Min	Max		Dimensionless Depth Ratios	Mean	Min	Max
Riffle Depth (d_{rif})	1.1	1.0	1.2	ft	Riffle Depth/Mean Riffle Depth (d_{rif}/d_{bkf})	1.4	1.25	1.5
Run Depth (d_{run})	0.96	0.88	1.04	ft	Run Depth/Mean Riffle Depth (d_{run}/d_{bkf})	1.2	1.1	1.3
Pool Depth (d_p)	1.7	1.4	1.9	ft	Pool Depth/Mean Riffle Depth (d_p/d_{bkf})	2.1	1.8	2.4
Glide Depth (d_g)	0.6	0.48	0.64	ft	Glide Depth/Mean Riffle Depth (d_g/d_{bkf})	0.7	0.6	0.8

Channel Materials

	Reach[b]	Riffle[c]	Bar		Reach[b]	Riffle[c]	Bar	Protrusion Height[d]
% Silt/Clay				D_{16}	0.2	1.0	1.0	mm
% Sand	30	25		D_{35}	10.0	8.0	2.0	mm
% Gravel	65	75		D_{50}	19.0	18.0	4.0	mm
% Cobble	5	10		D_{84}	87.0	90.0	14.0	mm
% Boulder				D_{95}	124.0	120.0	20.0	mm
% Bedrock				D_{100}	130.0	135.0	26.0	mm

a Min, max, mean depths are the average mid-point values except pools, which are taken at deepest part of pool.
b Composite sample of riffles and pools within the designated reach.
c Active bed of a riffle.
d Height of roughness feature above bed.
*Pattern is not measured on braided (D) steam types. Bed features are formed by convergence/divergence rather than riffle/pool features as in the C stream types.

Cross-section 45+56 (C4)

Cross-section 3B-1 (D4)

Figure 7-32. Comparison of C4 and D4 stream type reaches.

WATERSHED ASSESSMENT OF RIVER STABILITY AND SEDIMENT SUPPLY

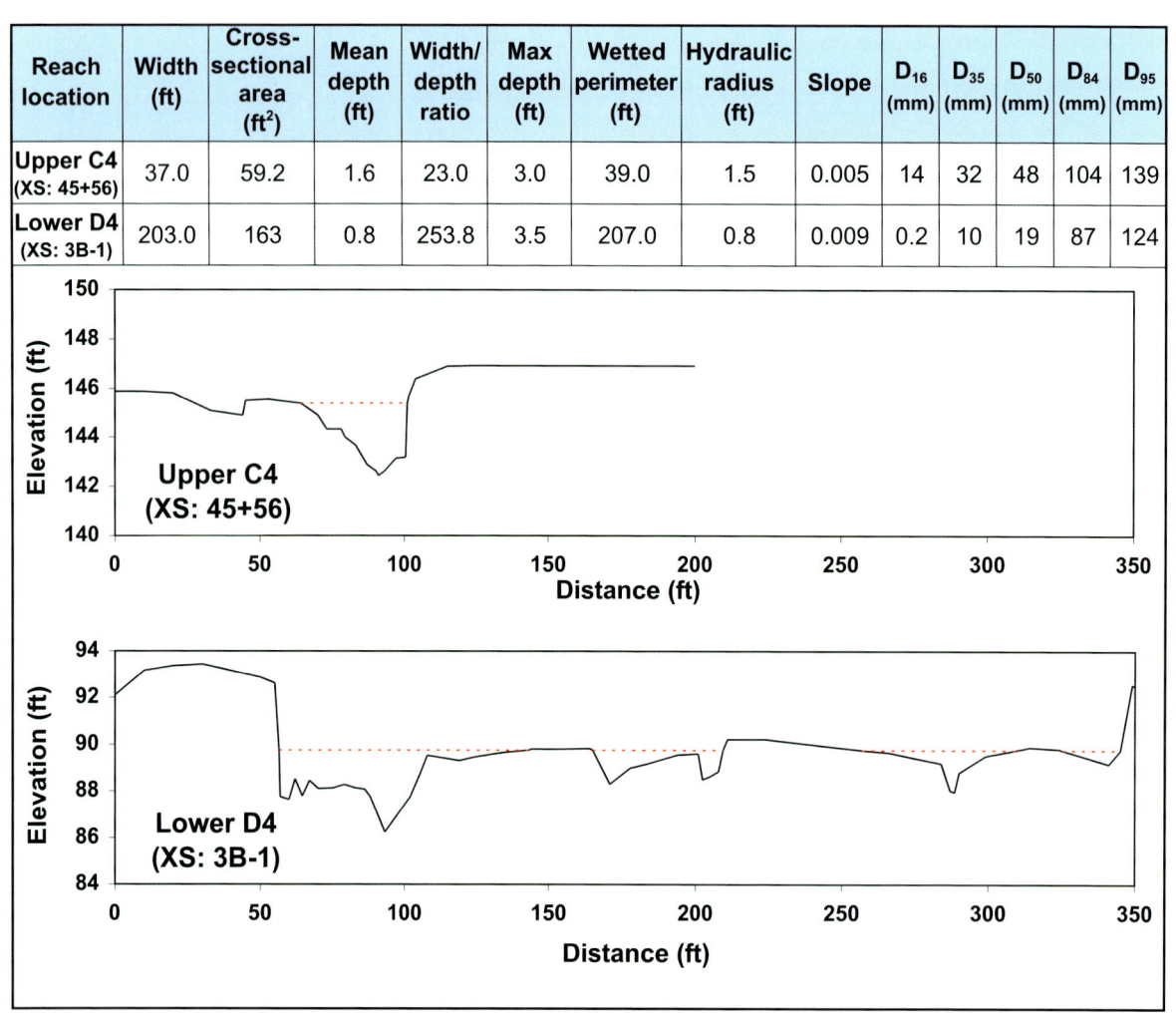

Reach location	Width (ft)	Cross-sectional area (ft²)	Mean depth (ft)	Width/ depth ratio	Max depth (ft)	Wetted perimeter (ft)	Hydraulic radius (ft)	Slope	D_{16} (mm)	D_{35} (mm)	D_{50} (mm)	D_{84} (mm)	D_{95} (mm)
Upper C4 (XS: 45+56)	37.0	59.2	1.6	23.0	3.0	39.0	1.5	0.005	14	32	48	104	139
Lower D4 (XS: 3B-1)	203.0	163	0.8	253.8	3.5	207.0	0.8	0.009	0.2	10	19	87	124

Figure 7-33. Wolf Creek cross-section comparison: Willow eradication area, upper C4 (XS: 45+56) vs. lower D4 (XS: 3B-1) reach, as depicted in the photographs in **Figure 7-32**.

Particle size		D_{16}	D_{35}	D_{50}	D_{84}	D_{95}
Reach	**Untreated (C4)**	14	32	48	104	139
	Treated (D4)	0.2	10	19	87	124

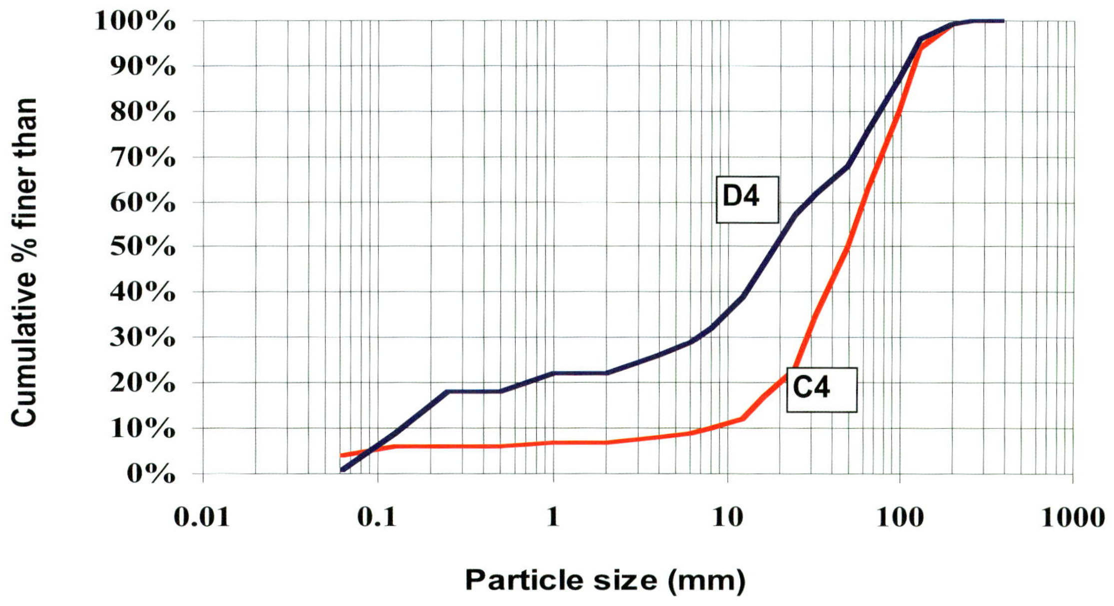

Figure 7-34. Comparison of bed-material size in the C4 (XS: 45+56) stream type and the D4 (XS: 3B-1) stream type, as shown in **Figures 7-32** and **7-33**.

Watershed Assessment of River Stability and Sediment Supply

Step 7: Stream channel stability indices (See Flowchart 7-6)

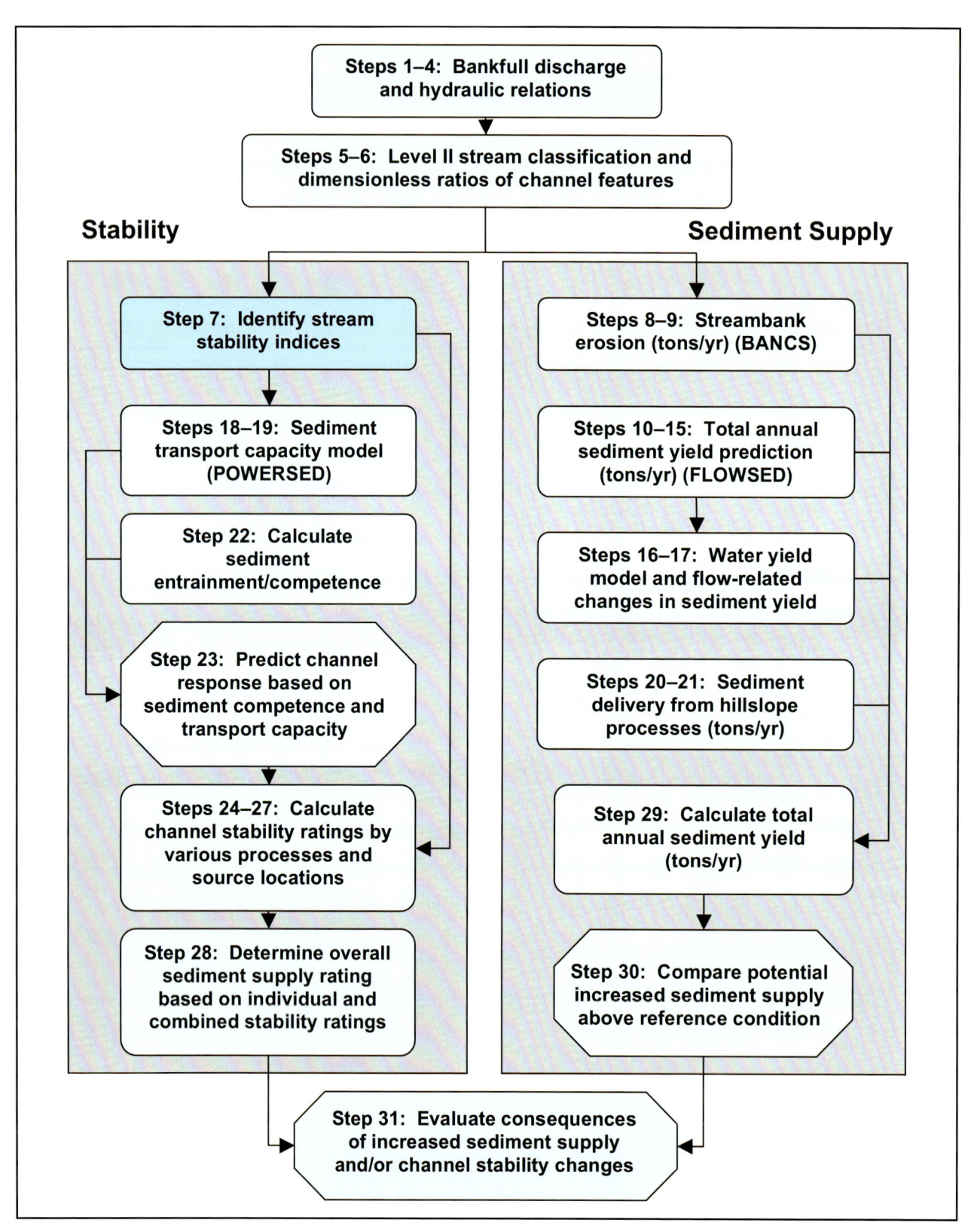

```
Steps 1–4: Bankfull discharge and hydraulic relations
        ↓
Steps 5–6: Level II stream classification and dimensionless ratios of channel features
```

Stability | **Sediment Supply**

Stability branch:
- **Step 7: Identify stream stability indices**
- Steps 18–19: Sediment transport capacity model (POWERSED)
- Step 22: Calculate sediment entrainment/competence
- Step 23: Predict channel response based on sediment competence and transport capacity
- Steps 24–27: Calculate channel stability ratings by various processes and source locations
- Step 28: Determine overall sediment supply rating based on individual and combined stability ratings

Sediment Supply branch:
- Steps 8–9: Streambank erosion (tons/yr) (BANCS)
- Steps 10–15: Total annual sediment yield prediction (tons/yr) (FLOWSED)
- Steps 16–17: Water yield model and flow-related changes in sediment yield
- Steps 20–21: Sediment delivery from hillslope processes (tons/yr)
- Step 29: Calculate total annual sediment yield (tons/yr)
- Step 30: Compare potential increased sediment supply above reference condition

Step 31: Evaluate consequences of increased sediment supply and/or channel stability changes

Flowchart 7-3d (Flowchart 5-3d). The sequential steps to determine stream stability and sediment supply, highlighting **Step 7**.

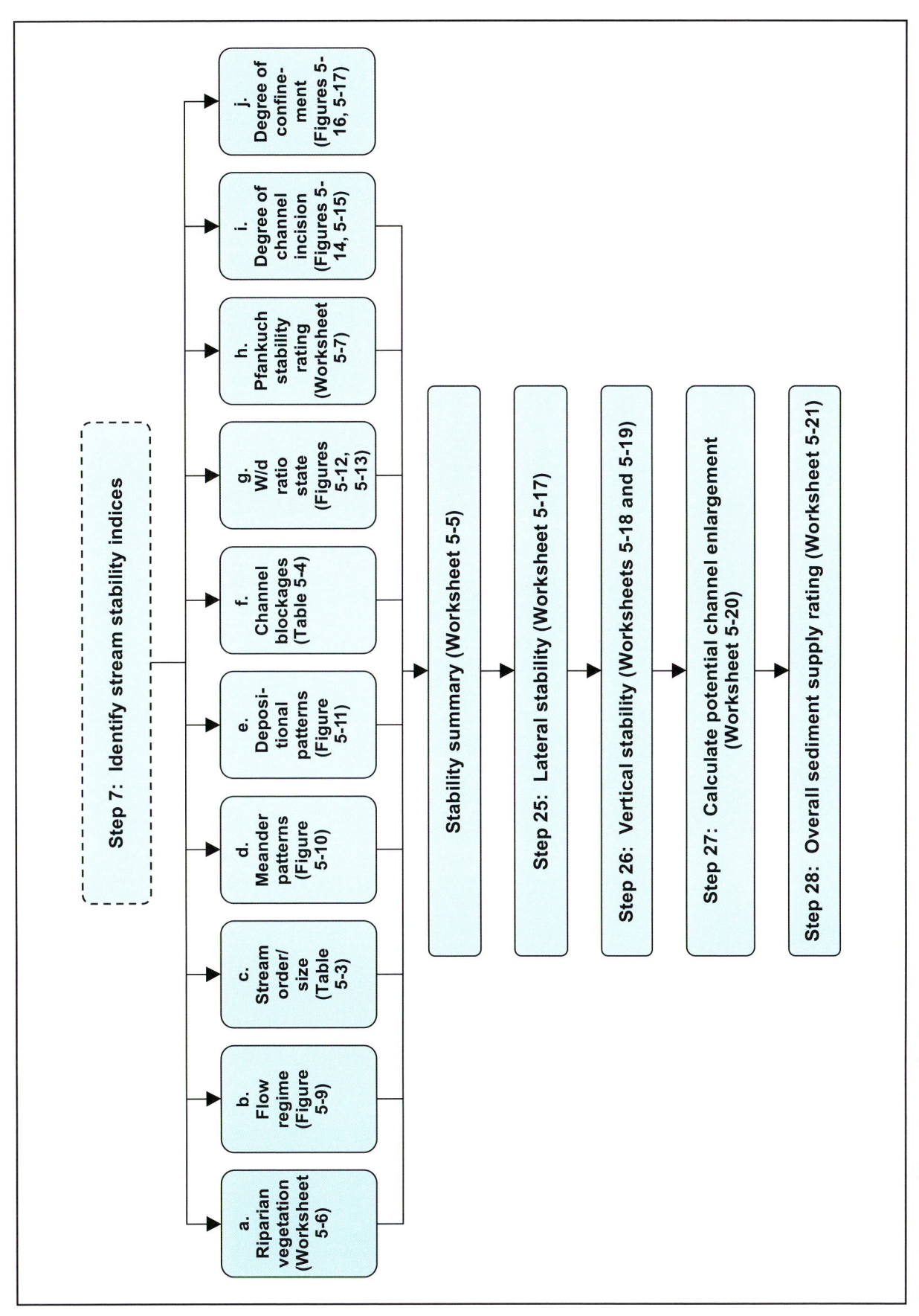

Chapter 7 — PLA   The WARSSS Methodology: A Case Study of Wolf Creek, Colorado

Step 7: Identify stream stability indices

a. Riparian vegetation (Worksheet 5-6)
b. Flow regime (Figure 5-9)
c. Stream order/size (Table 5-3)
d. Meander patterns (Figure 5-10)
e. Depositional patterns (Figure 5-11)
f. Channel blockages (Table 5-4)
g. W/d ratio state (Figures 5-12, 5-13)
h. Pfankuch stability rating (Worksheet 5-7)
i. Degree of channel incision (Figures 5-14, 5-15)
j. Degree of confinement (Figures 5-16, 5-17)

→ Stability summary (Worksheet 5-5)
→ Step 25: Lateral stability (Worksheet 5-17)
→ Step 26: Vertical stability (Worksheets 5-18 and 5-19)
→ Step 27: Calculate potential channel enlargement (Worksheet 5-20)
→ Step 28: Overall sediment supply rating (Worksheet 5-21)

Flowchart 7-6 (Flowchart 5-6). The multiple variables and procedures (a.–j.) used for stream stability prediction in **Steps 25–28**.

7-105

Step 7. Identify stream stability indices

A comparison between the C4 reference reach and the D4 impaired reach predicts stability consequences due to willow spraying and poor grazing practices. The photographs compare time-trend changes in stream morphology and stability (**Figures 7-4** and **7-5**; see **Step 3**, *RLA*) and demonstrate the importance of woody riparian vegetation in maintaining a stable C4 stream type on Wolf Creek. Oblique aerial photographs (**Figure 7-35**) and the contrasting reach images (**Figure 7-31**) provide visual evidence of extreme channel change due to poor land use practices.

The following variables were analyzed for both the reference and impacted reaches:

a. Riparian vegetation (**Worksheets 7-17a** and **7-17b**);

b. Flow regime (**Figure 7-37**);

c. Stream order/size (**Table 7-30**);

d. Meander patterns (**Figure 7-38**);

e. Depositional patterns (**Figure 7-39**);

f. Channel blockages (**Table 7-31**);

g. Width/depth ratio state (**Figure 7-40**);

h. Pfankuch channel stability rating (**Worksheets 7-18a** and **7-18b**);

i. Degree of channel incision (**Figure 7-41**); and

j. Degree of channel confinement (lateral containment) (**Figure 7-42**).

These variables are recorded in the stream stability summary worksheet and are used, along with sediment source data, in **Steps 25–28**. The completed summary worksheet appears at the end of **Step 7**.

Figure 7-35. Willow eradication and the resulting stream type change from a C4 to D4, Wolf Creek, Colorado.

a. Riparian vegetation

The riparian vegetation assessment was completed for both the C4 reference and D4 impacted reaches. **Figure 7-36** depicts the bank erosion on the D4 due to the willow eradication. The reference reach composition of cottonwood overstory, willow and alder understory and a sedge (*Carex* spp.) grass community (**Worksheet 7-17a**) was in stark contrast to the grass/forb composition of the impacted reach (**Worksheet 7-17b**). The species composition (existing and potential) and the percent of cover are recorded in Column (3) in the stream stability summary worksheet.

Figure 7-36. Typical eroding bank on the D4 (braided) reach of Wolf Creek following willow removal due to herbicide spraying.

Worksheet 7-17a (Worksheet 5-6). Riparian vegetation variables that influence channel stability for the C4 reference reach.

Stream:	Wolf Creek, C4 Stream Type		Location:	≈ 5000 ft above conf. of WF San Juan River	
Observers:	D. Rosgen	Reference reach: X Disturbed (impacted reach):		Date: 7/04	
Existing species composition:	Cottonwood/willow/sedge		Potential species composition:	Same	

	Riparian cover categories	Percent aerial cover *	Percent of site coverage **	Species composition	Percent of total species composition
1. Overstory	Canopy layer	60	30	Populus (Cottonwood) Picea (Spruce)	90 10
					100%
2. Understory	Shrub layer		40	Salix (Willow) Alnus (Alder) Acer (Maple)	80 10 10
					100%
3. Ground level	Herbaceous		25	Carex (Sedge) Poa (Bluegrass) Deschampsia (Hairgrass) Bromus (Bromegrass)	50 20 20 10
					100%
	Leaf or needle litter		5	**Remarks:** Condition, vigor and/or usage of existing reach: **High vigor, excellent condition, zero utilization**	
	Bare ground		0		

*Based on crown closure.
** Based on basal area to surface area.

Column total = 100%

Worksheet 7-17b (Worksheet 5-6). Riparian vegetation variables that influence channel stability for the D4 impaired reach.

Stream:	Wolf Creek, D4 Stream Type		Location:	≈ 4000 ft above conf. of WF San Juan River	
Observers:	D. Rosgen	Reference reach	Disturbed (impacted reach) X	Date: 7/04	
Existing species composition:	Grass/forb		Potential species composition:	Cottonwood/willow/sedge	

	Riparian cover categories	Percent aerial cover*	Percent of site coverage**	Species composition	Percent of total species composition
1. Overstory	Canopy layer	0	0		
					100%
2. Understory	Shrub layer		5	Salix (Willow)	90
				Alnus (Alder)	10
					100%
3. Ground level	Herbaceous		75	Poa (Bluegrass)	50
				Bromus (Bromegrass)	20
				Annual Forbs	30
					100%
	Leaf or needle litter		0	**Remarks:** Condition, vigor and/or usage of existing reach:	
	Bare ground		20	**Poor vigor, grass/forb in fair condition, heavy utilization**	

*Based on crown closure.
** Based on basal area to surface area.

Column total = 100%

b. Flow regime

Both the reference and impaired stream reaches are perennial stream channels with seasonal flow variations dominated by snowmelt and stormflow runoff (**Figure 7-37**). This information is recorded in Column (4) in the stream stability summary worksheet.

General Category	
E	Ephemeral stream channels: flows only in response to precipitation. Often used in conjunction with intermittent.
S	Subterranean stream channel: flows parallel to and near the surface for various seasons - a sub-surface flow that follows the stream bed.
I	Intermittent stream channel: one that flows only seasonally or sporadically. Surface sources involve springs, snowmelt, artificial controls, etc. Often this term is associated with flows that reappear along various locations of a reach then run subterranean.
(P)	Perennial stream channels: surface water persists yearlong.

Specific Category	
(1)	Seasonal variation in streamflow dominated primarily by snowmelt runoff.
(2)	Seasonal variation in streamflow dominated primarily by stormflow runoff.
3	Uniform stage and associated streamflow due to spring-fed condition, backwater, etc.
4	Streamflow regulated by glacial melt.
5	Ice flows/ice torrents from ice dam breaches.
6	Alternating flow/backwater due to tidal influence.
7	Regulated streamflow due to diversions, dam release, dewatering, etc.
8	Altered due to development, such as urban streams, cut-over watersheds or vegetation conversions (forested to grassland) that change flow response to precipitation events.
9	Rain-on-snow generated runoff.

Figure 7-37 (Figure 5-9). Flow regime variables that influence channel characteristics, sediment regime and biological interpretations for the C4 and D4 stream types.

c. Stream order and stream size

The C4 reference reach is a fourth-order stream with a bankfull width of 37 ft, categorizing it as S-5(4) using **Table 7-30**. The D4 impaired reach is also a fourth-order stream, but its bankfull width is 203 ft, making it an S-9(4). This information is recorded in Column (5) in the stream stability summary worksheet.

Table 7-30 (Table 5-3). Stream size categories for the C4 and D4 stream types (Rosgen, 1996).

Category	STREAM SIZE: Bankfull width meters	STREAM SIZE: Bankfull width feet	
S-1	<0.3	<1	
S-2	0.3 – 1.5	1 – 5	
S-3	1.5 – 4.6	5 – 15	
S-4	4.6 – 9	15 – 30	
S-5	9 – 15	30 – 50	← C4 Stream Type
S-6	15 – 22.8	50 – 75	
S-7	22.8 – 30.5	75 – 100	
S-8	30.5 – 46	100 – 150	
S-9	46 – 76	150 – 250	← D4 Stream Type
S-10	76 – 107	250 – 350	
S-11	107 – 150	350 – 500	
S-12	150 – 305	500 – 1000	
S-13	>305	>1000	

d. Meander patterns

The C4 reference reach displays regular meanders in the M1 category in **Figure 7-38**, while the D4 impaired reach, a braided channel, displays both M5 and M8 meanders. The shift in channel pattern from M1 for the reference C4 stream type to M5 and M8 for the D4 stream type is interpreted as lateral instability and is further evaluated and summarized in **Step 25**. The meander pattern ratings are recorded in Column (6) in the stream stability summary worksheet.

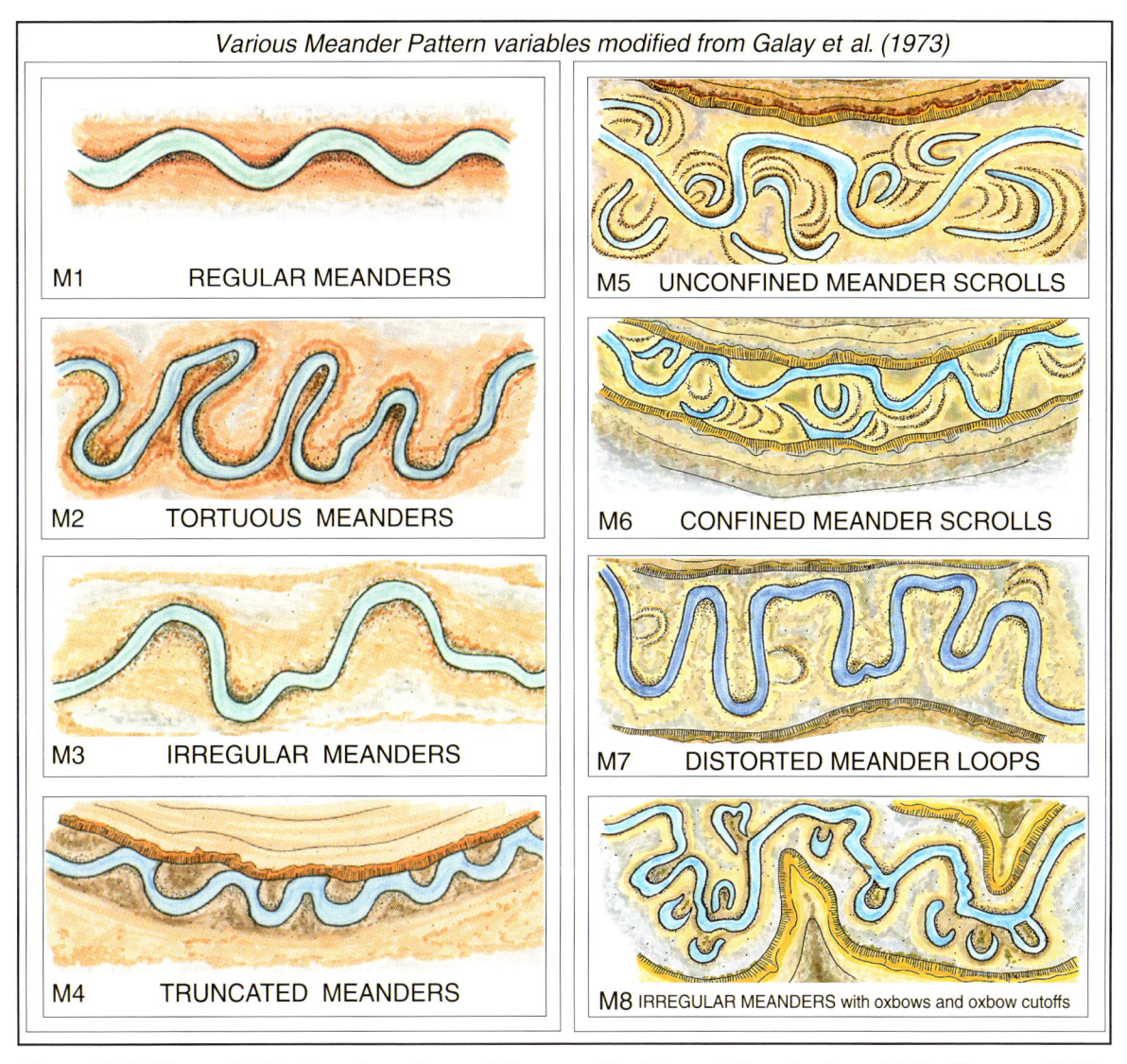

Figure 7-38 (Figure 5-10). Meander pattern relations used for interpretations for river stability (Rosgen, 1996).

e. Depositional patterns

The C4 reference reach contains point bars and mid-channel bars, referring to depositional categories B1 and B2 in **Figure 7-39**. The D4 impaired reach contains a variety of channel bars, relating to depositional categories B5, B6 and B7. These categories are indicative of very high sediment supply, lateral and vertical instability and channel enlargement. Large widths and shallow depths are associated with these depositional patterns, and loss of fish habitat is also evident as pools fill. The depositional pattern ratings are recorded in Column (7) in the stream stability summary worksheet. These depositional patterns will be used in the assessment of lateral and vertical stability ratings in **Steps 25** and **26**.

Various Depositional Features modified from Galay et al. (1973)

B1 POINT BARS

B2 POINT BARS with Few MID-CHANNEL BARS

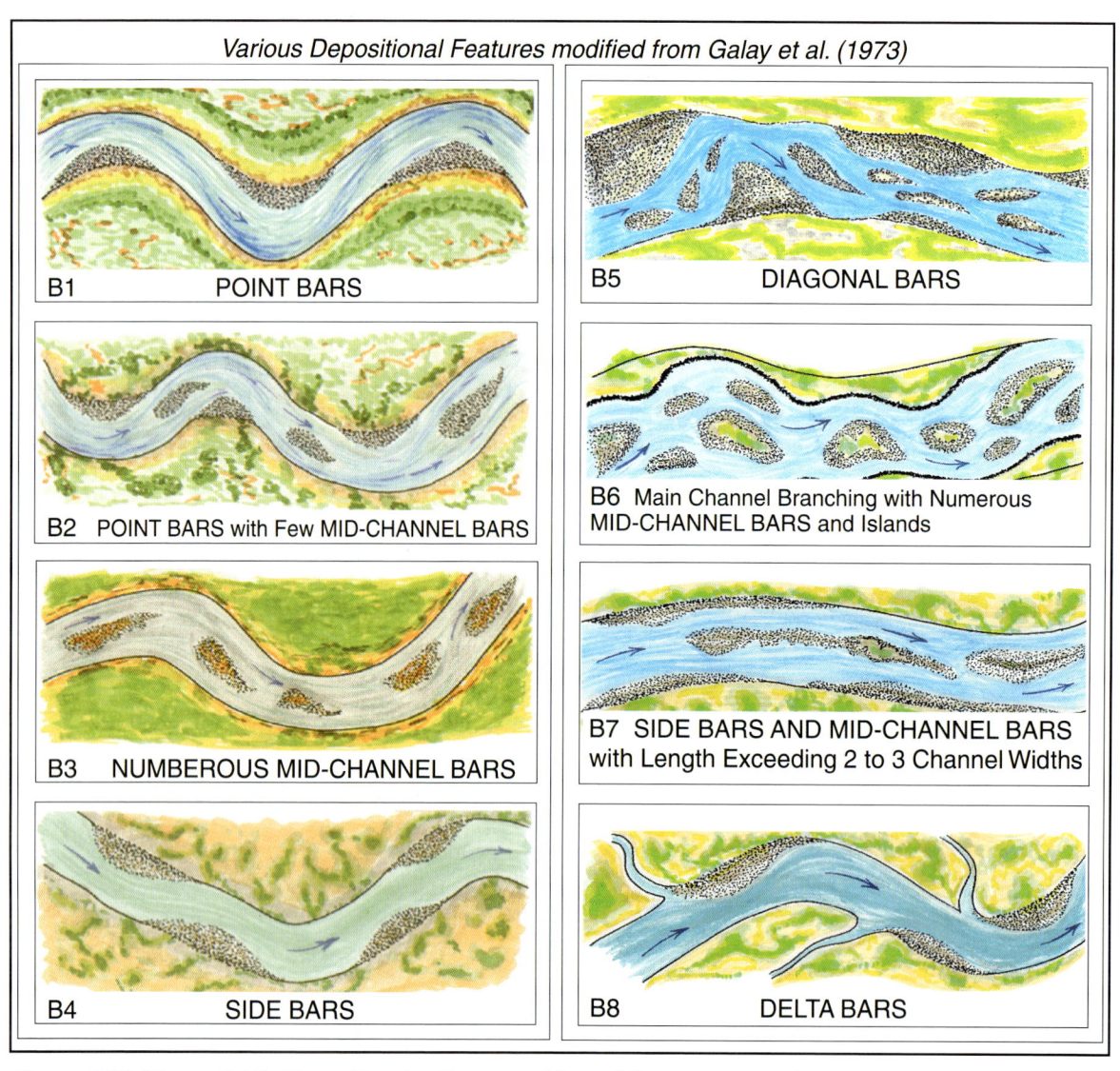

B3 NUMBEROUS MID-CHANNEL BARS

B4 SIDE BARS

B5 DIAGONAL BARS

B6 Main Channel Branching with Numerous MID-CHANNEL BARS and Islands

B7 SIDE BARS AND MID-CHANNEL BARS with Length Exceeding 2 to 3 Channel Widths

B8 DELTA BARS

Figure 7-39 (Figure 5-11). Depositional patterns used for stability assessment interpretations (Rosgen, 1996).

f. Channel blockages

Both the C4 reference and the D4 impaired reaches have moderate in-channel debris, corresponding to a D3 rating in **Table 7-31**. The moderate rating for large woody debris represents a low risk to river stability and function for these stream types. The rating reflects an actual benefit for fish habitat. The D3 ratings for the C4 and D4 reaches are reflected in the overall vertical stability summary by stream type in **Step 26**. This information is recorded in Column (8) in the stream stability summary worksheet.

Table 7-31 (Table 5-4). Various categories of in-channel debris, dams and/or channel blockages used to evaluate channel stability (Rosgen, 1996).

Description/extent		Materials, which upon placement into the active channel or flood-prone area, may cause adjustments in channel dimensions or conditions due to influences on the existing flow regime.
D1	None	Minor amounts of small, floatable material.
D2	Infrequent	Debris consists of small, easily moved, floatable material, e.g., leaves, needles, small limbs and twigs.
D3	Moderate	Increasing frequency of small- to medium-sized material, such as large limbs, branches and small logs, that when accumulated, affect 10% or less of the active channel cross-section area.
D4	Numerous	Significant build-up of medium- to large-sized materials, e.g., large limbs, branches, small logs or portions of trees that may occupy 10–30% of the active channel cross-section area.
D5	Extensive	Debris "dams" of predominantly larger materials, e.g., branches, logs and trees, occupying 30–50% of the active channel cross-section area, often extending across the width of the active channel.
D6	Dominating	Large, somewhat continuous debris "dams," extensive in nature and occupying over 50% of the active channel cross-section area. Such accumulations may divert water into the flood-prone areas and form fish migration barriers, even when flows are at less than bankfull.
D7	Beaver dams: Few	An infrequent number of dams spaced such that normal streamflow and expected channel conditions exist in the reaches between dams.
D8	Beaver dams: Frequent	Frequency of dams is such that backwater conditions exist for channel reaches between structures where streamflow velocities are reduced and channel dimensions or conditions are influenced.
D9	Beaver dams: Abandoned	Numerous abandoned dams, many of which have filled with sediment and/or breached, initiating a series of channel adjustments, such as bank erosion, lateral migration, avulsion, aggradation and degradation.
D10	Human influences	Structures, facilities or materials related to land uses or development located within the flood-prone area, such as diversions or low-head dams, controlled by-pass channels, velocity control structures and various transportation encroachments that have an influence on the existing flow regime, such that significant channel adjustments occur.

g. Width/depth ratio state

The C4 reference reach has a width/depth ratio of 23, while the D4 impaired reach has a width/depth ratio of 254. If the reference reach is used as the point of departure with a value of 1, the D4 reach is so unstable that it is off the stability rating chart (**Figure 7-40**) with a value of 11. A width/depth ratio increase greater than one order of magnitude above the reference condition indicates extreme instability. The instability is due to excess sediment deposition from a reduction in both sediment competence and sediment transport capacity. Typical of very high width/depth ratio departures are a bi-modal particle size distribution trending to finer sediment and filling of pools. Stream temperature and fish habitat are all adversely affected due to the very high width/depth ratio. The width/depth ratio states for the C4 and D4 stream types are recorded in Column (9) in the stream stability summary worksheet and are used in **Steps 25** and **26** to assess lateral and vertical stability.

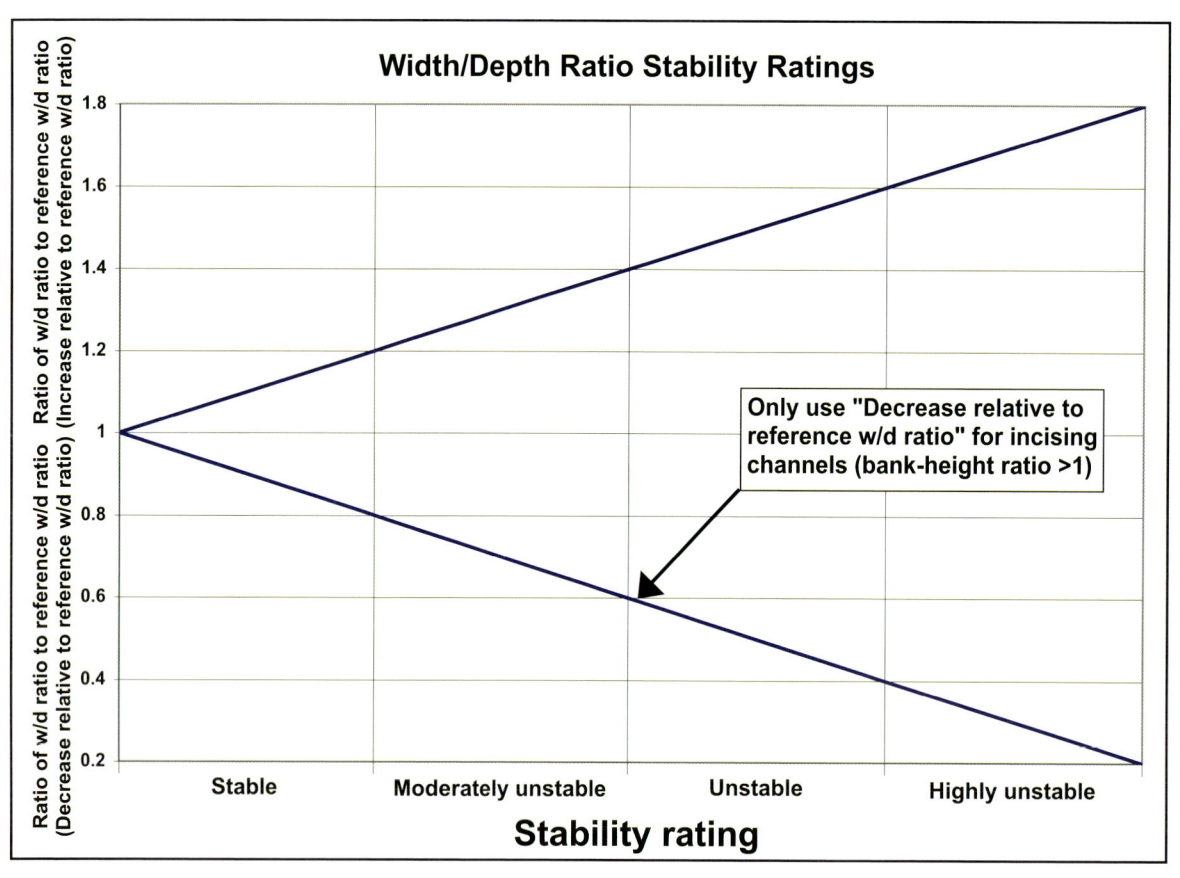

Figure 7-40 (Figure 5-13). Stability ratings based on departure of width/depth ratio from reference condition (Rosgen, 2001b).

h. Modified Pfankuch stability rating

The numerical channel stability rating summary for the modified Pfankuch method was 59 for the C4 reference reach (**Figures 7-29**, **7-31** and **7-32**, cross-section 45 + 56). The numerical score of 59 converts to an adjective rating of *good* or *stable* by stream type as shown in **Worksheet 7-18a**. This indicates that the sediment supply from channel sources at this reach is relatively low. This rating includes bank stability as well as depositional bars. Streambank erosion rates will be calculated separately in **Steps 8** and **9**.

The modified Pfankuch channel stability rating for the D4 stream type, however, was 133 at the typical reach shown in **Figures 7-29**, **7-31** and **7-32**. Because the potential stream type is a C4 for the impaired reach, the modified Pfankuch stability rating is *poor* or *unstable* (**Worksheet 7-18b**). The *poor* or *unstable* rating indicates a disproportionately high sediment supply, both from the streambanks and stored sediment due to excess deposition evident in the stability evaluation. The photographs in **Figures 7-31** and **7-32** depict the channel characteristics of the D4 reach as a result of spraying willows and converting to a grass/forb riparian vegetative community.

The Pfankuch stability ratings of *good* or *stable* for the C4 stream type and *poor* or *unstable* for the D4 stream type are reflected in the selection of dimensionless sediment rating curves by stream type and/or stability rating in **Step 12** to calculate annual sediment yield using the FLOWSED model. The Pfankuch stability ratings are also used in the sediment supply rating in **Step 28**. The *poor* or *unstable* D4 reach continues to show major impairment due to channel processes.

The user should use caution to realize that not all D4 stream types rate *poor* using the Pfankuch channel stability rating procedure adjusted by stream type. Many D4 stream types are the stable form in certain valley types. A rating up to 107 points for a D4 would be a *good* or *stable* condition, whereas a rating of 107 for a C4 would be on the borderline between *fair* and *poor*.

WATERSHED ASSESSMENT OF RIVER STABILITY AND SEDIMENT SUPPLY

Worksheet 7-18a (Worksheet 5-7). Pfankuch (1975) stream channel stability rating procedure for the C4 reference reach, as modified by Rosgen (1996, 2001b).

Stream:	Wolf Creek		Location: C4 Reference		Valley Type: VIII		Observers: D. Rosgen		Date: 7/04
Location	Key	Category	Excellent Description	Rating	Good Description	Rating	Fair Description	Rating	Poor Description	Rating
Upper banks	1	Landform slope	Bank slope gradient <30%.	2	Bank slope gradient 30–40%.	4	Bank slope gradient 40–60%.	6	Bank slope gradient >60%.	8
	2	Mass erosion	No evidence of past or future mass erosion.	3	Infrequent. Mostly healed over. Low future potential.	6	Frequent or large, causing sediment nearly yearlong.	9	Frequent or large, causing sediment nearly yearlong OR imminent danger of same.	12
	3	Debris jam potential	Essentially absent from immediate channel area.	2	Present, but mostly small twigs and limbs.	4	Moderate to heavy amounts, mostly larger sizes.	6	Moderate to heavy amounts, predominantly larger sizes.	8
	4	Vegetative bank protection	>90% plant density. Vigor and variety suggest a deep, dense soil-binding root mass.	3	70–90% density. Fewer species or less vigor suggest less dense or deep root mass.	6	50–70% density. Lower vigor and fewer species from a shallow, discontinuous root mass.	9	<50% density plus fewer species & less vigor indicating poor, discontinuous and shallow root mass.	12
Lower banks	5	Channel capacity	Bank heights sufficient to contain the bankfull stage. Width/depth ratio departure from reference width/depth ratio = 1.0. Bank-Height Ratio (BHR) = 1.0.	1	Bank heights sufficient to contain the bankfull stage. Width/depth ratio departure from reference width/depth ratio = 1.0–1.2. Bank-Height Ratio (BHR) = 1.0–1.1	2	Bankfull stage is not contained. Width/depth ratio departure from reference width/depth ratio = 1.2–1.4. Bank-Height Ratio (BHR) = 1.1–1.3.	3	Bankfull stage is not contained; over-bank flows are common with flows less than bankfull. Width/depth ratio departure from reference width/depth ratio > 1.4. Bank-Height Ratio (BHR) >1.3.	4
	6	Bank rock content	>65% w/ large angular boulders. 12"+ common.	2	40–65%. Mostly boulders and small cobbles 6–12".	4	20–40%. Most in the 3-6" diameter class.	6	<20% rock fragments of gravel sizes, 1-3" or less.	8
	7	Obstructions to flow	Rocks and logs firmly imbedded. Flow pattern w/o cutting or deposition. Stable bed.	2	Some present causing erosive cross currents and minor pool filling. Obstructions fewer and less firm.	4	Moderately frequent, unstable obstructions move with high flows causing bank cutting and pool filling.	6	Frequent obstructions and deflectors cause bank erosion yearlong. Sediment traps full, channel migration occurring.	8
	8	Cutting	Little or none. Infrequent raw banks <6".	4	Some, intermittently at outcurves and constrictions. Raw banks may be up to 12".	6	Significant. Cuts 12–24" high. Root mat overhangs and sloughing evident.	12	Almost continuous cuts, some over 24" high. Failure of overhangs frequent.	16
	9	Deposition	Little or no enlargement of channel or point bars.	4	Some new bar increase, mostly from coarse gravel.	8	Moderate deposition of new gravel and coarse sand on old and some new bars.	12	Extensive deposit of predominantly fine particles. Accelerated bar development.	16
Bottom	10	Rock angularity	Sharp edges and corners. Plane surfaces rough.	1	Rounded corners and edges. Surfaces smooth and flat.	2	Corners and edges well rounded in 2 dimensions.	3	Well rounded in all dimensions, surfaces smooth.	4
	11	Brightness	Surfaces dull, dark or stained. Generally not bright.	1	Mostly dull, but may have <35% bright surfaces.	2	Mixture dull and bright, i.e., 35–65% mixture range.	3	Predominantly bright, >65%, exposed or scoured surfaces.	4
	12	Consolidation of particles	Assorted sizes tightly packed or overlapping.	2	Moderately packed with some overlapping.	4	Mostly loose assortment with no apparent overlap.	6	No packing evident. Loose assortment, easily moved.	8
	13	Bottom size distribution	No size change evident. Stable material 80–100%.	4	Distribution shift light. Stable material 50–80%.	8	Moderate change in sizes. Stable materials 20–50%.	12	Marked distribution change. Stable materials 0–20%.	16
	14	Scouring and deposition	<5% of bottom affected by scour or deposition.	6	5–30% affected. Scour at constrictions and where grades steepen. Some deposition in pools.	12	30–50% affected. Deposits and scour at obstructions, constrictions and bends. Some filling of pools.	18	More than 50% of the bottom in a state of flux or change nearly yearlong.	24
	15	Aquatic vegetation	Abundant growth moss-like, dark green perennial. In swift water, too.	1	Common. Algae forms in low velocity and pool areas. Moss here, too.	2	Present but spotty, mostly in backwater. Seasonal algae growth makes rocks slick.	3	Perennial types scarce or absent. Yellow-green, short-term bloom may be present.	4
			Excellent total =	16	Good total =	30	Fair total =	5	Poor total =	8

Stream type	A1	A2	A3	A4	A5	A6	B1	B2	B3	B4	B5	B6	C1	C2	C3	C4	C5	C6	D3	D4	D5	D6
Good (Stable)	38-43	38-43	54-90	60-95	60-95	50-80	38-45	38-45	40-60	40-64	48-68	40-60	38-50	38-50	60-85	70-90	70-90	60-85	85-107	85-107	85-107	67-98
Fair (Mod. unstable)	44-47	44-47	91-129	96-132	96-142	81-110	46-58	46-58	61-78	65-84	69-88	61-78	51-61	51-61	86-105	91-110	91-110	86-105	108-132	108-132	108-132	99-125
Poor (Unstable)	48+	48+	130+	133+	143+	111+	59+	59+	79+	85+	89+	79+	62+	62+	106+	111+	111+	106+	133+	133+	133+	126+

Stream type	DA3	DA4	DA5	DA6	E3	E4	E5	E6	F1	F2	F3	F4	F5	F6	G1	G2	G3	G4	G5	G6		
Good (Stable)	40-63	40-63	40-63	40-63	40-63	50-75	50-75	40-63	60-85	60-85	85-110	85-110	90-115	80-95	40-60	40-60	85-107	85-107	90-112	85-107		
Fair (Mod. unstable)	64-86	64-86	64-86	64-86	64-86	76-96	76-96	64-86	86-105	86-105	111-125	111-125	116-130	96-110	61-78	61-78	108-120	108-120	113-125	108-120		
Poor (Unstable)	87+	87+	87+	87+	87+	97+	97+	87+	106+	106+	126+	126+	131+	111+	79+	79+	121+	121+	126+	121+		

Grand total = 59
Existing stream type = C4
*Potential stream type = C4
Modified channel stability rating = Stable

*Rating should be adjusted to potential stream type, not existing.

7-118

Chapter 7 — PLA The WARSSS Methodology: A Case Study of Wolf Creek, Colorado

Worksheet 7-18b (Worksheet 5-7). Pfankuch (1975) stream channel stability rating procedure for the D4 impaired reach, as modified by Rosgen (1996, 2001b)

Stream:	Wolf Creek		Location: D4 impaired				Valley Type: VIII		Observers: D. Rosgen		Date: 7/04	
Location	Key	Category	**Excellent** Description	Rating	**Good** Description	Rating	**Fair** Description	Rating	**Poor** Description	Rating
Upper banks	1	Landform slope	Bank slope gradient <30%.	(2)	Bank slope gradient 30–40%.	4	Bank slope gradient 40–60%.	6	Bank slope gradient >60%.	8
	2	Mass erosion	No evidence of past or future mass erosion.	(3)	Infrequent. Mostly healed over. Low future potential.	6	Frequent or large, causing sediment nearly yearlong.	9	Frequent or large, causing sediment nearly yearlong OR imminent danger of same.	12
	3	Debris jam potential	Essentially absent from immediate channel area.	2	Present, but mostly small twigs and limbs.	4	Moderate to heavy amounts, mostly larger sizes.	6	Moderate to heavy amounts, predominantly larger sizes.	8
	4	Vegetative bank protection	>90% plant density. Vigor and variety suggest a deep, dense soil-binding root mass.	3	70–90% density. Fewer species or less vigor suggest less dense or deep root mass.	6	50–70% density. Lower vigor and fewer species from a shallow, discontinuous root mass.	9	<50% density plus fewer species & less vigor indicating poor, discontinuous and shallow root mass.	(12)
Lower banks	5	Channel capacity	Bank heights sufficient to contain the bankfull stage. Width/depth ratio departure from reference width/depth ratio = 1.0. Bank-Height Ratio (BHR) = 1.0.	1	Bankfull stage is contained within banks. Width/depth ratio departure from reference width/depth ratio = 1.0–1.2. Bank-Height Ratio (BHR) = 1.0–1.1	2	Bankfull stage is not contained. Width/depth ratio departure from reference width/depth ratio = 1.2–1.4. Bank-Height Ratio (BHR) = 1.1–1.3.	3	Bankfull stage is not contained; over-bank flows are common with flows less than bankfull. Width/depth ratio departure from reference width/depth ratio > 1.4. Bank-Height Ratio (BHR) >1.3.	4
	6	Bank rock content	>65% w/ large angular boulders. 12"+ common.	2	40–65%. Mostly boulders and small cobbles 6–12".	4	20–40%. Most in the 3–6" diameter class.	6	<20% rock fragments of gravel sizes, 1–3" or less.	8
	7	Obstructions to flow	Rocks and logs firmly imbedded. Flow pattern w/o cutting or deposition. Stable bed.	2	Some present causing erosive cross currents and minor pool filling. Obstructions fewer and less firm.	4	Moderately frequent, unstable obstructions move with high flows causing bank cutting and pool filling.	6	Frequent obstructions and deflectors cause bank erosion yearlong. Sediment traps full, channel migration occurring.	8
	8	Cutting	Little or none. Infrequent raw banks <6".	4	Some, intermittently at outcurves and constrictions. Raw banks may be up to 12".	6	Significant. Cuts 12–24" high. Root mat overhangs and sloughing evident.	12	Almost continuous cuts, some over 24" high. Failure of overhangs frequent.	16
	9	Deposition	Little or no enlargement of channel or point bars.	4	Some new bar increase, mostly from coarse gravel.	8	Moderate deposition of new gravel and coarse sand on old and some new bars.	12	Extensive deposit of predominantly fine particles. Accelerated bar development.	16
Bottom	10	Rock angularity	Sharp edges and corners. Plane surfaces rough.	1	Rounded corners and edges. Surfaces smooth and flat.	2	Corners and edges well rounded in 2 dimensions.	3	Well rounded in all dimensions, surfaces smooth.	4
	11	Brightness	Surfaces dull, dark or stained. Generally not bright.	1	Mostly dull, but may have <35% bright surfaces.	2	Mixture dull and bright, i.e., 35–65% mixture range.	3	Predominantly bright, >65%, exposed or scoured surfaces.	4
	12	Consolidation of particles	Assorted sizes tightly packed or overlapping.	2	Moderately packed with some overlapping.	4	Mostly loose assortment with no apparent overlap.	6	No packing evident. Loose assortment, easily moved.	8
	13	Bottom size distribution	No size change evident. Stable material 80–100%.	4	Distribution shift light. Stable material 50–80%.	8	Moderate change in sizes. Stable materials 20–50%.	12	Marked distribution change. Stable materials 0–20%.	16
	14	Scouring and deposition	<5% of bottom affected by scour or deposition.	6	5–30% affected. Scour at constrictions and where grades steepen. Some deposition in pools.	12	30–50% affected. Deposits and scour at obstructions, constrictions and bends. Some filling of pools.	18	More than 50% of the bottom in a state of flux or change nearly yearlong.	24
	15	Aquatic vegetation	Abundant growth moss-like, dark green perennial. In swift water, too.	1	Common. Algae forms in low velocity and pool areas. Moss here, too.	2	Present but spotty, mostly in backwater. Seasonal algae growth makes rocks slick.	3	Perennial types scarce or absent. Yellow-green, short-term bloom may be present.	4
			Excellent total =	5	Good total =	4	Fair total =	0	Poor total =	124

Stream type	A1	A2	A3	A4	A5	A6	B1	B2	B3	B4	B5	B6	C1	C2	C3	C4	C5	C6	D3	D4	D5	D6
Good (Stable)	38-43	38-43	54-90	60-95	60-95	50-80	38-45	38-45	40-60	40-64	48-68	40-60	38-50	38-50	60-85	70-90	70-90	60-85	85-107	85-107	85-107	67-98
Fair (Mod. unstable)	44-47	44-47	91-129	96-132	96-142	81-110	46-58	46-58	61-78	65-84	69-88	61-78	51-61	51-61	86-105	91-110	91-110	86-105	108-132	108-132	108-132	99-125
Poor (Unstable)	48+	48+	130+	133+	143+	111+	59+	59+	79+	85+	89+	79+	62+	62+	106+	111+	111+	106+	133+	133+	133+	126+

Stream type	DA3	DA4	DA5	DA6	E3	E4	E5	E6	F1	F2	F3	F4	F5	F6	G1	G2	G3	G4	G5	G6		
Good (Stable)	40-63	40-63	40-63	40-63	40-63	50-75	50-75	40-63	60-85	60-85	60-85	60-85	90-115	80-95	40-60	40-60	85-107	85-107	90-112	85-107		
Fair (Mod. unstable)	64-86	64-86	64-86	64-86	64-86	76-96	76-96	64-86	86-105	86-105	86-105	86-105	116-130	96-110	61-78	61-78	108-120	108-120	113-125	108-120		
Poor (Unstable)	87+	87+	87+	87+	87+	97+	97+	87+	106+	106+	106+	106+	131+	111+	79+	79+	121+	121+	126+	121+		

Grand total =	133
Existing stream type =	D4
*Potential stream type =	C4
Modified channel stability rating =	Unstable

*Rating should be adjusted to potential stream type, not existing.

7-119

i. Degree of channel incision

Both the C4 and D4 reaches have a Bank-Height Ratio (BHR) of 1. Using **Figure 7-41**, river reaches with a BHR of 1 are not incised. The adjacent floodplain is still accessible with flows greater than the bankfull discharge. This information is recorded in Column (11) in the stream stability summary worksheet and is used for degradation potential in **Step 26**.

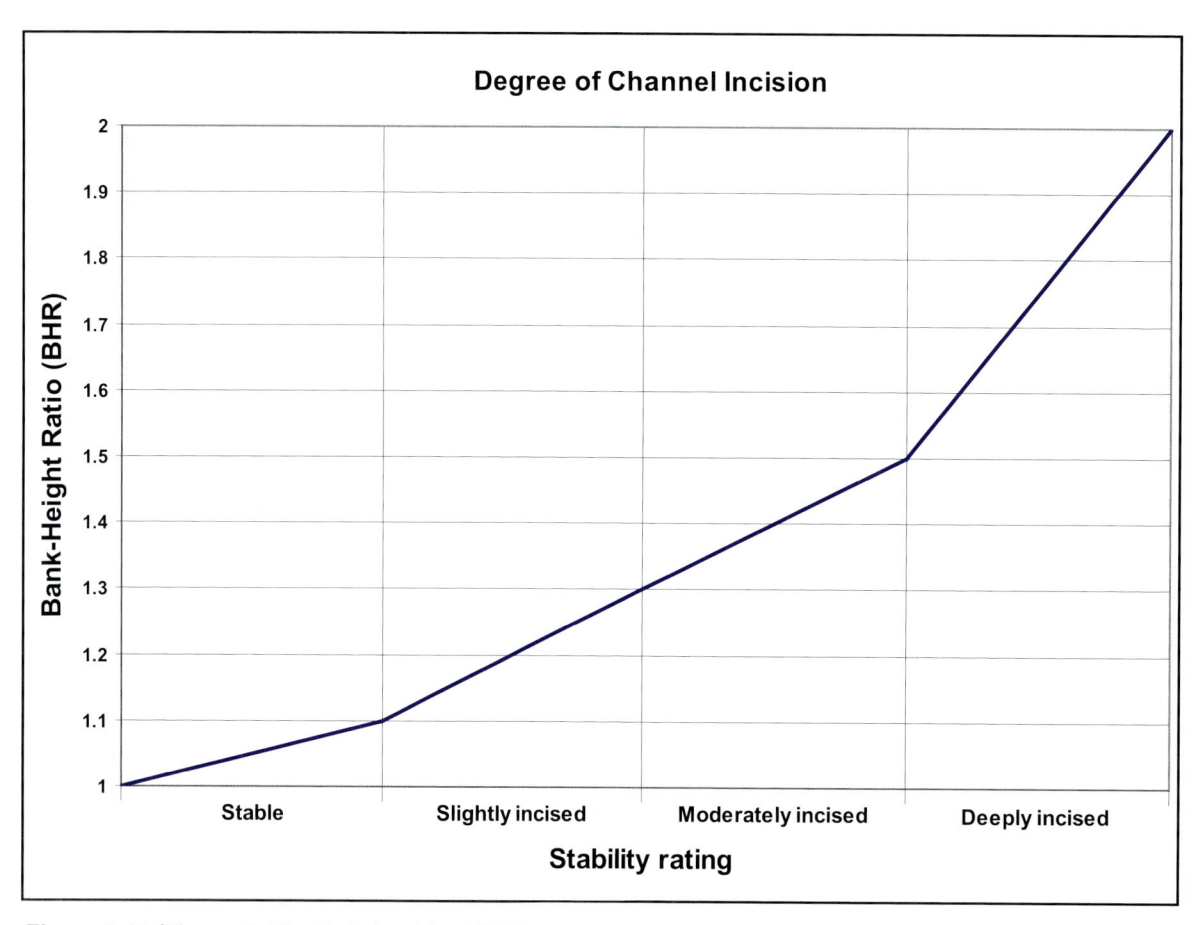

Figure 7-41 (Figure 5-15). Relationship of BHR ranges to corresponding stream stability ratings (Rosgen, 2001b).

j. Degree of channel confinement (lateral containment)

The Meander Width Ratio (MWR) for the C4 stream type is 4.8. For the D4 stream type, the belt width is the same as the bankfull width; thus, the MWR for the D4 stream type is 1.0. The departure ratio of the D4 stream type is 0.2 (MWR/MWR$_{ref}$). Using **Figure 7-42**, the C4 stream type is *unconfined*, while the D4 stream type rates as *confined*. This information is recorded in Column (12) in the stream stability summary worksheet and is used in **Steps 25** and **26** to predict lateral and vertical stability.

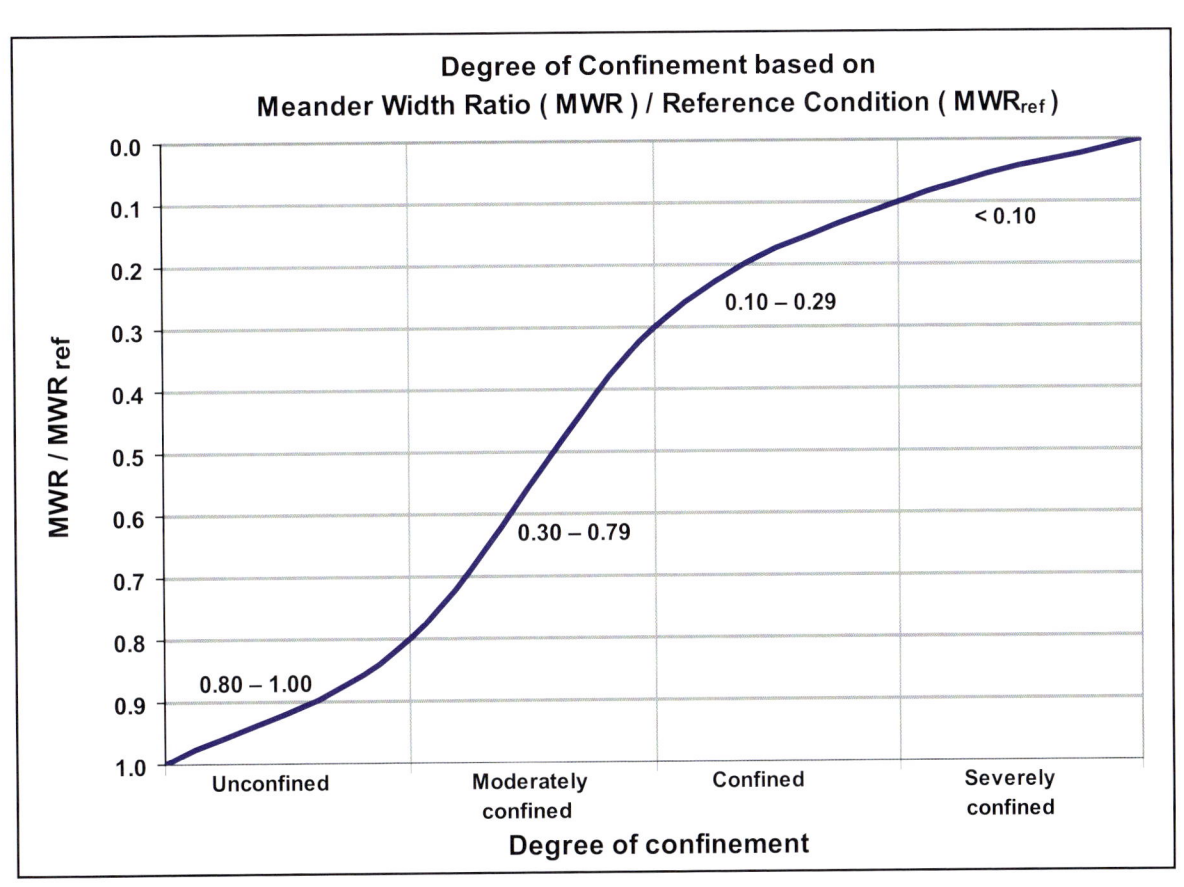

Figure 7-42 (Figure 5-17). Degree of confinement based on Meander Width Ratio (MWR) divided by reference condition Meander Width Ratio (MWR$_{ref}$).

WATERSHED ASSESSMENT OF RIVER STABILITY AND SEDIMENT SUPPLY

Worksheet 7-19 (Worksheet 5-5). Completed stream stability ratings summary worksheet for the C4 and D4 reaches (**Step 7**).

Stream: **Wolf Creek** Location: **above conf. of WF San Juan River** Observers: **D. Rosgen** Date: **7/04**

(1) Reach location	(2) Stream type (Worksheet 5-3)	(3) a. Riparian vegetation (Worksheet 5-6) Existing species composition	(3) Potential species composition	(4) b. Flow regime (Fig. 5-9)	(5) c. Stream order/size (Table 5-3)	(6) d. Meander patterns (Fig. 5-10)	(7) e. Depositional patterns (Fig. 5-11)	(8) f. Channel blockages (Table 5-4)	(9) g. W/d ratio (Worksheet 5-2)	(9) W/d stability rating (Fig. 5-13)	(10) h. Pfankuch channel stability rating (Worksheet 5-7)	(11) i. Bank-Height Ratio (BHR) (Worksheet 5-4)	(11) Stability rating (Fig. 5-15)	(12) j. MWR divided by MWR_ref	(12) Degree of confinement (Fig. 5-17)
1. Wolf Creek ref.	C4	Cottonwood/Willow/Sedge	Cottonwood/Willow/Sedge	P1, 2	S-5(4)	M1	B1, B2	D3	23	stable	59, stable	1.0	stable	1.0	unconfined
2. Wolf Creek impaired	D4	Grass/Forb	Cottonwood/Willow/Sedge	P1, 2	S-9(4)	M5, M8	B5, B6, B7	D3	254	highly unstable	133, unstable	1.0	stable	0.2	confined
3.															
4.															
5.															
6.															
7.															
8.															
9.															
10.															
11.															
12.															
13.															
14.															
15.															

Chapter 7 — PLA *The WARSSS Methodology: A Case Study of Wolf Creek, Colorado*

Steps 8–9: Streambank erosion (BANCS model) (See Flowchart 7-7)

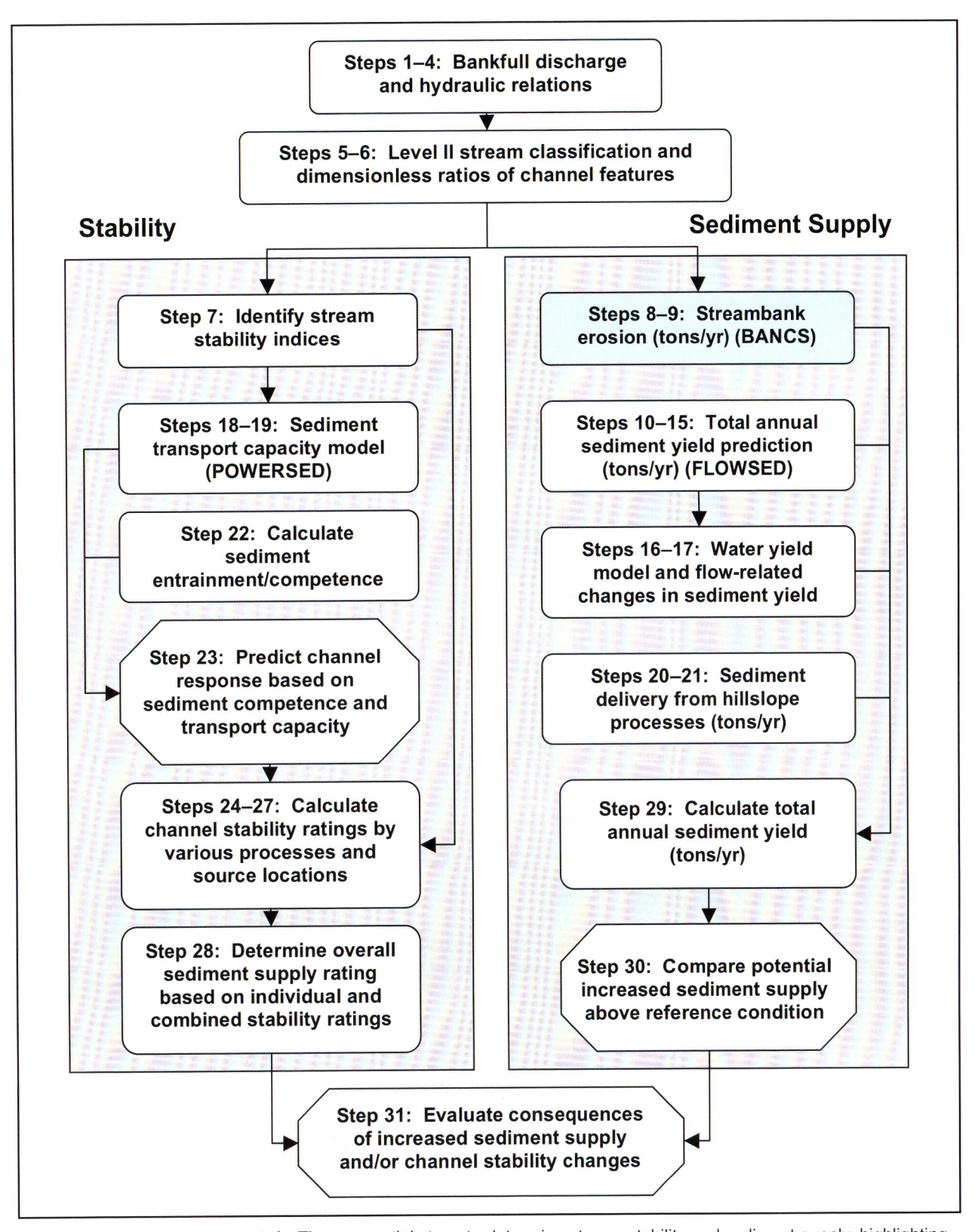

Steps 1–4: Bankfull discharge and hydraulic relations

Steps 5–6: Level II stream classification and dimensionless ratios of channel features

Stability

- **Step 7:** Identify stream stability indices
- **Steps 18–19:** Sediment transport capacity model (POWERSED)
- **Step 22:** Calculate sediment entrainment/competence
- **Step 23:** Predict channel response based on sediment competence and transport capacity
- **Steps 24–27:** Calculate channel stability ratings by various processes and source locations
- **Step 28:** Determine overall sediment supply rating based on individual and combined stability ratings

Sediment Supply

- **Steps 8–9:** Streambank erosion (tons/yr) (BANCS)
- **Steps 10–15:** Total annual sediment yield prediction (tons/yr) (FLOWSED)
- **Steps 16–17:** Water yield model and flow-related changes in sediment yield
- **Steps 20–21:** Sediment delivery from hillslope processes (tons/yr)
- **Step 29:** Calculate total annual sediment yield (tons/yr)
- **Step 30:** Compare potential increased sediment supply above reference condition

Step 31: Evaluate consequences of increased sediment supply and/or channel stability changes

Flowchart 7-3e (Flowchart 5-3e). The sequential steps to determine stream stability and sediment supply, highlighting Steps 8–9.

WATERSHED ASSESSMENT OF RIVER STABILITY AND SEDIMENT SUPPLY

Steps 8–9: Streambank erosion (BANCS model)

Step 8: Calculate Bank Erosion Hazard Index (BEHI) and Near-Bank Stress (NBS) ratings using the BANCS model

BEHI

Bank Erosion Hazard Index (BEHI) (Figure 5-19 and Worksheet 5-8)

- Bank height/bankfull height
- Rooting depth/bank height
- Weighted root density
- Bank angle (slope steepness)
- Surface protection
- Bank material
- Material stratification

BEHI rating (Worksheet 5-8)

NBS

Near-Bank Stress (NBS) (Methods 1–7 in Worksheet 5-9)

- (1) Transverse bar or split channel creating NBS
- (2) Ratio of radius of curvature to bankfull width
- (3) Ratio of pool slope to average water surface slope
- (4) Ratio of pool slope to riffle slope
- (5) Ratio of near-bank max depth to bankfull mean depth
- (6) Ratio of near-bank shear stress to bankfull shear stress
- (7) Velocity gradient

NBS rating (Worksheet 5-9)

Step 9: Predict annual streambank erosion rate (ft/yr) using BEHI and NBS ratings (use Figure 5-38 or 5-39 and Worksheet 5-10)

Step 25: Lateral stability (Worksheet 5-17)

Summary of sediment sources in tons/yr (*PLA* summary Worksheet 5-22)

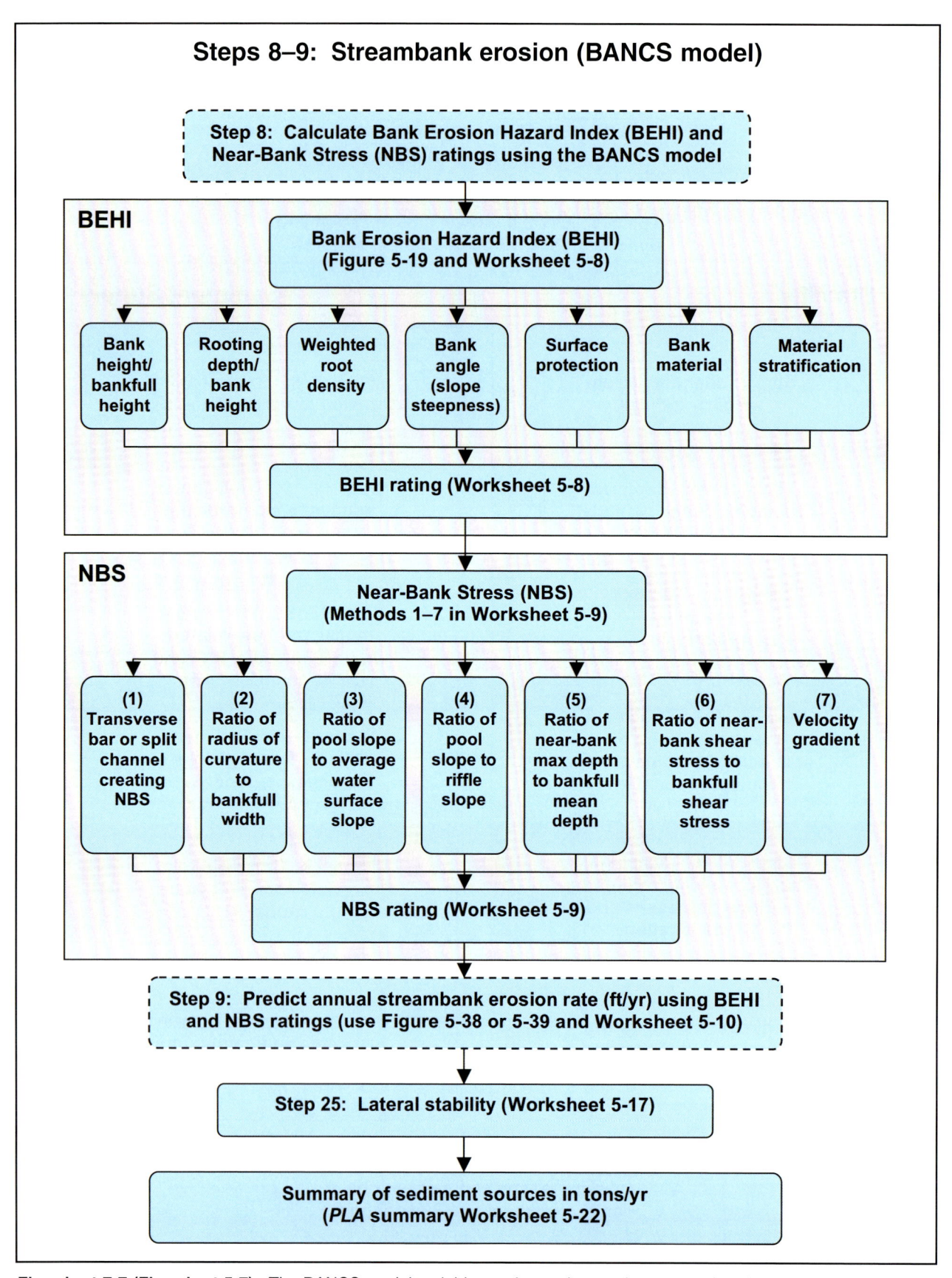

Flowchart 7-7 (Flowchart 5-7). The BANCS model variables, ratios and procedures associated with the Bank Erosion Hazard Index (BEHI) and Near-Bank Stress (NBS) to predict annual streambank erosion.

Step 8. Calculate Bank Erosion Hazard Index (BEHI) and Near-Bank Stress (NBS) ratings

The Bank Erosion Hazard Index (BEHI), (**Figure 7-43**), which evaluates the streambank characteristics susceptible to various erosional processes, was completed for both the reference reach and the impaired reach. The Near-Bank Stress (NBS) ratings were also completed for the reference reach and the impaired reach. The combined BEHI and NBS ratings are used in **Step 9** to predict lateral streambank erosion rates.

The C4 reference reach BEHI and NBS methods are documented in detail for station 9+80 to 11+50 in **Worksheets 7-20a** and **7-21a**. The D4 impaired reach BEHI and NBS methods for station 0+00 to 0+90 are documented in **Worksheets 7-20b** and **7-21b**. The BEHI and NBS ratings for the remaining stations for the C4 and D4 reaches are documented in **Step 9**, **Worksheets 7-22a** and **7-22b**, where the ratings were determined using the field calibration methods discussed in Chapter 5 and by measuring study bank heights and lengths for each station. In addition to the C4 and D4 reach assessments, the streambank erosion rates for the contributing reaches of Wolf Creek were also assessed using the field calibration methods and by measuring study bank heights and lengths.

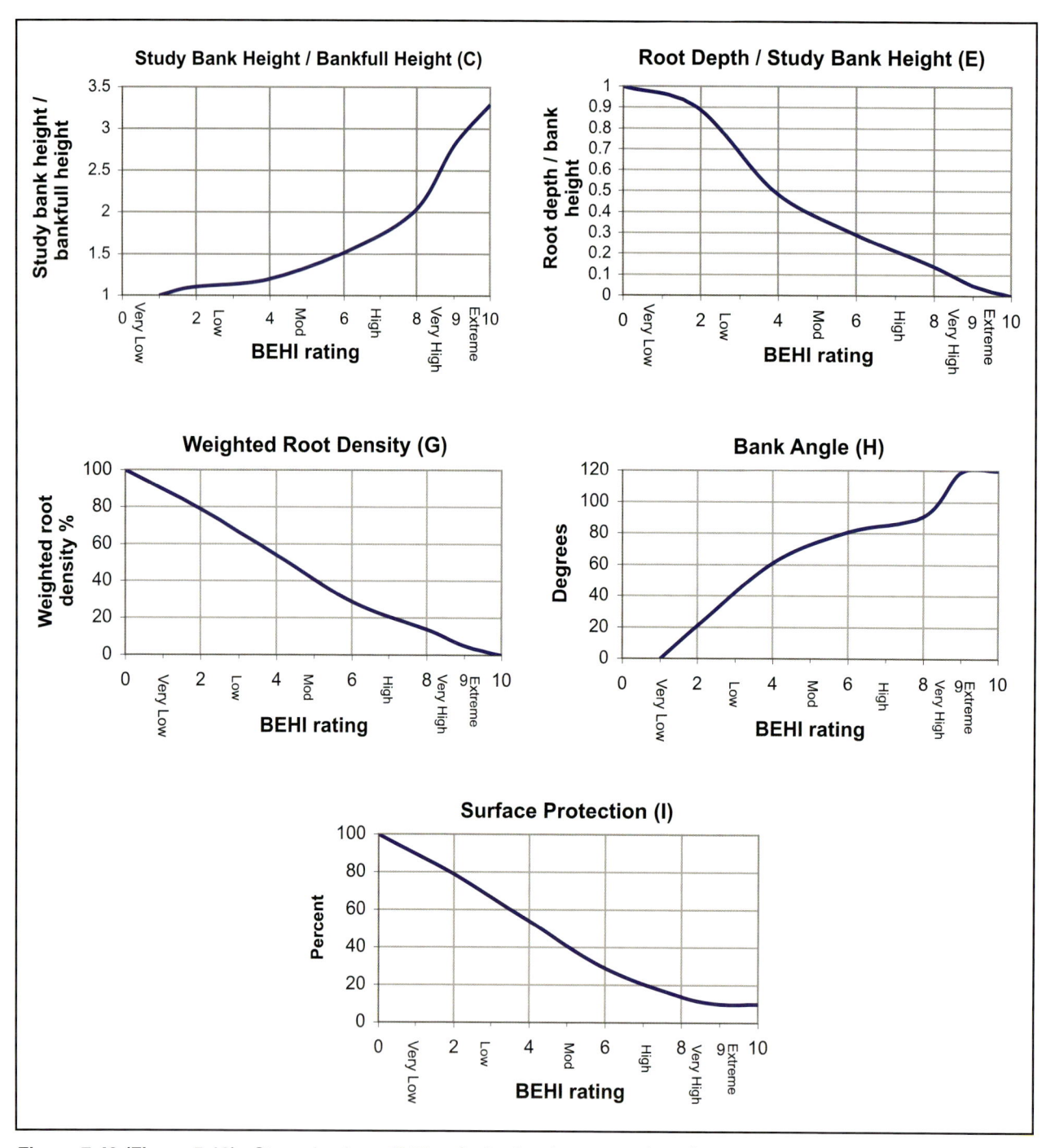

Figure 7-43 (Figure 5-19). Streambank erodibility criteria showing conversion of measured ratios and bank variables to a BEHI rating (Rosgen, 1996, 2001a). Use **Worksheet 5-8** variables to determine BEHI score.

Chapter 7 — PLA The WARSSS Methodology: A Case Study of Wolf Creek, Colorado

Worksheet 7-20a (Worksheet 5-8). Form to calculate Bank Erosion Hazard Index (BEHI) variables and overall BEHI rating for the C4 reference reach (Rosgen, 1996, 2001a).

Stream:	**Wolf Creek**	Location: ≈ **5000 ft above conf. WF San Juan R.**
Station:	**9+80 to 11+50**	Observers: **D. Rosgen**
Date:	**7/04** Stream Type: **C4**	Valley Type: **VIII**

BEHI Score (Fig. 5-19)

Study Bank Height / Bankfull Height (C)

| Study Bank Height (ft) = | 3.3 (A) | Bankfull Height (ft) = | 3 (B) | (A)/(B) = | 1.1 (C) | 1.9 |

Root Depth / Study Bank Height (E)

| Root Depth (ft) = | 3.3 (D) | Study Bank Height (ft) = | 3.3 (A) | (D)/(A) = | 1 (E) | 0 |

Weighted Root Density (G)

| Root Density as % = | 79% (F) | (F) × (E) = | 79% (G) | 2 |

Bank Angle (H)

| Bank Angle as Degrees = | 81 (H) | 6 |

Surface Protection (I)

| Surface Protection as % = | 79% (I) | 2 |

Bank Material Adjustment:
- **Bedrock** (Overall Very Low BEHI)
- **Boulders** (Overall Low BEHI)
- **Cobble** (Subtract 10 points if uniform medium to large cobble)
- **Gravel or Composite Matrix** (Add 5–10 points depending on percentage of bank material that is composed of sand)
- **Sand** (Add 10 points)
- **Silt/Clay** (no adjustment)

→ **Bank Material Adjustment**: 5

Stratification Adjustment
Add 5–10 points, depending on position of unstable layers in relation to bankfull stage: 0

Very Low	Low	Moderate	High	Very High	Extreme
5 – 9.5	⟨10 – 19.5⟩	20 – 29.5	30 – 39.5	40 – 45	46 – 50

Adjective Rating and Total Score: **Low** **16.9**

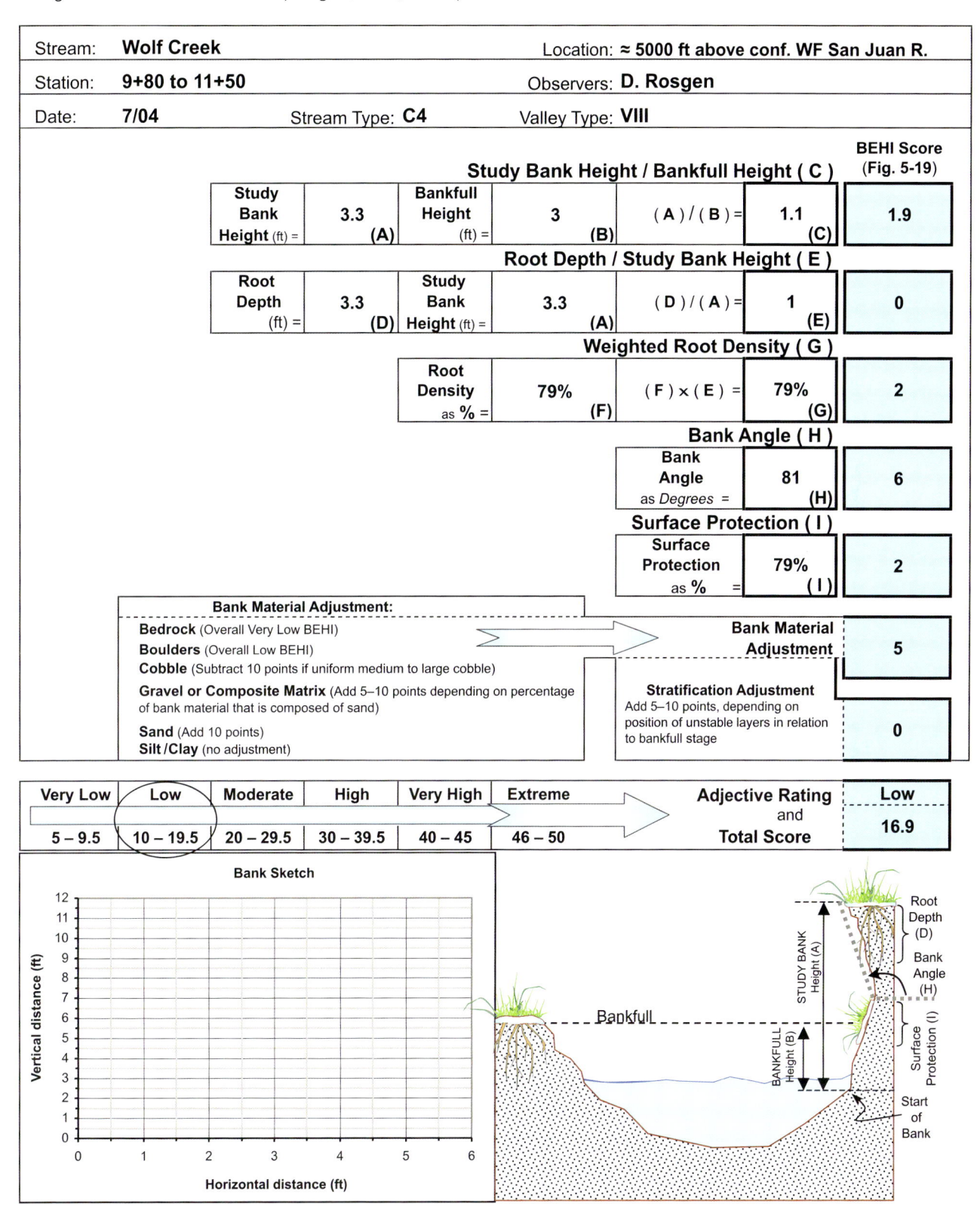

Bank Sketch

Watershed Assessment of River Stability and Sediment Supply

Worksheet 7-20b (Worksheet 5-8). Form to calculate Bank Erosion Hazard Index (BEHI) variables and overall BEHI rating for the D4 impaired reach (Rosgen, 1996, 2001a).

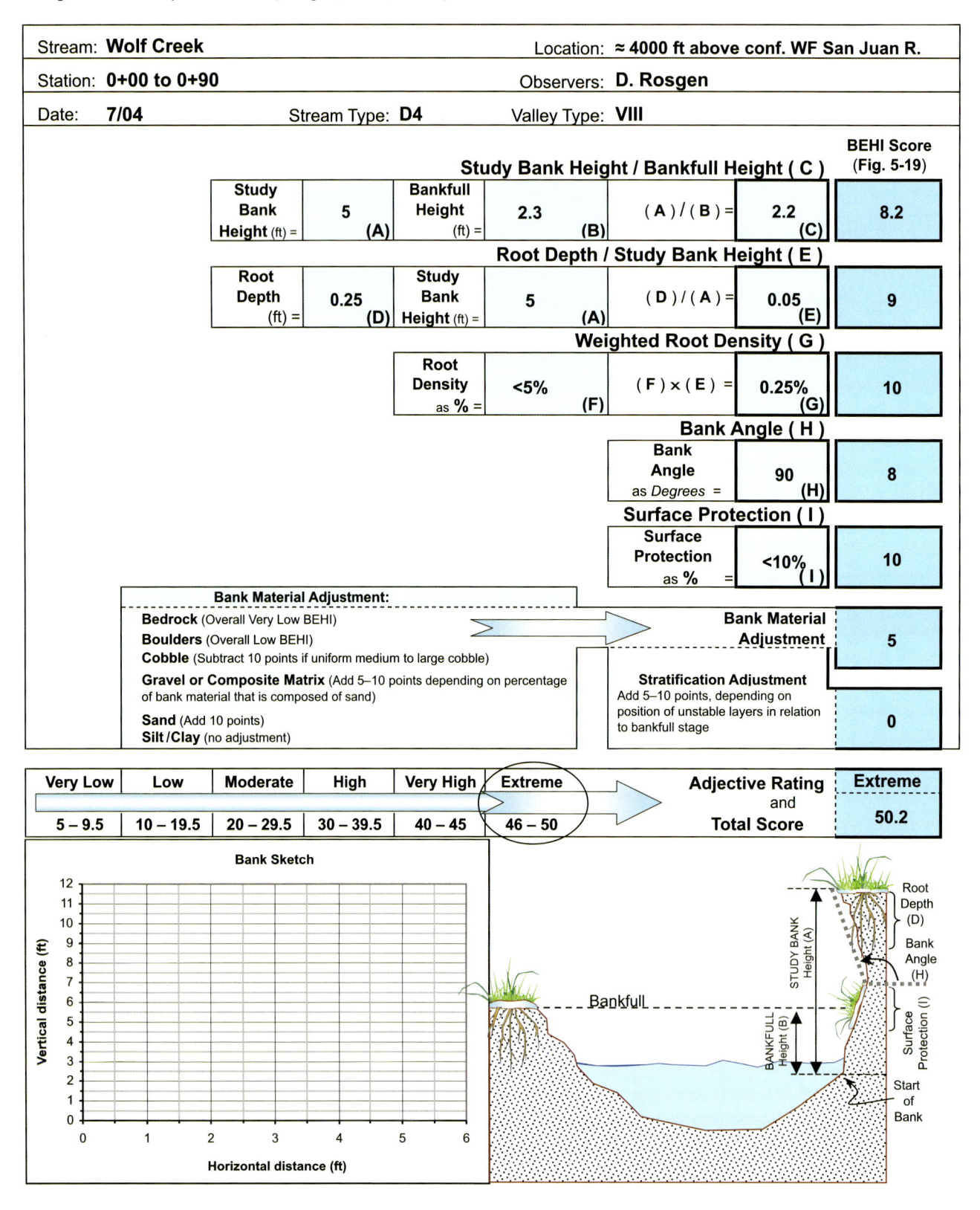

Stream: **Wolf Creek**	Location: ≈ **4000 ft above conf. WF San Juan R.**
Station: **0+00 to 0+90**	Observers: **D. Rosgen**
Date: **7/04** Stream Type: **D4**	Valley Type: **VIII**

Study Bank Height / Bankfull Height (C)

| Study Bank Height (ft) = | 5 (A) | Bankfull Height (ft) = | 2.3 (B) | (A)/(B) = | 2.2 (C) | BEHI Score (Fig. 5-19): **8.2** |

Root Depth / Study Bank Height (E)

| Root Depth (ft) = | 0.25 (D) | Study Bank Height (ft) = | 5 (A) | (D)/(A) = | 0.05 (E) | **9** |

Weighted Root Density (G)

| Root Density as % = | <5% (F) | (F) × (E) = | 0.25% (G) | **10** |

Bank Angle (H)

| Bank Angle as Degrees = | 90 (H) | **8** |

Surface Protection (I)

| Surface Protection as % = | <10% (I) | **10** |

Bank Material Adjustment:
- **Bedrock** (Overall Very Low BEHI)
- **Boulders** (Overall Low BEHI)
- **Cobble** (Subtract 10 points if uniform medium to large cobble)
- **Gravel or Composite Matrix** (Add 5–10 points depending on percentage of bank material that is composed of sand)
- **Sand** (Add 10 points)
- **Silt/Clay** (no adjustment)

Bank Material Adjustment: **5**

Stratification Adjustment Add 5–10 points, depending on position of unstable layers in relation to bankfull stage: **0**

Very Low	Low	Moderate	High	Very High	Extreme
5 – 9.5	10 – 19.5	20 – 29.5	30 – 39.5	40 – 45	46 – 50

Adjective Rating and Total Score: **Extreme 50.2**

Chapter 7 — PLA The WARSSS Methodology: A Case Study of Wolf Creek, Colorado

Worksheet 7-21a (Worksheet 5-9). Various field methods of estimating Near-Bank Stress (NBS) risk ratings to calculate erosion rate for the C4 reference reach.

Estimating Near-Bank Stress (NBS)

Stream: **Wolf Creek** Location: ≈ 5000 ft above conf. WF San Juan R.
Station: **9+80 to 11+50** Stream Type: **C4** Valley Type: **VIII**
Observers: **D. Rosgen** Date: **7/04**

Methods for estimating Near-Bank Stress (NBS)

(1)	Channel pattern, transverse bar or split channel/central bar creating NBS............	Level I	Reconaissance
(2)	Ratio of radius of curvature to bankfull width (R_c / W_{bkf})........................	Level II	General prediction
(3)	Ratio of pool slope to average water surface slope (S_p / S)........................	Level II	General prediction
(4)	Ratio of pool slope to riffle slope (S_p / S_{rif}).............................	Level II	General prediction
(5)	Ratio of near-bank maximum depth to bankfull mean depth (d_{nb} / d_{bkf}).........	Level III	Detailed prediction
(6)	Ratio of near-bank shear stress to bankfull shear stress (τ_{nb} / τ_{bkf})............	Level III	Detailed prediction
(7)	Velocity profiles / Isovels / Velocity gradient..............................	Level IV	Validation

Level I (1):
- Transverse and/or central bars-short and/or discontinuous........................... NBS = High / Very High
- Extensive deposition (continuous, cross-channel)................................ NBS = Extreme
- Chute cutoffs, down-valley meander migration, converging flow..................... NBS = Extreme

Level II

(2)
Radius of Curvature R_c (ft)	Bankfull Width W_{bkf} (ft)	Ratio R_c / W_{bkf}	Near-Bank Stress (NBS)
70	32	2.2	Moderate

(3)
Pool Slope S_p	Average Slope S	Ratio S_p / S	Near-Bank Stress (NBS)

Dominant Near-Bank Stress: Moderate

(4)
Pool Slope S_p	Riffle Slope S_{rif}	Ratio S_p / S_{rif}	Near-Bank Stress (NBS)

Level III

(5)
Near-Bank Max Depth d_{nb} (ft)	Mean Depth d_{bkf} (ft)	Ratio d_{nb} / d_{bkf}	Near-Bank Stress (NBS)
2.5	1.4	1.79	Moderate

(6)
Near-Bank Max Depth d_{nb} (ft)	Near-Bank Slope S_{nb}	Near-Bank Shear Stress τ_{nb} (lb/ft²)	Mean Depth d_{bkf} (ft)	Average Slope S	Bankfull Shear Stress τ_{bkf} (lb/ft²)	Ratio τ_{nb} / τ_{bkf}	Near-Bank Stress (NBS)

Level IV

(7)
Velocity Gradient (ft / sec / ft)	Near-Bank Stress (NBS)

Converting values to a Near-Bank Stress (NBS) rating

Near-Bank Stress (NBS) ratings	Method number						
	(1)	(2)	(3)	(4)	(5)	(6)	(7)
Very Low	N/A	> 3.00	< 0.20	< 0.40	< 1.00	< 0.80	< 0.50
Low	N/A	2.21 – 3.00	0.20 – 0.40	0.41 – 0.60	1.00 – 1.50	0.80 – 1.05	0.50 – 1.00
Moderate	N/A	(2.01 – 2.20)	0.41 – 0.60	0.61 – 0.80	(1.51 – 1.80)	1.06 – 1.14	1.01 – 1.60
High	See	1.81 – 2.00	0.61 – 0.80	0.81 – 1.00	1.81 – 2.50	1.15 – 1.19	1.61 – 2.00
Very High	(1)	1.50 – 1.80	0.81 – 1.00	1.01 – 1.20	2.51 – 3.00	1.20 – 1.60	2.01 – 2.40
Extreme	Above	< 1.50	> 1.00	> 1.20	> 3.00	> 1.60	> 2.40

Overall Near-Bank Stress (NBS) rating: Moderate

Worksheet 7-21b (Worksheet 5-9). Various field methods of estimating Near-Bank Stress (NBS) risk ratings to calculate erosion rate for the D4 impaired reach.

Estimating Near-Bank Stress (NBS)

Stream: **Wolf Creek**	Location: **≈ 4000 ft above conf. WF San Juan R.**	
Station: **0+00 to 0+90**	Stream Type: **D4**	Valley Type: **VIII**
Observers: **D. Rosgen**		Date: **7/04**

Methods for estimating Near-Bank Stress (NBS)

(1)	Channel pattern, transverse bar or split channel/central bar creating NBS.................	Level I	Reconnaissance
(2)	Ratio of radius of curvature to bankfull width (R_c / W_{bkf})................................	Level II	General prediction
(3)	Ratio of pool slope to average water surface slope (S_p / S)...........................	Level II	General prediction
(4)	Ratio of pool slope to riffle slope (S_p / S_{rif})...	Level II	General prediction
(5)	Ratio of near-bank maximum depth to bankfull mean depth (d_{nb} / d_{bkf})...........	Level III	Detailed prediction
(6)	Ratio of near-bank shear stress to bankfull shear stress (τ_{nb} / τ_{bkf}).............	Level III	Detailed prediction
(7)	Velocity profiles / Isovels / Velocity gradient..	Level IV	Validation

Level I (1): Transverse and/or central bars-short and/or discontinuous..........................NBS = High (**Very High**)
Extensive deposition (continuous, cross-channel)..NBS = Extreme
Chute cutoffs, down-valley meander migration, converging flow..........................NBS = Extreme

Level II

(2)
Radius of Curvature R_c (ft)	Bankfull Width W_{bkf} (ft)	Ratio R_c / W_{bkf}	Near-Bank Stress (NBS)

(3)
Pool Slope S_p	Average Slope S	Ratio S_p / S	Near-Bank Stress (NBS)

Dominant Near-Bank Stress: Very High

(4)
Pool Slope S_p	Riffle Slope S_{rif}	Ratio S_p / S_{rif}	Near-Bank Stress (NBS)

Level III

(5)
Near-Bank Max Depth d_{nb} (ft)	Mean Depth d_{bkf} (ft)	Ratio d_{nb} / d_{bkf}	Near-Bank Stress (NBS)
1.7	0.9	1.9	High

(6)
Near-Bank Max Depth d_{nb} (ft)	Near-Bank Slope S_{nb}	Near-Bank Shear Stress τ_{nb} (lb/ft²)	Mean Depth d_{bkf} (ft)	Average Slope S	Bankfull Shear Stress τ_{bkf} (lb/ft²)	Ratio τ_{nb} / τ_{bkf}	Near-Bank Stress (NBS)

Level IV

(7)
Velocity Gradient (ft / sec / ft)	Near-Bank Stress (NBS)

Converting values to a Near-Bank Stress (NBS) rating

Near-Bank Stress (NBS) ratings	Method number						
	(1)	(2)	(3)	(4)	(5)	(6)	(7)
Very Low	N/A	> 3.00	< 0.20	< 0.40	< 1.00	< 0.80	< 0.50
Low	N/A	2.21 – 3.00	0.20 – 0.40	0.41 – 0.60	1.00 – 1.50	0.80 – 1.05	0.50 – 1.00
Moderate	N/A	2.01 – 2.20	0.41 – 0.60	0.61 – 0.80	1.51 – 1.80	1.06 – 1.14	1.01 – 1.60
High	See	1.81 – 2.00	0.61 – 0.80	0.81 – 1.00	(**1.81 – 2.50**)	1.15 – 1.19	1.61 – 2.00
Very High	(**V.High**)	1.50 – 1.80	0.81 – 1.00	1.01 – 1.20	2.51 – 3.00	1.20 – 1.60	2.01 – 2.40
Extreme	Above	< 1.50	> 1.00	> 1.20	> 3.00	> 1.60	> 2.40

Overall Near-Bank Stress (NBS) rating: Very High

Step 9. Predict annual streambank erosion rate using BEHI and NBS ratings

The combined BEHI and NBS ratings determined in **Step 8** are used in this step to predict annual streambank erosion rates using the Yellowstone curve (**Figure 7-44**). This curve was used because the Wolf Creek Watershed is associated with volcanism and alpine glaciation similar to the geology of the empirically derived relations from Yellowstone National Park.

Worksheets 7-22a and **7-22b** were used to predict streambank erosion rates for the C4 and D4 reaches. The BEHI and NBS ratings were completed using the ratings calculated in **Worksheets 7-20a, 7-20b, 7-21a** and **7-21b** and by using the field calibration procedures discussed in Chapter 5, **Steps 8** and **9**. Study bank heights and lengths were also measured for all stations to determine lateral erosion rates. The final streambank erosion prediction rates are expressed in tons/yr per foot of lineal channel distance for a given BEHI and NBS rating.

The streambank erosion rates from the C4 reference reach were estimated at 5.5 tons/yr, or 0.006 tons/yr/ft for approximately 910 ft of stream length (**Worksheet 7-22a**). This is a natural stable streambank erosion rate for this stream and soil type. In contrast, the bank erosion prediction rate of the D4 impacted reach for 970 ft of assessed length generated approximately 429 tons/yr, or 0.44 tons/yr/ft (**Worksheet 7-22b**). It is important to note that not all of the eroded bank material will be transported out of the watershed as sediment. Much of the bank material may be stored in depositional features of the Wolf Creek channel.

For model validation purposes, a comparison of measured annual erosion rates on the D4 impaired reach is shown in **Figure 7-45**, where the average annual lateral erosion rate measured 1.6 ft. The BEHI and NBS ratings for this reach are both Very High. The predicted value using **Figure 7-44** (Yellowstone curve) is 1.7 ft, a very close agreement to the observed data from Wolf Creek. The predicted range of lateral erosion rate values for the same stream type and similar bank conditions at the D4 impaired reach, shown in **Worksheet 7-22b**, varied from 1.2 to 2.3 ft/yr with a weighted average annual rate of 1.87 ft/yr. The observed values fall well within the range of predicted values, indicating that the model is valid for this location. Additional independent data comparisons of predicted-to-observed bank erosion values used to test the bank erosion rates for rivers in Southwestern Colorado are summarized in Rosgen (2001a).

The upper Wolf Creek Watershed lateral erosion rates were also calculated using the field calibration methods and by measuring study bank heights and lengths. The streambank erosion rates for the existing conditions of the upper Wolf Creek Watershed represent 196 tons/yr. The streambank erosion rate for the D4 stream type as a comparison is 429 tons/yr. It is obvious that there is a short distance of reach that makes a significant disproportionate contribution of sediment. This streambank erosion information is used to compare various sources of sediment supply contributing to total annual sediment yield in **Steps 29–31**.

The resultant streambank erosion rates by stream type and reach location are shown in **Figure 7-46**. The color codes on this map indicate the streambank erosion rate in tons/yr per foot of channel length. This map is used to prioritize the highest sediment contribution in tons/yr/ft for site-specific mitigation, stabilization and/or restoration recommendations. This map also shows cumulative contributions of sediment by process and by location. Any mitigation design and priority can directly relate to sediment source reduction.

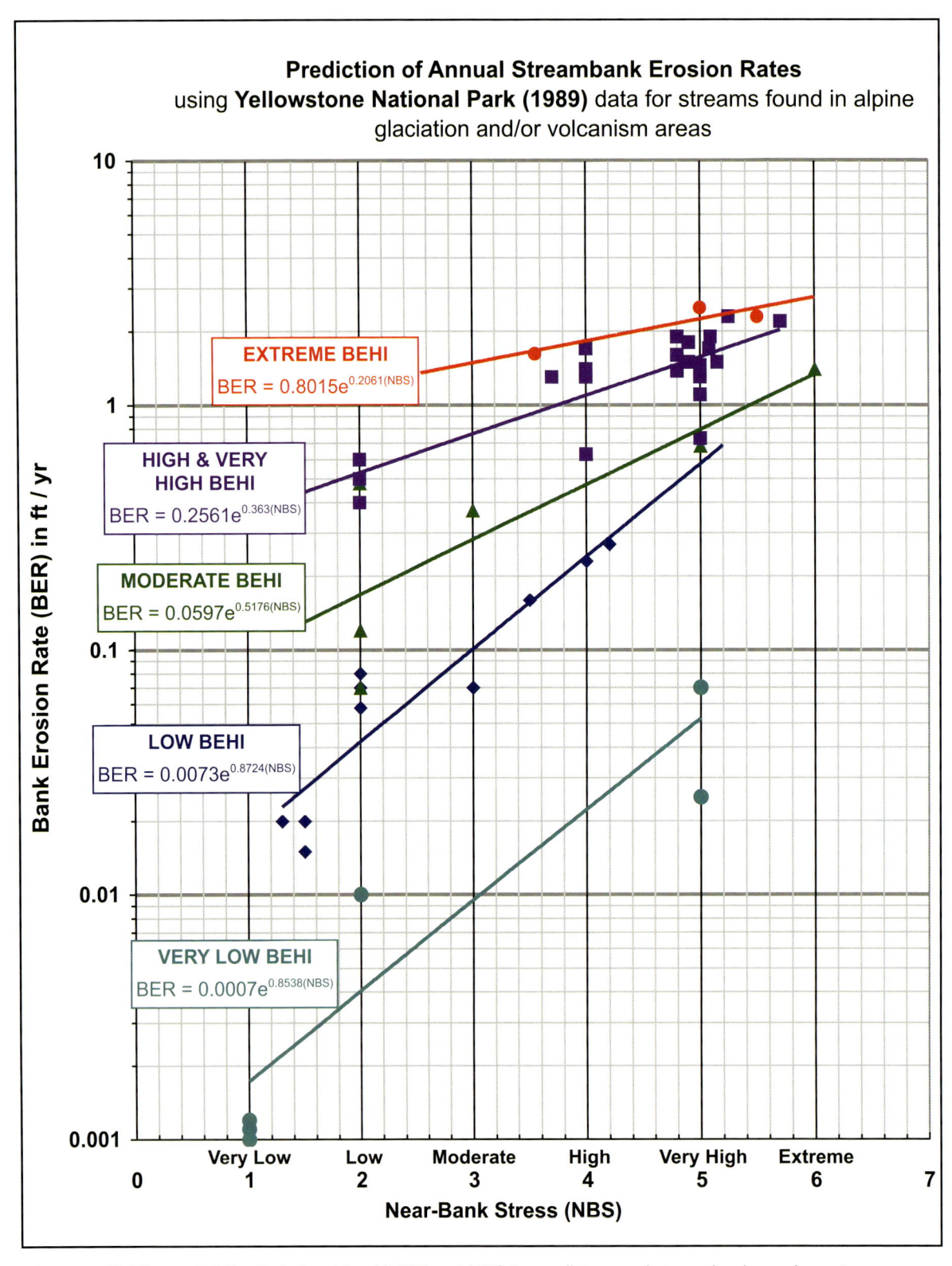

Figure 7-44 (Figure 5-39). Relationship of BEHI and NBS to predict annual streambank erosion rates, Yellowstone National Park data, 1989 (Rosgen, 1996, 2001a).

Worksheet 7-22a (Worksheet 5-10). Summary form of annual streambank erosion estimates for the C4 reference study reach.

Stream:	Wolf Creek			Location: above conf. WF San Juan		
Graph Used:	Yellowstone	Stream Type: C4		Total Bank Length (ft): 910'		
Observers:	D. Rosgen		Valley Type: VIII		Date: 7/04	

	(1) Station (ft)	(2) BEHI rating (Worksheet 5-8) (adjective)	(3) NBS rating (Worksheet 5-9) (adjective)	(4) Bank erosion rate (Figure 5-38 or 5-39) (ft/yr)	(5) Length of bank (ft)	(6) Study bank height (ft)	(7) Erosion subtotal [(4)X(5)X(6)] (ft³/yr)
1.	9+80 to 11+50	Low	Mod	0.09	170	3.3	50.5
2.	11+50 to 11+90	Very Low	Low	0.009	40	3.3	1.2
3.	11+90 to 12+20	Low	Very Low	0.02	30	3.3	2
4.	12+20 to 13+50	Very Low	Mod	0.005	130	3.3	2.1
5.	13+50 to 15+80	Low	Low	0.05	230	3.3	38
6.	15+80 to 18+90	Low	Very Low	0.02	310	3.3	20.5
7.							
8.							
9.							
10.							
11.							
12.							
13.							
14.							
15.							

Sum erosion subtotals in Column (7) for each BEHI/NBS combination	Total erosion (ft³/yr)	114
Convert erosion in ft³/yr to yds³/yr {divide Total erosion (ft³/yr) by 27}	Total erosion (yds³/yr)	4.22
Convert erosion in yds³/yr to tons/yr {multiply Total erosion (yds³/yr) by 1.3}	Total erosion (tons/yr)	5.5
Calculate erosion per unit length of channel {divide Total erosion (tons/yr) by total length of stream (ft) surveyed}	Total erosion (tons/yr/ft)	0.006

Worksheet 7-22b (Worksheet 5-10). Summary form of annual streambank erosion estimates for the D4 impaired study reach.

Stream:	Wolf Creek			Location: above conf. WF San Juan		
Graph Used:	Yellowstone	Stream Type: D4		Total Bank Length (ft): 970'		
Observers:	D. Rosgen		Valley Type: VIII	Date: 7/04		
(1)	(2)	(3)	(4)	(5)	(6)	(7)
Station (ft)	BEHI rating (Worksheet 5-8) (adjective)	NBS rating (Worksheet 5-9) (adjective)	Bank erosion rate (Figure 5-38 or 5-39) (ft/yr)	Length of bank (ft)	Study bank height (ft)	Erosion subtotal [(4)X(5)X(6)] (ft^3/yr)
1. 0+00 to 0+90	Extreme	Very High	2.3	90	5	1,035
2. 0+90 to 1+45	Very High	High	1.2	55	5	330
3. 1+45 to 3+80	Very High	Very High	1.7	235	5	1,997.5
4. 3+80 to 4+20	Extreme	Very High	2.3	40	5	460
5. 4+20 to 7+45	Extreme	Very High	2.3	325	5	3,737.5
6. 7+45 to 9+70	Very High	High	1.2	225	5	1,350
7.						
8.						
9.						
10.						
11.						
12.						
13.						
14.						
15.						
Sum erosion subtotals in Column (7) for each BEHI/NBS combination					Total erosion (ft^3/yr)	8,910
Convert erosion in ft^3/yr to yds^3/yr {divide Total erosion (ft^3/yr) by 27}					Total erosion (yds^3/yr)	330
Convert erosion in yds^3/yr to tons/yr {multiply Total erosion (yds^3/yr) by 1.3}					Total erosion (tons/yr)	429
Calculate erosion per unit length of channel {divide Total erosion (tons/yr) by total length of stream (ft) surveyed}					Total erosion (tons/yr/ft)	0.44

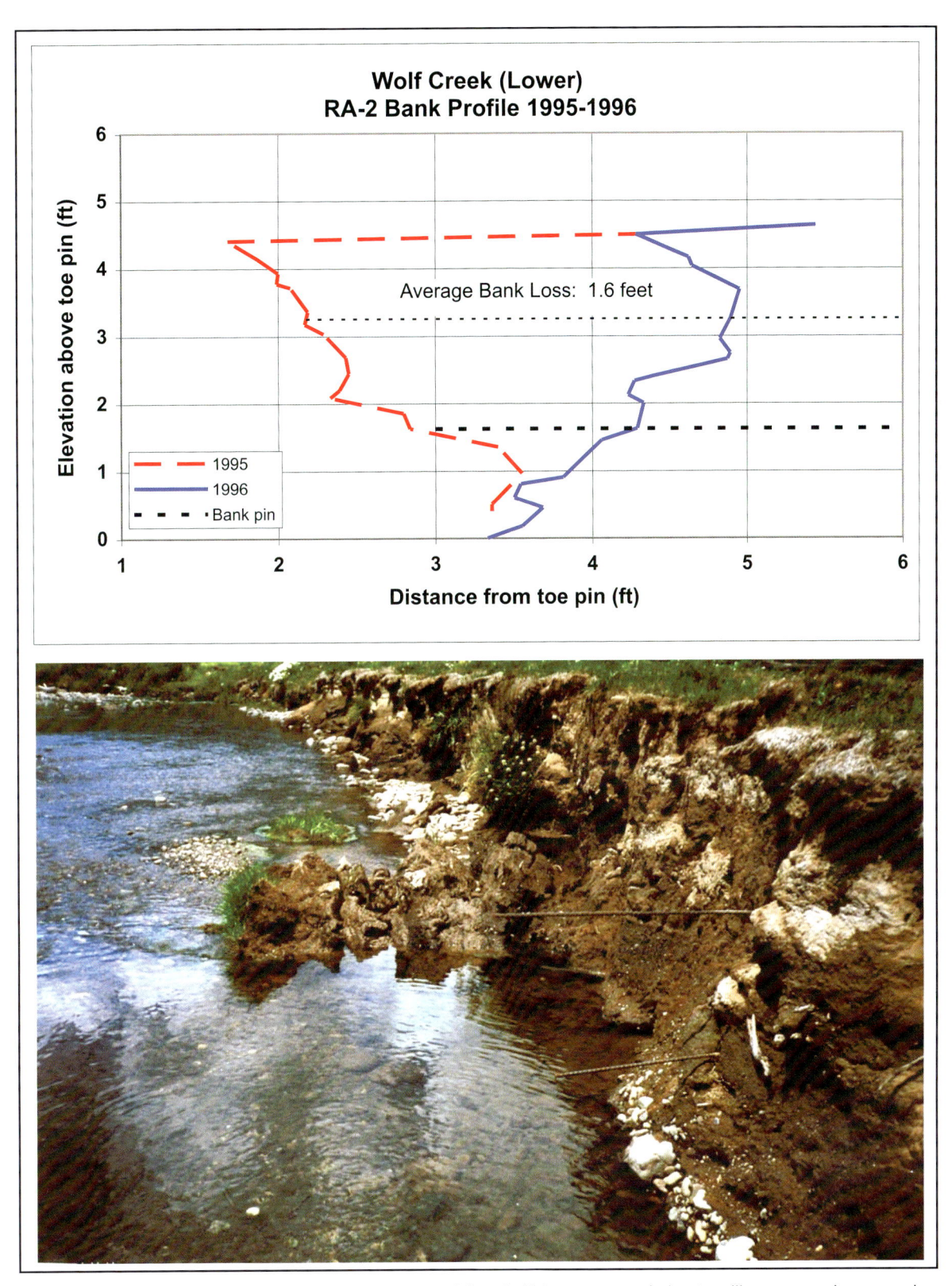

Chapter 7 — PLA   The WARSSS Methodology: A Case Study of Wolf Creek, Colorado

Figure 7-45. Streambank erosion rate on lower Wolf Creek (D4 stream type) due to willow removal; exposed pins represent one year of lateral erosion of 1.6 ft. The BEI II and NBS ratings are both Very High.

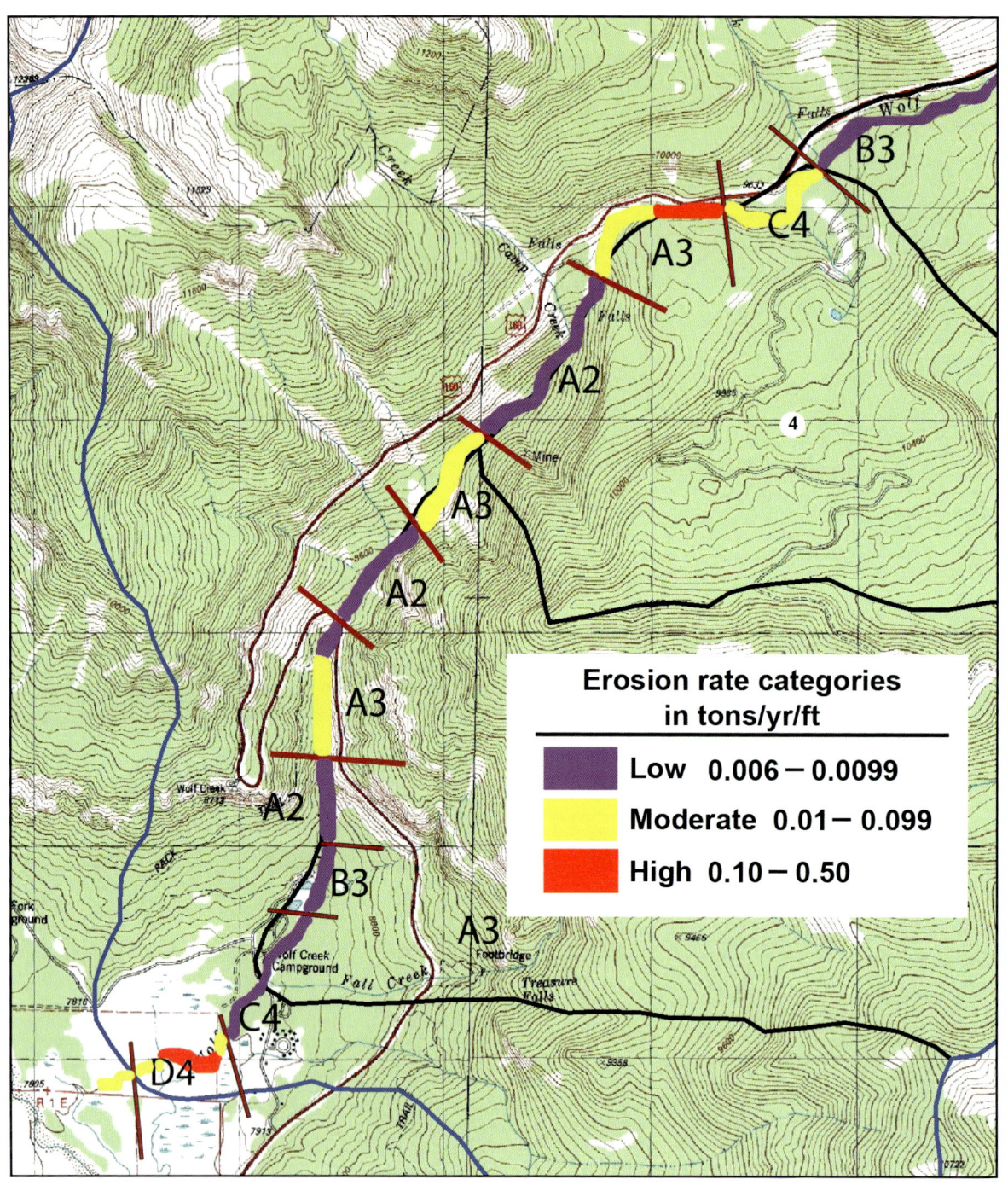

Figure 7-46. Streambank erosion rates by stream reach location and stream type.

Steps 10–17: Total annual sediment yield prediction and flow-related increases in sediment using water yield and FLOWSED models (See Flowchart 7-8)

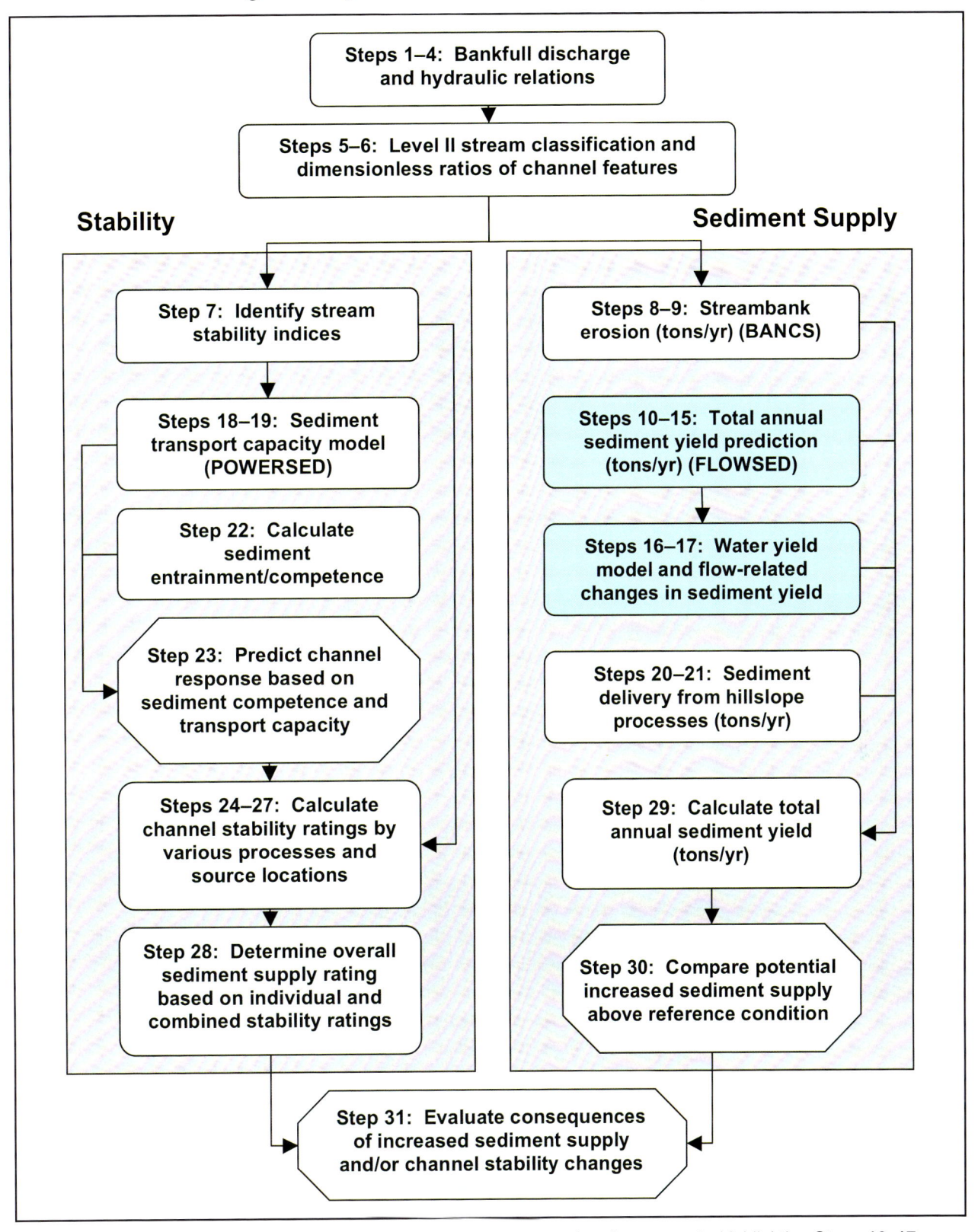

Stability | **Sediment Supply**

- Steps 1–4: Bankfull discharge and hydraulic relations
- Steps 5–6: Level II stream classification and dimensionless ratios of channel features

Stability:
- Step 7: Identify stream stability indices
- Steps 18–19: Sediment transport capacity model (POWERSED)
- Step 22: Calculate sediment entrainment/competence
- Step 23: Predict channel response based on sediment competence and transport capacity
- Steps 24–27: Calculate channel stability ratings by various processes and source locations
- Step 28: Determine overall sediment supply rating based on individual and combined stability ratings

Sediment Supply:
- Steps 8–9: Streambank erosion (tons/yr) (BANCS)
- Steps 10–15: Total annual sediment yield prediction (tons/yr) (FLOWSED)
- Steps 16–17: Water yield model and flow-related changes in sediment yield
- Steps 20–21: Sediment delivery from hillslope processes (tons/yr)
- Step 29: Calculate total annual sediment yield (tons/yr)
- Step 30: Compare potential increased sediment supply above reference condition

- Step 31: Evaluate consequences of increased sediment supply and/or channel stability changes

Flowchart 7-3f. The sequential steps to determine stream stability and sediment supply, highlighting **Steps 10–17**.

WATERSHED ASSESSMENT OF RIVER STABILITY AND SEDIMENT SUPPLY

Steps 11–17: Calculate flow-related changes in annual sediment yield using water yield and FLOWSED models

Step 16: Water yield model

Calculate sediment yield using FLOWSED Steps 11–15
- Pre-treatment condition
- Post-treatment condition

Determine pre-treatment bankfull discharge (Q_{bkf}) from water yield model

Step 12: Obtain or establish dimensionless sediment rating curves by stream type/stability

Step 11: Collect field data by stream type/valley type
- Bankfull discharge (Q_{bkf})
- Bankfull bedload sediment (kg/s)
- Bankfull suspended sediment (mg/l)

Step 14 (modified): Flow-duration curves from water yield model
- Pre-treatment condition
- Post-treatment condition

Step 13: Convert dimensionless sediment rating curves to dimensioned curves
- Pre-treatment condition
- Post-treatment condition

Step 15: Calculate total annual bedload and suspended sediment yield
- Pre-treatment condition
- Post-treatment condition

Step 17: Calculate flow-related changes in annual sediment yield

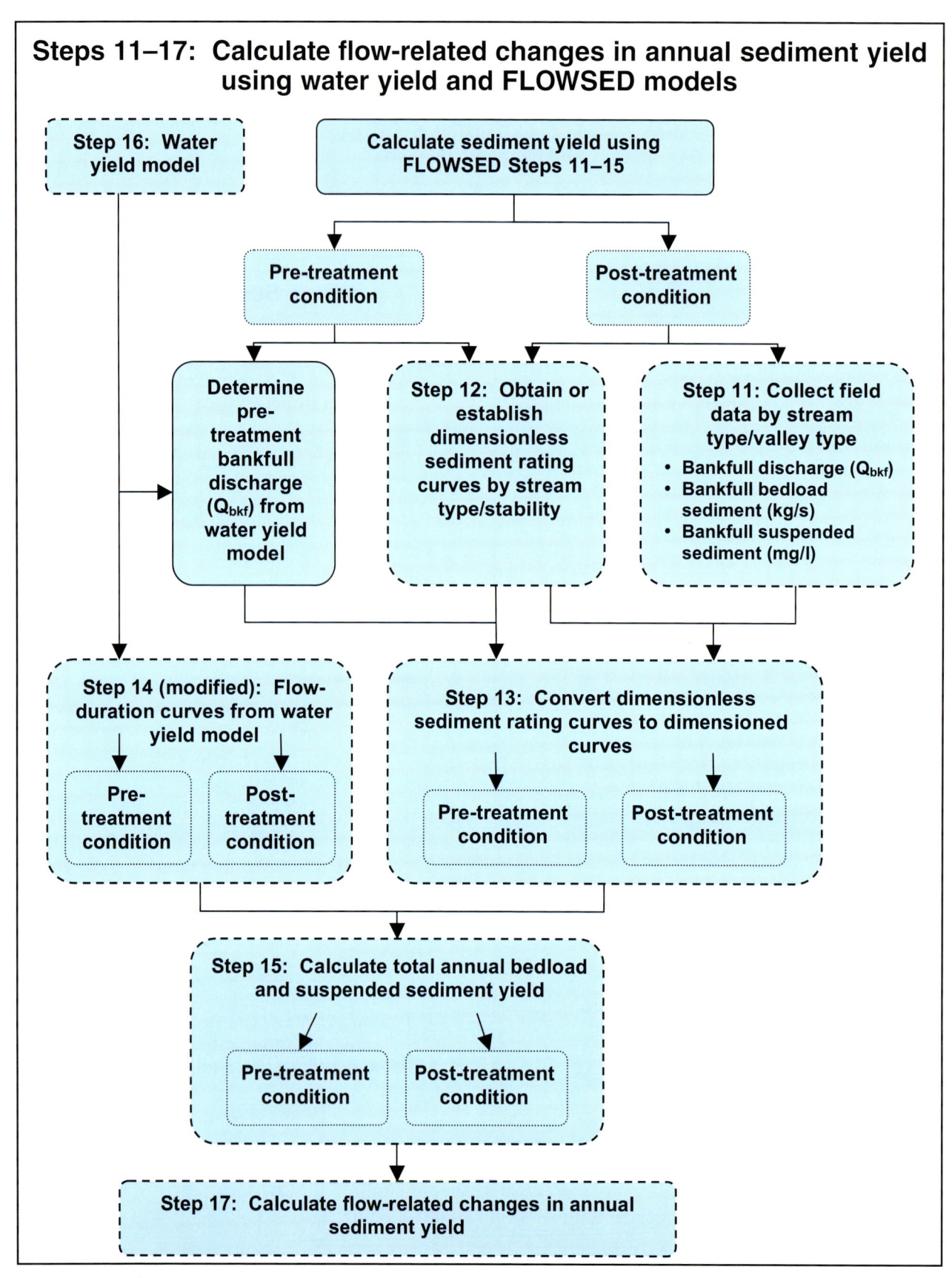

Flowchart 7-8 (Flowchart 5-12). Procedure to calculate total annual sediment yield increase by comparing pre-treatment with post-treatment water yield conditions (combining water yield and FLOWSED models).

Total annual sediment yield prediction and flow-related sediment yield increases using water yield and FLOWSED models: Steps 10–17

The objectives for the Wolf Creek Watershed were to calculate total annual sediment yield and flow-related increases in sediment yield. Therefore, the FLOWSED model was used in conjunction with a water yield model to determine flow-related changes in sediment yield. The procedure involves using pre- and post-treatment flow-duration curves from the water yield model to input directly into the FLOWSED model in **Step 14**, rather than developing dimensionless flow-duration curves from a gaging station (**Step 10**). The procedure combining the water yield and FLOWSED models is outlined in **Flowchart 7-8** and is discussed in detail in Chapter 5, **Steps 10–17**.

Both Wolf Creek above the lower C4 reference stream type and lower Wolf Creek below the D4 impaired stream type were analyzed using the FLOWSED and water yield models. Pre- and post-treatment flow-duration curves were developed for both the C4 reference and D4 impaired conditions.

Water yield model (Step 16)

A snowmelt water yield model modified by Troendle and Swanson (in press) based on the water yield model in WRENSS (USEPA, 1980) was used for the Wolf Creek case study. The use of the updated WRENSS model in conjunction with the FLOWSED model is shown in Appendix A. The Wolf Creek Watershed was delineated into five sub-watersheds (**Figures 7-3** and **7-28**). Aerial photographs showed extensive timber harvest clearcuts and roads in each sub-watershed (**Figure 7-2**). The drainage area simulated was 13,016 acres (20 mi^2). Timber harvest data was obtained from the San Juan National Forest. The treated stands were stratified by aspect, elevation and basal area reduction. The year of harvest was also recorded, and the appropriate evapo-transpiration and stand interception height recovery curve was applied. Road location and acres of cut bank, road surface and fill slope acres were used in the model as clearcuts with no hydrologic recovery. The roads included forest access roads and State Highway 160. Road sediment source prediction using the road impact index, however, separated forest access roads from State Highway 160. This separation was necessary because the sediment recovery is substantial on the forest access roads compared to State Highway 160 (**Steps 20** and **21**). According to the modified WRENSS snowmelt model, the potential increase in streamflow is approximately 2,778 acre ft for both the C4 reference and D4 impaired conditions. This represents an increase in 7% of mean annual discharge for Wolf Creek. The corresponding increase in water yield output from this snowmelt model was presented in pre- and post-treatment flow-duration curves. These curves were used directly in the FLOWSED model in **Step 14**.

The various pre- and post-treatment scenarios to predict total annual sediment yield for the C4 reference and D4 impaired conditions are depicted in **Flowchart 7-9**. **Worksheets 7-23a** through **7-23d** display the results of the water yield and FLOWSED models. Details of the individual steps to calculate total annual sediment yield and flow-related increases in sediment are discussed.

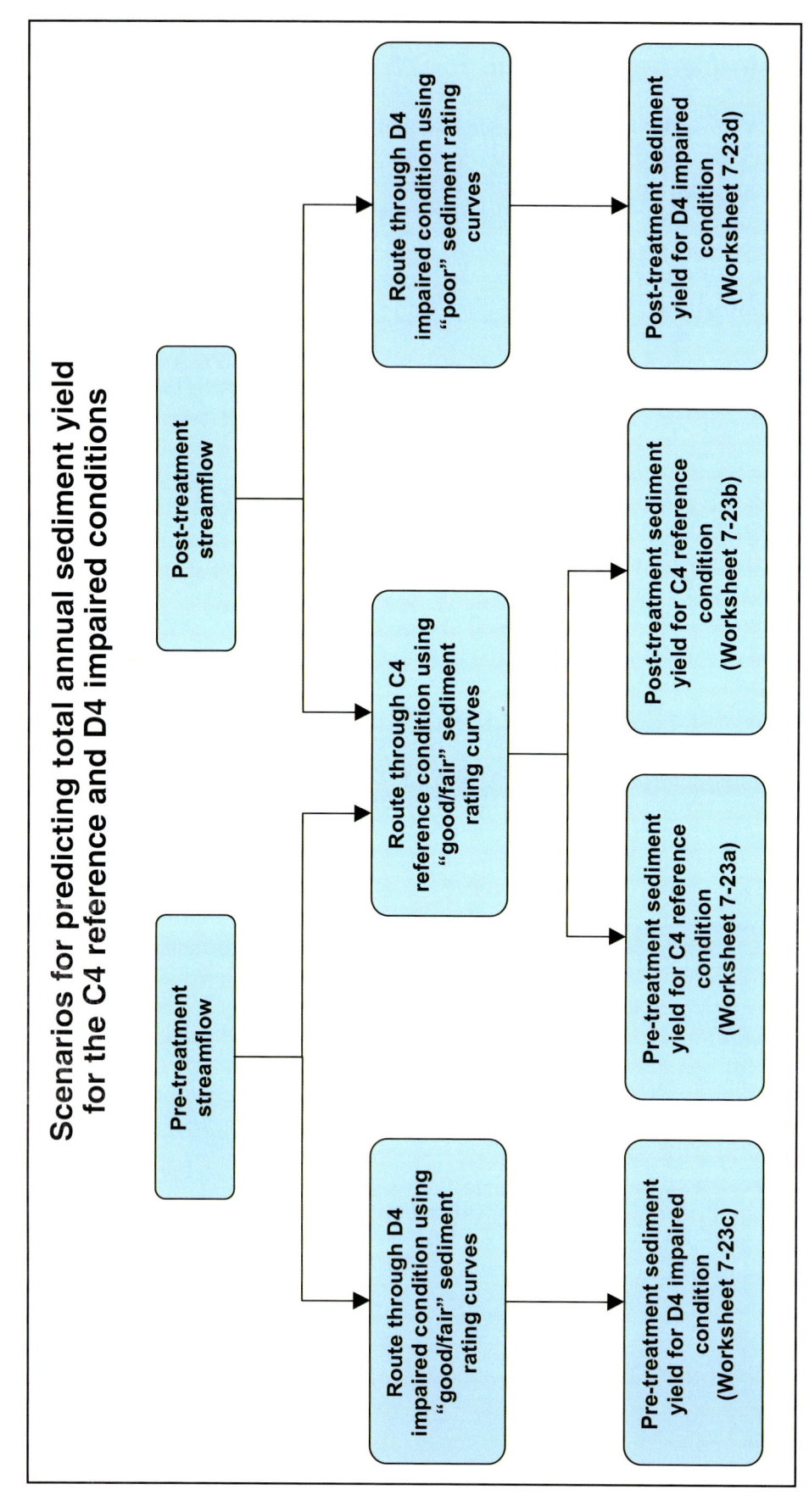

Flowchart 7-9. Various pre- and post-treatment scenarios for the C4 reference and D4 impaired conditions to predict total annual sediment yield.

Worksheet 7-23a (Worksheet 5-11). Pre-treatment total annual sediment yield for the C4 reference condition.

Stream: Wolf Creek C4 reference condition (pre-treatment) **Location:** ≈ 5000 ft above confluence of WF San Juan River **Date:** 7/04
Observers: D. Rosgen **Gage Station #:** 09341300 **Stream Type:** C4 **Valley Type:** VIII

Equation type	Intercept	Coefficient	Exponent	Form (e.g., linear, non-linear, etc.)	Equation name	Bankfull discharge (cfs)	Bankfull bedload (kg/s)	Bankfull suspended (mg/l)
1. Bedload (dimensionless)	-0.0113	1.0139	2.1929	nonlinear	"good/fair" Pagosa	259	0.093	27.3
2. Suspended sediment (dimensionless)	0.0636	0.9326	2.4085	nonlinear	"good/fair" Pagosa			
3. User-defined relations (bedload)								
4. User-defined relations (suspended sediment)								

Notes: Pre-treatment flow using 1985–1987 flow-duration curve from WRENSS output

			From dimensioned flow-duration curve				From sediment rating curves				Calculate		Calculate sediment yield	
(1)	(2)	(3)	(4)	(5)	(6)	(7)	(8)	(9)	(10)	(11)	(12)	(13)	(14)	(15)
Flow exceedence	Daily mean discharge	Mid-ordinate	Time increment (percent)	Time increment (days)	Mid-ordinate streamflow	Dimensionless streamflow	Dimensionless suspended sediment discharge	Suspended sediment discharge	Dimensionless bedload discharge	Bedload	Time adjusted streamflow	Suspended sediment [(5)×(9)]	Bedload sediment [(5)×(11)]	Suspended + bedload [(13)+(14)]
(%)	(cfs)	(%)	(%)	(days)	(cfs)	(Q/Q_{bkf})	(S/S_{bkf})	(tons/day)	(b_s/b_{bkf})	(tons/day)	(cfs)	(tons)	(tons)	(tons)
0%	368.7													
1%	285.4	0.5%	1%	3.65	327.0	1.26	1.70	40.80	1.67	14.86	3.27	148.92	54.23	203.15
2%	265.0	0.5%	1%	3.65	275.2	1.06	1.10	23.10	1.14	10.14	2.75	84.32	37.02	121.33
3%	250.1	0.5%	1%	3.65	257.6	0.99	1.00	18.60	0.99	8.76	2.58	67.89	31.97	99.86
4%	242.7	0.5%	1%	3.65	246.4	0.95	0.90	16.10	0.89	7.94	2.46	58.77	28.99	87.76
5%	237.2	0.5%	1%	3.65	240.0	0.93	0.80	14.80	0.84	7.49	2.40	54.02	27.32	81.34
10%	172.7	2.5%	5%	18.25	204.9	0.79	0.60	8.90	0.59	5.27	10.25	162.43	96.10	258.53
20%	77.7	5%	10%	36.50	125.2	0.48	0.20	2.10	0.19	1.72	12.52	76.65	62.81	139.46
30%	36.7	5%	10%	36.50	57.2	0.22	0.10	0.40	0.03	0.23	5.72	14.60	8.27	22.87
40%	23.7	5%	10%	36.50	30.2	0.12	0.10	0.20	0.00	0.00	3.02	7.30	0.00	7.30
50%	16.5	5%	10%	36.50	20.1	0.08	0.10	0.10	0.00	0.00	2.01	3.65	0.00	3.65
60%	10.6	5%	10%	36.50	13.6	0.05	0.10	0.10	0.00	0.00	1.36	3.65	0.00	3.65
70%	8.1	5%	10%	36.50	9.4	0.04	0.10	0.00	0.00	0.00	0.94	0.00	0.00	0.00
80%	6.0	5%	10%	36.50	7.1	0.03	0.10	0.00	0.00	0.00	0.71	0.00	0.00	0.00
90%	4.9	5%	10%	36.50	5.5	0.02	0.10	0.00	0.00	0.00	0.55	0.00	0.00	0.00
100%	3.1	5%	10%	36.50	4.0	0.02	0.10	0.00	0.00	0.00	0.40	0.00	0.00	0.00
											Annual totals:	682 (tons/yr)	347 (tons/yr)	1029 (tons/yr)

WATERSHED ASSESSMENT OF RIVER STABILITY AND SEDIMENT SUPPLY

Worksheet 7-23b (Worksheet 5-11). Post-treatment total annual sediment yield for the C4 reference condition.

Stream: **Wolf Creek C4 reference condition (post-treatment)** Location: **≈ 5000 ft above confluence of WF San Juan River** Date: **7/04**

Observers: **D. Rosgen** Gage Station #: **09341300** Stream Type: **C4** Valley Type: **VIII**

Equation type	Intercept	Coefficient	Exponent	Form (e.g., linear, non-linear, etc.)	Equation name	Bankfull discharge (cfs)	Bankfull bedload (kg/s)	Bankfull suspended (mg/l)
1. Bedload (dimensionless)	−0.0113	1.0139	2.1929	nonlinear	"good/fair" Pagosa	280	0.104	31.3
2. Suspended sediment (dimensionless)	0.0636	0.9326	2.4085	nonlinear	"good/fair" Pagosa			
3. User-defined relations (bedload)								
4. User-defined relations (suspended sediment)								

Notes: Post-treatment flow using 1985–1987 flow-duration curve from WRENSS output

(1)	(2)	(3)	(4)	(5)	(6)	(7)	(8)	(9)	(10)	(11)	(12)	(13)	(14)	(15)
Flow exceedence	Daily mean discharge	Mid-ordinate	Time increment (percent)	Time increment (days)	Mid-ordinate streamflow	Dimensionless streamflow	Dimensionless suspended sediment discharge	Suspended sediment discharge	Dimension-less bedload discharge	Bedload	Time adjusted streamflow	Suspended sediment [(5)×(9)]	Bedload sediment [(5)×(11)]	Suspended + bedload [(13)+(14)]
(%)	(cfs)	(%)	(%)	(days)	(cfs)	(Q/Q_{bkf})	(S/S_{bkf})	(tons/day)	(b_s/b_{bkf})	(tons/day)	(cfs)	(tons)	(tons)	(tons)
0%	398.0													
1%	308.0	0.5%	1%	3.65	353.0	1.26	1.70	50.40	1.67	16.50	3.53	183.96	60.22	244.18
2%	286.0	0.5%	1%	3.65	297.0	1.06	1.10	28.50	1.14	11.26	2.97	104.03	41.11	145.13
3%	270.0	0.5%	1%	3.65	278.0	0.99	1.00	23.00	0.99	9.73	2.78	83.95	35.50	119.45
4%	262.0	0.5%	1%	3.65	266.0	0.95	0.90	19.90	0.89	8.82	2.66	72.64	32.19	104.83
5%	256.0	0.5%	1%	3.65	259.0	0.93	0.80	18.30	0.84	8.31	2.59	66.80	30.34	97.13
10%	188.0	2.5%	5%	18.25	222.0	0.79	0.60	11.20	0.60	5.90	11.10	204.40	107.60	312.00
20%	84.0	5%	10%	36.50	136.0	0.49	0.20	2.60	0.20	1.94	13.60	94.90	70.80	165.70
30%	39.0	5%	10%	36.50	61.5	0.22	0.10	0.50	0.03	0.25	6.15	18.25	9.07	27.32
40%	25.0	5%	10%	36.50	32.0	0.11	0.10	0.20	0.00	0.00	3.20	7.30	0.00	7.30
50%	17.0	5%	10%	36.50	21.0	0.08	0.10	0.10	0.00	0.00	2.10	3.65	0.00	3.65
60%	11.0	5%	10%	36.50	14.0	0.05	0.10	0.10	0.00	0.00	1.40	3.65	0.00	3.65
70%	8.5	5%	10%	36.50	9.8	0.03	0.10	0.10	0.00	0.00	0.98	3.65	0.00	3.65
80%	6.3	5%	10%	36.50	7.4	0.03	0.10	0.00	0.00	0.00	0.74	0.00	0.00	0.00
90%	5.0	5%	10%	36.50	5.7	0.02	0.10	0.00	0.00	0.00	0.57	0.00	0.00	0.00
100%	3.1	5%	10%	36.50	4.1	0.01	0.10	0.00	0.00	0.00	0.41	0.00	0.00	0.00
										Annual totals:		847 (tons/yr)	387 (tons/yr)	1234 (tons/yr)

Worksheet 7-23c (Worksheet 5-11). Pre-treatment total annual sediment yield for the D4 impaired condition.

Stream: Wolf Creek D4 impaired condition (pre-treatment) Location: ≈ 4000 ft above confluence of WF San Juan River Date: 7/04
Observers: D. Rosgen Gage Station #: 09341300 Stream Type: D4 Valley Type: VIII

Equation type	Intercept	Coefficient	Exponent	Form (e.g., linear, non-linear, etc.)	Equation name	Bankfull discharge (cfs)	Bankfull bedload (kg/s)	Bankfull suspended (mg/l)
1. Bedload (dimensionless)	-0.0113	1.0139	2.1929	nonlinear	"good/fair" Pagosa	259	0.146	34.1
2. Suspended sediment (dimensionless)	0.0636	0.9326	2.4085	nonlinear	"good/fair" Pagosa			
3. User-defined relations (bedload)								
4. User-defined relations (suspended sediment)								

Notes: Pre-treatment flow using 1985–1987 flow-duration curve from WRENSS output

		From dimensioned flow-duration curve					From sediment rating curves				Calculate	Calculate sediment yield	
(1)	(2)	(3)	(4)	(5)	(6)	(7)	(8)	(9)	(10)	(11)	(12)	(13)	(14)	(15)
Flow exceedence	Daily mean discharge	Mid-ordinate	Time increment (percent)	Time increment (days)	Mid-ordinate streamflow	Dimensionless streamflow	Dimensionless suspended sediment discharge	Suspended sediment discharge	Dimensionless bedload discharge	Bedload	Time adjusted streamflow	Suspended sediment [(5)×(9)]	Bedload sediment [(5)×(11)]	Suspended + bedload [(13)+(14)]
(%)	(cfs)	(%)	(%)	(days)	(cfs)	(Q/Q_{bkf})	(S/S_{bkf})	(tons/day)	(b_s/b_{bkf})	(tons/day)	(cfs)	(tons)	(tons)	(tons)
0%	368.7													
1%	285.4	0.5%	1%	3.65	327.0	1.26	1.70	50.90	1.67	23.29	3.27	185.79	85.02	270.80
2%	265.0	0.5%	1%	3.65	275.2	1.06	1.10	28.80	1.14	15.90	2.75	105.12	58.03	163.15
3%	250.1	0.5%	1%	3.65	257.6	0.99	1.00	23.20	0.99	13.73	2.58	84.68	50.12	134.80
4%	242.7	0.5%	1%	3.65	246.4	0.95	0.90	20.10	0.89	12.45	2.46	73.37	45.45	118.81
5%	237.2	0.5%	1%	3.65	240.0	0.93	0.80	18.50	0.84	11.73	2.40	67.53	42.83	110.36
10%	172.7	2.5%	5%	18.25	204.9	0.79	0.20	11.20	0.59	8.26	10.25	204.40	150.66	355.06
20%	77.7	5%	10%	36.50	125.2	0.48	0.10	2.60	0.19	2.70	12.52	94.90	98.46	193.36
30%	36.7	5%	10%	36.50	57.2	0.22	0.10	0.50	0.03	0.36	5.72	18.25	12.96	31.21
40%	23.7	5%	10%	36.50	30.2	0.12	0.10	0.20	0.00	0.00	3.02	7.30	0.00	7.30
50%	16.5	5%	10%	36.50	20.1	0.08	0.10	0.10	0.00	0.00	2.01	3.65	0.00	3.65
60%	10.6	5%	10%	36.50	13.6	0.05	0.10	0.10	0.00	0.00	1.36	3.65	0.00	3.65
70%	8.1	5%	10%	36.50	9.4	0.04	0.10	0.10	0.00	0.00	0.94	3.65	0.00	3.65
80%	6.0	5%	10%	36.50	7.1	0.03	0.10	0.00	0.00	0.00	0.71	0.00	0.00	0.00
90%	4.9	5%	10%	36.50	5.5	0.02	0.10	0.00	0.00	0.00	0.55	0.00	0.00	0.00
100%	3.1	5%	10%	36.50	4.0	0.02	0.10	0.00	0.00	0.00	0.40	0.00	0.00	0.00
										Annual totals:		852 (tons/yr)	544 (tons/yr)	1396 (tons/yr)

Worksheet 7-23d (Worksheet 5-11). Post-treatment total annual sediment yield for the D4 impaired condition.

Stream:	Wolf Creek D4 impaired condition (post-treatment)					Location: ≈ 4000 ft above confluence of WF San Juan River								Date: 7/04
Observers: D. Rosgen						Gage Station #: 09341300				Stream Type: D4			Valley Type: VIII	

Equation type	Intercept	Coefficient	Exponent	Form (e.g., linear, non-linear, etc.)	Equation name	Bankfull discharge (cfs)	Bankfull bedload (kg/s)	Bankfull suspended (mg/l)
1. Bedload (dimensionless)	0.0718	1.0218	2.3772	nonlinear	"poor" Pagosa	280	0.16	37.7
2. Suspended sediment (dimensionless)	0.0989	0.9213	3.659	nonlinear	"poor" Pagosa			
3. User-defined relations (bedload)					Notes: Post-treatment flow using 1985-1987 flow-duration curve from WRENSS output			
4. User-defined relations (suspended sediment)								

				From dimensioned flow-duration curve			From sediment rating curves				Calculate	Calculate sediment yield		
(1)	(2)	(3)	(4)	(5)	(6)	(7)	(8)	(9)	(10)	(11)	(12)	(13)	(14)	(15)
Flow exceedence	Daily mean discharge	Mid-ordinate	Time increment (percent)	Time increment (days)	Mid-ordinate streamflow	Dimensionless streamflow	Dimensionless suspended sediment discharge	Suspended sediment discharge	Dimensionless bedload discharge	Bedload	Time adjusted streamflow	Suspended sediment [(5)×(9)]	Bedload sediment [(5)×(11)]	Suspended + bedload [(13)+(14)]
(%)	(cfs)	(%)	(%)	(days)	(cfs)	(Q/Q$_{bkf}$)	(S/S$_{bkf}$)	(tons/day)	(b$_s$/b$_{bkf}$)	(tons/day)	(cfs)	(tons)	(tons)	(tons)
0%	398.0													
1%	308.0	0.5%	1%	3.65	353.0	1.26	2.25	80.80	1.33	20.31	3.53	294.92	74.13	369.05
2%	286.0	0.5%	1%	3.65	297.0	1.06	1.24	37.60	1.08	16.50	2.97	137.24	60.21	197.45
3%	270.0	0.5%	1%	3.65	278.0	0.99	1.00	28.20	1.00	15.20	2.78	102.93	55.49	158.42
4%	262.0	0.5%	1%	3.65	266.0	0.95	0.86	23.40	0.94	14.39	2.66	85.41	52.51	137.92
5%	256.0	0.5%	1%	3.65	259.0	0.93	0.79	20.90	0.91	13.91	2.59	76.29	50.77	127.06
10%	188.0	2.5%	5%	18.25	222.0	0.79	0.49	11.10	0.75	11.39	11.10	202.58	207.87	410.44
20%	84.0	5%	10%	36.50	136.0	0.49	0.16	2.30	0.36	5.54	13.60	83.95	202.06	286.01
30%	39.0	5%	10%	36.50	61.5	0.22	0.10	0.60	0.31	0.46	6.15	21.90	16.94	38.84
40%	25.0	5%	10%	36.50	32.0	0.11	0.10	0.30	0.00	0.00	3.20	10.95	0.00	10.95
50%	17.0	5%	10%	36.50	21.0	0.08	0.10	0.20	0.00	0.00	2.10	7.30	0.00	7.30
60%	11.0	5%	10%	36.50	14.0	0.05	0.10	0.10	0.00	0.00	1.40	3.65	0.00	3.65
70%	8.5	5%	10%	36.50	9.8	0.03	0.10	0.10	0.00	0.00	0.98	3.65	0.00	3.65
80%	6.3	5%	10%	36.50	7.4	0.03	0.10	0.10	0.00	0.00	0.74	3.65	0.00	3.65
90%	5.0	5%	10%	36.50	5.7	0.02	0.10	0.10	0.00	0.00	0.57	3.65	0.00	3.65
100%	3.1	5%	10%	36.50	4.1	0.01	0.10	0.00	0.00	0.00	0.41	0.00	0.00	0.00
										Annual totals:		1038 (tons/yr)	720 (tons/yr)	1758 (tons/yr)

Step 10. Develop dimensionless flow-duration curve

Dimensionless flow-duration curves were not required for the Wolf Creek Watershed analysis. The flow-duration curves used for the FLOWSED model were obtained from the modified WRENSS water yield model (Troendle and Swanson, in press) and are used in **Step 14**.

Step 11. Collect bankfull discharge, bedload sediment and suspended sediment

Bankfull discharge and bedload and suspended sediment data were collected to calculate changes in sediment yield and flow relations using the water yield and FLOWSED models for the C4 and D4 reaches (**Worksheets 7-23a** through **7-23d**).

The individual bankfull values for discharge and bedload and suspended sediment are shown in **Worksheets 7-23a** through **7-23d**. The determined bankfull discharge of 280 cfs represents the existing (post-treatment) conditions for the C4 and D4 conditions; thus, using the water yield model, bankfull discharge was determined for the pre-treatment conditions. The water yield model showed an increase in approximately 7% of mean annual discharge for Wolf Creek; consequently, the pre-treatment bankfull discharge was determined to be approximately 259 cfs for the C4 and D4 conditions.

The measured bankfull values of bedload and suspended sediment for the post-treatment conditions on the C4 were 0.104 kg/s and 31.3 mg/l. The measured bankfull values of bedload and suspended sediment for the D4 impaired condition were 0.16 kg/s and 37.7 mg/l. These values represent the existing conditions after timber harvest. Pre-treatment bankfull values of bedload and suspended sediment are generally not available to compare to the post-treatment condition. The assumption of this approach is to compare the change in flow through the best indicators of sediment supply. Often, it is not possible to obtain pre-treatment sediment measurements after the fact. However, if a stable reference condition has been verified to be stable for many years, measured sediment data can be used as a surrogate for pre-treatment sediment supply. For the Wolf Creek case study, a stable C4 stream type downstream of the USGS streamgage on Wolf Creek, 2,500 ft upstream of the lower extent of the C4 stream type, was used to obtain reference "pre-treatment" sediment supply. Time-trend data from aerial photos indicated no apparent change in this reach from the 1976 to present day condition. Measured bankfull sediment data for this reach was 0.093 kg/s for bedload and 27.3 mg/l for suspended sediment.

Furthermore, the suspended sediment collected for the C4 reference and D4 impaired conditions was divided in the lab to separate the sand portion from the silt/clay (< 0.062 mm). This allows additional interpretation of depositional suspended sand load, and the suspended sand data is used in the POWERSED model that deals only with the hydraulically controlled sediment yield in **Steps 18** and **19**.

Step 12. Obtain or establish dimensionless bedload and suspended sediment rating curves

The dimensionless bedload and suspended rating curves that were used for both the reference and impaired conditions are shown in Chapter 2 in **Figures 2-52** and **2-53** for "good/fair" stability and in **Figures 2-58** and **2-59** for "poor" stability. The user has the option of selecting the appropriate dimensionless sediment rating curves for both the pre-treatment and post-treatment conditions. This is often done to isolate the effects of flow-related changes for the same sediment supply and to compare the same flow change on a higher supply, "poor" condition reach. This was done in the Wolf Creek example. The pre- and post-treatment C4 reference conditions and the pre-treatment D4 impaired condition used the "good/fair" sediment rating curves (**Flowchart 7-9**). The post-treatment D4 impaired condition used the "poor" sediment rating curves.

Step 13. Convert dimensionless bedload and suspended sediment rating curves to dimensioned sediment rating curves

The dimensionless bedload and suspended sediment rating curves for the pre- and post-treatment conditions for the C4 reference and D4 impaired conditions were converted to dimensioned sediment rating curves. This was completed by multiplying the bankfull discharge and sediment values obtained in **Step 11** by each of the ratios appropriate for the dimensionless relations selected in **Step 12**. For example, the measured bankfull discharge was 280 cfs for the post-treatment D4 condition and the bedload was 0.16 kg/s and the suspended sediment was 37.7 mg/l (**Step 11**). The post-treatment D4 condition used the "poor" dimensionless sediment rating curves (**Step 12**); thus, the dimensionless discharge values on the "poor" dimensionless sediment rating curves were multiplied by 280 and the dimensionless sediment values were multiplied by the measured values of bedload and suspended sediment. The values for the pre- and post-treatment conditions for the C4 and D4 impaired conditions are shown in **Worksheets 7-23a** through **7-23d**.

Step 14. Convert dimensionless flow-duration curve to dimensioned flow-duration curve

The need to convert a dimensionless flow-duration curve to a dimensioned flow-duration curve was negated at this step as the modified WRENSS water yield model generated pre- and post-treatment flow-duration curves for the C4 reference and D4 impaired conditions. The results of the flow-duration curves for pre- and post-treatment conditions are documented in **Worksheets 7-23a** through **7-23d**.

Step 15. Calculate total annual sediment yield for both bedload and suspended sediment

Table 7-32 summarizes the pre- and post-treatment total annual sediment yield values for the C4 reference and D4 impaired conditions using the modified WRENSS and FLOWSED models (**Worksheets 7-23a** through **7-23d**).

Table 7-32. Summary of total annual sediment yield for the C4 reference and D4 impaired conditions.

Condition	Bedload sediment (tons/yr)	Suspended sediment (tons/yr)	Total sediment (tons/yr)
C4 reference condition			
Pre-treatment (**Worksheet 7-23a**)	347	682	1029
Post-treatment (**Worksheet 7-23b**)	387	847	1234
D4 impaired condition			
Pre-treatment (**Worksheet 7-23c**)	544	852	1396
Post-treatment (**Worksheet 7-23d**)	720	1038	1758

Step 16. Select and run a water yield model

The selected water yield model was discussed previously in the discussion for **Steps 10–17** where a modified WRENSS snowmelt model (Troendle and Swanson, in press) was run for the Wolf Creek Watershed. The detailed results of the flow model applied to the Wolf Creek Watershed are included in a summary report by J. Nankervis in **Appendix A**.

The FLOWSED model utilized the flow-duration curve data generated by the modified WRENSS model to predict total annual bedload and suspended sediment for the pre- and post-treatment C4 reference condition (**Worksheets 7-23a** and **7-23b**) and the D4 impaired condition (**Worksheets 7-23c** and **7-23d**).

Step 17. Calculate flow-related changes in annual sediment yield

The increase in sediment due to flow-related sediment increase for the C4 reference condition is 205 tons/yr for total bedload and suspended sediment (**Table 7-33**). Bedload sediment yield is increased by 40 tons/yr and suspended sediment yield is increased by 165 tons/yr.

Table 7-33. Predicted flow-related annual sediment yield increase for upper Wolf Creek, representing the C4 reference stream reach.

C4 reference reach, upper Wolf Creek	Bedload sediment (tons/yr)	Suspended sediment (tons/yr)	Total sediment (tons/yr)
Pre-treatment annual sediment yield (**Worksheet 7-23a**)	347	682	1029
Post-treatment annual sediment yield (**Worksheet 7-23b**)	387	847	1234
Flow-related annual sediment yield increase	40	165	205

The increase of flow-related total sediment for the D4 impaired condition is approximately 362 tons/yr, of which 176 tons/yr are bedload and 186 tons/yr are suspended sediment (**Table 7-34**). A total of 362 tons can be thought of as 28, 10-yard, end-dump truck loads/yr, relating to a 7% increase in streamflow.

Table 7-34. Predicted flow-related annual sediment yield increase for lower Wolf Creek, representing the D4 impaired stream reach.

D4 impaired reach, lower Wolf Creek	Bedload sediment (tons/yr)	Suspended sediment (tons/yr)	Total sediment (tons/yr)
Pre-treatment annual sediment yield (**Worksheet 7-23c**)	544	852	1396
Post-treatment annual sediment yield (**Worksheet 7-23d**)	720	1038	1758
Flow-related annual sediment yield increase	176	186	362

Steps 18–19: Sediment transport capacity model (POWERSED)
(See Flowchart 7-10)

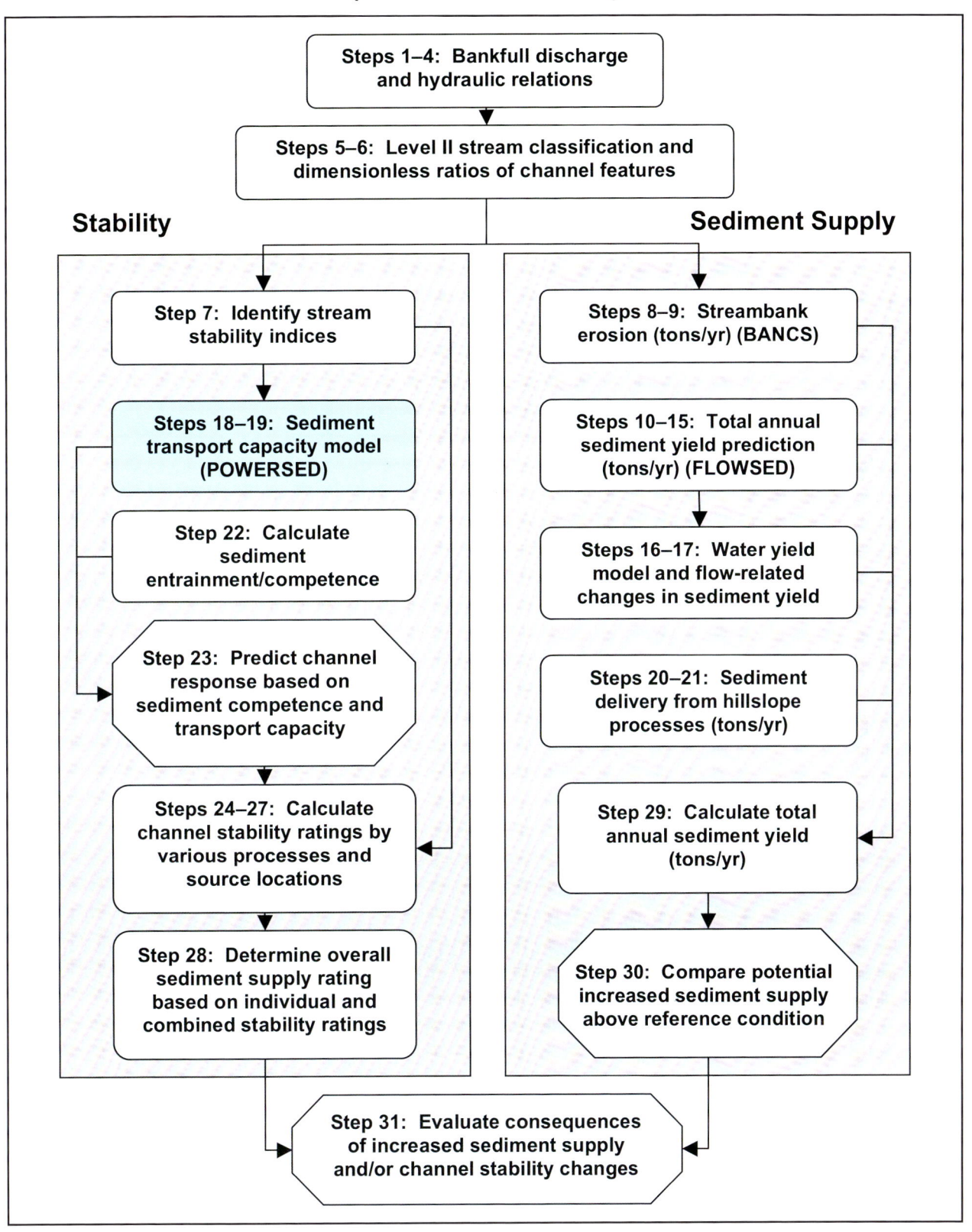

```
Steps 1–4: Bankfull discharge and hydraulic relations
        ↓
Steps 5–6: Level II stream classification and dimensionless ratios of channel features
```

Stability | **Sediment Supply**

Stability branch:
- Step 7: Identify stream stability indices
- **Steps 18–19: Sediment transport capacity model (POWERSED)**
- Step 22: Calculate sediment entrainment/competence
- Step 23: Predict channel response based on sediment competence and transport capacity
- Steps 24–27: Calculate channel stability ratings by various processes and source locations
- Step 28: Determine overall sediment supply rating based on individual and combined stability ratings

Sediment Supply branch:
- Steps 8–9: Streambank erosion (tons/yr) (BANCS)
- Steps 10–15: Total annual sediment yield prediction (tons/yr) (FLOWSED)
- Steps 16–17: Water yield model and flow-related changes in sediment yield
- Steps 20–21: Sediment delivery from hillslope processes (tons/yr)
- Step 29: Calculate total annual sediment yield (tons/yr)
- Step 30: Compare potential increased sediment supply above reference condition

Step 31: Evaluate consequences of increased sediment supply and/or channel stability changes

Flowchart 7-3g (Flowchart 5-3h). The sequential steps to determine stream stability and sediment supply, highlighting **Steps 18–19**.

WATERSHED ASSESSMENT OF RIVER STABILITY AND SEDIMENT SUPPLY

Steps 18–19: Sediment transport capacity model (POWERSED)

Step 18: Evaluate channel characteristics that change hydraulic and morphological variables and develop hydraulic geometry relations for a wide range of flows for the upstream and impaired reaches

→ Width | Depth | Slope | Velocity | Discharge

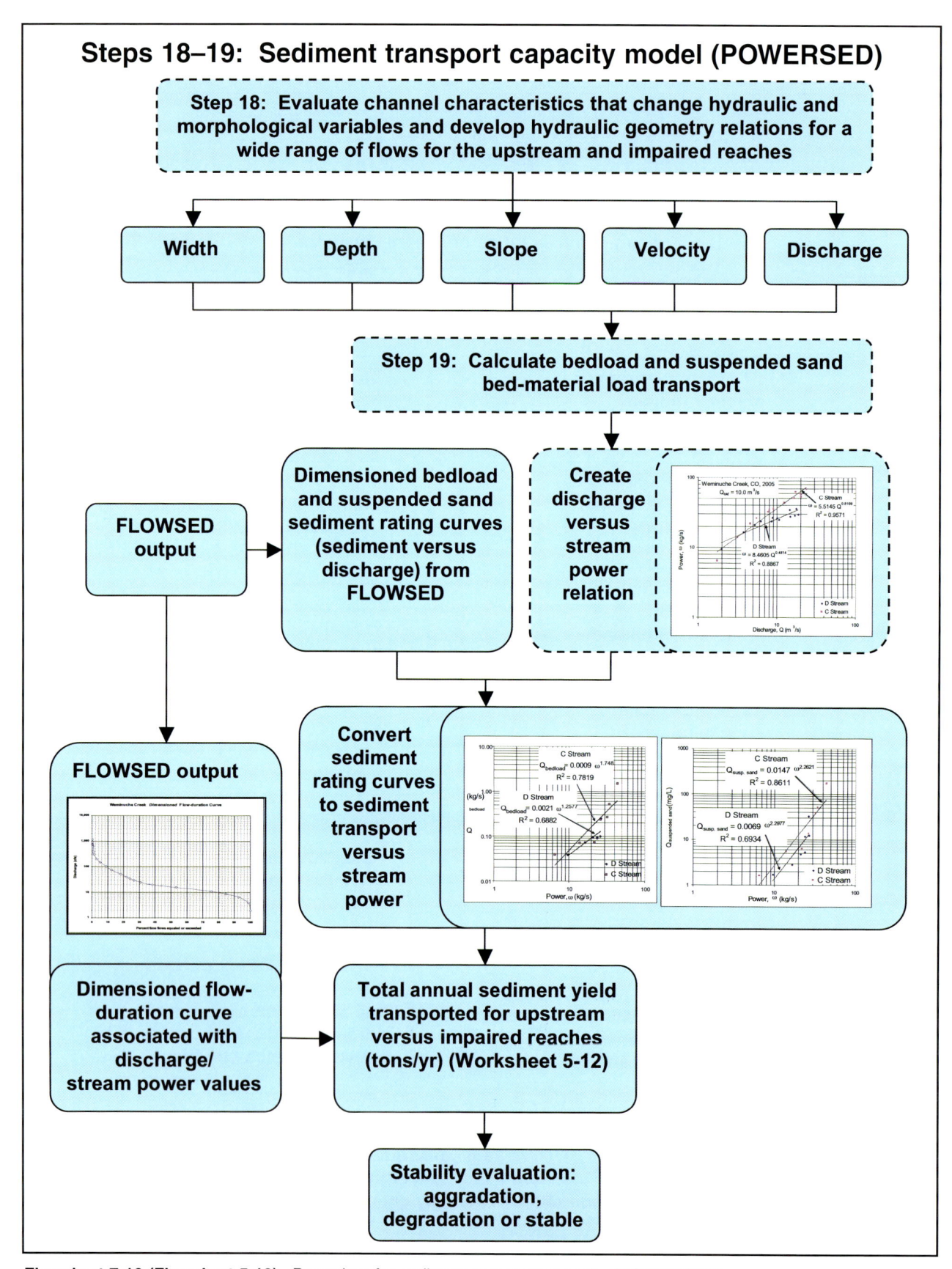

Step 19: Calculate bedload and suspended sand bed-material load transport

- FLOWSED output
- Dimensioned bedload and suspended sand sediment rating curves (sediment versus discharge) from FLOWSED
- Create discharge versus stream power relation

- FLOWSED output
- Dimensioned flow-duration curve associated with discharge/stream power values
- Convert sediment rating curves to sediment transport versus stream power
- Total annual sediment yield transported for upstream versus impaired reaches (tons/yr) (Worksheet 5-12)
- Stability evaluation: aggradation, degradation or stable

Flowchart 7-10 (Flowchart 5-13). Procedure for sediment transport capacity (POWERSED).

Step 18. Evaluate channel characteristics that change hydraulic and morphological variables and develop hydraulic geometry relations for a wide range of flows

The channel changes occurring in the lower reaches of Wolf Creek are of major concern given their high instability ratings. The following steps start to quantify and isolate various sediment and stability consequence relations related to specific processes and land uses.

Hydraulic relations are an important aspect of channel change. For example, in **Step 7** (channel stability analysis), the width/depth ratio (**Figure 7-40**) is rated as *highly unstable* for the C4 to D4 stream type conversion. An increase in the width/depth ratio on Wolf Creek results in a depth reduction for the same discharge. The reduced depth decreases velocity by increasing relative roughness (depth of flow over a D_{84}-size bed particle). The net effect is a reduction in shear stress, velocity and unit stream power (shear stress multiplied by velocity).

For a bankfull discharge of 280 cfs, the velocity of the Wolf Creek C4 stream type is approximately 4.7 ft/sec, with a depth of 1.6 ft and a slope of 0.005. The velocity of the D4 stream type for the same flow, however, is 1.7 ft/sec with a depth of 0.8 ft and a slope of 0.009. Shear stress is 0.5 lbs/ft^2 for the C4 stream type and 0.45 lbs/ft^2 for the D4 stream type. Although there is an increase in sediment supply from bank erosion in the D4 stream type, the difference in unit stream power for the same bankfull discharge is 2.6 lbs/ft/sec for the C4 versus 0.85 lbs/ft/sec for the D4 stream type. This indicates the potential loss of sediment transport capacity and associated stream channel instability of the D4 stream type.

Step 19. Calculate bedload and suspended sand bed-material load transport capacity (stream power)

Bedload models evaluate the ability of the river to transport the sediment load for various flows. This step calculates sediment transport capacity. Plotting the relationship between unit stream power and stream discharge for the C4 and D4 stream types for Wolf Creek reveals a stark contrast between the two types (**Figure 7-47**). Although the discharge increases from baseflow to bankfull, there is a very small increase in unit stream power with the D4 stream type as compared to the C4 stream type. Sediment rating curves relate sediment transport versus discharge. The rating curves using the Bagnold model show that sediment transport is increased for the D4 stream type, even though the river has a decrease in unit stream power. The sediment increase, however, may also be associated with an over-prediction of bedload transport. The model results are tested against observed values and with the POWERSED model. This condition indicates a high sediment supply that may be associated with streambank erosion. Sediment transport capacity is essential in securing river stability. The high width/depth ratio and depositional patterns previously identified would indicate potential aggradation.

A relationship for Wolf Creek bedload transport is also presented in **Figure 7-48**. The author modified the Bagnold equation (Chapter 2) to predict bedload transport for the C4 stream type. Model prediction values were compared to measured bedload transport for a wide range of C4 stream type stages. Fairly good agreement was shown between the model and observed values of unit stream power and unit bedload transport rate.

When the same model was applied to the D4 stream type for the same range of streamflows, the results indicated a major reduction in transport capacity, although the modified Bagnold model over-predicted bedload transport (**Figure 7-49**). The result of a decrease in sediment capacity and competence is reflected in the reduced size of bed-material gradations in the D4 stream type compared to the upstream C4 stream type (**Figure 7-34**; see **Step 6**).

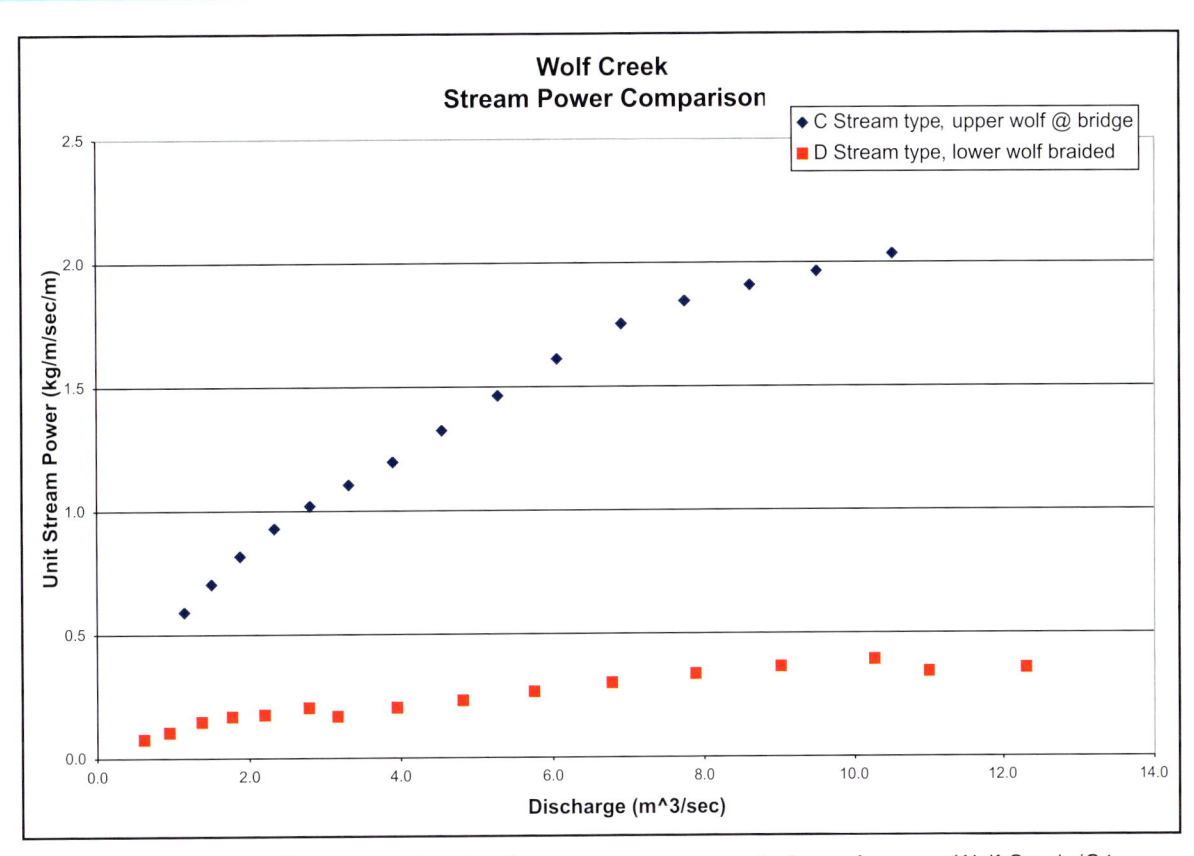

Chapter 7 — PLA The WARSSS Methodology: A Case Study of Wolf Creek, Colorado

Figure 7-47 (Figure 2-61). Relationship of unit stream power versus discharge for upper Wolf Creek (C4 stream type) and lower Wolf Creek (D4 stream type).

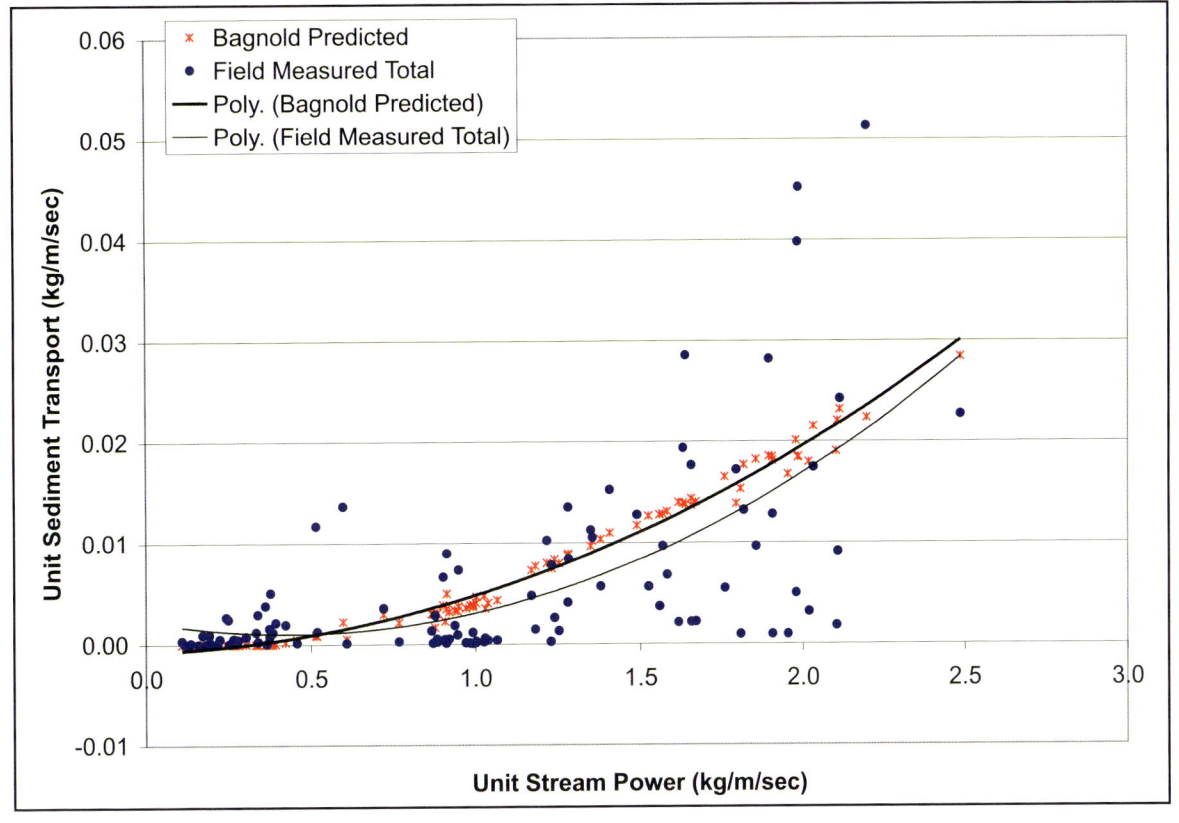

Figure 7-48 (Figure 2-62). Comparison of measured bedload transport versus predicted (Bagnold, 1980) for upper Wolf Creek, Colorado.

7-153

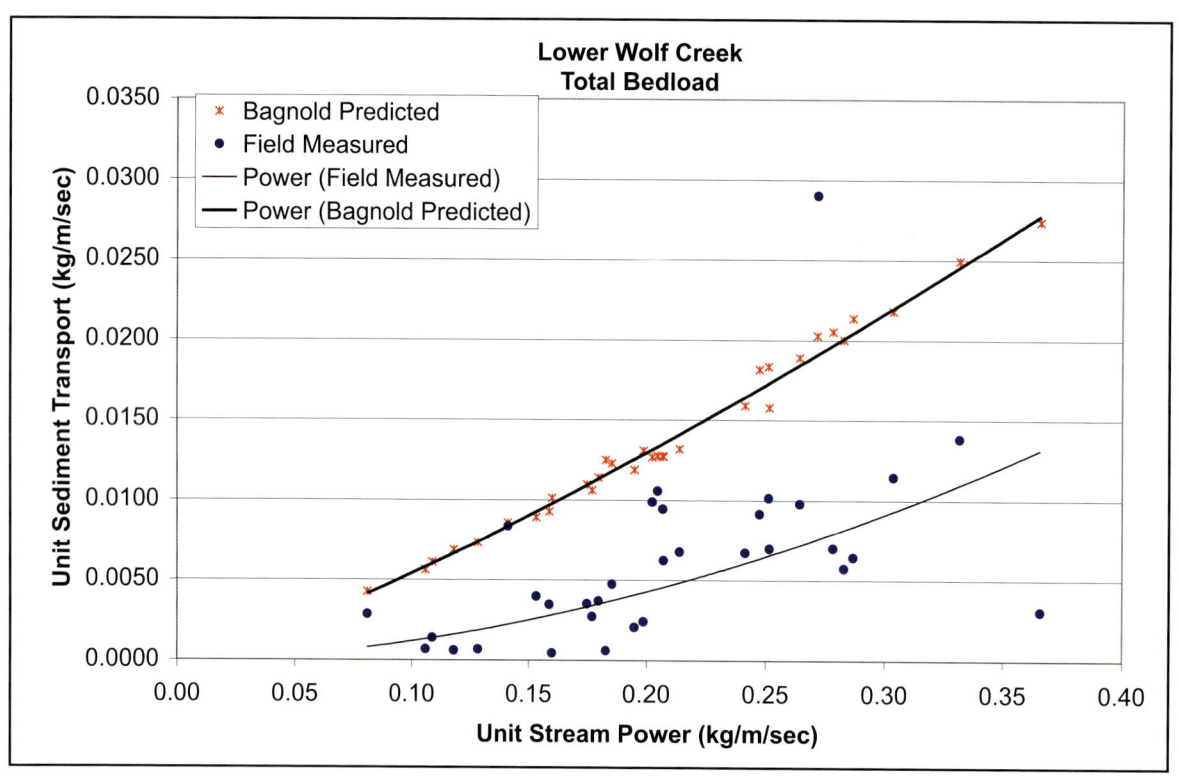

Figure 7-49 (Figure 2-63). Comparison of measured versus predicted (Bagnold, 1980) bedload transport for lower Wolf Creek (D4 stream type).

The Bagnold model over-predicted actual transport; thus, POWERSED using RIVERMorph™ was used to predict channel response. The POWERSED model incorporates unit stream power and local sediment supply. The model was used for sediment transport on Wolf Creek to compare differences between the C4 reference reach and the lower D4 impaired reach for the same flows. This model integrates dimensionless sediment rating curves, dimensionless flow-duration curves and hydraulic geometry by stages (**Flowchart 7-10**). The procedures for the POWERSED model are outlined in Chapter 5 and **Worksheet 5-12**. Analysts, however, should use models that they have calibrated and in which they have confidence and experience.

The predicted combined streamflow-related post-treatment bedload and suspended sediment yield values are higher for the lower Wolf Creek, D4 stream type than for the immediately upstream Wolf Creek, C4 stream type by 524 tons/yr (**Table 7-32**). This indicates a higher sediment supply as supported by the streambank erosion prediction (**Steps 8** and **9**). The sediment yield calculation of streambank erosion for the lower reach was 429 tons/yr for approximately 970 ft of stream reach in the D4 stream type. This could help explain the increase in sediment supply. If the increase in streamflow-related sediment yield was a major contributor, then the POWERSED model would indicate whether the D4 braided reach would have the capacity to route the increased sediment supply.

The POWERSED runs for the D4 stream type are summarized in **Table 7-35** and **Worksheets 7-24a, 7-24b** and **7-24c** where three cells are isolated and computed separately. The objective of POWERSED is to determine sediment transport capacity as part of a channel stability analysis. Because the hydraulic geometry changes drastically corresponding to a change from a C4 to D4 stream type, the sediment transport consequences must be ascertained. The suspended sediment excluded the washload (silt/clay for each size fraction) from the analysis as it is not hydraulically controlled. Hydraulic geometry by stage is computed to calculate stream power and corresponding sediment transport.

POWERSED was also used to compare the hydraulic character and sediment supply of the upstream, reference C4 reach. The results of the POWERSED prediction for the reference C4 stream type are shown in **Table 7-35**. These results are compared to measured values obtained in 1997 (at flows 18% above the bankfull discharge).

Furthermore, the results presented in **Table 7-35** show the reduction in both bedload and suspended sand sediment for the D4 impaired reach compared to the C4 reference reach. This reduction in both bedload and suspended sand sediment is due to the reduction in unit stream power associated with the D4 stream type. The POWERSED model shows a reduction from 1,418 tons/yr to 988 tons/yr, or a reduction of 430 tons/yr (**Table 7-35**), indicating *aggradation*.

Table 7-35. Summary of the predictions of annual sediment yield (tons/yr) using POWERSED from RIVERMorph™ for the C4 and D4 stream types.

C4 above the D4	Predicted values (tons/yr)	Measured values (tons/yr)*
Bedload	510	512
Suspended sand	908	1,194
Total annual sediment yield	1,418	1,706

D4 impaired reach	Predicted values (tons/yr)			
	Cell 1	Cell 2	Cell 3	Totals
Bedload	493	0	0	**493**
Suspended sand	464	21	10	**495**
Total annual sediment yield	957	21	10	**988**

*Based on 1997 streamflows (18% higher than bankfull).

Watershed Assessment of River Stability and Sediment Supply

Worksheet 7-24a (Worksheet 5-12). Bedload and suspended sand bed-material load transport prediction using the POWERSED model for cell 1 of the D4 impaired reach.

Stream: **Wolf Creek impaired (braided) reach (Cell 1)**   Location: **≈ 4000 ft above conf. of WF San Juan River**   Date: **7/04**
Observers: **D. Rosgen**   Stream Type: **D4**   Valley Type: **VIII**   Gage Station #: **09341300**

Flow-duration curve		Calculate	Hydraulic geometry			Measure					Calculate					
(1)	(2)	(3)	(4)	(5)	(6)	(7)	(8)	(9)	(10)	(11)	(12)	(13)	(14)	(15)	(16)	(17)	(18)
Exceedance probability	Daily mean discharge	Mid-ordinate stream-flow	Area	Width	Depth	Velocity	Slope	Shear stress	Stream power	Unit power	Time increment	Time increment	Daily mean bedload transport	Daily mean suspended sand transport	Time adjusted bedload transport [(13)×(14)]	Time adjusted suspended sand transport [(13)×(15)]	Time adjusted total transport [(16)+(17)]
(%)	(cfs)	(cfs)	(ft²)	(ft)	(ft)	(ft/s)	(ft/ft)	(lb/ft²)	(lb/s)	(lb/ft/s)	(%)	(days)	(tons/day)	(tons/day)	(tons)	(tons)	(tons)
100%	2.08																
90%	4.16	3.12	3.51	8.47	0.41	0.89	0.009	0.23	1.75	0.21	10%	36.50	0.00	0.00	0.00	0.00	0.00
80%	4.16	4.16	4.29	9.07	0.47	0.97	0.009	0.26	2.34	0.26	10%	36.50	0.00	0.00	0.00	0.00	0.00
70%	6.24	5.20	5.02	9.59	0.52	10.30	0.009	0.28	2.92	0.30	10%	36.50	0.00	0.00	0.00	0.00	0.00
60%	8.32	7.28	6.40	10.60	0.60	1.14	0.009	0.33	4.09	0.39	10%	36.50	0.00	0.00	0.00	0.00	0.00
50%	12.49	10.41	8.52	12.60	0.68	1.22	0.009	0.36	5.85	0.46	10%	36.50	0.00	0.10	0.00	3.65	3.65
40%	18.73	15.61	11.87	15.62	0.76	1.32	0.009	0.41	8.77	0.56	10%	36.50	0.00	0.10	0.00	3.65	3.65
30%	29.13	23.93	22.10	39.78	0.56	1.08	0.009	0.30	13.44	0.34	10%	36.50	0.00	0.20	0.00	7.30	7.30
20%	62.43	45.78	34.58	45.92	0.75	1.32	0.009	0.41	25.71	0.56	10%	36.50	0.00	0.30	0.00	10.95	10.95
10%	141.51	101.97	56.68	47.06	1.20	1.80	0.009	0.65	57.27	1.22	10%	36.50	2.20	1.90	80.30	69.35	149.65
5%	189.37	165.44	76.61	48.06	1.59	2.16	0.009	0.86	92.91	1.93	5%	18.25	8.60	9.10	156.95	166.08	323.03
2%	212.26	200.82	86.50	48.55	1.78	2.32	0.009	0.95	112.78	2.32	3%	10.95	21.20	15.40	232.14	168.63	400.77
1%	230.99	221.63	107.40	72.83	1.47	2.06	0.009	0.80	124.47	1.71	1%	3.65	6.50	9.50	23.73	34.68	58.40
															493	**464**	**957**

Total annual sediment yield (bedload and suspended sand bed-material load) (tons/yr):

Worksheet 7-24b (Worksheet 5-12). Bedload and suspended sand bed-material load transport prediction using the POWERSED model for cell 2 of the D4 impaired reach.

Stream:	Wolf Creek impaired (braided) reach (Cell 2)							Location: ≈ 4000 ft above conf. of WF San Juan River								Date: 7/04	
Observers:	D. Rosgen							Stream Type: D4	Valley Type: VIII			Gage Station #: 09341300					
Flow-duration curve		Calculate		Hydraulic geometry				Measure						Calculate			
(1)	(2)	(3)	(4)	(5)	(6)	(7)		(8)	(9)	(10)	(11)	(12)	(13)	(14)	(15)	(16)	(17)	(18)
Exceedance probability	Daily mean discharge	Mid-ordinate stream-flow	Area	Width	Depth	Velocity		Slope	Shear stress	Stream power	Unit power	Time increment	Time increment	Daily mean bedload transport	Daily mean suspended sand transport	Time adjusted bedload transport [(13)×(14)]	Time adjusted suspended sand transport [(13)×(15)]	Time adjusted total transport [(16)+(17)]
(%)	(cfs)	(cfs)	(ft²)	(ft)	(ft)	(ft/s)		(ft/ft)	(lb/ft²)	(lb/s)	(lb/ft/s)	(%)	(days)	(tons/day)	(tons/day)	(tons)	(tons)	(tons)
100%	0.52																	
90%	1.04	0.78	1.44	7.54	0.19	0.53		0.009	0.11	0.44	0.06	10%	36.50	0.00	0.00	0.00	0.00	0.00
80%	1.04	1.04	1.79	8.42	0.21	0.57		0.009	0.12	0.58	0.07	10%	36.50	0.00	0.00	0.00	0.00	0.00
70%	1.57	1.31	2.12	9.16	0.23	0.61		0.009	0.13	0.74	0.08	10%	36.50	0.00	0.00	0.00	0.00	0.00
60%	2.09	1.83	2.70	10.47	0.26	0.66		0.009	0.15	1.03	0.10	10%	36.50	0.00	0.00	0.00	0.00	0.00
50%	3.13	2.61	3.61	12.16	0.30	0.72		0.009	0.17	1.47	0.12	10%	36.50	0.00	0.00	0.00	0.00	0.00
40%	4.70	3.92	4.91	14.28	0.34	0.80		0.009	0.19	2.20	0.15	10%	36.50	0.00	0.00	0.00	0.00	0.00
30%	7.30	6.00	6.77	16.83	0.40	0.88		0.009	0.22	3.37	0.20	10%	36.50	0.00	0.00	0.00	0.00	0.00
20%	15.65	11.48	11.17	22.24	0.50	1.03		0.009	0.28	6.45	0.29	10%	36.50	0.00	0.10	0.00	3.65	3.65
10%	35.47	25.56	21.55	34.60	0.62	1.19		0.009	0.35	14.35	0.41	10%	36.50	0.00	0.20	0.00	7.30	7.30
5%	47.47	41.47	31.47	43.19	0.73	1.32		0.009	0.41	23.29	0.54	5%	18.25	0.00	0.30	0.00	5.48	5.48
2%	53.21	50.34	36.57	47.02	0.78	1.38		0.009	0.44	28.27	0.60	3%	10.95	0.00	0.30	0.00	3.29	3.29
1%	57.91	55.56	45.47	70.00	0.65	1.22		0.009	0.36	31.20	0.45	1%	3.65	0.00	0.40	0.00	1.46	1.46
											Total annual sediment yield (bedload and suspended sand bed-material load) (tons/yr):					0	21	21

Watershed Assessment of River Stability and Sediment Supply

Worksheet 7-24c (Worksheet 5-12). Bedload and suspended sand bed-material load transport prediction using the POWERSED model for cell 3 of the D4 impaired reach.

Stream:	Wolf Creek impaired (braided) reach (Cell 3)				Location: ≈ 4000 ft above conf. of WF San Juan River												Date: 7/04
Observers:	D. Rosgen						Stream Type: D4		Valley Type: VIII			Gage Station #: 09341300					

Flow-duration curve		Calculate		Hydraulic geometry			Measure						Calculate				
(1)	(2)	(3)	(4)	(5)	(6)	(7)	(8)	(9)	(10)	(11)	(12)	(13)	(14)	(15)	(16)	(17)	(18)
Exceedance probability	Daily mean discharge	Mid-ordinate stream-flow	Area	Width	Depth	Velocity	Slope	Shear stress	Stream power	Unit power	Time increment	Time increment	Daily mean bedload transport	Daily mean suspended sand transport	Time adjusted bedload transport [(13)×(14)]	Time adjusted suspended sand transport [(13)×(15)]	Time adjusted total transport [(16)+(17)]
(%)	(cfs)	(cfs)	(ft²)	(ft)	(ft)	(ft/s)	(ft/ft)	(lb/ft²)	(lb/s)	(lb/ft/s)	(%)	(days)	(tons/day)	(tons/day)	(tons)	(tons)	(tons)
100%	0.36																
90%	0.72	0.54	1.22	8.56	0.14	0.43	0.009	0.08	0.30	0.04	10%	36.50	0.00	0.00	0.00	0.00	0.00
80%	0.72	0.72	1.55	9.97	0.16	0.46	0.009	0.08	0.40	0.04	10%	36.50	0.00	0.00	0.00	0.00	0.00
70%	1.08	0.90	1.79	10.44	0.17	0.49	0.009	0.09	0.51	0.05	10%	36.50	0.00	0.00	0.00	0.00	0.00
60%	1.44	1.26	2.28	11.40	0.20	0.54	0.009	0.11	0.71	0.06	10%	36.50	0.00	0.00	0.00	0.00	0.00
50%	2.16	1.80	2.95	12.61	0.23	0.60	0.009	0.13	1.01	0.08	10%	36.50	0.00	0.00	0.00	0.00	0.00
40%	3.24	2.70	3.97	14.28	0.28	0.68	0.009	0.15	1.52	0.11	10%	36.50	0.00	0.00	0.00	0.00	0.00
30%	5.04	4.14	5.41	16.32	0.33	0.76	0.009	0.18	2.33	0.14	10%	36.50	0.00	0.00	0.00	0.00	0.00
20%	10.80	7.92	8.93	21.60	0.41	0.88	0.009	0.22	4.45	0.21	10%	36.50	0.00	0.00	0.00	0.00	0.00
10%	24.48	17.64	16.47	30.00	0.55	1.07	0.009	0.30	9.91	0.33	10%	36.50	0.00	0.10	0.00	3.65	3.65
5%	32.75	28.62	25.21	42.33	0.60	1.13	0.009	0.33	16.07	0.38	5%	18.25	0.00	0.20	0.00	3.65	3.65
2%	36.71	34.73	28.44	42.79	0.66	1.22	0.009	0.36	19.50	0.46	3%	10.95	0.00	0.20	0.00	2.19	2.19
1%	39.95	38.33	36.58	69.95	0.52	1.04	0.009	0.29	21.53	0.31	1%	3.65	0.00	0.20	0.00	0.73	0.73

Total annual sediment yield (bedload and suspended sand bed-material load) (tons/yr): 0 | 10 | 10

Chapter 7 — PLA   The WARSSS Methodology: A Case Study of Wolf Creek, Colorado

Steps 20–21: Sediment delivery from hillslope processes
(See Flowchart 7-11)

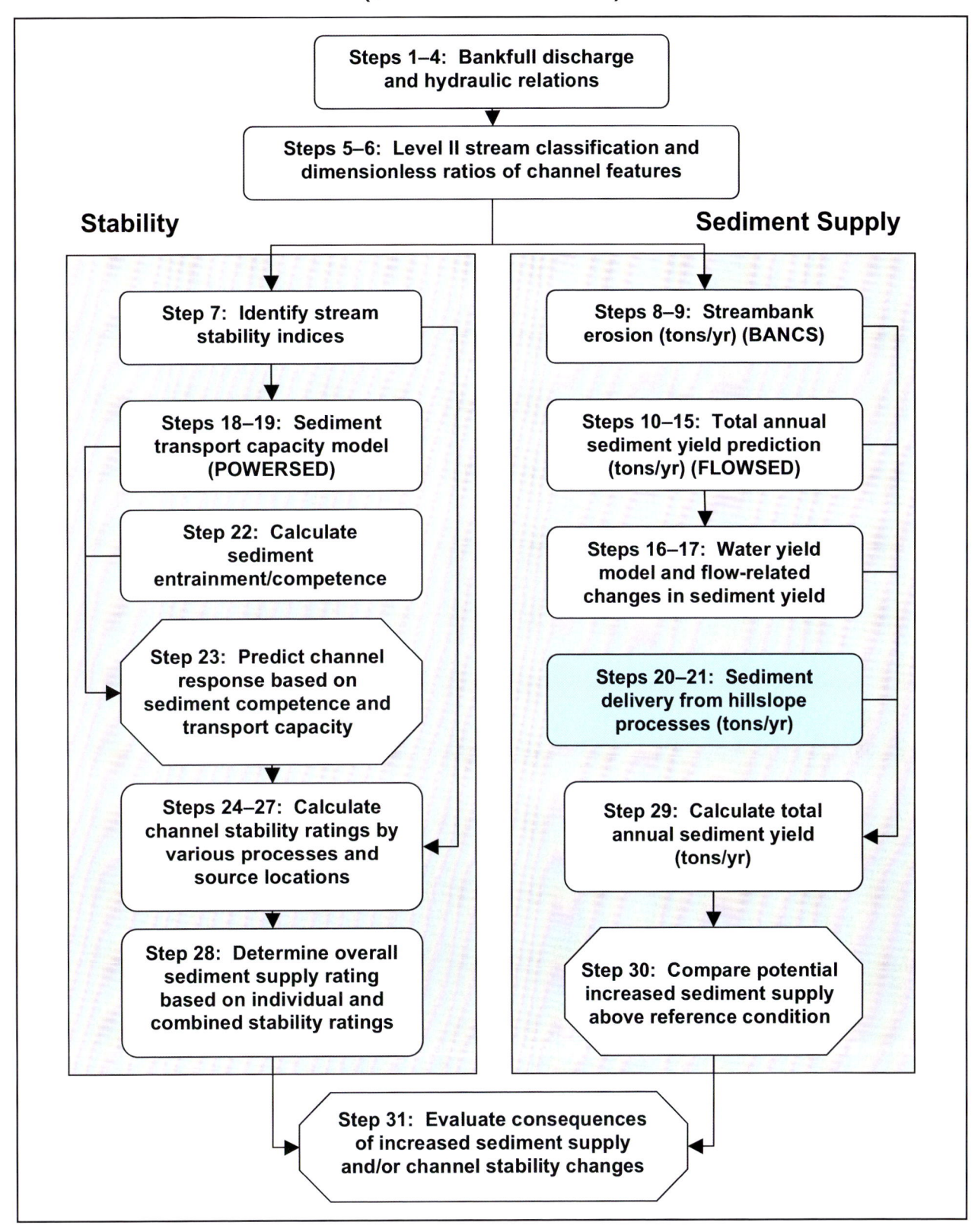

```
                Steps 1–4: Bankfull discharge
                   and hydraulic relations
                             │
                             ▼
         Steps 5–6: Level II stream classification and
              dimensionless ratios of channel features

   Stability                                    Sediment Supply

   Step 7: Identify stream            Steps 8–9: Streambank
      stability indices                erosion (tons/yr) (BANCS)

  Steps 18–19: Sediment             Steps 10–15: Total annual
  transport capacity model          sediment yield prediction
       (POWERSED)                      (tons/yr) (FLOWSED)

     Step 22: Calculate               Steps 16–17: Water yield
         sediment                    model and flow-related
   entrainment/competence           changes in sediment yield

  Step 23: Predict channel           Steps 20–21: Sediment
    response based on                delivery from hillslope
 sediment competence and              processes (tons/yr)
    transport capacity

  Steps 24–27: Calculate            Step 29: Calculate total
  channel stability ratings by        annual sediment yield
    various processes and                   (tons/yr)
       source locations

   Step 28: Determine overall        Step 30: Compare potential
     sediment supply rating          increased sediment supply
     based on individual and          above reference condition
    combined stability ratings

              Step 31: Evaluate consequences
               of increased sediment supply
              and/or channel stability changes
```

Flowchart 7-3h (Flowchart 5-3i). The sequential steps to determine stream stability and sediment supply, highlighting **Steps 20–21**.

7-159

WATERSHED ASSESSMENT OF RIVER STABILITY AND SEDIMENT SUPPLY

Steps 20–21: Sediment delivery from hillslope processes (tons/yr)

Step 20: Determine potential sediment delivery from roads, surface erosion and mass erosion

- Sediment delivery from roads → Road Impact Index (RII) (Table 5-14, Figure 5-46) → Preliminary annual sediment yield (Worksheet 5-13) → Total annual sediment yield due to roads (tons/yr)
  - Recovery adjustment (Figure 5-47) → Preliminary annual sediment yield
  - Mitigation adjustment (Table 5-14) → Preliminary annual sediment yield

- Sediment delivery from surface erosion → User's choice in model → e.g. RUSLE / e.g. WRENSS procedure / e.g. WEPP → Sediment delivery index (USEPA, 1980) (annual sediment yield in tons/yr)

- Sediment delivery from mass erosion → Soil mass movement (tons/yr) (Chapter V, USEPA, 1980) → Sediment delivery

Step 21: Summarize total annual sediment yield (tons/yr) from hillslope processes

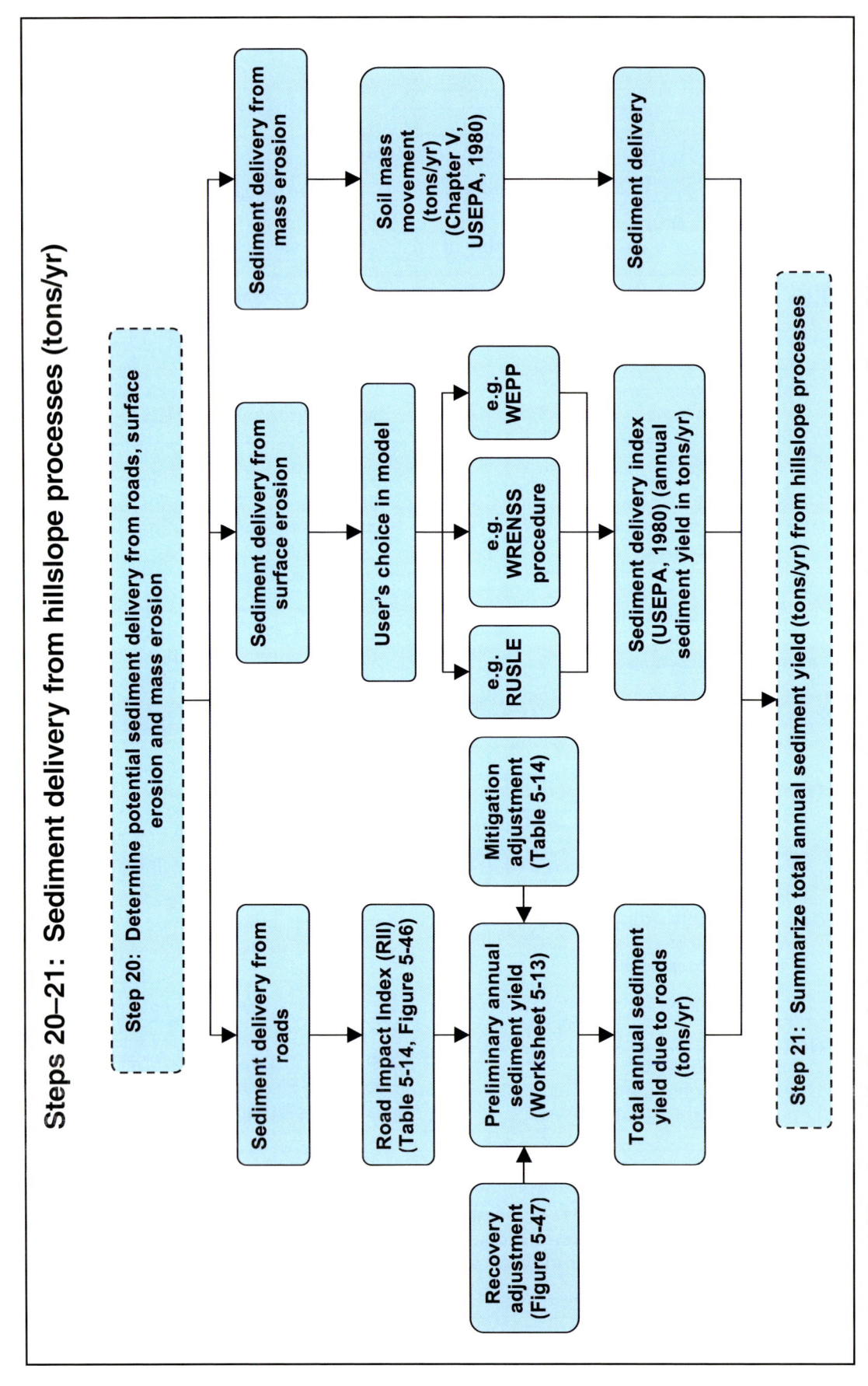

Flowchart 7-11 (Flowchart 5-15). Procedural sequence of hillslope erosional processes contributing to total annual sediment yield.

Step 20. Determine potential sediment delivery from roads, surface erosion and mass erosion

Sediment delivery from roads

This analysis identifies and calculates disproportionate contributions of sediment from road sources to the total sediment yield using the procedure in **Table 7-36**.

The Road Impact Index (RII) (**Figure 7-50**) was used for both the main State Highway 160 and for the older, secondary logging roads on national forest lands. To adjust the sediment yield predicted from road sources, an erosion recovery relation was applied (**Figure 7-51**).

The results, summarized in **Worksheet 7-25**, indicate 107 tons/yr from all secondary logging roads and 495 tons/yr from State Highway 160, for a total road sediment contribution of 602 tons/yr. Much of the sediment is from ditch-line and road fill erosion. Stream encroachment on road fills at stream crossings provides a significant sediment source.

Table 7-36 (Table 5-14). Procedure to calculate Road Impact Index (RII) and corresponding annual sediment yield from roads (record data in **Worksheet 5-13**).

Annual sediment yield calculation based on Road Impact Index (RII)

1. Divide watershed into 200–5,000 acre sub-watersheds (Column (2) in **Worksheet 5-13**).
2. Locate roads on topographic map and/or aerial photos.
3. Calculate total acres of road (Column (3) in **Worksheet 5-13**).
4. Count number of stream crossings (Column (4) in **Worksheet 5-13**).
5. Calculate sediment delivered by Road Impact Index (RII) (Column (5) in **Worksheet 5-13**):

    $$RII = \frac{(\text{acres of road})}{(\text{acres of sub-watershed})} \times (\text{number of stream crossings})*$$

    *if no stream crossings, multiply by 1.

6. Determine dominant (most representative) slope position: lower, mid or upper 1/3 slope (Column (6) in **Worksheet 5-13**).
7. Obtain sediment yield (tons/acre of road) by using **Equations 5-2** or **5-3** and comparing with **Figure 5-46** (Column (7) in **Worksheet 5-13**).

    $$y = 1.7 + 40x \quad \text{(lower 1/3 slope position)} \quad \textbf{(Equation 5-2)}$$

    $$y = -0.1595 + 3.0913x \quad \text{(mid or upper 1/3 slope position)} \quad \textbf{(Equation 5-3)}$$

    where:
    y = Sediment yield (tons/acre of road)
    x = Road Impact Index

8. Obtain potential sediment yield, assuming road construction is new. Use the recovery potential or other adjustments, below, to reduce sediment yield based on road age and/or quality of construction.

9. Calculate recovery potential using Megahan's (1974) negative exponential recovery relation, **Equation 5-4**:

    $$\varepsilon_t = 3.52 + 523.9e^{-0.00956t} \quad \textbf{(Equation 5-4)}$$

    where:
    ε_t = Erosion rate recovery (tons/mi^2/day)
    e = Natural logarithm
    t = Elapsed time since disturbance (days)

    See **Figure 5-47**. Because this empirical relation calculates erosion rate recovery for an Idaho site, to use this equation for recovery, a conversion of the elapsed time versus erosion rate reduction needs to be converted to *percent reduction* (**Figure 5-47**). This percent reduction can then be applied to the initial sediment yield as calculated from the Road Impact Index (RII). Insert erosion rate recovery in Column (9) in **Worksheet 5-13**.

    For recovery rate for durations greater than 1,200 days, a reduction of 95% of the initial sediment delivery is recommended for use. Rate of recovery beyond 1,200 days is relatively low.

10. Mitigation adjustments. Use 95% reduction in delivered sediment from calculated rates if roads are surfaced, with stable ditch and/or out-sloped road grade (Column (11) in **Worksheet 5-13**).

    a. If road fills are less than 200 ft from stream, treat as surface erosion. Decrease sediment yields if mitigation by surfacing, out-sloping, etc.; use with sediment delivery ratio relations.
    b. If road is involved in mass erosion or debris torrent, use mass erosion procedure from WRENSS (USEPA, 1980, Chapter V).
    c. If a road encroaches on the stream and subsequently alters dimension, pattern and/or profile, use channel process analysis to quantify impacts.

Chapter 7 — PLA   The WARSSS Methodology: A Case Study of Wolf Creek, Colorado

$y = 1.7 + 40x$

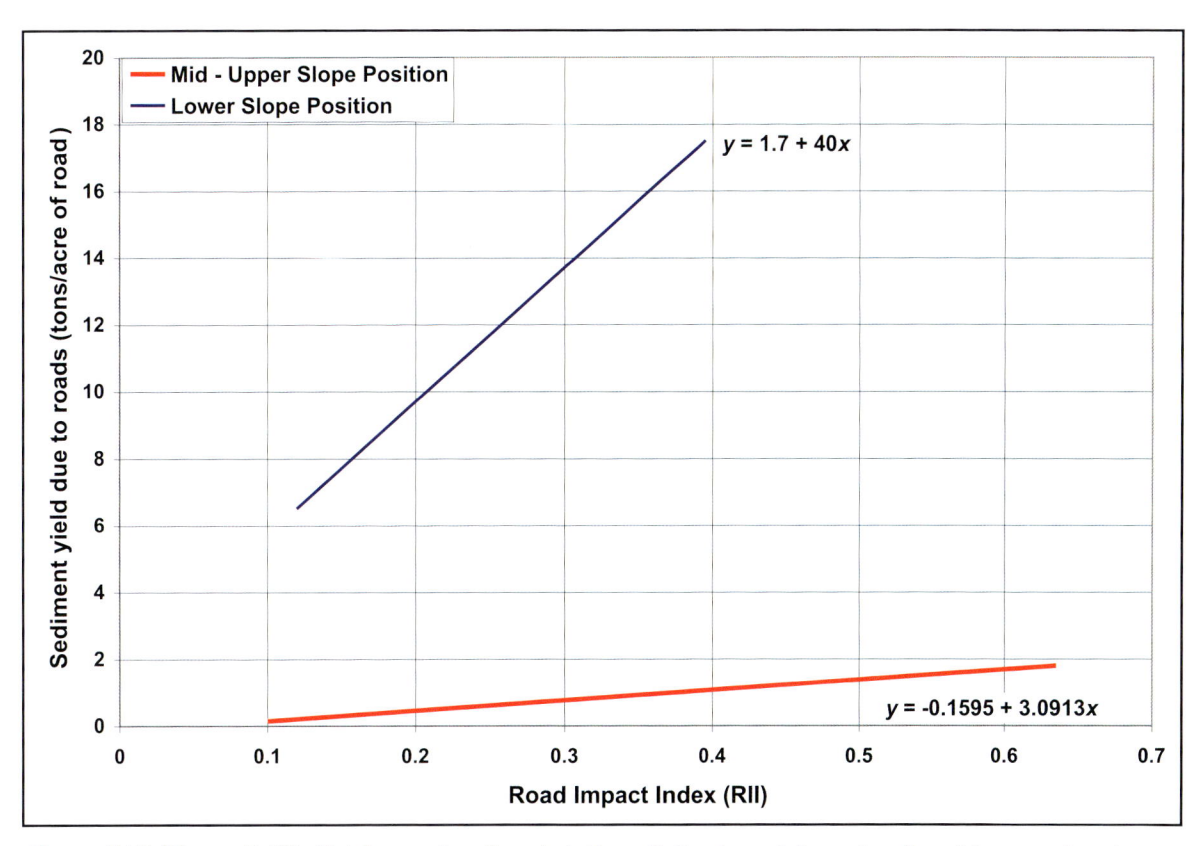

$y = -0.1595 + 3.0913x$

Figure 7-50 (Figure 5-46). Total annual sediment yield prediction from delivered sediment from roads using the Road Impact Index (RII) model (basic data from USDA Forest Service, Horse Creek Watershed, Idaho and Fool Creek, Colorado).

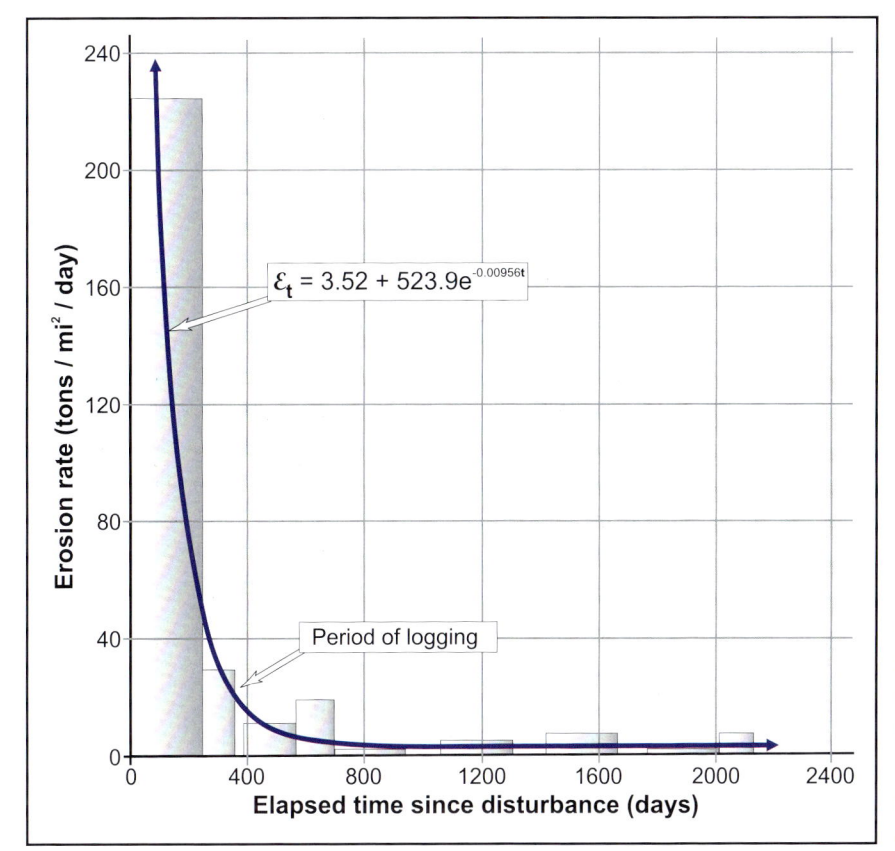

$\varepsilon_t = 3.52 + 523.9e^{-0.00956t}$

Figure 7-51 (Figure 5-47). Erosion rate recovery over time (Megahan, 1974). Use the relations to calculate percent reduction from initial erosion based on elapsed time.

7-163

WATERSHED ASSESSMENT OF RIVER STABILITY AND SEDIMENT SUPPLY

Worksheet 7-25 (Worksheet 5-13). Road Impact Index (RII) and corresponding annual sediment yield from roads using the steps in **Table 7-36** for the Wolf Creek Watershed.

Stream: **Wolf Creek**　　　　　　　　　　　　　　　　　　　　Location: **Sub-watersheds 1–5 and Highway 160**

Observers: **D. Rosgen**　　　　　　　　　　　　　　　　　　　　Date: **7/04**

	Sub-watershed location ID# (1)	Total acres of sub-watershed (Step 1) (2)	Total acres of road (Step 3) (3)	Number of stream crossings (Step 4) (4)	Road impact index [(3)/(2)X(4)] (Step 5) (5)	Dominant slope position (lower, mid or upper slope) (Step 6) (6)	Sediment yield (tons/acre of road) (Fig. 5-46, Step 7) (7)	Total tons [(3)X(7)] (8)	Erosion rate recovery (% from Fig. 5-47, Step 9) (convert to decimal) (9)	Total tons/yr [(8) − (8)×(9)] (10)	Mitigation adjustments (Step 10) (11)
1.	1	355	34	4	0.38	lower	16.90	574.60	95% = .95	28.73	
2.	2	1024	144	3	0.42	upper	1.14	164.16	95% = .95	8.21	
3.	3	861	27	8	0.25	lower	11.70	315.90	95% = .95	15.80	
4.	4	1013	88	3	0.26	lower	12.10	1064.80	95% = .95	53.24	
5.	5	2649	124	2	0.09	upper	0.12	14.88	95% = .95	0.74	
6.	Totals	5902	417					Total for sub-watershed roads:		107	
7.											
8.	Highway 160	3975	157	27	1.07	upper	3.15	494.55	0*	495	
9.											
10.											
11.											
12.											
13.											
14.											
15.											

*No recovery as cut banks, fill and ditch line are actively eroding

Total road sediment yield (tons/year): **602 tons/yr**

Sediment delivery from surface erosion

As the data indicated a low probability for surface erosion processes, no surface erosion data was gathered for the Wolf Creek Watershed. Surface erosion from compacted road surfaces is integrated in the road impact index relations.

Sediment delivery from mass erosion

Active landslides caused by Highway 160 were mapped in **Figure 7-6** (see **Step 5**, *RLA*). A site investigation calculated a landslide toe impinging on Wolf Creek. The length of the debris slide was 600 ft, the depth of erosion source was 1.48 ft and the height of the bank slope was 13.4 ft. The potential, available mass erosion sediment source totaled 12,004 ft^3 (444.6 yds^3 or 578 tons). The sediment delivery ratio was obtained from **Figures 7-52** and **7-53**, which were developed in WRENSS (USEPA, 1980). The relations indicate a sediment delivery of 50% based on slope position, shape and gradient. Therefore, the potential delivered sediment is 289 tons/yr.

Furthermore, because the landslide existed prior to the road construction, the assumption was made that half of the delivered sediment (144 tons/yr) is due to a natural geologic rate, while the other half (145 tons/yr) is due to Highway 160.

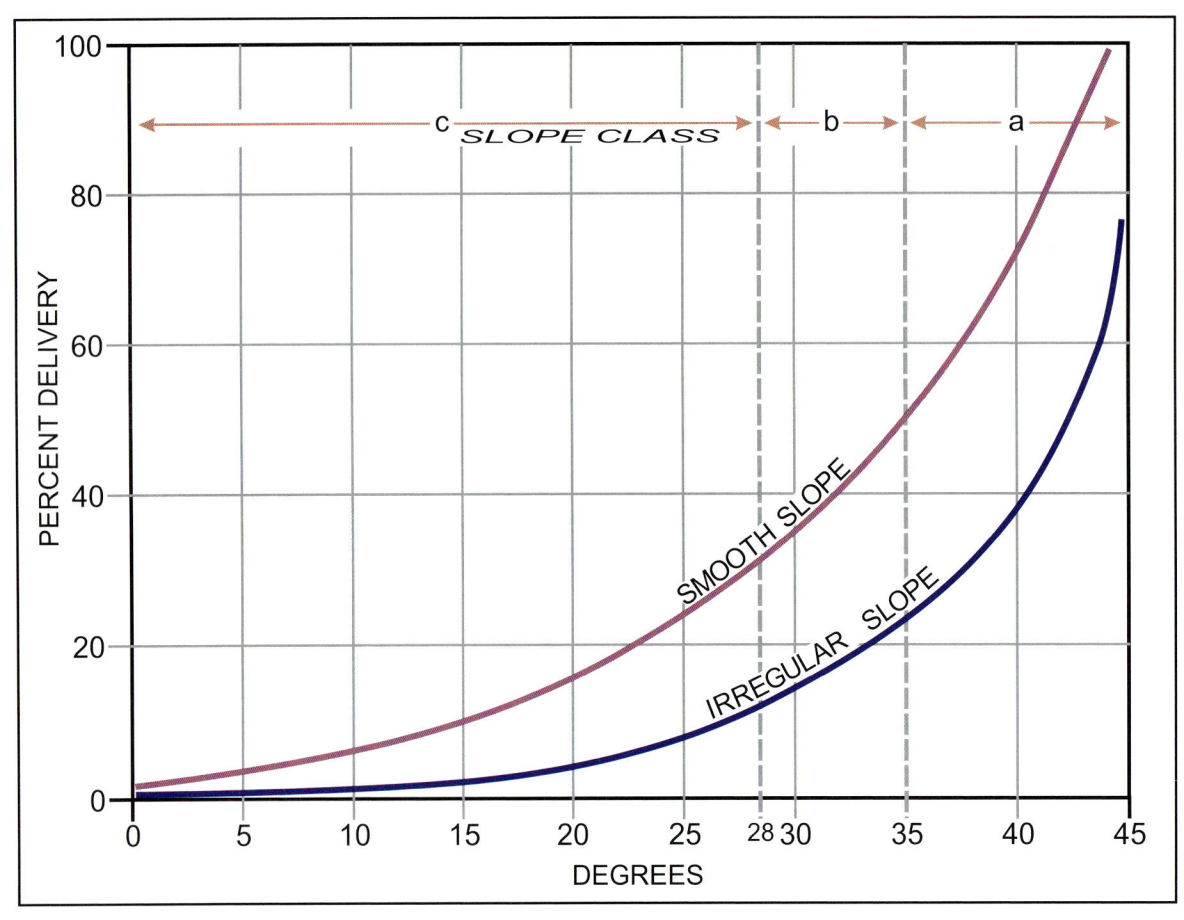

Figure 7-52 (Figure 5-51). Delivery potential of debris avalanche/debris torrent material to closest stream (USEPA, 1980).

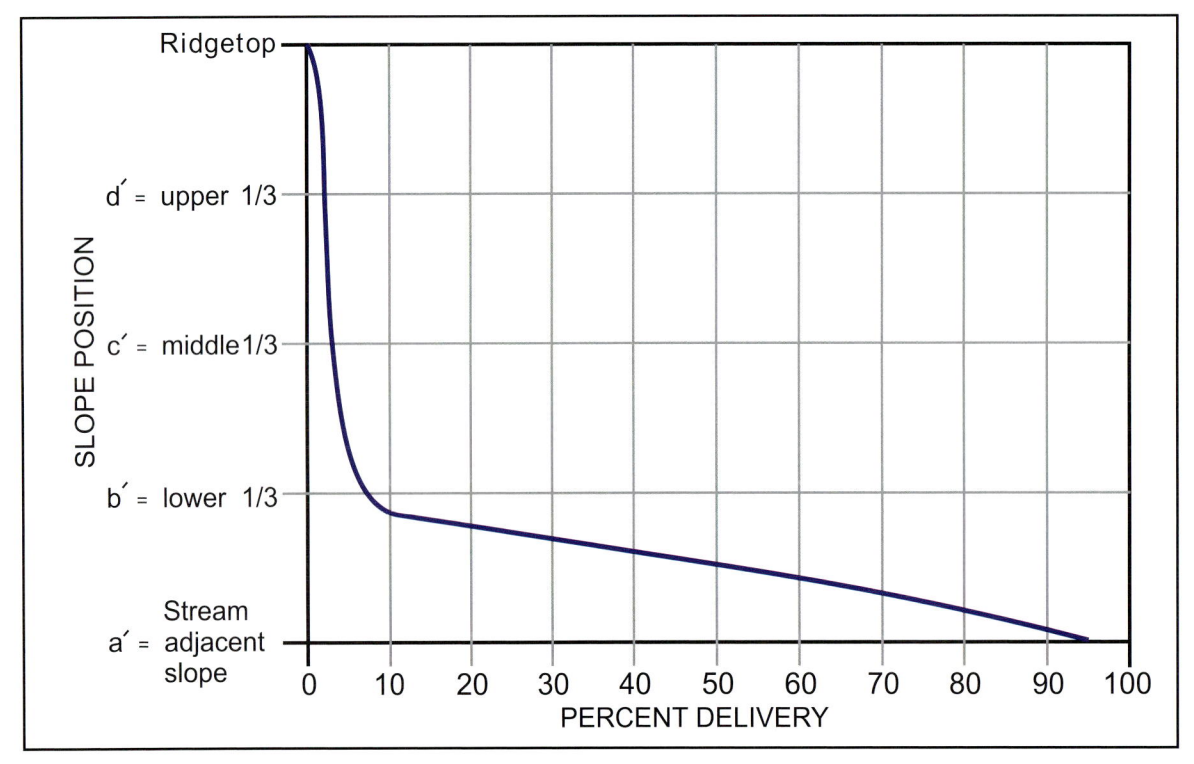

Figure 7-53 (Figure 5-52). Delivery potential of slump/earthflow material to closest stream (USEPA, 1980).

Step 21. Summarize total annual sediment yield (tons/year) from hillslope processes

Table 7-37 summarizes the predicted sediment yield from roads, mass erosion and surface erosion.

Table 7-37. Summary of sediment yield prediction from hillslope processes in the Wolf Creek Watershed.

Predicted sediment yield from hillslope processes
Roads:
Secondary logging roads: **107 tons/yr**
Highway 160: **495 tons/yr**
Total sediment from roads: **602 tons/yr**
Surface Erosion:
Conditions (except for roads) for Hortonian overland flow are rare. Road sediment is calculated separately based on the road impact index and a time-trend recovery relationship.
Mass erosion:
Landslide: **289 tons/yr** (145 tons/yr due to Highway 160)

WATERSHED ASSESSMENT OF RIVER STABILITY AND SEDIMENT SUPPLY

Step 22: Sediment competence prediction (See Flowchart 7-12)

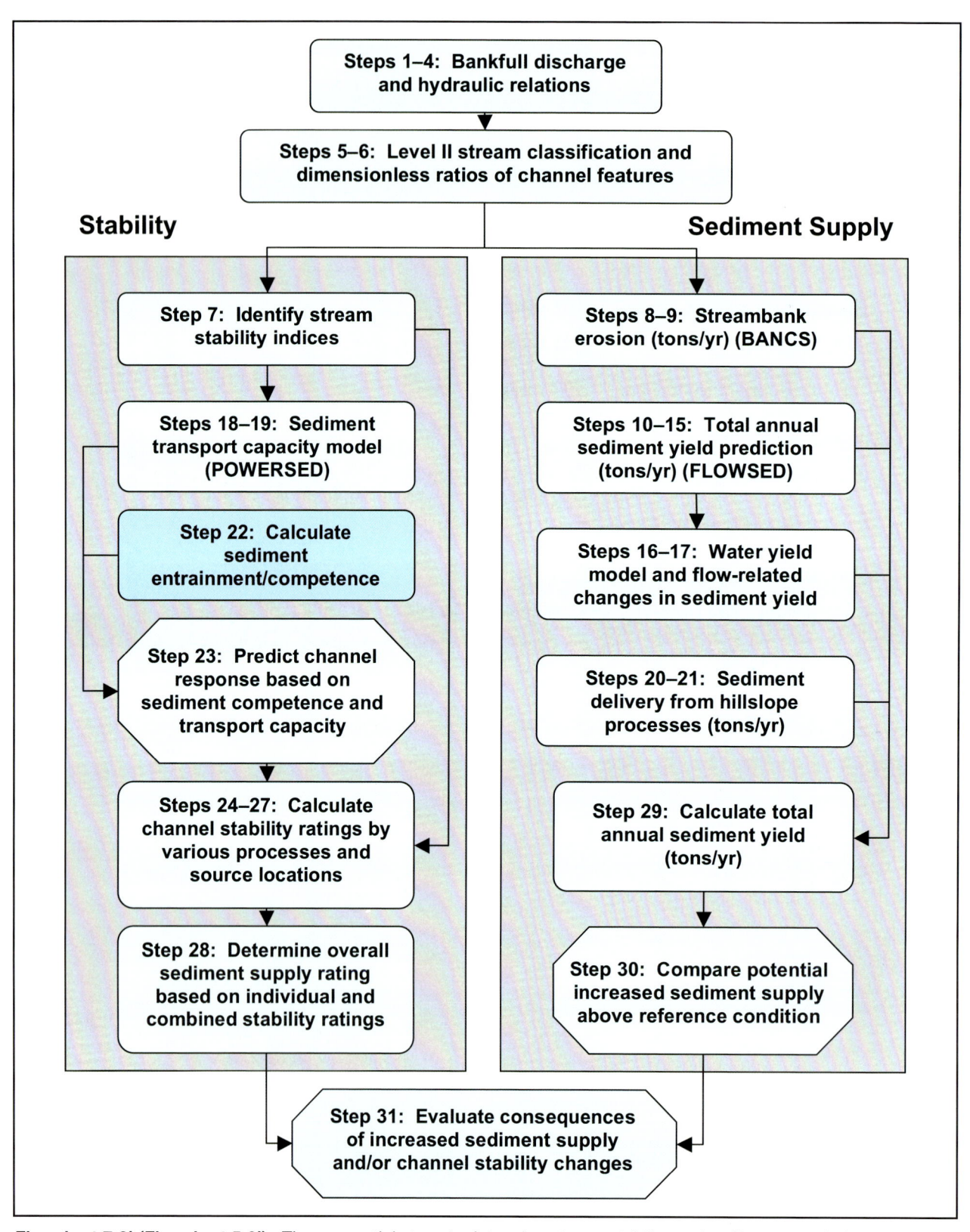

```
                Steps 1–4: Bankfull discharge
                    and hydraulic relations
                              ↓
             Steps 5–6: Level II stream classification and
                dimensionless ratios of channel features

   Stability                                    Sediment Supply

   Step 7: Identify stream          Steps 8–9: Streambank
      stability indices              erosion (tons/yr) (BANCS)
            ↓                                   ↓
   Steps 18–19: Sediment             Steps 10–15: Total annual
   transport capacity model          sediment yield prediction
        (POWERSED)                      (tons/yr) (FLOWSED)
            ↓                                   ↓
     Step 22: Calculate              Steps 16–17: Water yield
        sediment                     model and flow-related
   entrainment/competence            changes in sediment yield
            ↓                                   ↓
    Step 23: Predict channel          Steps 20–21: Sediment
       response based on             delivery from hillslope
   sediment competence and              processes (tons/yr)
      transport capacity                        ↓
            ↓                         Step 29: Calculate total
   Steps 24–27: Calculate             annual sediment yield
   channel stability ratings by            (tons/yr)
    various processes and                      ↓
       source locations               Step 30: Compare potential
            ↓                         increased sediment supply
   Step 28: Determine overall         above reference condition
     sediment supply rating
    based on individual and
   combined stability ratings

              Step 31: Evaluate consequences
                of increased sediment supply
               and/or channel stability changes
```

Flowchart 7-3i (Flowchart 5-3j). The sequential steps to determine stream stability and sediment supply, highlighting **Step 22**.

Chapter 7 — PLA   The WARSSS Methodology: A Case Study of Wolf Creek, Colorado

```
Step 22: Calculate sediment entrainment/competence
```

Collect field data:
- Bed material, riffle bed (D_{50})
- Bar samples (D_{max}, $\hat{D}_{50}$) (Figure 5-53)
- Average water surface slope (bankfull)
- Cross-section (mean bankfull depth)

Obtain ratio of D_{max}/D_{50}
(Worksheet 5-15)

Ratio outside range of 1.3–3.0

Ratio within range of 1.3–3.0

Calculate ratio
$D_{50}/\hat{D}_{50}$

Calculate dimensionless shear stress:
$\tau^* = 0.0384 \, (D_{max}/D_{50})^{-0.887}$

Ratio outside range of 3.0–7.0

Ratio within range of 3.0–7.0

Calculate dimensional shear stress:
$\tau = \gamma RS$
(Figure 5-54, Worksheet 5-15)

Calculate dimensionless shear stress:
$\tau^* = 0.0834 \, (D_{50}/\hat{D}_{50})^{-0.872}$

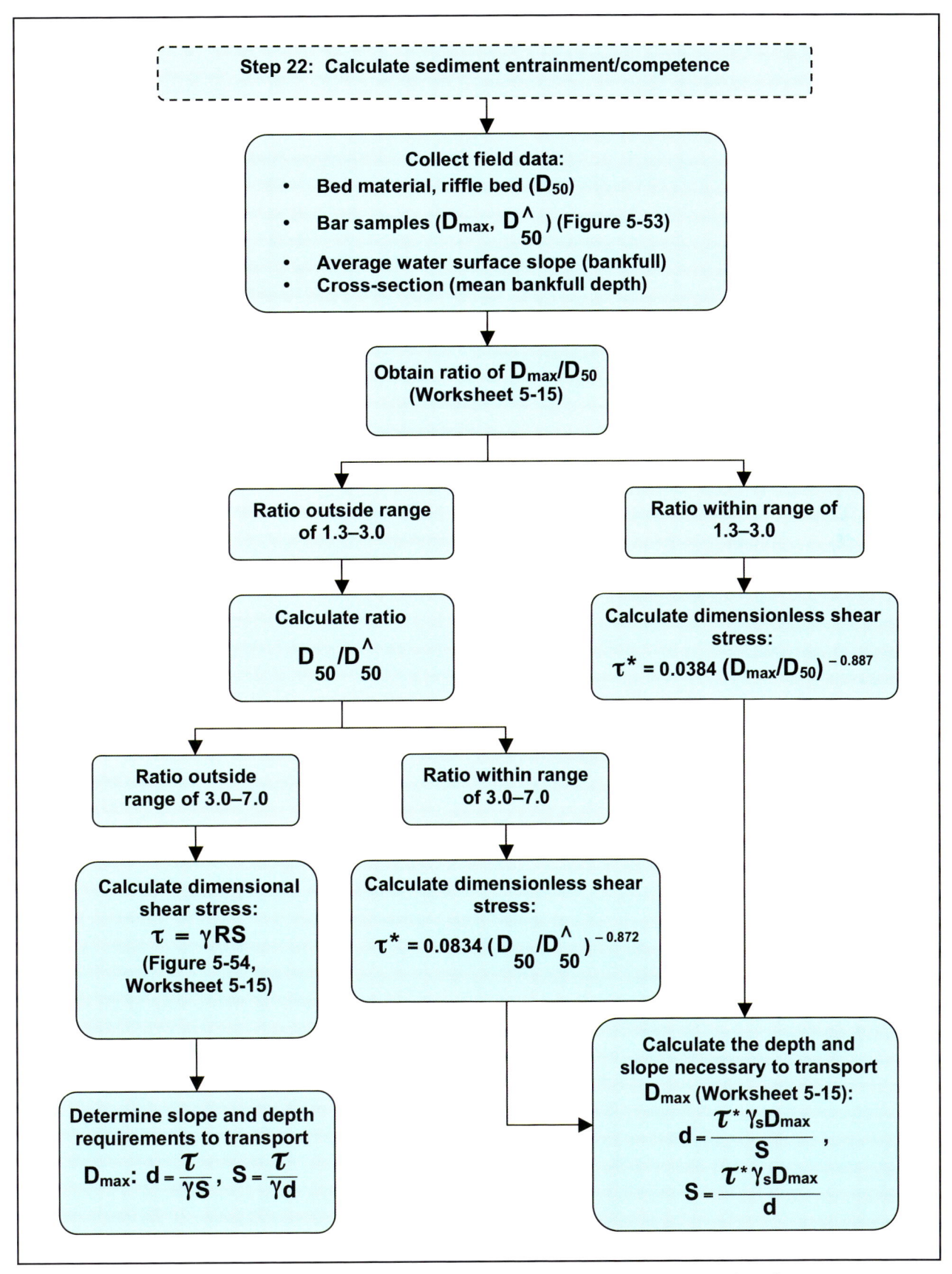

Determine slope and depth requirements to transport
D_{max}: $d = \dfrac{\tau}{\gamma S}$, $S = \dfrac{\tau}{\gamma d}$

Calculate the depth and slope necessary to transport D_{max} (Worksheet 5-15):
$d = \dfrac{\tau^* \gamma_s D_{max}}{S}$,
$S = \dfrac{\tau^* \gamma_s D_{max}}{d}$

Flowchart 7-12 (Flowchart 5-16). Generalized procedure to calculate sediment competence/entrainment.

Step 22. Calculate sediment entrainment/competence

The key reaches or receiving streams that may have an entrainment/competence problem due to an increased supply of larger sediment sizes are the lower C4 and D4 stream types. These two stream types were evaluated independently for competence. The results of the C4 reference reach entrainment relation indicate that the required depth of 1.5 ft matches the existing depth of 1.6, and the slope of 0.005 also matches the required slope of 0.0046. This C4 stream type reach is evaluated as *stable* (**Worksheet 7-26a**).

The impacted D4 stream type, however, has *aggradation* and *unstable* indicators. The required minimum depth is 1.0 ft, but the existing depth is 0.8 ft. The required slope to move the sediment size supplied (78mm) is 0.012, steeper than the existing valley slope. Thus, the D4 reach rates as ***not competent*** to move the sediment size from the C4 stream type reach immediately upstream, and is evaluated as *aggrading* (**Worksheet 7-26b**). This computation agrees with the actual reduction in bed-material size of the D4 stream type, as shown in **Figure 7-34**.

The relationships using dimensional shear stress in **Figure 5-54** in Chapter 5 overestimated particle entrainment compared to observed values. The empirically derived relations shown in **Figure 5-54** are developed from single-thread channels; thus, the C4 stream type matched predicted values close to observed values. However, the D4 stream type (braided, multiple channels) did not predict very well; consequently, the use of the dimensionless shear stress values were used where the predicted values matched closer to the observed values.

Worksheet 7-26a (Worksheet 5-15). Sediment competence calculation form to assess bed stability for the C4 reference reach.

Stream:	**Wolf Creek**		Stream Type:	**C4**	
Location:	**≈ 5000 ft above conf. WF San Juan R.**		Valley Type:	**VIII**	
Observers:	**D. Rosgen**		Date:	**7/04**	

Enter required information

48	D_{50}	Riffle bed material D_{50} (mm)			
8	$\hat{D}_{50}$	Bar sample D_{50} (mm)			
0.26	D_{max}	Largest particle from bar sample (ft)	78	(mm)	304.8 mm/ft
0.005	S	Existing bankfull water surface slope (ft/ft)			
1.6	d	Existing bankfull mean depth (ft)			
1.65	γ_s	Submerged specific weight of sediment			

Select the appropriate equation and calculate critical dimensionless shear stress

6	$D_{50}/\hat{D}_{50}$	Range: 3 – 7	Use EQUATION 1: $\tau^* = 0.0834 \, (D_{50}/\hat{D}_{50})^{-0.872}$
1.6	D_{max}/D_{50}	Range: 1.3 – 3.0	Use EQUATION 2: $\tau^* = 0.0384 \, (D_{max}/D_{50})^{-0.887}$
0.017	τ^*	Bankfull Dimensionless Shear Stress	EQUATION USED: **1**

Calculate bankfull mean depth required for entrainment of largest particle in bar sample

1.5	d	Required bankfull mean depth (ft)	$d = \dfrac{\tau^* \gamma_s D_{max}}{S}$

Check ✓: ☒ Stable   ☐ Aggrading   ☐ Degrading

Calculate bankfull water surface slope required for entrainment of largest particle in bar sample

0.0046	S	Required bankfull water surface slope (ft/ft)	$S = \dfrac{\tau^* \gamma_s D_{max}}{d}$

Check ✓: ☒ Stable   ☐ Aggrading   ☐ Degrading

Sediment competence using dimensional shear stress

0.5	Bankfull shear stress $\tau = \gamma d S$ (lbs/ft²) (substitute hydraulic radius, R, with mean depth, d)
35	Moveable particle size (mm) at bankfull shear stress **(Figure 5-54)**
0.9	Predicted shear stress required to initiate movement of D_{max} (mm) **(Figure 5-54)**
1.6	Predicted mean depth required to initiate movement of D_{max} (mm) $d = \dfrac{\tau}{\gamma S}$
0.005	Predicted slope required to initiate movement of D_{max} (mm) $S = \dfrac{\tau}{\gamma d}$

Worksheet 7-26b (Worksheet 5-15). Sediment competence calculation form to assess bed stability for the D4 impaired reach.

Stream:	**Wolf Creek**		Stream Type:	**D4**	
Location:	≈ **4000 ft above conf. WF San Juan R.**		Valley Type:	**VIII**	
Observers:	**D. Rosgen**		Date:	**7/04**	

Enter required information

19	D_{50}	Riffle bed material D_{50} (mm)			
4	$\hat{D}_{50}$	Bar sample D_{50} (mm)			
0.26*	D_{max}	Largest particle from bar sample (ft)	78*	(mm)	304.8 mm/ft
0.009	S	Existing bankfull water surface slope (ft/ft)			
0.8	d	Existing bankfull mean depth (ft)			
1.65	γ_s	Submerged specific weight of sediment			

Select the appropriate equation and calculate critical dimensionless shear stress

4.75	$D_{50}/\hat{D}_{50}$	Range: 3 – 7	Use EQUATION 1: $\tau^* = 0.0834 \, (D_{50}/\hat{D}_{50})^{-0.872}$
4.11	D_{max}/D_{50}	Range: 1.3 – 3.0	Use EQUATION 2: $\tau^* = 0.0384 \, (D_{max}/D_{50})^{-0.887}$
0.0214	τ^*	Bankfull Dimensionless Shear Stress	EQUATION USED: 1

Calculate bankfull mean depth required for entrainment of largest particle in bar sample

1.02	d	Required bankfull mean depth (ft)	$d = \dfrac{\tau^* \gamma_s D_{max}}{S}$

Check ✓: ☐ Stable ☑ Aggrading ☐ Degrading

Calculate bankfull water surface slope required for entrainment of largest particle in bar sample

0.012	S	Required bankfull water surface slope (ft/ft)	$S = \dfrac{\tau^* \gamma_s D_{max}}{d}$

Check ✓: ☐ Stable ☑ Aggrading ☐ Degrading

Sediment competence using dimensional shear stress

0.45	Bankfull shear stress $\tau = \gamma dS$ (lbs/ft²) (substitute hydraulic radius, R, with mean depth, d)
25	Moveable particle size (mm) at bankfull shear stress **(Figure 5-54)**
0.95	Predicted shear stress required to initiate movement of D_{max} (mm) **(Figure 5-54)**
1.7	Predicted mean depth required to initiate movement of D_{max} (mm) $d = \dfrac{\tau}{\gamma S}$
0.019	Predicted slope required to initiate movement of D_{max} (mm) $S = \dfrac{\tau}{\gamma d}$

*Required to be moved based on upstream C4 stream type.

Step 23: Channel response based on sediment competence and/or transport capacity

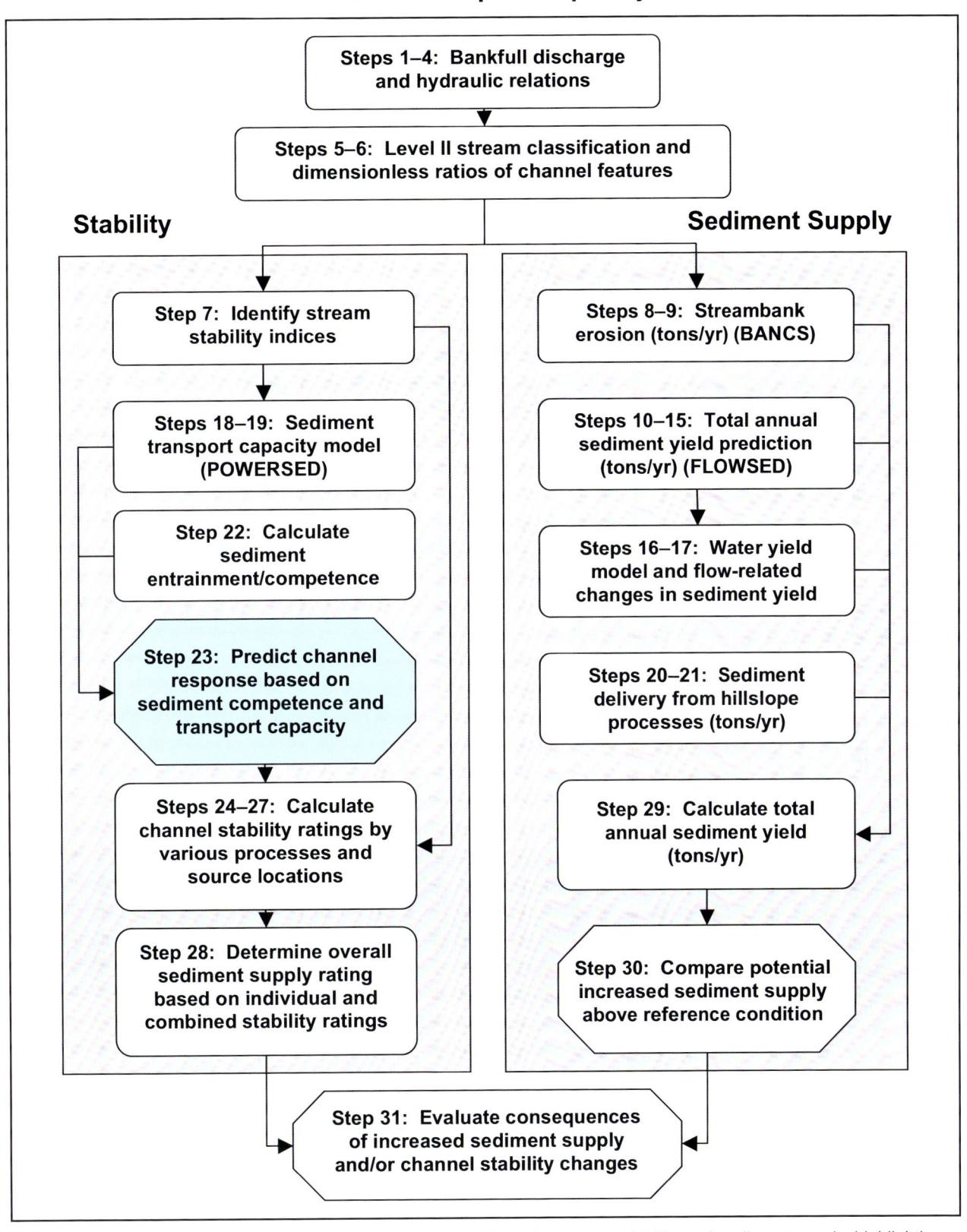

```
                    Steps 1–4: Bankfull discharge
                       and hydraulic relations
                                 │
                                 ▼
                  Steps 5–6: Level II stream classification and
                   dimensionless ratios of channel features
                                 │
      Stability                                          Sediment Supply
           │                                                    │
           ▼                                                    ▼
   Step 7: Identify stream                          Steps 8–9: Streambank
      stability indices                             erosion (tons/yr) (BANCS)
           │                                                    │
           ▼                                                    ▼
   Steps 18–19: Sediment                           Steps 10–15: Total annual
   transport capacity model                         sediment yield prediction
         (POWERSED)                                   (tons/yr) (FLOWSED)
           │                                                    │
           ▼                                                    ▼
      Step 22: Calculate                            Steps 16–17: Water yield
          sediment                                   model and flow-related
     entrainment/competence                         changes in sediment yield
           │                                                    │
           ▼                                                    ▼
     Step 23: Predict channel                         Steps 20–21: Sediment
     response based on                                 delivery from hillslope
     sediment competence and                             processes (tons/yr)
       transport capacity
           │                                                    │
           ▼                                                    ▼
   Steps 24–27: Calculate                           Step 29: Calculate total
   channel stability ratings by                       annual sediment yield
     various processes and                                  (tons/yr)
       source locations
           │                                                    │
           ▼                                                    ▼
   Step 28: Determine overall                       Step 30: Compare potential
    sediment supply rating                          increased sediment supply
    based on individual and                         above reference condition
    combined stability ratings
                          \                       /
                           ▼                     ▼
                    Step 31: Evaluate consequences
                     of increased sediment supply
                     and/or channel stability changes
```

Flowchart 7-3j (Flowchart 5-3k). The sequential steps to determine stream stability and sediment supply, highlighting **Step 23**.

Step 23. Predict channel response based on sediment competence and transport capacity

The sediment capacity reduction of 430 tons/yr of annual sediment yield (**Table 7-35** in **Step 19**) indicates *aggradation* or insufficient capacity. The competence calculations also indicate stream *aggradation* in the D4 stream type (**Worksheet 7-26b** in **Step 22**). The channel continues to widen due to the accelerated bank erosion, making the hydraulic conditions insufficient to accommodate the watershed's sediment supply. The receiving streams below the excess sediment supply (high energy A2 and B3 stream types) have the competence and capacity to transport the high sediment loads. Unfortunately, the unstable D4 impaired reach has a higher potential rate of aggradation/excess deposition as demonstrated in the competence and capacity steps.

Chapter 7 — PLA   The WARSSS Methodology: A Case Study of Wolf Creek, Colorado

Steps 24–27: Channel stability ratings by various processes and source locations (See Flowchart 7-13)

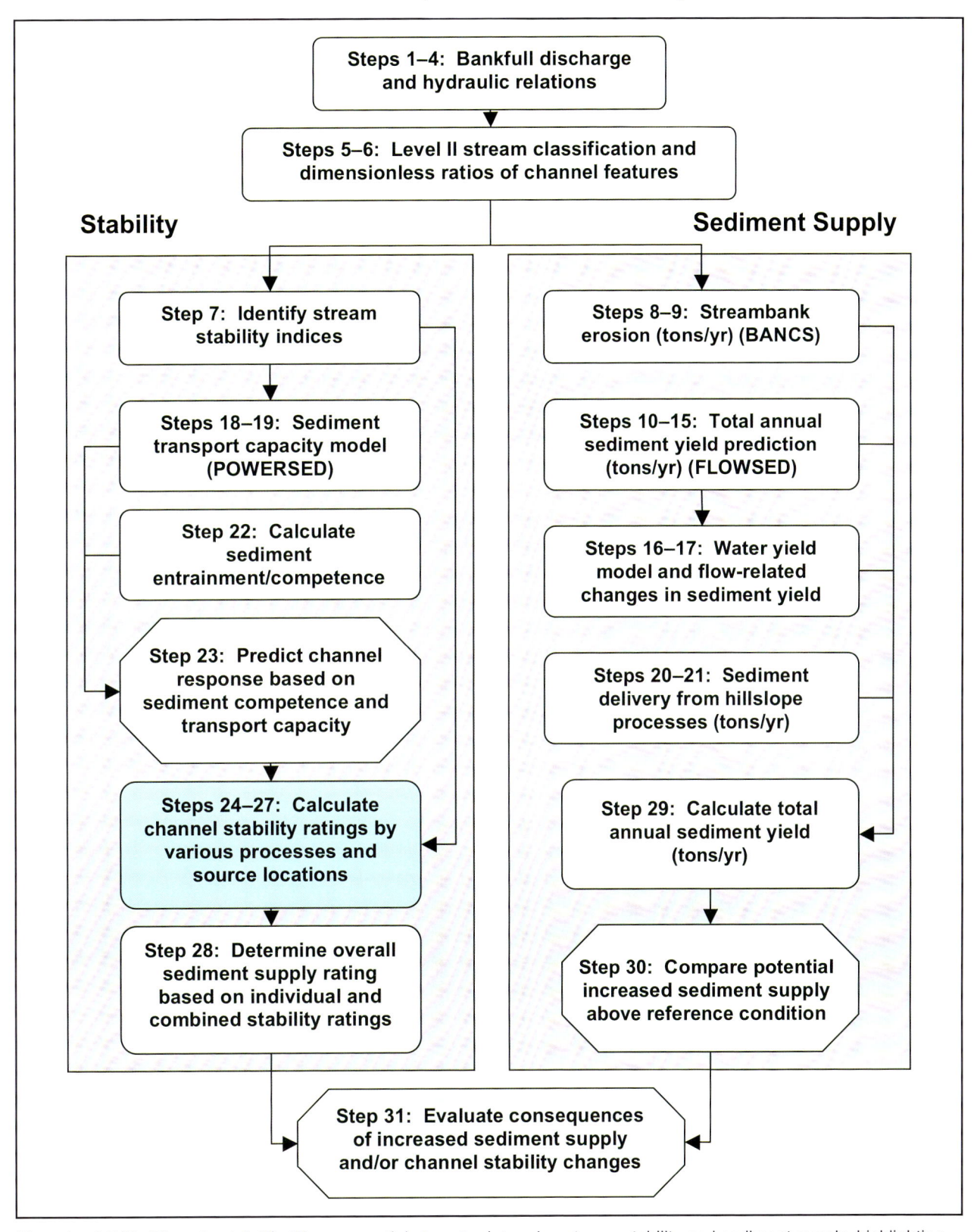

```
                    Steps 1–4: Bankfull discharge
                        and hydraulic relations
                                 │
                                 ▼
                Steps 5–6: Level II stream classification and
                 dimensionless ratios of channel features

   Stability                                      Sediment Supply

   Step 7: Identify stream               Steps 8–9: Streambank
      stability indices                  erosion (tons/yr) (BANCS)
             │                                       │
             ▼                                       ▼
   Steps 18–19: Sediment                 Steps 10–15: Total annual
   transport capacity model              sediment yield prediction
        (POWERSED)                         (tons/yr) (FLOWSED)
             │                                       │
             ▼                                       ▼
      Step 22: Calculate                  Steps 16–17: Water yield
         sediment                         model and flow-related
   entrainment/competence                 changes in sediment yield
             │                                       │
             ▼                                       ▼
   Step 23: Predict channel               Steps 20–21: Sediment
      response based on                   delivery from hillslope
   sediment competence and                   processes (tons/yr)
      transport capacity                            │
             │                                       ▼
             ▼                            Step 29: Calculate total
   Steps 24–27: Calculate                 annual sediment yield
   channel stability ratings by                 (tons/yr)
    various processes and                          │
       source locations                            ▼
             │                            Step 30: Compare potential
             ▼                            increased sediment supply
   Step 28: Determine overall             above reference condition
    sediment supply rating
    based on individual and
    combined stability ratings
                          │            │
                          ▼            ▼
              Step 31: Evaluate consequences
                of increased sediment supply
               and/or channel stability changes
```

Flowchart 7-3k (Flowchart 5-3l). The sequential steps to determine stream stability and sediment supply, highlighting **Steps 24–27**.

7-175

Watershed Assessment of River Stability and Sediment Supply

Steps 24–27: Channel stability ratings by various processes and source locations

Step 24: Calculate potential stream channel succession stage shift (Worksheet 5-16)

Step 25: Calculate lateral stability rating (Worksheet 5-17)
- W/d ratio state
- Depositional patterns
- Meander pattern
- BEHI/NBS
- Meander width ratio (confinement)

Step 26: Calculate vertical stability rating

Aggradation/excess sediment supply (Worksheet 5-18)
- Sediment competence
- Sediment capacity
- W/d ratio state
- Stream succession state
- Depositional patterns
- Debris blockages

Degradation/channel incision (Worksheet 5-19)
- Sediment competence
- Sediment capacity
- Bank-height ratio
- Stream succession state
- Meander width ratio

Step 27: Calculate potential channel enlargement (Worksheet 5-20)

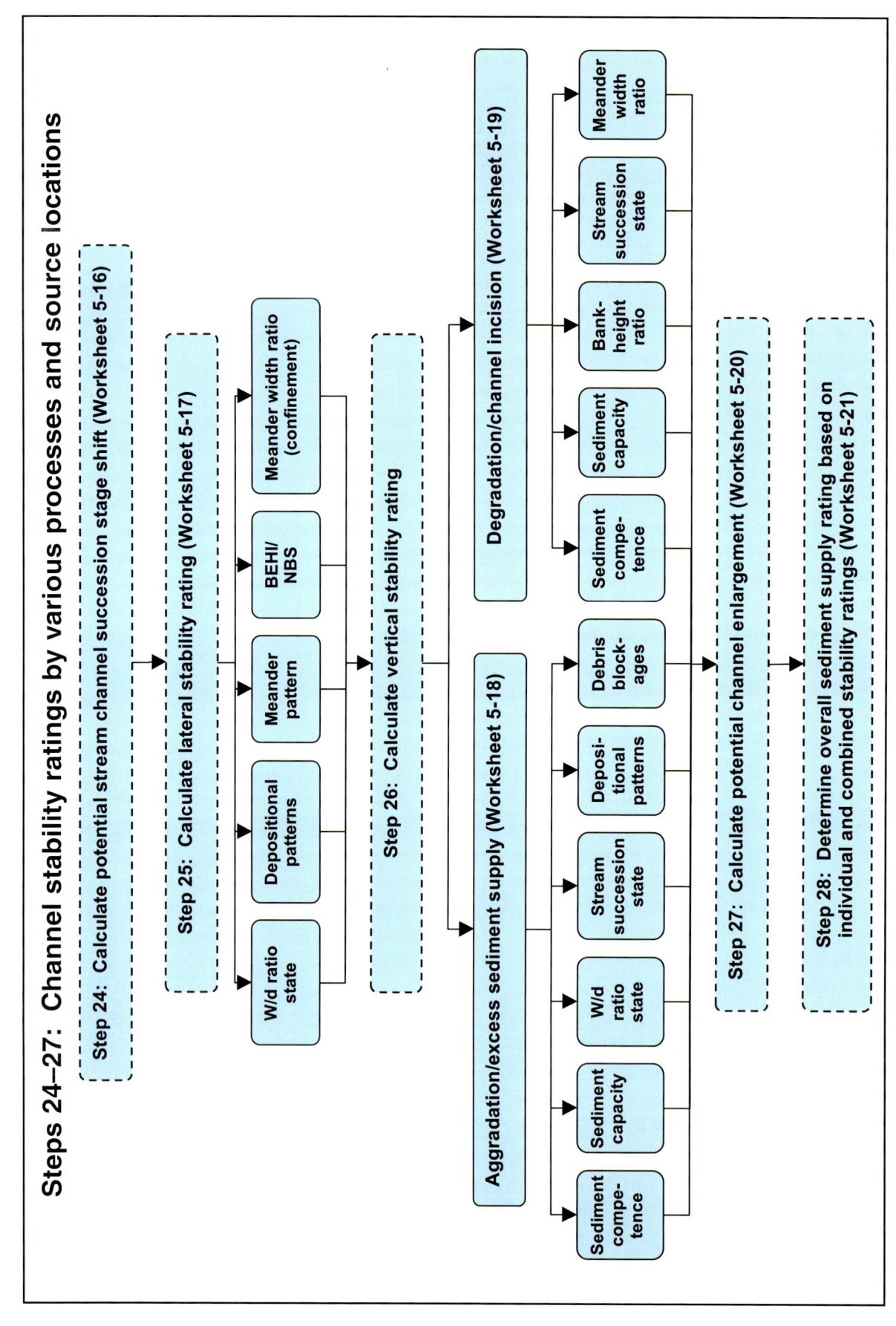

Step 28: Determine overall sediment supply rating based on individual and combined stability ratings (Worksheet 5-21)

Flowchart 7-13 (Flowchart 5-17). The procedural steps and variables used in the assessment of channel stability.

Chapter 7 — PLA The WARSSS Methodology: A Case Study of Wolf Creek, Colorado

Step 24. Calculate potential stream channel successional stage shift

The only reach that rated as a high risk for successional stage shifts and therefore evaluated in detail was the D4 impaired reach, which shifted from a C4 to a D4 stream type (scenario #2 in **Figure 7-54**). The stability ratings for the C4 reference and D4 impaired reaches are shown in **Worksheets 7-27a** and **7-27b**, respectively. These stability ratings are used in **Steps 26** and **27** to assess vertical stability and channel enlargement.

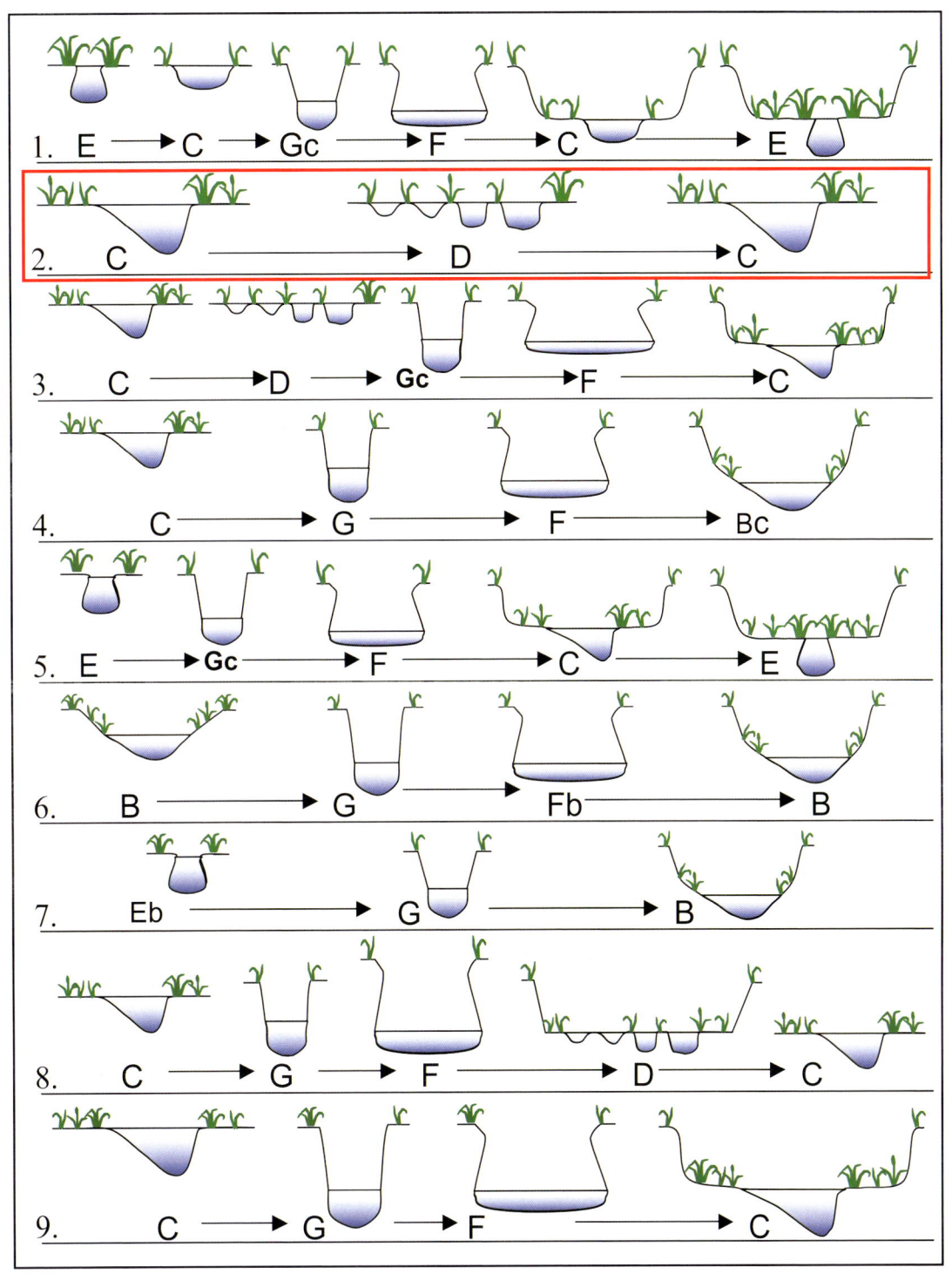

Figure 7-54 (Figure 2-38). Various channel evolution scenarios involving stream type succession, indicating scenario #2 for lower Wolf Creek (Rosgen, 1999, 2001b).

Worksheet 7-27a (Worksheet 5-16). Stability ratings for corresponding successional stage shifts of stream types for the C4 reference reach.

Stream: **Wolf Creek**	Stream Type: **C4**
Location: **above conf. WF San Juan R.**	Valley Type: **VIII**
Observers: **D. Rosgen**	Date: **7/04**
Stream type changes due to successional stage shifts (Figure 5-55)	**Stability rating (check (✓) appropriate rating)**
Stream type at potential, (C→E), (F_b→B), (G→B), (F→B_c), (F→C), (D→C) ⟵ *circled*	☑ Stable
(E→C)	☐ Moderately unstable
(G→F), (F→D)	☐ Unstable
(C→D), (B→G), (D→G), (C→G), (E→G)	☐ Highly unstable

Worksheet 7-27b (Worksheet 5-16). Stability ratings for corresponding successional stage shifts of stream types for the D4 impaired reach.

Stream: **Wolf Creek**	Stream Type: **D4**
Location: **above conf. WF San Juan R.**	Valley Type: **VIII**
Observers: **D. Rosgen**	Date: **7/04**
Stream type changes due to successional stage shifts (Figure 5-55)	**Stability rating (check (✓) appropriate rating)**
Stream type at potential, (C→E), (F_b→B), (G→B), (F→B_c), (F→C), (D→C)	☐ Stable
(E→C)	☐ Moderately unstable
(G→F), (F→D)	☐ Unstable
(C→D) ⟵ *circled*, (B→G), (D→G), (C→G), (E→G)	☑ Highly unstable

Step 25. Calculate lateral stability rating

The following five field stability indicators are used to calculate lateral stability ratings for the C4 reference and D4 impaired reaches:

1. W/d ratio state: W/d ratio divided by reference w/d ratio (**Step 7**, **Worksheet 7-19**);

2. Depositional pattern (**Step 7**, **Worksheet 7-19**);

3. Meander pattern (**Step 7**, **Worksheet 7-19**);

4. Dominant BEHI/NBS (**Step 9**, **Worksheets 7-22a** and **7-22b**); and

5. Degree of confinement: Meander width ratio divided by reference meander width ratio) (**Step 7**, **Worksheet 7-19**).

The lateral stability ratings are shown in **Worksheet 7-28a** for the C4 reference reach and in **Worksheet 7-28b** for the D4 impaired reach. The results indicate *stable* lateral stability for the C4 reference reach and *highly unstable* for the D4 impaired reach. These lateral stability ratings are also used in **Step 27** to determine potential channel enlargement and to calculate the overall sediment supply rating in **Step 28**. The *highly unstable* rating for the D4 impaired reach is interpreted in **Step 31**, and mitigation recommendations are offered.

Watershed Assessment of River Stability and Sediment Supply

Worksheet 7-28a (Worksheet 5-17). Lateral stability prediction summary for the C4 reference reach.

Stream:	Wolf Creek		Stream Type:	C4
Location:	≈ 5000 ft above conf. of WF San Juan River		Valley Type:	VIII
Observers:	D. Rosgen		Date:	7/04

Lateral stability criteria (choose one stability category for each criterion 1–5)	Stable	Moderately unstable	Unstable	Highly unstable	Selected points (from each row)
1. W/d ratio state (Worksheet 5-5)	< 1.2 1 **(2)**	1.2 – 1.4 (4)	1.4 – 1.6 (6)	> 1.6 (8)	2
2. Depositional pattern (Worksheet 5-5)	B1, B2 **B1, B2 (1)**	B4, B8 (2)	B3 (3)	B5, B6, B7 (4)	1
3. Meander pattern (Worksheet 5-5)	M1, M3, M4 **M1 (1)**		M2, M5, M6, M7, M8 (3)		1
4. Dominant BEHI / NBS (Worksheet 5-10)	L/VL, L/L, L/M, L/H, L/VH, M/VL **L/M (2)**	M/L, M/M, M/H, L/Ex, H/L (4)	M/VH, M/Ex, H/L, H/M, H/H, VH/VL, Ex/VL (6)	H/H, H/Ex, Ex/M, Ex/H, Ex/VH, VH/VH, Ex/Ex (8)	2
5. Degree of confinement (MWR / MWR$_{ref}$) (Worksheet 5-5)	0.8 – 1.0 **1.0 (1)**	0.3 – 0.79 (2)	0.1 – 0.29 (3)	< 0.1 (4)	1
				Total points	**7**

Overall lateral stability category (use total points and check (✓) stability rating)	Stable 7 – 9 ✓	Moderately unstable 10 – 12 ☐	Unstable 13 – 21 ☐	Highly unstable > 21 ☐

Worksheet 7-28b (Worksheet 5-17). Lateral stability prediction summary for the D4 impaired reach.

Stream:	Wolf Creek		Stream Type:	D4
Location:	≈ 4000 ft above conf. WF San Juan River		Valley Type:	VIII
Observers:	D. Rosgen		Date:	7/04

Lateral stability criteria (choose one stability category for each criterion 1–5)	Lateral stability categories				Selected points (from each row)
	Stable	Moderately unstable	Unstable	Highly unstable	
1. W/d ratio state (Worksheet 5-5)	< 1.2 (2)	1.2 – 1.4 (4)	1.4 – 1.6 (6)	> 1.6 **11** (8)⃝	8
2. Depositional pattern (Worksheet 5-5)	B1, B2 (1)	B4, B8 (2)	B3 (3)	B5, B6, B7 **B5, B6, B7** (4)⃝	4
3. Meander pattern (Worksheet 5-5)	M1, M3, M4 (1)		M2, M5, M6, M7, M8 **M5, M8** (3)⃝		3
4. Dominant BEHI / NBS (Worksheet 5-10)	L/VL, L/L, L/M, L/H, L/VH, M/VL (2)	M/L, M/M, M/H, L/Ex, H/L (4)	M/VH, M/Ex, H/L, H/M, H/H, VH/VL, Ex/VL (6)	H/H, H/Ex, Ex/M, Ex/H, Ex/VH, VH/VH, Ex/Ex **Ex/VH** (8)⃝	8
5. Degree of confinement (MWR / MWR_ref) (Worksheet 5-5)	0.8 – 1.0 (1)	0.3 – 0.79 (2)	0.1 – 0.29 **0.2** (3)⃝	< 0.1 (4)	3
				Total points	26

Overall lateral stability category (use total points and check (✓) stability rating)	Lateral stability category point range			
	Stable 7 – 9 ☐	Moderately unstable 10 – 12 ☐	Unstable 13 – 21 ☐	Highly unstable > 21 ☑

Step 26. Calculate vertical stability ratings

The vertical stability ratings are separated by aggradation or excess deposition (**Worksheets 7-29a** and **7-29b**) and degradation/channel incision (**Worksheets 7-30a** and **7-30b**).

Aggradation or excess deposition

The variables and individual stability ratings that are used to assess vertical stability for aggradation/excess deposition include the following six categories:

1. Sediment competence (**Step 22**, **Worksheets 7-26a** and **7-26b**);
2. Sediment capacity (**Steps 18** and **19**, **Table 7-35** and **Worksheets 7-24a**, **7-24b** and **7-24c**);
3. W/d ratio state (**Step 7**, **Worksheet 7-19**);
4. Stream succession states (**Step 24**, **Worksheets 7-27a** and **7-27b**);
5. Depositional patterns (**Step 7**, **Worksheet 7-19**); and
6. Debris blockages (**Step 7**, **Worksheet 7-19**).

The six categories are used to complete the vertical stability assessment for the C4 reference and D4 impaired reaches. The C4 reference reach indicates *no deposition* (**Worksheet 7-29a**). However, the D4 impaired reach indicates *aggradation* (**Worksheet 7-29b**). Interpretations for the *aggradation* rating are discussed in **Step 31**, along with mitigation recommendations.

Degradation/channel incision

The following are the five categories that are used to evaluate the overall vertical stability for degradation/channel incision:

1. Sediment competence (**Step 22**, **Worksheets 7-26a** and **7-26b**);
2. Sediment capacity (**Steps 18** and **19**, **Table 7-35** and **Worksheets 7-24a**, **7-24b** and **7-24c**);
3. Bank-Height Ratio (BHR) (degree of channel incision) (**Step 7**, **Worksheet 7-19**);
4. Stream succession states (**Step 24**, **Worksheets 7-27a** and **7-27b**); and
5. Channel confinement (Meander Width Ratio (MWR)) (**Step 7**, **Worksheet 7-19**).

All of the categories for the C4 reference and D4 impaired reaches rated as *not incised* for degradation/channel incision potential (**Worksheets 7-30a** and **7-30b**).

The vertical stability ratings for aggradation/excess sediment and degradation/channel incision are also used to determine potential channel enlargement in **Step 27** and to calculate overall sediment supply ratings in **Step 28**.

Worksheet 7-29a (Worksheet 5-18). Vertical stability prediction for excess deposition/aggradation for the C4 reference reach.

Stream:	Wolf Creek		Stream Type:	C4
Location:	≈ 5000 ft above conf. WF San Juan River		Valley Type:	VIII
Observers:	D. Rosgen		Date:	7/04

Vertical stability criteria (choose one stability category for each criterion 1–6)	Vertical stability categories for excess deposition / aggradation				Selected points (from each row)
	No deposition	Moderate deposition	Excess deposition	Aggradation	
1. Sediment competence (Worksheet 5-15)	Sufficient depth and/or slope to transport largest size available	Trend toward insufficient depth and/or slope-slightly incompetent	Cannot move D_{35} of bed material and/or D_{100} of bar material	Cannot move D_{16} of bed material and/or D_{100} of bar or subpavement size	2
	(2)	**(4)**	(6)	(8)	
2. Sediment capacity (POWERSED) (Worksheet 5-12)	Sufficient capacity to transport annual load	Trend toward insufficient sediment capacity	Reduction up to 25% of annual sediment yield of bedload and/or suspended	Reduction over 25% of annual sediment yield for bedload and/or suspended	2
	(2)	(4)	(6)	(8)	
3. W/d ratio state (Worksheet 5-5)	1.0 – 1.2	1.2 – 1.4	1.4 – 1.6	>1.6	2
	1.0 **(2)**	(4)	(6)	(8)	
4. Stream sucession states (Worksheet 5-16)	Current stream type at potential	(E→C)	(C→High w/d C), (B→High w/d B)	(C→D), (F→D)	2
	(2)	(4)	(6)	(8)	
5. Depositional patterns (Worksheet 5-5)	B1	B2, B4	B3, B5	B6, B7, B8	2
	(1)	B1, B2 **(2)**	(3)	(4)	
6. Debris / blockages (Worksheet 5-5)	D1, D2, D3	D4, D7	D5, D8	D6, D9, D10	1
	D3 **(1)**	(2)	(3)	(4)	
				Total points	11

	Vertical stability category point range for excess deposition / aggradation				
Vertical stability for excess deposition / aggradation (use total points and check (✓) stability rating)	No deposition 10 – 14 ☑	Moderate deposition 15 – 20 ☐	Excess deposition 21 – 30 ☐	Aggradation > 30 ☐	

Worksheet 7-29b (Worksheet 5-18). Vertical stability prediction for excess deposition/aggradation for the D4 impaired reach.

Stream:	Wolf Creek	Stream Type:	D4
Location:	≈ 4000 ft above conf. WF San Juan River	Valley Type:	VIII
Observers:	D. Rosgen	Date:	7/04

Vertical stability criteria (choose one stability category for each criterion 1–6)	Vertical stability categories for excess deposition / aggradation				Selected points (from each row)
	No deposition	**Moderate deposition**	**Excess deposition**	**Aggradation**	
1. Sediment competence (Worksheet 5-15)	Sufficient depth and/or slope to transport largest size available	Trend toward insufficient depth and/or slope-slightly incompetent	Cannot move D_{35} of bed material and/or D_{100} of bar material	Cannot move D_{16} of bed material and/or D_{100} of bar or subpavement size	8
	(2)	(4)	(6)	**(8)**	
2. Sediment capacity (POWERSED) (Worksheet 5-12)	Sufficient capacity to transport annual load	Trend toward insufficient sediment capacity	Reduction up to 25% of annual sediment yield of bedload and/or suspended	Reduction over 25% of annual sediment yield for bedload and/or suspended	8
	(2)	(4)	(6)	**(8)**	
3. W/d ratio state (Worksheet 5-5)	1.0 – 1.2	1.2 – 1.4	1.4 – 1.6	>1.6	8
	(2)	(4)	11 (6)	**(8)**	
4. Stream succession states (Worksheet 5-16)	Current stream type at potential	(E→C)	(C→High w/d C), (B→High w/d B)	(C→D), (F→D)	8
	(2)	(4)	(6)	**C→D (8)**	
5. Depositional patterns (Worksheet 5-5)	B1	B2, B4	B3, B5	B6, B7, B8	4
	(1)	(2)	(3)	**B5, B6, B7 (4)**	
6. Debris / blockages (Worksheet 5-5)	D1, D2, D3	D4, D7	D5, D8	D6, D9, D10	1
	D3 (1)	(2)	(3)	(4)	
				Total points	37

	Vertical stability category point range for excess deposition / aggradation				
Vertical stability for excess deposition / aggradation (use total points and check (✓) stability rating)	No deposition 10 – 14 ☐	Moderate deposition 15 – 20 ☐	Excess deposition 21 – 30 ☐	Aggradation > 30 ☑	

Worksheet 7-30a (Worksheet 5-19). Vertical stability prediction for channel incision/degradation for the C4 reference reach.

Stream:	Wolf Creek		Stream Type:	C4
Location:	≈ 5000 ft above conf. WF San Juan River		Valley Type:	VIII
Observers:	D. Rosgen		Date:	7/04

Vertical stability criteria (choose one stability category for each criterion 1–5)	Vertical stability categories for channel incision / degradation				Selected points (from each row)
	Not incised	Slightly incised	Moderately incised	Degradation	
1 Sediment competence (Worksheet 5-15)	Does not indicate excess competence	Trend to move larger sizes than D_{100} of bar or > D_{84} of bed	D_{100} of bed moved	Particles much larger than D_{100} of bed moved	2
	② (2)	(4)	(6)	(8)	
2 Sediment capacity (POWERSED) (Worksheet 5-12)	Does not indicate excess capacity	Slight excess energy: up to 10% increase above reference	Excess energy sufficient to increase load up to 50% of annual load	Excess energy transporting more than 50% of annual load	2
	② (2)	(4)	(6)	(8)	
3 Degree of channel incision (BHR) (Worksheet 5-5)	1.00 – 1.10	1.11 – 1.30	1.31 – 1.50	> 1.50	2
	1.0 ② (2)	(4)	(6)	(8)	
4 Stream sucession states (Worksheets 5-5 and 5-16)	Does not indicate incision or degradation	If BHR > 1.1 and stream type has w/d between 5–10	If BHR > 1.1 and stream type has w/d less than 5	(B→G), (C→G), (E→G), (D→G)	2
	② (2)	(4)	(6)	(8)	
5 Confinement (MWR / MWR$_{ref}$) (Worksheet 5-5)	0.80 – 1.00	0.30 – 0.79	0.10 – 0.29	< 0.10	1
	1.0 ① (1)	(2)	(3)	(4)	
				Total points	**9**

	Vertical stability category point range for channel incision / degradation				
Vertical stability for channel incision/ degradation (use total points and check (✓) stability rating)	Not incised 9 – 11 ☑	Slightly incised 12 – 18 ☐	Moderately incised 19 – 27 ☐	Degradation > 27 ☐	

WATERSHED ASSESSMENT OF RIVER STABILITY AND SEDIMENT SUPPLY

Worksheet 7-30b (Worksheet 5-19). Vertical stability prediction for channel incision/degradation for the D4 impaired reach.

Stream:	Wolf Creek		Stream Type:	D4
Location:	≈ 4000 ft above WF San Juan River		Valley Type:	VIII
Observers:	D. Rosgen		Date:	7/04

Vertical stability criteria (choose one stability category for each criterion 1–5)	Vertical stability categories for channel incision / degradation				Selected points (from each row)
	Not incised	**Slightly incised**	**Moderately incised**	**Degradation**	
1. Sediment competence (Worksheet 5-15)	Does not indicate excess competence (2) ◯	Trend to move larger sizes than D_{100} of bar or > D_{84} of bed (4)	D_{100} of bed moved (6)	Particles much larger than D_{100} of bed moved (8)	2
2. Sediment capacity (POWERSED) (Worksheet 5-12)	Does not indicate excess capacity (2) ◯	Slight excess energy: up to 10% increase above reference (4)	Excess energy sufficient to increase load up to 50% of annual load (6)	Excess energy transporting more than 50% of annual load (8)	2
3. Degree of channel incision (BHR) (Worksheet 5-5)	1.00 – 1.10 1.0 (2) ◯	1.11 – 1.30 (4)	1.31 – 1.50 (6)	> 1.50 (8)	2
4. Stream succession states (Worksheets 5-5 and 5-16)	Does not indicate incision or degradation (2) ◯	If BHR > 1.1 and stream type has w/d between 5–10 (4)	If BHR > 1.1 and stream type has w/d less than 5 (6)	(B→G), (C→G), (E→G), (D→G) (8)	2
5. Confinement (MWR / MWR_ref) (Worksheet 5-5)	0.80 – 1.00 N/A* (1) ◯	0.30 – 0.79 (2)	0.10 – 0.29 (3)	< 0.10 (4)	1
				Total points	9

	Vertical stability category point range for channel incision / degradation				
Vertical stability for channel incision/ degradation (use total points and check (✓) stability rating)	**Not incised** 9 – 11 ☑	**Slightly incised** 12 – 18 ☐	**Moderately incised** 19 – 27 ☐	**Degradation** > 27 ☐	

*MWR is not applicable to infer vertical stability for D4 braided stream types.

Step 27. Calculate potential channel enlargement

Channel enlargement is a key stability summary. The consequences of channel enlargement include land loss, flood-level increases with less than flood discharges and loss of fish habitat. The C4 reference and D4 impaired reaches were evaluated for potential channel enlargement in **Worksheets 7-31a** and **7-31b** using the successional stage shifts from **Step 24** and the lateral and vertical stability ratings from **Steps 25** and **26**.

The results of the channel enlargement prediction indicate *no increase* for the C4 reference reach. The D4 impaired reach, however, indicates *extensive* channel enlargement. This *extensive* enlargement rating is due to the successional stage shift from a C4 to D4 stream type, the *highly unstable* lateral stability rating and the *aggradation* rating for the vertical stability category (**Worksheet 7-31b**). The *extensive* channel enlargement rating is explained further in **Step 31**, and mitigation recommendations are offered.

Worksheet 7-31a (Worksheet 5-20). Channel enlargement prediction for the C4 reference reach.

Stream:	Wolf Creek		Stream Type:	C4
Location:	≈ 5000 ft above conf. WF San Juan River		Valley Type:	VIII
Observers:	D. Rosgen		Date:	7/04

Channel enlargement prediction criteria (choose one stability category for each criterion 1–4)	Channel enlargement prediction categories				Selected points (from each row)
	No increase	**Slight increase**	**Moderate increase**	**Extensive**	
1. Successional stage shift (Worksheet 5-16)	Stream type at potential, (C→E), (F$_b$→B), (G→B), (F→B$_c$), (F→C), (D→C)	(E→C)	(G→F), (F→D)	(C→D), (B→G), (D→G), (C→G), (E→G)	2
	at potential **(2)**	(4)	(6)	(8)	
2. Lateral stability (Worksheet 5-17)	Stable	Moderately unstable	Unstable	Highly unstable	2
	(2)	(4)	(6)	(8)	
3. Vertical stability excess deposition/ aggradation (Worksheet 5-18)	No deposition	Moderate deposition	Excess deposition	Aggradation	2
	(2)	(4)	(6)	(8)	
4. Vertical stability incision/ degradation (Worksheet 5-19)	Not incised	Slightly incised	Moderately incised	Degradation	2
	(2)	(4)	(6)	(8)	
				Total points	8

	Category point range				
Channel enlargement prediction (use total points and check (✓) stability rating)	No increase 8 – 10 ☑	Slight increase 11 – 16 ☐	Moderate increase 17 – 24 ☐	Extensive > 24 ☐	

Worksheet 7-31b (Worksheet 5-20). Channel enlargement prediction for the D4 impaired reach.

Stream:	Wolf Creek		Stream Type:	D4	
Location:	≈ 4000 ft above conf. WF San Juan River		Valley Type:	VIII	
Observers:	D. Rosgen		Date:	7/04	

Channel enlargement prediction criteria (choose one stability category for each criterion 1–4)	Channel enlargement prediction categories				Selected points (from each row)
	No increase	Slight increase	Moderate increase	Extensive	
1. Successional stage shift (Worksheet 5-16)	Stream type at potential, (C→E), (F_b→B), (G→B), (F→B_c), (F→C), (D→C)	(E→C)	(G→F), (F→D)	(C→D), (B→G), (D→G), (C→G), (E→G)	8
	(2)	(4)	(6)	(C→D) **(8)**	
2. Lateral stability (Worksheet 5-17)	Stable	Moderately unstable	Unstable	Highly unstable	8
	(2)	(4)	(6)	**(8)**	
3. Vertical stability excess deposition/ aggradation (Worksheet 5-18)	No deposition	Moderate deposition	Excess deposition	Aggradation	8
	(2)	(4)	(6)	**(8)**	
4. Vertical stability incision/ degradation (Worksheet 5-19)	Not incised	Slightly incised	Moderately incised	Degradation	2
	(2)	(4)	(6)	(8)	
				Total points	26

	Category point range				
Channel enlargement prediction (use total points and check (✓) stability rating)	No increase 8 – 10 ☐	Slight increase 11 – 16 ☐	Moderate increase 17 – 24 ☐	Extensive > 24 ☑	

Step 28: Overall sediment supply rating by individual and combined stability ratings (See Flowchart 7-14)

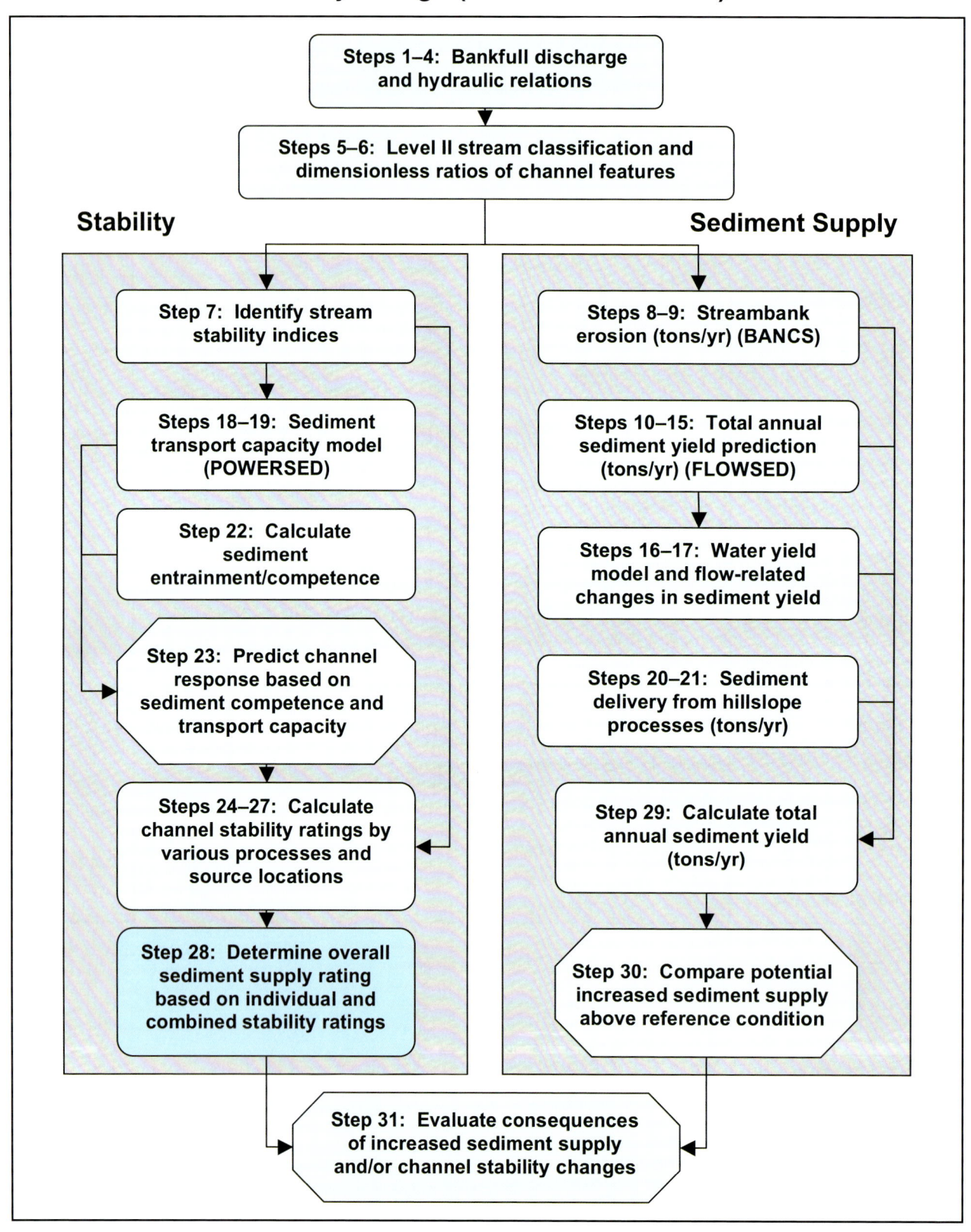

Flowchart 7-3l (Flowchart 5-3m). The sequential steps to determine stream stability and sediment supply, highlighting **Step 28**.

```
┌─────────────────────────────────────────────────────────────────────────┐
│   ┌─────────────────────────────────────────────────────────┐           │
│   │ Step 28: Determine overall sediment supply rating based on│          │
│   │   individual and combined stability ratings (Worksheet 5-21)│        │
│   └─────────────────────────────────────────────────────────┘           │
│         │         │         │         │         │                        │
│         ▼         ▼         ▼         ▼         ▼                        │
│    Lateral    Vertical  Vertical  Channel   Pfankuch                    │
│   stability  stability: stability: enlargement stability                │
│  (Worksheet  aggradation/ degradation/ (Worksheet rating                │
│    5-17)     excess     channel     5-20)    (Worksheet                 │
│              sediment   incision              5-7)                      │
│              supply     (Worksheet                                      │
│             (Worksheet   5-19)                                          │
│              5-18)                                                      │
│                                                                          │
│              PLA summary Worksheet 5-22                                 │
└─────────────────────────────────────────────────────────────────────────┘
```

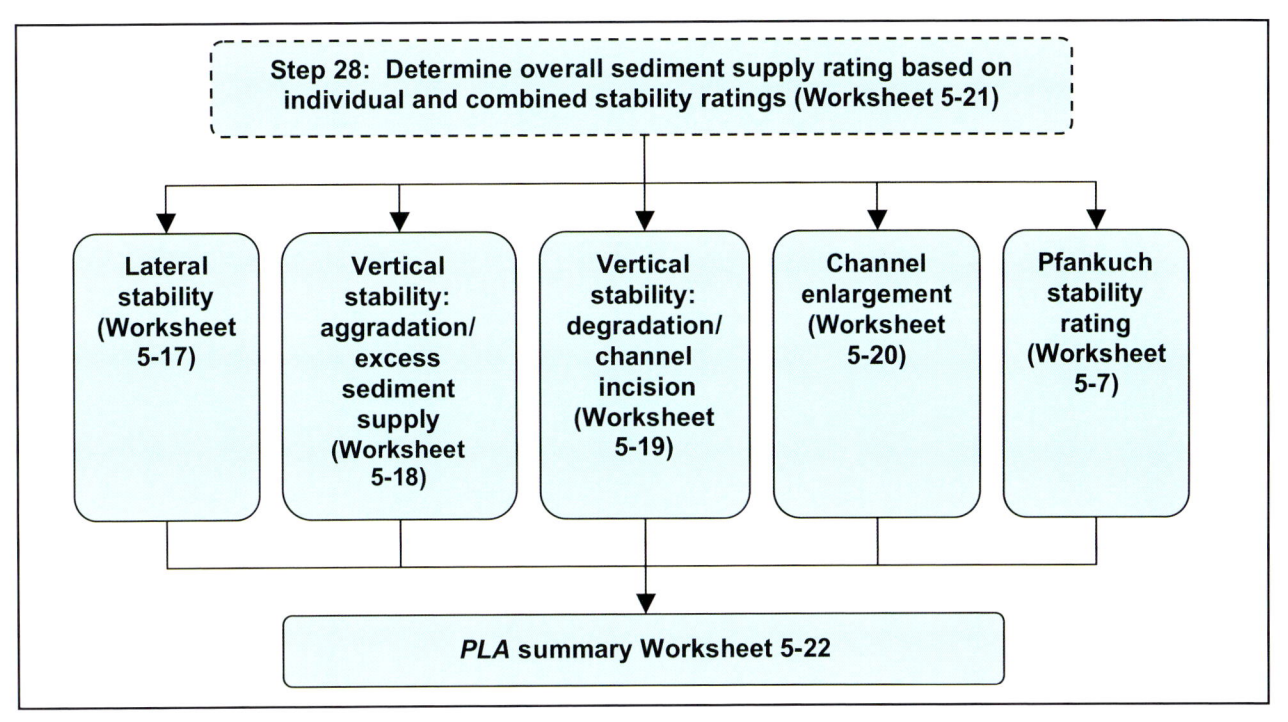

Flowchart 7-14 (Flowchart 5-18). The combined influence of individual stability ratings on sediment supply.

Step 28. Determine overall sediment supply rating from individual and combined stability ratings

The stability categories from **Steps 24–27** and the Pfankuch stability rating procedures (**Step 7**, **Worksheets 7-18a** and **7-18b**) are used to determine overall channel source sediment supply ratings in **Worksheet 7-32**. The overall rating of sediment supply allows the user to "back track" to identify the sources and causes related to disproportionate sediment supply (**Step 31**). This procedure allows recommendations for mitigation to specifically target processes, land uses and locations responsible for the high sediment supply.

The overall sediment supply rating for the C4 reference reach is *low*, as shown in **Worksheet 7-32**. On the contrary, the D4 stream type reach rates *very high*. These overall sediment supply ratings for the C4 reference and D4 impaired reaches are recorded and evaluated in the *PLA* summary **Steps 29–31**.

Watershed Assessment of River Stability and Sediment Supply

Worksheet 7-32 (Worksheet 5-21). Overall sediment supply ratings for the C4 reference and D4 impaired reaches.

Stream: Wolf Creek **Location:** above conf. West Fork of San Juan River **Observers:** D. Rosgen **Date:** 7/04

(1) Reach location	(2) Successional stage shift stability rating (Worksheet 5-16)*	(3) Lateral stability rating (Worksheet 5-17)		(4) Vertical stability for excess deposition/aggradation (Worksheet 5-18)		(5) Vertical stability for channel incision/degradation (Worksheet 5-19)		(6) Channel enlargement prediction (Worksheet 5-20)		(7) Pfankuch channel stability (Worksheet 5-7)		(8) Add points from Columns (3) to (7)	(9) Overall sediment supply rating; use column (8) points to determine adjective rating
	Stability rating:	Stability rating:	Points:	Stability rating:	Points:	Stability rating:	Points:	Stability rating:	Points:	Stability rating:	Points:		Sediment supply rating:
	Stable	Stable	1	No deposition	1	Not incised	1	No increase	1	Good: stable	1		5 = Low
	Mod. unstable	Mod. unstable	2	Mod. deposition	2	Slightly incised	2	Slight increase	2	Fair: mod. unstable	2		6–10 = Moderate
	Unstable	Unstable	3	Excess deposition	3	Mod. incised	3	Mod. increase	3				11–15 = High
	Highly unstable	Highly unstable	4	Aggradation	4	Degradation	4	Extensive	4	Poor: unstable	4		16–20 = Very high
1. C4 Reference	Stable	Stable	1	No deposition	1	Not incised	1	No increase	1	Good: stable	1	5	Low
2. D4 Impaired	Highly unstable	Highly unstable	4	Aggradation	4	Not incised	1	Extensive	4	Poor: unstable	4	17	Very High
3.													
4.													
5.													
6.													
7.													
8.													
9.													
10.													
11.													
12.													
13.													
14.													
15.													

*Successional stage shift stability rating is not used to determine overall sediment supply rating.

Steps 29–31: Summary evaluations

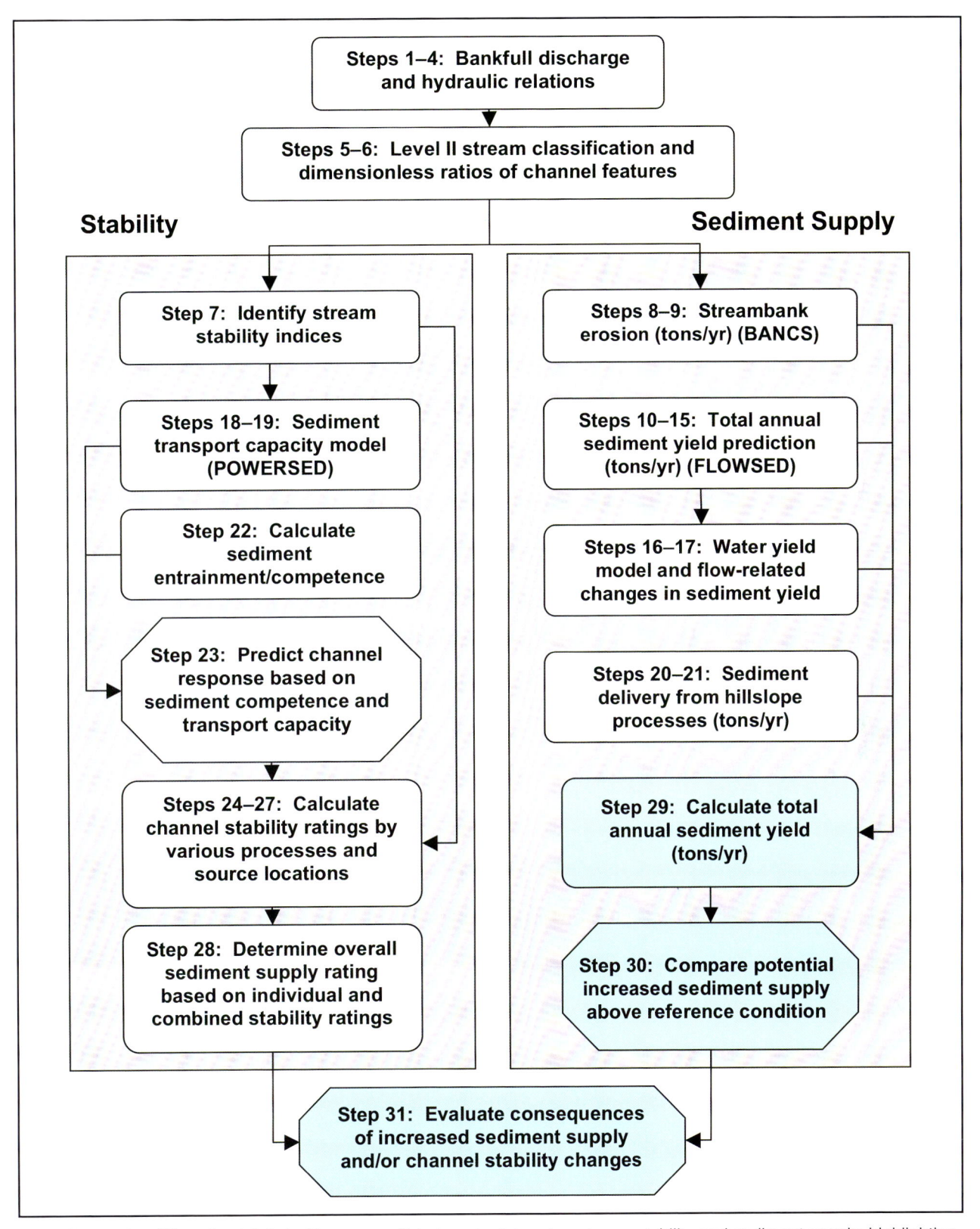

```
┌─────────────────────────────────────┐
│ Steps 1–4: Bankfull discharge       │
│ and hydraulic relations             │
└─────────────────────────────────────┘
                  ↓
┌─────────────────────────────────────┐
│ Steps 5–6: Level II stream          │
│ classification and dimensionless    │
│ ratios of channel features          │
└─────────────────────────────────────┘
```

Stability — **Sediment Supply**

- **Step 7:** Identify stream stability indices
- **Steps 18–19:** Sediment transport capacity model (POWERSED)
- **Step 22:** Calculate sediment entrainment/competence
- **Step 23:** Predict channel response based on sediment competence and transport capacity
- **Steps 24–27:** Calculate channel stability ratings by various processes and source locations
- **Step 28:** Determine overall sediment supply rating based on individual and combined stability ratings

- **Steps 8–9:** Streambank erosion (tons/yr) (BANCS)
- **Steps 10–15:** Total annual sediment yield prediction (tons/yr) (FLOWSED)
- **Steps 16–17:** Water yield model and flow-related changes in sediment yield
- **Steps 20–21:** Sediment delivery from hillslope processes (tons/yr)
- **Step 29:** Calculate total annual sediment yield (tons/yr)
- **Step 30:** Compare potential increased sediment supply above reference condition

Step 31: Evaluate consequences of increased sediment supply and/or channel stability changes

Flowchart 7-3m (Flowchart 5-3n). The sequential steps to determine stream stability and sediment supply, highlighting Steps 29–31.

Summary evaluations: Steps 29–31

Step 29. Calculate total sediment yield (all sources)

The following summary of the total annual sediment yield for the Wolf Creek Watershed is presented for the following conditions:

- Baseline (340 tons/yr);
- Reference, existing (1,234 tons/yr); and
- Impaired, existing (1,758 tons/yr).

The various erosional processes and locations responsible for the total annual sediment yields are shown in **Worksheet 7-33**. The sediment yield source and stability consequence summaries are shown by their step where the prediction was conducted in **Worksheet 7-33**. In addition to the summary in **Worksheet 7-33**, the secondary or forest access roads contributed 107 tons/yr compared to 495 tons/yr for State Highway 160 (**Step 20, Worksheet 7-25**). Highway 160 also contributed to the road-related mass erosion of 145 tons/yr. Thus, the combined influence of Highway 160 contributed directly to 640 tons/yr (36% of the total annual sediment yield). Accelerated streambank erosion contributed 429 tons/yr (24% of the total sediment yield) from 970 ft of the impaired reach (**Worksheet 7-22b**).

Departure of the D4 impaired reach from the baseline and reference condition sediment yields is presented in **Step 30**.

Worksheet 7-33 (Worksheet 5-22). PLA summary of sediment sources and stability ratings for multiple locations in the Wolf Creek Watershed.

Stream: Wolf Creek Location: Upper San Juan River Basin Observers: D. Rosgen Date: 7/04

(1) Sub-watershed or reach location	(2) Step 9: Streambank erosion (Worksheet 5-10) (tons/yr)	(3) Step 15: Total annual sediment yield (Worksheet 5-11) (tons/yr)	(4) Step 17: Flow-related sediment (Worksheet 5-11) (tons/yr)	(5) Step 19: Sediment transport capacity stability rating (Worksheet 5-12b) stable/aggrading/degrading	Roads (total sediment yield) (Worksheet 5-13) (tons/yr)	(6) Step 20: Surface erosion (total sediment yield) (tons/yr)	Mass erosion (total sediment yield) (tons/yr)	(7) Step 21: Hillslope (total sediment yield) (tons/yr)	(8) Step 22: Sediment competence/entrainment (Worksheet 5-15) stable/aggrading/degrading	(9) Step 23: Overall channel response due to sediment competence and capacity stable/aggradation/degradation	(10) Step 28: Overall channel source sediment supply rating (Worksheet 5-21) low/moderate/high/very high	(11) Step 29: Total sediment yield (Worksheet 5-11) Bedload (tons/yr)	Suspended (tons/yr)	Total (tons/yr)	(12) Step 30: Difference in sediment from baseline (Worksheet 5-23) Bedload (tons/yr)	Suspended (tons/yr)	Total (tons/yr)	(13) Step 31: Potential consequence for overall stability
1. Baseline condition	196*	340	n/a	n/a	n/a	n/a	144	144	n/a	n/a	n/a	116	224	340	n/a	n/a	n/a	stable
2. C4 reference reach	5.5**	1,234	205	stable	602	n/a	289	891	stable	stable	low	387	847	1,234	271	623	894	stable
3. D4 impaired reach	429**	1,758	362	aggrading	602	n/a	289	891	aggrading	aggradation	very high	720	1,038	1,758	604	814	1,418	aggradation

*includes steambank erosion for the A, B, upper C4 and lower C4 reaches.
**for approximately 910 ft of stream length for the C4 reference reach and approximately 970 ft of stream length for the D4 impaired reach

Step 30. Compare potential accelerated or increased sediment supply above a reference or baseline condition

Total annual sediment yield (Step 15)

Comparisons of total annual bedload and suspended sediment yield from various sources among the baseline, reference and impaired conditions are summarized in **Worksheet 7-33**, **Table 7-38** and **Worksheets 7-34a, 7-34b** and **7-34c**. Baseline total annual sediment yield (340 tons/yr) was obtained from "pre-disturbance" hillslope sediment source (144 tons/yr) and the channel processes of streambank erosion (196 tons) as shown in **Worksheets 7-33** and **7-34a**.

Increases in sediment supply above the baseline condition for the C4 reference and D4 impaired reaches are summarized in **Table 7-38**. The total annual sediment increase above baseline for the existing (post-treatment) condition for the C4 reference reach near the mouth of Wolf Creek is 894 tons/yr, of which 271 tons/yr is bedload and 623 tons/yr is suspended sediment (**Worksheet 7-34b**). This represents all sources following disturbance on the C4 stream type. The flow-related increase due to timber harvest and roads is 205 tons/yr, of which 40 tons/yr is bedload and 165 tons/yr is suspended sediment.

The total annual sediment yield departure from baseline for the D4 impaired reach is 1,418 tons/yr, which is 524 tons/yr greater than the upstream C4 reference reach (**Worksheet 7-34c**). This increase in sediment supply is related to 604 tons/yr of bedload and 814 tons/yr of suspended sediment. The flow-related sediment increase of the D4 impaired reach resulted in 362 tons/yr. This is 157 tons/yr greater than the C4 reference reach due to channel source sediment response.

Table 7-38. General summary of sediment yield changes above baseline following timber harvest, grazing and road construction activities for the Wolf Creek Watershed.

Location/ condition	Annual sediment yield (tons/yr)			Increase in sediment yield above baseline			Percent increase from baseline total annual sediment yield
	Bedload sediment (tons/yr)	Suspended sediment (tons/yr)	Total sediment (tons/yr)	Bedload sediment (tons/yr)	Suspended sediment (tons/yr)	Total sediment (tons/yr)	
Baseline	116	224	340	—	—	—	—
C4 reference reach*	387	847	1,234	271	623	894	263%
D4 impaired reach*	720	1,038	1,758	604	814	1,418	417%

*Sediment yields represent existing (post-treatment) condition.

The increases in sediment yield of 263% and 417% above baseline for the C4 reference and D4 reaches, respectively, are substantial (**Table 7-38**). The summaries shown in **Worksheets 7-33, 7-34a, 7-34b** and **7-34c** assist in site-specific and process-specific mitigation/sediment reduction recommendations.

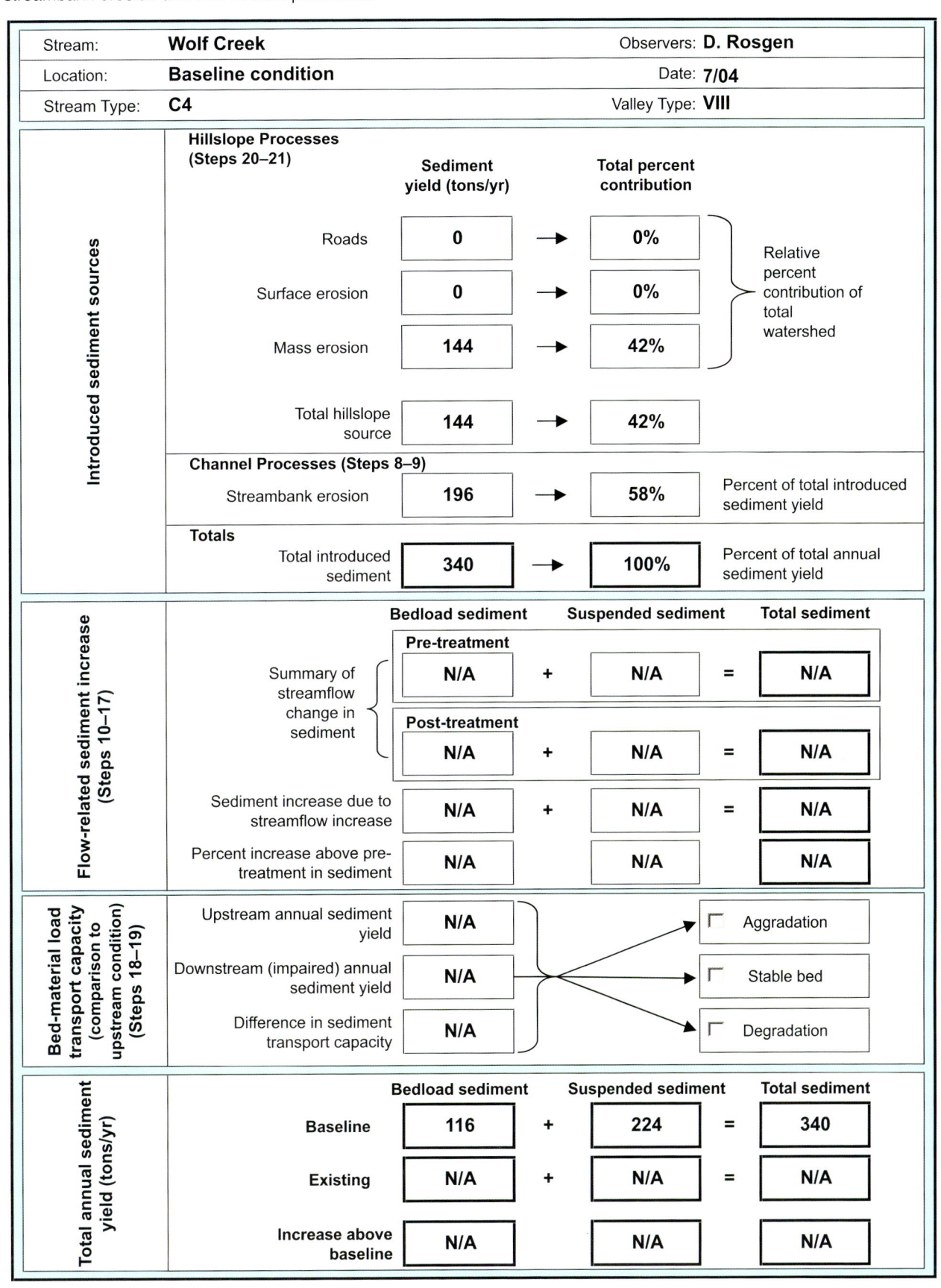

Chapter 7 — PLA   The WARSSS Methodology: A Case Study of Wolf Creek, Colorado

Worksheet 7-34a (Worksheet 5-23). Annual sediment yield by sources for the baseline condition, including hillslope, streambank erosion and flow-related processes.

Stream:	Wolf Creek	Observers: **D. Rosgen**
Location:	**Baseline condition**	Date: **7/04**
Stream Type:	**C4**	Valley Type: **VIII**

Introduced sediment sources

Hillslope Processes (Steps 20–21)

	Sediment yield (tons/yr)	Total percent contribution	
Roads	0	→ 0%	Relative percent contribution of total watershed
Surface erosion	0	→ 0%	
Mass erosion	144	→ 42%	
Total hillslope source	144	→ 42%	

Channel Processes (Steps 8–9)

Streambank erosion	196	→ 58%	Percent of total introduced sediment yield

Totals

Total introduced sediment	340	→ 100%	Percent of total annual sediment yield

Flow-related sediment increase (Steps 10–17)

	Bedload sediment	Suspended sediment	Total sediment
Summary of streamflow change in sediment — **Pre-treatment**	N/A	+ N/A	= N/A
Summary of streamflow change in sediment — **Post-treatment**	N/A	+ N/A	= N/A
Sediment increase due to streamflow increase	N/A	+ N/A	= N/A
Percent increase above pre-treatment in sediment	N/A	N/A	N/A

Bed-material load transport capacity (comparison to upstream condition) (Steps 18–19)

Upstream annual sediment yield	N/A	☐ Aggradation
Downstream (impaired) annual sediment yield	N/A	☐ Stable bed
Difference in sediment transport capacity	N/A	☐ Degradation

Total annual sediment yield (tons/yr)

	Bedload sediment	Suspended sediment	Total sediment
Baseline	116	+ 224	= 340
Existing	N/A	+ N/A	= N/A
Increase above baseline	N/A	N/A	N/A

7-197

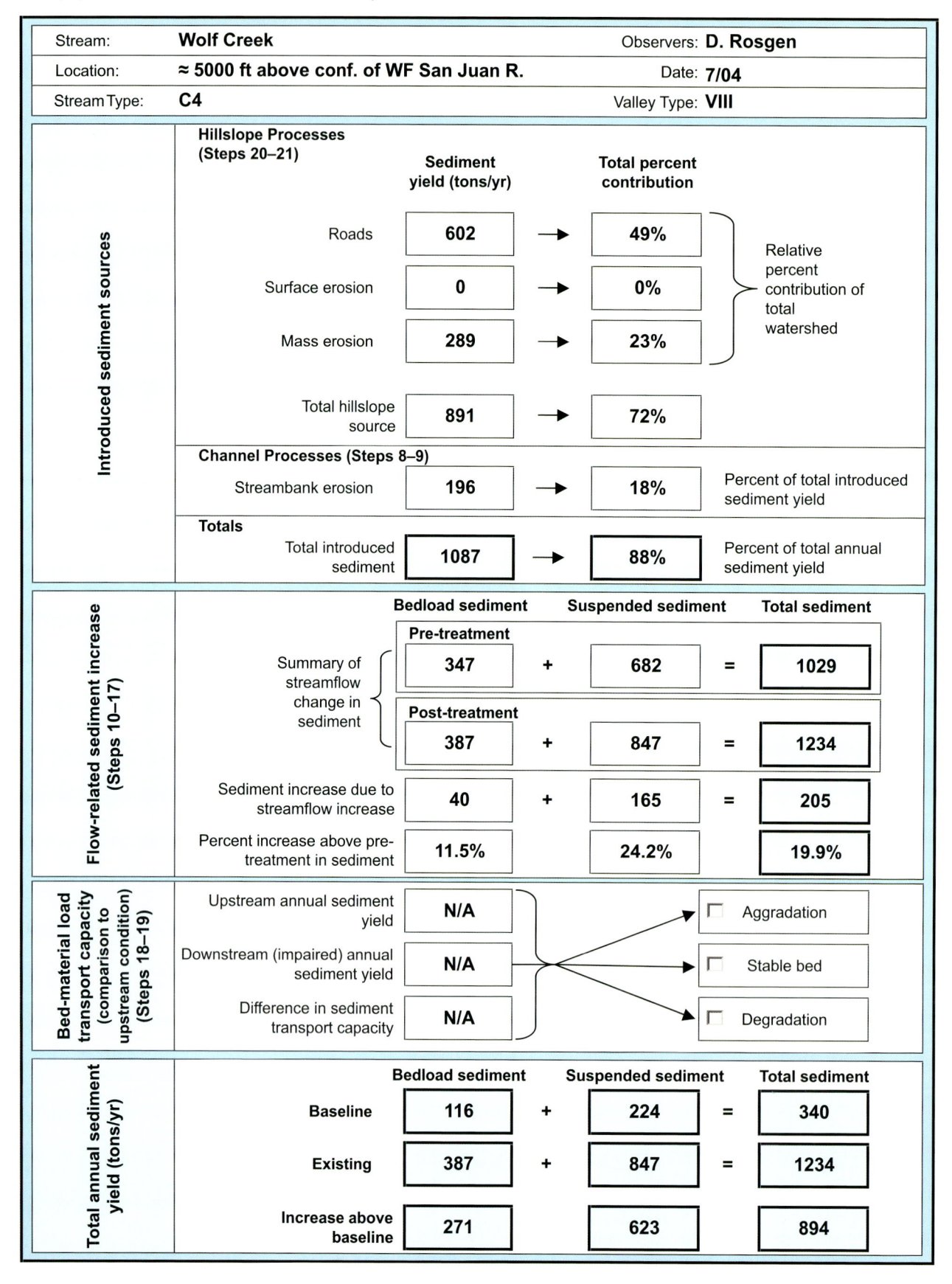

WATERSHED ASSESSMENT OF RIVER STABILITY AND SEDIMENT SUPPLY

Worksheet 7-34b (Worksheet 5-23). Annual sediment yield by sources for the C4 reference condition, including hillslope, streambank erosion and flow-related processes.

Stream:	**Wolf Creek**
Location:	**≈ 5000 ft above conf. of WF San Juan R.**
Stream Type:	**C4**
Observers:	**D. Rosgen**
Date:	**7/04**
Valley Type:	**VIII**

Introduced sediment sources

Hillslope Processes (Steps 20–21)

	Sediment yield (tons/yr)	Total percent contribution	
Roads	602	→ 49%	Relative percent contribution of total watershed
Surface erosion	0	→ 0%	
Mass erosion	289	→ 23%	
Total hillslope source	891	→ 72%	

Channel Processes (Steps 8–9)

Streambank erosion	196	→ 18%	Percent of total introduced sediment yield

Totals

Total introduced sediment	1087	→ 88%	Percent of total annual sediment yield

Flow-related sediment increase (Steps 10–17)

	Bedload sediment		Suspended sediment		Total sediment
Pre-treatment Summary of streamflow change in sediment	347	+	682	=	1029
Post-treatment	387	+	847	=	1234
Sediment increase due to streamflow increase	40	+	165	=	205
Percent increase above pre-treatment in sediment	11.5%		24.2%		19.9%

Bed-material load transport capacity (comparison to upstream condition) (Steps 18–19)

Upstream annual sediment yield	N/A	☐ Aggradation
Downstream (impaired) annual sediment yield	N/A	☐ Stable bed
Difference in sediment transport capacity	N/A	☐ Degradation

Total annual sediment yield (tons/yr)

	Bedload sediment		Suspended sediment		Total sediment
Baseline	116	+	224	=	340
Existing	387	+	847	=	1234
Increase above baseline	271		623		894

7-198

Worksheet 7-34c (Worksheet 5-23). Annual sediment yield by sources for the D4 impaired condition, including hillslope, streambank erosion and flow-related processes.

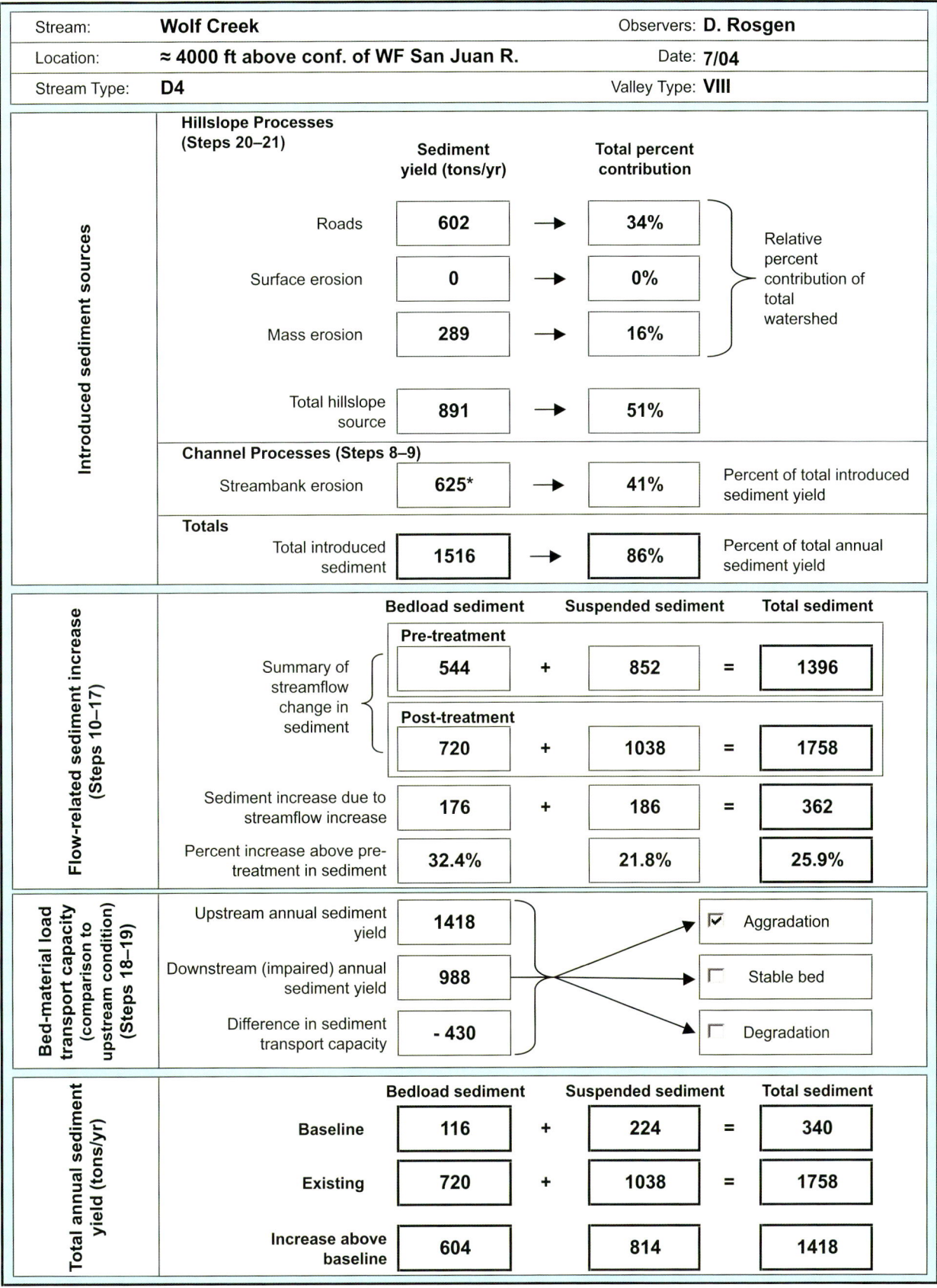

Stream:	Wolf Creek	Observers:	D. Rosgen
Location:	≈ 4000 ft above conf. of WF San Juan R.	Date:	7/04
Stream Type:	D4	Valley Type:	VIII

Introduced sediment sources

Hillslope Processes (Steps 20–21)

	Sediment yield (tons/yr)	Total percent contribution	
Roads	602	→ 34%	Relative percent contribution of total watershed
Surface erosion	0	→ 0%	
Mass erosion	289	→ 16%	
Total hillslope source	891	→ 51%	

Channel Processes (Steps 8–9)

Streambank erosion	625*	→ 41%	Percent of total introduced sediment yield

Totals

Total introduced sediment	1516	→ 86%	Percent of total annual sediment yield

Flow-related sediment increase (Steps 10–17)

	Bedload sediment	Suspended sediment	Total sediment
Pre-treatment	544 +	852 =	1396
Post-treatment	720 +	1038 =	1758
Sediment increase due to streamflow increase	176 +	186 =	362
Percent increase above pre-treatment in sediment	32.4%	21.8%	25.9%

Bed-material load transport capacity (comparison to upstream condition) (Steps 18–19)

Upstream annual sediment yield	1418	☑ Aggradation
Downstream (impaired) annual sediment yield	988	☐ Stable bed
Difference in sediment transport capacity	-430	☐ Degradation

Total annual sediment yield (tons/yr)

	Bedload sediment	Suspended sediment	Total sediment
Baseline	116 +	224 =	340
Existing	720 +	1038 =	1758
Increase above baseline	604	814	1418

*625 tons/yr includes 429 tons/yr from the D4 reach plus 196 tons/yr from upstream.

Sediment transport capacity (Step 19)

The sediment transport capacity using POWERSED (**Steps 18** and **19**) showed a major reduction in sediment transport. The upstream reference transport (using POWERSED for the C4 reference reach) was 1,418 tons/yr compared to the D4 stream type of 988 tons/yr, a reduction of 430 tons/yr (**Worksheet 7-34c**). This reduction total is made up of bedload and suspended sand, both hydraulically controlled. This analysis indicates aggradation/excess deposition for the D4 stream type.

Streambank erosion (Steps 8–9)

Approximately 196 tons/yr are contributed as a baseline sediment source (**Worksheet 7-33**). The streambank erosion makes up approximately 58% of the total baseline sediment yield. The C4 reference reach produced 5.5 tons/yr for 910 ft of stream length versus 429 tons/yr (970 ft) for the D4 impaired reach. The difference in erosion rate is 0.006 tons/yr/ft for the C4 reference reach versus 0.44 tons/yr/ft for the D4 impaired reach. The total streambank erosion source is 625 tons/yr, representing 36% of the total sediment supply (**Worksheet 7-34c**).

Hillslope processes (Steps 20–21)

The summary of sediment yield is summarized in **Step 29**. **Worksheets 7-25** and **7-33** indicate that State Highway 160 contributed 495 tons/yr where forest access roads, showing recovery since construction, contributed 107 tons/yr. The total road contribution is 602 tons/yr. In addition, mass erosion related to the widening of Highway 160 contributed an additional 145 tons/yr (above baseline). The road sediment total of 602 tons/yr plus 145 tons/yr from mass erosion equals a total of 747 tons/yr, or 42% of the existing total annual sediment yield relating to roads.

Sediment competence (Step 22)

The competence calculation resulted in an *aggrading* stability rating for the D4 stream type (**Worksheet 7-26b**). The calculation using dimensionless shear stress indicates that a bankfull depth of 1.02 ft and a slope of 0.012 are necessary to move a 78 mm particle. Using dimensional shear stress, the depth necessary to move a 78 mm particle is 1.7 ft for the existing slope of 0.009, or a slope of 0.019 for the existing depth of 0.8 ft (**Worksheet 7-26b**). It is obvious by both approaches that the existing depth of 0.8 ft and slope of 0.009 are not sufficient to move the 78 mm particle for the D4 stream type. The depth and slope of the upstream C4 reference reach are 1.6 ft and 0.005, respectively, to move the 78 mm particle (**Worksheet 7-26a**).

The photographs in **Figure 7-32** show the differences in the C4 reference reach compared to the D4 impaired reach. The field data shows a reduction in bed-material size gradation comparing the C4 reference reach to the D4 reach (**Figure 7-34**), where the D_{50} was reduced from 48 mm to 19 mm. The stability rating of *aggradation* is supported by the additional field monitoring data.

Channel stability and sediment supply summaries (Steps 24–28)

The overall integrative summary of the individual ratings are shown in **Worksheet 7-32**. The corresponding ratings compare the C4 reference stream type to the D4 impaired stream type. In addition, a listing of the criteria used to obtain the overall ratings in **Worksheet 7-32** is shown in **Table 7-39**. This table identifies the specific variables that may be addressed for mitigation or restoration in the final assessment interpretation in **Step 31**. **Worksheet 7-32** and **Table 7-39** provide a better understanding of the "limiting factors" leading to impairment.

Table 7-39. Summary ratings for various criteria contributing to sediment supply.

Criteria related to sediment supply	Rating summary — C4 reference reach	Rating summary — D4 impaired reach
Width/depth ratio state	Stable	Unstable
Stream succession	Stable	Unstable
Depositional patterns	Stable	Unstable
Debris blockages	Stable	Stable
Bank-height ratio	Stable	Stable
Meander width ratio	Stable	Unstable
Meander pattern	Stable	Unstable
Dominant BEHI/NBS	Stable	Unstable
Sediment capacity	Stable	Unstable
Sediment competence	Stable	Unstable

Step 31. Evaluate potential consequences of increased sediment supply and/or channel stability changes

A. Channel stability considerations and summary interpretations

Table 7-40 identifies the criteria or variables from **Steps 28** and **30** and from the *RRISSC* assessment that influenced the instability of the D4 impaired, the upper C4 (above landslide) and the A3 stream type reaches.

Table 7-40. Summary of the variables that influenced the instability of the D4 impaired, the upper C4 and the A3 stream type reaches.

D4 impaired reach (Worksheet 7-32)	Rating
A1. Successional stage shift	Highly unstable
A2. Lateral stability rating	Highly unstable
A3. Vertical stability (aggradation/excess sediment)	Aggradation
A4. Vertical stability (degradation/channel incision)	N/A
A5. Channel enlargement	Extensive
A6. Pfankuch channel stability rating	Poor: unstable
A7. Overall sediment supply	Very high
A3 stream type (Worksheet 7-13)	***RRISSC* rating**
A3. Vertical stability (aggradation/excess sediment)	Very high
A4. Vertical stability (degradation/channel incision)	Very high
A5. Channel enlargement	Very high
C4 upper stream type (above landslide) (Worksheet 7-13)	***RRISSC* rating**
A3. Vertical stability (aggradation/excess sediment)	High
A4. Vertical stability (degradation/channel incision)	Very high

Some of the discussions and recommendations dealing with channel stability issues also relate to the sediment supply concerns. For example, sediment transport capacity is a serious problem for bed stability and aggradation. It is also important as a calculation of sediment transport and supply. For this procedure, it will be appropriately discussed in each major category of channel stability and sediment supply.

The following discussion and analysis is designed to identify the specific variables associated with the unstable ratings in **Table 7-40**.

A1. Successional stage shift (Worksheet 7-27b)

The rating of *highly unstable* reflects a change from a C4 to a D4 braided stream type. The dimension, pattern and profile of the C4 reference stream type serves as the blueprint for the D4 stream type because it is stable. It is unlikely that the D4 stream type would restore itself in the presence of the geologically high sediment supply in the watershed; thus, restoration of the D4 reach is recommended as a mitigation priority. The restoration would convert the D4 reach to a C4 stream type. This restoration would improve the morphological, hydraulic, sedimentological and biological functions of the reach.

The stream classification data for the C4 reference stream type (**Worksheet 7-15a**) sets the initial stage for the stable dimension, pattern and profile criteria of the potential, stable stream type. An extension of this morphological data is shown in **Worksheet 7-16a** where dimensionless ratios of bed features for pools, glides, runs and riffles are described. Values of slope and depth of the bed features provide additional design criteria for restoration. An improvement in fish habitat would occur due to deeper baseflows, instream and overhead cover, lower temperatures and higher dissolved oxygen.

Furthermore, the hydraulics of the C4 stream type is reflected in **Worksheet 7-14**. The channel depth, slope and channel materials are reflected in the required bankfull velocity of 4.7 ft/sec for a bankfull discharge of 280 cfs. This data assists in the stability solution for the D4 impaired reach.

A2. Lateral stability (Worksheet 7-28b)

The rating of *highly unstable* for the D4 impaired reach is due to the following:

　　a. Width/depth ratio state;

　　b. Depositional patterns;

　　c. Meander patterns;

　　d. Dominant BEHI and NBS; and

　　e. Degree of confinement (meander width ratio).

The data and assessment of these variables help the user to focus on the change in the variables to reduce the impairment.

A2a. Width/depth ratio state

The width/depth ratio state rated *highly unstable* for the D4 impaired reach (**Worksheet 7-28b**). To correct this criterion, the C4 reference width of 37 ft and depth of 1.6 ft are used (**Worksheet 7-15a**). For the D4 impaired reach to be stable, the width of 203 ft needs to be reduced to 37 ft, and the depth of 0.8 ft needs to be increased to a mean bankfull depth of 1.6 ft. A comparison of the w/d ratio of the C4 and D4 reaches is shown in **Figure 7-33**. Morphological data including w/d ratio is presented with each cross-section.

The acceptable range of the w/d ratio for a stable C4 stream type can range in magnitude between 0.8 and 1.2 times the average w/d ratio. This means that the restored D4 w/d ratio can range from 18 to 28 (current C4 w/d = 23). The corresponding range of widths is calculated to be between 33 ft and 41 ft using:

$$W_{bkf} = [(w/d)(A)]^{1/2} \quad \text{Equation 7-1}$$

Thus, the range of depths can vary between 1.4 ft and 1.8 ft using:

$$d = W_{bkf} / (w/d) \quad \text{Equation 7-2}$$

or

$$d = A / W_{bkf} \quad \text{Equation 7-3}$$

The channel also requires a slope of 0.005 to maintain the hydraulic conditions with the associated width, depth and cross-sectional area for velocity, shear stress, stream power and sedimentological requirements.

A slope reduction from 0.009 for the braided D4 reach to a 0.005 slope is accomplished by increasing sinuosity and meander width ratio as the pattern matches the stable morphology of the C4 stream type.

A2b. Depositional patterns

The B5, B6 and B7 depositional patterns (**Figure 7-39**) indicate a *highly unstable* rating and accelerated lateral adjustment for the D4 impaired reach (**Worksheet 7-28b**). The stable patterns are the B1 and B2 categories representing typical point bars with few mid-channel bars, associated with the C4 reference reach (**Worksheet 7-28a**).

A2c. Meander patterns

The M5 and M8 categories (**Figure 7-38**) of unconfined meander scrolls and irregular meanders add to lateral instability. Stable reaches reflect the M1, M3 or M4 meander patterns for C4 stream types.

A2d. Dominant BEHI and NBS

The dominant BEHI and NBS combination for the D4 impaired reach is Extreme and Very High. This lateral instability was initially caused by spraying the riparian willows, cottonwoods and alders with 2, 4-D and 2,4,5-T foliar sprays. It is necessary to identify the variables within the BANCS model of BEHI and NBS to offset the unstable and high erosion consequence variables. The BEHI variables of study bank-height ratio, rooting depth, rooting density, bank angle and percent of surface protection can be altered by restoration stabilization or a change in grazing management strategies. Reduction of the extensive bars and establishment of a single-thread channel will reduce the Very High NBS. Often structures can be installed, such as J-hook vanes, using logs and/or boulders (Rosgen, 2001d) to decrease NBS and buy time to re-establish the plant community.

In addition to the rooting depth and density data, the assessment data in **Worksheet 7-17a** (**Step 7**) can provide the composition and density for the potential riparian community. The user is advised to study this valuable assessment step.

A2e. Degree of confinement

The degree of confinement rated *unstable* for the D4 impaired reach (**Worksheet 7-28b**). The degree of confinement is based on the Meander Width Ratio (MWR) of the impaired reach divided by the reference MWR_{ref} (**Figure 7-42**). To correct the degree of confinement requires reducing the bankfull width of the D4 reach from 203 ft to 37 ft. The existing width of 203 ft would become the belt width of the new C4 stream type. The MWR would then be converted to 203 ft divided by 37 ft, which equals 5.5. The reference ratios for the C4 reference reach range from 3.5 to 6.0 with a mean value of 4.8. The MWR value of 5.5 falls within this range.

A3. Vertical stability for excess sediment/aggradation (Worksheet 7-29b)

The D4 impaired reach rated as *aggradation* due to the following:

   a. Sediment competence;

   b. Sediment capacity;

   c. Width/depth ratio state;

   d. Stream succession states; and

   e. Depositional patterns.

Debris/blockages is another criterion to assess vertical stability for excess sediment/aggradation; however, this criterion was not contributing to the aggradation and rated as *no deposition* for the D4 impaired reach.

A3a. Sediment competence

Sediment competence rated as *aggradation* for the D4 impaired reach in **Worksheet 7-29b**. The largest particle made available from the C4 upstream reach is 78 mm. For the D4 impaired reach to successfully transport the 78 mm particle size, the required depth needs to be 1.6 ft and the slope needs to be 0.005 (**Worksheet 7-26a**).

A3b. Sediment capacity

The D4 impaired reach rated as *aggradation* for sediment capacity. The POWERSED model predicted that approximately 430 tons/yr of bedload and suspended sand would be deposited in the D4 impaired reach, which is approximately 970 ft long (**Worksheet 7-34c**). This represents approximately a quarter of the entire annual sediment yield of the watershed. The solution to this dilemma is to create the morphological variables (**Worksheets 7-15a** and **7-16a**) and the corresponding hydraulic relations (**Worksheet 7-14**) of the stable C4 reference reach. Converting the D4 to a C4 stream type will create the conditions to transport the sediment load as shown in **Table 7-35**.

A3c. Width/depth ratio state
See section *A2a. Width/depth ratio state*.

A3d. Stream succession states
See section *A1. Successional stage shift*.

A3e. Depositional patterns
See section *A2b. Depositional patterns*.

Discussion for the A3 stream types

For the A3 stream types that rated as *very high* risk in the *RRISSC* assessment (**Worksheet 7-13**), the following suggestions are offered.

A3 stream types are geologically very active. They have unlimited sediment supply (incised in glacial tills and/or landslide debris) and are high energy (low w/d ratios and steep slopes). Detailed worksheets to evaluate aggradation/excess sediment and degradation/channel incision always rate A3 stream types as *unstable*. However, the unlimited sediment supply and high energy are the natural, geologic conditions of the A3 stream types; thus, grade control structures, such as check dams and streambank stabilization methods, are generally discouraged. Although the A3 stream types contribute excess sediment from channel processes (bed and banks), the sediment is of a geologic origin. The increased streamflow has some adverse effects, but they are not sufficient to implement restoration or stabilization measures.

The exception to this is to stabilize the toe of a major landslide, as discussed in more detail in the summary section **B5. Mass erosion** related to Highway 160.

Discussion for the upper C4 stream type (above landslide)

The upper C4 stream type (**Figure 7-28**) rated as *high* risk for aggradation/excess sediment in the *RRISSC* assessment (**Worksheet 7-13**). The high rating is due to the following reasons:

- The landslide that blocked the valley flattened the slope of the upstream reach;
- The decrease in slope increased the width/depth ratio, which decreased the velocity and depth and resulted in a reduction in channel competence and capacity; and

- The riparian vegetation was eliminated due to constant inundation from the backwater condition. Mortality of mature spruce also occurred.

Mitigation recommendations include a bankfull notch to be cut in the slide debris to drain the channel and floodplain of backwater and to re-establish a stable C4 stream type.

A4. Vertical stability for degradation/channel incision (Worksheet 7-30b)

Both the C4 reference and D4 impaired reaches rated as *not incised*; therefore, discussion of the criteria for this stability rating is not required. However, for the A3 stream types that rated *very high* in the *RRISSC* assessment, see the **Discussion for the A3 stream type** in the summary section *A3.* **Vertical stability for aggradation/excess sediment** as this discussion relates to both categories.

Discussion for the upper C4 stream type (above landslide)

The upper C4 stream type rated as *very high* risk for degradation/channel incision in the *RRISSC* assessment (**Worksheet 7-13**). The *PLA* procedure would not rate this C4 stream type as degradation due to its present condition in backwater deposition. However, the risk of a potential headcut exists if the debris from the landslide were to be eroded or the base level lowered. Because this backwater reach was created by the landslide debris, the reach can be altered by changes in the same landslide debris. If the local base level were to lower by three to four feet, a headcut gully would advance, creating a G4 stream type. This would be a serious impairment and would cause an increase in sediment supply. The mitigation recommendation described in the summary section *A3.* **Vertical stability for aggradation/excess sediment** for the upper C4 stream type would also apply to prevent degradation/channel incision.

A5. Channel enlargement (Worksheet 7-31b)

Channel enlargement is a serious instability for the D4 impaired reach due to corresponding land loss, high sediment consequence from streambanks and bed, loss of fish habitat and water quality problems. To offset this instability, mitigation recommendations must look at the criteria responsible for the impairment, which include the following criteria:

A5a. Successional stage shift
See section *A1.* **Successional stage shift**.

A5b. Lateral stability
See section *A2.* **Lateral stability**.

A5c. Vertical stability for excess sediment/aggradation
See section *A3.* **Vertical stability for excess sediment/aggradation**.

Discussion for the A3 stream types

The A3 stream types rated as *very high* risk for channel enlargement in the *RRISSC* assessment (**Worksheet 7-13**). The worksheet variables for this stream type always generate *unstable* ratings. The enlargement process is very active in A3 stream types due to the active streambank rejuvenation and shift in bed levels. The nature of the heterogeneous, unconsolidated soil materials and the steep slopes with low w/d ratios make these stream types susceptible to channel enlargement. However, these are natural, geologic conditions and therefore mitigation prescriptions are not recommended on the majority of A3 stream types. The exception is the short distance of the A3 stream type that is adjacent to an active landslide (see the recommendations for Highway 160 in the summary section *B5. Mass erosion*).

A6. Pfankuch stability rating (Worksheet 7-18b)

The D4 impaired reach rated as *poor* or *unstable*. A comparison between the Pfankuch channel stability ratings for the C4 reference reach and the D4 impaired reach shows major departure. The *poor* or *unstable* rating is primarily due to:

- Poor riparian vegetation;
- Loss of channel capacity;
- Erodible nature of the streambanks;
- Obvious cutting (scour) and excess deposition; and
- A shift to a bi-modal bed-material size distribution (increase in sand-sized particles in the substrate).

These ratings for the D4 impaired reach can be studied in detail in **Worksheet 7-18b**. Converting this stream back to its previous morphological C4 stream type that existed prior to the spraying of the woody riparian vegetation is a strong recommendation based on these assessments. This stream type conversion would shift the majority of the other categories discussed to a *stable* and *low* sediment supply condition. Additionally, fish habitat would be greatly improved.

A7. Overall sediment supply rating (Worksheet 7-32)

The D4 impaired reach rated as *very high* risk for overall sediment supply. The combined individual stability rating categories are used to evaluate the overall sediment supply. The rating of *very high* necessitates a review of the criteria for each category contributing to the rating. In such a manner, specific recommendations can be made to reduce the sediment supply based on the various sources and causes. The ratings for each category are summarized in **Worksheet 7-32**. The following are the criteria responsible for the *very high* risk rating:

A7a. Lateral stability
See section *A2. Lateral stability*.

A7b. Vertical stability for excess sediment/aggradation
See section *A3. Vertical stability for excess sediment/aggradation*.

A7c. Channel enlargement
See section *A5. Channel enlargement*.

A7d. Pfankuch stability rating
See section *A6. Pfankuch stability rating*.

This concludes the discussion of the criteria for the channel stability ratings. Many of the ratings discussed are estimates of stability. The indicators that are used reflect the various erosional and depositional processes that contribute to impairment and high sediment supply. The following section deals with a detailed, quantitative evaluation of the processes that influence sediment supply.

B. Sediment supply considerations and summary interpretations

The objective of this discussion on sediment supply and corresponding impairment is to review the summaries to:

- Identify specific locations and land uses that contribute disproportionate sources of sediment;
- Identify erosional and depositional processes of high sediment supply;
- Prescribe mitigation that reflects the contributing process; and
- Set priorities for mitigation based on the magnitude and consequence of sediment-related impairment.

A review of **Worksheets 7-33**, **7-34a**, **7-34b**, and **7-34c**, the summaries in **Steps 29** and **30** and the channel stability summary in **Step 31** is recommended prior to starting this last phase of *PLA*. Discussion will be organized by the following subjects:

1. Sediment competence;
2. Sediment capacity;
3. Sediment supply from roads;
4. Surface erosion;
5. Mass erosion;
6. Direct channel impacts;
7. Streambank erosion;
8. Flow-related sediment yield; and
9. Total annual sediment yield.

B1. Sediment competence (Worksheets 7-26a and 7-26b)

In a review of **Worksheet 7-26a** and the discussion in **Step 31** section *A3a. Sediment competence*, the D4 impaired reach did not have the competence to move the largest particle (78 mm) made available from immediately upstream (the C4 reference reach). The impaired reach needs to have a bankfull mean depth of 1.6 ft and a slope of 0.005 to move the 78 mm particle. The existing bankfull mean depth of 0.8 ft and the slope of 0.009 indicates *aggradation*.

Converting the D4 to a C4 stream type would provide the morphological and hydraulic conditions to regain the sediment competence of the stream. Stream restoration is recommended to shape the dimension, pattern and profile to a C4 inside the D4 stream type. The numbers for dimension, pattern and profile are obtained from **Worksheets 7-15a** and **7-16a**, which represent a stable, reference C4 stream type. This would reduce the extensive bars as shown in **Figure 7-32**. The particle size distribution would coarsen to match the C4 stream type particle size data shown in **Figure 7-34**. Furthermore, this restoration recommendation would help maintain stability and would enhance the fishery resources.

It is doubtful that the D4 impaired reach would self-stabilize by fencing the stream and/or planting woody riparian species. In the presence of the high sediment supply from upstream and the low bankfull shear stress of the D4 stream type, the excess bar deposition will likely continue. This deposition will continue to distribute stress on the near-bank region, contributing to the increase in width/depth ratio and channel enlargement.

B2. Sediment capacity

To many observers, interpretation of the values of total annual sediment yield in tons/yr does not evoke a clear understanding of consequence. It often generates more questions, such as:

- How much sediment is too much sediment?
- When are thresholds exceeded?
- What is the relationship between sediment yield and stream channel impairment?
- How effective is mitigation on reducing sediment supply and improving river health?
- How do we structure (design) mitigation to be the most effective?

The following discussion using the *PLA* assessment for the Wolf Creek Watershed helps answer the aforementioned questions.

It is often the case where a stream has sufficient shear stress to transport the largest size made available (competence) but the stream is aggrading with small gravel and sand. These small grain sizes appear to be able to move at the bankfull stage; however, the stream does not have the sediment transport capacity to move its own load. The question of how much is too much has posed a difficult computational challenge. The POWERSED model, however, has predicted actual aggradation resulting from loss of channel capacity. The most common problem associated with loss of capacity is an increase in width/depth ratio and decreases in mean bankfull velocity and mean bankfull depth. Slope changes are also sensitive to sediment transport capacity. The product of velocity and shear stress equals unit stream power (**Equation 2-11**). If the stream enlarges due to instability, it continues to lose its capacity to transport its own load.

The POWERSED model was run for both the C4 reference and D4 impaired reaches. The results are summarized in **Table 7-35** in **Step 19**. The POWERSED model indicated that the D4 stream type would only transport 988 tons/yr (493 tons/year bedload and 464 tons/yr suspended sand), a reduction of 430 tons/yr (or a 30% reduction of capacity). The FLOWSED model using flow-duration curves and dimensionless sediment rating curves indicate a higher transport. The POWERSED model, however, evaluates the hydraulic geometry changes due to the altered channel's dimension, pattern and/or profile. In comparing the C4 to the D4 stream type, bankfull width increased from 37 to 203 ft, bankfull depth decreased from 1.6 to 0.8 ft, mean velocity decreased from 4.7 to 1.7 ft/sec and unit stream power decreased from 2.1 lbs/sec/ft to 0.76 lbs/sec/ft.

The result of the change in morphology and hydraulics for the D4 impaired reach indicates that 430 tons/yr would deposit on an average runoff year. The predicted annual sediment yield value of 1,758 tons/yr is similar to the measured annual sediment yield of 1,780 tons/yr for the D4 impaired reach.

An interesting observation is the continued channel enlargement in the presence of decreased sediment transport capacity (depositing 430 tons/yr in a 970 ft reach). The estimated streambank erosion for the same stream length is 429 tons/yr; i.e., the sediment that went into storage put stress on the banks from the bars and accelerated streambank erosion rates. The aerial photo time-trend studies (**Figure 7-35**) verified this process.

The solution to increase the sediment transport capacity is to convert the D4 to the C4 reference stream type using the morphological, hydraulic and sedimentological relations in **Worksheets 7-14**, **7-15a** and **7-16a**. If the reference reach is not immediately upstream from the impaired reach and is located where the bankfull discharge is much different, the use of dimensionless ratios are used to obtain the bankfull dimensions to match the design bankfull discharge. The POWERSED model would then be run *to evaluate the sediment transport capacity of the new channel design*. If the sediment transport capacity is restored, the stream will not aggrade. Monitoring by annual resurveying of cross-sections and longitudinal profiles will verify the sediment transport capacity.

B3. Sediment supply from roads (Worksheet 7-25)

The contribution of sediment supply in the Wolf Creek Watershed is substantial. The secondary logging or forest access roads proportionately contribute the least direct sediment input at approximately 107 tons/yr (6% of the total annual sediment yield). This accounts for the vegetal recovery over time and the low use of these roads. However, State Highway 160 contributes approximately 495 tons/yr (28% of the total annual sediment yield). Combined, the roads directly contribute 602 tons/yr (34% of the total annual sediment yield). Site-specific stabilization and sediment reduction mitigation would be warranted.

The road impact index that is used to predict sediment yield shows exponentially larger sediment delivery on the lower slope position. As a result, mitigation on those roads located in the lower slope position would be most cost-efficient and effective at reducing sediment. Highway 160 is also a priority due to its higher proportion of sediment contribution. The following list of stabilization/sediment reduction recommendations would be effective based on the assessment:

- Disperse water concentrations down the ditch line by installing more drop-outlet cross-drains;
- Rock-line the ditch lines to prevent ditch-line erosion;
- Cleaning ditch lines over-steepens cut banks leading to active cut bank erosion. Allow a berm or bench between the ditch line and cut bank and revegetate. This helps flatten the slope and reduces the sediment delivery to the ditch line;
- Dissipate the energy below the "shotgun" culverts on the road fills with step/pool dissipaters; also consider installing downspouts below the culverts; and
- Revegetate cut banks and fill slopes.

The forest access roads that are not being used and maintained should be out-sloped and the culverts should be removed with back-sloping and revegetating the exposed surfaces, including the road fill at the crossing. Additional recommendations for treatment of these roads due to flow-related sediment are discussed.

A debris, revetment catch, crib wall or similar structure would reduce an extensive source of sediment from directly entering the surface waters at the toe of the fill slope at the road crossings of drainageways and where Highway 160 encroaches on Wolf Creek two-thirds of the way down from Wolf Creek pass.

The debris slides created by the drainage from the road near the "switchbacks" should be carried down the ditch line and drained off the bedrock cliffs, rather than the fill on the east side of the road. The annual sediment contributions from this source is extensive and the maintenance is high on the lower road position due to the slide debris.

As approximately one-third of the total sediment yield is from roads, the aforementioned mitigation is warranted and should prove effective at overall sediment reduction.

B4. Surface erosion

Normally, surface erosion can be a significant process in need of mitigation due to compaction from surface disturbance activities and vegetation removal/exposed bare soil. However, in the assessment of the Wolf Creek Watershed, the sites that had been disturbed for forestry and recreation-related activities had a higher infiltration rate than precipitation rate, which means that the potential for overland flow is insignificant. The exception to this is the roads that were evaluated separately. No mitigation is recommended for any surface disturbance activity other than roads.

B5. Mass erosion

The watershed has several geologically active slides due to the volcanism and alpine glaciation geology. In addition, Highway 160 had undercut a major slump/earthflow area as located in **Figure 7-14** with the label "landslide." The debris from this landslide caused headward deposition of sediment, killing the riparian vegetation and spruce trees. Downstream, the toe of the landslide caused a landslide in glacial till across from the slide debris due to convergence. The toe of both slides creates extensive sediment (145 tons/yr) at this location, associated with an A3 stream type. The recommendation for this reach and landslide is to construct a large boulder step/pool channel off-set from the toe of both slides. This would reduce the slope rejuvenation (over-steepening). The *PLA* evaluation indicates that this work is warranted to potentially reduce approximately 145 tons/yr (8% of the total annual sediment yield).

B6. Direct channel impacts

Direct channel impacts often are associated with past activities of straightening, levees, lining and raising, lowering or changing the dimension, pattern, profile and materials of natural channels. In the Wolf Creek Watershed, the direct impacts are associated with road encroachment, culvert crossings and the spraying of riparian vegetation (discussed in the following ***B7. Streambank erosion*** section).

The location in need of mitigation or stabilization is the fill encroachment on Highway 160 below treasure falls on Fall Creek, a tributary to lower Wolf Creek. The potential exists for a major fill failure at this location due to a poor design in the existing fill. Overflow pipes should be installed to handle the flood debris from this tributary at an elevation above the bankfull stage. Elevated pipes are presently eroding the fill and contributing excess sediment to Fall Creek. The toe of the fill needs stabilization, and a dissipater, step/pool is needed below the existing culvert.

B7. Streambank erosion

The natural, geologic rate of 0.006 tons/yr/ft from an alluvial channel (excludes many of the B, A1 and A2 stream types in the watershed) is not uncommon. The recommendation for sediment reduction is to reduce the accelerated streambank erosion. The map in **Figure 7-46** indicates the location of various streambank erosion rates and is helpful for this discussion. The location of naturally high erosion rates on the steep A3 stream types (incised in heterogeneous, unconsolidated glacial tills) are not candidates for stabilization. Although increases in streamflow can increase channel-source sediment from these A3 stream types, there is not much that can be done to stabilize these channels. In fact, the sediment supply from streambank erosion for baseline conditions (196 tons/yr) comes from these A3 stream types.

The accelerated streambank erosion rate of the 970 ft reach of the D4 impaired stream type is 429 tons/yr. This reach is a candidate for stabilization to reduce the high, disproportionate sediment supply. This reach is contributing 24% of the total annual sediment yield from the Wolf Creek Watershed. The cause of this accelerated erosion rate is due to the spraying of the willows, cottonwoods and alders. The streambank erosion rate caused channel enlargement from 37 ft to 203 ft over a period of 16 years. The time-tend aerial photo analysis verifies that the stream continues to widen rather than self-stabilize. A stream restoration project would be proposed to reduce the streambank erosion.

The BEHI values for the D4 impaired reach in **Step 8** of *PLA* indicate that reducing the study bank-height ratios, flattening the bank angles and creating greater surface protection and rooting depth and density would improve the BEHI values. The NBS values rated Very High to Extreme due to the excess sediment supply from the bedload and suspended sand deposition due to loss of sediment transport capacity (see section ***B2. Sediment capacity***). The bars create redistribution of energy and local slope changes that put excessive boundary stress on the erodible streambanks. To reduce the NBS, reshaping of the channel to match the stable C4 reference reach would reduce the bar-related sediment deposition problems.

Of further interest is that the prediction with the POWERSED model of the amount of coarse sediment deposition (bedload and suspended sand) in this 970 ft reach is 430 tons/yr. The amount of streambank erosion is 429 tons/yr predicted from the BANCS model. This indicates that the stream is seeking sediment continuity by balancing the sediment supply through channel enlargement (streambank erosion).

The accuracy of prediction is always a concern due to the complexity and uncertainty of natural systems. Measurement of actual erosion rates in the braided, D4 impaired reach are shown in **Figure 7-45**. The BEHI and NBS ratings are both Very High. The corresponding measured annual erosion rate using bank pins and a toe pin bank profile resurvey from 1995 to 1996 indicated a loss of 1.6 ft/yr (lateral erosion rate). The prediction using the Yellowstone curve (**Figure 7-44**) for the same Very High BEHI and NBS ratings was 1.55 ft/yr, very close to the measured values. The predicted verses measured values using the Yellowstone curve have been in close agreement using data collected in the same geology (Rosgen, 2001a). The estimates of total annual yield of streambank contributions are reasonable estimates based on this work. Restoring 970 ft of this reach to potentially reduce the sediment yield by 24% is a realistic prediction.

Only "hardening" or stabilizing the streambank would not regain the natural stability or offset the impairment of the physical and biological function of the D4 impaired reach. The aggradation process, for example, would continue to build the bars, losing channel capacity and loss of fish habitat. The recommendation to change the dimension, pattern and profile to match the C4 reference stream type as discussed in section ***B2. Sediment capacity*** would help reduce the impairment of this reach. The criteria for restoration include the data from **Worksheet 7-16a** (reference reach) to restore the unstable D4 reach comparing the corresponding data shown in **Worksheet 7-16b**. See also the discussion in ***A1. Successional stage shift***.

B8. Flow-related sediment yield

The increase in water yield contributed to an increase of 205 tons/yr for the C4 reference reach (**Worksheet 7-34b**) and 362 tons/yr for the D4 impaired reach (**Worksheet 7-34c**). Thus, the cumulative increase in flow-related sediment is 567 tons/yr. Streambank erosion is included in the channel-source increase due to flow increases. Of the 567 tons/yr, 429 tons/yr are related to the sediment source from the streambank erosion for 970 ft. This indicates that approximately 138 tons/yr are related to flow increases if the impaired reach is restored.

The streamflow increase is due to timber harvest clearcut blocks from 30 to 45 years ago. In high-elevation, snowmelt-dominated spruce and fir forests, it often takes up to 90 years to offset the increase in streamflow from timber harvest. This is due to the interception and snow deposition during storms, as influenced by adjacent stand heights and stand density (basal area). It is not reasonable to assume that evapo-transpiration, increased snow deposition and interception loss will be recovered in short time periods.

Roads, however, that add to increased streamflow by sub-surface flow interception and snowpack accumulation create increased routing of ditch-line water to the drainageways. Roads also increase the time of concentration and drainage density. The recommended procedure to restore the roads is to place the fill back into the cut bank using a large angle blade dozer or excavator. This creates an out-sloped road and regains the original slope allowing for revegetation. This also puts these road acres back into production and stops the interception of sub-surface water and reduces snow deposition. The overall effect of the road obliteration/restoration is to reduce the potential increased flow and sediment routing efficiency.

In the future, if selected roads are needed, then the material can be excavated from the cut bank and replaced to the road surface. In the intervening years, those roads would not contribute to the flow increase, sediment yields and maintenance challenges for the agencies or individuals managing recreational use.

B9. Total annual sediment yield

This is the final major item discussed in *PLA*. The summaries presented in **Worksheets 7-33** and **7-34c** and **Table 7-40** indicate the relative contributions of the various processes to the total annual sediment yield. The prediction of total annual sediment yield at the lowest end of the D4 stream type is related to the measured values in **Table 7-41**. The predicted value is within 1.2% of measured total annual sediment yield comparing measured data from 1997 to the FLOWSED prediction. Thus, the predicted values are reasonable or likely to be observed.

Table 7-41. A comparison of predicted-to-measured sediment yield values for lower Wolf Creek.

Lower Wolf Creek D4 impaired reach	Predicted values (FLOWSED)	Measured values (1997)
Bedload sediment	720	534
Suspended sediment	1,038	1,246
Total sediment	1,758	1,780

In review of the data summaries, it is recommended to evaluate those processes that are evidently contributing to high sediment supply and causing impairment. Many of the discussions have been previously presented by individual process and location to understand that the Wolf Creek Watershed has an annual increase in sediment of 1,418 tons/yr compared to 340 tons/yr for the baseline condition (**Worksheet 7-34c**). The resultant change in yield should place a priority emphasis for mitigation in this watershed. Of this increase, the relative contributions and percent departures for the major activities/processes are evident. For example, using **Worksheet 7-34c**, the 602 tons/yr from roads are contributing 34% of the total sediment yield, while streambank erosion on the lower 970 ft is contributing 429 tons/yr (24% of the total sediment yield). Thus, approximately 58% of the accelerated sediment can be potentially reduced if mitigation was emphasized on just these processes and locations.

This data helps put priorities on potential effective mitigation. The flow-related sediment increases based on an increment of increase of 567 tons/yr can also be minimized by reducing the streambank erosion of 429 tons/yr. This proposed restoration would reduce the 567 tons/yr to 138 tons/yr for flow-related sediment increase by restoring the D4 stream type to a C4 stream type.

Any predictions of sediment yield are not free of the uncertainty and complexity inherent in the processes driving actual sediment values. Monitoring (Chapter 6) is a solution to help ease the uncertainty of prediction and increase one's confidence in the ability to make mitigation effective to offset the predicted impairments.

The responsible manager must know the cause, consequence and hopefully the correction of the impairment of river systems due to the cumulative effects of past land uses. With a thorough analysis of documented observations of past impacts, one can be more effective at preventing similar problems in the future.

APPENDIX A

Water Yield Simulation Using a Modified WRENSS Procedure, Version 1.01

by:
Jim Nankervis, Berthoud, CO
Blue Mountain Consultants

A summary of the hydrologic simulations of Wolf Creek, a tributary of the San Juan River near Pagosa Springs, Colorado, and its sub-watersheds.

Objective

Model change in water yield as a result of forest management practices (e.g., timber harvest) for the Wolf Creek Watershed using a modification of WRENSS (Troendle and Swanson, in press) and San Juan National Forest (SJNF) data to determine maximum potential (complete hydrologic utilization) versus current (2000) conditions.

The forest data

In 1998, a request was made to the SJNF to provide data necessary to simulate hydrologic response with the WRENSS (Troendle and Leaf, 1980) model for the area of interest. In November 2001, with the help of Andrew Peavy (USFS Inventory and Monitoring Institute, Ft. Collins, Colorado), an electronic dataset was compiled from the SJNF's data describing homogeneous polygons of different cover type, size class, aspect, elevation, precipitation, harvest and the associated area within each sub-watershed. A summary of the Wolf Creek Watershed cover type acreages by harvest and size is described in **Table 1**. The SJNF defined cover type as Forbes, Grass, Non-Forest Land, Private, Shrub, Forest (by specie) or Water. Harvest information was simply denoted as yes or no. The SJNF defines the size of the timber within the forested polygons as Non-stocked (N), Seedling (E), Sapling (S), Pole (M) or Sawtimber (L and V). For purposes of this analysis, size classes were grouped into four categories; N (Non-stocked), S (Seedling/Sapling), P (Pole), or W (Sawtimber).

The precipitation data

Andrew Peavy assisted in the generation of precipitation data by superimposing the Oregon State University Climate Center Precipitation Map for Colorado over the DEM's for the Wolf Creek Watershed to estimate mean annual precipitation in 500' elevation zones. Mean monthly data for the Upper San Juan precipitation gage (#06M03S-located within the watershed) was used to estimate percentage of annual precipitation occurring each month. These percentages were then used to estimate mean monthly precipitation for each elevation zone (**Table 2**) and were applied to each forested polygon for the purpose of running the modified WRENSS water yield model.

Table 1. Summary of cover type, harvest, size class and acreage for the Wolf Creek Watershed.

Cover	Harvest	Size class	Acres
FOR			740.33
GRA			547.30
NFL			1720.60
PVT			240.45
SHR			167.60
TAA	No Harvest	P	102.46
TAA	No Harvest	W	471.72
		Subtotal	574.18
TDF	No Harvest	P	6.97
TDF	No Harvest	W	1295.08
		Subtotal	1302.05
TSF	No Harvest	N	17.74
TSF	No Harvest	P	98.01
TSF	No Harvest	S	75.75
TSF	No Harvest	W	5263.85
		Subtotal	5455.35
TDF	Harvest	W	36.37
		Subtotal	36.37
TSF	Harvest	N	398.17
TSF	Harvest	P	22.71
TSF	Harvest	S	524.63
TSF	Harvest	W	1284.91
		Subtotal	2230.43
WAT			1.75
		Total	13016.40

Appendix A   Water Yield Simulation Using a Modified WRENSS Procedure, Version 1.01

Table 2. Estimated mean monthly precipitation (in) by elevation zone (ft), Wolf Creek Watershed.

Elev	Jan	Feb	Mar	Apr	May	Jun	Jul	Aug	Sep	Oct	Nov	Dec
8000	2.260	2.290	3.118	2.907	1.395	0.628	1.796	2.124	2.188	3.469	3.750	2.035
8500	2.288	2.318	3.156	2.942	1.412	0.636	1.818	2.150	2.214	3.511	3.796	2.060
9000	2.439	2.471	3.364	3.137	1.505	0.678	1.938	2.292	2.360	3.743	4.046	2.196
9500	2.656	2.691	3.664	3.417	1.639	0.738	2.111	2.496	2.571	4.077	4.407	2.392
10000	2.952	2.990	4.071	3.796	1.821	0.820	2.346	2.774	2.856	4.530	4.897	2.657
10500	3.239	3.281	4.468	4.166	1.999	0.900	2.575	3.044	3.135	4.971	5.374	2.916
11000	3.454	3.499	4.765	4.443	2.131	0.960	2.746	3.246	3.343	5.301	5.731	3.110
11500	3.521	3.566	4.856	4.528	2.172	0.978	2.798	3.309	3.407	5.403	5.841	3.170
12000	3.560	3.606	4.911	4.579	2.197	0.989	2.830	3.346	3.446	5.464	5.907	3.205
12500	3.636	3.683	5.016	4.677	2.244	1.010	2.890	3.417	3.519	5.581	6.033	3.274
13000	3.597	3.644	4.962	4.627	2.220	1.000	2.859	3.381	3.482	5.521	5.968	3.239

Water yield

After the precipitation and forest data are merged, polygons with similar attributes can be aggregated (i.e., cover type, harvest, elevation, aspect and age class) reducing the number of observations from 3500+ to 141 forested data records. Once aggregated, these data need to be conditioned for the purpose of simulating 1) current hydrologic function (year 2000), and 2) complete hydrologic utilization (theoretical). An electronic version of the WRENSS procedure was used (Troendle and Leaf, 1980 v1.01; Troendle and Swanson, in press). There are two parameters that need to be adjusted when using WRENSS: basal area and precipitation.

Basal area

Hydrologic utilization is a function of stand density or basal area. Specific basal area data were not provided with the SJNF dataset so the relationship (and assumptions) developed by Troendle and Nankervis (2000) to predict basal as a function of age (mean of each specie size class) was used. A summary of basal area by specie and size class for current conditions is listed in **Table 3**. For harvested stands, the assumption is that size classes N and S were clearcut and that P and W were shelterwood (partial) cut and under the assumption that the mean basal area of N and S size classes equaled 30 and 40 $ft^2 acre^{-1}$, respectively, and the shelterwood cuts of P and W were cut back to 80 $ft^2 acre^{-1}$ (Troendle and Nankervis, 2000). To estimate water yield for complete hydrologic utilization conditions, all size classes in all polygons are set to Sawtimber so that basal area for all polygons are ≥ basal area for complete hydrologic utilization.

III

Table 3. Summary of basal area used for current conditions in the Wolf Creek Watershed.

Cover type	Size class	Basal Area (ft² acre⁻¹) Complete[a]	NonHarvest	Harvest
TAA	P	110	120	
TAA	W	110	140	
TDF	P	110	120	
TDF	W	110	140	
TDF	W	110		80
TSF	N	140	30	
TSF	S	140	40	
TSF	P	140	120	
TSF	W	140	140	
TSF	N	140		30
TSF	S	140		40
TSF	P	140		80
TSF	W	140		80

[a] Basal area where hydrologic utilization is 100%.

Precipitation

The electronic WRENSS procedure used does not account for the current understanding of snow interception loss (rather than redistribution) for timber stands reduced in density. To account for these situations, winter monthly (November–April) precipitation values are modified by specie, size class and aspect (see Troendle and Nankervis, 2000, for description and assumptions). **Table 4** lists the modifier coefficients used by specie, harvest and size class. These values are multiplied by mean monthly precipitation to adjust for interception loss. Note that for nonharvest conditions, only N and S size classes are modified because we assume that basal area for P and S size classes are at 100 percent hydrologic utilization. For harvest conditions, P and S are modified because of the assumption that current mature stands were partially harvested to a basal area of 80 ft²acre⁻¹.

Table 4. Summary of precipitation modifier coefficients for the Wolf Creek Watershed.

Aspect	TAA NonHarvest P	TAA NonHarvest W	TDF NonHarvest P	TDF NonHarvest W	TDF Harvest W	TSF NonHarvest N	TSF NonHarvest S	TSF NonHarvest P	TSF NonHarvest W	TSF Harvest N	TSF Harvest S	TSF Harvest P	TSF Harvest W
EW	N/A	N/A	N/A	N/A	1.065	1.35	1.195	N/A	N/A	1.35	1.195	1.09	1.12
NO	N/A	N/A	N/A	N/A	1.095	1.48	1.264	N/A	N/A	1.48	1.264	1.13	1.18
SO	N/A	N/A	N/A	N/A	1.040	1.20	1.110	N/A	N/A	1.20	1.110	1.05	1.07

WRENSS output

The water yield predicted by the two simulations is 20.5 inches for the complete hydrologic utilization and 21.9 inches for the current year 2000, a change of 1.4 inches or about a 7% increase in water yield.

The flow-duration curve

By making some assumptions and correcting for changes in flow at certain quantiles based on observed flow changes in the Coon Creek pilot project (Troendle et al., 2001), current and baseline flow-duration curves were generated.

First, the predicted water yield (in inches) was distributed over time intervals for each of the simulations using the equations (inches water x area in acres)/(12 x 1.98 x days in interval), where days in interval equals 36.5 accumulated (36.6, 73.0, 109.5, ... 365) to get streamflow in ft^3s^{-1}. The change per interval and sum to get total change was calculated. Next, from Coon Creek (**Table 2**), percent change per quantile was calculated (note Coon Creek flow is measured seasonally, about 183 days/year). Because no maximum flow can be estimated from the WRENSS model, the upper 2 quantiles (percentages) are combined at this time. For each quantile, multiply the percent change in flow observed at Coon Creek by the sum of the change in flow from WRENSS to distribute over each quantile except 1.0. Convert these values to $ft^3s^{-1}mi^{-2}$ to normalize for area by dividing by 15 (the forested area simulated in the WRENSS procedure). Now multiply each value by 18 (watershed area in mi^2 above the Wolf Creek gage) to make values comparable to the observed flows at the gage. Next, the largest flow value (in the smallest interval) is apportioned between the 1.0 and 0.95 quantiles based on the observed changes at Coon Creek, 0.5152 of the value to 1.0 and 0.4848 to 0.95. We now have the adjusted seasonal flow changes, per quantile (0.5, 0.55, ..., 1.0) for the Wolf Creek gage (**Table 5**). The two big assumptions here are that timber harvest has no effect on low or baseflows (October–March) and that the observed percent changes in flow/quantile at Coon Creek are appropriate for simulating a response at Wolf Creek.

The daily mean flow records for the Wolf Creek gage (USGS# 09341300) for water years 1985–87 and water years 1997–2000 were used to develop flow-duration curves. Subtracting the estimated flow values for each quantile (between 0.5 and 1.0) provides an estimate of what the flow-duration curve would be when the entire forested area above the Wolf Creek gage is at complete hydrologic utilization (**Table 6**).

WATERSHED ASSESSMENT OF RIVER STABILITY AND SEDIMENT SUPPLY

Table 5. Calculation of change in flow by quantile (seasonal) for Wolf Creek gage.

Interval (days)	Base WRENSS (ft^3s^{-1})	2000 WRENSS (ft^3s^{-1})	Change in flow (ft^3s^{-1})	Percent change of total change Coon Creek[a]	Corrected change from Coon Creek (ft^3s^{-1})	Area (15) corrected ($ft^3s^{-1}mi^{-2}$)	Area (18) corrected Wolf Creek gage (ft^3s^{-1})	Wolf Creek gage with upper quantile (ft^3s^{-1})	Quantile
				0.2698				14.696	1.00
36.5	226.889	242.384	15.495	0.254	23.773[b]	1.585	28.527	13.831	0.95
73.0	113.444	121.192	7.747	0.190	8.645	0.576	10.373	10.373	0.90
109.5	75.630	80.795	5.165	0.079	3.602	0.240	4.322	4.322	0.85
146.0	56.722	60.596	3.874	0.079	3.602	0.240	4.322	4.322	0.80
182.5	45.378	48.477	3.099	0.032	1.441	0.096	1.729	1.729	0.75
219.0	37.815	40.397	2.582	0.016	0.720	0.048	0.864	0.864	0.70
255.5	32.413	34.626	2.214	0.032	1.441	0.096	1.729	1.729	0.65
292.0	28.361	30.298	1.937	0.032	1.441	0.096	1.729	1.729	0.60
328.5	25.210	26.932	1.722	0.016	0.720	0.048	0.864	0.864	0.55
365.0	22.689	24.238	1.549	0.000	0.000	0.000	0.000	0.000	0.50

[a] From Troendle et al., 2001.
[b] Value includes percent for 1.0 and 0.95 quantile.

Table 6. Current and estimated baseline flow-duration curves, Wolf Creek.

Quantile	Current (ft^3s^{-1})	Adjustment (ft^3s^{-1})	Baseline (ft^3s^{-1})
1.00	398.0	14.7	383.3
0.95	227.0	13.8	213.2
0.90	162.0	10.4	151.6
0.85	108.9	4.3	104.6
0.80	69.0	4.3	64.7
0.75	45.0	1.7	43.3
0.70	34.8	0.9	33.9
0.65	28.0	1.7	26.3
0.60	22.0	1.7	20.3
0.55	16.0	0.9	15.1
0.50	12.0	0.0	12.0
0.45	11.0	0.0	11.0
0.40	9.6	0.0	9.6
0.35	8.5	0.0	8.5
0.30	7.5	0.0	7.5
0.25	6.8	0.0	6.8
0.20	6.0	0.0	6.0
0.15	5.4	0.0	5.4
0.10	4.8	0.0	4.8
0.05	3.9	0.0	3.9
0.00	2.0	0.0	2.0

References

Ackers, P., & White, W.R. (1973). Sediment transport: New approach and analysis. *Journal of the Hydraulics Division, ASCE, 99*(11), 2041–2060.

Andrews, E.D. (1980). Effective and bankfull discharges of streams in the Yampa River Basin, Colorado and Wyoming. *Journal of Hydrology, 46,* 311–330.

Andrews, E.D. (1983). Entrainment of gravel from naturally sorted riverbed material. *Geological Society of America Bulletin, 94,* 1225–1231.

Andrews, E.D. (1984). Bed-material entrainment and hydraulic geometry of gravel-bed rivers in Colorado. *Geological Society of America Bulletin, 95,* 371–378.

Andrews, E.D., & Erman, D.C. (1986). Persistence in the size distribution of surficial bed material during an extreme snowmelt flood. *Water Resources Research, 22*(2), 191–197.

Andrews, E.D., & Nankervis, J.M. (1995). Effective discharge and the design of channel maintenance flows for gravel-bed rivers. In J.E. Costa, A.J. Miller, K.W. Potter, & P.R. Wilcock (Eds.), *Natural and anthropogenic influences in fluvial geomorphology* (pp. 151–164). Washington, DC: American Geophysical Union Monograph Series 89.

Annable, W.K. (1995). *Morphological relationships of rural water courses in Southwestern Ontario for use in natural channel designs.* Unpublished master's thesis, University of Guelph, Guelph, Ontario, Canada.

Annable, W.K. (1996). *Database of morphological characteristics of watercourses in Southern Ontario.* Ontario: Ministry of Natural Resources.

Ashworth, P.J., & Ferguson, R.I. (1989). Size-selective entrainment of bed load in gravel bed streams. *Water Resources Research, 25*(4), 627–634.

Bagnold, R.A. (1966). An approach to the sediment transport problem from general physics (USGS Professional Paper 422–I). Reston, VA: U.S. Geological Survey.

Bagnold, R.A. (1980). An empirical correlation of bedload transport rates in flumes and natural rivers. *Proceedings of the Royal Society, London, 372,* 452–473.

Barbour, M.T., Stribling, J.B., & Karr, J.R. (1991). *Biological criteria: Streams- fourth draft* (EPA Contract No. 68-CO-0093). Washington, DC: U.S. Environmental Office of Science and Technology.

Benda, L., & Dunne, T. (1987). Sediment routing by debris flows. In R.L. Beschta, T. Blinn, G.E. Grant, G.G. Ice and F.J Swanson (Eds.), *Erosion and sedimentation in the Pacific Rim* (Pub. 165) (pp. 213–223). Wallingford, Oxfordshire, UK: International Association of Hydrological Sciences.

Bliss, C.I. (1935). The calculation of the dosage mortality curve. *Annals of Applied Biology, 22*(1), 134–167.

Bosch, J.M., & Hewlett, J.D. (1982). A review of catchment experiments to determine the effect of vegetation changes on water yield and evapotranspiration. *Journal of Hydrology, 55,* 2–23.

Bunte, K., & Abt, S.R. (2001). *Sampling surface and subsurface particle-size distributions in wadable gravel and cobble-bed streams for analyses in sediment transport, hydraulics, and streambed monitoring* (Gen. Tech. Rep. RMRS-GTR-74). Fort Collins, CO: U.S. Department of Agriculture, Forest Service, Rocky Mountain Research Station.

Burroughs, E.R. (1984). Landslide hazard rating for portions of the Oregon Coast Range. In C.L. O'Loughlin, & A.J. Pearce (Eds.), *Proceedings of a symposium on effects of forest land use on erosion and slope stability* (pp. 264–274). Honolulu: University of Hawaii.

Caine, N. (1980). The rainfall intensity-duration control of shallow landslides and debris flows. *Geografiska Annaler, 62A*(1-2), 23–27.

Charlton, F.G., Brown, P.M., & Benson, R.W. (1978). *The hydraulic geometry of some gravel rivers in Britain* (Report IT 180). Wallingford, UK: Hydraulics Research Station.

Copeland, R.R., McComas, D.N., Thorne, C.R., Suar, P.J., Jones, M.M., & Fripp, J.B. (2001). *Hydraulic design of stream restoration projects* (ERDC/CHL, TR-01-28). Vicksburg, MS: U.S. Army Corps of Engineers, Engineer Research and Development Center, Coastal and Hydraulics Laboratory.

Costa, J.E. (1983). Paleohydraulic reconstruction of flash flood peaks from boulder deposits in the Colorado Front Range. *Geological Society of America Bulletin, 94,* 986–1004.

Davis, W.M. (1902). Base level, grade, and peneplain. *Journal of Geology, 10,* 77–111.

DeBano, L.F., & Rice, R.M. (1975). Water-repellant soils: Their implications in forestry. *Journal of Forestry, 71*(4), 222–223.

Devore, B. (1998). The Stream Team. *The Minnesota Conservation Volunteer, November-December,* 10–19.

Dietrich, W.E., & Dunne, T. (1978). Sediment budget for a small catchment in mountainous terrain. *Zeitschrift fur Geomorphologie Neue Folge Supplementband, 29,* 191–206.

References

Diplas, P. (1987). Bedload transport in gravel-bed streams. *Journal of Hydraulic Engineering, 113*(3), 277–291.

Dunne, T., & Leopold, L.B. (1978). *Water in environmental planning*. San Francisco, CA: W.H. Freeman and Company.

Einstein, H.A. (1950). *The bedload function for sediment transport in open channel flow* (Tech. Bull. No. 1026). Washington, DC: U.S. Department of Agriculture, Soil Conservation Service.

Elliot, W.J., & Foltz, M. (2001, July-August). *Validation of the FS WEPP interfaces for forest roads and disturbances*. Presentation at the American Society of Agricultural Engineers Annual International Meeting, Sacramento, California.

Elliot, W.J., & Hall, D.E. (1997). Water Erosion Prediction Project (WEPP) forest applications (Gen. Tech. Rep. Draft). Odgen UT: U.S. Department of Agriculture, Forest Service, Intermountain Research Station.

Emmett, W.W. (1972). *The hydraulic geometry of some Alaskan streams south of the Yukon River* (Open File Report 72–108). Reston, VA: U.S. Geological Survey.

Emmett, W.W. (1975). *The channels and waters of the upper Salmon River Area, Idaho* (USGS Professional Paper 870A). Reston, VA: U.S. Geological Survey.

Emmett, W.W. (1980). A field calibration of the sediment-trapping characteristics of the Helley-Smith Bedload Sampler (USGS Professional Paper 1139). Reston, VA: U.S. Geological Survey.

Emmett, W.W. (2001). U.S. Geological Survey sediment data for Idaho. Personal Communication.

Erman, D.C., Andrews, E.D., & Yoder-Williams, M. (1988). Effects of winter floods on fishes in the Sierra Nevada. *Canadian Journal of Fisheries and Aquatic Sciences, 45*(12), 2195–2200.

Federal Interagency Stream Restoration Working Group. (1998). *Stream corridor restoration: Principles, processes, and practices* (GPO Item No. 0120-A). Washington, DC: Author.

Flanagan, D.C., & Nearing, M.A. (1995). *USDA-Water Erosion Prediction Project: Hillslope profile and water-shed model documentation* (NSERL Report No. 10). West Lafayette, IN: U.S. Department of Agriculture-Agriculture Research Service, National Soil Erosion Research Laboratory.

Galay, V.J., Kellerhals, R., & Bray, D.I. (1973). Diversity of river types in Canada. In *Fluvial Process and Sedimentation* (Proceedings of Hydrology Symposium No. 9) (pp. 217–250). Ottowa, Ontario: National Research Council of Canada.

Gomez, B., & Church, M. (1989). An assessment of bedload sediment transport formulae for gravel bed rivers. *Water Resources Research, 25*(6), 1161–1186.

Graf, W.H. (1971). *Hydraulics of sediment transport*. New York: McGraw-Hill.

Griffiths, G.A. (1981). Stable channels with mobile gravel beds. *Journal of Hydrology, 52*, 291–305.

Haan, C.T., Barfield, B.J., & Hayes, J.C. (1994). *Design hydrology and sedimentology for small catchments*. San Diego, CA: Academic Press.

Hack, J.T. (1960). Interpretations of erosional topography in humid temperate regions. *American Journal of Science, 258A*, 80–97.

Harmon, W., & Jessup, A. (1999). Personal communication summarizing research findings on the Mitchell River streambank erosion studies in North Carolina. North Carolina State University and Natural Resources Conservation Service, respectively.

Harrelson, C.C., Rawlins, C.L, & Potyondy, J.P. (1994). *Stream channel reference sites: An illustrated guide to field technique* (Gen. Tech. Rep. RM-245). Fort Collins, CO: U.S. Department of Agriculture, Forest Service, Rocky Mountain Forest and Range Experiment Station.

Hey, R.D. (1997). Stable river morphology. In C.T. Thorne, R.D. Hey and M.D. Newson (Eds.), *Applied fluvial geomorphology for river engineering and management* (pp. 223–236). New York: John Wiley and Sons, Inc.

Hey, R.D., & Thorne, C.R. (1986). Stable channels with mobile gravel beds. *Journal of Hydraulic Engineering, 112*, 671–689.

Howes, D.E. (1987). A method for predicting terrain susceptible to landsliding following timber harvesting: A case study from the Southern Coast mountains of British Columbia. In R.H. Swanson, P.Y. Bernier, and P.D. Woodward (Eds.), *Forest hydrology and watershed management* (Pub.167, pp. 143–154). Wallingford, Oxfordshire, UK: International Association of Hydrological Sciences.

James, L.A. (1991). Incision and morphologic evolution of an alluvial channel recovering from hydraulic mining sediment. *Geological Society of American Bulletin, 103*, 723–736.

Jarrett, R.D., & Costa, J.E. (1986). *Hydrology, geomorphology, and dam-break modeling of the July 15, 1982 Lawn Lake Dam and Cascade Lake Dam failures, Larimer County, Colorado* (USGS Professional Paper 1369). Washington, DC: U.S. Government Printing Office.

Kelsey, H.M. (1982). Hillslope evolution and sediment movement in a forested headwater basin, Van Duzen River, North Coastal California. In *Sediment budgets and routing in forested drainage basins* (Gen. Tech. Rep. PNW-141, pp. 86–96). Washington, DC: U.S. Government Printing Office.

References

Kluckhohn, B. (Photographer) (1998). In B. Devore (author), The stream team. *The Minnesota Conservation Volunteer, November-December.* Reprinted with permission from Bruce Kluckhohn.

Komar, P.D. (1987a). Selective gravel entrainment and the empirical evaluation of flow competence. *Sedimentology, 34,* 1165–1176.

Komar, P.D., & Karling, P.A. (1991). Grain sorting in gravel-bed streams and the choice of particle sizes for flow-competence evaluations. *Sedimentology, 38*(3), 489–502.

Kondolf, M.G. (1997). Hungry water: Effects of dams and gravel mining on river channels. *Environmental Management, 21*(4), 533–551.

Kurz, J.D. (Photographer) (1999). E4 Stream Type, South Fork of the South Platte River, CO.

Laflen, J.M., Elliot, W.J., Flanagan, D.C., Meyer, C.R., & Nearing, M.A. (1997). WEPP- predicting water erosion using a process-based model. *Journal of Soil and Water Conservation, 52,* 96–102.

Lane, E.W. (1955). Design of stable channels. *Transactions of the American Society of Civil Engineers, 120,* 1234–1279.

Lane, L.J., & Nearing, M.A. (Eds.) (1989). USDA-Water Erosion Prediction Project: Hillslope profile model documentation (NSERL Rep. No. 2). West Lafayette, IN: U.S. Department of Agriculture, Agricultural Research Service, National Soil Erosion Research Laboratory.

Langbein, W. B. (1964). Geometry of river channels. *Journal of the Hydraulics Division American Society of Civil Engineers, 90*(HY2), 301–312.

Lehre, A.K., Collins, B.D., & Dunne, T. (1983). Post-eruption sediment budget for the North Fork Toutle River Drainage, June 1980-June 1981. *Zeitschrift fur Geomorphologie Neue Folge Supplementband, 46,* 143–163.

Leopold, L.B. (1994). *A View of the river.* Cambridge, MA: Harvard University Press.

Leopold, L.B., & Emmett, W.W. (1997). *Bedload and river hydraulics- Inferences from the East Fork River, Wyoming* (USGS Professional Paper 1583). Reston, VA: U.S. Geological Survey.

Leopold, L.B., & Maddock, T. (1953). *The hydraulic geometry of stream channels and some physiographic implications* (USGS Professional Paper 252). Reston, VA: U.S. Geological Survey.

Leopold, L.B., Rosgen, D.L., & Silvey, H.L. (2000). *The river field book.* Pagosa Springs, CO: Wildland Hydrology Books.

Leopold, L.B., Wolman, G.M., & Miller, J.P. (1964). *Fluvial processes in geomorphology*. San Francisco, CA: W.H. Freeman.

Leven, R. (1977). Suspended sediment rating curves from USGS and USDA Forest Service-Redwood Creek. Six Rivers N.F., Arcata, CA. In *An approach to water resources evaluation of non-point silvicultural sources* (EPA -600/8-80-012, p. VI.40). Athens, GA: U.S. Environmental Protection Agency.

Lillesand, T.M., & Kiefer, R.W. (1999). *Remote sensing and image interpretation* (4th ed.). New York: John Wiley and Sons, Inc.

Limerinos, J.T. (1970). *Determination of the Manning Co-Efficient from measured bed roughness in natural channels.* (Water Supply Paper 1898B). Reston, VA: U.S. Geological Survey.

Lopes, V.L., Osterkamp, W.R., Espinosa, M.B. (2001a). Evaluation of selected bedload equations under transport- and supply-limited conditions. *Proceedings of the Seventh Interagency Sedimentation Conference, 1*, I-192–I-198.

Lopes, V.L., Folliott, P.F., & Baker, M.B. (2001b). Impacts of vegetative practices on suspended sediment from watersheds in Arizona. *Journal of Water Resources Planning and Management, 127*(1), 41–47.

López, J.L.S. (2004). Channel response to gravel mining activities in mountain rivers. *Journal of Mountain Science, 1*(3), 264–269.

Lowham, H.W. (1976). *Techniques for estimating flow characteristics of Wyoming streams* (Water Resources Investigations 76–112). Cheyenne, WY: U.S. Geological Survey, Water Resources Division.

MacDonald, L., Smart, A., & Wissmar, R. (1991). *Monitoring guidelines to evaluate effects of forestry activities on streams in the Pacific Northwest and Alaska* (EPA/910/9-91-001). Washington, DC: U.S. Environmental Protection Agency.

Mackin, J.H. (1948). Concept of the graded river. *Geological Society of America Bulletin, 59*, 463–512.

Meeuwig, R.D. (1970). Infiltration and soil erosion as influenced by vegetation and soil in Northern Utah. *Journal of Range Management, 23*(3), 185–188.

Megahan, W.F. (1974). *Erosion over time on severely disturbed granitic soil: A model* (Res. Pap. INT-156). Ogden, UT: U.S. Department of Agriculture, Forest Service, Intermountain Forest and Range Experiment Station.

Miller, M.C., McCave, I.N., & Komar, P.D. (1977). Threshold of sediment motion in unidirectional currents. *Sedimentology, 24*, 507–528.

National Research Council, Water Science and Technology Board, Commission on Geosciences, Environment, and Resources. (1992). *Restoration of aquatic ecosystems, science, technology, and public policy*. Washington, DC: National Academy Press.

Nearing, M.A., Foster, G.R., Lane, L.J., & Finkner, S.C. (1989). A process-based soil erosion model for USDA-Water Erosion Prediction Project Technology. *Transactions of the American Society of Agricultural Engineers, 32*(5), 1587–1593.

Neibling, W.H., & Foster, G.R. (1977). Estimating deposition and sediment yield from overland flow processes. In *Proceedings of international symposium on urban hydrology, hydraulics, and sediment control* (pp. 75–86). Lexington: University of Kentucky.

Nilsen, T.H., Taylor, F.A., & Brabb, E.E. (1976). *Recent landslides in Atameda County, California (1940-71): An estimate of economic losses and correlations with slope, rainfall and ancient landslide deposits* (USGS Bulletin 1398). Reston, VA: U.S. Geological Survey.

Nixon, M. (1959). A study of bankfull discharges of rivers in England and Wales. *Proceedings of the Institution of Civil Engineers, 12*, 157–175.

Osterkamp, W. (1994). Personal communication on techniques of bankfull determination for step/pool channel types. Denver, CO: U.S. Geological Survey.

Osterkamp, W.R., Day, T.J., & Parker, R.S. (1992). A sediment monitoring program for North America. In *Erosion and sediment transport monitoring programs in river basins* (IAHS Publication 210, pp. 391–396). Wallingford, Oxfordshire, UK: International Association of Hydrological Sciences.

Pacific Southwest Inter-Agency Committee, Water Management Subcommittee, Sedimentation Task Force. (1968). *Report on factors affecting sediment yield in the Pacific Southwest area*.

Parker, G. (1979). Hydraulic geometry of active gravel rivers. *Journal of the Hydraulics Division of the American Society of Civil Engineers, 105*(HY 9), 1185–1201.

Parker, G. (1990). Surface-based bedload transport relation for gravel rivers. *Journal of Hydraulics Research, 28*(4), 417–436.

Parker, G., Klingeman, P.C., & McLean, D.C. (1982). Bedload and size distribution in paved gravel-bed streams. *Journal of Hydraulics Division, American Society of Civil Engineers, 108*, 545–571.

Pfankuch, D.J. (1975). *Stream reach inventory and channel stability evaluation* (USDAFS No. R1-75-002, GPO No. 696-260/200). Washington, DC: U.S. Government Printing Office.

Pickup, G. (1988). Hydrology and sediment models. In M.G. Anderson (Ed.), *Modelling geomorphological systems* (pp. 153–215). Chichester, UK: John Wiley and Sons, Inc.

Pickup, G., Higgins, R.J., & Grant, I. (1983). Modelling sediment transport as a moving wave – The transfer and deposition of mining waste. *Journal of Hydrology, 60*, 281–301.

Piest, R.F., Bradford, J., & Spomer, R.G. (1975). Mechanisms of erosion and sediment movement from gullies. In *Proceedings present and prospective technology for predicting sediment yields and sources* (ARS-S-40). Oxford, MS: U.S. Department of Agriculture, National Sedimentation Laboratory.

Pitlick, J. (1985). *The effect of a major sediment influx on Fall River, Colorado.* Unpublished master's thesis, Colorado State University, Fort Collins, CO.

Pizzuto, J.E., & Mecklenburg, T.S. (1989). Evaluation of a linear bank erosion equation. *Water Resources Research, 25*(5), 1005–1013.

Porth, L. (2006). Personal communication on hydrologic recovery in snowmelt-dominated watersheds. Fort Collins, CO: U.S. Department of Agriculture, Forest Service, Rocky Mountain Forest and Range Experiment Station.

Reid, L.M. (1981). *Sediment production from gravel-surfaced forest roads, Clearwater Basin, Washington* (Publication FRI-UW-8108). Seattle: University of Washington, Fisheries Research Institute.

Reid, L.M., & Dunne, T. (1984). Sediment production from forest road surfaces. *Water Resources Research, 20*, 1753–1761.

Reid, L.M., & Dunne, T. (1996). *Rapid evaluation of sediment budgets; Geo-Ecology Texts.* Reiskirchen, Germany: Catena Verlag.

Renard, K.G., Foster, G.R., Weesies, D.K., & Yoder, D.C. (1997). *Predicting soil erosion by water: A guide to conservation planning with the Revised Universal Soil Loss Equation* (RUSLE) (Agriculture Handbook No. 703). Tucson, AZ: U.S. Department of Agriculture, Agriculture Research Service, Southwest Watershed Research Center.

Rice, R.M., & Pillsbury, N.H. (1982). Predicting landslides in clearcut patches. In *Recent developments in the explanation and prediction of erosion and sediment yield* (Pub. No.137, pp. 303–311). Wallingford, Oxfordshire, UK: International Association of Hydrological Sciences.

Riefenberger, E., & Baird, D. (1991). Dump Creek: A man made ecological disaster. In *Proceedings of the fifth Federal Interagency Sedimentation Conference*, (pp. 10-57–10-60.). Washington DC: Federal Energy Regulatory Committee.

Rosgen, D.L. (1973). The use of color infrared photography for the determination of sediment production. In *Fluvial process and sedimentation* (pp. 381–402). Ottowa, Ontario, Canada: National Research Council of Canada.

References

Rosgen, D.L. (1976). The use of color infrared photography for the determination of suspended sediment concentrations and source areas. In *Proceedings of the third Federal Interagency Sedimentation Conference* (pp. 30–42). Reston, VA: U.S. Geological Survey, Sedimentation Committee of the Water Resources Council.

Rosgen, D.L. (1994). A classification of natural rivers. *Catena, 22,* 169–199.
Figure 2-13, Table 2-2, Figure 2-14, Figure 2-22 and Table 2-4 Reprinted from *Catena, 22,* Rosgen, D.L., A classification of natural rivers, 169–199, Copyright 1994, with permission from Elsevier.

Rosgen, D.L. (1996). *Applied river morphology.* Pagosa Springs, CO: Wildland Hydrology Books.

Rosgen, D.L. (1997). A geomorphological approach to restoration of incised rivers. In S.S.Y Wang, E.J Langendoen, & F.D. Shields (Eds.), *Proceedings of the conference on management of landscapes disturbed by channel incision.* Oxford: University of Mississippi.

Rosgen, D.L. (1998). The reference reach-A blueprint for natural channel design. In D.F. Hayes (Ed.), *Engineering approaches to ecosystem restoration (Proceedings of the wetlands engineering and river restoration conference).* Reston, VA: American Society of Civil Engineers.

Rosgen, D.L. (1999). Development of a river stability index for clean sediment TMDL's. In D.S. Olsen & J.P. Potyondy (Eds.), *Proceedings of Wildland Hydrology* (pp. 25–36). Herndon, VA.: American Water Resources Association.

Rosgen, D.L. (2001a). A practical method of computing streambank erosion rate. In *Proceedings of the seventh Federal Interagency Sedimentation Conference: Vol. 1.* (pp. II-9–II-15). Reno, NV: Subcommittee on Sedimentation.

Rosgen, D.L. (2001b). A stream channel stability assessment methodology. In *Proceedings of the seventh Federal Interagency Sedimentation Conference: Vol. 1.* (pp. II-18–II-26). Reno, NV: Subcommittee on Sedimentation.

Rosgen, D.L. (2001c). A hierarchical river stability/watershed-based sediment assessment methodology. In *Proceedings of the seventh Federal Interagency Sedimentation Conference: Vol. 1.* (pp. II-97–II-106). Reno, NV: Subcommittee on Sedimentation.

Rosgen, D.L. (2001d). The Cross-Vane, W-Weir and J-Hook vane structures … their description, design and application for stream stabilization and river restoration. In D.F. Hayes (Ed.), *Wetlands engineering and river restoration 2001 (Proceedings of the wetlands engineering and river restoration conference) (Chapter 3).* Reston, VA: American Society of Civil Engineers.

Rosgen, D.L. (2003). *Critical review of a classification of natural rivers.* Doctoral dissertation, University of East Anglia, Norwich, Norfolk.

Rosgen, D.L. (2006). FLOWSED-POWERSED: Prediction models for suspended and bedload transport. In *Proceedings of the eighth Federal Interagency Sediment Conference*. In Press.

Rosgen, D.L. & Silvey, H.L. (2005). *The reference reach field book*. Fort Collins, CO: Wildland Hydrology Books.

Schumm, S.A., Harvey, M.D., & Watson, C.C. (1984). *Incised channels: Morphology, dynamics and control*. Littleton, CO: Water Resources Publications.

Schumm, S.A., & Lichty, R.W. (1963). *Channel widening and floodplain construction along the Cimarron River in Southwestern Kansas* (USGS Professional Paper, 352D, pp. 71–88). Reston, VA: U.S. Geological Survey.

Schoklitsch, A. (1962). *Handbuch des Wasserbaues* (3rd ed.). Vienna: Springer-Verlag.

Seaber, P.R., Kapinos, F.P., & Knapp, G.L. (1987). *Hydrologic unit maps* (USGS Water Supply Paper, 2294). Denver, CO: U.S. Geological Survey.

Sheppard, W.D., Troendle, C.E., & Edminster, C.B. (1991). Linking water and growth and yield models to evaluate management alternatives in subalpine ecosystems. In *Getting to the future through silviculture. Proceedings of the national silviculture workshop* (Gen. Tech. Rep. INT-291, pp. 42-49). Ogden, UT: U.S. Department of Agriculture, Forest Service, Rocky Mountain Forest and Range Experiment Station.

Shields, A. (1936). Application of similarity principles and turbulence research to bedload movement. In W.P. Ott, and J.C. Unchelen (Translators), *Mitteilungen der preussischen Nersuchsanstalt fur Wassed Brau end Schiffbuar* (Rep. No. 167). Pasedena, CA: California Institute of Technology.

Simon, A. (1989a). A model of channel response in disturbed alluvial channels. *Earth Surface Processes and Landforms, 14*, 11–26.

Simon, A. (1989b). The discharge of sediment in channelized alluvial streams. *Water Resources Bulletin, 25*(6), 1177–1188.

Simon, A. (1992). Energy, time, and channel evolution in catastrophically disturbed fluvial systems. *Geomorphology, 5*, 345–372.

Simon, A., & Hupp, C.R. (1986). Channel evolution in modified Tennessee streams. In *Proceedings of the fourth Federal Interagency Sedimentation Conference* (pp. 5.71–5.82). Washington, DC: U.S. Government Printing Office.

Simon, A., Curini, A., Darby, S., & Langenden, E. (1999). Streambank mechanics and the role of bank and near-bank processes. In S. Darby, & A. Simon (Eds.), *Incised river channels* (pp. 123–152). New York: John Wiley and Sons, Inc.

References

Simon, A., Dickerson, W., & Heins, A. (2004). Suspended-sediment transport rates at the 1.5 year recurrence interval for ecoregions of the United States: Transport conditions at the bankfull and effective discharge. *Geomorphology, 58*, 243–262.

Strahler, A.N. (1957). Quantitative analysis of watershed geomorphology. *American Geophysical Union Transactions, 38*, 913–920.

Swanson, R.H. (2004). The complete WRENSS hydrologic model. *Proceedings of the 72nd Western Snow Conference, 72*, 99–104.

Swanson, D.N., & Swanson, F.J. (1976). Timber harvesting, mass erosion and steep land geomorphology in the Pacific Northwest. In D.R. Coates (Ed.), *Geomorphology and Engineering* (pp. 199–221). Stroudsburg, PA: Dowden, Hutchinson and Ross, Inc.

Thorne, C. (1999). Bank processes and channel evolution in North-central Mississippi. In S. Darby, & A. Simon, (Eds.), *Incised river channels* (pp. 97–121). New York: John Wiley and Sons, Inc.

Tollner, E.W., Barfield, B.J., Haan, C.T., & Kao, T.Y. (1976). Suspended sediment filtration capacity of simulated vegetation. *Transactions of the American Society of Agricultural Engineers, 19*(4), 678–682.

Toy, J.T., & Foster, G.R. (1998). *Guidelines for the use of the Revised Universal Soil Loss Equation (RUSLE) Version 1.06 on mined lands, construction sites and reclaimed lands*. Denver, CO: Office of Surface Mining.

Troendle, C.A., & Leaf, C.F. (1980). Hydrology. In *An approach to Water Resource Evaluation of Non-point Silvicultural Sources (A Procedural Handbook)* (EPA 600/8-80-012) (pp. III-1–III-173). Athens, GA: Environmental Research Laboratory.

Troendle, C.A. & Nankervis, J.M. (2000). *Estimating additional water yield from changes in management of national forests in the North Platte Basin*. Final Report submitted to the Platte River Office, Bureau of Reclamation. March 2000. 51pp.

Troendle, C.A., & Olsen, W.K. (1994). Potential effects of timber harvest and water management on streamflow dynamics and sediment transport. In W.W. Covington & L.F. DeBano (Tech. coordinators), *Sustainable ecological systems: Implementing an ecological approach to land management* (Gen. Tech. Report RM-247) (pp. 34–41). Fort Collins, CO: U.S. Department of Agriculture, Rocky Mountain Forest and Range Experimentation Station.

Troendle, C.A., & Swanson, R. (in press). User's guide of the hydrologic component of the WRENSS procedure (Gen. Tech. Rep.). Fort Collins, CO: U.S. Forest Service, Rocky Mountain Research Station.

Troendle, C.A., Rosgen, D.L., Ryan, S., Porth, L., & Nankervis, J.M. (2001). Developing a "reference" sediment transport relationship. In *Proceedings of the seventh Federal Interagency Sedimentation Conference: Vol. 1.* (pp. II-73–II-80). Reno, NV: Subcommittee on Sedimentation.

U.S. Army Corps of Engineers. (1998). *Flood Hydrograph Package: User's Manual,* Retrieved from http://www.hec.usace.army.mil/software/legacysoftware/hec1/ documentation/hec1user.pdf

U.S. Department of Agriculture, Forest Service. (2003). *Identifying bankfull stage in the Eastern and Western United States [DVD-ROM].* Fort Collins, CO: Author.

U.S. Environmental Protection Agency. (1997a). *New policies for establishing and implementing Total Maximum Daily Loads (TMDLs).* Washington, DC: Author.

U.S. Environmental Protection Agency. (1997b). *Monitoring guidance for determining the effectiveness of nonpoint source controls* (EPA-841-B-96-004). Washington, DC: Author.

U.S. Environmental Protection Agency. (1980). *(WRENSS) An approach to Water Resource Evaluation of Non-point Silvicultural Sources (A Procedural Handbook)* (EPA-600/8-80-012). Athens, GA: Environmental Research Laboratory.

U.S. Environmental Protection Agency, Office of Water. (1999). *Protocol for developing sediment TMDLs* (EPA 841-B-99-004). Washington, DC: Author. Retrieved from http://www.epa.gov/owow/tmdl/sediment/pdf/sediment.pdf

U.S. Geological Survey. (1980). *Surface water investigations measurement of stream discharge.* Reston, VA: Author.

U.S. Geological Survey. (1999). *Field methods for measurement of fluvial sediment.* Reston, VA: Author.

Vanoni, V.A. (Ed). (1975). *Sedimentation engineering (Manuals and reports on engineering practice, No. 54).* New York: American Society of Civil Engineers.

Wildland Hydrology (1986). [Pagosa Springs, CO data set]. Unpublished raw data.

Williams, G.P. (1978). Bankfull discharges of rivers. *Water Resources Research, 14* (6), 1141–1154.

Williams, G.P. (1983). Improper use of regression equations in earth sciences. *Geology, 11,* 195–197.

Williams, G.P., & Rosgen, D.L. (1989). *Measured total sediment loads (suspended and bedloads) for 93 United States streams* (Open File Rep. 89-67). Denver, CO: U.S. Geological Survey.

Williams, J.R. (1975). Sediment yield prediction with universal equation using runoff energy factor. In *Present and prospective technology for predicting sediment yields and sources: Proceedings of sediment yield workshop* (ARS-S-40) (pp. 244–252). Oxford, MS: U.S. Department of Agriculture, National Sedimentation Laboratory.

Williams, J.R., & Berndt, H.D. (1972). Sediment yield computed with universal equation. *Journal of the Hydraulics Division, American Society of Civil Engineers, 98*(HY12), 2087–2097.

Wilson, D. (1985). Subjective techniques for identification and hazard assessment of unstable terrain. In D. Swanston (Ed.), *Proceedings of the workshop on slope stability: Problems and solutions in forest management* (Gen. Tech. Rep. PNW-180, pp. 36–42). Portland, OR: U.S. Department of Agriculture, Forest Service, Pacific Northwest Forest and Range Experiment Station.

Winward, A.H. (2000). *Monitoring the vegetation resources in riparian areas* (Gen. Tech. Rep. RMRS-GTR-47). Ogden, UT: U.S. Department of Agriculture, Forest Service, Rocky Mountain Research Station.

Wischmeier, W.H., & Smith, D.D. (1978). *Predicting rainfall erosion losses- A guide to conservation planning* (Agriculture Handbook 537). Washington, DC: U.S. Department of Agriculture-Science and Education Administration.

Wolman, M. G. (1955). *The natural channel of Brandywine Creek, Pennsylvania* (USGS Professional Paper 271). Washington, DC: U.S. Geological Survey.

Wolman, M.G., & Miller, J.P. (1960). Magnitude and frequency of forces in geomorphic processes. *Journal of Geology, 68*, 54–74.

Yang, C.T. (1979). Unit stream power equations for total load. *Journal of Hydrology, 40*(1/2), 123–138.

Yang, C.T. (1996). *Sediment transport: Theory and practice*. New York: McGraw-Hill.

Yang, C.T., Trevino, M.A., & Simoes, F.J. (1998). GSTARS 2.0 [Manual]. Denver, CO: U.S. Department of the Interior, Bureau of Reclamation, Sedimentation and River Hydraulics Group, Technical Service Center.

Subject Index

A

Abandoned floodplain: 2-56, 5-14
— see also terrace
Accelerated streambank erosion: 1-8, 2-38, 2-47, 2-48, 2-77, 2-89, 2-90, 2-91, 4-62, 4-69, 5-43, 5-162, 6-5, 7-21, 7-47, 7-51, 7-97, 7-194, 7-211, 7-213
Acres cleared: 4-36, 7-49
Aerial photograph: 3-1, 4-7, 5-68, 7-3, 7-4, 7-8, 7-85
Aggradation: 1-7, 1-8, 2-30, 2-35, 2-37 thru 2-39, 2-42, 2-43, 2-44, 2-47, 2-55, 2-76, 2-87, 3-8, 3-17, 3-18, 4-11, 4-21, 4-62 thru 4-66, 4-68, 5-4, 5-41 thru 5-43, 5-106, 5-111, 5-114, 5-129, 5-134, 5-138, 5-141, 5-142, 5-146, 5-147, 5-159, 5-162, 6-5, 6-7, 7-10, 7-16, 7-17, 7-32, 7-65, 7-67, 7-68, 7-71, 7-79, 7-83, 7-97, 7-115, 7-152, 7-155, 7-170, 7-174, 7-182 thru 7-184, 7-187, 7-200, 7-203, 7-205 thru 7-207, 7-209, 7-210, 7-214
Agricultural applications: 2-3, 4-5, 4-6, 7-24, 7-25
Alder: 5-17, 7-108
Alluvial channels: 5-9, 5-14
Alluvial fans: 2-26
Alluvial valley fills: 2-26
Alluvium: 7-4, 7-12
Alnus (spp.): 5-17
— see also alder
Alpine glaciation: 2-92, 4-16, 5-78, 5-80, 7-4, 7-36, 7-131, 7-213
Altered channel: 4-4, 4-6, 7-7, 7-23, 7-25, 7-221
Altered channel dimension, pattern, profile and materials:
— see channel dimension; channel pattern; profile; channel materials
Anastomosed channels: 2-22
Annual sediment yield: 1-11, 2-4, 2-6, 2-35, 2-58, 4-15, 4-49, 5-54, 5-82, 5-83, 5-85, 5-88, 5-94, 5-97 thru 5-102, 5-111, 5-113 thru 5-118, 5-126, 5-138, 5-156, 5-158, 5-160 thru 5-162, 6-13, 7-40, 7-57, 7-77, 7-117, 7-131, 7-137, 7-138, 7-140, 7-141 thru 7-144, 7-147, 7-148, 7-155, 7-160, 7-162 thru 7-164, 7-167, 7-174, 7-194, 7-196 thru 7-200, 7-206, 7-209, 7-210, 7-211, 7-213, 7-215
Anthropogenic: 1-1, 1-8, 2-1, 4-67, 5-34, 6-4
Aquatic ecosystem: 1-8, 6-5
Average slope: 2-3, 5-69, 5-72
Avulsion: 2-30, 2-38, 2-76, 2-82, 2-83, 2-85 thru 2-87, 3-18, 5-42, 7-17, 7-115

B

Bagnold equation: 2-73, 2-75, 5-111, 7-152
Bagnold relation: 2-73, 2-75, 5-11, 7-152
BANCS model: 2-87, 2-89, 2-92, 5-52, 5-53, 5-78, 5-161, 6-13, 7-123, 7-124, 7-205, 7-214
Bank
— Bank angle: 5-6, 5-55, 5-61, 5-161, 7-205
— Bank Assessment for Non-point source Consequences of Sediment (BANCS): 2-37, 2-87, 2-92, 5-54, 7-78
— see also BANCS model
— Bank erodibility factors:
— see BEHI
— Bank erosion: 2-35, 2-37, 2-40, 2-42, 2-44, 2-47, 2-57, 2-85, 2-87, 2-92, 3-4, 4-11, 4-43, 4-49, 4-68, 4-74, 5-6, 5-36, 5-38, 5-50, 5-53 thru 5-56, 5-68, 5-78, 5-161, 6-6, 6-13 thru 6-15, 7-32, 7-47, 7-78, 7-108, 7-124, 7-125, 7-127, 7-128, 7-131, 7-151
— see also streambank erosion
— Bank Erosion Hazard Index (BEHI): 2-37, 5-53 thru 5-56, 5-78, 5-161, 6-6, 7-124, 7-125, 7-127, 7-128
— see also BEHI
— Bank erosion pins:
— see bank pins
— Bank erosion processes:
— see streambank erosion processes
— Bank erosion rates: 2-35, 2-57, 2-92, 4-49, 5-36, 5-50, 5-54, 7-47, 7-131
— see also streambank erosion rates
— Bank-Height Ratio (BHR): 2-40, 4-44, 4-47, 4-49, 4-69, 4-72, 5-6, 5-44, 5-47, 5-55, 5-58, 5-146, 5-159, 5-161, 6-7, 7-53, 7-72, 7-120, 7-182, 7-205
— Bank material adjustment: 5-63
— Bank pins: 4-74, 6-5, 6-7, 6-13, 6-16, 7-214
— Bank profile: 6-7, 6-13, 6-15, 7-214
— Bank stratification: 5-64
see also incision, degree of; channel incision; belt width

Bankfull
— Bankfull cross-sectional area: 5-18, 5-20, 5-25, 7-90
— Bankfull dimensionless shear stress: 2-13, 2-14, 5-131, 5-134
— Bankfull dimensions: 4-44, 5-8 thru 5-10, 5-14, 5-16, 7-211
— Bankfull dimensional shear stress: 2-72
— Bankfull discharge: 1-7, 1-9, 2-18, 2-28, 2-29, 2-35, 2-40, 2-55 thru 2-58, 2-61, 2-71, 2-83, 2-87, 4-7, 4-8, 4-39, 4-40, 5-7 thru 5-9, 5-11, 5-14, 5-17 thru 5-21, 5-25, 5-47, 5-86, 5-87, 5-89, 5-90, 5-92, 5-94 thru 5-96, 5-100 thru 5-102, 5-106, 5-129 thru 5-131, 5-134, 6-7, 7-87 thru 7-91, 7-120, 7-145, 7-146, 7-151, 7-155, 7-203, 7-211
— see also effective discharge; dominant discharge
— Bankfull discharge return period:
— see return period
— Bankfull height: 5-55, 5-58
— Bankfull maximum depth: 2-40, 5-18
— Bankfull mean depth: 2-20, 5-18, 5-20, 5-65, 5-71, 5-72, 7-90, 7-210
— Bankfull mean velocity: 5-20, 5-21
— Bankfull shear stress: 5-65, 5-72, 5-73, 7-210
— Bankfull stage: 2-13, 2-14, 2-18, 2-40, 2-55 thru 2-57, 2-61, 2-67, 2-70, 2-72, 2-73, 4-65, 4-72, 5-6, 5-9, 5-14 thru 5-20, 5-47, 5-58, 5-69, 5-86, 5-92, 5-129 thru 5-131, 5-134, 5-138, 6-6, 6-9, 7-210, 7-213
— Bankfull stage indicators: 5-14 thru 5-17
— see also field indicators of bankfull stage
— Bankfull calibration: 5-13 thru 5-25
Bar sample: 2-13, 2-70, 2-72, 2-73, 2-75, 5-6, 5-130, 5-131, 5-133, 5-134, 6-7, 6-9, 6-16
Bare ground: 3-17, 4-4, 4-26, 4-29, 7-16, 7-23
Bars:
— Central bars: 5-14, 5-67, 5-161
— Delta bars: 4-66, 5-41, 7-67, 7-114
— Diagonal bars: 4-66, 5-41, 7-67, 7-114

XXI

Subject Index

— Mid-channel bars: 4-66, 5-41, 7-67, 7-114, 7-204
— Point bars: 2-13, 2-70, 4-66, 5-14, 5-41, 5-130, 7-67, 7-114, 7-204
— Side bars: 4-66, 5-41, 7-67
— Transverse bars: 5-66, 5-67, 7-129, 7-130
— see also depositional features
Basal area: 7-139, 7-215
Base level: 1-7, 2-40, 2-83, 3-19, 4-21, 5-47, 5-146, 7-71, 7-79, 7-207
Baseflow: 4-43, 4-50, 4-69, 6-3, 7-152
Beaver dams: 5-42, 7-115
Bed features: 2-24, 2-55, 5-69, 6-10, 6-16, 7-203
Bed scour: 2-40, 2-42, 7-18
— see also degradation; excess shear stress; critical depth computation; entrainment
Bed-material load: 2-19, 2-74, 5-107 thru 5-109, 5-111, 7-152, 7-156 thru 7-158
Bed-material size distribution: 2-84, 4-41, 7-97, 7-208
— see also pebble count
Bedload: 2-6 thru 2-8, 2-10, 2-12, 2-14 thru 2-19, 2-33, 2-34, 2-38, 2-58, 2-60 thru 2-62, 2-64 thru 2-70, 2-72 thru 2-75, 2-87, 2-88, 4-43, 5-6, 5-14, 5-38, 5-45, 5-85 thru 5-87, 5-92 thru 5-95, 5-97, 5-101, 5-102, 5-106 thru 5-109, 5-111, 5-114, 5-130, 5-131, 5-134, 5-138, 5-154, 5-156, 6-2, 6-3, 6-6, 6-7, 6-9, 7-145 thru 7-148, 7-152 thru 7-158, 7-196, 7-200, 7-206, 7-211, 7-214
Bedload rating curves: 2-18, 2-61, 2-67, 2-68
Bedload sediment: 2-6, 2-18, 2-33, 2-34, 2-58, 2-61, 2-62, 2-67, 5-6, 5-85, 5-87, 5-92 thru 5-94, 5-95, 5-106, 5-130, 5-131, 5-134, 6-3, 6-6, 7-145, 7-148
Bedload sediment yield: 5-85, 7-148
Bedrock: 2-26, 4-34, 5-23, 5-63, 7-4, 7-212
Bedrock-controlled valleys: 2-26
BEHI: 2-37, 2-87, 2-89 thru 2-92, 5-6, 5-53 thru 5-64, 5-72, 5-74, 5-78 thru 5-80, 5-93, 5-144, 5-161, 6-6, 6-13, 7-124 thru 7-128, 7-131, 7-132, 7-135, 7-179, 7-203 thru 7-205, 7-214
Belt width: 5-18, 5-50, 7-121, 7-205
— see also Bank-Height Ratio (BHR); incision, degree of
Benthic: 6-5
Berm: 7-212

Best management practices: 2-3, 4-26, 5-164
Betula: 5-17
— see also birch
Biological function: 1-7, 7-214
Biological monitoring: 6-1, 6-5
Birch: 5-17
Boulders: 1-9, 5-63, 7-205
Boundary shear stress: 4-48, 5-68
Braided channel: 2-35, 2-42, 7-113
Bridges: 2-40, 2-42, 3-17, 4-4, 4-21, 4-62, 4-69, 4-71, 7-16, 7-17, 7-23, 7-71, 7-73
— see also stream crossing impacts; culverts
Broad-level classification: 2-22
— see also river classification

C

Calibration monitoring: 6-6
Canopy: 5-37, 7-109, 7-110
— see also riarian overstory
Cantilever failure: 4-46, 5-58, 5-61
Canyons: 2-26
Carex (spp.): 7-108
— see also sedges
Central bars: 5-14, 5-67, 5-161
— see also bars
Channel:
— Channel bed: 2-43, 5-138, 6-10
— Channel blockages: 4-57, 4-58, 5-34, 5-42, 7-106, 7-115
— Channel capacity: 2-38, 7-208, 7-210, 7-214
— Channel confinement: 5-34, 5-50, 5-146, 7-106, 7-121, 7-182
— see also confinement; lateral containment; Meander Width Ratio (MWR)
— Channel containment: 1-8
— see also channel entrenchment; vertical containment
— Channel dimension: 2-58, 4-4, 4-6, 4-58, 5-110, 6-7, 7-7, 7-18, 7-23, 7-25
— Channel enlargement: 1-8, 2-38, 2-42, 2-47, 3-8, 4-10, 4-11, 4-34, 4-42, 4-52, 4-59 thru 4-63, 4-65, 4-68, 5-41, 5-50, 5-142, 5-144, 5-146, 5-149, 7-10, 7-32, 7-40, 7-62 thru 7-64, 7-79, 7-83, 7-114, 7-177, 7-179, 7-182, 7-187 thru 7-189, 7-207 thru 211, 7-214
— Channel entrenchment: 2-40
— see also vertical containment
— Channel evolution: 2-44 thru 2-46, 2-58, 3-6, 3-18, 4-61, 4-67, 4-69, 5-6, 5-143, 7-17, 7-69, thru 7-71, 7-79, 7-177

— see also stream channel succession; morphological shift
Channel hydraulics: 5-106, 5-138
Channel incision: 1-7, 2-40, 2-42, 4-43, 4-68, 4-69, 4-71, 4-72, 5-6, 5-34, 5-44, 5-47, 5-48, 5-146, 5-148, 5-159, 7-73, 7-83, 7-106, 7-120, 7-182, 7-185, 7-186, 7-206, 7-207
— see also incision, degree of; Bank-Height Ratio (BHR)
Channel length: 4-11, 4-50, 4-53, 4-55, 4-56, 7-32, 7-58, 7-59, 7-131
Channel materials: 1-7, 2-20, 2-24, 2-43, 2-58, 3-13, 4-6, 4-7, 5-18, 5-106, 5-110, 5-111, 5-130, 5-131, 7-12, 7-25, 7-90, 7-203
Channel morphology: 2-38, 2-85, 3-18, 4-43
Channel pattern: 4-43, 5-65, 5-67, 7-113
Channel processes: 2-6, 3-18, 4-43, 7-17
Channel restoration:
— see restoration
Channel roughness:
— see roughness
Channel shape: 5-85
Channel slope: 2-38, 5-6, 5-18
Channel stability: 1-6, 1-7, 2-24, 2-33, 2-57, 2-58, 2-60, 2-67, 2-68, 2-87, 3-1, 3-6, 4-2, 4-15, 4-25, 4-43, 4-50, 4-52, 5-1, 5-6, 5-32, 5-34, 5-37, 5-46, 5-87, 5-106, 5-107, 5-111, 5-139 thru 5-141, 5-151, 5-154, 5-156, 5-159, 5-164, 7-7, 7-23, 7-51, 7-75, 7-104, 7-106, 7-109, 7-110, 7-117 thru 7-119, 7-151, 7-154, 7-175, 7-176, 7-201 thru 7-203, 7-208, 7-209
— see also river stability; equilibrium
Channel stability rating: 2-58, 5-6, 5-34, 5-46, 5-87, 5-107, 7-106, 7-117 thru 7-119
Channel stabilization: 7-48
Channelization: 1-9, 2-57, 2-58, 3-19, 4-11, 4-56, 4-62, 4-69, 7-32
— see also land uses
Check dam: 4-6, 5-42, 7-25, 7-206
Chute: 5-41, 5-67
Classification: 2-20 thru 2-24, 2-26, 2-29, 2-30, 2-33, 2-43, 2-44, 3-10, 3-13, 4-7, 4-9, 5-6, 5-17, 5-26 thru 5-29, 5-87, 7-5, 7-12, 7-26 thru 7-30, 7-47, 7-48, 7-84, 7-92, 7-93, 7-95, 7-96, 7-203
— see also river classification; valley classification

XXII

Subject Index

Clay: 2-19, 3-13, 4-16, 5-106, 7-36, 7-145, 7-155
Clean sediment: 1-2, 1-11
Clean Water Act: 1-1
Clear water discharge: 1-8, 2-40, 3-17, 4-5, 4-10, 4-59, 7-16, 7-24
— *see also hungry water*
Clearcut condition: 3-13, 7-14, 7-48
Climate: 1-7, 2-2, 2-38, 4-67, 5-14, 6-1
Cobble: 1-9, 2-39, 3-13, 4-55, 5-20, 5-63, 5-129, 5-134, 5-138, 6-13
Cohesive bank material: 5-56, 5-63, 7-127, 7-128
Colloidal sediments: 5-106
Colluvial slopes: 5-14
Colluvial valley: 2-26
Colluvium: 4-16, 7-4, 7-36
Color-infrared photography: 1-11, 3-5
Competence: 2-6, 2-12, 2-14, 2-38, 2-42, 2-70, 2-71, 2-73, 2-77, 4-11, 4-57, 4-62, 5-4, 5-6, 5-114, 5-127 thru 5-129, 5-134, 5-135, 5-137, 5-138, 5-146, 5-156, 5-159, 5-160, 6-7, 6-9, 7-32, 7-116, 7-152, 7-168 thru 7-174, 7-182, 7-200, 7-205, 7-206, 7-209, 7-210
Concentration: 2-19, 2-37, 2-58, 3-5, 4-5 thru 4-7, 4-16, 4-21, 4-35, 4-39, 7-12, 7-24, 7-25, 7-36, 7-78, 7-215
— *see also suspended sediment*
Confined meander scrolls: 7-204
Confinement: 1-8, 1-9, 2-43, 4-11, 5-34, 5-50, 5-51, 5-144, 5-146, 7-32, 7-106, 7-121, 7-179, 7-182, 7-203, 7-205
— *see also lateral containment; Meander Width Ratio (MWR)*
Continuous slopes: 2-3, 3-16, 7-15, 7-35
Continuum: 2-20
Convergence: 7-213
Cornus (spp.): 5-17
— *see also dogwood*
Cottonwood: 2-76, 5-17, 7-108
Crest gage: 6-7, 6-10
Critical depth computation: 2-42
Critical dimensionless shear stress: 2-8, 2-11, 2-70
Critical shear stress: 2-9, 2-70, 5-136
Cross-sectional area: 2-8, 4-58, 5-9, 5-18, 5-20, 5-25, 5-47, 7-90, 7-204
Crown cover:
— *see riparian overstory; canopy*
Culverts: 2-40 thru 2-42, 3-17, 4-4, 4-21, 4-69, 4-71, 5-159, 7-16, 7-17, 7-23, 7-71, 7-73, 7-80, 7-212
— *see also stream crossing impacts; bridges*
Cumulative effects: 1-1, 2-54, 4-42, 5-162, 7-40, 7-51, 7-77, 7-216

Cut bank: 2-6, 4-4, 7-7, 7-15, 7-21, 7-78, 7-139, 7-212, 7-215

D

Dams: 2-6, 4-5, 4-6, 5-42, 5-101, 7-24, 7-25, 7-115, 7-206
Darcy-Weisbach friction factor: 5-23
Debris avalanche: 3-16, 4-14 thru 4-17, 5-125, 7-15, 7-35, 7-36, 7-71, 7-166
Debris blockage: 4-57, 5-42, 7-60
Debris cones: 2-26
Debris dams: 5-42, 7-115
— *see also channel blockages*
Debris flow: 3-16, 7-15, 7-35, 7-71
Debris management: 4-49, 4-58, 7-78
Debris torrents: 1-8, 1-9, 2-3, 2-4, 3-16, 4-14, 7-15
Degradation: 1-7, 1-8, 2-30, 2-35, 2-37, 2-40 thru 2-44, 2-48, 2-55, 3-8, 3-17, 3-18, 4-7, 4-10, 4-11, 4-21, 4-42, 4-52, 4-69, 4-70 thru 4-72, 5-4, 5-42, 5-106, 5-111, 5-129, 5-134, 5-138, 5-141, 5-142, 5-146, 5-148, 5-159, 5-162, 6-5, 6-7, 7-10, 7-16, 7-17, 7-25, 7-32, 7-71 thru 7-74, 7-79, 7-83, 7-115, 7-120, 7-182, 7-185, 7-186, 7-206, 7-207
Degree of channel incision: 2-40, 5-6, 5-34, 5-44, 5-47, 5-48, 5-146, 5-159, 7-106, 7-120, 7-182
Degree of impairment: 5-1
Delivery distance: 5-119
Delta: 2-26, 2-38, 5-106
Departure: 1-6, 1-7, 2-20, 2-24, 2-26, 2-30, 2-57, 2-58, 2-61, 4-63, 4-65, 4-69, 4-73, 4-74, 5-1, 5-4, 5-28, 5-30, 5-34, 5-36, 5-43, 5-44, 5-106, 5-154, 5-156, 5-157, 5-160, 6-3 thru 6-5, 7-66, 7-75, 7-83, 7-97, 7-116, 7-121, 7-194, 7-196, 7-208
Deposition: 1-1, 1-7, 1-8, 2-15, 2-37, 2-38, 2-42, 2-75, 2-77, 2-87, 3-17, 3-18, 4-11, 4-16, 4-41, 4-43, 4-62 thru 4-66, 4-68, 5-6, 5-14, 5-41, 5-43, 5-67, 5-106, 5-114, 5-119, 5-138, 5-146, 5-147, 5-159, 6-10, 7-16, 7-17, 7-32, 7-36, 7-65, 7-67, 7-68, 7-79, 7-116, 7-117, 7-174, 7-182 thru 7-184, 7-200, 7-205, 7-207, 7-208, 7-210, 7-213 thru 7-215
Depositional features: 2-13, 4-52, 4-66, 5-14, 5-67, 7-1, 7-67, 7-131
— *see also bars*
Depositional patterns: 3-16, 4-63, 4-66, 5-34, 5-41, 5-146, 5-159, 7-67, 7-106, 7-114, 7-152, 7-182, 7-203 thru 7-206

Digital terrain models: 2-5, 4-15
Dimension: 7-18
— *see also channel dimension*
Dimensional shear stress: 2-72, 7-170
Dimensioned flow-duration curve: 5-96, 5-97, 5-102, 7-146
Dimensioned sediment rating curve: 2-19, 5-92 thru 5-94, 5-106, 7-146
Dimensionless bedload rating curve: 2-69, 5-87
Dimensionless flow-duration curve: 5-85, 5-87, 5-89 thru 5-91, 5-96, 5-102, 7-145, 7-146
Dimensionless ratios: 2-61, 5-6, 5-26 thru 5-28, 5-30, 5-31, 5-39, 7-92, 7-93, 7-97, 7-203, 7-211
Dimensionless sediment rating curve: 2-61, 5-94, 5-102
Dimensionless shear stress: 2-8 thru 2-11, 2-13, 2-14, 2-16, 2-70, 2-73, 2-75, 5-131, 5-134, 7-170
Direct channel impacts: 1-9, 3-19, 4-11, 4-53, 4-54, 4-57, 4-58, 4-61, 5-161, 7-7, 7-18, 7-19, 7-32, 7-33, 7-51, 7-57, 7-61, 7-71, 7-78, 7-83, 7-209, 7-213
Discontinuous slopes: 3-16
Disequilibrium: 1-7, 2-30, 2-38, 2-47, 2-82, 7-71
— *see also stream channel instability*
Dissected slopes: 3-14, 3-17, 4-26, 4-35, 7-14, 7-16
Dissolved oxygen: 6-5, 7-203
Diversions
— Depletions: 4-41, 5-101
— Imported water: 2-42, 2-55, 3-17, 4-5, 4-39, 4-40, 5-101, 7-16, 7-24
— *see also operational hydrology; land uses*
Dogwood: 5-17
Dominant discharge: 2-55
— *see also bankfull discharge; effective discharge*
Drainage area: 2-2, 2-43, 2-71, 4-4, 4-7, 4-8, 4-44, 5-9, 5-12, 5-20, 5-25, 5-47, 5-94, 5-100, 7-3, 7-23, 7-89, 7-90, 7-139
Drainage basin: 2-37
Drainage density: 2-3, 2-26, 3-16, 4-5, 4-6, 4-21, 4-26, 4-30, 4-69, 7-4, 7-14, 7-24, 7-25, 7-215
Dredging: 2-42, 3-4, 3-19, 4-11, 4-53, 4-69, 7-32
— *see also land uses*
Dry ravel: 2-2, 4-47
Dynamic equilibrium: 2-37, 2-38

XXIII

Subject Index

E

Effective discharge: 2-56
— see also bankfull discharge; dominant discharge
Effectiveness monitoring: 1-4, 4-74, 6-2 thru 6-4, 6-18, 7-81
Einstein bedload function: 2-10
Encroachment: 3-19, 4-5 thru 4-7, 5-50, 7-7, 7-24, 7-25, 7-51, 7-57, 7-161, 7-213
Energy dissipation: 4-11, 7-32
Energy distribution: 2-87, 5-65, 5-67
Engineering: 1-6, 4-53, 5-4
Enlargement: 2-42, 2-47, 2-56, 2-75, 2-87, 4-10, 4-11, 4-34, 4-42, 4-52, 4-59, 4-61, 4-62, 5-36, 5-41, 5-142, 5-144, 5-146, 5-149, 6-5, 7-10, 7-17, 7-32, 7-40, 7-62, 7-63, 7-79, 7-83, 7-179, 7-182, 7-187 thru 7-189, 7-207, 7-208, 7-211, 7-214
— see also channel enlargement
Entrainment: 2-4, 2-8, 2-12, 2-35, 2-38, 2-70 thru 2-72, 2-77, 4-15, 4-46, 5-55, 5-59, 5-114, 5-128, 5-129, 5-131, 5-134, 6-7, 6-10, 7-169, 7-170
— see also Shields relation; fluvial entrainment
Entrenchment: 1-9, 2-20, 2-40, 4-7, 5-6, 5-18, 5-36
— see also channel entrenchment; vertical containment
Entrenchment ratio: 1-9, 2-20, 4-7, 5-6, 5-18, 5-36
Eolian deposition: 2-26
Ephemeral gully: 2-43
Ephemeral stream: 5-38, 7-111
Equilibrium: 1-7, 1-8, 2-37, 2-38, 2-47, 2-57, 2-58, 3-16, 4-11, 7-32
Erodibility: 2-3, 2-35, 2-43, 2-87, 4-11, 4-14, 4-26, 5-57, 7-32, 7-65, 7-126
Erosion:
— see bed scour; dry ravel; fluvial entrainment; freeze-thaw; streambank erosion; mass erosion; surface erosion; roads; hillslope processes; channel processes
Erosion rate recovery: 5-116, 7-163
Evapo-transpiration: 2-53, 3-16, 3-17, 4-5, 4-21, 5-100, 7-15, 7-16, 7-24, 7-78, 7-139, 7-215
Evolutionary stages: 5-94
— see also stream succession; morphological shift; channel evolution
Excess sediment supply: 2-39, 4-10, 4-62, 4-63, 5-41, 7-7, 7-65, 7-71, 7-79, 7-174, 7-214
— see also aggradation

Excess shear stress: 2-40, 2-55, 5-129
— see also critical depth computation
Existing condition (watershed): 4-65

F

Factor of safety approach: 2-37
Field calibration: 5-14, 5-17, 5-54, 6-6, 7-90, 7-125, 7-131
Field indicators of bankfull stage: 5-14 thru 5-17
Field-sieve: 2-72, 5-129, 6-9
Fire: 5-119
— see also land uses
Fish habitat: 1-12, 2-6, 2-47, 2-77, 2-84, 4-44, 4-69, 7-97, 7-114 thru 7-116, 7-187, 7-203 thru 7-208, 7-214
Flood control: 2-57, 4-6, 4-53, 7-25
— see also land uses
Flood frequency: 2-55, 5-17 thru 5-19, 5-101
Flood-prone area: 2-20, 5-6, 5-18, 5-42, 7-17, 7-78, 7-115
Flood-prone area width: 2-20, 5-6, 5-18
Floodplain: 1-9, 2-26, 2-35, 2-51, 2-55 thru 2-58, 2-82, 3-17, 3-18, 4-21, 4-68, 5-14, 5-18, 5-19, 5-159, 7-16 thru 7-18, 7-120, 7-207
Floods: 1-10, 2-38, 2-42, 5-38, 5-47, 7-3
Flow-duration curve: 2-54, 5-85, 5-87, 5-89 thru 5-91, 5-96, 5-97, 5-102, 5-107, 5-111, 7-145 thru 7-147
Flow modification: 5-101
— see also operational hydrology
Flow regime: 2-40, 2-57, 4-61, 5-6, 5-34, 5-38, 5-42, 5-85, 5-100, 7-35, 7-106, 7-111, 7-115
Flow-related sediment: 2-67, 3-6, 3-17, 4-34, 4-35, 4-38, 4-39, 5-85, 5-101, 5-102, 5-154, 5-156, 5-161, 5-162, 7-21, 7-48, 7-49, 7-51, 7-139, 7-148, 7-196, 7-209, 7-212, 7-214, 7-216
FLOWSED: 2-67, 2-74, 5-45, 5-82 thru 5-88, 5-92, 5-94, 5-98 thru 5-100, 5-102, 5-106, 5-107, 5-111, 5-138, 5-162, 6-6, 7-117, 7-137 thru 7-139, 7-145, 7-147, 7-211, 7-215
Fluvial-dissected landscapes: 2-26
Fluvial entrainment: 2-4, 2-35, 2-77, 4-15, 4-46, 5-55, 5-59
Forest ecosystem: 5-120
Four "C's" of river assessment: 1-12
Frame and pin: 6-7 thru 6-9
Freeze-thaw: 2-2, 4-47, 5-58
Friction factor: 5-23

G

Gaging station: 5-11, 7-139
Geologic erosion rate: 1-1
Geologic hazard map: 5-6, 5-125
Geologic map: 5-6, 5-125
Geology: 1-6, 2-2, 2-4, 2-5, 2-92, 3-4, 3-13, 4-7, 4-8, 4-12, 4-14 thru 4-16, 5-4, 5-78 thru 5-80, 7-4, 7-12, 7-26, 7-36, 7-131, 7-213, 7-214
Geomorphic characterization: 2-20, 2-26
Geomorphology: 1-1, 1-6, 5-4
GIS: 1-11, 3-4, 3-10, 4-4, 5-12
Glacial melt: 5-38, 7-111
Glacial moraine:
— see glacial till
Glacial outwash: 2-26, 3-18, 7-17
Glacial till: 1-9, 4-16, 7-36, 7-213
Glacial-trough valley: 2-26
Glacio-lacustrine: 2-26
Grade control structures: 2-43, 7-206
Graded river: 2-38
Gradient: 1-9, 2-3, 2-6, 2-26, 2-28, 3-10, 3-14, 4-16, 4-17, 4-19, 4-24, 4-26, 4-30, 4-48, 5-65, 5-67, 5-74, 5-77, 5-106, 5-119, 5-123, 5-125, 6-8, 7-36, 7-37, 7-165
Gravel: 1-9, 2-16, 2-25, 2-28, 2-29, 2-32, 2-35, 2-37, 2-39, 2-42, 3-13, 4-41, 4-43, 4-55, 5-20, 5-63, 5-65, 5-129, 5-134, 5-138, 6-5, 6-13, 7-90, 7-210
Gravitational acceleration: 2-8, 2-16, 5-23
Grazing: 2-32, 2-33, 2-76, 2-79, 2-84, 3-4, 4-6, 4-49, 4-53, 4-55, 4-62, 4-68, 4-74, 5-159, 6-4, 7-3, 7-6, 7-7, 7-25, 7-57, 7-106, 7-196, 7-205
— see also land uses
Ground cover: 2-3, 4-11, 4-26, 4-31, 5-36, 5-119, 5-123, 6-8, 7-32
Guidance criteria: 3-2, 3-13 thru 3-20, 4-4, 4-14, 4-34, 7-14 thru 7-20, 7-35
Gully: 2-40, 2-41, 2-43, 2-57, 3-8, 4-11, 4-26, 7-32, 7-207
Gully erosion: 2-43, 2-57, 3-8, 4-11, 4-26, 7-32

H

Headcut: 2-40, 2-43, 2-76, 2-82, 7-207
Helley-Smith bedload sampler: 2-72, 5-87, 6-6, 6-7
High width/depth ratio: 2-38, 2-58, 2-87, 7-116, 7-152
Hillslope processes: 2-1, 2-6, 3-13, 4-4, 4-5, 4-14, 4-42, 4-62, 5-112 thru 5-114, 5-126, 5-154, 5-156, 7-14,

Subject Index

7-23, 7-24, 7-35, 7-65, 7-159, 7-160, 7-167, 7-200
Hortonian overland flow: 2-3, 4-26, 7-12, 7-19, 7-46
Hungry water: 1-8, 4-41
— *see also clear water discharge*
Hydraulic geometry: 2-24, 2-25, 2-27, 2-29, 2-30, 2-37, 5-18, 5-25, 5-106, 5-110, 5-111, 7-151, 7-154, 7-155
Hydraulic mean depth: 5-23, 5-111
Hydraulic radius: 2-8, 5-23, 5-111, 5-130
Hydrograph: 2-53 thru 2-55, 5-89, 5-100, 7-3, 7-16, 7-21
Hydrologic processes: 2-53, 3-16, 4-5, 4-34, 5-38, 7-15, 7-24, 7-47
Hydrology: 2-1, 2-55, 3-17, 4-8, 4-10, 4-11, 4-34, 4-35, 4-39 thru 4-41, 5-1, 5-14, 5-47, 5-100, 5-101, 7-16, 7-32, 7-89, 7-90
Hydropower dams: 4-5, 7-24

I

Implementation monitoring: 5-164, 6-1, 6-4
Imported water: 2-42, 2-55, 3-17, 4-5, 4-39, 4-40, 5-101, 7-16, 7-24
In-channel mining: 4-50, 4-51, 4-52, 4-58, 4-59, 4-61, 4-69, 7-57, 7-71
— *see also land uses*
Incised channel: 2-43
Incision, degree of: 2-40, 5-6, 5-34, 5-44, 5-47, 5-48, 5-146, 5-159, 7-106, 7-120, 7-182
— *see also Bank-Height Ratio (BHR); channel incision; belt width*
Increased sediment supply: 2-42, 2-57, 2-61, 2-75, 4-42, 5-154, 5-157, 5-159, 7-154, 7-196, 7-202
Infiltration: 2-3, 4-5, 5-119, 5-121, 7-24, 7-46, 7-212
Infiltration capacity: 5-121
Initiation of particle movement: 2-8
— *see also particle movement*
Instability: 1-1, 1-7, 1-10, 2-1, 2-5, 2-15, 2-30, 2-35, 2-38, 2-48, 2-58, 2-68, 2-77, 2-82, 2-89, 2-93, 3-17, 3-20, 4-12, 4-15, 4-21, 4-35, 4-39, 4-43, 4-53, 4-55, 4-61, 4-69, 4-73, 4-74, 5-1, 5-4, 5-50, 5-85, 5-118, 5-151, 6-5, 7-21, 7-51, 7-57, 7-58, 7-63, 7-80, 7-81, 7-113, 7-114, 7-116, 7-151, 7-202, 7-204, 7-207
— *see also stream channel instability; disequilibrium*
Instream cover: 1-1, 4-68
Inter-gorge valley: 2-26
Interception: 2-6, 2-53, 3-16, 3-17, 4-5, 4-6, 4-11, 4-17, 4-21, 4-40, 7-6, 7-7, 7-15, 7-16, 7-24, 7-25, 7-32, 7-40, 7-47, 7-78, 7-80, 7-139, 7-215
Intermittent flow: 5-38, 7-111
Irregular meanders: 7-204
Irregular slopes:
— *see discontinuous slopes*

L

Lacustrine valleys: 2-26
Land clearing: 3-13, 3-14, 4-11, 4-17, 4-20, 4-26, 7-32
— *see also land uses*
Land drainage: 2-57, 2-58, 3-4, 3-19, 4-6, 4-53, 7-25
Land uses:
— Channelization: 1-9, 2-57, 2-58, 3-19, 4-11, 4-56, 4-62, 4-69, 7-32
— Diversions: 2-42, 2-55, 3-17, 4-5, 4-6, 4-8, 4-11, 4-34, 4-35, 4-39, 4-40, 4-66, 5-38, 5-42, 5-100, 5-101, 7-16, 7-24, 7-25, 7-32, 7-115
— depletions: 4-41, 5-101
— imported water: 2-42, 2-55, 3-17, 4-5, 4-39, 4-40, 5-101, 7-16, 7-24
— Dredging: 2-42, 3-4, 3-19, 4-11, 4-53, 4-69, 7-32
— Fire: 5-119
— Flood control: 2-57, 4-6, 4-53, 7-25
— Grazing: 2-32, 2-33, 2-76, 2-79, 2-84, 3-4, 4-6, 4-49, 4-53, 4-55, 4-62, 4-68, 4-74, 5-159, 6-4, 7-3, 7-6, 7-7, 7-25, 7-57, 7-106, 7-196, 7-205
— In-channel mining: 4-50 thru 4-52, 4-58, 4-59, 4-61, 4-69, 7-57, 7-71
— Land clearing: 3-13, 3-14, 4-11, 4-17, 4-20, 4-26, 7-32
— Levees: 3-4, 3-19, 4-11, 4-53, 4-55, 4-56, 7-32, 7-213
— Mining: 1-8, 4-4, 4-5, 4-6, 4-8, 4-26, 4-50 thru 4-52, 4-58, 4-59, 4-61, 4-69, 7-6, 7-23 thru 7-25, 7-57, 7-71
— Ramping flow: 2-42, 3-17, 4-5, 4-11, 4-35, 4-39, 7-16, 7-24, 7-32
— Reservoirs: 1-7, 1-8, 2-40, 2-55, 3-17, 4-5, 4-8, 4-11, 4-34, 4-35, 4-39, 4-40, 5-38, 5-100, 5-106, 7-16, 7-24, 7-32
— Roads: 1-7, 2-3, 2-4, 2-6, 2-54, 3-13, 3-16, 3-17, 3-19, 4-4 thru 4-8, 4-11, 4-12, 4-14 thru 4-17, 4-20 thru 4-22, 4-24, 4-25, 4-42, 4-62, 4-69, 5-6, 5-101, 5-114 thru 5-118, 5-154, 5-156, 5-157, 5-160, 5-161, 6-7, 7-3, 7-4, 7-6, 7-7, 7-12, 7-16, 7-19, 7-21, 7-23 thru 7-25, 7-32, 7-33, 7-35, 7-36, 7-40 thru 7-43, 7-45 thru 7-49, 7-57, 7-65, 7-77, 7-79, 7-80, 7-83, 7-139, 7-161 thru 7-164, 7-167, 7-194, 7-196, 7-200, 7-209, 7-211, 7-212, 7-215, 7-216
— Silviculture: 4-26, 7-7
— Urban development: 3-16, 4-5, 4-7, 4-8, 4-39, 5-17, 7-24, 7-25
Landform: 2-55, 4-15, 4-16, 7-36
Landscape overview: 3-10, 7-12
Landslide: 2-5, 2-6, 3-16, 3-17, 4-15, 4-17, 7-3, 7-12, 7-14 thru 7-16, 7-71, 7-165, 7-202, 7-206 thru 7-208, 7-213
Large woody debris: 4-6, 4-53, 4-57, 4-58, 5-62, 7-25, 7-57, 7-60, 7-78, 7-115
Lateral accretion: 2-42, 3-18, 5-50, 6-7, 7-17
Lateral adjustment: 5-40, 7-204
Lateral bank erosion rate: 2-35
Lateral containment: 5-50
— *see also confinement; Meander Width Ratio (MWR)*
Lateral migration: 5-40, 5-42
Lateral stability: 5-40, 5-51, 5-78, 5-142, 5-144, 5-145, 6-5, 7-179, 7-180, 7-181, 7-187, 7-203, 7-207, 7-209
Levees: 3-4, 3-19, 4-11, 4-53, 4-55, 4-56, 7-32, 7-213
LIDAR: 1-11, 3-4
Limiting factor analysis: 6-5
Lining: 3-19, 4-6, 4-55, 4-56, 7-25, 7-213
Lithology: 5-9, 5-94, 6-4
Livestock: 2-32, 2-33, 3-6, 3-19, 4-53, 4-58, 6-3, 7-7
Local base level: 1-7, 2-40, 4-21, 5-47, 5-146, 7-71, 7-79, 7-207
— *see also base level*
Longitudinal profile: 2-40, 5-15, 5-18, 5-69, 5-107, 5-131, 6-3
Low bank height:
— *see Bank-Height Ratio (BHR)*
Low width/depth ratio: 1-9, 2-58

M

Management prescriptions: 4-2, 4-74, 5-1, 5-160, 6-2
Manning's "n": 5-23, 5-24, 7-90
— *see also roughness coefficient*
Mass erosion: 1-8, 2-1, 2-3, 2-4, 2-35, 3-8, 3-13, 3-16, 4-11, 4-12, 4-14 thru 4-21, 4-46, 4-47, 4-62, 5-55, 5-58, 5-61, 5-114, 5-125, 5-126, 5-154, 5-156, 5-160, 5-161, 6-7,

XXV

Subject Index

7-4, 7-6, 7-10, 7-12, 7-15, 7-19, 7-21, 7-32, 7-33, 7-35 thru 7-39, 7-65, 7-71, 7-77, 7-83, 7-161, 7-165, 7-167, 7-194, 7-200, 7-206, 7-208, 7-209, 7-213
Mass wasting: 3-16
— *see also mass erosion*
Maximum anticipated precipitation: 5-121
Maximum depth: 2-40, 4-72, 5-18, 5-65, 5-71, 5-72, 5-130, 6-3
Mean bankfull depth: 4-65, 5-6, 5-44, 7-204, 7-211
Mean daily bankfull discharge: 5-87, 5-89, 5-90
Mean depth: 2-8, 2-20, 2-57, 5-6, 5-9, 5-18, 5-20, 5-23, 5-65, 5-71, 5-72, 5-111, 7-90, 7-210
Meander arc angle: 5-18
Meander geometry: 5-17 thru 5-19
Meander length: 6-16
Meander length ratio: 5-31
— *see also patern*
Meander pattern: 5-6, 5-40, 5-144, 7-113, 7-179
Meander scrolls: 7-204
Meander wavelengths: 6-3
Meander Width Ratio (MWR): 5-18, 5-50, 5-51, 5-144, 5-146, 7-121, 7-179, 7-182, 7-203 thru 7-205
— *see also confinement; lateral containment*
Meanders: 1-9, 2-55, 7-113, 7-204
Metamorphic geology: 2-92, 5-78, 5-79
Migration barrier: 5-42, 7-115
Mining: 1-8, 4-4 thru 4-6, 4-8, 4-26, 4-50 thru 4-52, 4-58, 4-59, 4-61, 4-69, 7-6, 7-23 thru 7-25, 7-57, 7-71
— *see also land uses*
Mitigation: 1-2, 1-4, 1-7, 1-12, 2-3, 2-6, 2-43, 2-68, 4-1, 4-2, 4-8, 4-10, 4-14, 4-20, 4-21, 4-25, 4-26, 4-33 thru 4-35, 4-42, 4-44, 4-49, 4-52, 4-56, 4-58, 4-61, 4-63, 4-66, 4-69, 4-73, 4-74, 5-1, 5-4, 5-6, 5-34, 5-36, 5-54, 5-86, 5-93, 5-97, 5-118, 5-126, 5-141, 5-151, 5-154, 5-156, 5-157, 5-159 thru 5-162, 5-164, 6-1 thru 6-6, 6-13, 6-18, 7-1, 7-40, 7-48, 7-51, 7-57, 7-62, 7-65, 7-71, 7-75, 7-77 thru 7-81, 7-83, 7-131, 7-179, 7-182, 7-187, 7-191, 7-196, 7-201, 7-203, 7-207 thru 7-213, 7-216
Mobilize bed material:
— *see initiation of particle movement; particle entrainment*

Momentary maximum value:
— *see dimensionless flow-duration curve*
Monitoring: 1-4, 1-7, 3-2, 4-2, 4-42, 4-52, 4-58, 4-73, 4-74, 5-1, 5-4, 5-164, 6-1 thru 6-9, 6-11, 6-13, 6-15, 6-17, 6-18, 7-1, 7-75, 7-77, 7-78, 7-80, 7-81, 7-200, 7-211, 7-216
— Calibration: 6-6
— Effectiveness: 6-2
— Implementation: 6-3
— Methods: 6-6
— Objectives: 6-2
— Validation: 6-6
Moraine:
— *see glacial till*
Morphological description: 2-20, 5-30, 7-97
Morphological shift: 2-30, 2-44, 5-159
— *see also stream channel succession; channel evolution*
Morphology: 2-20, 2-37, 2-38, 2-85, 3-16, 3-18, 4-42, 4-43, 4-67, 5-1, 5-6, 5-28, 5-30, 5-34, 5-44, 7-106, 7-204, 7-211
— *see also dimensionless ratios*
Multiple-thread channel:
— *see braided channel*

N

Natural channel design: 2-24, 4-68
Near-bank maximum depth: 5-65, 5-71, 5-72
Near-bank region: 2-77, 5-65, 5-67, 7-210
Near-bank shear stress: 5-65, 5-72 thru 5-74, 5-76, 5-77
Near-Bank Stress (NBS): 2-37, 2-87, 5-6, 5-53 thru 5-55, 5-65 thru 5-67, 5-69, 5-74, 5-76, 5-77, 5-78, 5-161, 6-3, 6-6, 7-124, 7-125, 7-129, 7-130
Negative exponential recovery relation: 5-115, 5-116
Non-point source pollution: 1-2, 3-2

O

Operational hydrology: 2-55, 3-17, 4-8, 4-10, 4-34, 4-35, 4-39 thru 4-41, 5-100, 5-101, 7-16
— *see also reservoir releases; diversions; flow modification*
Ordinary high water: 2-57
Over-steepened: 4-69
Over-wide channels: 4-65, 7-66
Overhead cover: 6-5, 7-203

Overland flow: 2-1 thru 2-3, 4-26, 5-118, 5-119, 5-121, 7-12, 7-19, 7-46, 7-77, 7-212
Oxbow cutoffs: 5-40, 7-113

P

Paired watershed: 1-6, 6-4
Particle detachment: 5-59
Particle entrainment: 2-8, 2-72, 7-170
Particle movement: 2-8
— *see also initiation of particle movement*
Particle size distribution: 2-84, 5-14, 5-19, 6-10, 7-116, 7-210
Pattern: 1-7, 1-9, 2-20, 2-24, 2-26, 2-30, 2-37, 2-38, 2-40, 2-43, 2-47, 2-57, 2-58, 3-19, 4-4, 4-6, 4-7, 4-43, 4-50, 4-58, 4-62, 4-67, 5-6, 5-30, 5-40, 5-41, 5-50, 5-65, 5-67, 5-110, 5-138, 5-144, 5-160, 5-161, 6-7, 7-7, 7-18, 7-23, 7-25, 7-51, 7-67, 7-83, 7-97, 7-99, 7-100, 7-113, 7-114, 7-179, 7-203, 7-204, 7-210, 7-211, 7-213, 7-214
— *see also channel pattern*
Pavement sample: 2-13, 5-131, 5-134
Peak flow: 2-56, 2-57, 6-7, 7-48
Pebble count: 3-13, 4-7, 5-18, 5-107, 5-130, 6-16
— *see also bed-material size distribution; Wolman pebble count*
Percent impervious: 4-5, 4-8, 4-35, 4-39, 5-17, 5-100, 7-24
Perennial flow: 5-38, 7-111
Permanent cross-section: 5-72, 6-2, 6-7, 6-10, 6-13
Pfankuch channel stability rating: 5-6, 5-34, 5-87, 5-107, 7-106, 7-117
Piping: 5-64
Plant science: 1-6
— *see also riparian vegetation*
Point bar: 2-13, 2-14, 2-42, 2-70, 5-130, 5-138, 6-9
— *see also bars, depositional features*
Polygon area: 5-121
Pool slope: 5-18, 5-65, 5-69, 5-70
Pool to pool spacing: 5-31
Pools: 2-6, 2-15, 2-24, 2-38, 2-77, 4-68, 5-16, 5-69, 5-70, 6-3, 6-16, 7-114, 7-116, 7-203
Populus (spp.): 5-17
Post-treatment condition: 5-100, 5-101, 5-102, 7-145
POWERSED: 2-74, 5-6, 5-25, 5-85, 5-86, 5-92, 5-103 thru 5-111, 5-114, 5-138, 5-160, 6-6, 7-145, 7-149, 7-150, 7-152, 7-154 thru 7-158, 7-200, 7-206, 7-210, 7-211, 7-214

Subject Index

Pre-treatment condition: 5-101, 5-102
Precipitation: 2-3, 4-14 thru 4-16, 5-6, 5-9, 5-17, 5-100, 5-101, 5-121, 6-7, 7-3, 7-36, 7-212
Precipitation data: 5-100, 5-101
Precipitation isohyetal map: 5-6
Process-specific mitigation: 1-4, 4-2, 4-14, 4-25, 4-74, 5-1, 5-34, 5-151, 5-154, 5-162, 5-164, 6-2, 7-1, 7-65, 7-79, 7-80, 7-196
Process-specific monitoring: 5-4
Profile: 1-7, 1-9, 2-20, 2-24, 2-26, 2-30, 2-37, 2-38, 2-40, 2-43, 2-47, 2-55, 2-57, 2-58, 3-19, 4-4, 4-6, 4-7, 4-50, 4-58, 4-67, 5-6, 5-15, 5-18, 5-30, 5-50, 5-69, 5-74, 5-107, 5-110, 5-131, 5-138, 5-160, 5-161, 6-3, 6-7, 6-13, 6-15, 7-7, 7-18, 7-23, 7-25, 7-51, 7-83, 7-97, 7-99, 7-100, 7-203, 7-210, 7-211, 7-213, 7-214
— see also longitudinal profile
Proposed condition (watershed): 5-100, 5-101, 5-154
Proposed restoration: 7-216
Protrusion height: 5-23
— see also relative roughness
Protrusion ratio: 2-72
Pyroclastic soils: 4-16, 7-4, 7-36

Q

Quasi-equilibrium: 2-37, 2-38, 2-47, 2-57

R

Radius of curvature: 4-8, 4-44, 4-48, 4-49, 5-6, 5-18, 5-65, 5-68, 5-161, 7-54
Rain-on-snow: 5-38, 5-100
Ramping flow: 2-42, 3-17, 4-5, 4-11, 4-35, 4-39, 7-16, 7-24, 7-32
— see also land uses
Rearing habitat: 1-1, 6-5
Receiving stream reach: 4-42
Recovery potential: 2-33, 4-43, 4-73, 7-75
Recurrence interval: 2-55, 2-57, 5-14, 5-17, 5-18
Redds: 2-6
Reference reach: 1-9, 2-24, 2-26, 2-30, 2-33, 2-47, 2-82, 4-7, 4-8, 4-65, 5-1, 5-6, 5-13, 5-15, 5-19, 5-28, 5-34, 5-39, 5-106, 5-134, 6-4, 7-83, 7-97, 7-99, 7-106, 7-108, 7-109, 7-112 thru 7-114, 7-116 thru 7-118, 7-125, 7-127, 7-129, 7-131, 7-154, 7-155, 7-170, 7-171, 7-178 thru 7-180, 7-182, 7-183, 7-185, 7-187, 7-188, 7-191, 7-196, 7-200, 7-204 thru 7-206, 7-208, 7-210, 7-211, 7-214
Regime channel: 2-37
Regional curves: 2-57, 4-7, 4-44, 5-9 thru 5-11, 5-16, 5-94, 7-89, 7-90
Regular meanders: 7-113
Regulated flow: 4-34
Rejuvenated: 2-26
Relative roughness: 5-6, 5-20, 5-22, 7-151
Reservoir releases: 2-42, 4-11, 4-35, 4-40, 7-32
— see also operational hydrology
Reservoirs: 1-7, 1-8, 2-40, 2-55, 3-17, 4-5, 4-8, 4-11, 4-34, 4-35, 4-39, 4-40, 5-38, 5-100, 5-106, 7-16, 7-24, 7-32
— see also land uses
Resistance: 5-20, 5-22, 5-24, 5-106, 7-90
— see also Manning's "n"; friction factor
Restoration: 1-2, 1-7, 1-8, 1-12, 2-43, 2-55, 2-68, 4-42, 4-49, 4-52, 4-68, 4-69, 4-73, 4-74, 5-54, 5-93, 5-97, 5-138, 5-151, 5-156, 5-159 thru 5-161, 5-164, 6-1, thru 6-4, 6-18, 7-48, 7-75, 7-77, 7-79, 7-80, 7-131, 7-201, 7-203, 7-205, 7-206, 7-210, 7-214, 7-216
Return period: 2-57, 5-15, 5-17, 5-19, 5-101
Revised Universal Soil Loss Equation (RUSLE): 2-2, 5-118
Riffle: 2-13, 2-14, 5-6, 5-15, 5-16, 5-18, 5-20, 5-65, 5-68 thru 5-71, 5-130, 5-131
Riffle slope: 5-18, 5-65, 5-70
Rill erosion: 4-26, 4-47, 5-119, 5-160, 7-46
Riparian grazing: 4-49, 4-68
Riparian impacts: 5-1, 7-6, 7-21, 7-57
Riparian overstory: 5-36, 5-37, 7-109, 7-110
Riparian species conversion: 3-19, 4-31, 4-32, 4-46
— see also riparian vegetation change
Riparian vegetation: 1-9, 1-10, 2-24, 2-30, 2-32, 2-35, 2-47, 2-48, 2-77, 3-19, 4-6, 4-8, 4-11, 4-43, 4-46, 4-49, 4-53, 4-55, 4-56, 4-58, 4-62, 4-68, 4-69, 4-74, 5-6, 5-34, 5-36 thru 5-38, 6-1, 6-3, 7-7, 7-10, 7-18, 7-25, 7-32, 7-33, 7-51, 7-57, 7-58, 7-62, 7-97, 7-106, 7-108 thru 7-110, 7-207, 7-208, 7-213
Riparian vegetation change: 2-77, 4-6, 4-49, 4-55, 4-58, 7-7, 7-25, 7-58

Risk rating system: 4-1, 4-12, 4-43, 7-33
River:
— River classification: 2-20, 3-13
— River entrenchment:
— see channel entrenchment; vertical containment
— River inventory: 4-7, 7-26
— River management: 3-1, 3-6, 4-4 thru 4-7, 7-23 thru 7-25
— River morphology: 2-37, 5-1, 5-28, 5-30, 5-34, 5-44
— see also morpholoy
— River pedestals: 2-35, 2-36
— River restoration: 1-2, 4-42, 4-68, 5-160
— River stability: 1-1, 1-5, 1-7, 1-9, 2-37, 2-38, 3-5, 3-6, 3-10, 3-19, 4-34, 4-41, 5-1, 5-6, 5-40, 5-159, 5-160, 6-1, 6-5, 7-23, 7-83, 7-113, 7-115, 7-152
— see also equilibrium
River instability: 1-10, 2-1, 4-44, 4-73, 5-1, 5-85, 5-118, 6-5, 7-80
— see also stream channel instability; disequilibrium
RIVERMorph™: 2-74, 5-4, 5-20, 5-25, 5-86, 5-90, 5-100, 5-102, 5-107, 5-110, 6-13, 7-154, 7-155
Road Impact Index (RII): 4-21, 4-23, 5-115 thru 5-117, 5-160, 7-41, 7-161 thru 7-164
Road sediment yield prediction:
— see Road Impact Index (RII)
Roads: 1-7, 2-3, 2-4, 2-6, 2-54, 3-13, 3-16, 3-17, 3-19, 4-4 thru 4-8, 4-11, 4-12, 4-14 thru 4-17, 4-20 thru 4-22, 4-24, 4-25, 4-42, 4-62, 4-69, 5-6, 5-101, 5-114 thru 5-118, 5-154, 5-156, 5-157, 5-160, 5-161, 6-7, 7-3, 7-4, 7-6, 7-7, 7-12, 7-16, 7-19, 7-21, 7-23 thru 7-25, 7-32, 7-33, 7-35, 7-36, 7-40 thru 7-43, 7-45, 7-46 thru 7-49, 7-57, 7-65, 7-77, 7-79, 7-80, 7-83, 7-139, 7-161 thru 7-164, 7-167, 7-194, 7-196, 7-200, 7-209, 7-211, 7-212, 7-215, 7-216
— see also land uses
Root density: 5-55, 5-60
— see also weighted root density
Root depth: 5-55, 5-59, 5-60
Rooting depth: 2-77, 4-16, 4-46, 4-69, 5-6, 5-36, 5-59, 5-60, 5-161, 7-36, 7-205, 7-214
Roughness: 2-37, 5-6, 5-20, 5-22, 5-23, 5-24, 5-111, 5-119, 5-123, 7-151
Roughness coefficient: 5-24
— see also manning's "n"

XXVII

Subject Index

Runoff: 1-11, 2-1, 2-2, 2-6, 2-35, 2-42, 2-53 thru 2-55, 2-61, 3-5, 3-16, 4-5, 4-6, 4-11, 4-21, 4-35, 4-39, 4-40, 4-42, 5-101, 5-119, 5-122, 5-125, 6-2, 6-3, 6-8, 6-10, 6-13, 7-7, 7-24, 7-25, 7-32, 7-78, 7-111, 7-211
Rural watershed: 4-38, 4-39, 7-49
RUSLE: 2-2, 5-118

S

Salix (spp.): 5-17 — see also willows
Sand: 1-9, 2-19, 2-26, 2-39, 2-57, 2-58, 2-74, 2-77, 3-13, 5-23, 5-63, 5-85 thru 5-87, 5-92, 5-106 thru 5-109, 5-111, 5-129 thru 5-131, 5-134, 5-138, 7-97, 7-145, 7-152, 7-155 thru 7-158, 7-200, 7-206, 7-208, 7-210, 7-211, 7-214
Sapping: 5-64
Saturation: 2-4, 3-16, 4-11, 7-15, 7-32, 7-46
Scour: 2-37, 2-40, 2-42, 2-55, 2-72, 4-11, 4-21, 4-69, 4-71, 5-14, 5-69, 5-102, 5-134, 6-7, 6-10 thru 6-12, 6-16, 7-18, 7-32, 7-72, 7-73, 7-208
Scour chains: 6-7, 6-10, 6-11, 6-16
Sedges: 7-52
Sediment:
    — Sediment competence: 2-14, 2-70, 2-71, 2-73, 4-62, 5-6, 5-127, 5-128, 5-134, 5-135, 5-137, 5-138, 5-146, 5-156, 5-159, 5-160, 6-7, 6-9, 7-116, 7-168, 7-169, 7-171 thru 7-174, 7-182, 7-200, 7-205, 7-209, 7-210
        — see also sediment entrainment
    — Sediment concentration: 2-58, 2-87, 3-5, 5-38, 5-106
    — Sediment delivery index: 5-118, 5-120, 5-121, 5-123
    — Sediment delivery potential: 2-1, 2-3, 2-6, 4-14, 4-17, 4-19, 4-20, 4-21, 4-25, 4-28 thru 4-30, 5-160, 5-161, 7-33, 7-35
    — Sediment delivery ratio: 2-2, 2-3, 7-165
    — Sediment entrainment: 2-8, 5-127, 5-129, 7-170
        — see also sediment competence
    — Sediment load: 2-6, 2-19, 2-37, 2-38, 2-73, 4-11, 4-66, 5-101, 5-126, 7-32, 7-152, 7-206
    — Sediment particle size distribution: — see particle size distribution
    — Sediment rating curve: 2-17, 2-18, 2-33, 2-57, 2-58, 2-61, 2-62, 2-68, 2-69, 5-87, 5-94, 5-95, 5-97, 5-102, 6-3, 6-7
    — Sediment sources: 1-2, 1-4, 1-6, 1-11, 2-48, 3-1, 3-2, 3-4, 4-1, 4-49, 4-74, 5-1, 5-4, 5-118, 5-126, 5-151, 5-154 thru 5-156, 5-159, 5-160, 6-7, 7-3, 7-12, 7-21, 7-40, 7-48, 7-77, 7-81, 7-195
    — Sediment standards: 1-11
    — Sediment storage: 2-75, 5-4
    — Sediment supply rating: 5-141, 5-144, 5-146, 5-149 thru 5-151, 7-117, 7-179, 7-190, 7-191, 7-208
    — Sediment transport capacity: 2-77, 2-87, 4-11, 5-4, 5-25, 5-42, 5-103, 5-104, 5-106, 5-111, 5-129, 5-138, 5-156, 5-160, 7-32, 7-116, 7-149 thru 7-152, 7-154, 7-200, 7-203, 7-210, 7-211, 7-214
    — Sediment yield: 1-2, 1-11, 2-3, 2-4, 2-6, 2-35, 2-57, 2-58, 2-87, 4-1, 4-15, 4-17, 4-49, 5-1, 5-54, 5-82, 5-83, 5-85 thru 5-88, 5-93, 5-94, 5-97 thru 5-102, 5-111, 5-113 thru 5-118, 5-125, 5-126, 5-138, 5-154, 5-156, 5-158, 5-160 thru 5-162, 6-7, 6-13, 7-40, 7-77, 7-117, 7-131, 7-137 thru 7-145, 7-147, 7-148, 7-154, 7-155, 7-160 thru 7-164, 7-167, 7-174, 7-194, 7-196 thru 7-200, 7-206, 7-209 thru 7-216
    — see also bedload sediment yield; suspended sediment yield; annual sediment yield
Sediment entrainment: 2-8, 5-129, 7-170
Spring-fed: 5-38, 5-100
Sedimentary rock: 5-78, 5-79
Seepage pressure: 5-64
Sensitivity to disturbance: 2-31, 2-33
Shear stress: 2-7 thru 2-11, 2-13, 2-14, 2-16, 2-38, 2-40 thru 2-42, 2-55, 2-70, 2-72, 2-73, 2-75, 2-77, 2-87, 4-11, 4-48, 4-65, 4-66, 4-68, 4-69, 5-44, 5-47, 5-65, 5-68, 5-72 thru 5-74, 5-76, 5-77, 5-106, 5-111, 5-129 thru 5-131, 5-134, 5-136, 5-138, 7-32, 7-151, 7-170, 7-210, 7-211
Shear velocity: 5-23
Shields relation: 2-8, 5-134, 5-136
    — see also entrainment
Silvicultural activities: 4-5, 4-6, 7-6, 7-7, 7-23, 7-25
Silvicultural impacts: 2-57
Silviculture: 4-26, 7-7
    — see also land uses
Single-thread channel: 7-205
Sinuosity: 1-9, 4-7, 5-6, 5-18, 5-41

Site-specific mitigation: 4-2, 4-25, 4-33, 5-157, 5-164, 7-131
Skewed probit transformations: 5-120
Slip failure: 5-64
Slope: 1-6, 1-9, 2-1 thru 2-4, 2-6 thru 2-8, 2-14, 2-15, 2-24, 2-26, 2-37, 2-38, 2-40, 2-42, 2-57, 2-70, 2-72, 2-73, 2-76, 3-14, 3-16, 3-17, 4-4, 4-7, 4-8, 4-10, 4-11, 4-15 thru 4-17, 4-19 thru 4-21, 4-23 thru 4-26, 4-30, 4-62, 5-6, 5-14, 5-18, 5-19, 5-23, 5-36, 5-41, 5-42, 5-65, 5-67, 5-69, 5-70, 5-72, 5-85, 5-106, 5-110, 5-111, 5-119, 5-121 thru 5-123, 5-125, 5-129, 5-131, 5-134, 5-138, 5-159, 5-160, 6-3, 6-5, 6-8, 7-7, 7-14, 7-16, 7-21, 7-32, 7-36 thru 7-38, 7-41, 7-43, 7-44, 7-71, 7-77, 7-78, 7-90, 7-139, 7-151, 7-165, 7-170, 7-200, 7-203, 7-205, 7-206, 7-210 thru 7-214
Slope shape: 2-3, 4-16, 4-17, 4-19, 5-119, 5-123, 5-125, 7-36, 7-37
Slump/earthflow: 2-3, 2-5, 3-16, 4-14, 4-15, 4-16, 4-17, 5-125, 5-126, 7-13, 7-15, 7-35, 7-36, 7-166, 7-213
Snow deposition: 3-17, 7-16, 7-215
Snowmelt: 1-7, 1-11, 2-35, 2-53 thru 2-55, 2-61, 3-17, 4-16, 5-36, 5-89, 5-100, 5-119, 5-121, 6-2, 6-3, 7-3, 7-16, 7-36, 7-46, 7-48, 7-111, 7-139, 7-147, 7-215
Snowmelt model: 2-54, 5-100, 7-139, 7-147
    — see also WRENSS
Snowmelt runoff: 1-11, 2-35, 2-61, 6-2
Soil erosion: 5-118, 5-126, 6-8
    — see also surface erosion
Soil matrix: 5-64
Soil science: 1-6
Spawning habitat: 1-1
Stability: 1-1, 1-2
Step/pool channel: 7-213
Stiff diagram: 5-120, 5-121
Stormflow: 1-7, 2-54, 2-55, 4-40, 4-42, 5-38, 5-89, 5-96, 5-100, 5-101, 6-2, 6-3, 7-111
Stormflow model: 5-100, 5-101
    — see also unit hydrograph model; TR-55
Stormflow runoff: 4-40, 6-2, 6-3, 7-111
Strahler stream order: 5-39
Straightening: 2-40, 3-4, 3-19, 4-6, 4-11, 4-53, 4-55, 4-56, 4-69, 5-161, 7-24, 7-25, 7-32, 7-213
Stratification: 2-24, 2-27, 3-10, 5-39, 5-55, 5-64, 6-5
Stratification adjustment: 5-64

XXVIII

Subject Index

Stream:
- Stream buffers: 7-7
- Stream channel instability: 1-1, 1-2, 1-7, 1-8, 2-35, 2-38, 4-12, 7-151
  - *see also disequilibrium*
- Stream channel succession: 2-44, 5-142
  - *see also morphological shift; channel evolution*
- Stream classification: 2-20, 2-21, 2-23, 2-24, 2-26, 2-29, 2-33, 2-43, 2-44, 3-10, 4-7, 4-9, 5-6, 5-17, 5-26 thru 5-29, 5-87, 7-5, 7-26 thru 7-30, 7-47, 7-48, 7-84, 7-92, 7-93, 7-95, 7-96, 7-203
- Stream condition: 6-5
- Stream crossing impacts:
  - Culverts: 2-40 thru 2-42, 3-17, 4-4, 4-21, 4-69, 4-71, 5-159, 7-16, 7-17, 7-23, 7-71, 7-73, 7-80, 7-212
  - Bridges: 2-40, 2-42, 3-17, 4-4, 4-21, 4-62, 4-69, 4-71, 7-16, 7-17, 7-23, 7-71, 7-73
- Stream enhancement :
  - *see also stream restoration; restoration*
- Stream order: 3-10, 5-6, 5-34, 5-39, 7-12, 7-106, 7-112
  - *see also stream size*
- Stream power: 2-14 thru 2-16, 2-38, 2-40 thru 2-42, 2-55, 2-73, 2-74, 2-77, 2-87, 4-10, 4-11, 4-65, 4-66, 4-68, 5-6, 5-44, 5-106, 5-110, 5-111, 5-138, 7-7, 7-32, 7-151 thru 7-155, 7-211
- Stream restoration: 2-55, 4-49, 5-156, 5-159, 7-210, 7-214
  - *see also stream enhancement; restoration*
- Stream size: 5-39, 7-112
  - *see also stream order*
- Stream stability: 2-6, 2-57, 4-69, 5-4, 5-5, 5-7, 5-26, 5-32 thru 5-36, 5-38, 5-47, 5-49, 5-50, 5-52, 5-82, 5-93, 5-98, 5-103, 5-112, 5-127, 5-137, 5-139, 5-150, 5-153, 6-5, 7-17, 7-84, 7-86, 7-87, 7-92, 7-104 thru 7-106, 7-108, 7-111 thru 7-116, 7-120 thru 7-123, 7-137, 7-149, 7-159, 7-168, 7-173, 7-175, 7-190, 7-193
- Stream succession state: 5-146, 5-159, 7-182, 7-205, 7-206
- Stream temperature: 7-116
- Stream type classification: 5-28, 7-26
  - *see also river classification*

Streambank:
- Streambank erosion: 1-8, 2-2, 2-30, 2-35, 2-37, 2-38, 2-43, 2-47, 2-48, 2-58, 2-76, 2-77, 2-78, 2-87, 2-89, 2-90, 2-91, 2-92, 2-93, 3-8, 3-17, 3-19, 4-11, 4-44, 4-45, 4-46, 4-47, 4-48, 4-49, 4-50, 4-59, 4-61, 4-62, 4-63, 4-69, 4-74, 5-4, 5-6, 5-36, 5-43, 5-52, 5-53, 5-54, 5-55, 5-65, 5-69, 5-70, 5-72, 5-74, 5-78, 5-79, 5-80, 5-81, 5-154, 5-156, 5-158, 5-161, 5-162, 6-5, 6-6, 6-7, 6-13, 7-3, 7-10, 7-15, 7-16, 7-21, 7-32, 7-47, 7-51 thru 7-56, 7-62, 7-78, 7-79, 7-83, 7-97, 7-117, 7-123 thru 7-125, 7-131, 7-132 thru 7-136, 7-152, 7-154, 7-194, 7-196 thru 7-200, 7-209, 7-211, 7-213 thru 7-216
- Streambank erosion processes: 2-35
- Streambank erosion rates: 2-35, 2-37, 2-78, 2-87, 2-89, 2-92, 2-93, 5-43, 5-54, 5-55, 5-74, 5-78, 5-79, 5-80, 5-161, 7-3, 7-83, 7-117, 7-125, 7-131, 7-132, 7-136, 7-211, 7-213
- Streambank stabilization: 4-34, 4-49, 7-206

Streamflow:
- Duration: 1-6, 2-3, 2-6, 2-19, 2-53, 2-54, 2-57, 2-67, 2-74, 3-16, 4-5, 4-10, 4-11, 4-16, 4-34, 4-39, 4-41, 5-6, 5-85, 5-86, 5-87, 5-89, 5-90 thru 5-92, 5-94, 5-96, 5-97, 5-100, 5-101, 5-102, 5-106, 5-107, 5-111, 7-24, 7-32, 7-36, 7-47, 7-139, 7-145 thru 7-147, 7-154, 7-211
- Magnitude: 1-6, 1-9, 2-1, 2-4, 2-6, 2-33, 2-36, 2-53, 2-54, 2-56 thru 2-58, 2-61, 2-73, 2-84, 3-16, 3-18, 4-5, 4-10, 4-11, 4-14, 4-34, 4-39, 4-41, 4-74, 5-42, 5-85, 5-97, 5-100, 5-114, 5-120, 5-121, 5-123, 5-134, 5-156 thru 5-157, 5-159, 6-2 thru 6-5, 7-17, 7-24, 7-32, 7-47, 7-80, 7-116, 7-208, 7-209
- Timing: 2-38, 2-53, 3-16, 3-17, 4-5, 4-10, 4-11, 4-34, 4-39, 4-41, 5-85, 5-100, 7-16, 7-24, 7-32, 7-47
- Streamflow change: 3-8, 3-17, 4-35, 4-40, 4-41, 4-42, 5-85, 7-10, 7-16, 7-19, 7-47, 7-62, 7-71, 7-78, 7-83

Streamgage: 2-55, 2-71, 5-20, 5-86, 7-90, 7-145
Structural control: 7-4
Study bank height: 5-6, 5-54, 5-55, 5-58 thru 5-60

Study bank-height ratio: 5-55, 5-58, 5-161, 7-205
Sub-pavement sample: 2-13, 5-131, 5-134
Sub-surface flow interception: 2-6, 3-16, 3-17, 7-6, 7-15, 7-16, 7-40, 7-47, 7-78, 7-215
Substrate: 2-13, 4-68, 7-208
Subterranean flow: 5-38, 7-111
Surface disturbance: 1-7, 3-13, 3-14, 4-4 thru 4-6, 4-17, 4-21, 4-27, 4-49, 7-12, 7-14, 7-23 thru 7-25, 7-46, 7-212
Surface erosion: 1-8, 2-1 thru 2-3, 3-14, 4-11, 4-12, 4-14, 4-26 thru 4-33, 4-62, 5-55, 5-58, 5-114, 5-118, 5-120, 5-154, 5-156, 5-160, 6-7 thru 6-9, 7-12, 7-14, 7-19, 7-32, 7-35, 7-46, 7-77, 7-161, 7-165, 7-167, 7-209, 7-212
Surface protection: 2-3, 5-55, 5-62, 5-161, 7-205, 7-214
Surface roughness: 5-119
Surface runoff: 2-6, 4-6, 4-11, 4-35, 5-119, 7-25, 7-32
Suspended sediment: 1-11, 2-17, 2-19, 2-33, 2-35, 2-58 thru 2-69, 2-87, 2-88, 5-6, 5-38, 5-45, 5-85 thru 5-87, 5-92 thru 5-95, 5-97, 5-101, 5-102, 5-106, 5-154, 5-156, 6-6, 6-7, 7-145, 7-146 thru 7-148, 7-154, 7-155, 7-196
Suspended sediment rating curve: 2-17, 2-19, 2-33, 2-58 thru 2-60, 2-62, 2-68, 2-69, 2-87, 2-88, 5-87, 5-95, 5-97, 6-7
Suspended sediment yield: 2-35, 5-85 thru 5-87, 5-97, 7-148, 7-154, 7-196

T

Temperature:
- *see stream temperature*

Terrace: 2-35, 2-36, 2-56, 2-57, 3-13, 5-14, 5-18, 5-19
- *see also abandoned floodplain*

Texture of eroded material: 5-119, 5-123
Thalweg: 5-130, 5-131, 6-9
Timber harvest: 2-5, 2-53, 2-54, 3-4, 3-20, 4-6, 5-101, 7-3, 7-4, 7-6, 7-15, 7-17, 7-19, 7-21, 7-25, 7-33, 7-40, 7-48, 7-139, 7-145, 7-196, 7-215
Time-trend recovery: 5-100
Toe pins: 4-74, 6-5
Topographic maps: 2-20, 3-4, 3-10, 3-14, 4-7, 4-8, 4-12, 5-6, 5-125, 7-26
Tortuous meanders: 5-40, 7-113

Subject Index

Total annual sediment yield: 1-11, 2-6, 2-35, 4-49, 5-82, 5-83, 5-85, 5-88, 5-97, 5-99, 5-111, 5-113, 5-114, 5-116, 5-126, 5-156, 5-160 thru 5-162, 7-131, 7-137 thru 7-144, 7-147, 7-160, 7-163, 7-167, 7-194, 7-196, 7-200, 7-209 thru 7-211, 7-213, 7-215
— *see also annual sediment yield*
Total bed-material load: 2-19, 2-74
Total load: 5-156
Total maximum daily load (TMDL): 1-2
TR-55: 2-54, 5-100
— *see also stormflow model*
Transport: 1-1, 1-7, 1-9, 1-10, 2-1, 2-2, 2-6 thru 2-8, 2-10, 2-12, 2-14 thru 2-19, 2-33, 2-37, 2-38, 2-40, 2-42, 2-54 thru 2-56, 2-61, 2-67, 2-70, 2-73 thru 2-75, 2-77, 2-87, 3-4, 4-11, 4-21, 4-43, 4-57, 4-58, 4-66, 5-4, 5-25, 5-38, 5-42, 5-50, 5-102 thru 5-104, 5-106 thru 5-109, 5-111, 5-114, 5-119, 5-121, 5-129 thru 5-131, 5-134, 5-137, 5-138, 5-156, 5-160, 6-7, 6-8, 7-12, 7-32, 7-116, 7-149, 7-150, 7-152 thru 7-158, 7-173, 7-174, 7-200, 7-203, 7-205, 7-206, 7-210, 7-211, 7-214
Transport agent: 5-119, 5-121
Transverse bar: 5-65, 5-67, 5-70
— *see also diagonal bars*
Truncated meanders: 5-40, 7-113

U

Unconfined meander scrolls: 7-204
Unconsolidated: 1-9, 3-13, 7-208, 7-213
Uniform slopes:
— *see continuous slopes*
Unit hydrograph approach: 5-100
— *see also stormflow model*
Unit stream power: 2-14 thru 2-16, 2-38, 2-40, 2-73, 2-77, 2-87, 4-65, 5-6, 5-44, 5-106, 5-110, 5-111, 7-151 thru 7-155, 7-211
Urban development: 3-16, 4-5, 4-7, 4-8, 4-39, 5-17, 7-24, 7-25
— *see also land uses*
Urban hydrology: 5-100
Urban watershed: 3-16, 4-35

V

Validation monitoring: 1-4, 1-7, 5-1, 5-4, 6-2, 6-6
Valley classification: 2-26
Valley type: 1-6, 1-9, 2-24, 2-26 thru 2-28, 2-30, 2-48, 2-76, 3-10, 3-12, 3-18, 4-7, 4-67, 4-74, 5-28, 5-30, 5-39, 7-17, 7-26, 7-69
Vegetation:
— Alnus, spp. (Alder:)
— *see alder*
— Carex, spp. (Sedges):
— *see sedges*
— Populus, spp. (Cottonwood):
— *see cottonwood*
— Salix, spp. (Willow):
— *see willow*
— Vegetation alteration: 3-19, 5-6, 5-36, 5-100, 7-18, 7-57
— Vegetation canopy:
— *see canopy*
— Vegetation composition: 4-4, 4-6, 4-8, 4-44, 4-46, 4-55, 5-37, 7-18, 7-23, 7-25, 7-52
— Vegetation-controlling influence: 4-43
— Vegetation density: 4-4, 7-23
— Vegetation overstory: 5-36, 5-37, 7-109, 7-110
— Vegetation understory: 5-36, 5-37, 7-109, 7-110
Velocity distribution: 5-75
Velocity estimation: 5-20, 5-21
Velocity gradient: 4-48, 5-65, 5-67, 5-74, 5-77
— *see also velocity profiles/ isovels/velocity gradient*
Velocity isovels: 5-74, 5-75, 5-77
— *see also velocity profiles/ isovels/velocity gradient*
Velocity profiles: 5-65, 5-74
Velocity profiles/isovels/velocity gradient: 5-65, 5-74
Vertical bank profile:
— *see bank profile*
Vertical containment: 2-20, 2-40, 4-72, 5-47
— *see also channel entrenchment*
Vertical stability: 5-41, 5-42, 5-44, 5-45, 5-47, 5-50, 5-142, 5-146 thru 5-148, 5-159, 6-5, 7-114 thru 7-116, 7-121, 7-177, 7-182 thru 7-187, 7-205, 7-207, 7-209
Vertically contained: 2-20, 2-42, 4-34
Volcanic rock: 4-16, 7-36
Volcanism: 5-78, 5-80, 7-4, 7-12, 7-131, 7-213

W

Washload: 2-19, 5-85, 5-106, 7-155
Water:
— Water availability values: 5-122
— Water quality: 1-1, 1-11, 3-2, 3-8, 3-17, 4-4, 4-5, 7-16, 7-24, 7-207
— Water Resource Evaluation of Non-point Source pollution from Silvicultural activities (WRENSS):
— *see WRENSS*
— Water surface slope: 2-8, 2-14, 5-18, 5-23, 5-65, 5-69, 5-70, 5-72, 5-111, 5-131
— Water temperature: 4-5, 6-5, 7-24
— Water yield model: 2-54, 4-42, 5-6, 5-85, 5-100, 5-102, 5-107, 7-139, 7-145 thru 7-147
Water Erosion Prediction Procedure (WEPP): 2-2
— *see also WEPP*
Wavelength: 5-18
— *see also meander wavelength*
Weak link stream: 4-35 thru 4-42, 7-48, 7-49, 7-65
Weak zones: 5-64
— *see also bank stratification; BEHI*
Wedge of incision:
— *see incision, degree of*
Weighted root density: 5-55, 5-60
WEPP: 2-2, 2-3, 5-118
Wet-sieve: 5-130, 5-131
— *see also field-sieve*
Width/depth ratio: 1-9, 2-15, 2-20, 2-27 thru 2-29, 2-35, 2-38, 2-40, 2-43, 2-52, 2-57, 2-58, 2-76, 2-87, 4-7, 4-8, 4-58, 4-62, 4-63, 4-65, 4-68, 4-69, 4-72, 5-9, 5-34, 5-41, 5-43 thru 5-45, 5-71, 5-138, 5-159, 7-65, 7-66, 7-72, 7-106, 7-116, 7-151, 7-152, 7-203 thru 7-206, 7-210, 7-211
Willow eradication: 2-35, 2-47, 7-98, 7-102, 7-107, 7-108
Willows: 1-9, 5-14, 5-17, 7-6, 7-7, 7-18, 7-97, 7-117, 7-204, 7-213
Wolman Pebble Count: 5-130
— *see also pebble count*
WRENSS: 2-2, 2-3, 2-5, 2-6, 2-54, 2-61, 4-15, 5-100, 5-114, 5-118, 5-125, 7-139, 7-145 thru 7-147, 7-165
— *see also snowmelt model; mass erosion; surface erosion*
9-207 data: 5-20